This item must be returned or renewed by the last date shown below. The loan period may be shortened if it is reserved by another reader. A fine will be due if it is not returned on time.

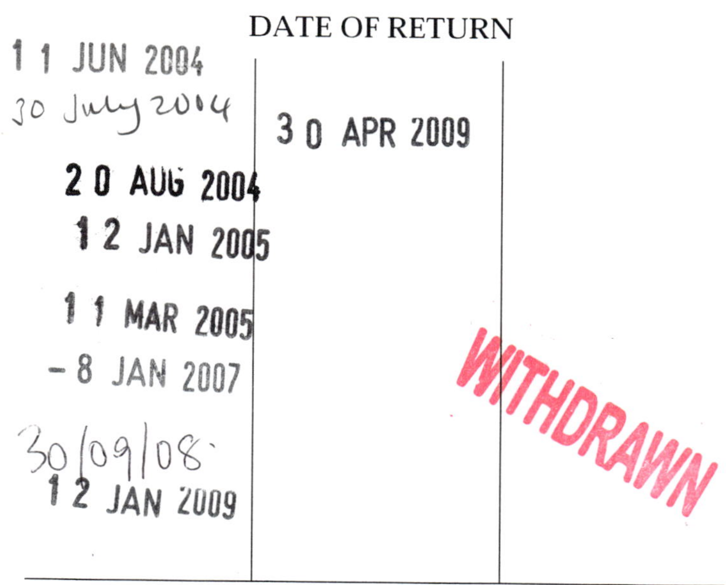

UNIVERSITY COLLEGE LONDON
Gower Street London WC1E 6BT

LF4D

RL15425.26.8.03

DRUG METABOLIZING ENZYMES

Cytochrome P450 and Other Enzymes in Drug Discovery and Development

DRUG METABOLIZING ENZYMES

Cytochrome P450 and Other Enzymes in Drug Discovery and Development

Edited by
Jae S. Lee, R. Scott Obach
and Michael B. Fisher

FONTIS
MEDIA

Although great care has been taken to provide accurate and current information, neither the author(s) nor the publisher, nor anyone else associated with this publication, shall be liable for any loss, damage, or liability directly or indirectly caused or alledged by this book. The material contained herein is not intended to provide specific advice or recommendation for any specific situation.

Trademark notice: Product or corporate names may be trademarks or registered trademarks and are used only for identification and explanation without intent to infringe.

Library of Congress Cataloging-in-Publication Data
A catalog record for this book is available from the Library of Congress.

ISBN: 0-8247-4293-1

This book is printed on acid-free paper.

This book is a joint publication of FontisMedia S.A. and Marcel Dekker, Inc.

FONTISMEDIA S.A.
FontisMedia S.A.
Avenue Vinet 19
CH-1004 Lausanne, Switzerland
tel: 41-21-648-3971; fax: 41-21-648 3975
WWW: http://www.fontismedia.com

MARCEL DEKKER INC.
Headquarters
Marcel Dekker, Inc.
270 Madison Avenue, New York, NY 10016
tel: 212-696-9000; fax: 212-685-4540

Eastern Hemisphere Distribution
Marcel Dekker AG
Hutgasse 4, Postfach 812, CH-4001 Basel, Switzerland
tel: 41-61-261-8482; fax: 41-61-261-8896

World Wide Web
http://www.dekker.com

The publisher offers discounts on this book when ordered in bulk quantities. For more information, write to Special Sales/Professional Marketing at the headquarters address above.

Current printing (last digit):
10 9 8 7 6 5 4 3 2 1

PRINTED IN THE NETHERLANDS

DEDICATION

This book is dedicated to Michael R. Rowley (1969 - 2002),
who passed away January 19, 2002, after a long battle with cancer.
Mike, you were a good friend, and you are sorely missed.

M. B. Fisher

Contents

Editors' Preface   xiii
Foreword   xv
    Anthony Y. H. Lu

Chapter 1

DIOXYGEN ACTIVATION BY CYTOCHROMES P450: A ROLE FOR MULTIPLE OXIDANTS IN THE OXIDATION OF SUBSTRATES   1
Alfin D. N. Vaz

1	INTRODUCTION	1
	1.1 Ligands to heme in CYP enzymes	2
	1.2. Reaction cycle of CYP enzymes	3
2	ACTIVE OXIDANT(S) IN CYP REACTIONS	6
	2.1 Heme-oxene as oxidant	6
	2.2 Heme-peroxo as an oxidant	11
	2.3 Multiple Oxidant Hypothesis in CYP-catalyzed oxidations	17
3	HEME-HYDROPEROXO AS AN OXIDANT	19
4	HEME-SUPEROXO AS ACTIVE OXIDANT	24
5	THE TWO-STATE THEORY	25
	CONCLUSIONS	27
	REFERENCES	27

Chapter 2

APPLICATION OF LC/MS, LC/NMR, NMR AND STABLE ISOTOPES IN IDENTIFYING AND CHARACTERIZING METABOLITES   33
A. E. Mutlib and John P. Shockcor

1	INTRODUCTION	33
2	SEPARATION OF METABOLITES FROM ENDOGENOUS COMPONENTS	36
	2.1 Sample Clean-up on Solid Phase Cartridges	37
3	LIQUID CHROMATOGRAPHY/MASS SPECTROMETRY	39

4   NUCLEAR MAGNETIC RESONANCE (NMR)   46
     4.1   Continuous-Flow LC-NMR   46
     4.2   Time-Slice LC-NMR   47
     4.3   Stop-Flow LC-NMR   47
     4.4   Loop-Storage   49
     4.5   LC-NMR-MS   49
     4.6   Recent Advances in NMR   50
5   SPECIFIC EXAMPLES   51
     5.1   Metabolism of Efavirenz (DPC 266) and Renal Toxicity   51
     5.2   Characterization of Unusual Metabolites   54
     5.3   Pharmacologically Active Metabolites   80
6   CONCLUSIONS   81
     REFERENCES   83

Chapter 3

BIOACTIVATION   87

Jack Uetrecht

1   INTRODUCTION   87
2   INVOLVEMENT OF REACTIVE METABOLITES IN
     ADVERSE REACTIONS   88
     2.1   Types of Adverse Drug Reactions   88
     2.2   The Hapten Hypothesis   89
     2.3   The Danger Hypothesis   91
     2.4   Immune Response in the Absence of Covalent Binding   92
3   ENZYME SYSTEMS RESPONSIBLE FOR
     METABOLIC ACTIVATION   93
4   TYPES OF REACTIVE METABOLITES   94
     4.1   General Types of Reactive Metabolites   94
     4.2   Alkyl Halides   96
     4.3   Primary Aryl Amines   97
     4.4   Aromatic Nitro Drugs   101
     4.5   Hydrazines   104
     4.6   Other Nitrogen-Containing Aromatic Compounds   106
     4.7   Acyl Glucuronides and Co-A Esters   107
     4.8   Quinone-type Reactive Metabolites   110
     4.10   Metabolites, such as Epoxides, that are Reactive
            because of Ring Strain   121
     4.11   Furans and Thiophenes   125
     4.12   Sulfhydryl-Containing Drugs   126
     4.13   Thiono Sulfur Compounds   127
     4.14   Isocyanates   129
     4.15   Acetylenes   131

ix

    4.16   Methylenedioxyphenyl compounds   131
    4.17   Free Radicals   132
5   THE USE OF BIOACTIVATION SCREENS FOR LEAD
DRUG SELECTION   135
6   SUMMARY AND CONCLUSIONS   138
    REFERENCES   139

Chapter 4

CHEMICALLY REACTIVE METABOLITES IN DRUGDISCOVERY AND DEVELOPMENT

Thomas A. Baillie

CHEMICALLY REACTIVE METABOLITES IN DRUGDISCOVERY AND DEVELOPMENT   147

REFERENCES   154

Chapter 5

CYTOCHROME P450 AND ITS PLACE IN DRUG DISCOVERY AND DEVELOPMENT

Dennis Smith

CYTOCHROME P450 AND ITS PLACE IN DRUG DISCOVERY AND DEVELOPMENT   155

    INTRODUCTION   155
1   FROM SINGLE ENZYME TO SUPERFAMILY OF HUMAN ISOFORMS   155
2   INHIBITION OF HUMAN ISOFORMS   160
3   INDUCTION OF HUMAN ISOFORMS   163
4   DIVERSITY OF HUMAN ISOFORMS   166
5   RESPONSE TO P450 METABOLISM IN DRUG DESIGN   168
6   P450 AS A DRUG TARGET   173
7   SUMMARY   176
    REFERENCES   176

Chapter 6

CYTOCHROME P450 IN LABORATORY ANIMAL SPECIES

Margit Spatzenegger, Stephanie L. Born and James R. Halpert

CYTOCHROME P450 IN LABORATORY ANIMAL SPECIES   179

1   INTRODUCTION   179
2   LABORATORY ANIMAL MODELS AS PREDICTORS OF HUMAN METABOLISM   179
    2.1   Basic Approaches   180
    2.2   Interspecies Scaling   182
3   FROM ANIMAL TO MAN: COMPARISON OF CATALYTIC SELECTIVITY BY ONE SUBFAMILY   185

x

3.1	CYP1A	185
3.2	CYP2A	186
3.3	CYP2B	187
3.4	CYP2C	188
3.5	CYP2D	189
3.6	CYP2E	189
3.7	CYP3A	190
3.8	CYP4A	191
4	REGULATION OF LABORATORY ANIMAL CYTOCHROME P450	192
4.1	Regulation of CYP1 genes	192
4.2	Regulation of CYP2B genes	193
4.3	Regulation of CYP3A genes	195
4.4	Regulation of CYP4A genes	196
5	NON-NATIVE ANIMAL MODELS	197
5.1	The Gene Knockout Mouse Model	197
5.2	The Humanized Mouse Model	200
6	CONCLUSIONS	201
	REFERENCES	202

Chapter 7

	TYPICAL AND ATYPICAL ENZYME KINETICS	211
	J. Brian Houston, Kathryn E. Kenworthy and Aleksandra Galetin	211
1	INTRODUCTION	212
1.1	General considerations of *In Vitro* Experimentation	213
1.2	Scope of this chapter	215
2	THE MICHAELIS-MENTEN APPROACH FOR ANALYSIS OF ENZYME KINETIC DATA	215
2.1	Enzyme Kinetic Studies in Drug Metabolism	216
2.2	Assumptions in Enzyme Kinetics	216
2.3	Methodological Considerations	217
2.4	Analysis of Substrate Kinetic Data	218
2.5	Analysis of Enzyme Inhibition Data	223
3	MULTISITE APPROACH FOR ANALYSIS OF A TYPICAL KINETIC DATA	227
3.1	Homotropic Effects	229
3.2	Heterotropic Effects	235
3.3	Criteria for Selection of an Appropriate Multisite Kinetic Model in Prediction of Drug Interactions	246
4	*IN VIVO* RELEVANCE OF *IN VITRO* ATYPICAL KINETICS AND CURRENT PERSPECTIVE	248
	REFERENCES	251

Chapter 8

CYTOCHROME P450 REACTION PHENOTYPING
Larry C. Wienkers and Jeffrey C. Stevens

1	INTRODUCTION	256
2	EFFECT OF INCUBATION CONDITIONS ON REACTION PHENOTYPING	262
3	ENZYME KINETICS AND REACTION PHENOTYPING	266
4	CORRELATION ANALYSIS	274
5	EXPRESSED P450S AND REACTION PHENOTYPING	281
6	CHEMICAL INHIBITORS AND REACTION PHENOTYPING	285
7	ANTIBODIES	294
8	CONCLUDING REMARKS	299
9	ACKNOWLEDGEMENTS	300
	REFERENCES	300

Chapter 9

DRUG-DRUG INTERACTIONS AND THE CYTOCHROMES P450
Kenneth A. Bachmann, Barbara J. Ring and Steven A. Wrighton

1		INTRODUCTION	311
	1.1	The Numerology of Drug-Drug Interactions	311
	1.2	The Centrality of the Cytochromes P450 in Drug-Drug Interactions	312
	1.3	Clinical Ramifications of CYP-Based Drug-Drug Interactions	314
	1.4	Electronic Sources of Information on CYP-Based Drug-Drug Interactions	315
2		*IN VITRO* DRUG-DRUG INTERACTIONS: DETERMINING THE POTENTIAL OF AN NCE TO INHIBIT THE CYPS	317
	2.1	*In Vitro* to *In Vivo* Extrapolations	319
	2.2	Irreversible or Quasi-Irreversible Inhibition	321
3		*IN VIVO* DRUG-DRUG INTERACTION STUDIES	322
	3.1	CYP-Based Interactions with Nutraceuticals and Food	327
	3.2	Transporter-Mediated Drug-Drug Interactions Masquerading as CYP-Based Interactions	330
	3.3	Pharmacogenetics and CYP-Based Drug-Drug Interactions	331
4		CONCLUSION	332
		REFERENCES	332

Chapter 10

CYP GENE INDUCTION BY XENOBIOTICS AND DRUGS 337

Academic authors: Jean-Marc Pascussi, Sabine Gerbal-Chaloin, Martine Daujat, Lionel Drocourt, Lydiane Pichard-Garcia, Marie-José Vilarem and Patrick Maurel
Drug Industry authors: Sylvie Klieber, François Torreilles, Martine Bourrié, François Guillou and Gérard Fabre

1	INTRODUCTION	337
2	CLINICAL AND PHARMACOLOGICAL CONSEQUENCES OF CYP1-3 GENE INDUCTION IN MAN	338
3	MECHANISMS OF CYP GENE INDUCTION	339
	3.1 Induction of the CYP1A family	339
	3.2 Cross-talks between the AhR pathway and other transcriptional factors and signal transduction pathways	340
	3.3 Induction of CYP2 and CYP3 families	344
	3.4 Cross-talk between PXR and CAR	349
	3.5 Cross-talk between PXR and CAR and other nuclear receptors	350
4	*IN VITRO* SYSTEMS TO SCREEN CYP INDUCERS	354
	4.1 Direct receptor-ligand binding assays	355
	4.2 Enhanced green fluorescent protein (EGFP)-based recombinant cell bioassay	356
	4.3 Ligand-induced receptor-coactivator interaction assays	356
	4.4 Cell lines for the direct screening of CYP1A inducers	357
	4.5 Cell line co-transfections: gene reporter assays	358
	4.6 Primary human hepatocytes for the direct screening of CYP inducers	359
5	*IN VIVO* EVALUATION OF CYP INDUCTION IN MAN	360
	5.1 Induction of CYP1A2 isoform	361
	5.2 Induction of CYP3A4 isoform	362
6	CONCLUSION	364
	REFERENCES	365

Chapter 11

CYTOCHROME P450 PHARMACOGENETICS 375

Robert L. Haining and Aiming Yu

1	INTRODUCTION	375
	1.1 Chapter Introduction	375
	1.2 Brief History of Pharmacogenetics	375
	1.3 Fundamentals of Pharmacogenetic Mutations	376
	1.4 *In Vivo* versus *In Vitro* Considerations	380
	1.5 Key principles of pharmacogenetics	380

2   P450 PHARMACOGENETICS   381
     2.1   CYP2C9 and Warfarin   381
     2.2   The CYP2D6 Story   385
     2.3   Other Human Cytochrome P450s of Real and/or
            Potential Pharmacogenetic Interest   401
3   PHARMACOGENETICS: PROMISE VERSUS PRACTICE   412
4   CONCLUSIONS   414
     ACKNOWLEDGEMENT   415
     REFERENCES   415

Chapter 12

ROLE OF INTESTINAL CYTOCHROMES P450 IN DRUG DISPOSITION

ROLE OF INTESTINAL CYTOCHROMES P450 IN DRUG
DISPOSITION   421

Mary F. Paine and Kenneth E. Thummel   421

1   INTRODUCTION   422
     1.1   Absorption and the First-Pass Effect   422
     1.2   Drug Movement Through the Gastrointestinal Barrier   423
     1.3   Anatomic and Physiologic Considerations   424
     1.4   Clinical Relevance of Intestinal First-Pass Metabolism   426
     1.5   Overview of Intestinal Drug Metabolizing Enzymes   428
2   INTESTINAL CYTOCHROMES P450   429
     2.1   Individual Subfamilies/Isoforms   430
3   SUMMARY AND PERSPECTIVE   445
     REFERENCES   446

Chapter 13

PREDICTION OF HEPATIC CLEARANCE IN HUMANS FROM EXPERIMENTAL ANIMALS AND *IN VITRO* DATA

PREDICTION OF HEPATIC CLEARANCE IN HUMANS FROM
EXPERIMENTAL ANIMALS AND *IN VITRO* DATA   453

Masato Chiba, Yoshihiro Shibata, Hiroyuki Takahashi, Yasuyuki Ishii and Yuichi Sugiyama

1   INTRODUCTION   454
2   STRATEGY FOR THE PREDICTION OF HEPATIC
     CLEARANCE IN HUMANS   458
     2.1   Allometric Scaling Methods to predict Clearance in Humans 458
     2.2   Direct Prediction of Hepatic Clearance in Humans and
            Experimental Animals from In-vitro Liver Microsomal Data   463
     2.3   Empirical Prediction of Hepatic Clearance in Humans by
            In Vitro-In Vivo Relationship in Experimental Animals   468
     2.4   Empirical Prediction of Hepatic Clearance in Humans by
            In Vitro-In Vivo Calibration of Human Hepatocytes Data   470

xiv

3   OPTIMIZATION OF METABOLIC CLEARANCE BY
STRUCTURE MODIFICATION   474
4   CONCLUSION   477
    REFERENCES   478

Chapter 14

NON-P450 MEDIATED OXIDATIVE METABOLISM OF XENOBIOTICS

483

Dieter Lang and Amit S. Kalgutkar

1   FLAVIN-CONTAINING MONOOXYGENASES (FMOS)   484
    1.1   Nomenclature and Molecular Characteristics of FMOs   484
    1.2   Tissue Distribution and Species Differences of FMOs   485
    1.3   Regulation of FMO Expression   486
    1.4   Catalytic Mechanism   488
    1.5   Characteristics of FMO Catalyzed Reactions   489
    1.6   Diagnostic Tools to Distinguish P450- and FMO-Catalyzed
    Reactions   494
    1.7   Diagnostic FMO Substrates *In Vitro* and *In Vivo*   496
    1.8   Polymorphism of Human FMOs and Relevance of FMOs to
    Drug Metabolism   498
2   MONOAMINE OXIDASES   499
    2.1   Multiplicity, Tissue Distribution, and Species Differences
    for the Mammalian Isozymes   499
    2.2   Crystal Structure   501
    2.3   Catalysis   501
    2.4   SAR Analysis of MAO Substrates   503
    2.5   MAO Inhibition   508
    2.6   MAO and Drug Metabolism   511
3   MOLYBDENUM HYDROXYLASES   511
    3.1   Tissue Distribution   513
    3.2   Species Distribution   514
    3.3   Catalysis   515
    3.4   Substrate Specificity of Molybdenum hydroxylases   516
    3.5   Antiviral Prodrug Activation by AO and XO   521
    3.6   Molybdenum Hydroxylase Mediated Reductions   522
    3.7   Inhibitors of Molybdenum Hydroxylases   525
    3.8   Molybdenum Hydroxylases and Drug Metabolism   526
4   CARBONYL REDUCTASES AND ALCOHOL- AND
    ALDEHYDE DEHYDROGENASES   527
    4.1   Alcohol Dehydrogenases   527
    4.2   Aldehyde Dehydrogenases   527
    4.3   Carbonyl Reductases   528
    REFERENCES   529

Chapter 15

THE ROLE OF SULFOTRANSFERASES (SULTS) AND UDP-GLUCURONOSYLTRANSFERASES (UGTS) IN HUMAN DRUG CLEARANCE AND BIOACTIVATION

Michael W.H. Coughtrie and Michael B. Fisher

THE ROLE OF SULFOTRANSFERASES (SULTS) AND UDP-GLUCURONOSYLTRANSFERASES (UGTS) IN HUMAN DRUG CLEARANCE AND BIOACTIVATION		541
GENERAL INTRODUCTION		541
INTRODUCTION TO THE SULFOTRANSFERASE ENZYME FAMILY		541
1	ASSAY OF SULFOTRANSFERASE ACTIVITY	543
2	MOLECULAR TOOLS AVAILABLE TO STUDY SULFOTRANSFERASES	545
3	ROLE OF SULFATION IN BIOACTIVATION OF DRUGS AND OTHER XENOBIOTICS	548
4	SULFOTRANSFERASE PHARMACOGENETICS	550
5	INTRODUCTION TO THE UDP-GLUCURONOSYLTRANSFERASE ENZYME FAMILY	551
6	REACTION MECHANISM/STRUCTURE	552
7	ENZYMOLOGY AND METHODOLOGIES	554
8	REGULATION	557
9	PHARMACOGENETICS	558
10	ROLE IN DRUG AND XENOBIOTIC METABOLISM	561
	10.1 Bioactivation.	561
	10.2 Structure-biotransformation relationships.	563
	10.3 Kinetics and scaling.	564
	10.4 First-pass metabolism.	565
11	CONCLUSIONS	566
	ACKNOWLEDGEMENTS	567
	REFERENCES	567
INDEX		577

Editors' Preface

Drug metabolism is a very interdisciplinary science, requiring knowledge and experience in areas as diverse as biochemistry, enzymology, biotransformation, physiology, toxicology, in vitro metabolism, and mathematical modelling. Moreover, the evolving role of the research scientist in drug metabolism departments in the pharmaceutical industry, especially here at Pfizer, is serving on interdisciplinary drug discovery and development project teams as the "drug metabolism expert". Recent examinations of the shortcomings of scientists in the progression from academia to industry emphasized the need for training in several disciplines within drug metabolism.[1] The trend of more new scientists entering industrial drug metabolism directly after graduate school results in fewer new scientists having postdoctoral opportunity to gain experience in other sub-disciplines. Also, scientists from other disciplines, such as medicinal chemistry, pharmacology, pharmaceutics, and toxicology require a working knowledge of drug metabolism science in order to be optimally effective in their roles in drug discovery and development.

For some time now a book has been needed that addresses the topic of drug metabolism from a practical viewpoint in order to address some of these shortcomings of the trained drug metabolism scientist. The message of the book should be accessible to those recently graduating with scientific degrees: both drug metabolism scientists and those in related disciplines. Although the book will certainly be useful in graduate-level pharmaceutical science courses, the principal aim is to provide a useful training text for the large number of scientists that require a basic understanding of the issues of drug metabolism in drug discovery and development. This book offers a solid foundation for all those entering or already working in the pharmaceutical industry that need to acquire a basic understanding of these critical issues. It was written to be accessible to the relative novice to pharmaceutical discovery and development, but also be sufficiently detailed, relevant, and up-to-date to have value for the more practiced scientist.

Our concept for the book was to assemble leading scientists in the drug metabolism field together for each of the areas to be covered in the book, and to maintain a balance where possible between academic and industrial scientists. The structure was based on impressions of the areas that constitute the foundation of the field. We owe gratitude to Dr. Anthony Lu for initial discussions of the purpose, scope, and content of this volume, as well as those numerous colleagues with whom we informally discussed this project. The areas covered in the book include such drug metabolism basics as P450 mechanism, structure-activity relationships, P450 reaction phenotyping, drug-drug interactions, induction, pharmacogenetics, and identification of metabolites; critical knowledge for the industrial drug metabolism scientist, such as P450 in laboratory animals, intestinal P450, in vitro-in vivo correlations and clearance projections, reactive metabolites, typical and atypical enzyme kinetics; and areas of growing importance, such as non-P450 oxidations and conjugating enzymes. Additionally, the authors were encouraged to bring their personal perspective to their contributions. However, some recent and very exciting findings, such as the first crystal structures described for mammalian P450 enzymes[2] and some computational approaches are notably absent. Their exclusion was not out of oversight or a lack of appreciation of these advances, but only because these areas are in such infancy that they have not yet made their way to the realm of universal application in applied drug research. There is no doubt that these areas are the focus of intensive research investigations and the fruits of these efforts will be harvested in the coming years—perhaps in time for an updated volume!

Finally, we extend our sincere gratitude to the contributing authors who have provided this volume with a set of high quality chapters. Writing up-to-date reviews of areas of scientific research, especially very active areas, takes a substantial amount of time and effort, and we are grateful to these scientists for their efforts and patience. The result is a systematic account of the latest findings in the field of cytochrome P450 and drug metabolism research and its impact on pharmaceutical research.

Editorial Team

Jae S. Lee, PhD
R. Scott Obach, PhD
Michael B. Fisher, PhD

Pharmacokinetics, Dynamics, and Metabolism
Groton Laboratories
Pfizer Global Research and Development

REFERENCE

1. Stevens, J.C.; Dean, D.C.; Preusch, P.C.; Correia, M.A. *Drug Metab. Dispos.* **2003**, *31*, 360.
2. Wester, M. R.; Johnson, E. F.; Marques-Soares, C.; Dansette, P. M.; Mansuy, D.; Stout, C. D. *Biochemistry* **2003**, *42,* 6370.

Foreword

In a paper reflecting his research career and celebrating the centenary of the *Journal of Biological Chemistry*, Professor Minor J. Coon of the University of Michigan made the following observation: "Those unfamiliar with basic research in biochemistry and related fields may assume that important discoveries are the result of brilliant ideas that are single mindedly pursued until, many years later, the answer is obtained, perhaps along with important biomedical applications. The progress of science is almost always more haphazard, as ambitious young scientists are influenced by their teachers, by the cooperative or competitive work of others, the availability of new techniques, and chance findings that may lead to different goals" (*J. Biol. Chem.* **2002**, *227*, 28351-28363).

This has been very true in the field of cytochrome P450 and drug metabolism research the last 40 years. The great debate in the 1960s over whether there was just one or more than one cytochrome P450 present in rat liver microsomes inspired investigators in the 1970s to conduct numerous studies that established the presence of multiple forms of cytochrome P450 in experimental animals. The significance of individual cytochrome P450s on the metabolism of therapeutic agents in humans is now well established. Research involving the inhibition of cytochrome P450 by either reversible inhibitors or mechanism-based inhibitors via MI-complex formation or covalent binding in the 1970s was aimed at elucidating the mechanism of cytochrome P450-mediated reactions; more recent studies have now established such inhibition to be the primary cause of drug-drug interactions in therapy. Discovering strain differences in mice on the induction of benzo(a)pyrene hydroxylase by 3-methylcholanthrene in the 1970s led to the demonstration of the presence of Ah receptor in the 1990s. Induction of various cytochrome P450s by drugs via receptor-mediated process represents another mechanism of drug-drug interactions in clinics. In the 1970s, the identification of slow metabolizers and extensive metabolizers in the population to metabolize certain drugs helped establish genetic polymorphism of various cytochrome P450s, a key factor responsible for individual variability in drug response and drug safety.

Few of us working in the field in the 1960s and 1970s ever dreamed that cytochrome P450 would play such a pivotal role in today's drug design and drug development. As a young and uninitiated scientist in the early 1970s, I once asked a noted clinical pharmacologist: "Do you think our studies on cytochrome P450 will eventually be useful in drug development?" "No way" he answered "After all, drug companies have successfully developed drugs for years without any knowledge of cytochrome P450". I was very disheartened after this conversation, worried that the kind of science I studied and worked on day in and day out actually had very little value. However, the situation changed in the 1980s when many talented young scientists started to purify, characterize and clone human cytochrome P450s. Heterologous expression of human cytochrome P450s, production of polyclonal and monoclonal antibodies against individual P450 enzymes, and development of selective probe substrates and inhibitors for each human cytochrome P450 made it possible for scientists to identify the cytochrome P450 form responsible for the metabolism of therapeutic agents. This information is the key factor in evaluating genetic polymorphism in metabolism and potential drug-drug interactions. The extent of metabolism of compounds by cytochrome P450 is also a major factor to consider in selecting drug development candidates to ensure adequate bioavailability.

The application of basic cytochrome P450 knowledge to drug design and development represents one of the most important advances in cytochrome P450 research in recent years. Since cytochrome P450 plays a pivotal role in eliminating drugs from the body, drug design must consider the unique properties of this versatile enzyme. In addition, changes in cytochrome P450 function in vivo (such as repression, inhibition and induction) have an impact on drug efficacy and safety in humans. This book addresses many of these important issues in drug development by some of the most talented young scientists in our field. My congratulations to Michael B. Fisher, R. Scott Obach, and Jae S. Lee for doing such a wonderful job. It is my hope that through understanding many of these important issues and continued progress in cytochrome P450 and drug metabolism research, the pharmaceutical industry can develop superior and safer therapeutic agents.

Anthony Y. H. Lu, Ph.D   May 10, 2003
Laboratory for Cancer Research
Department of Chemical Biology
College of Pharmacy, Rutgers University
Piscataway, New Jersey 08854

Chapter 1

Dioxygen Activation by Cytochromes P450: A Role for Multiple Oxidants in the Oxidation of Substrates

Alfin D. N. Vaz

Discovery Pharmacokinetics, Dynamics, and Metabolism, Pfizer Inc.
Eastern Point Rd, Groton, CT 06340

1 INTRODUCTION

Cytochromes P450 (CYPs) constitute a superfamily of heme proteins members of which are present in all branches of the phylogenetic tree from archaebacteria to higher mammals.[1] Over 2000 members have thus far been identified, a majority of them only known by DNA sequences extracted from genome databases. A significantly smaller number have either been purified by classical methods from mammalian, plant, yeast and bacterial sources, or, have been expressed as catalytically competent cytochrome P450 proteins in recombinant systems. These enzymes have diverse biological functions that include roles in the biosynthesis and regulation of cellular effectors, such as the synthesis of steroids and steroid hormones; hydroxy- and epoxy- arachidonic acid derivatives in hypertension; juvenile growth hormone in insects; absisic and salicylic acids in plant growth and wound healing; the homeostasis of signaling agents like retinoids, leucotrienes, prostaglandins, and epoxy eicosatrienoic acids; roles in inheritable diseases connected to mutations in specific CYP genes; the metabolic disposition of xenobiotics and pharmaceuticals; and the unique ability to activate molecular oxygen to oxidizing species with the capacity to effect oxidation reactions extending from the hydroxylation at un-activated carbon-hydrogen bond to the N- and S-oxidation of nitrogen and sulfur soft bases. This diversity of biological function and chemical capacity has attracted researchers with interests as diverse as human genetics and mechanistic organic chemistry.

CYP enzymes occur as soluble or membrane-associated forms. Soluble forms have thus far been found primarily in bacteria, whereas membrane-bound forms occur in yeast and higher organisms, usually in the endoplasmic reticulum and in mitochondrial membranes. For functional activity the CYP enzymes require reducing equivalents from the cofactors NADH or NADPH. These reducing equivalents are transferred to the CYP enzyme either by a flavoprotein reductase alone or in conjunction with an iron-sulfur protein. Currently there are known three well-defined classes of functional CYP enzyme systems. The first class is a system involving a ternary complex between the CYP enzyme (soluble or membrane bound), a flavoprotein reductase, and an iron-sulfur protein. This class is common with soluble CYPs in bacteria, and is also found in mammalian mitochondria. The second class is a system involving a binary complex between the CYP enzyme (membrane bound) and a flavoprotein reductase (membrane bound). This class is common to CYPs in the endoplasmic reticulum of eucaryotic cells. The third class is a self-contained system, in that the heme domain of the CYP and flavin domain of the reductase are on a single polypeptide chain. Only a few examples of this class have been identified in bacteria, mammals, and plants.[2] A fourth class has recently been described that involves a single polypeptide chain having domains for a flavoprotein reductase, an iron-sulfur cluster, and a cytochrome P450 type heme.[3]

1.1   Ligands to heme in CYP enzymes

The characteristic common to all CYP enzymes and unique to this family of heme proteins is a thiolate ligand to the heme iron from a cysteine residue located on a signature sequence near the carboxyl terminal of the protein sequence. This thiolate ligand is responsible for the unique absorbance at around 450 nm for the carbon monoxide complex of the ferrous form of these enzymes. This spectral characteristic distinguishes them from other heme proteins with histidine-derived imidazole ligands. The thiolate ligand imparts unique spectral signatures to the heme Soret absorbance when ligands bind within the active site of these enzymes either near, or at, the sixth co-ordination sphere of the heme iron. In the crystal structures of CYP enzyme in the resting state thus far determined, water occupies the 6th co-ordination site of the heme iron.[4] Ligands, such as substrates or inhibitors, that bind to the active site but do not coordinate directly to the heme iron may effect a spin state shift from low to high in the heme iron and a consequent loss of the water molecule coordinated at the 6th ligand site. This results in a characteristic blue shift of the Soret band from ~420 nm to ~390 nm, and is commonly used to establish binding of a ligand to the active site. Some compounds containing nitrogen, sulfur, or oxygen atoms that can co-ordinate to the heme iron may displace the coordinated water molecule showing a red shift in the Soret band characteristic of the heteroatom occupying the 6th coordination site.[5]

The cysteine thiolate ligand is also critical for functional activity. It has long been known that the P420 form of the CYP protein, where thiolate ligation is lost, is devoid of catalytic activity.[6] Studies with site specific mutants where the cysteine thiolate was replaced by the imidazole of histidine, or hydroxyl of serine resulted in proteins that are either very low in, or devoid of, catalytic activity. The heme was found to be weakly bound requiring reconstitution of the protein with heme.[7,8] The histidine mutant of CYP101 was found to have very low levels of camphor hydroxylase activity that was associated with a slow rate of reduction and a high rate of auto-oxidation. These results indicate the need of a thiolate ligand for functional activity of CYP101, and by analogy, other CYP proteins. Thus, the need of these enzymes for the thiolate ligand to play a critical role in the activation of molecular oxygen to the as yet confounding oxidant species capable of effecting the diversity of oxidation reactions that extend from the readily oxidizable soft bases such as N- and S-heteroatoms to the hydroxylation of hydrocarbons and its precise role in this process is as yet unclear and speculative.[9]

1.2. Reaction cycle of CYP enzymes

The majority of CYP proteins thus far examined function as oxidative catalysts. An atom from molecular oxygen is incorporated into the oxidized product and the other oxygen atom is reduced to water. However, exceptions are known where the oxygenated product derives its oxygen atom from water. A few CYP proteins function as reductases and directly transfer reducing equivalents from the cofactor to the substrate.[10] The generalized reaction cycle shown in Scheme 1 for the activation of molecular oxygen and substrate oxidation has evolved over the past 30 years from mechanistic studies on various CYP enzymes. Several steps of the reaction cycle and the intermediates formed in these steps have been well characterized for soluble as well as membrane bound forms and show common characteristics, thus allowing for a generalized formulation of the reaction cycle. Typically, substrates bind to the resting (Fe^{3+}) state of the enzyme, this is generally accompanied by a low to high spin-state shift in the heme iron resulting in displacement of the bound water molecule from the 6th co-ordination site and a change in the reduction potential of the heme iron.[11] Next, electron transfer occurs from the cofactor via either the flavoprotein reductase or the iron-sulfur protein (step 2). With some, but not all CYP enzymes, substrate binding is essential for efficient electron transfer to the heme from the electron transfer partner proteins of the CYP system. Oxygen binds rapidly (essentially limited by O_2 diffusion) only to the reduced (Fe^{2+}) CYP proteins to form the ferrous-dioxygen complex (step 3). Like hemoglobin and myoglobin the oxygen in the CYP-dioxygen complex is bound end-on at the heme iron. However, unlike these oxygen carrier proteins the electron density in the CYP-dioxygen complex resides in the oxygen ligand, having a ferric-superoxo anion character rather than the ferrous-dioxygen character of the oxygen carrier proteins. In some CYPs the ferric-superoxo complex is unstable, and dissociates to give the ferric resting enzyme and superoxide.[12,13] A second electron transfer from the electron transfer partners to the CYP-dioxygen complex results in the heme-peroxo complex (step 4). The heme-peroxo complex is also unstable, with many CYP enzymes it dissociates to give the resting ferric enzyme and hydrogen peroxide. Steps 3 and 4 can result in the apparent non-productive consumption of reducing equivalents and the release of reactive oxygen species (superoxide and hydrogen peroxide).

Step 1
Step2
Step 3
Fe^III + [S]
[S]-Fe^III
1e
Fe^II-[S]
O2
{Fe^II-O2}-[S]
[SOH]
O2•−
"Peroxide shunt"
•OH
{Fe^III-O2•−}-[S]
Step7
{FeIII}-[S•]
1e
Step4
Step6
OH
Step5
O
H+
H+
{Fe^IV•+}-[S]
{Fe^III}- [S]
{Fe^II-O2−}-[S]
H2O2

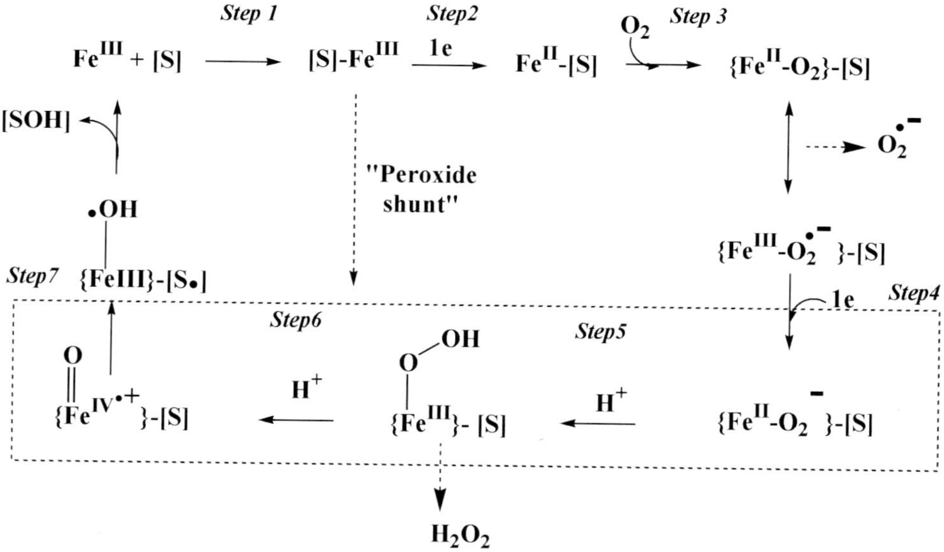

Scheme 1

The next three steps are critical for the activation of dioxygen to the putative terminal oxidant in all CYP-catalyzed reactions; yet, these steps have proven most refractory to clear elucidation and are the subject of ongoing mechanistic debate. The proton uptake steps (5 and 6) reflect the overall two protons consumed in the reaction cycle leading to substrate oxidation. Rates for these proton transfers have not been determined, and, except for the observation of the ferric-hydroperoxo complex in transient state cryo-crystallographic and spectroscopic studies it has not been established if the two-proton uptake occurs in a stepwise or concerted manner.[14-16] Theoretical studies suggested that protonation of the distal oxygen atom of the heme-hydroperoxo complex is isothermal, and results in the heterolysis of the peroxo bond to irreversibly generate the hypervalent iron-oxo complex termed the "iron-oxene" or "oxene."[17-19] This species was considered to be the ultimate oxidant responsible for the diverse oxidative reactions catalyzed by this unique class of heme proteins.

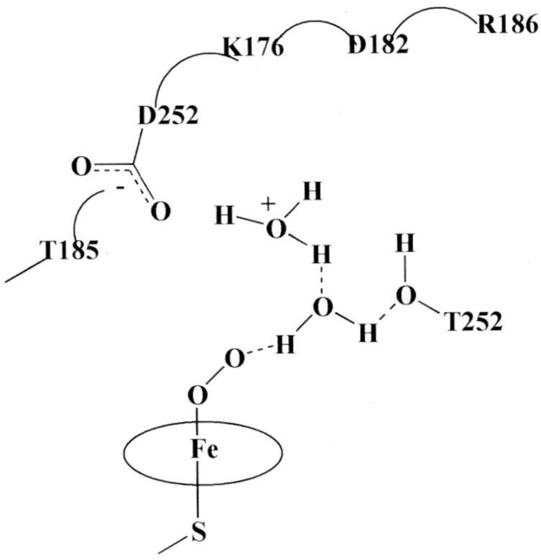

Scheme 2

In contrast to peroxidases where histidine serves as the proton donor for heterolysis of the peroxide bond, the crystal structures of CYP enzymes thus far solved do not reveal a distinct acidic group within close proximity of a putative hydroperoxo heme complex to serve such a function.[20,21] However, a significant number of structured water molecules have been identified within the active sites, for example, in the crystal structure of CYP101 the hydroxyl group of Thr252 was found to be hydrogen bonded to a water molecule that is within hydrogen bonding distance of a putative hydroperoxo heme complex, and Thr252 was proposed to serve the role of a protic residue within the active site (Scheme 2).[22,23] With a few exceptions, the threonine at this location in the I-helix, is highly conserved within the CYP superfamily.[24] Refinements in the crystal structure of CYP 101 have revealed a hydrogen bond network involving Thr252, Asp251, Lys178, Asp182, and Arg186, with Asp251 proposed as a "switch" between the solvent accessible residues (Lys178, Asp182 and Arg186) and water within the active site (Scheme 2).[23]

The oxidation of substrates (step 7) has generally been thought to be a stepwise process with electron or hydrogen atom abstraction from the substrate to give a transient enzyme-confined "caged" substrate radical with subsequent oxygen atom rebound to give the hydroxylated product. In the past decade this concept has been questioned and is undergoing a broad conceptual revision to account for many of the novel chemistries exhibited by this unique superfamily of heme proteins.

2   ACTIVE OXIDANT(S) IN CYP REACTIONS

2.1   *Heme-oxene as oxidant.*

The ability of CYP enzymes to introduce a molecular oxygen-derived oxygen atom into un-activated C-H bonds of a substrate indicated the need for a strong oxidant. An iron mono-oxygen species similar to Compound I of peroxidases was proposed to be this high valent oxidant.[25] This view was revised in the late 70's by the observations of Groves and others who demonstrated isotopic scrambling, loss of stereochemical integrity, allylic rearrangements and large intramolecular isotope effects in oxidations catalyzed by CYP enzymes.[26-30] For example, as shown in Scheme 3(a) the hydroxylation of exo-tetradeuteronorbornane by P450 LM2 (CYP2B4) resulted in a mixture of exo- and endo-norborneols, with the exo-norborneol retaining a significant (25%) amount of four deuterium atoms and the endo-norborneol containing 9% of molecules with three deuterium atoms. The hydroxylation of 1,1,4,4-tetradeutero cyclohexene and 1,4,5,6-tetrachloro-2-cyclohexene yielded the allylic alcohols with bond migration (Scheme 3(b) and 3(c)).   These studies, in conjunction with the well characterized oxidant species in the chemistry of peroxidases[31,32] and the exclusive incorporation of a molecular oxygen-derived oxygen atom at the site of hydroxylation, led to the generalized "oxygen rebound" mechanism.[28] In the 1980s and early 1990s all CYP-catalyzed reactions were viewed as having the mechanistic commonality of a stepwise process that involved an electron or hydrogen atom abstraction by a putative Compound I-like oxidant species to give a transient radical intermediate, oxygen rebound from a putative Compound II-like species to the radical intermediate yielded the hydroxylated products (Scheme 4). Rearrangement of the transient radical intermediate is in competition with oxygen rebound, and the extent to which rearrangement occurs depends on the ratio of the rate constants for oxygen rebound and radical rearrangement.

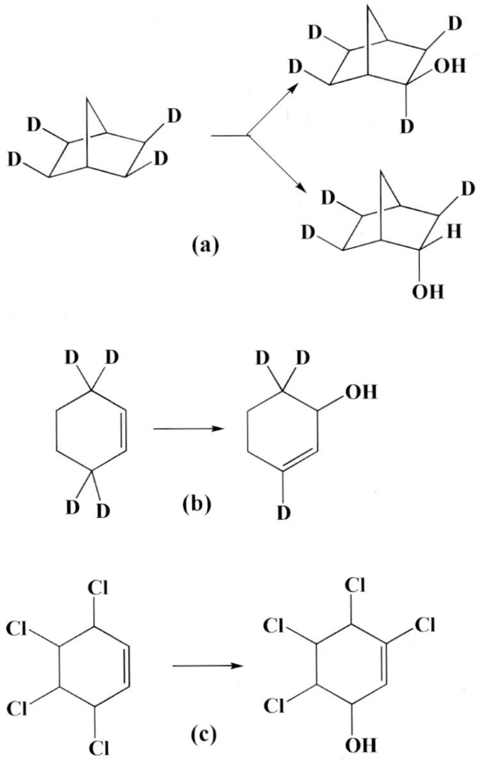

Scheme 3

The need for molecular oxygen and reducing equivalents can be bypassed by the use of oxidants such as alkylhydroperoxides, hydrogen peroxide, peracids, and iodosobenzenes. These oxidants are formally at the oxidation state of the heme-hydroperoxo intermediate. In reactions with CYP enzymes, commonly termed the "peroxy shunt," these oxidants have been shown to effect similar, but not identical, reactions as the molecular oxygen and reducing equivalents driven process.[33] Mechanisms for the "peroxy shunt" reaction have generally been depicted in a manner consistent with a Compound I-like species generated from the peroxy oxidant by the CYP enzyme followed by stepwise substrate oxidation as described above. Mechanistic studies with substituted alkyl or benzoyl hydroperoxides indicated that the alkyl components were involved in the transition state for the oxidations effected by these "shunt" oxidants, and a slight modification of the prevailing mechanism was proposed in that a three way complex between the substrate, the peroxy oxidant and the heme iron was considered as the transition state, homolysis of the peroxy bond yielded an alkoxy radical and a Compound II-like intermediate. Abstraction of a hydrogen atom by the alkoxy radical yielded the substrate radical that recombined with the heme-iron bound oxygen atom of the Compound II-like species.[34,35] Extensive efforts to identify a Compound I-like species in CYP-catalyzed reactions either in the molecular oxygen and reducing equivalents driven process, or the "peroxy shunt" reaction have been largely unsuccessful in connecting a distinct spectral species with the oxidation of a substrate.[36-39] Transient state cryo-crystallographic and spectroscopic studies of the CYP101-catalyzed hydroxylation of camphor have come the closest to "seeing" a Compound I-like intermediate. These studies have also distinctly demonstrated the intermediate complexes of the superoxo- and peroxo-anions, and of hydroperoxide with the heme iron.[14-16]

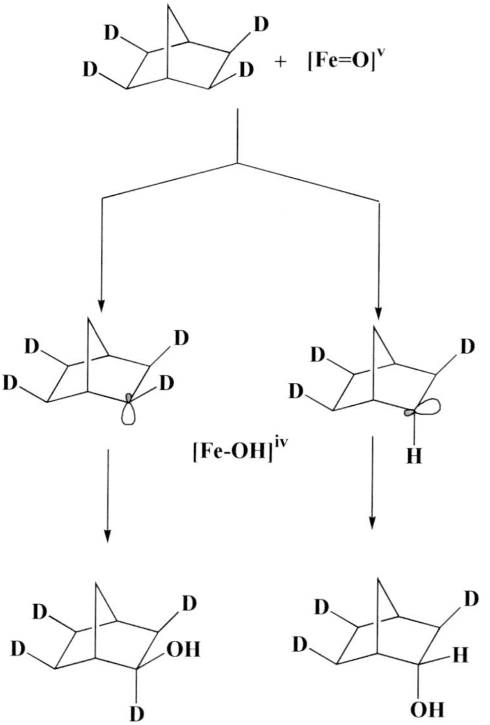

Scheme 4

Radicals generated adjacent to strained rings undergo ring opening with defined measurable rates.[40-43] When such radical intermediates are captured to give rearranged and un-rearranged products, the ratio of rearranged to un-rearranged products gives a measure of the capture rate (Scheme 5). This principle has been used to determine the oxygen rebound rate in the "oxygen rebound" mechanism. Early studies with bicyclo[2.1.0]pentane (Scheme 6 (a)) as probe substrate that forms a rearranged alcohol established an oxygen rebound rate constant of $\sim 1.4 \times 10^{10}$ s^{-1} corresponding to a radical lifetime of about 70 ps.[41,44] Subsequent studies with radical clocks designed to have faster rearrangement rates revealed that the oxygen rebound rate apparently increased with an increase in the rearrangement rate constants (Scheme 6(b) – (f)). This translates to a variation in activation energy from 4 kcal/mole for bicyclo[2.1.0]pentane (Scheme 6(a)) to zero kcal/mole for 1,1-diphenyl-2-methyl-cyclopentane (Scheme 6(e)).[40,42,45-48] The high rate for oxygen rebound with (**e**) was ascribed to enzyme active site constraints imposed on the biphenyl rings.[42] However, for substrate (f) where the radical intermediate has no flexibility, the rate constant for oxygen rebound is 1.3×10^{13} sec^{-1}, which is in the order of a bond vibration.[45] After a number of radical clocks were examined, a correlation diagram between the ratio of rearranged to un-rearranged product and the radical rearrangement rate constant showed essentially no correlation, and it became apparent that the results were inconsistent with the fundamental premise of a discrete radical intermediate in the "oxygen rebound" mechanism accounting for all the observed results.[48]

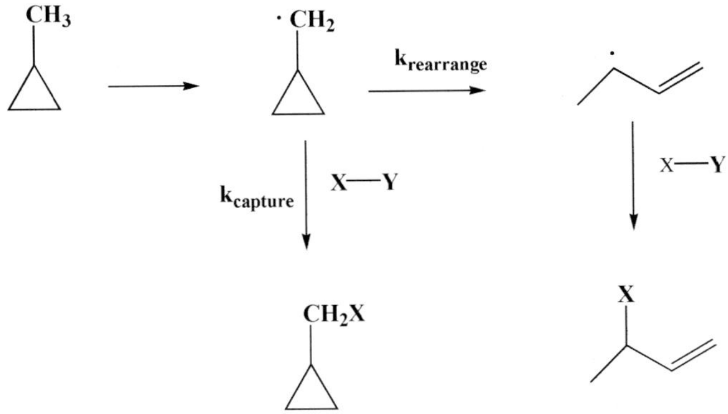

Scheme 5

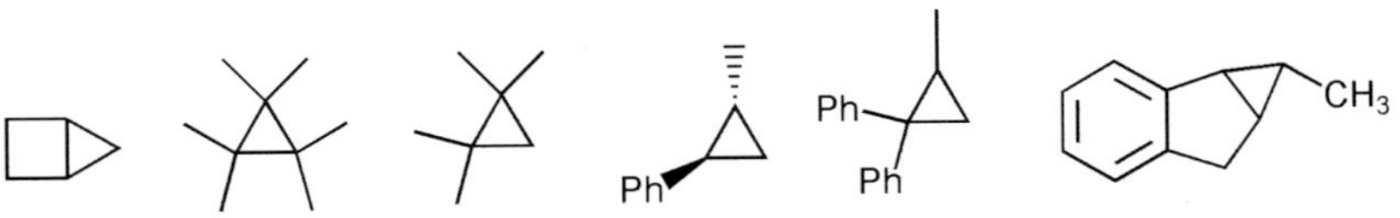

Scheme 6

$$k_{ox}\,(s^{-1}) = 1.4 \times 10^{10} \quad 2.3 \times 10^{11} \quad 2.5 \times 10^{11} \quad 1.5 \times 10^{12} \quad 7 \times 10^{12} \quad 1.3 \times 10^{13}$$

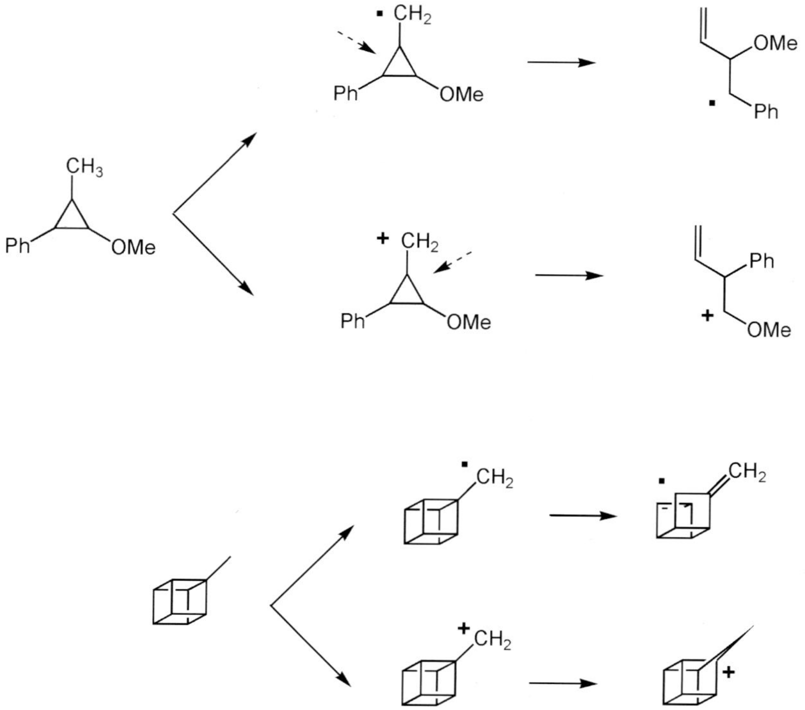

Scheme 7

Carbocations as intermediates were considered early in mechanistic studies of CYP catalysis. The differential rearrangement of radical or cation intermediates derived from norcarane can be used to distinguish between such intermediates in CYP-catalyzed reactions. Early studies with norcarane indicated no rearrangement products.[27] However, more recent studies with this substrate have identified small amounts of rearrangement products consistent with both radical and cationic intermediates.[47,49] Scheme 7 shows cationic and radical pathways for the rearrangement of such species derived from methyl cubane and the hypersensitive radical clocks 1-phenyl-2-methyl-3-alkoxy-cyclopropanes. The rearrangement products from these probes distinguish between radical and cationic intermediates. A cubylmethyl radical rearranges by cleavage of cubyl bonds, whereas the cubylmethyl cation rearranges by ring expansion to give the homocubyl cation.[50] The cyclopropylcarbinyl radical ring opens to give predominantly (>50:1) benzylic radical-derived products whereas the cyclopropylcarbinyl cation ring opens to give products derived only from the oxonium ion. Studies with various CYPs and methyl cubane gave the cation-derived product, homocubanol, to extents varying from 0 to 30% of the total oxidation at the methyl carbon.[43,51-53] The hypersensitive radical clock probes 1-phenyl-2-methyl-3-alkoxy-cyclopropanes gave small amounts of rearrangement products corresponding to a radical intermediate. The primary rearrangement product was derived from a cation-like intermediate. If the cation were derived by oxidation of the intermediate radical, the extent of radical rearrangement products would be expected to be significantly larger since the rearrangement rates of the radical intermediates from 1-phenyl-2-methyl-3-alkoxy-cyclopropanes are very fast ($>5 \times 10^{11}$ sec^{-1}). Accordingly, these probes indicate that their CYP catalyzed oxidation must occur by pathways not exclusively dictated by the "oxygen rebound" mechanism.[28] Thus, several lines of evidence now indicate that even for hydrocarbon hydroxylation the Oxygen Rebound mechanism is not the exclusive reaction pathway for CYP-catalyzed oxidations.

2.2   *Heme-peroxo as an oxidant.*

The first challenge to the prevailing view that a heme-oxene is the unique oxidant in all CYP-catalyzed reactions came from the seminal work by Akhtar and colleagues on the mechanism of the aromatase catalyzed demethylation of androgens to form estrogens.[54] As shown in Scheme 8 the 19-demethylation of androgens to estrogens by aromatase is a three step sequential process that involves an aliphatic type hydrocarbon hydroxylation; oxidation of a primary alcohol to an aldehyde; and, at the time, a unique terminal oxidative rearrangement of the aldehyde intermediate to form formic acid and aromatize the steroid A ring. This reaction was known to be stereochemically specific for loss of the β-hydrogens from C_1 and C_2. Multiple mechanisms were proposed for the terminal reaction on the assumption of an oxene species as the active oxidant.[55-63] Each of these mechanistic proposals failed to account for all the experimental observations. Akhtar and colleagues using isotopic label tracking techniques demonstrated that the formic acid formed in the terminal oxidative rearrangement of the aldehyde intermediate retained the 19-oxo androgen's hydrogen atom. In addition, when $^{18}O_2$ was used, the 19-oxo-androgen-derived formic acid had incorporated an atom of ^{18}O (Scheme 9). These researchers concluded correctly that an electrophilic Compound I-like species could not account for these results. They proposed the novel concept that the heme-peroxo species was the more likely oxidant in a nucleophilic type of reaction with the 19-oxo carbonyl group of the androgen to form a peroxyhemiacetal intermediate that rearranged to the estrogen and formic acid (Scheme 10).[54] This mechanistic hypothesis was consistent with all the experimental observations. Akhtar and colleagues subsequently demonstrated that the demethylation of lanosterol, the deacetylation of 17-hydroxy-progesterone to androstenedione, and the deacetylation of progesterone, by the respective steroid biosynthetic CYP enzymes were mechanistically identical to the aromatase-catalyzed reaction.[64-66] Cole and Robinson have provided chemical models in support of the heme-peroxo hypothesis by demonstrating that the peroxy adduct of the 19-oxo androgen intermediate could rearrange to the aromatic ring.[67-70]

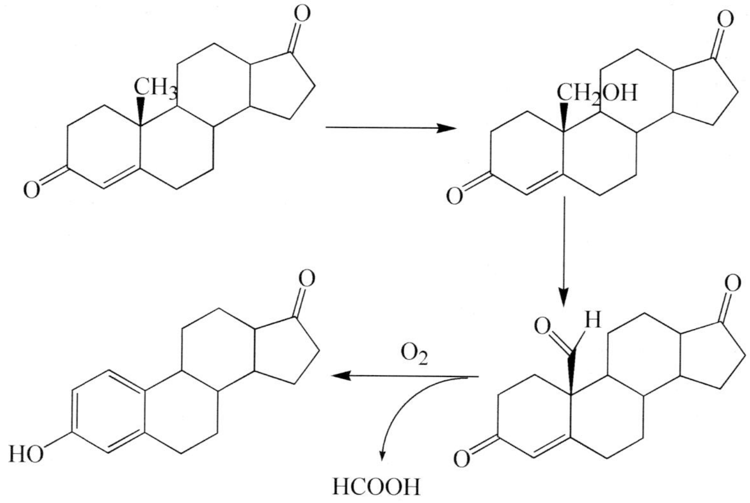

Scheme 8

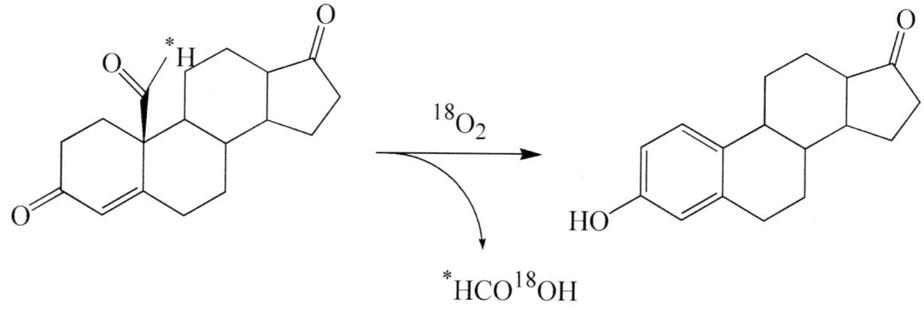

Scheme 9

The Akhtar hypothesis remained confined largely to the realm of steroid biosynthetic CYP enzymes, and found little acceptance as a generalized CYP mechanism throughout the 1980s. In the early 1990s evidence that this mechanistic hypothesis was applicable to xenobiotic metabolism as well was provided by studies on the oxidative deformylation of xenobiotic aldehydes to give olefins and formate (Scheme 11).[71,72] In the reaction of cyclohexanecarboxaldehyde, the formic acid was shown to retain the aldehyde hydrogen and incorporate an atom of oxygen from molecular oxygen. At a fundamental chemical level, this transformation is identical to the steroidogenic deformylation, and was shown to be common to a number of purified rabbit hepatic CYP enzymes.[72] To evaluate the role of a heme-peroxo intermediate in these deformylations, the ability of hydrogen peroxide, cumene hydroperoxide, meta-chloro-perbenzoic acid, and iodosobenzene to support oxidative deformylation, as surrogate oxygen donors in the peroxide shunt reaction was examined. As shown in Scheme 12 complexes of the heme iron with the peroxy oxidants indicate that only hydrogen peroxide complexed to the heme iron is nucleophilic. Additionally, iodosobenzene would either provide the electrophilic Compound I-like species or be non-nucleophilic. Of these oxidants examined, only hydrogen peroxide was effective in deformylation of the cyclohexanecarboxaldehyde.[71] This result was in complete agreement the nucleophilic peroxo oxidant proposed by Akhtar and colleagues.[54] Subsequently, 3-oxo-decalin-4-ene-9-carboxaldehyde, an analog of the A and B rings of the 19-oxo steroid intermediate in the aromatase reaction, was shown to be aromatized by CYP2B4 with stereochemistry comparable to that of the aromatase reaction (Scheme 13). This confirmed oxidative deformylation as a mechanistic model for the terminal step of the aromatase reaction.[73]

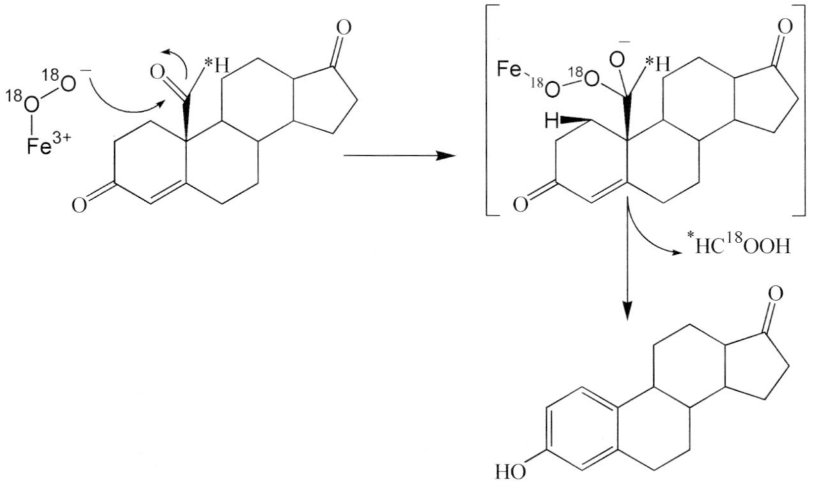

Scheme 10

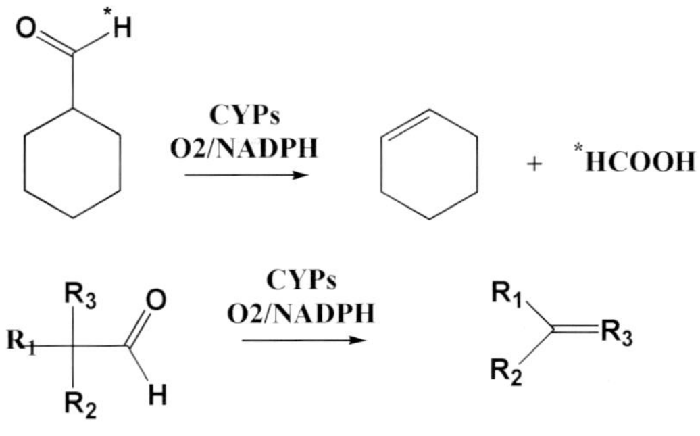

Scheme 11

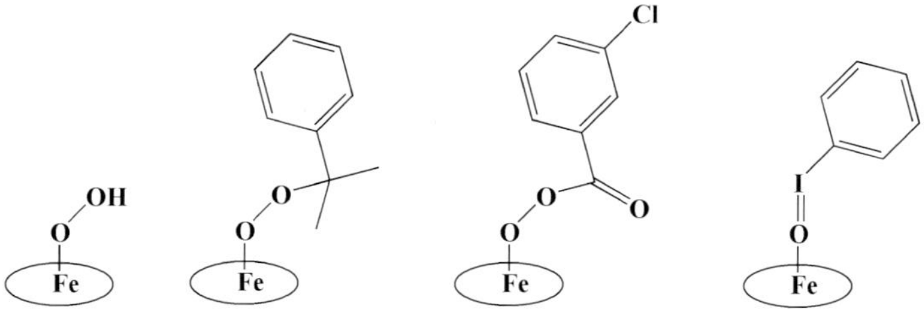

Scheme 12

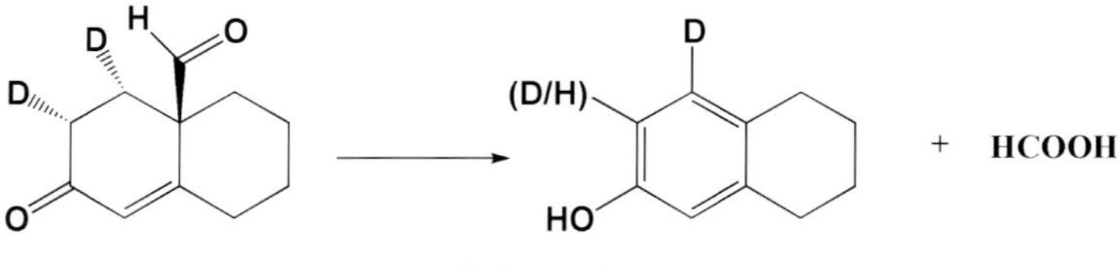

Scheme 13

Further support for a nucleophilic heme-peroxo as an active oxidant in CYP-catalyzed deformylations built on the studies of CYP101 by Poulos, Sligar, Ishimura and their co-workers.[22,74-76] From the crystal structure of CYP101 Poulos and colleagues proposed a role for the highly conserved Thr252 as the proton donor for heterolysis of the hydroperoxo heme complex. Sligar, Ishimura and their coworkers demonstrated that mutation of Thr252 to Ala resulted in complete loss of camphor hydroxylase activity.[74,75] However, this mutation had no effect on substrate binding, or on 1st and 2nd electron transfer rates to the heme-iron, and the enzyme stoichiometrically released hydrogen peroxide.[74] These authors concluded that Thr252 was indeed the active site proton donor for the activation of molecular oxygen to the putative "iron oxene". These observations were critical in the design of experiments to assess the role of a nucleophilic peroxo oxidant in the deformylation of xenobiotics. Since experiments with surrogate oxidants suggested the heme-peroxo complex to be the active oxidant in deformylation of aldehydes,[71] it was rationalized that disruption of proton delivery to the active site of CYP 2B4 should result in an increase in the heme-peroxo dependent deformylation of aldehydes, and a loss in hydroxylase activity. Sequence homology modeling shows threonine 302 in CYP2B4 to correspond to Thr252 in CYP101. Its mutation to alanine resulted in greater than 80% loss in the hydroxylation of cyclohexane and toluene and a 5-to-15-fold enhancement in the deformylation of aldehydes.[77] As predicted, this result was consistent with a nucleophilic heme-peroxo complex as the oxidant in the deformylation reaction. Akhtar and colleagues applied this rationale to the cleavage reaction catalyzed by CYP 17,20 lyase and obtained comparable results.[78]

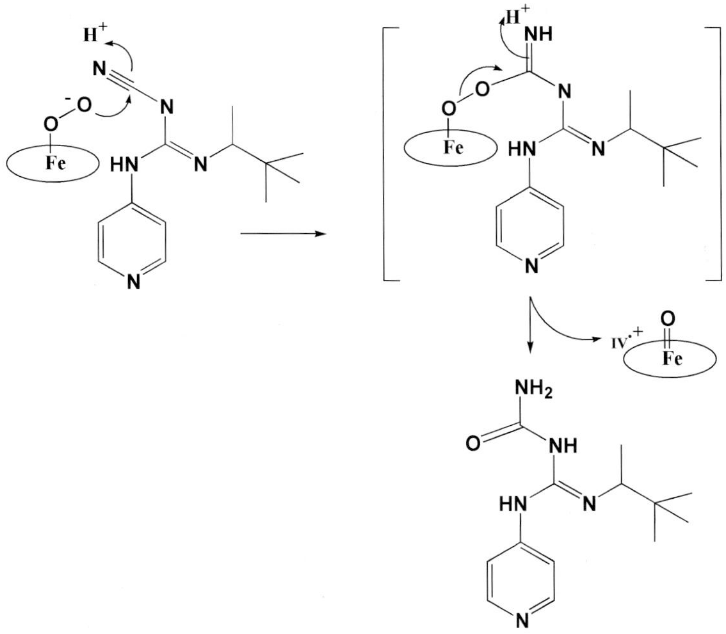

Scheme 14

The heme-peroxo as an nucleophilic oxidant species in certain types of reactions where an electrophilic reaction center is present in a substrate molecule is fairly well accepted and roles for this species have been proposed for the nitric oxide generating step of the nitric-oxide-synthase enzymes.[79-85] Recently the heme-peroxo complex has been proposed as a nucleophilic species for the hydrolysis of pinacidil by CYP3A4 (Scheme 14).[86] Studies with porphyrin biomimetics have demonstrated stable porphyrin-peroxo complexes capable of effecting nucleophilic reactions comparable to the steroidogenic and xenobiotic aldehyde deformylations.[87-91] Thus, evidence with several CYP enzymes and porphyrin models has concluded with a role for the heme-peroxo complex where electrophilic reaction centers undergo oxidative transformations via a nucleophilic heme-peroxo intermediate.

While the role for a nucleophilic heme-peroxo complex in some CYP reactions has gained acceptance, the mechanism by which the peroxyhemiacetal/ketal intermediates rearrange to form the olefin and carboxylate products remains unclarified. Oxidative deformylation / deacylation can be considered to proceed by the stepwise or concerted processes shown in Scheme 15. The stepwise process involves homolysis of the O-O bond in the peroxyhemiacetal/ketal to give an acyloxy radical intermediate and a heme-hydroxyl radical or compound II equivalent at the heme center (Scheme 15, path A). The unstable acyloxy radical intermediate undergoes rapid β-scission of the α-C-C bond to give a carbon centered radical and the carboxylate fragment. The carbon radical intermediate either undergoes hydrogen atom abstraction by the Compound II equivalent to yield the olefin, or, oxygen rebound from the Compound II equivalent to yield an alcohol. Akhtar and colleagues have argued in support of this mechanism for the cleavage reactions catalyzed by the steroidogenic CYP enzymes.[65,92] This view is based on precedence for homolytic activation of the peroxide bond by CYP enzymes as observed for peracids and hydroperoxides,[93,99] and, the observation of epitestosterone derived from progesterone in the CYP17,20 lyase reaction, a process that must derive from rebound of oxygen to the carbon radical intermediate.[92] A stepwise radical mechanism for the formation of olefins requires protein constraints on mobility of the 'caged' radical species to account for the high degree of stereochemical specificity observed in the aromatase, lanosterol-14a-demethylase, pregnenalone 17,20-lyase-16,17-desaturase, and deformylation reactions. The alternate mechanism for forming olefins from the peroxyhemiacetal/ketal intermediate favored by us involves a concerted rearrangement in a cyclic transition state with the heme-iron functioning as a Lewis acid (Scheme 15, path B). This mechanism accounts for the observed cis-stereochemical specificity observed in all the deformylation reactions without the need to invoke a protein constrained 'caged' radical intermediate. Further support for a concerted rearrangement is deduced from the deformylation of 2-methyl butraldehyde. The small size of a four carbon radical intermediate excludes significant protein constraints on orientation for hydrogen atom abstraction. Additionally, the differential reactivity of a methyl versus a methylene carbon for hydrogen atom abstraction should favor 2-butene as the reaction product. However, 1-butene was the only detected product.[72] We envision the concerted and radical pathways to be in competition, leading independently to the olefinic product and to the cleaved alcohol products, inactivation of CYPs, and heme alkylation that has been observed in the reaction of aldehydes with some CYP enzymes.[100-104] This distinction has yet to be confirmed, and the extent to which protein structure has an influence on the competing pathways remains to be established.

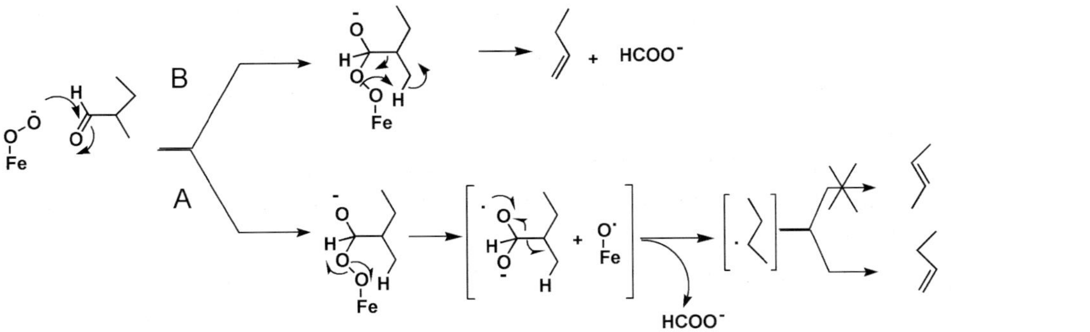

Scheme 15

2.3 *Multiple Oxidant Hypothesis in CYP-catalyzed oxidations*

Since studies had established deformylation of xenobiotic aldehydes to be common to several CYP isoforms[72] it was of interest to examine if the role of the conserved threonine was universal. Accordingly, the Thr303Ala mutant of CYP2E1, as an N-terminal truncated form, was examined in the deformylation and hydrocarbon hydroxylation reactions. As with the CYP2B4T302A mutant, the CYPD2E1T303A mutant was more effective than the wild type enzyme in deformylations. However, in contrast to CYP2B4T302A, CYP2E1T303A was not deficient in the hydroxylation of hydrocarbons.[106] These results were in apparent contradiction to the earlier results with CYP101 and CYP2B4, and indicated that at least with CYP2E1, the conserved threonine did not serve as a unique proton donor within the active site. However, the ability of CYP2E1T303A to significantly enhance oxidative deformylation when presented with a carbonyl function suggested that threonine 303 did have a role in modulating the steady state concentration of the nucleophilic heme-peroxo species.

An alternate conceptualization for the role of the active site threonine is to consider it as a regulator of proton delivery within the active site. In this capacity, and depending on the active site characteristics of the CYP isoform involved, it may function either to modulate the pKa of the heme-peroxo/ hydroperoxo species and/or regulate the rate of proton delivery to the heme-hydroperoxo species (Scheme 16). As such it would be expected that the steady state levels of the various heme-oxygen complexes would be affected. Thus, for reactions such as hydrocarbon hydroxylation, where a strong oxidant such as the terminal iron-oxene is essential, and the rate-limiting step is C-H bond cleavage, mutation of the conserved threonine may have little or no effect on the reaction rate. Whereas, for reactions where intermediates such as the heme-peroxo or heme-hydroperoxo species play a role, mutation could be expected to show significant effects. This reconsideration of the role of the conserved I-helix threonine led to the novel, and as yet controversial, Multiple Oxidant Hypothesis which considered the oxidants in CYP-catalyzed oxidations to be a continuum extending from the iron-superoxo to the iron-oxene complexes, and the CYP enzymes to be capable of using these different heme iron-bound dioxygen intermediates as oxidants, the choice in oxidant was both CYP isoform and substrate dependent (Scheme 17).[106,112]

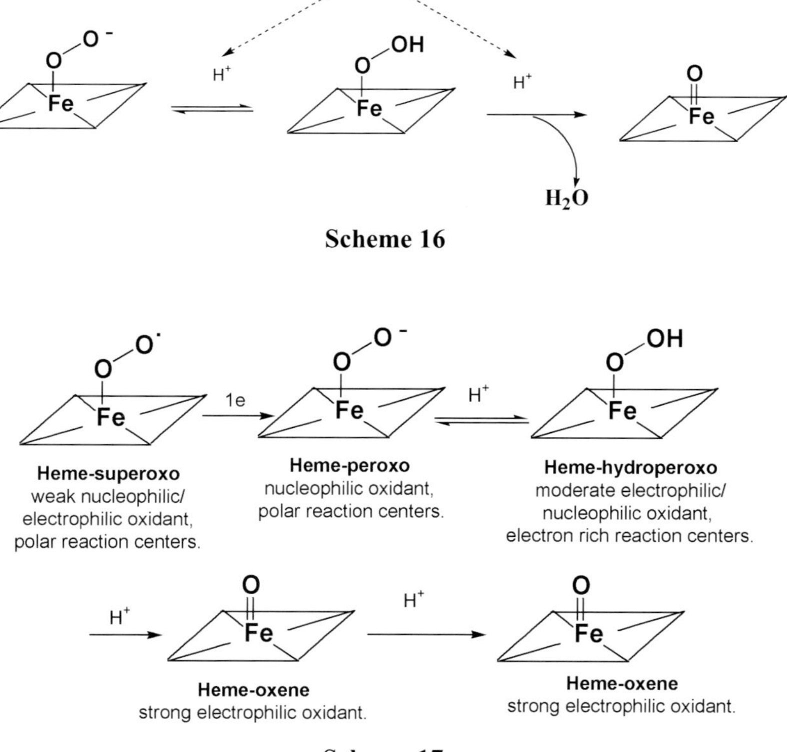

3   HEME-HYDROPEROXO AS AN OXIDANT

While the nucleophilic and electrophilic reactivities of a hydroperoxo species may be readily distinguished by the type of substrate that is being oxidized, the distinction between the electrophilic hydroperoxo and the oxene oxidants is less obvious. This is particularly true with non-polar electrophilic reaction centers such as olefins that could be oxidized by either oxidant. A kinetic approach based on the isotope sensitive branching theory developed to determine isotope effects in CYP-catalyzed reactions was used to distinguish between electrophilic oxidations effected by the heme-hydroperoxo and heme-oxene complexes.[105] These kinetic approaches are shown in Scheme 18. As shown in the scheme if two oxidation products (P1 and P2) are formed from a common oxidant species [ES], then the ratio of products ([P1]/[P2]) will be independent of the concentration of the oxidant effecting the two reactions. The product ratio depends only on the intrinsic reactivity of the oxidant towards either reaction product. Alternatively, if the two products are derived from distinct oxidant species (ES and ES*), then the ratio of products will depend on the effective concentration of the respective oxidants and their respective intrinsic reactivity towards the reaction they effect.[106] As discussed earlier, mutation of the conserved threonine may alter proton transfer rates or the pK_a of the heme-peroxo species, thus altering the equilibrium concentrations of the oxidant species. Therefore, if a single oxidant is responsible for both products, the wild type and mutant enzymes should produce the same ratio of these products, since this is controlled only by the intrinsic reactivity of the single oxidant species towards the two reactions. If however, the two products are derived from different oxidant species, then the ratio of products should be different for the wild type and mutant proteins. The direction of shift in the product ratio from wild type to mutant enzyme will depend on the equilibrium concentration of the dominant oxidant species in the steady state.

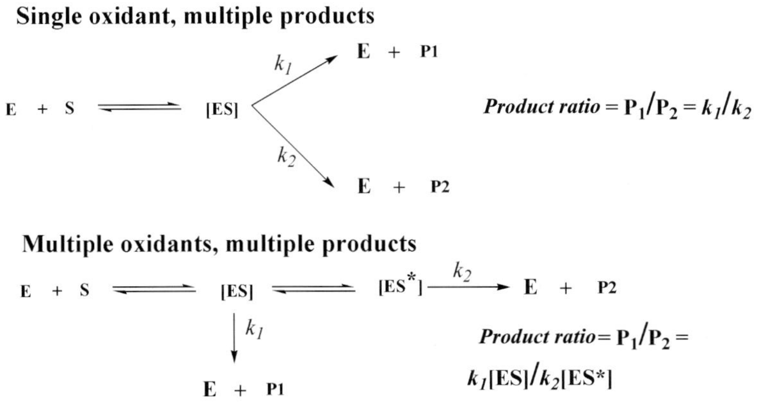

Single oxidant, multiple products

$$E + S \rightleftharpoons [ES] \quad\Big\langle\quad \begin{matrix} \xrightarrow{k_1} E + P1 \\ \xrightarrow{k_2} E + P2 \end{matrix}$$

Product ratio $= P_1/P_2 = k_1/k_2$

Multiple oxidants, multiple products

$$E + S \rightleftharpoons [ES] \rightleftharpoons [ES^*] \xrightarrow{k_2} E + P2$$

$$[ES] \xrightarrow{k_1} E + P1$$

Product ratio $= P_1/P_2 = k_1[ES]/k_2[ES^*]$

Scheme 18

2-Methylbutane, *cis-* and, *trans*-2-butenes, and cyclohexene are substrates that provided unique characteristics to examine this hypothesis. As shown in Scheme 19- 2-methylbutane can only undergo hydroxylation reactions. This molecule provides nine primary, two secondary, and one tertiary C-H sites that can undergo hydroxylation. When this substrate was examined with CYP2E1 and CYP2E1T303A, the ratio of primary to secondary pentanols was not influenced by mutation of the conserved active site threonine. This result is consistent with a single oxidant species effecting the same reaction in wild type and mutant enzymes.[106] Cyclohexene, *cis-* and *trans*-2-butenes, provide allylic C-H and olefinic bonds that have distinct reactivities within the same molecule. Additionally, *cis-* and *trans*-2-butenes can provide an insight into the geometry of the transition state for epoxidation. Since the energy barrier for rotation about a single bond is ~3kcal/mole, disruption of the olefinic bond to a radical or radical-cation type intermediate in the transition state should result in epimerization and loss of geometrical integrity in the epoxide product. Furthermore, these molecules are small with no polar functional groups that may allow for protein constraints within the active site. When these molecules were examined with CYP2E1 and CYP2E1T303A, the mutant enzyme showed a dramatic preference for epoxidation over allylic hydroxylation.[106] Also, *cis-* and *trans*-2-butenes yielded exclusively *cis* and *trans* 2,3-epoxybutanes with both enzymes (Scheme 20). Taken together, these results strongly suggest that the epoxidation and allylic hydroxylation reactions are affected by more than one oxidant, and that the transition state for epoxidation does not involve disruption of the double bond to a radical intermediate. The heme-hydroperoxo intermediate was proposed to be this oxidant (Scheme 21).[106] With CYP2B4 and CYP2B4T303A the ratio of allylic alcohol to epoxide was comparable for cyclohexene whereas for both 2-butenes the allylic alcohol was detectable but below quantitative limits. Hence the role of a hydroperoxo species as the oxidant for epoxidation could not be distinctly established for CYP2B4. By the kinetic principle used to distinguish oxidants in this analysis, the absolute rates for epoxidation or allylic hydroxylation are not correct measures for the role of any oxidant species.[4] In studies on the sulfoxidation of p-tolyl-methyl sulfide, CYP2E1T303A was shown to be significantly more effective in sulfoxidation than benzylic hydroxylation when compared to CYP2E1 (McGinnity and Vaz, unpublished). This reaction is comparable to the epoxidation and allylic hydroxylation discussed above and suggestive of a similar role for the heme hydroperoxo species in sulfoxidation reactions.

Recent studies with CYP biomimetics have revealed that porphyrin-based models also use multiple electrophilic oxidants. Collman and colleagues reported that in competitive reactions the ratio of oxidation products from

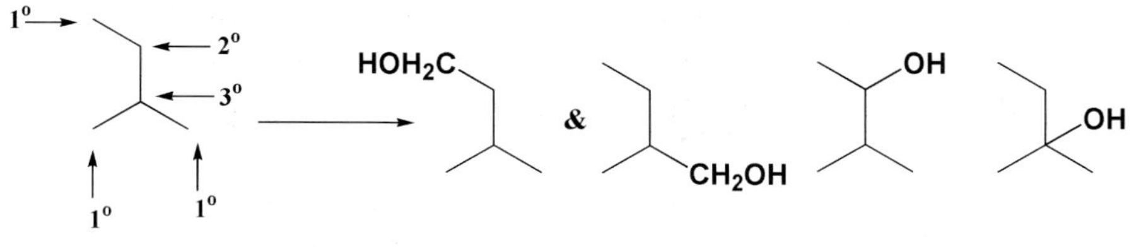

2-Methyl butane

Scheme 19

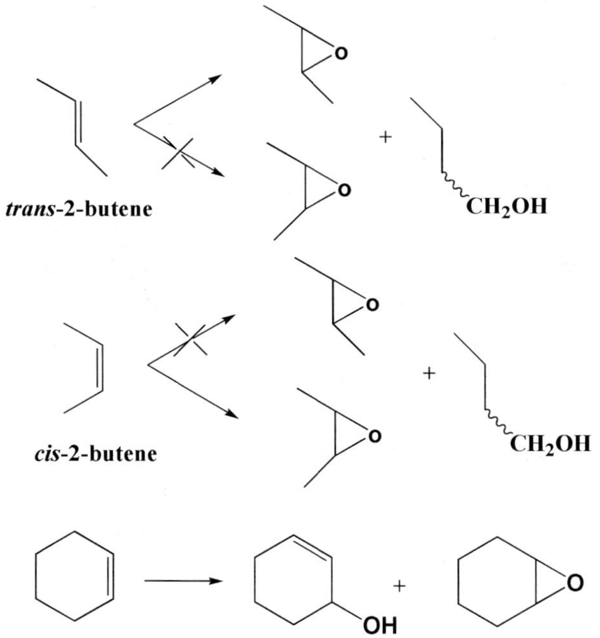

trans-2-butene

cis-2-butene

Scheme 20

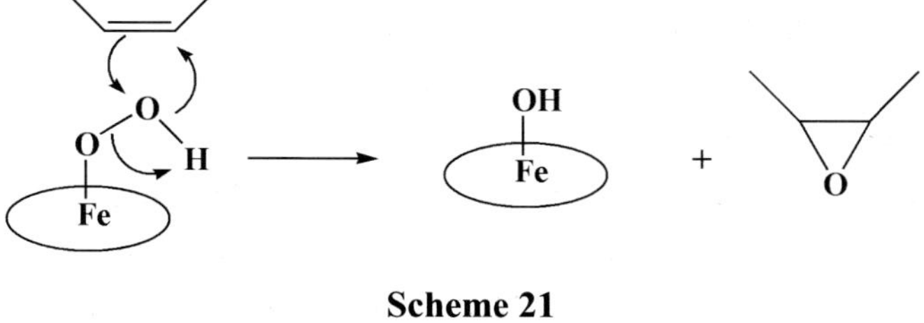

Scheme 21

different substrates was dependent on the oxidant used, suggesting that a unique oxidant was not responsible.[107] Nam and colleagues using porphyrin complexes in aprotic media have concluded that a hydroperoxo porphyrin complex was the intermediate responsible for epoxidation of cyclohexene.[108] More recently Suzuki and colleagues used thiolate ligated porphyrin models of CYP enzymes with a series of *para* substituted peroxybenzoic acids as oxygen donors in the hydroxylation/epoxidation of cyclooctane and cyclooctene.[109] Their results indicate that the rate ratios of alkane hydroxylation to alkene epoxidation were dependent on the electronics of the *para*-substituent on the oxidant. These authors concluded that substrate and oxidant interact with each other during the oxygen atom transfer reaction, i.e., the oxidation reaction occurs before O-O bond cleavage. Volz and colleagues have examined kinetic isotope effects in the N-dealkylation and sulfoxidation of *para*-N,N-dimethylamino-phenyl-methyl-sulfide by P450BM3 and its conserved threonine to alanine mutant (BM3T268A), and have essentially concluded that multiple electrophilic oxidants, that are not interchangeable, must be responsible for the sulfoxidation and N-dealkylation reactions.[110] However, these authors have refrained from attributing the sulfoxidation to the heme hydroperoxo complex. In a joint report the Dawson and Sligar laboratories have reported that the heme-hydroperoxo species of CYP101T252A is the species responsible for epoxidizing an olefinic camphor derivative.[111] This quintessential CYP protein has been the basis for much of our own work with the rabbit liver CYP isoforms that led to the "multiple oxidant hypothesis."[77,106,112] Thus, growing evidence in the literature now points to the heme-hydroperoxo species as a competent oxidant in electrophilic reactions effected by CYP enzymes. The scope of this oxidant in relation to the extensive body of literature on electrophilic oxidations affected by CYP proteins has yet to be evaluated.

A reaction relatively recently identified to be catalyzed by CYP enzymes is the *ipso* substitution of *para*- substituted phenols and anilines.[113,114] This reaction is of particular significance to drug metabolism since the metabolites can be quinonoid species that can react with cellular components, or redox cycle, resulting in oxidative stress. In preliminary studies on the *ipso* substitution of *para*- substituted phenols we reported the CYP2E1(T303A) mutant to be more effective than the wild type CYP2E1, a result we interpreted to involve the heme-hydroperoxo complex as the active oxidant (Scheme 22).[115] Follow-up studies by Vatsis and Coon have substantiated the preliminary findings, and established that the enhancement by the CYP2E1T303A mutant is common to a series of *para*- substituted phenols.[116] Interestingly pH dependence of the reactions between the wild type and mutant CYP2E1 enzymes were different and dependent on the substrate examined.

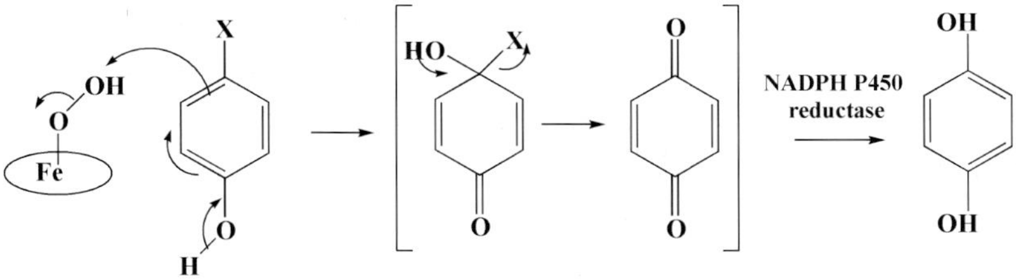

Scheme 22

In a recent study on the *para* hydroxylation of 3,5-dideuterio-aniline by CYP2E1 and CYP2E1T303A, the mutant enzyme was found to be more active than the wild type enzyme, and with both systems deuterium was completely retained in the hydroxylated product. These results indicate that no NIH shift occurs in the formation of the hydroxylated product and suggests a role for the heme hydroperoxo intermediate as the oxidant with activated aromatic rings as shown in Scheme 23 (Vaz, unpublished results).

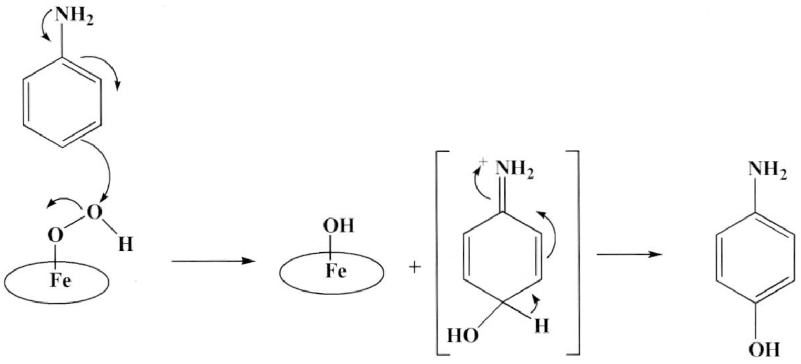

Scheme 23

An expansion for the role of the heme-hydroperoxo species in the hydroxylation of C-H bonds has evolved from the work of Newcomb and colleagues with the use of "radical clock" substrates and CYPs 2B4, 2E1 and their respective active site conserved threonine to alanine mutants previously used to propose a role for the heme-hydroperoxo species in the epoxidation of olefins.[106] When the radical clocks *trans*1-phenyl-2-methyl-cyclopropane and its *para*-trifluoromethyl derivative (Scheme 24) were examined with CYPs 2B4 and 2E1 and their respective conserved threonine to alanine mutants, hydroxylation at the methyl carbon was significantly decreased with the mutants of both isoforms, and phenolic hydroxylation increased. The *para*-trifluoromethyl probe which cannot be ring hydroxylated gave methyl group hydroxylation and cyclopropyl ring-opened products. Cyclopropyl ring-opened products predominated with the mutant enzymes. A similar result was obtained with the hypersensitive alkoxyl cyclopropyl probes where reaction products were derived from a carbocation intermediate (Scheme 7). Taken together these results were interpreted to implicate the insertion of hydroxyl into the C-H bond by the heme-hydroperoxo group to yield a protonated alcohol.[46,53,117]

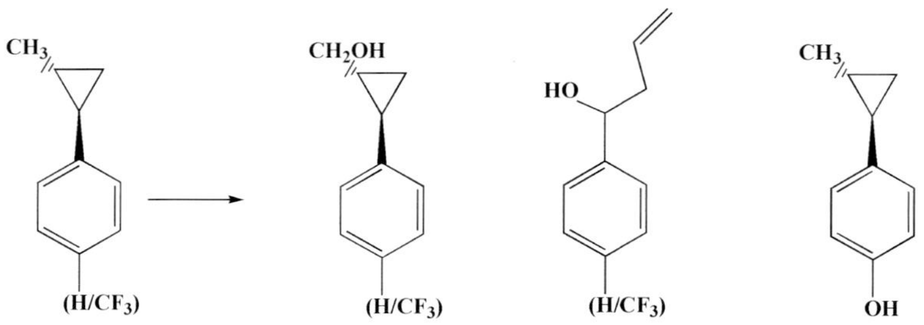

Scheme 24

4   HEME-SUPEROXO AS ACTIVE OXIDANT

The heme-superoxo complex is the first activated dioxygen species formed in the CYP reaction cycle. In contrast to hemoglobin and myoglobin where the complex is a ferrous-dioxygen species, calculations suggest that the CYP oxygen complex is a ferric-superoxide species.[118] This is corroborated by cryogenic studies of the oxygenated singly reduced CYP101 protein.[14-16] The radical anion nature of superoxide suggests it can function as a radical, a base, or a nucleophile. Nanni and colleagues proposed that some oxidation reactions attributed to superoxide in aprotic solvents are indirect, resulting from proton abstraction by superoxide anion and subsequent oxidation by the disproportionation products, H_2O_2.[119,120] Nucleophilic reactions of superoxide are generally seen in aprotic, non-hydrogen bonding solvents. The radical character of superoxide has been demonstrated to function in direct oxidations of compounds with readily abstractable hydrogen atoms.[119] The oxidation of N-hydroxyphentermine and N-(2-methyl-1-phenyl-2-propyl)hydroxylamine by superoxide to the respective nitroxide intermediates was shown to involve direct hydrogen atom abstraction by superoxide.[121,122] Superoxide dismutase inhibition of biological oxidations is generally used to implicate superoxide as the oxidant. However, when bound to a metal ion, as in the heme-superoxo complex of CYP enzymes, superoxide dismutase would not be expected to show any effect. The most relevant CYP-catalyzed oxidations that have been implicated to involve the heme-superoxide complex are the NOS-catalyzed oxidation of N-hydroxy-arginine to nitric oxide [79-81] and CYP-catalyzed cleavage of N-hydroxy guanidines, amidoximes and ketoximes.[84,85,123] Scheme 25 shows current thinking on the role of the heme-superoxo species in CYP- and nitric-oxide-synthase catalyzed reactions.

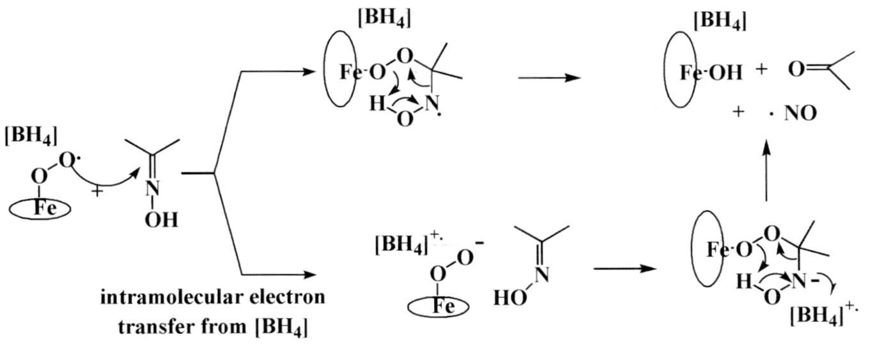

Scheme 25

5   THE TWO-STATE THEORY

Following the initial proposal that the heme-hydroperoxo complex was the likely oxidant in the epoxidation of olefins, Shaik and colleagues and others calculated the activation energy difference between a CYP-derived heme-hydroperoxo and a Compound-I-like species to effect epoxidation of ethylene is between 23-39 kcal in favor of the Compound-I-like species.[17] Shaik and colleagues have proposed the "two-state-reactivity" (TSR) theory that attempts to harmonize the apparent discrepancies in various experimental results that are inconsistent with the oxygen rebound mechanism.[17,124-128] According to this theory the heme-oxene exists in two closely lying electronic states, a quartet (high spin) and a doublet (low spin). In the oxygen rebound mechanism these states are competitive, and accordingly, product distributions are dependent on the relative reactivity of the two states. Hydroxylations effected by the high spin state are stepwise, giving a transition cluster of a substrate and heme-hydroxo radicals. This cluster collapses to the hydroxylated product. Hydroxylations effected by the low spin doublet state are effectively concerted, with the transition cluster undergoing essentially a barrierless collapse to the hydroxylated product.[126] Since the low spin state also involves a transition cluster, discreet radical intermediates cannot be ruled out by this oxidant state. However, the primary source of radical intermediates must be from the high spin (quartet state) of the oxidant. Un-rearranged products are possible from both states whereas rearranged products arise essentially from the high spin state. Thus, the failure to obtain a consistent measure of the oxygen rebound rate with radical clocks with radical rearrangement rates from 1×10^{10} to 1×10^{13} sec^{-1} may be accounted for by the relative contribution of the high and low spin states of the Compound I-like species to the oxidation of these substrates.

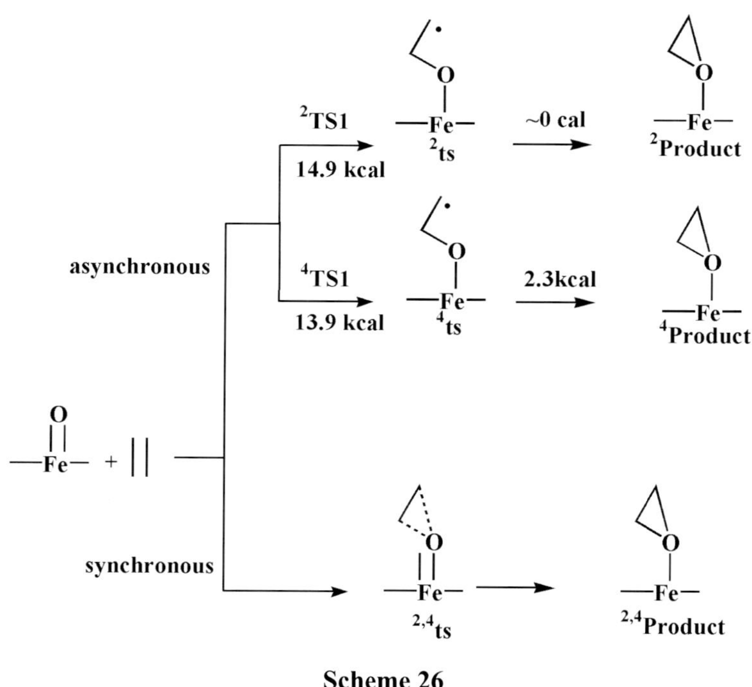

Scheme 26

This theory has also been applied to the epoxidation of olefins.[129] Scheme 26 shows the results of DFT calculations for the epoxidation of ethylene, which can be effected by either synchronous or asynchronous pathways. The synchronous (doublet or quartet states) pathway is found to be energetically less favorable by 7.3 kcal/mole than the more energetic of the asynchronous pathways (⁴TS1). The two spin state pathways for the asynchronous pathway have slightly different energetics. The activation energies for the low and high spin states (²TS1 and ⁴TS1) are 14.9 kcal mole^{-1} and 13.9 kcal mole^{-1}, respectively. The low spin transition state (^{2}ts) has an essentially barrier less collapse to the low spin heme iron coordinated product, whereas, the corresponding high spin transition state (^{4}ts) has a barrier of ~2.3 kcal mole^{-1}. The retention of stereochemistry in the epoxidation of olefins by CYP enzymes has been attributed to the very low energy for collapse of the transition states in the asynchronous processes thus making them stepwise but apparently concerted when considered from a stereochemical viewpoint.

The two-state theory provides a computational basis for many of the observed results that led to the multiple oxidant hypothesis. Recent experimental results with the conserved threonine to alanine mutant of CYP101 that is devoid of hydroxylase activity has been shown to be effective in the epoxidation of an olefinic derivative of camphor. This mutant while incapable of making a compound-I-like intermediate can make the heme-hydroperoxo intermediate and is now proposed as the oxidant in the epoxidation reaction.[111] Recent kinetic isotope effect studies with radical clock probes reveal inconsistencies between the experimental results and the computed values expected by the two-state theory.[130]

CONCLUSIONS

Studies conducted within the past decade on the mechanism of CYP-catalyzed oxidations suggest a complex process where mono-oxygenation reactions may be effected by all of the intermediates in the course of activation of molecular oxygen to the putative ultimate oxygenating species – the heme-oxene. This may in part account for the diversity of reactions effected by CYP enzymes. The extent to which these intermediates are involved in drug metabolism has yet to be determined. A few examples are known where the heme-superoxo and heme-peroxo intermediates are the oxidant species. The role of a heme-hydroperoxo species in electrophilic oxidations is controversial. A computational approach suggests that electrophilic oxidations are effected only by a compound-I-like species with two spin states and with distinct chemical reactivity, whereas several lines of experimental evidence point to the contrary. Clearly, further experimentation is necessary to explore these alternatives.

REFERENCES

1. Danielson, P. B. *Curr. Drug. Metab.* **2002**, *3*, 561-597.
2. De Mot, R.; Parret, A. H. A. *Trends in Microbiology* **2002**, *10*, 502-508.
3. Roberts, G. A.; Grogan, G.; Greter, A.; Flitsch, S. L.; Turner, N. J. *Journal of Bacteriology* **2002**, *184*, 3898-3908.

4. Ortiz de Montellano Paul, R.; De Voss James, J. *Natural Product Reports* **2002**, *19*, 477-493.

5. White, R. E.; Coon, M. J. *J. Biol. Chem.* **1982**, *257*, 3073-3083.

6. Wells, A. V.; Li, P.; Champion, P. M.; Martinis, S. A.; Sligar, S. G. *Biochem.* **1992**, *31*, 4384-4393.

7. Auclair, K.; Moeenne-Loccoz, P.; Ortiz de Montellano, P. R. *J. Amer. Chem. Soc.* **2001**, *123*, 4877-4885.

8. Vatsis, K. P.; Peng, H.-M.; Coon, M. J. *J. Inorg. Biochem.* **2002**, *91*, 542-553.

9. Ullrich, V. *Archives of Biochemistry and Biophysics* **2002**, *409*, 45-51.

10. Usuda, K.; Toritsuka, N.; Matsuo, Y.; Kim, D. H.; Shoun, H. *Appl. Environ. Microbio.* **1995**, *61*, 883-889.

11. Honeychurch, M. J.; Hill, H. A. O.; Wong, L.-L. *FEBS Letters* **1999**, *451*, 351-353.

12. Kuthan, H.; Tsuji, H.; Graf, H.; Ullrich, V.; Werringloer, J.; Estabrook, R. W. *FEBS Letters* **1978**, *91*, 343-345.

13. Puntarulo, S.; Cederbaum, A. I. *Free Radical Biology & Medicine* **1998**, *24*, 1324-1330.

14. Schlichting, I.; Berendzen, J.; Chu, K.; Stock, A. M.; Maves, S. A.; Benson, D. E.; Sweet, R. M.; Ringe, D.; Petsko, G. A.; Sligar, S. G. *Science* **2000**, *287*, 1615-1622.

15. Davydov, R.; Macdonald, I. D. G.; Makris, T. M.; Sligar, S. G.; Hoffman, B. M. *J. Amer. Chem. Soc.* **1999**, *121*, 10654-10655.

16. Denisov, I. G.; Makris, T. M.; Sligar, S. G. *J. Biol. Chem.* **2001**, *276*, 11648-11652.

17. Ogliaro, F.; de Visser, S. P.; Cohen, S.; Sharma, P. K.; Shaik, S. *J. Amer. Chem. Soc.* **2002**, *124*, 2806-2817.

18. Harris, D. L.; Loew, G. H. *J. Amer. Chem. Soc.* **1998**, *120*, 8941-8948.

19. Kamachi, T.; Shiota, Y.; Ohta, T.; Yoshizawa, K. *Bulletin of the Chemical Society of Japan* **2003**, *76*, 721-732.

20. Pond, A. E.; Ledbetter, A. P.; Sono, M.; Goodin, D. B.; Dawson, J. H. *Electron Transfer in Chemistry* **2001**, *3*, 56-104.

21. Poulos, T. L. *Porphyrin Handbook* **2000**, *4*, 189-218.

22. Raag, R.; Martinis, S. A.; Sligar, S. G.; Poulos, T. L. *Biochem.* **1991**, *30*, 11420-11429.

23. Vidakovic, M.; Sligar, S. G.; Li, H.; Poulos, T. L. *Biochem.* **1998**, *37*, 9211-9219.

24. Xiang, H.; Tschirret-Guth, R. A.; Ortiz De Montellano, P. R. *J. Biol. Chem.* **2000**, *275*, 35999-36006.

25. Hamilton, G. A.; Giacin, J. R.; Hellman, T. M.; Snook, M. E.; Weller, J. W. *Ann. N.Y. Academ. Sci.* **1973**, *212*, 4-12.

26. White, R. E.; Miller, J. P.; Favreau, L. V.; Bhattacharyya, A. *J. Amer. Chem. Soc.***1986**, *108*, 6024-6031.

27. White, R. E.; Groves, J. T.; McClusky, G. A. *Acta Biologica et Medica Germanica* **1979**, *38*, 475-482.

28. Groves, J. T.; McClusky, G. A.; White, R. E.; Coon, M. J. *Biochem. Biophys. Res. Comm.* **1978**, *81*, 154-160.

29. Groves, J. T.; Adhyam, D. V. *J. Amer. Chem. Soc.* **1984**, *106*, 2177-2181.
30. Hjelmeland, L. M.; Aronow, L.; Trudell, J. R. *Molec. Pharmacol.* **1977**, *13*, 634-639.
31. Chance, B.; Sies, H.; Boveris, A. *Physiological Reviews* **1979**, *59*, 527-605.
32. Gold, A.; Weiss, R. *Journal of Porphyrins and Phthalocyanines* **2000**, *4*, 344-349.
33. Ortiz de Montellano, P. R. *Cytochrome P450: Structure, Mechanism, Biochemistry.*; 2nd Edition ed.; Plenum , New York, 1995.
34. Blake, R. C., 2nd; Coon, M. J. *J. Biol. Chem.* **1980**, *255*, 4100-4111.
35. Blake, R. C., 2nd; Coon, M. J. *J. Biol. Chem.* **1981**, *256*, 12127-12133.
36. Schunemann, V.; Jung, C.; Trautwein, A. X.; Mandon, D.; Weiss, R. *FEBS Letters* **2000**, *479*, 149-154.
37. Sono, M.; Eble, K. S.; Dawson, J. H.; Hager, L. P. *J. Biol. Chem.* **1985**, *260*, 15530-15535.
38. Shimada, H.; Sligar, S. G.; Yeom, H.; Ishimura, Y. *Catalysis by Metal Complexes* **1997**, *19*, 195-221.
39. Blake, R. C., 2nd; Coon, M. J. *J. Biol. Chem.* **1989**, *264*, 3694-3701.
40. Bowry, V. W.; Lusztyk, J.; Ingold, K. U. *J. Amer. Chem. Soc.* **1991**, *113*, 5687-5698.
41. Bowry, V.; Lusztyk, J.; Ingold, K. U. *Pure and Applied Chemistry* **1990**, *62*, 213-216.
42. Atkinson, J. K.; Ingold, K. U. *Biochem.* **1993**, *32*, 9209-9214.
43. Newcomb, M.; Chestney, D. L. *J. Amer. Chem. Soc.* **1994**, *116*, 9753-9754.
44. Ortiz de Montellano, P. R.; Stearns, R. A. *J. Amer. Chem. Soc.* **1987**, *109*, 3415-3420.
45. Newcomb, M.; Le Tadic, M.-H.; Putt, D. A.; Hollenberg, P. F. *J. Amer. Chem. Soc.* **1995**, *117*, 3312-3313.
46. Newcomb, M.; Le Tadic-Biadatti, M.-H.; Chestney, D. L.; Roberts, E. S.; Hollenberg, P. F. *J. Amer. Chem. Soc.* **1995**, *117*, 12085-12091.
47. Newcomb, M.; Shen, R.; Lu, Y.; Coon, M. J.; Hollenberg, P. F.; Kopp, D. A.; Lippard, S. J. *J. Amer. Chem. Soc.* **2002**, *124*, 6879-6886.
48. Newcomb, M.; Toy, P. H. *Accounts of Chemical Research* **2000**, *33*, 449-455.
49. Auclair, K.; Hu, Z.; Little, D. M.; Ortiz de Montellano, P. R.; Groves, J. T. *J. Amer. Chem. Soc.* **2002**, *124*, 6020-6027.
50. Eaton, P. E.; Yip, Y. C. *J. Amer. Chem. Soc.* **1991**, *113*, 7692-7697.
51. Newcomb, M.; Shen, R.; Choi, S.-Y.; Toy, P. H.; Hollenberg, P. F.; Vaz, A. D. N.; Coon, M. J. *J. Amer. Chem. Soc.* **2000**, *122*, 2677-2686.
52. Tadic-Biadatti, M.-H. L.; Newcomb, M. *J. Chem. Soc., Perkin Trans. 2: Phys. Org. Chem.* **1996**, 1467-1473.
53. Newcomb, M.; Hollenberg, P. F.; Coon, M. J. *Arch. Biochem. Biophys.* **2002**, *409*, 72-79.
54. Akhtar, M.; Calder, M. R.; Corina, D. L.; Wright, J. N. *Biochem. J.* **1982**, *201*, 569-580.
55. Korzekwa, K. R.; Trager, W. F.; Smith, S. J.; Osawa, Y.; Gillette, J. R. *Biochem.* **1991**, *30*, 6155-6162.
56. Watanabe, Y.; Ishimura, Y. *J. Amer. Chem. Soc.* **1989**, *111*, 8047-8049.

57. Watanabe, Y.; Ishimura, Y. *J. Amer. Chem. Soc.* **1989**, *111*, 410-411.
58. Covey, D. F.; Hood, W. F. *Cancer Research* **1982**, *42*, 3327-3333.
59. Beusen, D. D.; Carrell, H. L.; Covey, D. F. *Biochem.* **1987**, *26*, 7833-7841.
60. Fishman, J. *Cancer Research* **1982**, *42*, 3277s-3280s.
61. Fishman, J.; Goto, J. *J. Biol. Chem.* **1981**, *256*, 4466-4471.
62. Goto, J.; Fishman, J. *Science* **1977**, *195*, 80-81.
63. Osawa, Y. *Excerpta Med. Int. Congr. Ser.* **1973**, *273*, 814-819.
64. Shyadehi, A. Z.; Lamb, D. C.; Kelly, S. L.; Kelly, D. E.; Schunck, W.-H.; Wright, J. N.; Corina, D.; Akhtar, M. *J. Biol. Chem.* **1996**, *271*, 12445-12450.
65. Akhtar, M.; Lee-Robichaud, P.; Akhtar, M. E.; Wright, J. N. *J. Steroid Biochem. and Molec. Bio.* **1997**, *61*, 127-132.
66. Akhtar, M.; Lee-Robichaud, P.; Wright, J. N.; Akhtar, M. E. *Biochem. Soc. Trans.* **1998**, *26*, 322-325.
67. Cole, P. A.; Robinson, C. H. *J. Amer. Chem. Soc.* **1988**, *110*, 1284-1285.
68. Cole, P. A.; Robinson, C. H. *Journal of Medic. Chem.* **1990**, *33*, 2933-2942.
69. Cole, P. A.; Robinson, C. H. *J. Chem. Soc., Perkin Trans. 1* **1990**, 2119-2125.
70. Cole, P. A.; Robinson, C. H. *J. Amer. Chem. Soc.* **1991**, *113*, 8130-8137.
71. Vaz, A. D. N.; Roberts, E. A.; Coon, M. J. *J. Amer. Chem. Soc.* **1991**, *113*, 5886-5887.
72. Roberts, E. S.; Vaz, A. D.; Coon, M. J. *Proceed. Nat. Acad. Sci. USA* **1991**, *88*, 8963-8966.
73. Vaz, A. D. N.; Kessell, K. J.; Coon, M. J. *Biochemistry* **1994**, *33*, 13651-13661.
74. Martinis, S. A.; Atkins, W. M.; Stayton, P. S.; Sligar, S. G. *J. Amer. Chem. Soc.* **1989**, *111*, 9252-9253.
75. Imai, M.; Shimada, H.; Watanabe, Y.; Matsushima-Hibiya, Y.; Makino, R.; Koga, H.; Horiuchi, T.; Ishimura, Y. *Proceed. Nat. Acad. Sci. USA* **1989**, *86*, 7823-7827.
76. Raag, R.; Poulos, T. L. *Biochem.* **1989**, *28*, 7586-7592.
77. Vaz, A. D. N.; Pernecky, S. J.; Raner, G. M.; Coon, M. J. *Proceed. Nat. Acad. Sci. USA* **1996**, *93*, 4644-4648.
78. Lee-Robichaud, P.; Akhtar, M. E.; Akhtar, M. *Biochem. J.* **1998**, *330*, 967-974.
79. Marletta, M. A.; Hurshman, A. R.; Rusche, K. M. *Curr. Opin. Chem. Bio.* **1998**, *2*, 656-663.
80. Marletta, M. A. *Advances in Experimental Medicine and Biology* **1993**, *338*, 281-284.
81. Marletta, M. A. *J. Biol. Chem.* **1993**, *268*, 12231-12234.
82. Pufahl, R. A.; Wishnok, J. S.; Marletta, M. A. *Biochem.* **1995**, *34*, 1930-1941.
83. Clague, M. J.; Wishnok, J. S.; Marletta, M. A. *Biochem.* **1997**, *36*, 14465-14473.
84. Mansuy, D.; Boucher, J.-L. *Drug Metabolism Reviews* **2002**, *34*, 593-606.
85. Mansuy, D.; Boucher, J. L.; Clement, B. *Biochimie* **1995**, *77*, 661-667.
86. Zhang, Z.; Li, Y.; Stearns, R. A.; Ortiz de Montellano, P. R.; Baillie, T. A.; Tang, W. *Biochem.* **2002**, *41*, 2712-2718.
87. Wertz, D. L.; Sisemore, M. F.; Selke, M.; Driscoll, J.; Valentine, J. S. *J. Amer. Chem. Soc.* **1998**, *120*, 5331-5332.
88. Wertz, D. L.; Valentine, J. S. *Structure and Bonding (Berlin)* **2000**, *97*, 37-60.
89. Selke, M.; Valentine, J. S. *J. Amer. Chem. Soc.* **1998**, *120*, 2652-2653.

90. Sisemore, M. F.; Selke, M.; Burstyn, J. N.; Valentine, J. S. *Inorg. Chem.* **1997**, *36*, 979-984.

91. Meunier, B.; Bernadou, J. *Structure and Bonding (Berlin)* **2000**, *97*, 1-35.

92. Lee-Robichaud, P.; Shyadehi, A. Z.; Wright, J. N.; Akhtar, M.; Akhtar, M. *Biochem.* **1995**, *34*, 14104-14113.

93. White, R. E. S., Stephen G.; Coon, Minor J *J. Biol. Chem.* **1980**, *255*, 11108-11111.

94. Barr, D. P. M., Martha V.; Guengerich, F. Peter; Mason, Ronald P. *Chem. Res. Toxicol.* **1996**, *9*, 318-325.

95. Correia, M. A. Y., Kunquan; Allentoff, Alban J.; Wrighton, Steven A.; Thompson, John A. *Arch. Biochem. Biophys.* **1995**, *317*, 471-478.

96. Shimizu, T. M., Yoshinori; Hatano, Masahiro. *J. Biol. Chem.* **1994**, *269*, 13296-13304.

97. White, R. E.; Sligar, S. G.; Coon, M. J. *J. Biol. Chem.* 10 **1980**, *255*, 11108-11111.

98. Shimizu, T.; Murakami, Y.; Hatano, M. *J. Biol. Chem.* **1994**, *269*, 13296-13304.

99. Barr, D. P.; Martin, M. V.; Guengerich, F. P.; Mason, R. P. *Chem. Res. Toxocol.* **1996**, *9*, 318-325.

100. Bestervelt, L. L.; Vaz, A. D. N.; Coon, M. J. *Proceed. Nat. Acad. Sci. USA* **1995**, *92*, 3764-3768.

101. Kuo, C. L.; Raner, G. M.; Vaz, A. D.; Coon, M. J. *Biochem.* **1999**, *38*, 10511-10518.

102. Kuo, C. L.; Vaz, A. D.; Coon, M. J. *J. Biol. Chem.* **1997**, *272*, 22611-22616.

103. Raner, G. M.; Vaz, A. D. N.; Coon, M. J. *Molecular Pharmacology* **1996**, *49*, 515-522.

104. Raner, G. M.; Chiang, E. W.; Vaz, A. D.; Coon, M. J. *Biochem.* **1997**, *36*, 4895-4902.

105. Korzekwa, K. R.; Trager, W. F.; Gillette, J. R. *Biochem.* **1989**, *28*, 9012-9018.

106. Vaz, A. D. N.; McGinnity, D. F.; Coon, M. J. *Proceed. Nat. Acad. Sci. USA* **1998**, *95*, 3555-3560.

107. Collman, J. P.; Chien, A. S.; Eberspacher, T. A.; Brauman, J. I. *J. Amer. Chem. Soc.* **2000**, *122*, 11098-11100.

108. Nam, W.; Lee, H. J.; Oh, S.-Y.; Kim, C.; Jang, H. G. *Journal of Inorganic Biochem.* **2000**, *80*, 219-225.

109. Suzuki, N.; Higuchi, T.; Nagano, T. *J. Amer. Chem. Soc.* **2002**, *124*, 9622-9628.

110. Volz, T. J.; Rock, D. A.; Jones, J. P. *J. Amer. Chem. Soc.* **2002**, *124*, 9724-9725.

111. Jin, S.; Makris, T. M.; Bryson, T. A.; Sligar, S. G.; Dawson, J. H. *J. Amer. Chem. Soc.* **2003**, *125*, 3406-3407.

112. Vaz, A. D. N. *Curr. Drug Metab.* **2001**, *2*, 1-16.

113. Ohe, T.; Mashino, T.; Hirobe, M. *Drug Metabolism and Disposition 25* **1997**, 116-122.

114. Ohe, T.; Mashino, T.; Hirobe, M. *Arch. Biochem. Biophys.* **1994**, *310*, 402-409.

115. Vaz, A. D. N.; Coon, M. J. In *12th International Symposium on Microsomes and Drug Oxidations,*: Montpellier, France, 1998, p Abstract #410.

116. Vatsis Kostas, P.; Coon Minor, J. *Arch. Biochem. Biophys.* **2002**, *397*, 119-129.

117. Toy, P. H.; Newcomb, M.; Coon, M. J.; Vaz, A. D. N. *J. Amer. Chem. Soc.* **1998**, *120*, 9718-9719.

118. Segall, M. D.; Payne, M. C.; Ellis, S. W.; Tucker, G. T.; Eddershaw, P. J. *Xenobiotica* **1999**, *29*, 561-571.

119. Nanni, E. J., Jr.; Sawyer, D. T. *J. Amer. Chem. Soc.* **1980**, *102*, 7591-7593.

120. Nanni, E. J., Jr.; Stallings, M. D.; Sawyer, D. T. *J. Amer. Chem. Soc.* **1980**, *102*, 4481-4485.

121. Duncan, J. D.; Di Stefano, E. W.; Miwa, G. T.; Cho, A. K. *Biochem.* **1985**, *24*, 4155-4161.

122. Fukuto, J. M.; Di Stefano, E. W.; Burstyn, J. N.; Valentine, J. S.; Cho, A. K. *Biochem.* **1985**, *24*, 4161-4167.

123. Jousserandot, A.; Boucher, J.-L.; Henry, Y.; Niklaus, B.; Clement, B.; Mansuy, D. *Biochem.* **1998**, *37*, 17179-17191.

124. Filatov, M.; Harris, N.; Shaik, S. *J. Chem. Soc., Perkin Trans. 2: Phys. Org. Chem.* **1999**, 399-410.

125. Filatov, M.; Shaik, S. *J. Phys. Chem. A* **1998**, *102*, 3835-3846.

126. Ogliaro, F.; Harris, N.; Cohen, S.; Filatov, M.; de Visser, S. P.; Shaik, S. *J. Amer. Chem. Soc.* **2000**, *122*, 8977-8989.

127. Ogliaro, F.; Cohen, S.; de Visser, S. P.; Shaik, S. *J. Amer. Chem. Soc.* **2000**, *122*, 12892-12893.

128. Schoneboom Jan, C.; Lin, H.; Reuter, N.; Thiel, W.; Cohen, S.; Ogliaro, F.; Shaik, S. *J. Amer. Chem. Soc.* **2002**, *124*, 8142-8151.

129. de Visser, S. P.; Ogliaro, F.; Harris, N.; Shaik, S. *J. Amer. Chem. Soc.* **2001**, *123*, 3037-3047.

130. Newcomb, M.; Aebisher, D.; Shen, R.; Chandrasena, R. E. P.; Hollenberg, P. F.; Coon, M. J. *J. Amer. Chem. Soc.* **2003**, *125*, 6064-6065.

Chapter 2

Application of LC/MS, LC/NMR, NMR and Stable Isotopes in Identifying and Characterizing Metabolites

A. E. Mutlib[1] and John P. Shockcor[2]

[1]*Pharmacokinetics, Dynamics and Metabolism, Pfizer Global Research and Development, Ann Arbor Laboratories, 2800 Plymouth Road, Ann Arbor, MI 48105*

[2]*Metabometrix Ltd., Prince Consort Rd., South Kensington, London SW7 2BP, United Kingdom*

1   INTRODUCTION

The metabolic disposition of xenobiotics is of considerable interest to scientists from several disciplines including medicinal chemists, toxicologists and pharmacologists. Historically one of the objectives of a medicinal chemist was to deliver useful therapeutic agents with desirable pharmacokinetic properties. However, often compounds exhibiting respectable pharmacokinetics have failed during development or after marketing because of undesirable side effects or unexpected toxicities. Often these side effects have been attributed to the metabolites formed from the parent compound. Because of this, the formation and subsequent characterization of metabolites of drugs has attracted considerable attention during the last few years. Consequently the metabolic disposition of compounds is seriously considered during the search for potentially useful therapeutic agents.

Recently, drug discovery and development efforts have been directed at elucidating the metabolic soft spots present on compounds. It is common to investigate the extent of metabolism of a new chemical entity (NCE) as well as the nature of the products formed. This is frequently achieved by carrying out *in vitro* metabolism studies on a NCE employing liver subcellular fractions (such as microsomal, cytosolic and S9), liver slices, hepatocytes and cDNA-expressed enzymes from various species. Hepatocytes are gaining popularity, especially with the availability of cryopreserved tissues, in studying biotransformation pathways and differences in the metabolism of xenobiotics among species. Despite this, liver microsomes remain the most widely used *in vitro* preparations to study the metabolic disposition of compounds for a number of reasons. The metabolic half-life and clearance of compounds are usually determined by incubations performed with liver microsomes.[1,2] A short half-life usually warrants further studies to elucidate the metabolic soft spot on the molecule. In the event that the short half-life is attributed to the production of a single major product, efforts are subsequently made to fully elucidate the identity of this metabolite. It becomes important to know the exact structure of the metabolite so that the medicinal chemist can make appropriate modifications to produce a more metabolically stable analog. If multiple products are formed during the biotransformation process then it becomes more difficult to characterize the structures of the metabolites. It is equally important to demonstrate that the NCE under investigation is not selectively metabolized extensively by only one particular species, metabolic system or enzyme class. An extensive conversion of a NCE to a metabolite by rodents does not necessarily mean that other species will dispose the compound in a similar manner. Hence the metabolism of a NCE by liver subcellular fractions or hepatocytes from various species is usually performed. Several studies have shown that compounds tend to be metabolized differently by various species.[3] In some instances, the species differences in the metabolic disposition of compounds have been utilized to demonstrate selective toxicities.[4]

The progress of a compound from discovery to development is often accompanied by safety studies in different species. It is during these safety evaluations that a number of compounds are eliminated due to their inherent toxicities. Toxic and carcinogenic responses for some drugs are produced solely by the parent compounds, whereas for others, the responses arise as a result of the formation of reactive metabolites. The formation of reactive intermediates from various compounds has been directly or indirectly linked to toxicities observed in various species.[5-11] Hence it becomes important for the drug metabolism chemist to investigate the likelihood that reactive metabolic intermediate(s) are formed from a NCE. It is desirable to minimize the risk of toxicities by decreasing the probability of forming reactive intermediates from a NCE. Attempts are in progress to characterize potential reactive intermediates formed by performing microsomal incubations of NCEs. Due to the instability and transient existence of the reactive intermediates, their presence is often demonstrated by trapping experiments with a nucleophile such as glutathione (GSH). The GSH adducts are usually fully characterized because knowledge of the structures of these conjugates provide valuable insight into the possible identity of the reactive chemical intermediates from which they are derived.

The possibility of a NCE being metabolized to a pharmacologically active compound also exists. For several years, biotransformation was considered synonymous with the inactivation of pharmacologically active compounds. However, it has been demonstrated that metabolites of some drugs are pharmacologically active. There are several known cases of marketed drugs that were found to be metabolized to an active moiety. At times the metabolites have been found to possess better pharmacological activity or a superior toxicity profile than the compounds from which they were derived. A classical example is terfenadine, an antihistamine agent, which caused prolongation of the cardiac QT interval.[12,13] The characterization of the major metabolite of terfenadine led to successful development of a compound (Allegra®) that is considered as a better alternative to terfenadine in treating allergies. Another example is acetaminophen, the *O*-dealkylated metabolite of phenacetin. Acetaminophen possesses better analgesic activity compared to phenacetin. Furthermore, acetaminophen does not produce methemoglobinemia and hemolytic anemia as phenacetin does.[14] Although phase I reactions can frequently give rise to pharmacologically active metabolites, phase II conjugation reactions can also produce biologically active metabolites. Formation of pharmacologically active morphine 6-glucuronide and minoxidil sulfate are two such examples.[15,16] Hence it would be desirable to characterize metabolites of NCEs considered for development. The isolated metabolites or synthetic standards of the metabolites could then be tested for pharmacological activity. This will also ensure that an adequate patent coverage is also obtained for the pharmacologically active metabolites, which could potentially be developed as therapeutic agents in the future.

During the drug discovery and development, it becomes vitally important to identify and characterize the structures of metabolites. Elucidating the structures of metabolites as early as possible during drug discovery appears logical. A considerable amount of time and resources can be saved if the role of metabolism was considered during the compound selection process. It is the authors' intent to illustrate, with examples, how the metabolites of various structurally diverse groups of compounds have been isolated and characterized. Furthermore, the implications of metabolite characterization will be discussed. It is pointless to identify and characterize metabolites if one does not make full use of the data obtained. One must question how the metabolites are formed from the parent molecules once the structures are identified. The pharmacological activity and the potential toxicity of the metabolites can also be investigated once the structures of metabolites are identified.

The overall biotransformation schematics for a compound can only be achieved by comprehensive characterization of all the metabolites produced. This is only accomplished by judicious selection of analytical techniques and a thorough knowledge of the potential chemical reactions that a molecule may undergo during the biotransformation processes. Likewise, the availability of synthetic standards and radio- and stable isotope labeled compounds can accelerate the identification of metabolites, especially in complex biological matrices. The chromatographic techniques that are frequently used for the purposes of isolating metabolites are often under emphasized. Frequently, obtaining metabolite profiles in complex biological matrices does not necessarily require complete separation of components. However, separation of various components is required when one needs to isolate metabolites for characterization by various spectroscopic techniques such as nuclear magnetic resonance (NMR). An attempt will be made to demonstrate the utility of the separation procedures used for metabolite identification purposes. Additionally, in this chapter the analytical techniques that have been extensively utilized in the past to elucidate the structures of metabolites will be discussed, followed by specific examples of compounds whose metabolism were elucidated utilizing these techniques.

2   SEPARATION OF METABOLITES FROM ENDOGENOUS COMPONENTS

Separating the metabolites of a compound from endogenous components present in a complex biological fluid can be a daunting and laborious task. However, with the advent of robust liquid chromatography/mass spectrometry (LC/MS) techniques, the task of physically separating each component is often not required. This is especially true if complete identification of the metabolites is not required. Such is usually the case in very early discovery stages whereby the LC/MS profiles of *in vitro* and *in vivo* extracts suffice. These LC/MS profiles are usually carried out to show the presence of compound related peaks in biological extracts. The mass spectrum for each peak usually shows the pseudomolecular ion (see below), which frequently suggests the biotransformation pathway that led to that metabolite. For example, an initial LC/MS profile of bile or urine obtained from an animal dosed with a compound would show if there were any metabolites formed by hydroxylation, glucuronidation, sulfation or other metabolic pathway based on the pseudomolecular ions alone. At this stage further MS/MS studies are performed to gain additional insight into the structural makeup of the metabolite. Frequently, there is a need to isolate a metabolite(s) for additional characterization by other techniques, such as NMR. A brief description of the separation techniques that we have used in our laboratories over the last few years has been provided. The overall impact of these techniques will be substantiated by specific examples later in the chapter.

2.1   Sample Clean-up on Solid Phase Cartridges

One of the most frequently used separation techniques for a variety of compounds of varyingious degrees of polarity, polarity has been solid phase cartridges packed with silica based C18 material available from several vendors. The following procedures have been frequently used to perform preliminary clean up or purification of biological samples containing metabolites. Cartridges packed with other polymeric materials such as an ion or cation exchanger, or with different carbon loading (C2, C8, C18 etc), have also been utilized to carry out purifications of samples.

Bile and Urine: Bile duct-cannulated rats have frequently been used for the purpose of isolating and characterizing metabolites of compounds. Rats are given the highest tolerable dose (non-toxic) of the compound followed by collection of bile and urine over 24 hr periods. It is highly desirable to administer the animals multiple doses over several days in the event that the metabolite profile changes with time (such as due to induction). The samples are collected over ice and stored refrigerated until used. An initial LC/MS profile of the bile or urine is done to get an idea on the polarity and retention of metabolites on a C18 column. In the case of very polar compounds alternate cartridge types are used for sample extraction and purification. A gradient HPLC system utilizing a linear ramp from 0% organic (such as acetonitrile) to 80-90% organic is performed on a C18 column to obtain the profile of metabolites excreted in urine or bile. The aqueous component of the HPLC mobile phase is usually water containing a small percentage of either an organic acid (e.g. 0.01% formic acid) or a volatile buffer (e.g., 10 mM ammonium formate). It is now common to use small diameter columns (e.g., 2.0 mm) available in various lengths (e.g., 5 – 25 cm) to perform LC/MS analyses of biological samples. The flow rate is typically 0.1 – 0.4 ml/min on these columns and the eluent from the HPLC column is directly introduced into the ion source of the mass spectrometer. The LC/MS profiling of samples provide useful information on the relative polarities of metabolites and their retention on the C18 column. This information is then used to judge the conditions needed to separate various metabolites from the endogenous components present in the biological samples. A common method used for pre-cleanup of samples using SPE cartridges is outlined below.

The urine and bile samples are usually loaded onto a preconditioned C18 cartridge (10 grams/60 cc) and eluted under gravity. If radiolabeled samples are available, the retention of the compound and its metabolites can be easily determined. However in many situations, studies are conducted with non-labeled compounds and care is needed to ensure that the cartridges are not overloaded which would result in the loss of potentially useful metabolites. After the samples have been eluted through the cartridge, a simple wash with deionized water (2 x column volume) is done to remove any inorganics and salts. Again, if radiolabeled samples are being used, one can determine the loss of compound related material during this step. The cartridges are subsequently eluted with 20-30 ml of solvents consisting of a mixture of an organic (such as methanol or acetonitrile) and an aqueous solvent (water or an acid solution). The percentage of the organic solvent is increased in an incremental manner (e.g., 5%, 10%, 15%, 20%, etc.). The fractions are collected and subsequently analyzed by LC/MS for the presence of metabolites. If sufficient quantities of the metabolites are present, flow injection analysis (FIA) of the fractions is adequate. Usually a 1:1 mixture of an organic and an aqueous solution (e.g., methanol : 0.01% formic acid) is used for the FIA. However, in certain cases, individual LC/MS runs are performed for each fraction to ensure that the presence of metabolites is not obscured due to ion suppression. This is especially important when dealing with low-level metabolites that co-elute with endogenous components present in bile or urine. Once the metabolites are located in the C18 fractions, the next step is to concentrate the samples and perform further purifications. The volumes of the solvents are usually reduced using nitrogen as the inert gas with one of the commercially available evaporators such as TurboVap (Zymark). The samples from the same fraction are pooled, dried and further purified using either a semi-preparative column or a packed cartridge (e.g., C18). If the fraction shows significant amounts of endogenous components, it is further purified on a C18 cartridge using different eluting solvent mixtures than the initial one. For example if the initial purification was done using mixtures of acetonitrile and 0.1% formic acid, the second clean-up on the C18 is done using mixtures of acetonitrile and 0.1% acetic acid. This usually leads to changes in the retention times of the various components in the mixture. Sometimes separation of components can be achieved using a packing material other than C18. The C18 fractionation leads to "zoning" of various metabolites in different fractions. The 5% methanol fraction will obviously contain more polar metabolites than the 50% fraction. This fractionation of metabolites then allows suitable isocratic HPLC methods to be developed quickly on a semi-preparative column that will allow adequate separation of metabolites from each other and from the endogenous components. For example the 5-10% methanol fraction, when dried and reconstituted in water, can be loaded onto a semi-preparative column (300 × 10 mm) using a mobile phase consisting of methanol:water (5:95 v/v) delivered at 3.5 ml/min. Again a slight variation in the mobile phase (acetonitrile instead of methanol) can lead to a fairly good separation of components in the mixture. One must be aware of the fact that acetonitrile will elute components earlier on a C18 column than methanol. The HPLC peaks are often collected using a fraction collector. Several injections are made so that adequate amounts of metabolites are collected for characterization purposes, especially for NMR analyses. If the levels of the metabolites are sufficiently high (at least 10 µg), the samples are usually suitable for a high-field NMR (500 or 600 MHz) analyses. The fractions, which are collected in clean 250 ml Erlenmeyer flasks, are dried in a tared conical centrifuge tube using a Speed-Vac concentrator. A fairly good idea of the amount of metabolite collected from the HPLC is obtained once the sample is dried. Frequently, one must remove the HPLC mobile phase salts or buffers after this stage. This is typically done by reconstituting the sample in water and extracting on a C18 cartridge (100 mg/3 cm^3) using only methanol and water as the eluting solvents. Absolute care in handling the samples is needed at this stage to ensure that the isolated metabolite is not contaminated during the final step of removing the HPLC salts. One must ensure that the extraction manifold is clean and free of contamination. Some manifolds have disposable polyethylene tips that can be replaced before the clean up of the sample takes place. The C18 cartridge is conditioned with methanol (2 × 2 ml) followed by a wash with water (2 × 2 ml) before loading the sample. The cartridges are then washed with water (2 × 2ml) to remove the salts before eluting with methanol. Again, the final elution can be done with much lower percentage of methanol if the metabolites are fairly polar in nature. Elution with 100% methanol is only recommended for very non-polar compounds because there is a possibility of column bleed leading to interferences in the NMR spectrum if a high percentage organic is used. The eluent from the cartridge is then dried thoroughly before NMR analyses.

3   LIQUID CHROMATOGRAPHY/MASS SPECTROMETRY

The ability to characterize metabolites has been greatly accelerated with the introduction of versatile analytical techniques such as liquid chromatography/mass spectrometry (LC/MS) and liquid chromatography/nuclear magnetic resonance (LC/NMR) spectroscopy. In the past, metabolites had to be isolated from biological matrices such as bile or urine so that information on the molecular weights and possible structures could be obtained by direct probe mass spectrometric techniques.[17-19] This was a tedious and laborious approach whereby several chromatographic runs were required before a metabolite could be subjected to mass spectral analysis. With the advent of more robust atmospheric pressure ionization techniques such a electrospray ionization for LC/MS,[20] the steps needed to purify samples for initial mass spectral analysis have been eliminated. It is routine these days to inject biological samples without any purification onto a HPLC column coupled to a mass spectrometer. This has been made possible with the design of better interfaces between the HPLC systems and the ion-inlet of the mass spectrometers. Furthermore, if one diverts the eluent from the HPLC column to a waste container for the first few minutes, the chances of successfully identifying metabolites are greatly improved. This procedure ensures that the salts in the sample, usually running with the solvent front, do not reach the ion source and hence suppress ionization. Again, it is wise to also divert the solvent to waste once all the metabolites have eluted from the column. This is especially important if a gradient run is performed and a lot of endogenous components are eluted from the column with the mobile phase consisting of a high percentage of organic component. Most mass spectrometers are now equipped with divert valves that can be activated at various times during the analysis of samples. The two most widely used interfaces for metabolite characterization purposes are pneumatically assisted electrospray [21-23] and atmospheric pressure chemical ionization (APCI) sources.[24-26] The application of these techniques in obtaining structural information of metabolites has been adequately demonstrated in the literature. Electrospray ionization (ESI) is a very gentle technique, which makes it ideal for analyzing fragile metabolites such as glutathione (GSH) or other peptide conjugates.[27,28] Similarly, APCI is suited for compounds that are not amenable to analysis by ESI technique. Because of the higher temperatures employed with APCI, it might not be a suitable technique to search for fragile metabolites. Nonetheless, in our experience, APCI has, in most cases, been able to provide the pseudomolecular ions of metabolite conjugates including GSH, glucuronide, sulfate and peptide adducts. However, because of its soft ionization, ESI-LC/MS is the method of choice in most cases. Both of these interfaces are interchangeable on the front end of most mass spectrometers and with the ability to switch ionization modes (positive and negative); at least four different options are available to test the appropriate mode of analysis of samples. Mass spectrometers are continuously improving in design, robustness and sensitivity so that they are able to handle diverse sample types and provide as much structural information as possible. Additionally, the software has been improved considerably so that it has become much easier to trace and track the metabolites of a compound in a complex mixture.

Three types of mass spectrometers are now considered complimentary in elucidating the structures of metabolites: the multi-stage quadrupole, the ion trap and the time-of-flight instruments. Each of these mass spectrometers exhibit different capabilities and limitations. A comprehensive discussion on the different types of the mass spectrometers and their capabilities is beyond the scope of this chapter. The reader is referred to other suitable publications, which discuss the techniques of mass spectrometry in greater detail.[29,30] Until recently, triple quadrupole mass spectrometers were the most widely used instruments to carry out LC/MS analysis of biological extracts. Most of these mass spectrometers can be fitted with either APCI or ESI sources and are used to provide pseudomolecular ions during the LC/MS analyses of samples. Frequently the MS/MS capability of these quadrupole instruments are used to get an idea of the potential site(s) of modifications on the molecule. At times, advantage is taken of the ability to fragment ions at the front-end of the instrument so that MS/MS/MS experiments may be performed. This provides additional structural information. The quadrupole instruments also allow neutral loss and parent ion scans which are useful techniques in detecting metabolites in complex mixtures. Ion trap instruments are gaining popularity in identifying metabolites in biological matrices. The ability of ion traps to handle multiple consecutive MS/MS experiments (MS^n) has accelerated their acceptance in laboratories dedicated to metabolite identification and characterization. The hybrid quadrupole-time of flight (Q-TOF) instruments provide accurate masses of metabolites, hence aiding in the structural elucidation process. The accurate masses can be used as a quasi-elemental analysis for metabolites in question. This is especially useful if a number of possible structures of metabolites exist for a given average monoisotopic mass. For example, one can use accurate mass measurements to deduce if a compound had been methylated (+14 amu) or oxidized to a keto-functional group (also +14 amu). The MS/MS capability also allows one to obtain accurate masses of the fragments, which enhances structural assignments. Hence a combination of these various types of mass spectrometers in the drug metabolism laboratory is extremely helpful in obtaining structural information as early as possible.

The mass spectrometers are used to provide the molecular weights of the metabolites. These are usually obtained as pseudomolecular ions, MH^+ or $[M-H]^-$, corresponding to positive or negative ion modes of operation, respectively. The softer ionization techniques allow the metabolism chemists to obtain these pseudomolecular ions with minimum fragmentation. The metabolites are easily distinguished from the substrate by observing characteristic increments or mass shifts from the parent pseudomolecular ion. Some of the metabolites produced by Phase I and II metabolic reactions are listed in Tables 1 and 2, respectively. While the lists are by no means exhaustive, these could be used by a novice metabolism chemist to investigate various metabolic pathways of a compound. A combination of metabolic reactions leading to secondary metabolites can lead to mass shifts that are not listed in these two tables. In addition to the expected pseudomolecular ions, one must be able to locate solvent (e.g. acetonitrile, +41 amu; NH_4^+, +17) and ion adducts (e.g., Na^+, +22) that could complicate an easy interpretation of mass spectral data. Sometimes analyzing samples using another eluting solvent (e.g. formic acid instead of ammonium formate) may simplify the mass spectra by eliminating the presence of ions produced by solvent adducts. Furthermore, one should try to avoid salts such as phosphates or NaCl when analyzing samples. This can be achieved by a preliminary "cleanup" of samples using solid phase cartridges (see above).

Table 1. Identification of Phase I metabolites by observing the shifts in MH$^+$ of the substrate*

Mass shift from the parent MH$^+$	Metabolic Reactions	Comments
+48	Addition of three oxygens (+16,+16,+16).	Rare but can detect these metabolites as very polar compounds. Further oxidation can lead to +46 and +44 mass shifts.
+34	Oxidation to form an epoxide which then hydrolyzes to dihydrodiol (+16, +18)	Can have two additions of oxygens (+16,+16) with a reduction (+2H) of a carbonyl group or a double bond.
+32	Addition of two oxygen (+16,+16) atoms such as the formation of sulfone from sulfide. Formation of dihydroxylated metabolites and catechols.	Can also explain other metabolic reactions such as ring openings by hydrolysis (+18) and further oxidation (+14); e.g. α-oxidation on an alicyclic amine with subsequent ring opening and further oxidation of aldehyde to carboxylic acid.
+30	Addition of two oxygen atoms with a loss of 2 hydrogens (by further oxidation); e.g. conversion of aliphatic methyl to an aldehyde followed by further oxidation to carboxylic acid.	Can add two oxygen atoms at two different positions on the molecule, followed by further oxidation of one of the hydroxyls (aliphatic) to a carbonyl functional group.
+28	Addition of two oxygen atoms followed by further oxidations leading to a loss of 4 hydrogens; e.g., oxidation of catechol to quinone	Formation of O-methyl catechols
+16	Hydroxylation to alcohols, phenols, epoxides, hydroxylamines. Oxidation to N-oxides, S-oxides and sulfones, oxidation of aldehyde to carboxylic acids	
+14	Oxidation at aliphatic position followed by further oxidation to a carbonyl; α-oxidation of cyclic amines to form lactam; oxidation of alcohol to carboxylic acid	N- and O-methylation
+12	Several possibilities: Can add +30 (as above) with a loss of water (-18). Oxidation of methyl to carboxylic acid (+30) followed by cyclization (-18) with amine or alcohol to form lactam or lactone, respectively.	
+3	Reductive cleavage of isoxazole followed by hydrolysis of imine to carbonyl	Nitrogen rule in getting the odd number of mass addition.
+2	Reduction of a carbonyl to a hydroxyl; reduction of aldehyde to an alcohol.	
+1	Hydrolysis of amide to form a carboxylic acid	Hydrolysis of imine to a carbonyl. Replacement of N with O by any metabolic process
-28	Loss of two methyls by oxidative demethylation; oxidative de-ethylation	
-18	Loss of water through dehydration, or intramolecular ester/amide formation	
-14	Reduction of carboxylic acid to alcohol (through the aldehyde intermediate), oxidative demethylation	
-2	Formation of an unsaturated bond through dehydrogenation reactions and by addition of oxygen followed by loss of water (+16, -18)	Oxidative defluorination can also lead to a net loss of 2 amu. Loss of water can be through some internal cyclization process such as lactam formation.

* This is not an exhaustive list. Other mass shifts can result from secondary metabolism, unusual metabolic pathways and combinations of metabolic reactions.

Table 2. Identification of Phase II metabolites by observing the shifts in MH^+ of the substrate[**]

Specific Conjugation Reactions		
Mass shift from the parent MH^+	Metabolic Reactions	Comments
+499	Formation of diconjugate-hydroxylation (+16), addition of GSH (+307) and glucuronide (+176).	Usually very polar metabolites. If GSH added to the molecule by substitution, should see +497 mass shift. If another hydroxylation, +515 observed.
+483	Formation of diconjugate-GSH (+307) and glucuronide (+176).	Michael addition or addition of GSH across a double bond leading to a total addition of 307 amu.
+481	Formation of diconjugate-GSH (+305) and glucuronide.	Substitution of a leaving group with GSH (+305)
+403	Hydroxylation (+16), addition of GSH (+307) and sulfation (+80).	Diconjugate formation. Additional hydroxylation can lead to +419 mass shift.

The metabolites described above can be degraded by mercapturic acid pathway leading to diconjugates showing cysteinylglycine, cysteine and N-acetylcysteine moieties. In such cases, substitute the mass value for the GSH (+307 or 305) with +177, 120 and 163 for cysteinylglycine, cysteine and N-acetylcysteine, respectively.

+323	Hydroxylation (+16), addition of GSH (+307)	Mass of 321 could be due to GSH substitution (+305)
+307	Addition of GSH (+307)	Substitution with GSH (+305)
+272	Diconjugate –formed by hydroxylation (+16), glucuronidation (+176) and sulfation (+80)	Very polar metabolites. Can be detected either in positive or negative ion modes
+256	Diconjugate – glucuronide (+176) and sulfate (+80)	
+220	Carbamyl glucuronide of amines (+44, +176)	Fragile and can be unstable. Loss of 44 amu seen with these conjugates. Can also produce ammonia adducts (+17 amu) masking the true MH^+ (net +237 shift)
+192	Hydroxylation (+16) and glucuronidation (+176)	Hydroxylation of substrates followed by glucuronidation is quite common. Aliphatic and aromatic compounds.
+176	Glucuronidation of phenols, alcohols (ethers), carboxylic acids (acyl glucuronide), N-glucuronides of 1°, 2°, 3° amines, S-glucuronides, C-glucuronides	C- and S- glucuronides are rare
+162	Glycosidation of alcohols and amines.	Glycosides of 1° amines seen in plasma of various species.
+129	Glutamates of amines or GSH conjugates (glutamic acid pathway)	Addition of 129 amu to GSH conjugates quite commonly seen as minor metabolites.
+143	Carnitine conjugates of carboxylic acids	Will observe M^+ due to the presence of tertiary amine.
+107	Taurine conjugate of carboxylic acids	
+96	Hydroxylation (+16) and sulfation (+80)	
+80	Sulfation of an existing hydroxyl functional group	
+57	Glycine conjugate of carboxylic acids	
+42	Acetylation 1°, 2° amines	
+14	O- and N-methylation	

[**] This is not an exhaustive list. Other mass shifts can result from unusual metabolic pathways and combinations of metabolic reactions.

Detecting metabolites in a complex biological matrix such as rat bile can be a challenge. However, as stated earlier, a softer ionization technique such as ESI has made it feasible to obtain the molecular weights of compounds relatively easily. Frequently the technique of neutral loss analyses is performed during the search of metabolites. These neutral loss studies are most easily performed with triple quadrupole instruments that are configured in such a way to monitor the losses of neutral fragments during collision induced fragmentations. For example, the presence of glucuronide, sulfate, taurine or GSH conjugates are easily detected by monitoring losses of ions at m/z 194 or 176, 80, 125 and 129, respectively. Neutral loss is commonly used to detect reactive metabolites (trapped by GSH) of compounds by monitoring the loss of glutamate (-129 amu) from the GSH adducts. Separation of components in a complex mixture is essential because ion suppression by endogenous components can lead to an important metabolite being overlooked. Hence it is reasonable to resolve all the components in a complex mixture when analyzing for metabolites for the first time. A high-throughput approach with a relatively short analysis time is not the best way of obtaining metabolite profiles for a compound. Often the HPLC systems linked to the mass spectrometers are equipped with variable wavelength detectors. Simultaneous monitoring of UV signals with total ion current (TIC) can provide very useful data, especially for metabolites with UV absorbing chromphores. Frequently, investigators have failed to utilize both of these detectors simultaneously to obtain metabolite profiles in biological samples. At times, the metabolites produced from a compound may not ionize as well as the parent substrate, and therefore can be easily overlooked if TIC was only used to monitor for the presence of metabolites. Furthermore, a small metabolite can also produce an exaggerated mass spectral response that can lead to over interpretation of data. Hence it is useful to look at the two sets of signals (UV and TIC) simultaneously during metabolite profiling. This technique is very useful for metabolite studies performed for compounds in discovery stages where radiolabeled compounds may not be available. Compounds that do not posses chromophores can be problematic if relying only on one of these two techniques to identify metabolites. In such cases, derivatization or use of radiolabeled compound may be needed to perform reliable metabolite profiling. The presence of natural isotopes of atoms such as chlorine in a molecule can also greatly assist in metabolite profiling. The metabolite profiling of a compound such as efavirenz was greatly accelerated by the presence of a chlorine atom in the molecule.[3] The metabolites were easily identified by observing the characteristic ion pattern in the mass spectra. The "twin ions' showed a ratio consistent with the natural abundance of ^{35}Cl and ^{37}Cl isotopes (67 and 33%, respectively). Stable isotope labeled (e.g. deuterium) compounds, if available in the discovery stages can also be used to detect metabolites in complex mixtures. Usually a 1:1 mixture of non-labeled and labeled compounds are mixed in equimolar quantities and metabolism studies performed. This allows the metabolism chemist to distinguish the drug related material from endogenous components in a complex biological extract by observing the appearance of "twin ions". These ions are usually separated from each other depending on the number of deuteriums incorporated in the labeled compound. If the mass spectrum of a metabolite shows a "twin ion" pattern different from the substrate, that would suggest a loss of label. This provides very useful information as to the possible site(s) of metabolism. For example, if a 1:1 mixture of toluene and deuterium labeled toluene (label on methyl group) was used for metabolism studies, one would expect to see MH^+ at m/z at 92 and 95 in a 1:1 ratio in the mass spectrum of parent compound. If a metabolite was detected that showed ions at m/z 108 and 111, that would suggest hydroxylation on the aromatic ring (since not labeled on the ring system). However, if a metabolite showed ions at m/z at 108 and 110, this would suggest hydroxylation on the methyl group (whereby one of the deuterium was lost during oxidative hydroxylation). Using this technique, one can get information on possible sites of modification without extensive scale-up and isolation of the metabolites. One can also get an idea of the deuterium isotope effect when the ratio of ions in the 'twin pair" show a dramatic change. Compounds labeled with ^{13}C can be used for metabolism studies as well. The ^{13}C-labeled compound can be useful for MS/MS and NMR studies. A combination of deuterium and ^{13}C-labeled compound can provide very useful structural information. However, obtaining such a dual labeled compound may not be easy, especially in the early discovery stages. The application of deuterium and ^{13}C- labeled compounds has been used to solve structures of complex metabolites as well as to aid us in understanding metabolic reactions and mechanisms leading to such products.

The use of deuterium labeled solvents these days to assist in structural elucidation of metabolites in discovery stages (where we are limited by the amount of compound available for metabolite isolation etc), is steadily growing in popularity. This technique has been found to be useful in distinguishing hydroxylations (e.g., formation of alcohols, phenols) from oxidations on heteroatoms (e.g., formation of sulfoxides, *N*-oxides) provided the molecule is capable of undergoing any of these reactions. In the presence of deuterated solvent (such as D_2O), the protons on functional groups such as OH, NH_2, or SH, will be exchanged with deuterium. This leads to mass shifts corresponding to the number of exchangeables in the molecule. Hence a metabolite formed by hydroxylation at an aliphatic position will show a net addition of 1 amu, while a product resulting from *N*-oxidation, for example, will not show this effect.

A combination of different mass spectrometers along with the availability of either stable or radiolabeled compounds has made metabolite identification much easier these days. Nonetheless the complete structural elucidation of metabolites often requires the use of synthetic standards or comprehensive NMR analyses of the unknowns.

4   NUCLEAR MAGNETIC RESONANCE (NMR)

As mentioned above, it is often necessary to more fully characterize a metabolite using NMR spectroscopy. However, isolation of even the relatively small quantities (1-10 µg) of a metabolite needed for NMR analysis often represents the rate-determining step in the process of metabolite structure elucidation. The application of the hyphenated techniques LC-NMR and LC-NMR-MS has made it possible to eliminate or reduce isolation steps and provided a new set of powerful tools for the study of drug metabolism.[27, 31-52] LC-NMR is also particularly useful in cases where the metabolites tend to be unstable and degrade during isolation or storage over a period of time. In the following sections the basic techniques of LC-NMR and LC-NMR-MS will be discussed.

4.1   Continuous-Flow LC-NMR

The simplest method of operation is on-flow detection. This mode of operation is generally only practical when using ^{1}H or ^{19}F NMR for detection unless enriched compounds are used. On-flow LC-NMR is a series of 1D spectra acquired for 16 to 32 transients into 2K to 8K data points. Total acquisition time for each transient is typically around 1 second. The data are multiplied by a line-broadening function of 1 to 3 Hz to improve the signal-noise ratio and zero-filled by a factor of two before Fourier transformation in the F2 domain only. This results in a contour plot of intensity with ^{1}H or ^{19}F NMR chemical shift on the horizontal axis and chromatographic retention time on the vertical axis (Figure 1).

If on-flow detection is required during a solvent gradient elution, the NMR resonance positions of the solvent peaks will shift as the solvent proportions change. For effective solvent suppression, it is therefore necessary to determine these solvent resonance frequencies as the chromatographic run proceeds. This is accomplished by measuring a single exploratory scan as soon as a chromatographic peak is detected in real time during the chromatographic run, and then applying solvent suppression irradiation at these frequencies as the peak elutes.

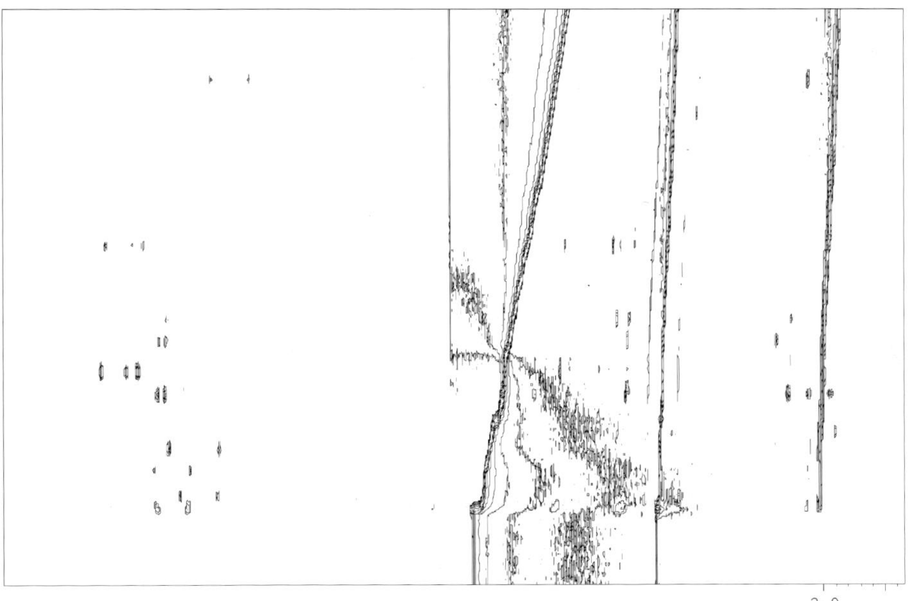

Figure 1. Typical on-flow ¹H LC-NMR data with chemical shift on the horizontal axis and retention-time on the vertical axis. These data were obtained using a capillary-LC flow probe and show the metabolites of phenacetin in human urine at submicrogram levels.

4.2   Time-Slice LC-NMR

Stopping the flow at short intervals over a chromatographic peak and collecting NMR data is referred to as "time-slicing". The time-slicing method may be useful if there is poor chromatographic separation, if the compounds under study have weak or no UV chromophores or if the exact chromatographic retention time is unknown. It is also possible to time-slice through an entire chromatographic run producing the equivalent of ana continuous-flow experiment with higher signal-to-noise. The data from such a time-slicing experiment is shown in Figure 2.

4.3   Stop-Flow LC-NMR

When the retention times of the compounds to be separated are known, or if they can be detected using mass spectrometers, UV (including diode arrays), radiochemical or fluorescence detectors, stop-flow LC-NMR becomes an option. Upon detection, the PC controlling the liquid chromatograph allows the pumps to continue running, moving the peak of interest into the correct position in the NMR probe. When the pumps have stopped, normal high-resolution NMR is possible. The long length of capillary tubing connecting the LC system to the

NMR probe does cause some loss of resolution in the separation. However, the NMR detection cell volume is typically 30-120 μl, and this represents the limiting factor in the chromatographic resolution. The practicality of the stop-flow approach has been amply demonstrated and although several separate stops are often made in each chromatographic run, the quality of the resulting NMR spectra are such that good structural information can be obtained. Even long 2-D experiments, which provide correlation between NMR resonances, based on mutual spin-spin coupling such as COSY or TOCSY and heteronuclear correlation studies such as HMQC or HSQC can be performed when sufficient amounts of the metabolite are in the probe. These 2-D experiments are extremely important when working on complex metabolites.

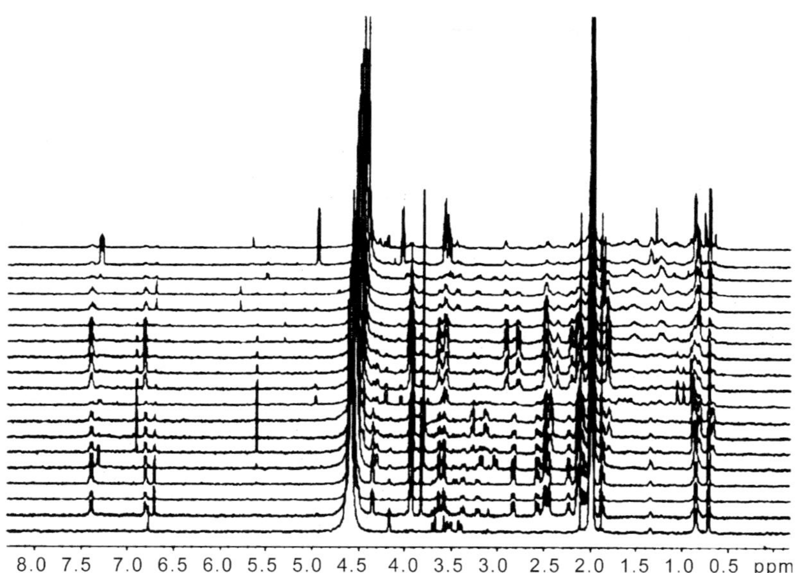

Figure 2. Typical time-slice LC-NMR data on a rat bile fraction. These data were used to assign, by inspection, 6 different metabolites.

TOCSY (TOtal Correlated SpectroscopY) is used to determine which signals arise from protons within a spin-system, especially when there are overlapping multiplets or there is extensive second order coupling. TOCSY provides long range as well as short-range correlations and is especially useful when coupling constants are small. However, not all correlations necessarily appear. Often multiple experiments, with variation to the mixing time parameter, are required to observe all correlations.

Heteronuclear correlation experiments (^{1}H to ^{13}C) like HSQC and HMQC provide ^{13}C chemical shift information from protonated carbons. These correlations arise from the large one-bond H-C couplings. ^{13}C chemical shifts are typically better resolved and more diagnostic than ^{1}H chemical shifts, and thus provide important information for assignment of the spectra of both nuclei. However, not all the ^{13}C nuclei are necessarily directly bound to protons. Long-range heteronuclear correlation (HMBC) can yield signals for these ^{13}C nuclei while suppressing the signals from those with directly bound protons. Data from HMBC experiments can thus provide the ^{13}C chemical shift of carbonyls and other quaternary carbons making total assignment of a metabolite possible.

4.4   Loop-Storage

Most LC-NMR hardware offers the option of moving eluting fractions to some type of device, typically capillary loops, for storage and later off-line NMR study. When there are several fractions of interest in a chromatographic run, it can be useful to move the first fraction in the NMR flow-probe without interrupting the chromatography and place the remaining fractions of interest in storage loops for subsequent analysis. Another advantage to loop-storage is that it avoids contamination of the flow-cell that often occurs in stop-flow mode when a high concentration component elutes prior to the peak of interest. The high concentration component will not be fully flushed from the NMR flow-cell resulting in contamination of the peak of interest. In loop-storage mode the peak of interest is directed to a clean storage-loop and the NMR flow-cell can be thoroughly washed prior to transfer. Typically, the loop temperature is controlled to ensure that peaks do not decompose while awaiting transfer for analysis. This method is now the most commonly used technique for metabolism studies and is particularly useful when LC-NMR-MS is being employed.

4.5   LC-NMR-MS

The extension of an LC-NMR system to include mass spectrometry has been in application for several years.[27,32,35,36] The feasibility of coupling a mass spectrometer to an LC-NMR was proposed as a way of obtaining complimentary sets of data for an unknown peak several years ago.[32] By directly-coupling a mass spectrometer to an LC-NMR system it is possible to obtain critical mass spectral data on a single injection of sample. Configuring the system to have the sample reach the mass spectrometer before it reaches the NMR flow cell enables the mass spectrometer to be employed as an experimental control device for analyzing complex mixtures. Mass spectrometry is an ideal detector, providing a high degree of selectivity in monitoring molecules of interest. It provides data rapidly and can thus yield valuable information on parent or daughter ion mass prior to initiating time-consuming NMR experiments. However one should be aware of the limitations of this technique, especially with compounds that do not ionize well under atmospheric pressure condition.

4.6   Recent Advances in NMR

Three major advances have occurred in NMR hardware recently, which will have a major impact on the way NMR is employed in metabolism studies. These advances all attempt to address major weaknesses and sensitivity with NMR technique.

4.6.1   Solid Phase Extraction-NMR

SPE-NMR uses a computer controlled cartridge based solid-phase extraction system, post-column, in place of the loop storage system. Chromatography is carried out using normal protonated solvents instead of the deuterated solvents typically used in LC-NMR. The selected peaks for one or more injections are stored in SPE cartridges. Prior to moving the peaks to the NMR probe for analysis they are dried with nitrogen and transferred with a small quantity of an appropriate deuterated solvent. This can result in a considerable increase the concentration of sample in the NMR probe when compared to a stop-flow or loop-storage experiment. This technique is considered suitable for analyzing metabolites present in plasma, urine and in other biological samples where the concentration of metabolites may not be high enough for easy purification.

4.6.2   Capillary-LC-NMR

Capillary LC-NMR is based on an NMR probe which has a very small (5 μl) flow-cell. The reduction in RF coil size results in a direct sensitivity enhancement which can be as high as a factor of five. However, because of the small flow-cell size and the need to couple them to capillary LC systems they are not the solution to all sensitivity problems. We have found that they excel in obtaining on-flow LC-NMR data as shown in Figure 1.

Combining a Capillary-LC probe with an SPE-NMR system designed for small volume work is the next logical step and would result in further improvements in overall sensitivity.

4.6.3   Cryogenic LC-NMR probes

The first cryogen LC-NMR probes have recently been demonstrated. By super cooling the NMR coils it is possible to achieve increases in sensitivity of up to a factor of 4. Using these probes with a normal LC-NMR system or a SPE-NMR system will be the standard LC-NMR configuration of the future. This will allow NMR data, especially critical 2D NMR experiments; to be collected much faster and thus reduce the time needed to elucidate important metabolite structures.

5   SPECIFIC EXAMPLES

The following examples will demonstrate the methods and techniques that were used in our laboratories to identify and characterize metabolites present in complex biological matrices such as bile, urine and plasma. Details are provided as to how some polar phase II metabolites (e.g., glucuronide, glutamate and GSH conjugates) of various compounds have been isolated and characterized. It is the authors' point of view that the methods described in these examples could easily be applied to a structurally diverse group of compounds. These examples will also illustrate how the data from these metabolite identification studies were used to address toxicity issues, active metabolites, metabolic soft spots for further modification by chemists and elucidation of novel metabolic pathways.

5.1   *Metabolism of Efavirenz (DPC 266) and Renal Toxicity*

The metabolism of a non-nucleoside reverse transcriptase inhibitor, efavirenz (DMP 266, Scheme 1) was investigated in various species.[3,4,53] DMP 266 possesses a chlorine substituent, which made it easy to identify drug-related components in a complex matrix such as rat urine or bile by using LC/MS. The identification of DPC 266 metabolites was initially investigated in rats. Urine samples were also obtained from cynomolgus monkeys, hamsters, guinea pigs and humans for metabolite profiling and to isolate metabolites. Bile and urine samples were extracted on the solid phase C18 cartridges as described above. Prior to the C18 extractions, LC/MS profiles were obtained so that the tentative structures of metabolites could be obtained. The C18 fractions of bile and urine were individually checked for the presence of metabolites before further purifications were carried out on semi-preparative HPLC columns. The metabolic pathways were confirmed by either further *in vitro* studies with isolated metabolites or by dosing the animals with synthetic standards of metabolites. The metabolic pathways obtained in rats, cynomolgus monkeys and humans are shown in Scheme 1. It was important to obtain the complete metabolic pathways in each species because of the search for a plausible explanation for species selective renal toxicity in rats.

As seen in Scheme 1, species differences in the formation of metabolites were observed. The major metabolite excreted by all the species was the glucuronide conjugate of 8-OH efavirenz. Interestingly, rats excreted a cysteinylglycine conjugate in the urine. The presence of cysteinylglycine conjugate (**M10**) in urine was unique because it has not been previously isolated and characterized as a urinary metabolite of a GSH adduct. However, it has been postulated that cysteinylglycines are intermediates in the catabolism of GSH conjugates to mercapturic acids.[54] The structure of the cysteinylglycine conjugate was unequivocally established by LC/MS and NMR analysis of the isolated metabolite. Furthermore the GSH adduct, **M9** (postulated precursor of **M10**), was isolated from rat bile and its structure confirmed by mass spectral analysis and by various NMR experiments.[3,28] Metabolites **M9** and **M10** are two examples of diconjugates that were found to be formed by an initial sulfation followed by enzyme catalyzed GSH conjugation. These diconjugates are usually produced in small quantities and are often difficult to isolate from endogenous components. Frequently, these diconjugates are ignored during the metabolite profiling of a compound as they are considered to have low importance. This is perhaps one example where the identification of a minor metabolite proved to be very useful in resolving a toxicity issue.

In vitro metabolism of efavirenz and its metabolites was conducted with liver and kidney sub cellular fractions (microsomes and hepatocytes) to confirm the formation of GSH conjugate, **M9** (precursor of **M10**), and the pathways involved in the catabolism of this metabolite. The *in vitro* studies suggested that rats could convert **M11** to **M9**, while humans and cynomolgus monkeys could not. The *in vivo* metabolite profiles showed cynomolgus monkeys produced a significant amount of **M11** (precursor to **M9**) but no **M9**. The human profile suggested formation of neither **M11** nor **M9**. The results from the metabolite identifications and subsequent metabolism studies suggested that the GSH conjugate produced by rats could contribute towards the selective renal toxicity. Further studies were conducted to confirm the contribution of the GSH conjugate towards renal toxicity. Experiments were conducted with the deuterated analog of efavirenz and with a specific enzyme inhibitor of γ-glutamyltranspeptidase, acivicin, that were found to greatly reduce the nephrotoxicity in carefully designed studies in rats.[4,47] Because the structures of the metabolites were well characterized as well as the metabolic pathways were very well defined, it was relatively easy to target a particular enzyme and a specific site on the molecule to test if the GSH adduct played a role in nephrotoxicity. Scheme 2 illustrates how the knowledge gained from metabolite characterization was used to design experiments to test the hypothesis that the GSH adduct formation could play a role in causing nephrotoxicity in rats. These studies confirmed that GSH adduct was responsible for causing species-specific nephrotoxicity in rats.[4] Mechanisms were proposed for the formation of reactive metabolites that could have been responsible for the renal toxicity observed in rats.

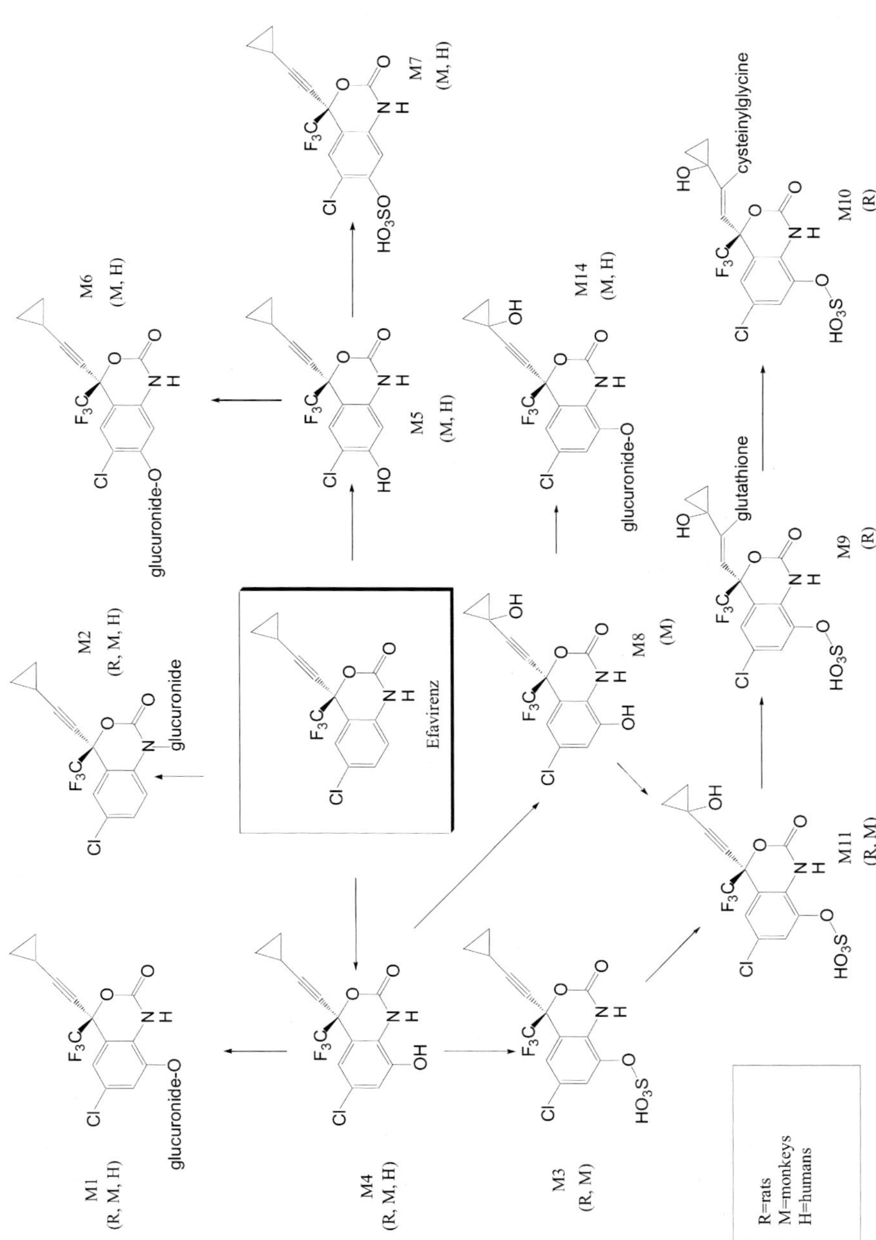

Scheme 1. Proposed metabolic pathways of efavirenz in rats, monkeys and humans.

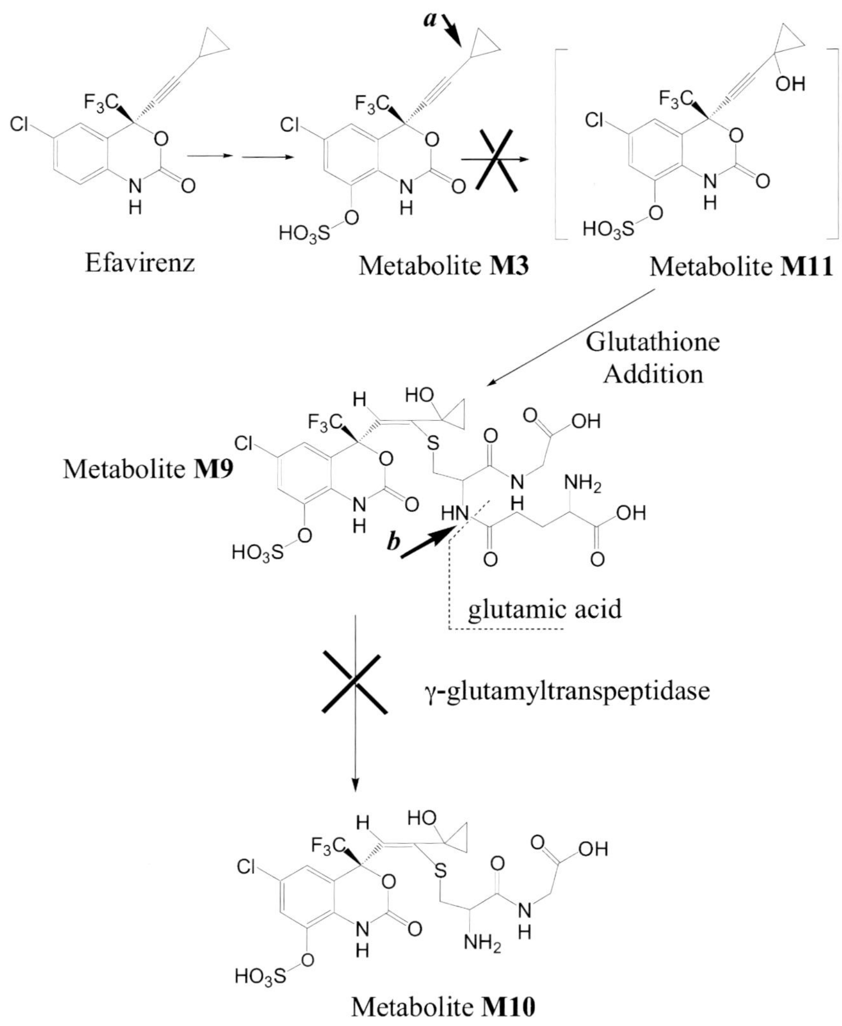

Efavirenz   Metabolite **M3**   Metabolite **M11**

Glutathione
Addition

Metabolite **M9**

glutamic acid

γ-glutamyltranspeptidase

Metabolite **M10**

Scheme 2. Metabolic sites affected by incorporating deuterium in the molecule (site *a*) as well as by administering acivicin, a γ-glutamyltranspeptidase inhibitor (site *b*) one hour prior to dosing rats with efavirenz. Complete identification of the metabolites enabled appropriate studies (such as synthesis of deuterium labeled analog) to be designed to address the renal toxicity observed in rats.

5.2   *Characterization of Unusual Metabolites*

5.2.1 *Metabolism of Acetaminophen: Formation of Peptide Conjugates.*

With the advent of LC/MS and LC/NMR it has become easier to characterize metabolites that were once difficult to isolate and identify. These techniques have enabled us to uncover the existence of an alternate pathway for the disposition of

GSH adducts of several structurally diverse compounds. Studies were conducted using acetaminophen as a model compound to investigate the role of the glutamic acid pathway in disposition of GSH adducts.[27] Although the mercapturic acid pathway was the major route of degradation of the GSH adducts, it was found that the conjugation of the glutathione, cysteinylglycine, and cysteine adducts of acetaminophen with the γ-carboxylic acid of the glutamic acid took place as well. The coupling of the GSH adduct and the products from the mercapturic acid pathway with the glutamic acid led to unusual peptide conjugates (Scheme 3). The identities of these adducts were confirmed unambiguously by LC/MS, NMR and by comparisons with synthetic standards. The addition of glutamic acids led to larger peptides, in contrast to the mercapturic acid pathway, in which the GSH adducts are broken down to smaller molecules.

Male Sprague-Dawley rats with cannulated bile ducts were dosed orally with acetaminophen once daily at 500 mg/kg for 2 days. Serial bile and urine samples were collected over ice and pooled from all of the animals on a daily basis and stored frozen at -20° C until analyzed. The isolation of GSH-derived conjugates from rat bile was done as described above using C18 cartridges (10 g/60 ml) and a semi-preparative HPLC column. Details of the purification processes for all of the acetaminophen metabolites are given elsewhere.[27] The LC/MS and LC/UV profiles of acetaminophen metabolites present in rat bile were obtained (Figure 3). LC/MS analyses of bile samples showed that the polar metabolites, including the glucuronide (**M1**), sulfate (**M2**), cysteine (**M3**), cysteinylglycine (**M4**), and GSH (**M5**) conjugates, were well resolved from each other and from the endogenous components. A comparison of the LC/MS (TIC) and HPLC/UV profiles clearly demonstrates a lack of correlation between TIC and the UV traces. For example, metabolite **M1** (the glucuronide conjugate) was the major metabolite as determined by HPLC/UV trace; but it showed a very weak MS response. Furthermore, Figure 3 also shows that a complete separation of metabolites is not necessary in order to obtain information on their possible molecular weights (by MH$^+$ ions etc). Figure 3 also illustrates that not all metabolites are polar and elute before the substrate on an HPLC system. Metabolites **M4-M7** eluted after acetaminophen on the reversed phase c18 column. Hence, one should always look for the presence of metabolites eluting after the parent compound by examining the TIC and UV chromatograms.

A number of unusual metabolites **M6-M9**, which showed losses of 129 amu during the MS/MS analyses on an ion trap, were found in the rat bile samples. These metabolites co-eluted with each other and with other metabolites of acetaminophen. The tetrapeptide conjugate (**M6**, Scheme 3) which co-eluted with the GSH adduct of acetaminophen was isolated on an analytical column (Waters Symmetry C18, 3.9 x 150 mm) using an isocratic mobile phase

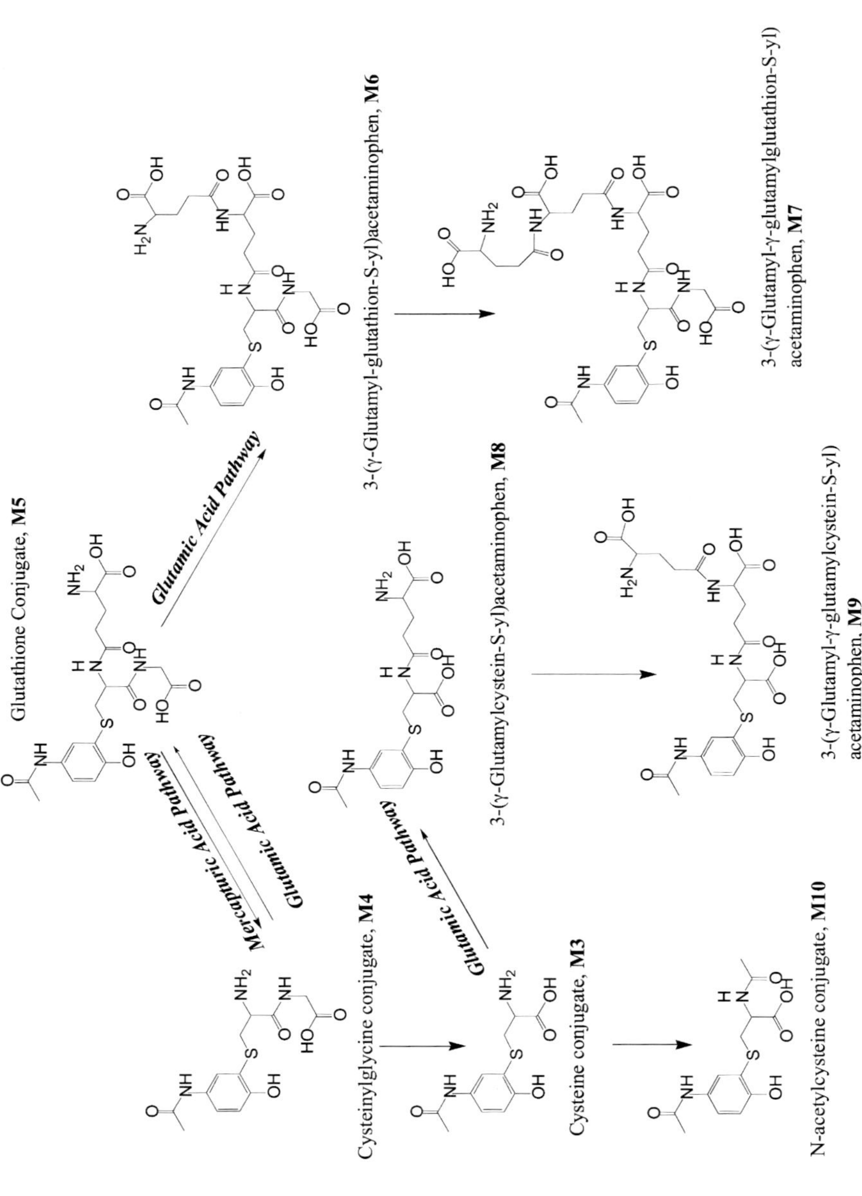

Scheme 3. Proposed disposition of GSH adducts of acetaminophen via the mercapturic acid and the glutamic acid pathways.

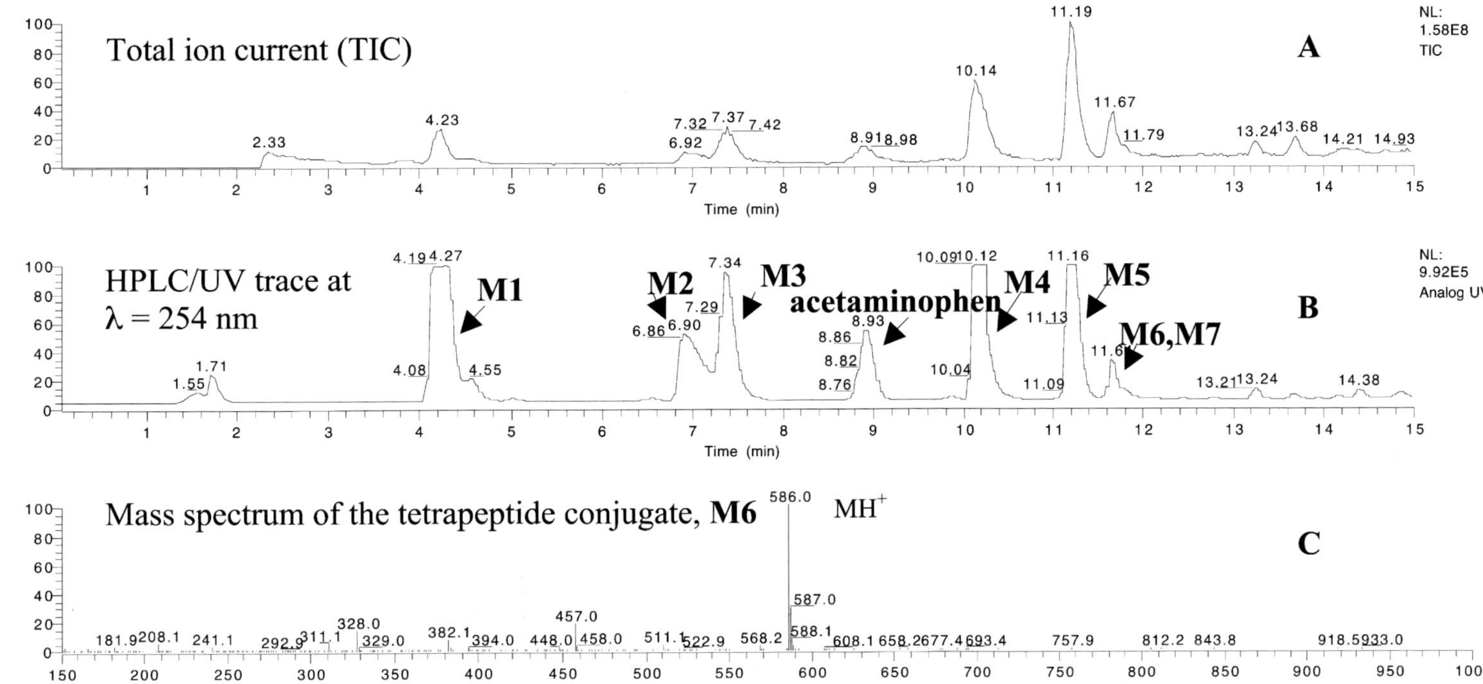

Figure 3. LC/MS and HPLC/UV profiles (A and B, respectively) of acetaminophen metabolites present in rat bile showing the presence of glucuronide conjugate (**M1**), sulfate conjugate (**M2**), cysteine (**M3**), cysteinylglycine (**M4**), GSH adduct (**M5**), and tetrapeptide and pentapetide conjugates (**M6** and **M7**, respectively). The extracted ion chromatograms of metabolites **M6** (*m/z* 586), **M7** (*m/z* 715), **M8** (*m/z* 400), and **M9** (*m/z* 529) showed that these metabolites co-eluted with each other or with other metabolites of acetaminophen.

consisting of a mixture of acetonitrile and 0.05% TFA (5:95 v/v) delivered at 0.7 ml/min. The peak corresponding to this metabolite was collected, pooled from several injections, dried and submitted for LC/NMR analyses. The MS/MS fragmentation patterns of the GSH adduct (**M5**) and the tetrapeptide conjugate (**M6**) are shown in Figure 4. The mass spectral data showed MH^+ at m/z 457 and 586 for metabolites **M5** and **M6**, respectively. The MH^+ at m/z 586 for **M6** was consistent with an addition of a glutamic acid onto the GSH adduct of acetaminophen. The structure of this metabolite was supported by MS^n experiments that showed that the glutamic acid was linked to the GSH adduct via the γ-carboxylic acid. The MS/MS fragmentation of the pseudomolecular ion (m/z 586) of the tetrapeptide produced ions at m/z 457 and 511 formed by losses of glutamate and glycine residues, respectively. Loss of glutamate from ion m/z 511 yielded a significant ion at m/z 382. The structure of this conjugate was confirmed by LC/NMR experiments and by comparison with a synthetic standard. The ^{1}H-LC/NMR and TOCSY of this metabolite are shown in Figures 5 and 6, respectively. While the ^{1}H-LC/NMR provided an idea of the relative purity for this metabolite, a 2-D TOCSY was found to be extremely useful in assigning the proton signals. TOCSY experiments are essential in elucidating the structures of unknown metabolites. However, other 2D-NMR experiments such as HMBC, whereby critical correlations between protons and carbons are established, often require isolation of metabolites.

The ^{1}H NMR spectrum clearly showed the presence of additional resonances for the extra glutamic acid at δ 4.28 (1H, glu α, m), 2.3 (2H, glu γ, m) and 1.90 – 2.00 (2H, glu β, m) in addition to the usual GSH proton signals. The correlation between the α, β, and γ protons of the two glutamic acids in the molecule is shown by TOCSY data (Figure 6). A number of other peptide conjugates were isolated from rat bile and characterized spectroscopically. The MS and NMR analyses of these conjugates confirmed the presence of additional glutamic acid moieties linked to the intact GSH, cysteinylglycine, and cysteine adducts of acetaminophen. The coupling of glutamic acid to other amino acids was demonstrated to occur via its γ-carboxylic acid group. This was confirmed by comparing the NMR and MS data of the isolated metabolites with those of synthetic standards. It is postulated that γ-glutamyltranspeptidase (GGT) could be involved in the formation of these conjugates. The discovery of these tetrapeptide and other peptide conjugates of acetaminophen represent a novel disposition of GSH conjugates of compounds. The catabolism of GSH adducts via the glutamic acid pathway has recently been described in the literature.[27] GSH adducts undergoing catabolism via mercapturic acid pathway lead to smaller molecules such as cysteinylglycine and cysteine conjugates. In this study, an alternate disposition pathway for GSH adducts (and for their breakdown products formed

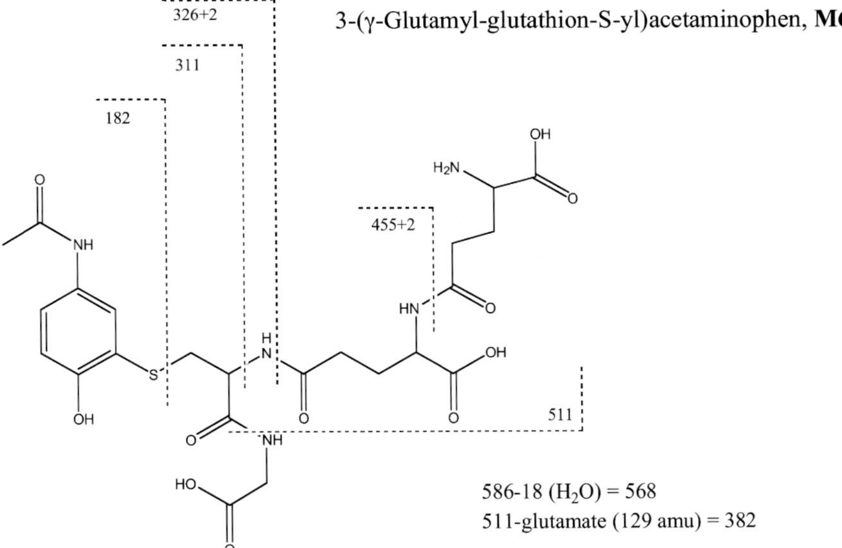

Figure 4. MS/MS fragmentation patterns of the GSH adduct (**M5**) and the tetrapeptide conjugate (**M6**). The fragment ions *a* (*m/z* 382) and *b* (*m/z* 182) are formed by an elimination of glutamic acid and by cleavage at the C-S bond of cysteine moiety, respectively. The mass spectral analysis was done in the positive ion mode, employing an ion trap instrument.

via the mercapturic acid pathway) was described in which the conjugates were taken as substrates by GGT that added glutamic acid sequentially. This pathway, termed as the glutamic acid pathway, seems to act in the opposite manner

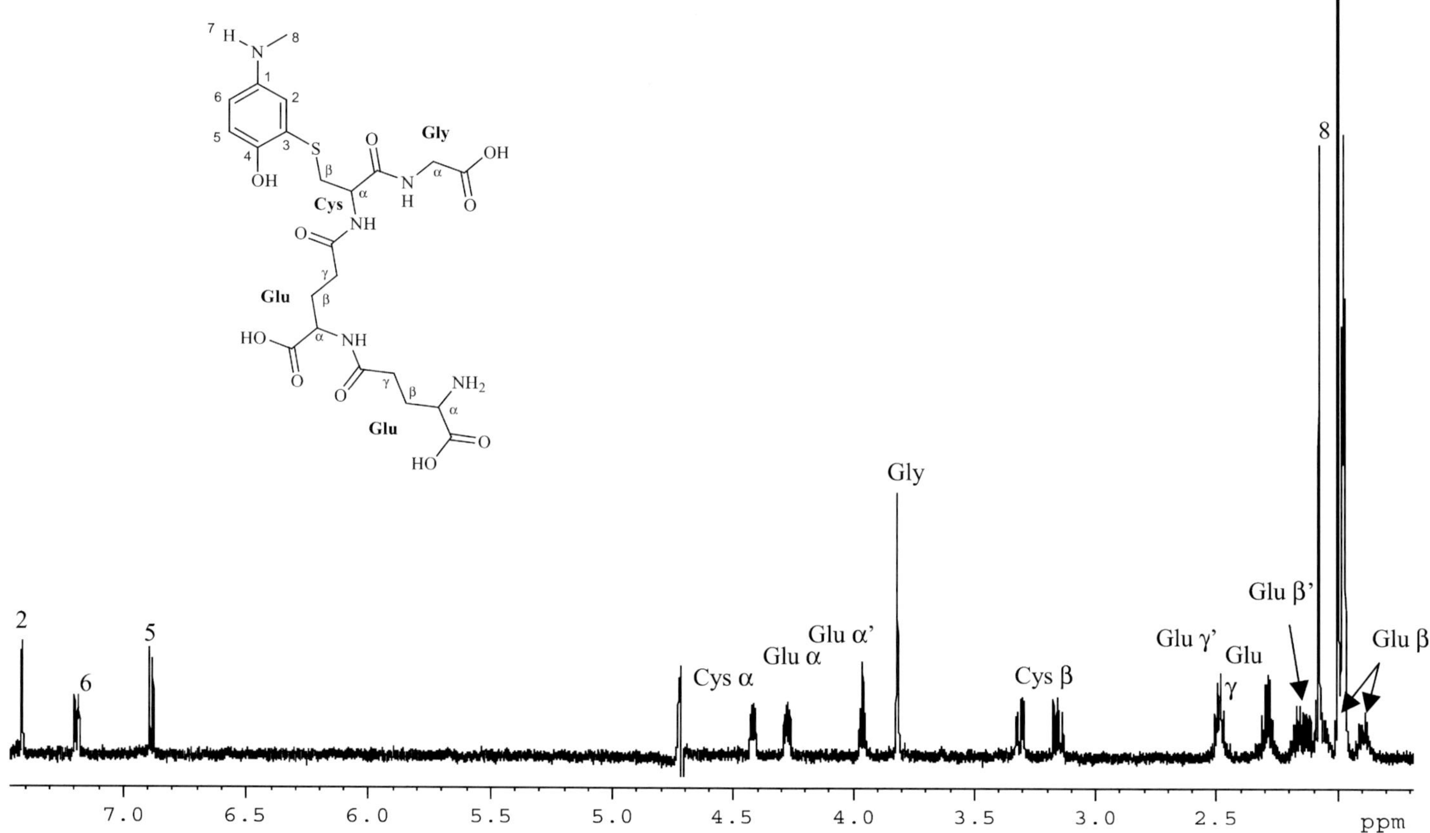

Figure 5. ^{1}H-LC/NMR of the tetrapeptide conjugate, 3-(γ-glutamyl-glutathion-S-yl)acetaminophen isolated from rat bile.

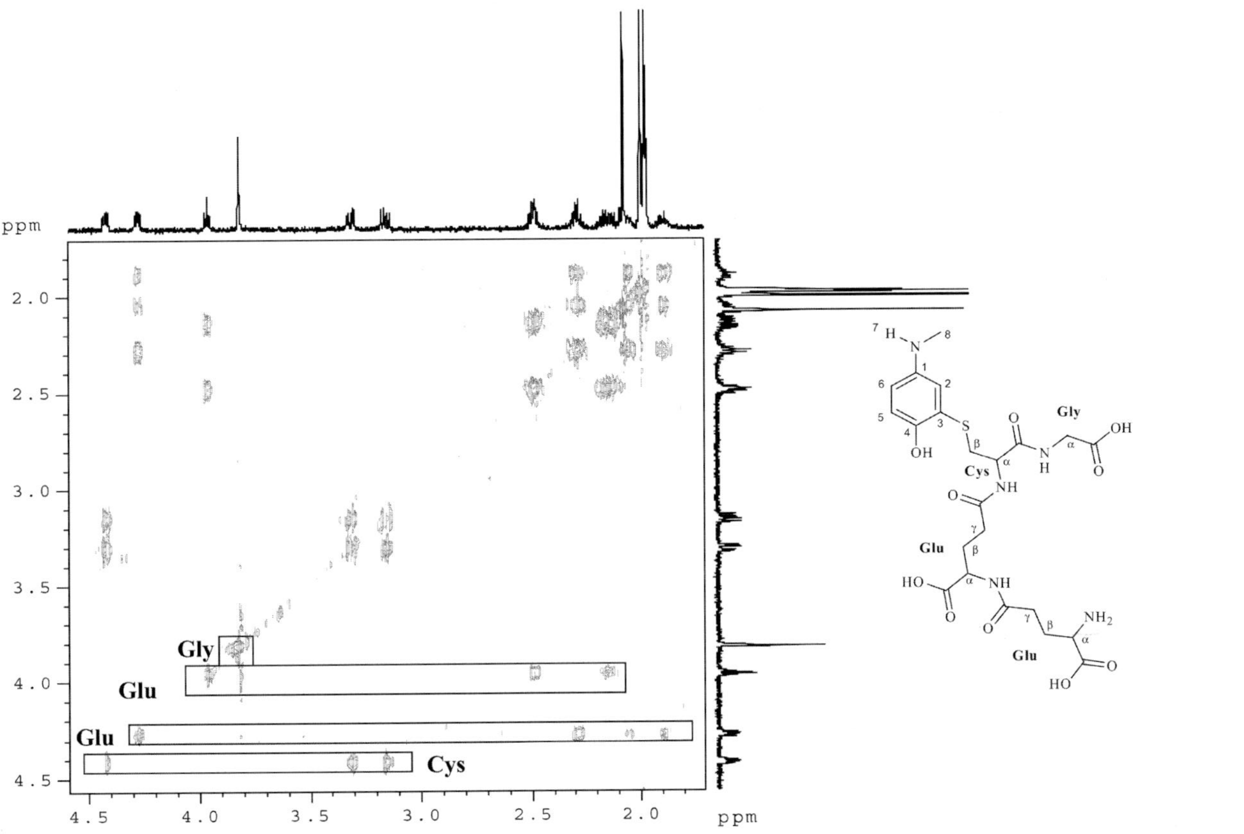

Figure 6. TOCSY of the tetrapeptide conjugate, 3-(γ-glutamyl-glutathion-S-yl)acetaminophen isolated from rat bile. The TOCSY data shows the correlation between the protons in the same spin systems. This enabled us to detect two sets of proton signals from two separate glutamate moieties in the molecule.

compared with the mercapturic acid pathway. This pathway has gone unnoticed in the past for a number of reasons. Analytical techniques such as LC/MS and LC/NMR were not available in the past to characterize these products that were once considered trivial and unidentifiable. Numerous metabolism studies in the past with radiolabeled compounds have shown the existence of radioactive peaks in radiochromatograms that were not characterized. It is quite possible that some of the radioactivity could have been due to products resulting from such undiscovered disposition pathways. With the introduction of LC/MS, LC/NMR and high field NMR it has become much easier to characterize compounds that were once considered unidentifiable by previously existing techniques.

5.2.2   *Metabolism of a non-nucleoside reverse transcriptase inhibitor (DPC 961): An Example of P450-mediated ring expansion.*

The metabolism of a disubstituted alkyne, (*S*)-6-chloro-4-(cyclopropylethynyl)-4-(trifluoromethyl)-3,4-dihydro-2(1*H*)-quinazolinone (DPC 961), was studied in rats. Bile-duct-cannulated Sprague-Dawley rats housed in metabolic cages were given an oral DPC 961 suspension twice daily at 30 mg/kg for 3 days before treatment with either unlabeled DPC 961 (50 – 500 mg/kg) or a ^{13}C-enriched compound (300 mg/kg). The metabolic pathways for DPC 961 in rats are shown in Scheme 4.

DPC 961 was metabolized extensively by rats to a number of metabolites, including the glucuronide and sulfate conjugates of the 8-hydroxylated metabolite. The glucuronide and sulfate conjugates were easily deduced from the corresponding addition of 176 and 80 amu, respectively, to the parent molecular weight (see Table 2). In addition to these metabolites, several GSH-derived adducts were found in bile and urine of rats treated with DPC 961.[55] These GSH adducts were detected in the bile samples by obtaining the expected pseudomolecular ions (+323 amu, see Table 2) during the LC/MS analyses. These metabolites were either excreted in rat bile or degraded to mercapturic acid conjugates and eliminated in urine. The existence of these catabolites was demonstrated by observing appropriate pseudomolecular ions corresponding to cysteinylglycine, cysteine and *N*-acetylcysteine conjugates (Table 2). In our previous studies with the structural analogue of DPC 961, efavirenz (see above), we had elucidated the metabolic pathways in different species. As with DPC 961, a GSH conjugate was detected in the bile of rats treated with efavirenz. This GSH conjugate (or its breakdown products) has been implicated as the agent responsible for selective nephrotoxicity observed in rats given high doses of efavirenz.[4,53] Due to a slight structural modification of DPC 961 as compared to efavirenz, we were interested if DPC 961 and efavirenz would produce similar

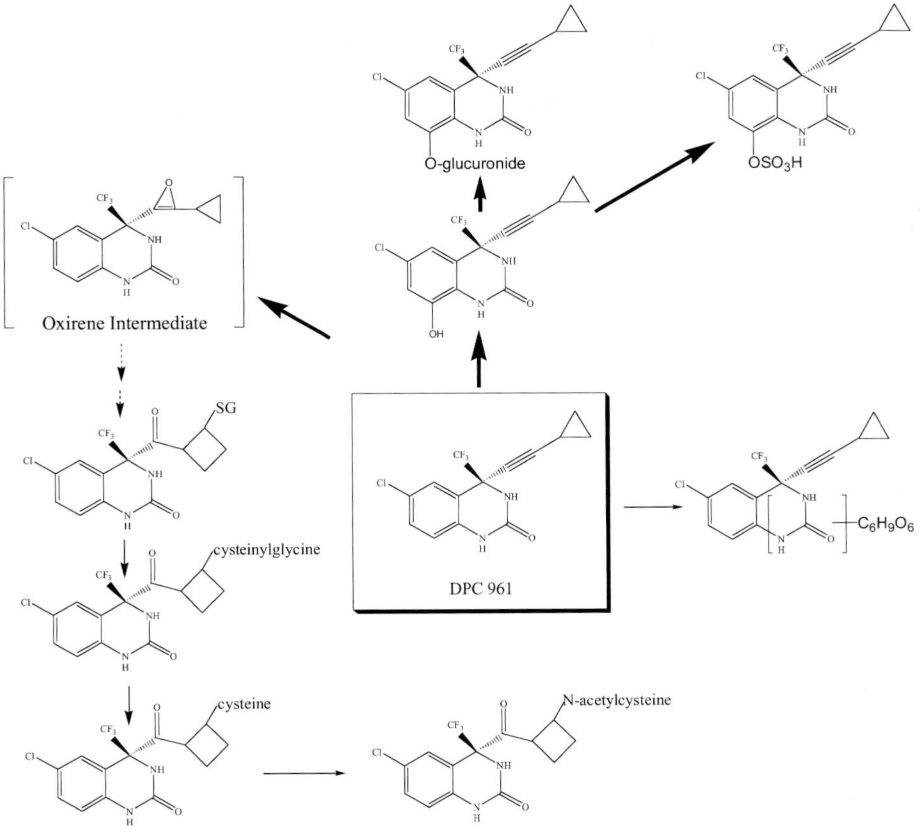

Scheme 4. Proposed metabolic pathways for DPC 961 in rats

GSH conjugates. The structure of GSH adducts from DPC 961 was found to be unique from that produced by efavirenz. In the case of efavirenz, the compound was bioactivated through hydroxylation on the cyclopropyl ring (methine carbon) before an enzyme-catalyzed addition of GSH could occur. For DPC 961, it appears that the oxidation occured preferentially on the alkyne, probably giving rise to an unstable oxirene metabolic intermediate which underwent rapid conversion to an α,β-unsaturated carbonyl compound. A postulated mechanism leading to the formation of the GSH adducts from the oxirene metabolite is shown in Scheme 5.

Rat P450 3A1 and 1A2 catalyzed the oxidation of the triple bond to the postulated oxirene intermediate, which rearranged to form a reactive cyclobutenyl ketone. GSH adds to this metabolic intermediate via 1,4-Michael addition, producing the two isomeric GSH adducts, GS-1 and GS-2. The structures of these two adducts are shown in Scheme 4. The LC/MS analysis of these adducts exhibited MH$^+$ at *m/z* 638 (corresponding to a net addition of 323 amu, see Table 2) with characteristic ion fragments at *m/z* 563, 509, 492, 406, and 331. The structures of these two GSH adducts were elucidated after a series of NMR experiments. The NMR data for one of the adducts are shown in Figures 7-10 (^{1}H-NMR, TOCSY, HSQC, and HMBC, respectively).

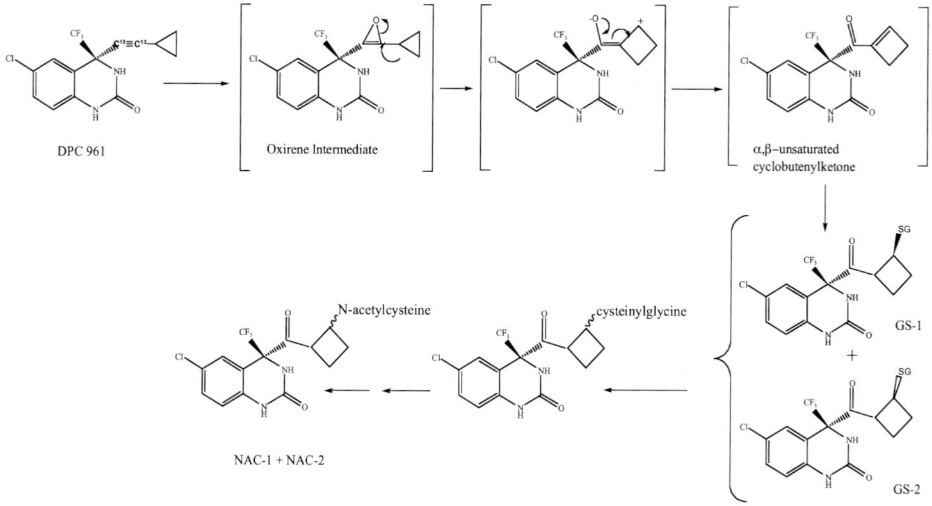

Scheme 5. Proposed mechanism for the formation of glutathione conjugates from the oxirene metabolite of DPC 961.

The addition of oxygen to one of the carbons of the triple bond in DPC 961 was demonstrated by characterizing the GSH adduct isolated from rats treated with ^{13}C-labeled DPC 961. The NMR studies confirmed that oxygen was attached to one of the labeled carbons. This was demonstrated by comparing the ^{13}C resonances of the alkyne carbons of ^{13}C-labeled DPC 961 with those of the isolated GSH adduct. The ^{13}C resonances of the labeled carbons had apparently changed due to the change in the spin states from sp-hybridized carbons to sp^2 and sp^3. Figure 11 shows the one-dimensional ^{13}C NMR spectra of the labeled DPC 961 (top) and the labeled metabolite (bottom). The spectrum of DPC 961 shows acetylene ^{13}C chemical shifts (68 and 92.5 ppm). The spectrum of the GSH conjugate shows a ^{13}C chemical shift (200 ppm) for carbonyl 2 and a methine ^{13}C chemical shift (49.5 ppm) for carbon d (Figure 11).

In this study, the combination of LC/MS, NMR, and stable labeled isotopes, enabled us to characterize unique GSH conjugates formed from a postulated

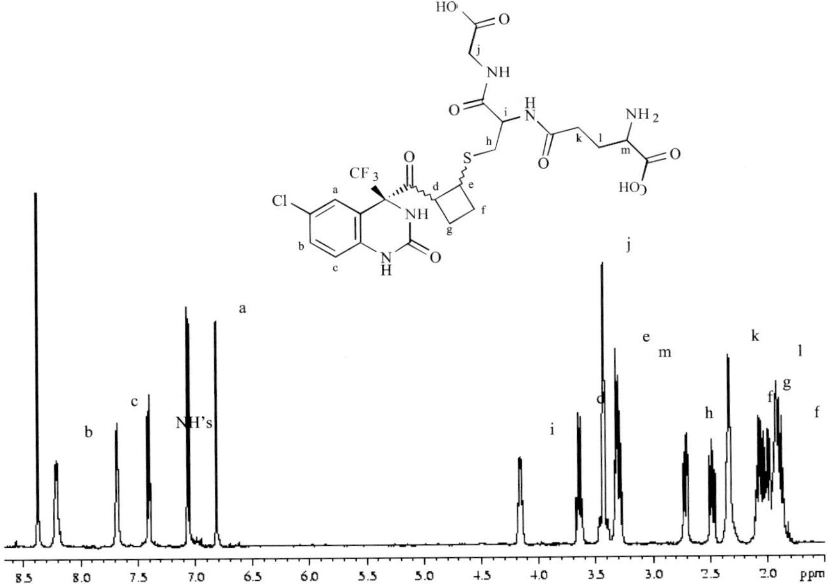

Figure 7. ^{1}H-NMR of one of the GSH adduct isomers isolated from rat bile.

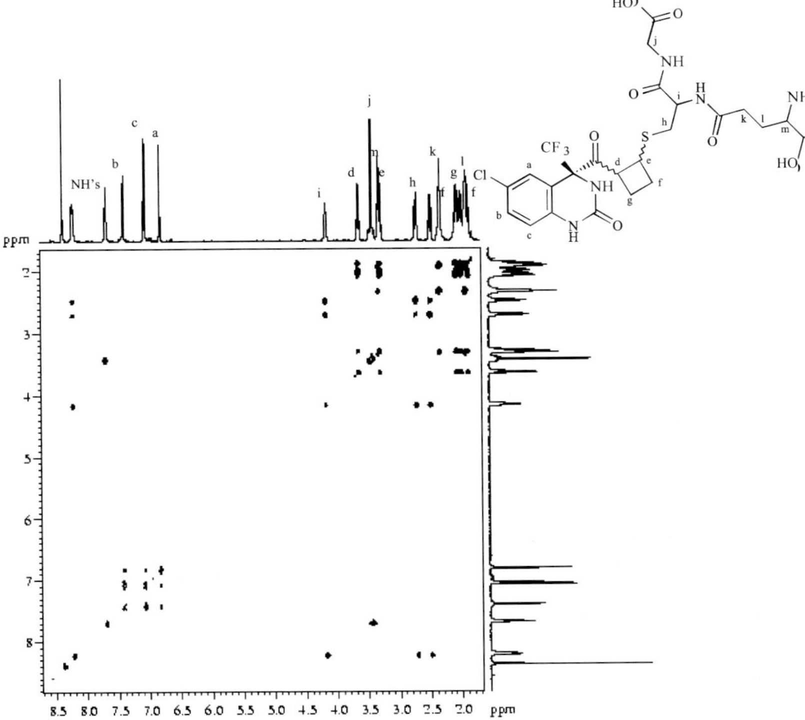

Figure 8. Total correlated spectroscopy (TOCSY) of the glutathione adduct (GS-1). The protons *d* - *g* of the cyclobutane ring are observed as part of one spin system.

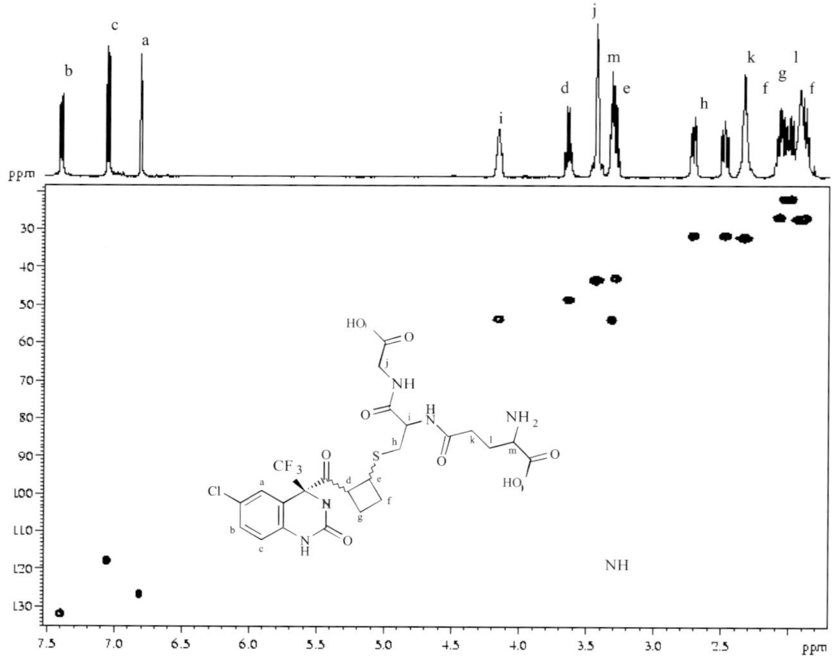

Figure 9. The HSQC spectrum of GS-1 showing the correlations between the carbons and the attached protons.

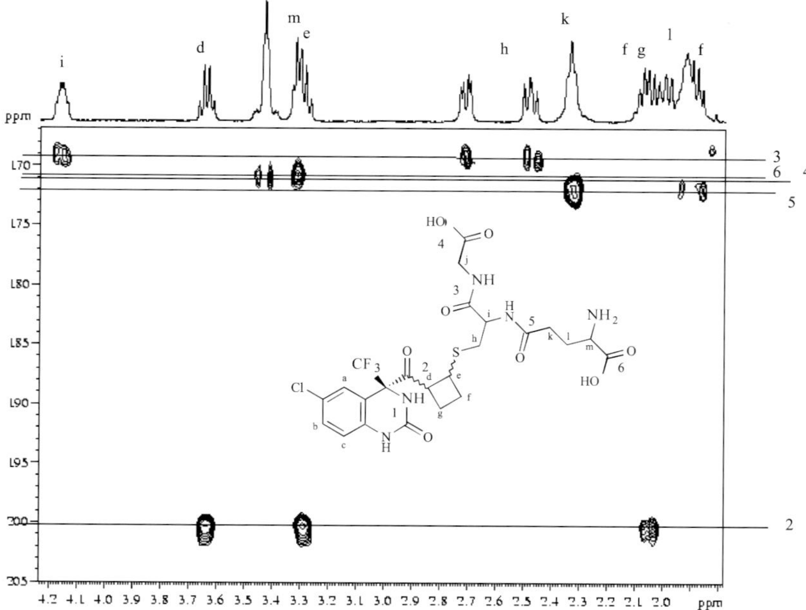

Figure 10. HMBC spectrum of GS-1 depicting the correlations between the carbons of the carbonyls (labeled 2-6) with adjacent protons (^2J and ^3J coupling).

oxirene intermediate. The stable labeled analogue of DPC 961 enabled us to provide spectroscopic evidence for the direct attachment of oxygen onto one of the carbons of the triple bond. The structure of the GSH adduct also showed that P450 was capable of mediating a reaction which ultimately led to an expansion of a cyclopropyl ring to a cyclobutyl ring.

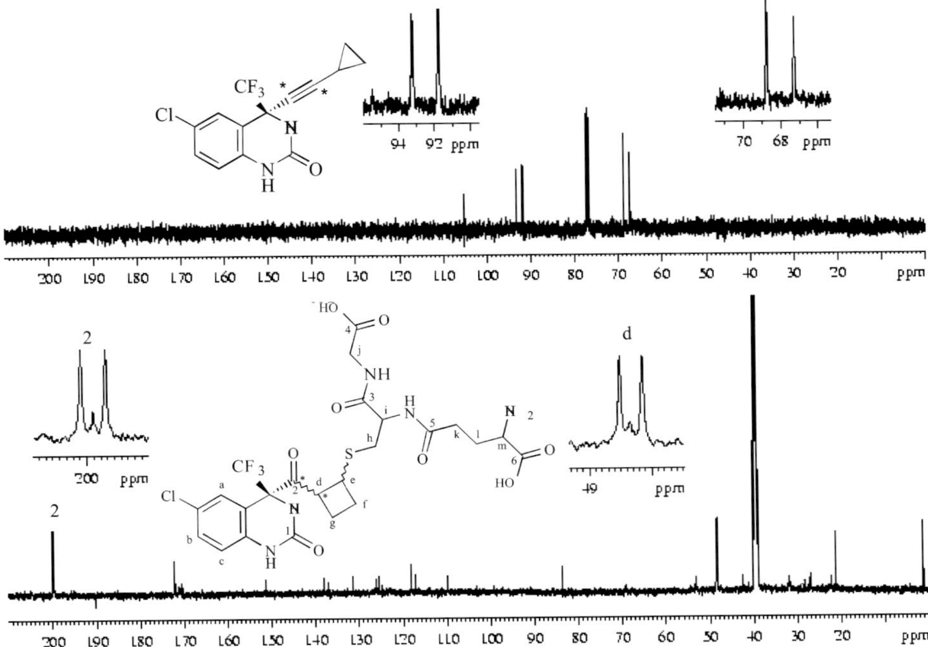

Figure 11. Carbon-13 spectra of the labeled DPC 961 (top) and the isolated glutathione adduct (bottom).

5.2.3   *Metabolism of (±)-N-methyl-N-(1-methyl-3,3-diphenylpropyl) formamide and 1-[3-(aminomethyl)phenyl]-N-[3-fluoro-2'-(methylsulfonyl)-[1,1'-biphenyl]-4-yl]-3-(trifluoromethyl)-1H-pyrazole-5-carboxamide (DPC 423). Characterization of glucuronide conjugates of unstable metabolic intermediates*

Intact glucuronides of a number of compounds have been characterized by MS.[56-59] Traditionally, the polar glucuronide conjugates have been analyzed as derivatized species by electron impact or chemical ionization MS[57,58,60] or by enzymatic hydrolysis, whereby the released aglycone is analyzed.[61] Underivatized glucuronides are now commonly analyzed by LC/MS instruments equipped with electrospray or atmospheric pressure chemical ionization interfaces. As most glucuronide conjugates are fairly stable, useful information regarding the exact nature of the aglycones can be obtained if the glucuronides are characterized spectroscopically in their intact forms. Described below are two examples of compounds, which formed unstable intermediates whose identities were established by identifying their glucuronide conjugates.

Formamide   N-hydroxymethylformamide   Glucuronide conjugate

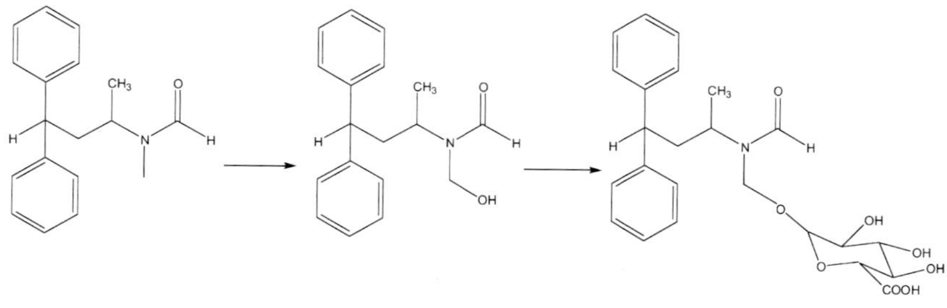

Figure 12. Metabolism of (±)-*N*-methyl-*N*-(1-methyl-3,3-diphenylpropyl)formamide.

    The metabolism of (±)-*N*-methyl-*N*-(1-methyl-3,3-diphenylpropyl)formamide (Figure 12) has been previously described.[62]

    A carbinolamide, *N*-hydroxymethyl-*N*-(1-methyl-3,3-diphenylpropyl)form-amide, was isolated from the β-glucuronidase-treated bile samples obtained from rats dosed with the (±)-*N*-methyl-*N*-(1-methyl-3,3-diphenylpropyl) formamide. This metabolite could only be characterized by GC/MS after its conversion to the volatile trimethylsilyl derivative. Carbinolamides and carbinolamines are generally unstable compounds decomposing rapidly by elimination of an aldehyde to yield the corresponding amide or amine.[63] However, if the electron density at the nitrogen is decreased by amide resonance, the stability of metabolically derived carbinolamides is increased sufficiently to form conjugates of glucuronic acid. The intact glucuronide conjugate of N-hydroxymethyl-*N*-(1-methyl-3,3-diphenylpropyl)formamide was isolated and characterized by combination of NMR and LC/MS/MS. Several male Sprague-Dawley rats were dosed with the formamide at 100 mg/kg (i.p.) and bile collected over 18 hours. Bile samples from several rats were pooled and diluted with 3 volumes of water before extracting with ethyl acetate to obtain the non-polar metabolites. The remaining aqueous phase was applied to a column (3.5 x 15 cm) of Amberlite XAD-2 resin (BDH Inc., Toronto, Canada) to concentrate the sample. The column was washed with several volumes of water and the conjugated metabolites then eluted with methanol.

The organic solvent was removed under vacuum and the gummy residue was chromatographed on silica gel dry packed under suction in a sintered glass funnel to form a column (6.5 x 4.5 cm). The column was first eluted with chloroform and followed by increasing percentages of methanol in chloroform to 80% v/v. The volume of fractions collected was 100 ml. Aliquots (10 ml) from these fractions were then evaporated in scintillation vials and the residues reconstituted in 0.5 M sodium acetate buffer (pH 5.0) before adding β-glucuronidase. After incubating for 24 hr at 37°C, the samples were extracted with ethyl acetate and derivatized with MSTFA for GC/MS analysis of TMS derivatives of the hydroxylated metabolites. The fractions containing the hydroxylated metabolites were pooled and the solvent removed in preparation for HPLC. The reconstituted residue from the fractions containing the glucuronide conjugates was further purified on a semi-preparative HPLC column (Whatman Partisil 10 ODS-2, 250 x 9 mm) using a mixture of methanol and 0.02 M sodium acetate (45:55, pH 4.0) delivered at 3.5 ml/min. After pooling appropriate fractions from several HPLC runs, the methanol was removed under vacuum and the remaining aqueous sample passed through a preconditioned C18 cartridge. After washing the cartridge with two volumes of water, the glucuronide conjugates were eluted with methanol. The organic solvent was removed by evaporation, and the sample submitted for ^{1}H-NMR, and LC/MS/MS analysis. Whereas the glucuronide of the carbinolamide and that of the aromatic ring hydroxylated formamide were readily separated from bile pigments on a silica column, HPLC was required to separate the two glucuronides from each other. The fraction from the silica column chromatography containing a mixture of glucuronide conjugates was analyzed by LC/MS/MS using an electrospray interface. The mobile phase consisting of methanol in 0.05 M ammonium acetate buffer (pH 4.0) was delivered at a rate of 1 ml/min with 20% of the eluent diverted to the mass spectrometer. The expected protonated molecular ion of the two conjugates, *m/z* 460, was selected and monitored during the LC/MS run. Three distinct peaks were observed based on retention times. Peaks A and B, eluting at 3-5 minutes, corresponded to phenolic glucuronides whereas peak C, eluting at approximately 10 minutes, corresponded to the carbinolamide conjugate. ^{1}H-NMR and COSY 2D-NMR confirmed the structure of the carbinolamide glucuronide conjugate corresponding to peak C.[62] The ^{1}H-NMR showed the presence of two rotamers present in the sample. A series of homonuclear decoupling experiments were conducted to assign the protons of the conjugate.

The metabolism of 1-[3-(aminomethyl)phenyl]-N-[3-fluoro-2'-(methyl-sulfonyl)-[1,1'-biphenyl]-4-yl]-3-(trifluoromethyl)-1*H*-pyrazole-5-carboxamide (DPC 423) (Figure 13) has been recently described.[52,64] DPC 423 was metabolized extensively to a number of metabolites (see Scheme 6) including a number of unusual metabolites.

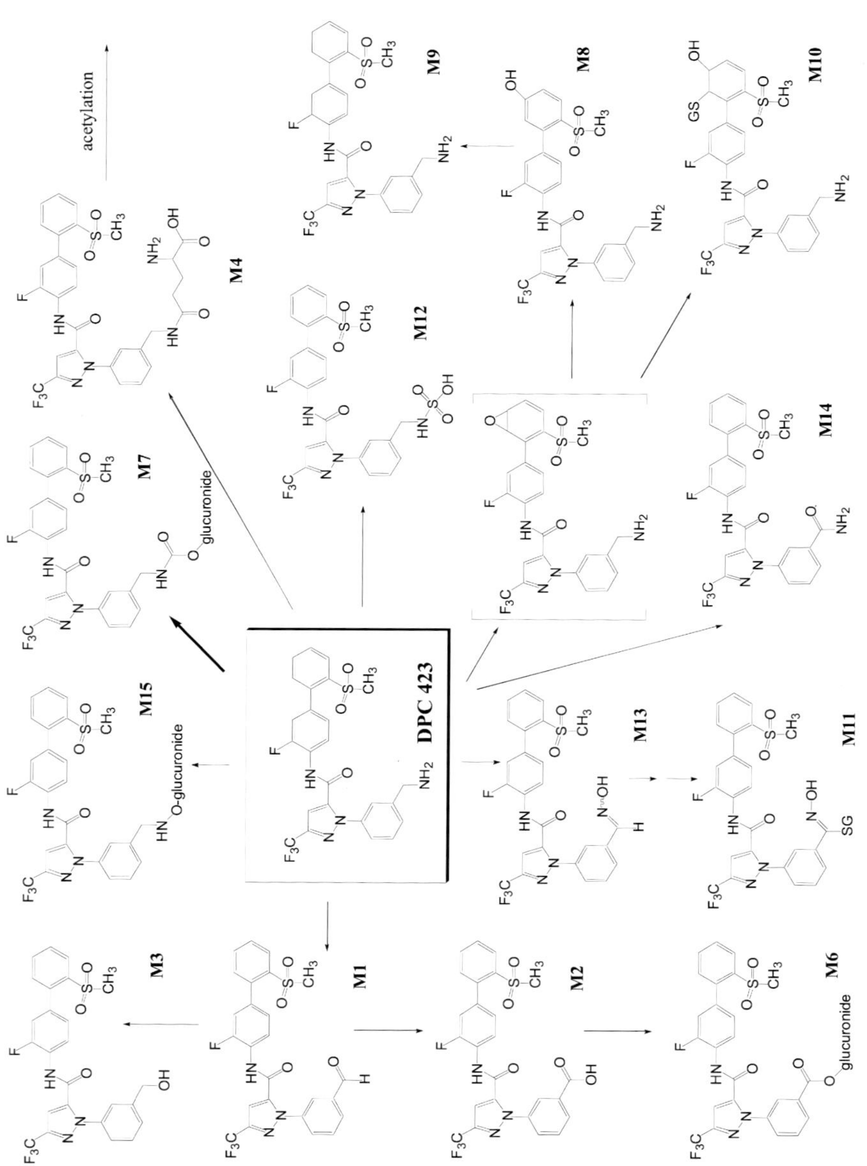

Scheme 6. Proposed metabolic pathways of DPC 423 in rats.

One of the metabolites was identified as the carbamyl glucuronide apparently formed after an initial formation of an unstable carbamic acid intermediate. This intermediate is too unstable to characterize by any of the available analytical techniques. However the glucuronide conjugate was very stable and was easily characterized by LC/MS and NMR. Male Sprague-Dawley rats with cannulated bile ducts were dosed with DPC 423 at 100 mg/kg and urine and bile collected over ice. The LC/MS profiling of bile and urine were completed by injecting an aliquot of sample directly onto the LC/MS column. The metabolites were separated on a Waters Symmetry C18 column (2.1 x 150 mm) by a gradient solvent system consisting of acetonitrile and 10 mM ammonium formate, pH 3.5. The percentage of acetonitrile was increased from 15 to 80 over 20 minutes with the solvent flow rate set at 0.4 ml/min. LC/MS was carried out by coupling the HPLC system to a Finnigan ion trap mass spectrometer. Liquid chromatography electrospray ionization mass spectrometry was performed in either positive or negative ion modes. MS^n of fragment ions on the ion trap mass spectrometer (LCQ, ThermoFinnigan) were obtained with 20-25% relative collision energy. The isolation of metabolites from rat bile was performed initially on 10 g/60 cc C18 cartridges as described in Section 2.1. It was found that most of the metabolites had eluted with 70-80% methanol in 0.1% acetic acid. These fractions were dried and reconstituted in 0.1% acetic acid and repurified on a 10-gram C18 cartridge, but this time using acetonitrile instead of methanol as the organic modifier. The fractions containing the metabolites were pooled, dried and further purified on an analytical column (Beckman C18, 4.6 x 250 mm) using acetonitrile and ammonium formate (10 mM, pH 3.5) as the mobile phase (28:72 v/v, pH 3.5). The peak corresponding to the carbamyl glucuronide was collected, dried under vacuum, and analyzed spectroscopically by NMR. The LC/MS analysis showed the carbamyl glucuronide was the major metabolite excreted in bile showing $[M+NH_4]^+$ at *m/z* 770 with the protonated ion appearing at *m/z* 753 (< 2% relative ion abundance). The LC/MS/MS spectrum of the carbamyl glucuronide showed a characteristic loss of 193 amu (elimination of glucuronic acid + NH_3) to give an ion at *m/z* 577. The ion fragment at *m/z* 577 was an addition of 44 amu to the parent compound, suggesting an incorporation of CO_2 in the molecule. Further MS/MS studies with the fragment ion at *m/z* 577 showed a loss of 44 amu ($-CO_2$) from each, producing ion at *m/z* 533 which corresponded to the MH^+ of DPC 423 aglycone, respectively. The ^{1}H-NMR of the carbamyl glucuronide conjugate showed that the proton signals of the metabolite had not changed significantly. However a new set of proton signals corresponding to the glucuronic acid moiety was observed. The formation of such carbamyl *O*-β-D-glucuronides from amines has been described before;[65-67] however, a carbamyl glucuronide of a benzylamine has not been described previously. Usually acid or enzymatic hydrolysis of these conjugates resulted in the formation of the parent compound, because the intermediate carabamic acid undergoes facile decarboxylation. The formation of carbamyl glucuronide conjugates through an intermediate carbamic acid may in fact be quite common with amines but would have gone unnoticed because of instability or because of enzyme hydrolysis back to the parent compound. The softer ionization techniques such as electrospray ionization (ESI) used during LC/MS has made identification of such conjugates possible.

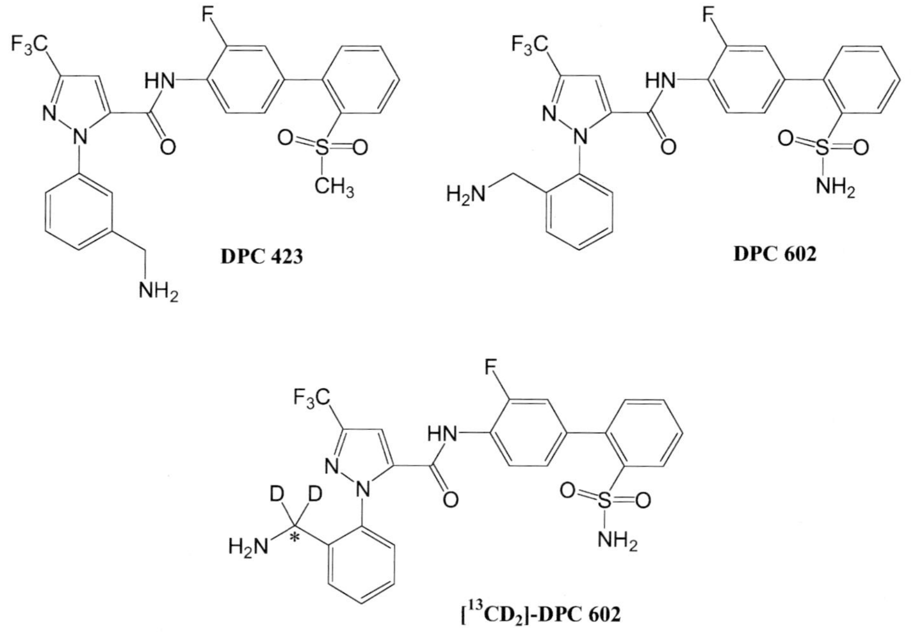

Figure 13. Structures of DPC 423, DPC 602, and ^{13}[CD$_2$]-DPC 602.

5.2.4 Metabolism of 1-[3-(aminomethyl)phenyl]-N-[3-fluoro-2'-(methylsulfonyl)-[1,1'-biphenyl]-4-yl]-3-(trifluoromethyl)-1H-pyrazole-5-carboxamide (DPC 423) and its analogues. An example of reactive metabolite formation and its detection and characterization by LC/MS and NMR.

Occasionally, a compound or a series of compounds is metabolized to products whose structures are not easily determined. In such instances, extensive use of NMR and stable labeled compounds can greatly assist in the elucidation of structures. As stated before the structures of these metabolites can very often provide additional information such as novel metabolic pathways and formation of reactive intermediates. The metabolism of DPC 423 and its analogues produced a number of unique metabolites whose structures were determined using such techniques.[52,64] Several metabolites, some of which were considered potentially reactive, were identified in rats. Bile-duct cannulated male Sprague-Dawley rats were dosed with DPC 423, DPC 602 and $[^{13}CD_2]$ DPC 602 (see Figure 13). A novel GSH adduct, was isolated from bile of rats dosed with DPC 423. Similar GSH conjugates of DPC 602 and $[^{13}CD_2]$ DPC 602; analogues of DPC 423 were isolated, characterized spectroscopically, and shown to have identical mass spectral fragmentation pathways. The bile samples were subjected to solid phase extraction as described above. The samples from C18 extraction (70% methanol fraction) containing the metabolites were pooled, dried under nitrogen, and reconstituted in water for further purification on a 2 g C18 cartridge. The metabolites were eluted from the cartridge with 5 ml aliquots of solvents consisting of different percentages of acetonitrile in 0.05% TFA. The 40% fraction which contained the GSH conjugate, was dried and further purified on a semi-preparative column (Beckman C18, 10 x 250 mm) using a mobile phase consisting of a mixture of acetonitrile and 0.05% TFA (2:3 v/v) delivered at 3.5 ml/min. The peaks corresponding to the GSH conjugates from DPC 423, DPC 602 and $[^{13}CD_2]$DPC 602 were collected from several injections, pooled, dried, and the dry powder analyzed by NMR. The structures of these GSH conjugates were initially suspected to be either an amide with N-S bond or a nitrogen–oxygen juxtaposed amide with a C-S bond. Rats dosed with DPC 423, DPC 602, and $[^{13}CD_2]$ DPC 602 excreted significant quantities of unique GSH conjugates, which showed net addition of 319 amu to the parent molecular weights. The pseudomolecular ions (MH^+) of these GSH conjugates from DPC 423, DPC 602, or $[^{13}CD_2]$ DPC 602 dosed rats were at *m/z* 852, 853, and 854, respectively. The high-resolution mass spectrum of DPC 423 GSH adduct showed MH^+ at *m/z* 852.1745 (852.1745 calculated). The MS/MS spectrum showed characteristic losses of pyroglutamate (-129) and glycine, giving ion fragments at *m/z* 723.1413 (calculated 723.1319) and 777.1505 (calculated 777.1424), respectively.

The major fragment formed by the neutral loss of hydroxylamine from ion at *m/z* 723 produced an ion at *m/z* 690.1138 (calculated 690.1104). Rats dosed with $[^{13}CD_2]$ DPC 602 excreted GSH conjugate with MH^+ at *m/z* 854. The MS/MS spectrum of the MH^+ ions of the conjugates from animals dosed with labeled and non-labeled DPC 602 is shown in Figure 14. The mass spectral data of the labeled DPC 602 conjugate suggested loss of the two deuteriums on the aminomethyl side chain, hence giving a net addition of only 1 amu to MH$^+$. An addition of GSH to a reactive metabolic intermediate usually leads to a net gain of either 305, 307, 321 or 323 amu to the parent mass (Table 2). The difference of 2 amu from the expected GSH conjugate of a hydroxylated metabolite (319 vs 321 amu) suggested the possibility of an unsaturated double bond. The loss of two deuteriums combined with a net increase of 1 amu (corresponding to the mass increment as a result of retention of C13) to the molecular weight of the GSH adduct of [^{13}CD$_2$] DPC 602 suggested a structure such as shown in Figure 15. However, the structures of these GSH conjugates could only be confirmed by NMR experiments. The ^{1}H-NMR and the assignments of the proton signals of the GSH adduct of DPC 602 are shown in Figure 15. In the NMR studies, similar correlation of the cysteine methylene (β, β') protons and the aromatic proton, *a*, with the quaternary imino carbon at 148 ppm was observed for the GSH conjugates of DPC 602 and [^{13}CD$_2$] DPC 602 (see Figures 16 and 17, respectively).

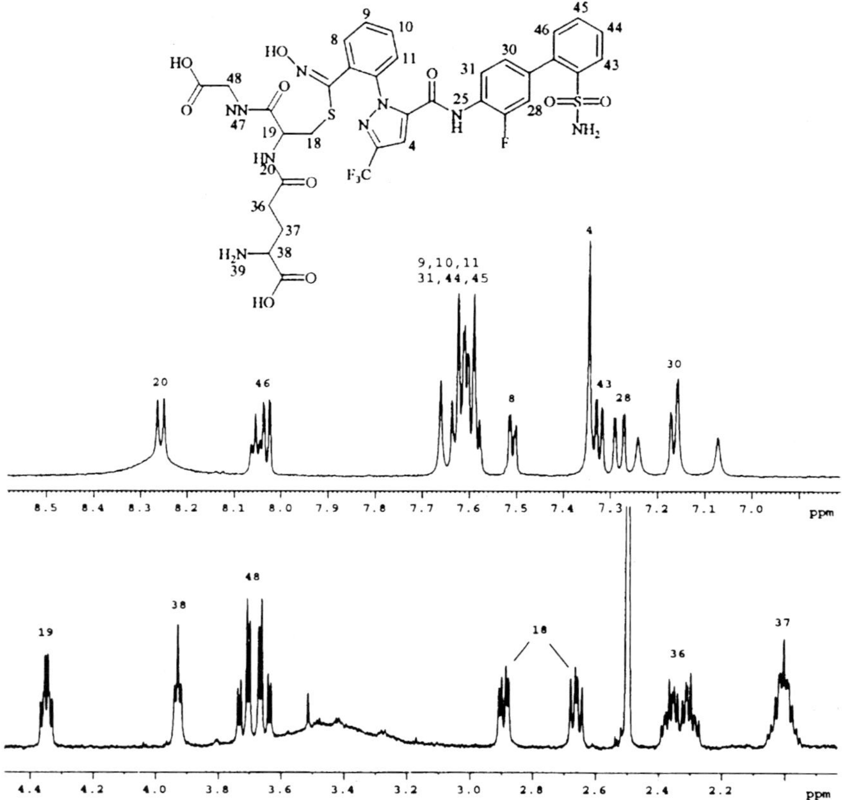

Figure 14. LC/MS/MS of GSH conjugates present in bile of rats dosed with DPC 602 (A) and [^{13}CD$_2$] DPC 602 (B). For the position of labeling see Figure 13.

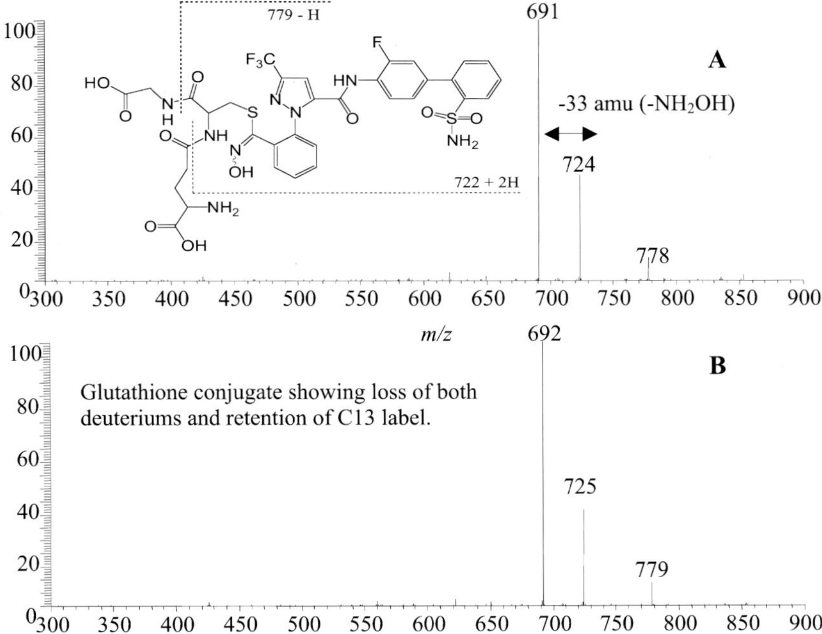

Figure 15. ^{1}H-NMR of GSH adduct isolated from bile of rats dosed with 100 mg/kg of DPC 602. The assignments for the proton signals were made from the TOCSY experiments.

The strong correlation observed in Figure 17 is due to the C13 label on the aminomethyl side chain. The presence of C13 label enabled the correlation data to be obtained in a relatively short period of time. The correlations clearly indicated that the benzyl carbon of the parent compound (DPC 602) was now a quaternary carbon with a chemical shift consistent with an imine. These correlations and the chemical shift of the quaternary carbon are consistent with the proposed structure. The GSH adduct of benzaldehyde oxime was synthesized to serve as a further confirmation for the proposed structure of the GSH adducts. The GSH conjugate of benzaldehyde oxime was characterized and compared with the GSH adduct of DPC 423 and DPC 602. The LC-ESI/MS showed MH$^+$ at *m/z* 427. MS/MS of MH$^+$ produced a characteristic base peak at *m/z* 265. This loss of the pyroglutamate moiety followed by the loss of hydroxylamine to produce a cyclic product with *m/z* 265 is similar to the fragmentation pattern observed for all the benzylamine GSH adducts isolated from *in vivo*. The HMBC spectrum of the synthetic GSH adduct showed a strong correlation of the cysteine methylene protons (β, β') and the aromatic protons (*a* and *b*) with the quaternary imino carbon at 150 ppm. These correlations were nearly identical

    A. E. Mutlib and John P. Shockcor

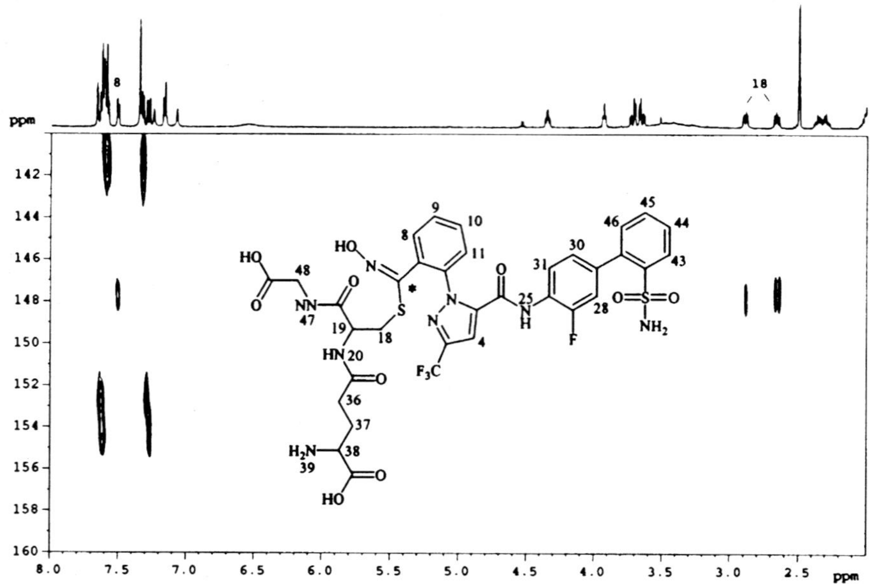

Figure 16. Partial HMBC spectrum of oxime derived GSH adduct isolated from bile of rats dosed with 100 mg/kg of DPC 602. Note the correlation between the cysteine protons (position 18) and aromatic proton (position 8) to the imino carbon at 148 ppm (marked with an asterix (*)).

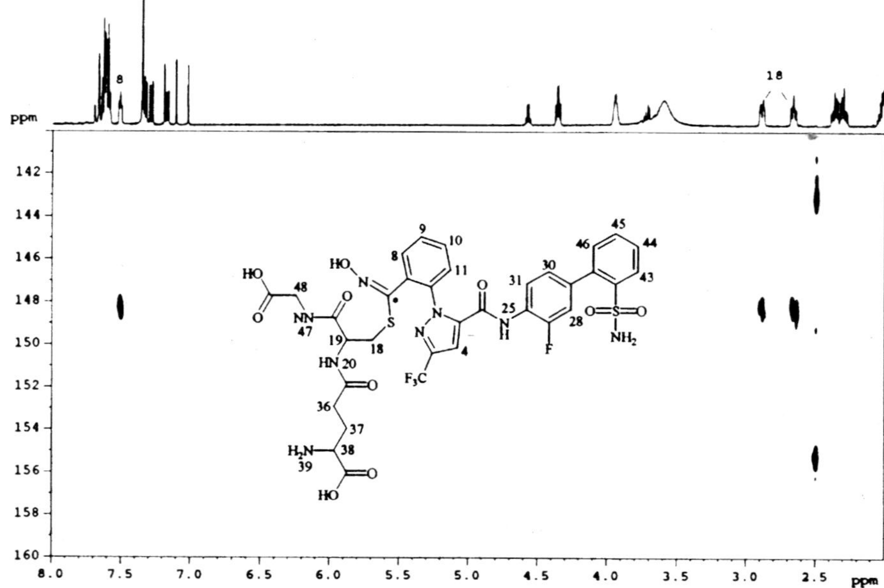

Figure 17. Partial HMBC spectrum of oxime derived GSH adduct isolated from bile of rats dosed with 100 mg/kg of $^{13}CD_2$-DPC 602. A strong correlation of the cysteine methylene protons (position 18) and the aromatic proton (*8*) with the imino carbon (*), also the position of ^{13}C label, at 148 ppm was observed. This result was obtained in a relatively short period of time due to the presence of stable labeled isotope (^{13}C).

to those seen in the spectra of the metabolites. *In vivo* and *in vitro* studies with synthetic oxime intermediate from DPC 423 showed an adduct identical to the one isolated from bile of rats dosed with DPC 423. This supported the intermediacy of an aldoxime as a precursor to the GSH adducts. It is postulated that the benzylamine moiety of DPC 423 and its analogue, DPC 602, was oxidized to a hydroxylamine, which was subsequently converted to a nitroso intermediate. Subsequent rearrangement of the nitroso leads to an aldoxime, which in turn is metabolized by P450 to a reactive intermediate. It was postulated that the aldoxime produces a radical or a nitrile oxide intermediate that reacts with GSH and hence produces this unusual GSH adduct.[64] In this study, a combination of LC/MS/MS, NMR and stable labeled analogues enabled us to demonstrate unequivocally the structures of unique GSH adducts formed from bioactivation of oximes. The structure of this GSH adduct enabled us to postulate plausible metabolic pathways leading to reactive metabolic intermediates.

5.5.5 *Metabolism of benzylamine. Application of stable isotope labeled compound in metabolism studies.*

The *in vivo* and *in vitro* disposition of benzylamine was investigated in rats.[68] Benzylamine was metabolized to only a small extent by rat liver subcellular fractions. In contrast, it was extensively metabolized *in vivo* in rats. *In vivo* studies performed with stable isotope labeled benzylamine enabled rapid mass spectrometric identification of metabolites present in rat bile and urine. The major metabolite of benzylamine was the hippuric acid (**M1**) formed by glycine conjugation of benzoic acid (Scheme 7). LC/MS analysis of bile and urine obtained from rats dosed with 1:1 equimolar mixture of either d_0:d_7 (deuterium labeled at all the positions)- or d_0:d_2 (deuterium labeled on the benzylic position only)-benzylamine, showed the presence of several GSH related adducts (**M2-M8**) in addition to the hippuric acid metabolite (Scheme 7). Rats that were dosed with 1:1 w/w mixture of d_0:d_2 benzylamine produced **M3** that showed MH$^+$ at *m/z* 316 only; i.e. both the deuteriums were lost due to sequential oxidation at the benzylic position. Studies with 1:1 w/w mixture of d_0:d_7 benzylamine showed loss of 2 deuteriums and retention of 5 labels (MH$^+$ at *m/z* 321), presumably on the aromatic ring of **M3**. The data suggested that an addition of oxygen at the benzylic position took place. Furthermore, because both of the deuteriums were lost, a further oxidation was postulated to occur at this position. The oxidation at this position was later confirmed by using benzamide as the substrate, which also produced **M3** as a metabolite. This is one example of how the labeled compounds can be used to accelerate structural elucidation of metabolites and also provide an insight into the possible mechanisms of metabolite formations.

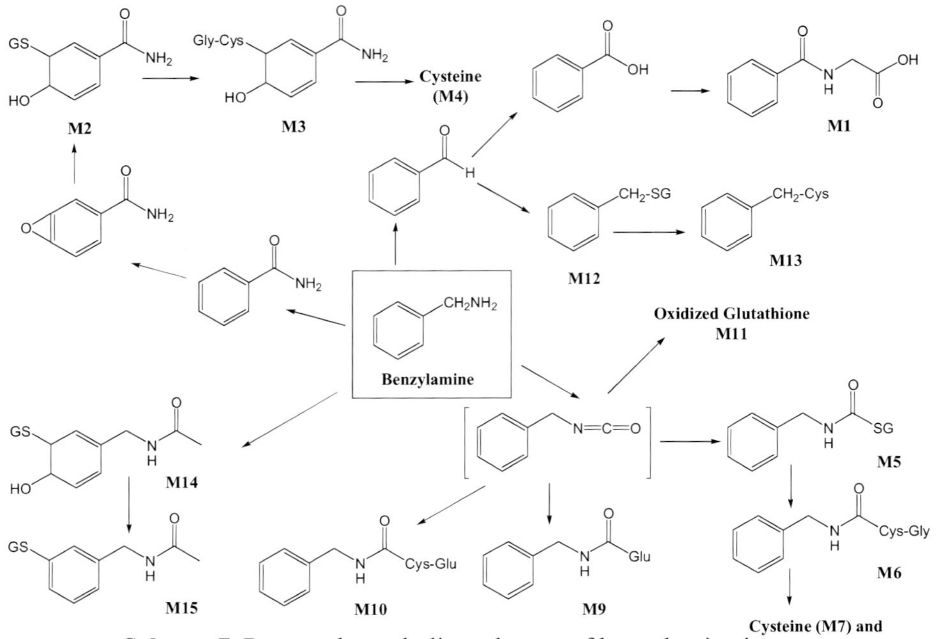

Scheme 7. Proposed metabolic pathways of benzylamine in rats.

A number of other unusual metabolites, including **M9**-**M11** were detected in bile samples by the use of deuterium labeled compound and LC/MS. The presence of various GSH adducts indicated that benzylamine was metabolized to a number of reactive intermediates. A previously undocumented pathway included the formation of a new carbon-nitrogen bond that led to a potentially reactive intermediate, Ar-CH$_2$-NH(CO)-X, capable of interacting with various nucleophiles. The origin of this reactive intermediate was postulated to occur via the formation of either a formamide or carbamic acid metabolites. Metabolites which were produced by the reaction of this intermediate, Ar-CH$_2$-NH(CO)-X are shown in Scheme 7. Bioactivation of amines via this pathway has not been previously described. The complete characterization of some of the metabolites was achieved by synthesizing appropriate standards since it was easier to synthesize than to isolate these compounds from bile. The mass spectral and chromatographic comparisons were made to confirm the identities of the metabolites. This provided conclusive evidence for the existence of reactive metabolic intermediates such as benzyl isocyanate. The formation of these reactive intermediates from benzylamine is novel and further studies are needed to define the metabolic pathways/enzymes responsible for their formations. The conversion of an amine to such reactive metabolites has been

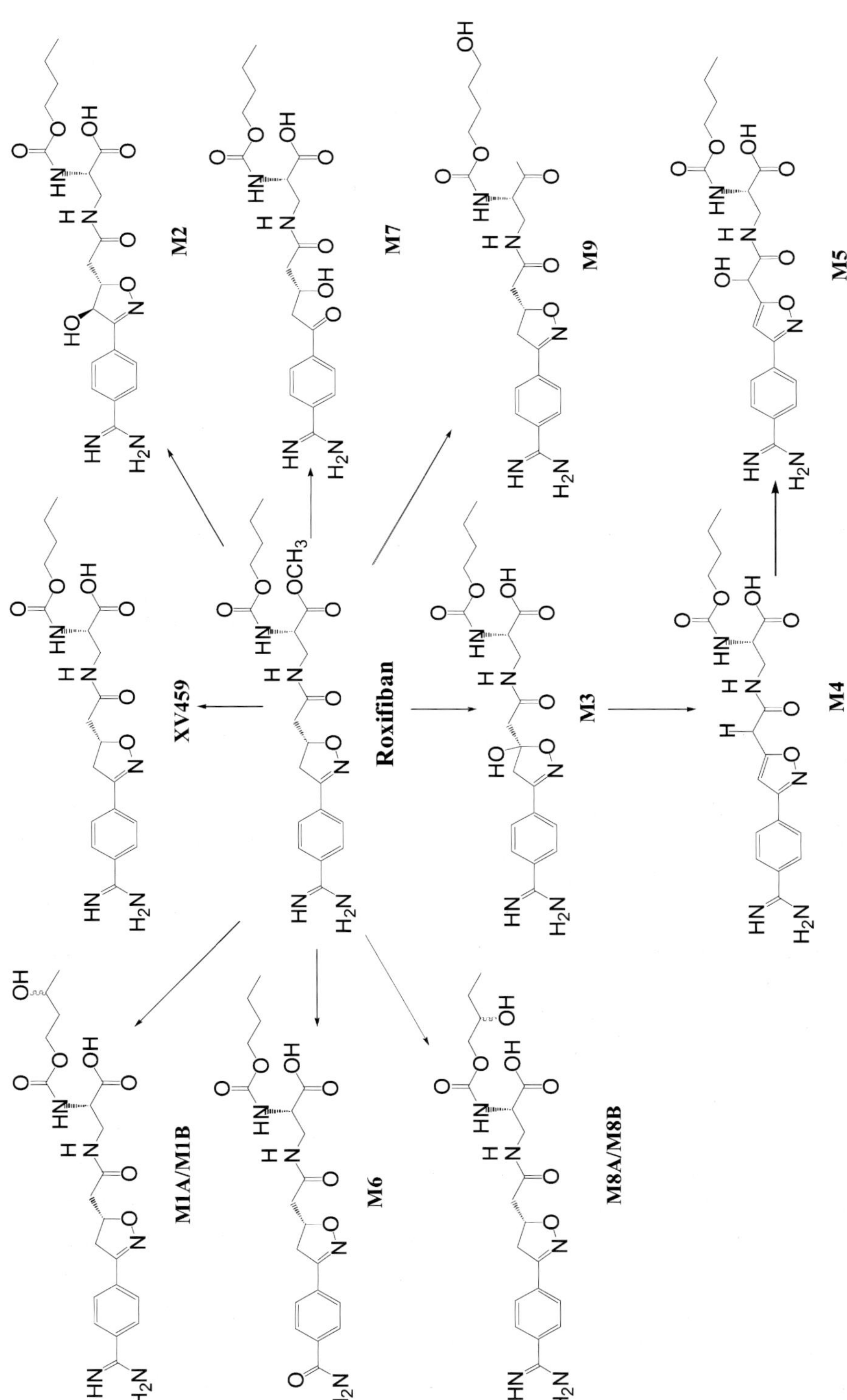

Scheme 8. Proposed metabolic pathways of roxifiban in rats.

overlooked in the past perhaps due to either difficulty in characterizing such polar metabolites or because of low levels of metabolites produced and detected. By the use of very sensitive analytical methods, we were able to demonstrate the existence of multiple metabolic pathways, which produced unusual metabolites. Characterization of some of the minor metabolites provided an insight into the possible bioactivation pathways for benzylamine. It is essential to elucidate the structures of detectable metabolites of a compound so that an appreciation of various metabolic pathways involved in its disposition (detoxification and bioactivation) can be realized.

5.3   Pharmacologically Active Metabolites

5.3.1 Metabolism of methyl N^3-[2-{3-(4-formamidino-phenyl)-isoxazolin-5(R)-yl}-acetyl]-N^2-(1-butyloxycarbonyl)-2,3-(S)-diaminopropionate (roxifiban).

The metabolism of roxifiban in rats, dogs and humans has been described previously.[43] Roxifiban is an ester prodrug that requires *in vivo* hydrolysis to the zwitterion XV459 (Scheme 8) producing a potent and selective antagonist of the platelet glycoprotein IIb/IIIa receptor.[69,70] Metabolism studies showed that XV459 was formed rapidly from roxifiban in the presence of liver microsomes, liver slices or intestinal segments. Because of the rapid hydrolysis of roxifiban, only the metabolites of the zwitterions were isolated and characterized. The conversion of roxifiban or its hydrolyzed product, XV459, to other metabolites was minimal and it was a challenge to characterize the hydroxylated products that were formed in small quantities. In this case, it was important to characterize all of the minor metabolites because subsequent clinical studies showed that pharmacodynamic response was perhaps due to the synergestic effects of XV459 and its hydroxylated metabolites. Using deuterium labeled roxifiban facilitated the identification of these metabolites in biological fluids. The metabolites were characterized by ESI-LC/MS, high-field NMR and LC/NMR experiments. The bile duct-cannulated Sprague-dawley rats were dosed with either unlabeled roxifiban or with equimolar mixture of $d_0 : d_4$-roxifiban (four unsubstituted positions on the aromatic ring labeled with deuterium) at 100 mg/kg and bile/urine collected at 0-8 and 8-24 hr intervals. LC/MS was carried out by coupling a HPLC system to a mass spectrometer. HPLC was carried out using a quaternary pump coupled in sequence to an autosampler and to a Beckman C18 column (250 x 4.6 mm). The metabolites were separated by an isocratic solvent system consisting of acetonitrile/10 mM ammonium formate, pH 3.5 (15:85 v/v) delivered at 1.0 ml/min. Aliquots of urine samples were injected directly onto the HPLC column. A post-column split (1:1) introduced $\cong$ 0.5 ml/min eluent to the mass spectrometer. LC/NMR was obtained for **M1** (see Scheme 7), which eluted as a polar metabolite. The chromatography was carried out on a Waters Symmetry C18 column (3.9 x 150 mm) using a gradient solvent system consisting of acetonitrile and D_2O (each containing 0.05% trifluoroacetic acid). The percentage of acetonitrile was increased linearly from 10 to 50% over 30 min. ^{1}H-NMR spectra were obtained in stop-flow mode using a Bruker Avance 500 MHz NMR spectrometer equipped with a ^{1}H/^{13}C LC/NMR flow-probe, with a cell volume of 120 μL. Suppression of the residual water and acetonitrile signals was carried out using WET solvent suppression method.[71] Data were acquired with 256 transients into 32 K data points with a pulse repetition time of 2.8 s. Chemical shifts were referenced to acetonitrile at δ 2.0 ppm. The NMR data for the rest of the metabolites were obtained in DMSO-d_6 on a Bruker Avance 500 MHz spectrometer. The structures of these metabolites were obtained using one- and two-dimensional NMR techniques. The metabolites of roxifiban excreted in rat urine are shown in Scheme 8. The levels of these metabolites in rat urine were very low compared to the excreted unchanged parent compound. However, each of these metabolites was isolated in sufficient quantities for their *in vitro* pharmacological activities to be determined. Interestingly, all of the hydroxylated metabolites exhibited similar pharmacological activity as the parent zwitterions, XV459. Subsequent studies in humans showed that these metabolites were produced in significant quantities by intestinal metabolism of roxifiban when the compound was administered at very low dose (1-2 mg). A discrepancy in pharmacokinetic profile of XV459 (active compound) and the pharmacological activity was explained by the presence of these hydroxylated metabolites. Hence, identification of metabolites played an important role in explaining this discrepancy in clinical trials. Furthermore, because the identities of these metabolites were well established during the preclinical studies, the demonstration of the existence of pharmacologically active entities in biological samples such as human plasma and urine (even at such low administered doses) was facile.

6   CONCLUSIONS

Identification of metabolites is a very critical step in succeeding at discovery and development of novel therapeutic agents. It is important to know the metabolic soft spots in a molecule, as well as the formation of any reactive or pharmacologically active metabolites. This information can be used to make decisions regarding chemistry efforts in modifying existing structures, performing further pharmacological tests on metabolites, and assessing potential liabilities associated with formation of reactive intermediates. With the analytical tools such as LC/MS, NMR and LC/NMR at our disposal, it has become easier to characterize metabolites in early stages of drug discovery. The structural information obtained from these early metabolism studies can be used to redirect synthetic efforts in obtaining better compounds for development. The sooner the metabolic soft spots are identified through metabolite identification, the better the success rate for making metabolically stable compounds. One must keep in mind the potential species differences in metabolism when investigating metabolic soft spots in a compound. Significant investment is currently being made in screening compounds that have the potential liability of forming reactive metabolites. Currently LC/MS is the method of choice in detecting such reactive intermediates (usually trapped by a nucleophile such as glutathione) generated *in vitro* using subcellular fractions. Although a lot of compounds can form GSH conjugates *in vitro* (suggesting formation of reactive metabolites), the significance of these findings can be debatable. However, considering the potential risks associated with the formation of reactive metabolites, it would be desirable to have compounds that do not form such intermediates. The ability to perform LC/MS^n experiments, notably with ion trap instruments, has made it feasible to better determine the structures of metabolites than in the past. The use of hybrid instruments such as the Q-TOF® that can provide accurate masses of metabolites, as well their fragment ions is making metabolite identification much easier than in past. However, in most cases, NMR is an indispensable tool in confirming the structures of metabolites. Ideally, an ion trap (capable of MS^n experiments), a triple quadrupole (for MS/MS, parent ion and neutral loss scans), a Q-TOF (for accurate mass measurements) and dedicated LC/NMR instruments will enable a metabolism chemist to elucidate structures of metabolites accurately and in a timely manner. LC/NMR techniques are now almost routinely used to profile and identify metabolites of compounds present in biological fluids such as bile and urine. The general stop-flow approach in LC/NMR is the most widely used method to characterize metabolites; although various hyphenated techniques such LC/NMR/MS and SPE/LC/NMR techniques are gaining popularity. However, not all drug metabolism studies will be amenable to the LC/NMR techniques. Frequently one is still faced with the challenge of isolating metabolites from biological samples, especially if larger quantities of compounds are needed for elaborate 2-D NMR experiments. Nevertheless, LC/ NMR will play an extremely important role in the determination of metabolite structures in drug metabolism laboratories. Mass spectrometry and NMR are complimentary techniques, with data needed from both to elucidate the structures of unknown metabolites. These techniques are important when the synthetic standards of metabolites are not available, especially in the discovery stages or if the metabolites possess unusual or unexpected structures. Along with these two analytical techniques, the use of stable labeled compounds has been frequently employed by the metabolism chemists to understand the metabolic disposition of xenobiotics. Compounds labeled with deuterium or ^{13}C have been frequently used to elucidate the structures of metabolites as well as the possible mechanisms leading to their formation. While the aim of the metabolism chemist is usually to identify the structures of major metabolites, it is often necessary to investigate the nature of so-called minor metabolites as well. These minor metabolites could possess inherent pharmacological activities or they may contribute toward the toxicity observed in a particular species. As illustrated within this chapter, identification of metabolites has led to the discovery of novel metabolites and metabolic pathways, identification of pharmacologically active entities, and elucidation of potentially toxic agents.

REFERENCES

1. Obach, R. S. *Drug Metab Dispos.* **1999**, *27*, 1350.
2. Obach, R. S. *Curr. Opin. Drug Disc. Dev.* **2001**, *4*, 36.
3. Mutlib, A. E.; Chen, H.; Nemeth, G. A.; Markwalder, J. A.; Seitz, S. P.; Gan, L. S.; and Christ, D. D. *Drug Metab. Dispos.* **1999**, *27*, 1319.
4. Mutlib, A. E.; Gerson, R. J.; Meunier, P. C.; Haley, P. J.; Chen, H.; Gan, L-S.; Davies, M. H.; Gemzik, B.; Christ, D. D.; Krahn, D. F.; Markwalder, J. A.; Seitz, S. P.; Robertson, R. T.; and Miwa, G. T. *Toxicol. Appl. Pharmacol.* **2000**, *169*, 102.
5. Gillette, J. R. In *Xenobiotic Metabolism and Disposition*, Ed. By R. Kato, R.W. Estabrook, and M.N. Cayen, **1989**, Taylor and Francis, London, UK.
6. Burek, D. J.; Nitsche, K. D.; Bell, T. J.; Wackerle, D. L.; Childs, R. C.; Beyer, J. E.; Dittenber, D. A.; Rampy, L. W.; and McKenna, M. J. *Fundam. Appl. Toxicol.* **1984**, *4*, 30.
7. Green, T. *Xenobiotica.* **1990**, 20, 1233.
8. Reitz, R. H.; Mandrala, A. L.; and Guengerich, F. P. *Toxicol. Appl. Pharmacol.* **1989**, *97*, 230.
9. Chinje, E.; Kentish, P.; Jarnot, B.; George, M.; and Gibson, G. *Toxicol. Lett.* **1994**, *71*, 69.
10. Bell, D. R.; Plant, N. J.; Rider, C. G.; Na, L.; Brown, S.; Ateitalla, I.; Acharya, S. K.; Davies, M. H.; Elias, E.; and Jenkins, R. *Biochem. J.* **1993**, *294*, 173.
11. Hard, G. C.; and Whysner, J. *Crit. Rev. Toxicol.* **1994**, *24*, 231.
12. Monahan, B. P.; Ferguson, C. L.; Killeavy, E. S.; Lloyd, B. K.; Troy, J.; and Cantilena, L. R. *JAMA.* **1990**, *264*, 2788.
13. Honig, P. K.; Wortham, D. C.; Zamani, K.; Conner, D. P.;Mullin, J. C.; and Cantilena, L. R. *JAMA,* **1993**, *269*, 1535.
14. Flower, R. J.; Moncada, S.; and Vane, J. R. In *Goodman and Gilman's The Pharmacological Basis of Therapeutics,* **1985**, Macmillan Publisher, Inc., New York.
15. Mulder, G. J. *Trends Pharmacol. Sci.* **1992**, *13*, 302.
16. Bray, K. and Quast, U. *Naunyn-Schmeidbergs arch. Pharmacol.* **1991**, *344*, 351.

17. Cotter, R. J. and Fenselau, C. *Biomed. Mass Spectrom.* **1981**, *6*, 287.
18. Schulten, H. R. and Games, D. E. *Biomed. Mass Spectrom.* **1974**, *1*, 120.
19. Bruins, A. *Biomed. Mass Spectrom.* **1981**, *8*, 31.
20. Covey, T. R.; Lee, E. D.; Bruins, A. P.; and Henion, J. D. *Anal. Chem.* **1986**, *58*, 1451A.
21. Bruins, A. P.; Covey, T. R.; and Henion, J. D. *Anal. Chem.* **1987**, *59*, 2642.
22. Lee, E. D.; Mueck, W.; Henion, J. D.; and Covey, T. R. *J. Am. Chem. Soc.* **1989**, *111*, 4600.
23. McLuckey, S. A.; Van Berkel, G. J.; Glish, G. L.; Huang, E. C.; and Henion, J. D. *Anal. Chem.* **1991**, *63*, 375.
24. Huang, C. H.; Wachs, T.; Conboy, J. J.; and Henion, J. D. *Anal. Chem.* **1990**, *62*, 713A.
25. Lee, E. D.; Muck, W.; and Henion, J. D. Biomed. *Environ. Mass Spectrom.* **1992**, *21*, 189.
26. Gilbert, J. D.; Hand, E. L.; Yuan, A. S.; Olah, T. V.; and Covey, T. R. *Biol. Mass Spectrom.* **1992**, *21*, 63.
27. Mutlib, A. E.; Shockcor, J.; Espina, R.; Graciani, N.; Du, A.; and Gan, L-S. *J. Pharmacol. Exp. Ther.* **2000**, *294*, 735.
28. Mutlib, A. E.; Chen, H. C.; Nemeth, G.; Gan, L-S., and Christ, D. D. *Drug Metab. Dispos.* **1999**, *27*, 1045.
29. Burlingame, A. L.; Baillie, T. A.; and Russell, D. *Anal. Chem.* **1992**, *64*, 467R and references cited therein
30. Capriolli, R. M., Malorni, A, and Sindona, G (eds) *Mass Spectrometry in Biomolecular Sciences* **1996**.
31. Lindon, J. C.; Nicholson, J.K.; Wilson, I.D.; *Adv. in Chromatogr.* **1995**, *36*, 315.
32. Mutlib, A. E.; Strupczewski, J.T.; and Chesson, S. M. *Drug Metab. Dispos.* **1995**, *23*, 951.
33. Shockcor, J. P.; Frick, L. W.; Wurm, R. M.; Sanderson, P. N.; Farrant, R. D.; Sweatman, B. C.; Lindon, J. C.; *Xenobiotica* **1996**, *26*, 189.
34. Shockcor, J. P.; Silver, I. S.; Wurm, R. M.; Sanderson, P, N.; Farrant, R. D.; Sweatman, B. C.; Lindon, J. C. *Xenobiotica* **1996**, *26*, 41.
35. Lindon, J.C.; Nicholson, J.K.; Wilson, I.D; *Progress in Nuclear Magnetic Resonance Spectroscopy* **1996**, *29*, 1.
36. Shockcor, J. P.; Unger, S. E.; Wilson, I. D.; Foxall, P. J. D.; Nicholson, J. K.; Lindon, J. C. *Anal. Chem.* **1996**, *68*, 4431.
37. Farrant, R. D.; Cupid, B. C.; Nicholson, J. K.; Lindon, J. C. *J. Pharm. Biomed. Anal.* **1997**, *16*, 1.
38. Ehlhardt, W.; Woodland, J. M.; Baughman, T. M.; Vandenbranden, M.; Wrighton, S. A.; Kroin, J. S.; Norman, B. H.; Maple, S. R.; *Drug Metab. Dispos.* **1998**, *26*, 42.
39. Scarfe, G. B.; Wright, B.; Clayton, E.; Taylor, S.; Wilson, I. D.; Lindon, J. C.; Nicholson, J. K. *Xenobiotica* **1998**, *28*, 373.
40. Nicholls, A. W.; Lindon, J. C.; Farrant, R. D.; Shockcor, J. P.; Wilson, I. D.; Nicholson, J. K. *J. Pharm. Biomed. Anal.* **1999**, *20*, 865.
41. Ismail, I.M.; Dear, G. J.; Mutch, P. J.; Davies, L. H.; Plumb, R. S.; Sweatman, B. C. *Xenobiotica* **1999**, *29*, 957.

42. Dachtler, M.; Handel, . H.; Glaser, T.; Lindquist, D.; Hawk, R. M.; Karson, C. N.; Komoroski, R. A.; Albert, K.; *Mag. Res. Chem.* **2000**, *38*, 951.

43. Mutlib, A.; Diamond, S.; Shockcor, J.; Way, R.; Nemeth, G.; Gan, L.; Christ, D. *Xenobiotica* **2000**, *30*, 1091.

44. Shockcor, J. P.; Unger, S. E.; Savina, P.; Nicholson, J. K.; Lindon, J. C. *J. Chromatogr. B* **2000**, *748*, 269.

45. Sidelmann, U. G.; Bjornsdottir, I.; Shockcor, J. P. Hansen, S. H.; Lindon, J. C.; Nicholson, J. K.; *J. Pharm. Biomed. Anal.* **2001**, *24*, 569.

46. Zhang, K. Y. E.; Hee, B.; Lee, C. A.; Liang, B. H.; Potts, B.C. M. *Drug Metab. Dispos.* **2001**, *29*, 729.

47. Mutlib, A.; Shockcor, J.; Chen, S.Y.; Espina, R.; Lin, J.R.; Graciani,N.; Prakash, S. *Drug Metab. Dispos.* **2001**, *29*, 1296.

48. Prueksaritanont, T.; Subramanian, R.; Fang, X.J.; Ma, B; Qiu, Y; Lin, J.H.; Pearson, P.G. Baillie, T.A. Drug Metab. Dispos. **2002**, *30*, 505.

49. Bao, D.H.; Thanabal, V; Pool, W.F. *J. Pharm. Biomed. Anal.* **2002**, *28*, 23.

50. Chen, H.; Shockcor, J.P.; Chen, W.Q.; Espina, R.; Gan, L.S.; Mutlib, A.E. Chem. Res. Toxicol. **2002**, *15*, 388.

51. Fisher, M.B.; Jackson, D.; Kaerner, A.; Wrighton, S.A.; Borel, A.G. *Drug Metab. Dispos.* **2002**, *30*, 270.

52. Mutlib, A.E.; Shockcor, J.P.; Chen, S.Y.; Espina, R.J.; Pinto, D.J.; Orwat, M.J.; Prakash, S.R.; Gan, L.S. Chem. Res. Toxicol. **2002**, *15*, 48.

53. Gerson, R. J.; Mutlib, A. E.; Meunier, P. C.; Haley, P. J.; Chen, H.; Gan, L-S.; Davies, M. H.; Gemzik, B.; Christ, D. D.; Krahn, D. F.; Markwalder, J. A.; Seitz, S. P.; Miwa, G. T.; and Robertson, R. T. *Toxicol. Sci.* **1999**, *48* (suppl. 1), 1833.

54. Commandeur, J. N. M.; Stijntjes, G. J.; and Vermeulen, N. P. E. *Pharamcol. Rev.* **1995**, *47*, 271.

55. Mutlib, A.E.; Chen, H.; Shockcor, J.P.; Chen, S.Y.; Espina, R.J.; Cao, K.; Du, A.; Nemeth, G.; Prakash, S.R.; Gan, L.S. Chem. Res. Toxicol. 2000, *13*, 775.

56. Lehman, J. P.; and Fenselau, C. *Drug Metab. Dispos.* **1982**, *10*, 446.

57. Fenselau, C. and Johnson, L. P. *Drug Metab. Dispos.* **1980**, *8*, 274.

58. Cairns, T. and Siegmund, E. G. *Anal. Chem.* **1982**, *54*, 2456.

59. Liberato, D. J.; Fenselau, C.; Vestel, M. L.; and Yergey, A. L. *Anal. Chem.* **1983**, *55*, 1742.

60. Mutlib, A. E.; Cheung, H. T. A.; and Watson, T. R. *J. Steroid. Biochem.* **1987**, *28*, 65.

61. Dutton, G. J. (ed). "Glucuronic acid, Free and Combined" **1966,** Academic Press, New York.

62. Mutlib, A. E. and Abbott, F. S. *Drug Metab. Dispos.* **1992**, *20*, 451.

63. Ross, D.; farmer, P. B.; Gescher, A.; Hickman, J. A.; and Threadgill, M. D. *Biochem. Pharmacol.* **1983**, *32*, 1773.

64. Mutlib, A. E.; Chen, S-Y.; Espina, R. J.; Shockcor, J.; Prakash, S. R.; and Gan, L-S. *Chem. Res. Toxicol.* **2002**, *15*, 63.

65. Straub, K., Davis, M. and Hwang, B. *Drug Metab. Dispos.* **1988**, *16*, 359.

66. Tremaine, L. M., Stroh, J. G., and Ronfeld, R. A. *Drug Metab. Dispos.* **1989**, *17*, 58.

67. Delbressine, L. P. C., Funke, C. W., van Tilborg, M., and Kaspersen. *Xenobiotica* **1990**, *20*, 133.

68. Mutlib, A.E.; Dickenson, P.; Chen, S-Y.; Espina, R. J.; Daniels, J. S.; and Gan, L-S. Chem. *Res. Toxicol.* **2002**, in press.

69. Xue, C. B.; Rafalski, M.; Roderik, J.; Eyermann, C. J.; Mousa, S.; Olson, R. E.; and DeGrado, W. F. *Bioorg. Med. Chem..* **1996**, *6*, 339.

70. Wityak, J.; Sielecki, T. M.; Pinto, D. J.; Emmett, G.; Sze, J. Y.; Liu, J.; Tobin, A. E.; Wang, S.; and Jiang, B. *J. Med. Chem.* **1997**, *40*, 50.

71. Smallcombe, S. H.; Patt, S. L.; and Keifer, P. J. Mag. Res. A. **1995**, *117*, 295.

Chapter 3

Bioactivation

Jack Uetrecht, M.D., Ph.D.

*Professor of Pharmacy and Medicine, Canada Research Chair in Adverse Drug Reactions,
University of Toronto, 19 Russell St., Toronto, Canada M5S 2S2*

1   INTRODUCTION

The term bioactivation will be used in this chapter to refer to metabolism of drugs to chemically reactive metabolites. In another context it could easily refer to the conversion of a prodrug to a pharmacologically active agent. Bioactivation is an important issue for drug development because many types of adverse reactions are believed to be due to chemically reactive metabolites rather than due to the parent drug. It has recently been reported that, of the new prescription drugs approved in the US during the period from 1975 to 2000, 10.2% acquired a new "black box" warning or had to be withdrawn from the market.[1] In addition, unknown numbers of potential drugs were withdrawn from development late in the process because of evidence to suggest that they might be associated with an unacceptable incidence of adverse reactions. This constitutes a major problem for drug development, not to mention for the patients who have adverse reactions. In order to deal effectively with the issue we need a much better understanding of the mechanisms involved in adverse drug reactions, especially idiosyncratic drug reactions. However, it is likely that, even at the present time, drug development could be made safer and more efficient by being more cognizant of the issue of bioactivation at a very early stage in drug development. In the past, the major reason that drugs failed was because of unfavorable pharmacokinetic characteristics; however, better analytical instrumentation and more attention to this issue have made this a much less common reason for drugs to fail and has increased the relative importance of adverse drug reactions.

2   INVOLVEMENT OF REACTIVE METABOLITES IN ADVERSE REACTIONS

The idea that bioactivation is important for toxicity was first proposed by the Millers to explain the mechanism of chemical-induced carcinogenesis,[2] and this is now a well-established principle. Soon thereafter it was proposed that acetaminophen toxicity is due to a reactive imidoquinone,[3] and there is now ample evidence to support this premise. Over time, this concept has been broadened to include other types of adverse reactions.

2.1   *Types of adverse drug reactions*

There are several types of adverse reactions. The most common type involves an extension of the pharmacological properties of the drug; these are sometimes referred to as type A reactions.[4] For example, opiates are analgesics that act on the central nervous system, and they also cause central nervous system depression leading to drowsiness, and at high doses, can lead to apnea and death. Such reactions are predictable based on the pharmacology of the drug, and similar effects are likely to occur in animals during preclinical testing. Most type A reactions are mediated by the parent drug and do not involve metabolic activation. One very worrisome adverse reaction of this type, which did not receive much attention until recently when it led to the withdrawal of several drugs, is prolongation of the QT interval of the electrocardiogram. This prolongation can lead to a potentially fatal cardiac arrhythmia, i.e. *torsade de pointes*. It is idiosyncratic in that it involves ion channels, which are genetically polymorphic, and some patients are much more susceptible than others to this arrhythmia. However, although difficult in practice, in principle this reaction is predictable in that it is possible to determine the effect of drugs on the QT interval, and many of the polymorphisms associated with an increased risk of a serious arrhythmia are known. Therefore, I prefer not to call this reaction idiosyncratic, although it would be perfectly reasonable to do so, and most people would likely call this an idiosyncratic reaction.

I prefer to reserve the term idiosyncratic drug reaction for those reactions that do not occur in most patients at any dose and do not involve the known pharmacological properties of the drug. These reactions are often referred to as dose independent; however, it is very important to emphasize that, although most patients (and animals) will not have an idiosyncratic reaction at any dose, in susceptible patients the reactions are dose dependent. All effects are dose dependent! The dose response relationship is, in some cases, shifted to the left of the dose response relationship for the desired pharmacological effect; however, in many cases the dose response relationship for an idiosyncratic reaction

- in susceptible patients - is in the same range as the pharmacological effects. These reactions are also often referred to as allergic reactions, hypersensitivity reactions or type B (bizarre) reactions. Although somewhat less common than type A reactions, in many ways idiosyncratic drug reactions pose a larger problem because they are impossible to accurately predict from animal testing. In fact, they are usually not even detected in clinical trials where the average number of patients that are exposed to a drug is about 3000. It is also impossible at the present time to predict which patient is at increased risk of such a reaction. Although the incidence of such reactions is low, considering the number of patients who are treated with drugs and the number of different drugs that can cause idiosyncratic reactions, the total number of idiosyncratic reactions is significant, and many such reactions are quite serious. It is the unpredictable and serious nature of these reactions that make them a common cause for removal of a drug from the market.

2.2   The hapten hypothesis

The characteristics of idiosyncratic drug reactions suggest that they are mediated by the immune system. In a few cases this has been demonstrated but in most cases it has not. It is widely believed that most idiosyncratic drug reactions are caused by reactive metabolites of drugs rather than by the drugs themselves. It must be acknowledged that this hypothesis has not been proven; however, there is a very large amount of circumstantial evidence to support it. The principle, discovered almost a century ago by Landsteiner,[5] is that low molecular mass compounds are not immunogenic and, in order to induce an immune response, they must bind to proteins or other macromolecules. A small molecule that modifies a macromolecule to make it immunogenic is referred to as a hapten. The best understood idiosyncratic drug reaction is penicillin-induced anaphylactic reactions. It is clear that IgE antibodies against penicillin-modified proteins mediate these reactions. Penicillin is one of the few examples where the parent drug is chemically reactive without bioactivation, but the principle of a small molecule covalently binding to a macromolecule is the same as for drugs that require metabolic activation in order to form haptens. Another good example is halothane-induced hepatotoxicity. Halothane is metabolized to a very reactive trifluoroacetyl chloride (Fig. 1), which reacts with amino groups and other types of nucleophiles on protein, and halothane-hepatitis is associated with antibodies against trifluoroacetylated protein.[6] These antibodies indicate an immune response directed against metabolite-modified protein. The presence of these antibodies correlates with toxicity. This strongly suggests that the toxicity is an immune-mediated reaction caused by the reactive metabolite, although it is possible that the liver damage is actually mediated by cytotoxic T cells rather than antibodies and the antibodies are merely a marker for the immune response. Furthermore, in a series of analogs - halothane, isoflurane and desflurane - there is a direct relationship between the fraction of the drug that is metabolized to the reactive metabolite (virtually the same reactive metabolite for all three agents, Fig. 1) and the incidence of idiosyncratic hepatotoxicity.[7]

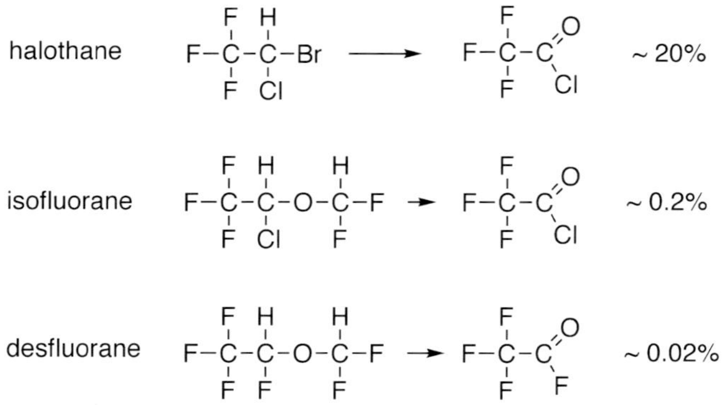

Figure 1. Differences in the degree of bioactivation of three similar gaseous anesthetics.

In many cases antibodies are produced against the enzyme that produced the reactive metabolite. A major antibody observed in halothane hepatitis is against CYP2E1, the major P450 responsible for the formation of the reactive metabolite.[8] Tienilic acid-induced hepatotoxicity is associated with antibodies against CYP2C9, the major enzyme responsible for the production of the reactive metabolite.[9] Dihydralazine-induced liver toxicity is associated with antibodies against CYP1A2,[10] and both CYP1A2 and CYP3A4 are responsible for the oxidation of dihydralazine.[11] Myeloperoxidase can also be a target for drug-induced antibodies. Several drugs are metabolized to reactive metabolites by myeloperoxidase/hydrogen peroxide and many of these same drugs are associated with the induction of a lupus-like syndrome.[12] Drug-induced lupus is often associated with antibodies referred to as p-ANCA, and many of these antibodies are against myeloperoxidase.[13] This data further supports the involvement of reactive metabolites in the induction of idiosyncratic drug reactions.

Most drugs that are associated with a relatively high incidence of idiosyncratic drug reactions are known to form reactive metabolites. On the other hand, most drugs probably form a reactive metabolite to some degree, and most are not associated with an unacceptable risk of idiosyncratic drug reactions. Two factors are likely to be important in the relevance of the reactive metabolites formed by a drug: the amount that actually binds to macromolecules and the specific macromolecules to which the reactive metabolites bind. Unfortunately, we know little about either factor. Quantification of binding is virtually impossible in humans in the target organ of the toxicity. Therefore, we do not have any precise data to determine if there is a good correlation between the amount of binding and the probability that a drug will be associated with a high incidence of idiosyncratic drug reactions. Any quantification that is done is either *in vitro* and/or in animals and usually only in hepatic tissue; therefore, we can only guess what is happening in other organs in humans. Furthermore, only a few proteins that are modified by reactive metabolites have been identified, and there is no clear picture of which ones are most relevant for idiosyncratic drug reactions.[14] One simple empirical observation is that drugs given at a dose of 10 mg/day or less are unlikely to be associated with a high incidence of idiosyncratic drug reactions.[15] There is nothing magic about 10 mg/day, but if the dose is sufficiently low, even if the drug efficiently forms a reactive metabolite, the amount formed appears to be too low to cause an idiosyncratic reaction. The converse is also true: if the dose of a drug is very high there is a higher probability that it will cause an idiosyncratic reaction. Two good examples are felbamate and procainamide, which are given in doses of greater than a gram a day. Both form reactive metabolites and both are associated with a relatively high incidence of idiosyncratic drug reactions.

2.3   *The Danger Hypothesis*

There may be other risk factors in addition to covalent binding that are important in the induction of an idiosyncratic drug reaction. It is know that when a foreign protein is injected into a person or animal there is little immune response unless an adjuvant is added. Janeway has referred to adjuvants as the immunologist's "dirty little secret".[16] Adjuvants upregulate costimulatory molecules on antigen presenting cells, and without costimulation, the usual response of the immune system to foreign protein is tolerance. Polly Matzinger has gone one step further and proposed that the "foreignness" of a protein is not an important factor in the induction of an immune response.[17-19] Unless the protein is associated with some sort of "danger signal" the response of the immune system is tolerance. Obviously, other sorts of cell damage may induce a danger signal analogous to that induced by adjuvant, which would upregulate costimulatory molecules, and so the Danger Hypothesis is not completely novel.

However, there are fundamental differences between the classical model in which the immune system responds to foreign proteins and the Danger Hypothesis.

There are many examples in which a factor that would be expected to act as a danger signal increases the risk of an idiosyncratic drug reaction. These include the increase in the incidence of adverse reactions to drugs, especially sulfonamides, in patients with AIDS.[20,21] AIDS is associated with immunosuppression and might be expected to decrease the risk of an immune-mediated reaction. However, it is also likely to produce a strong danger signal, and macrophages are activated in patients with AIDS. Surgery appears to increase the risk of procainamide-induced agranulocytosis and heparin-induced thrombocytopenia by a factor of about ten, and surgery is likely to provide a strong danger signal.[22,23] Even something as seemingly trivial as influenza vaccine appears to increase the risk of vesnarinone-induced agranulocytosis.[24] By their nature, reactive metabolites are likely to cause some degree of cell damage, and the ability of a drug to cause cell damage, in addition to acting as a hapten, may be an important factor in whether it is associated with a high incidence of idiosyncratic drug reactions. This hypothesis needs to be tested because it may make it possible to more accurately predict when formation of a reactive metabolite is likely to be associated with a high incidence of idiosyncratic reactions and, just as important, when it is less likely to be important. If drug-induced danger is an important factor then metabolites that are very reactive and only bind to the enzyme that formed them should not be associated with idiosyncratic drug reactions. It seems unlikely that modification of an enzyme, such as P450, would cause a danger signal, although such a reactive metabolite would still be undesirable because it could inhibit P450 and lead to drug interactions. One possible test is tienilic acid, which is said to only bind to P450.[25] If this is true it should not generate a danger signal, yet it is well known to cause idiosyncratic liver toxicity. However, the binding is decreased by the presence of glutathione (abbreviated G-SH in figures) and so some reactive metabolite must be able to escape and could potentially cause cell damage.

2.4   *Immune response in the absence of covalent binding*

Although there is a large amount of circumstantial evidence to support the involvement of reactive metabolites in the pathogenesis of idiosyncratic drug reactions, Pichler has found T cells from patients with idiosyncratic drug reactions that "recognize" drug in the absence of metabolism.[26,27] He proposes that antigen-presenting cells can present a drug in the absence of covalent binding and processing of a haptenized antigen. If this is true there would be no need for the formation of a reactive metabolite. This is a radical departure from the accepted model for the antigenicity of small molecules. If confirmed, it would not refute Landsteiner's work because he was dealing with agents that reproducibly induced contact sensitization, and the mechanism of idiosyncratic reactions could be different. On the other hand it is hard to understand the selective advantage of having an immune system that would respond to small molecules in this way. It also does not fit with the two examples of idiosyncratic drug reactions that we know the most about, i.e. penicillin-induced anaphylaxis or halothane-induced hepatotoxicity. However, many idiosyncratic drug reactions do not seem to fit well into the simple hapten model. More importantly, it is not clear that such presentation in the absence of covalent binding could induce an immune response. The T cells that have been detected may be an extension of an immune reaction that is initiated by covalent binding, but once initiated, the concentration of parent drug would be higher than hapten, and the immune system might shift over to a reaction directed against the patent drug. Another possibility is that, at least in some cases, the significance of metabolic activation is due to the requirement of a danger signal rather than for hapten formation.

3   ENZYME SYSTEMS RESPONSIBLE FOR METABOLIC ACTIVATION

Although the cytochrome P450 system is most commonly associated with metabolic activation, virtually any metabolic enzyme can form reactive metabolites. Oxidation usually makes a compound more electrophilic because oxygen is an electronegative element; however, this is not always the case. P450 is the enzyme most often involved because it has very broad substrate specificity, and so it is responsible for the oxidation of the widest range of drugs. Furthermore, it is also the dominant oxidative enzyme in the liver, the major site of drug metabolism. However, the liver is not the only target organ for toxicity, and many drugs are too reactive to escape the liver in significant concentrations. Other tissues contain varying concentrations of P450. However, it is unclear in tissues, such as the skin, to what degree metabolism in these tissues contributes to idiosyncratic reactions. The flavin-containing monooxygenase can also oxidize drugs, but the substrates for this enzyme are limited to certain easily oxidized nitrogen- and sulfur-containing drugs such as the bioactivation of methimazole (see Section 4.13). The major oxidation system in neutrophils is the combination of NADPH oxidase, which generates superoxide and ultimately hydrogen peroxide, and myeloperoxidase. Myeloperoxidase is oxidized by hydrogen peroxide and, in turn, oxidizes chloride to HOCl. This system is known to form reactive metabolites of most of the drugs that are associated with agranulocytosis (in which neutrophils or their precursors are the target of toxicity) as well as drugs associated with drug-induced lupus.[28] There is evidence that other peroxidases, such as prostaglandin synthase, can also oxidize drugs, but the significance of these enzymes is less clear.[29]

Reduction can also lead to reactive metabolites. The reduction of aromatic nitro groups leads to the same reactive metabolites as the oxidation of a primary aromatic amino group (see Section 4.4). P450 can be responsible for the reduction of nitro groups, but other enzymes, such as xanthine oxidase and microsomal NADPH-cytochrome C are also involved.[30] In addition, much reductive metabolism of orally administered drugs occurs in anaerobic bacteria in the gut before the drug ever reaches the liver.[31]

The conjugating enzymes are usually considered protective pathways, but glucuronidation of carboxylic acids leads to electrophilic acyl glucuronides (see section 4.7). Other conjugation pathways, such as sulfation and acetylation can also provide a better "leaving group" when conjugating OH groups and thereby facilitate the formation of an electrophilic species. Even conjugation with glutathione, a pathway responsible for detoxifying electrophiles, can sometimes lead to a more reactive species. This is less common with drugs; however, there are examples with other xenobiotics, such as dibromoethane, in which conjugation with glutathione leads to a reactive sulfur mustard.[32] Hydrolysis can also be involved in the formation of reactive metabolites. One example is felbamate in which the first step leading to the reactive metabolite is hydrolysis of a carbamate (see Section 4.9). Hydrolysis is also involved in enterohepatic cycling, which provides a drug with more than one chance to be metabolically activated.

One factor that can lead to an increased risk of an idiosyncratic drug reaction is polymorphism in the enzymes responsible for the bioactivation of drugs and detoxication of the reactive metabolites produced. There are major polymorphisms in several enzymes including P450 2D6, 2C19 and acetylation.[33] These polymorphisms play an important role in several idiosyncratic reactions, such as isoniazid-induced hepatitis, hydralazine-induced lupus and procainamide-induced lupus.[34-36] However, they are probably not the major variable responsible for the idiosyncratic nature of such reactions, and most attempts to link idiosyncratic reactions to polymorphisms in drug metabolism have failed.[37-39] If idiosyncratic drug reactions are immune-mediated, there are far more genes involved in the control of the immune response than there are in control of drug metabolism; therefore, it is likely that most of the genetic polymorphism contributing to the idiosyncratic nature of these reactions resides in the immune system.

4   TYPES OF REACTIVE METABOLITES

4.1   *General types of reactive metabolites*

Most reactive metabolites are electrophiles, i.e. they are electron deficient and react with nucleophiles, which are compounds that have electrons that can be donated to the electrophile to form a new bond. Electrophiles can be classed as "hard" or "soft" depending on how localized the electrophilic nature is. A localized positive charge would make an electrophile hard while, if the charge is delocalized, it would be soft. Nucleophiles can also be classed as hard or soft. A sulfur nucleophile is softer than a nitrogen nucleophile because the sulfur atom is larger and the bonding electrons are further away from the nucleus and more diffuse. In general, hard electrophiles tend to react with hard nucleophiles and soft electrophiles tend to react with soft nucleophiles. The approximate reactivity of an electrophile with specific amino acids in a protein can often be determined by chemical synthesis of the electrophile and direct reaction of the electrophile with the *N*-acetylated amino acid. *N*-Acetylated amino acids are used to block the nucleophilic amino group that would be involved in a peptide bond in a protein. However, the nucleophilicity of an amino acid in a protein can be considerably greater than that of the individual amino acid (unpublished observation). Probably the most toxic reactive metabolites are those with intermediate reactivity. If the reactive metabolite has a low reactivity, most of it is likely to be detoxified by systems such as glutathione/glutathione transferase. At the other extreme, if the metabolite is extremely reactive, it is more likely to react with water or the enzyme that formed it, and this is less likely to lead to an idiosyncratic drug reaction.

When a reaction between an electrophile and nucleophile occurs, a new bond is formed and this usually means that a bond must be broken in order to keep the appropriate number of bonds to the atom being "attacked". Therefore, one factor that can contribute to making a compound reactive is a good "leaving group". The leaving group usually leaves with a negative charge so that an approximation can be made as to how good a leaving group will be by looking at the pKa of the corresponding acid. For example, sulfuric acid is a strong acid because, when it loses a proton, the sulfate group easily accommodates the negative charge left behind. For the same reason, sulfate is a good leaving group. An example is tamoxifen, a synthetic antiestrogen, which is oxidized to an allylic alcohol, and this appears to form a sulfate conjugate (see Fig. 18). Loss of the sulfate forms a reactive carbocation, which is reactive with DNA.[40] Other good leaving groups are chloride, bromide, acetate, etc. Another factor that determines reactivity is how well the positive charge left behind can be stabilized; however, if it is an S_N2-type reaction (a reaction in which a new bond is being formed at the same time the bond to the leaving group is being broken) this is less of a factor.

Another general type of reactive metabolite is one with a polarized double bond that allows a new bond to be formed without the loss of a leaving group. A classic example is Michael acceptors, which have a carbonyl group conjugated with a carbon-carbon double bond. The double bond can be used to form a new bond with a nucleophile without a leaving group, and the carbonyl group makes the compound electrophilic. The simplest example is acrolein (see Fig. 30). A more common example is a quinone and related electrophiles as discussed in Section 4.8.

Another factor that can make a compound reactive is ring strain. The normal bond angle of a sp^3-hybidized carbon is 109° and that for a sp^2-hybridized carbon is 120°. Thus a 3- or 4-membered ring results in a significant amount of bond angle strain, and if the ring contains oxygen, nitrogen or sulfur, the compound will be electrophilic. Epoxides are a common example and specific examples of this type of reactive metabolite are provided in section 4.10.

The other major type of reactive metabolite is the free radical.[41] Free radicals are characterized by an unpaired electron. It is important to note that these reactive species are not usually electrophilic and they do not usually covalently bind to proteins or DNA, although they can add across an unsaturated bond, most commonly in a lipid.[42] The most common reaction of a free radical is to abstract a hydrogen atom from another molecule resulting in a new free radical (see Section 4.17). If the hydrogen atom comes from a molecule such as vitamin E, the resultant free radical is so stable that it prevents further reaction; however, in many cases such reactions initiate a free radical chain reaction. Free radicals usually cause a different type of cell damage than electrophiles; specifically, they are more likely to cause oxidative stress or damage to lipid membranes.

In the following sections a link will be made between the reactive metabolites produced by specific drugs and the types of idiosyncratic reactions associated with those drugs; however, this link has not been demonstrated to be causal. In many cases a drug can form several different types of reactive metabolite, and it is often very difficult to determine the relative contribution of each reactive metabolite to a specific type of toxicity.

4.2 Alkyl halides

With the exception of the gaseous anesthetics there are not many drugs that are alkyl halides. Although halides, such as chloride and bromide, are good leaving groups, simple alkyl halides react only very slowly with glutathione or protein sulfhydryl groups, although the reaction with glutathione can be significant in the presence of glutathione transferase. If the halide is at a benzylic or allylic position this increases the reactivity, but I can not think of any drugs with this type of structure. In contrast, a halogen attached to a sp^2 hybridized carbon, such as on a benzene ring, is even less reactive. However, when halogenated compounds are oxidized it can make them much more reactive. The classic example is halothane. The parent drug is relatively inert, but when it is oxidized by P450, it forms a very reactive acetyl halide as shown in Fig. 1. The carbonyl group makes the structure much more electrophilic and, in the case of halothane, the trifluoromethyl group further adds to the electrophilic nature of the metabolite. As mentioned above, other halogenated anesthetics are metabolized in a similar manner, and the degree to which the agent is metabolized by this pathway correlates with the risk that the drug will cause liver toxicity.[7]

4.3 Primary aryl amines

Aryl amine drugs are, in general, associated with a high incidence of idiosyncratic drug reactions. There are several aryl amine carcinogens whose reactive metabolites involve oxidation to a hydroxylamine followed by conjugation of the oxygen with a better leaving group, such as sulfate or acetate. These conjugates lead to a nitrenium ion as shown in Fig. 2. This is common in carcinogens but uncommon in drugs. Most drugs that are primary aryl amines have an electron withdrawing group *para* to the amine group, and this makes it much harder to form a nitrenium ion. The major reactive metabolite that appears to be responsible for covalent binding in these drugs is the nitroso metabolite,[43] which reacts with sulfhydryl groups to form sulfinamides as shown in Fig 2. The sulfinamide bond is very easily hydrolyzed back to the amine with very mild acid or base. This principle has been used to measure exposure to certain carcinogenic amines. Specifically, blood is taken from subjects, the red cells are isolated and lysed, and the hemoglobin is treated to release the amine (e.g. aminobiphenyl) that was bound to hemoglobin sulfhydryl groups due to a nitroso metabolite. Then the free amine is quantified.[44] Aromatic amines are also chlorinated by activated neutrophils to form reactive metabolites, and this may play a role in the observation that most drugs that are aromatic amines cause agranulocytosis.[28] In addition to covalent binding, the intermediate metabolites of aromatic amines undergo extensive redox cycling that can lead to methemoglobinemia and hemolytic anemia. This is a feature of dapsone and some of the sulfonamides; however, although procainamide may redox cycle, it does not lead to methemoglobinemia or hemolytic anemia. This appears to be both a function of the stronger electron withdrawing effect of the sulfone[45] and also the distribution of the dapsone hydroxylamine into red cells.[46] Redox cycling may also lead to significant cell stress.

Another metabolic pathway of aromatic amines is acetylation. In general this is a protective pathway because it decreases oxidation of the amine. The acetyl transferases are polymorphic,[47] and therefore patients who are slow acetylators can be at increased risk of toxicity as mentioned above. However, the hydroxylamine metabolite of aromatic amines can also be a substrate for acetyl transferases, and because acetate is a better leaving group than hydroxide, this increases reactivity (Fig. 2).

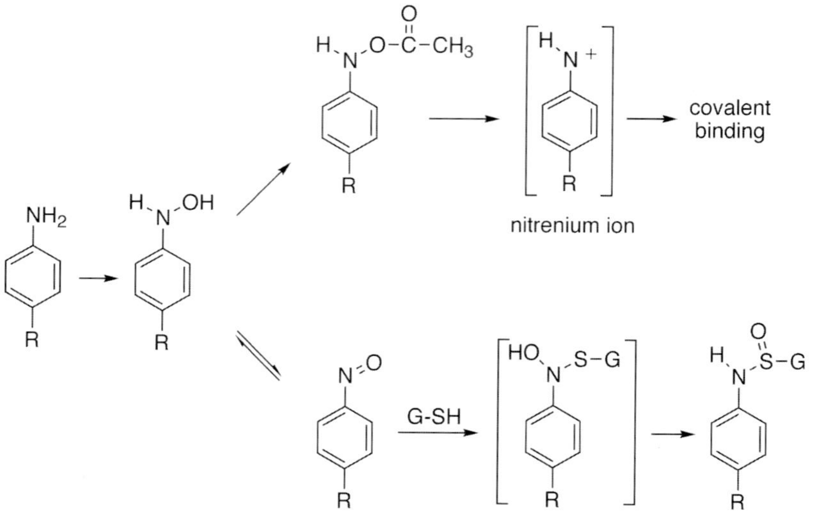

Figure 2. Bioactivation of aryl amines. (In this Fig. and subsequent Figs., G-SH will be used as an abbreviation for glutathione and R represents any other chemical group.)

Structures of representative aryl amine drugs are shown in Fig. 3. The most common aryl amine drugs are the sulfonamides. The only sulfonamides that are still available in North America are sulfamethoxazole and sulfisoxazole, and they are only available in combination with trimethoprim or erythromycin, respectively. Although they are widely used antibacterial agents, they are associated with a relatively high incidence of various types of idiosyncratic reactions. These reactions include generalized hypersensitivity reactions, hepatotoxicity, lupus, agranulocytosis and skin rashes. One skin rash in particular, toxic epidermal necrolysis, is associated with a 30-40% mortality rate.[48] These adverse reactions have led to severe restrictions on the use of sulfas in the United Kingdom. Sulfamethoxazole is known to be oxidized to a hydroxylamine and a nitroso metabolite,[49,50] and a specific antibody has been used to detect covalent binding to microsomes *in vitro*; however, covalent binding was not detected *in vivo*.[51] Most patients with delayed-type idiosyncratic reactions to sulfamethoxazole had autoantibodies against proteins of the endoplasmic reticulum, and only one had anti-drug antibodies.[52] The adverse reactions associated with sulfonamide antibiotics are almost surely due to the aromatic amine portion of the molecule. Yet, because of the name, other sulfonamides that are not primary aryl amines often carry the warning that they are contraindicated in patients with a history of an adverse reaction to sulfonamide antibiotics. The most recent example of this is celecoxib, which is a sulfonamide but not an aryl amine, and yet it carries this warning. This warning is not based on significant clinical data,[53] and many patients with a history of adverse reactions to sulfonamides have been challenged to celecoxib without incident (unpublished observation).

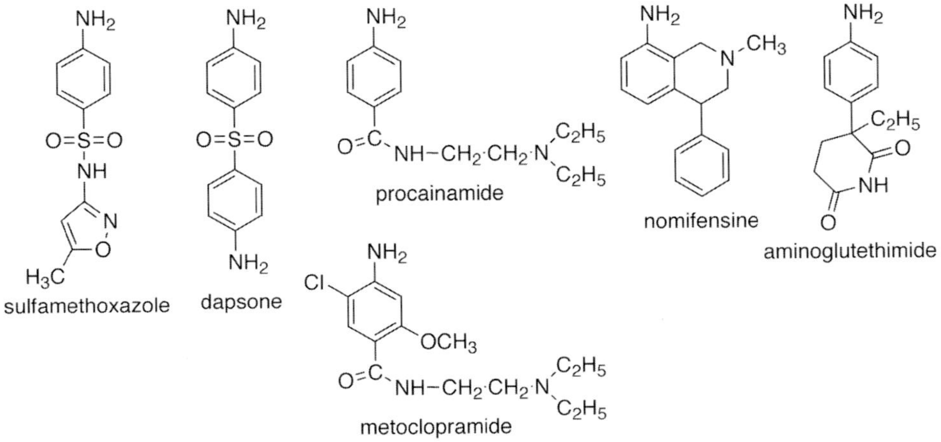

Figure 3. Structures of representative aryl amine drugs.

Dapsone is another aryl amine and it is associated with a relatively high incidence of agranulocytosis and a flu-like syndrome.[54,55] However, it is probably the only primary aryl amine given at high dose that does not appear to be associated with the induction of a lupus-like syndrome. It is also associated with hemolytic anemia, which is due to the hydroxylamine metabolite,[56] and methemoglobinemia. These last two reactions are not idiosyncratic in that they occur in virtually everyone at a sufficiently high dose of the drug. Dapsone is also metabolized to hydroxylamine and nitroso metabolites by P450[57] and neutrophils.[58]

Procainamide is an antiarrhythmic agent, which is associated with the highest incidence of drug-induced lupus of any known drug.[36] It is also associated with a relatively high incidence of agranulocytosis and various other adverse reactions, such as drug fever.[23,59] It is also oxidized to hydroxylamine and nitroso metabolites,[60,61] and covalent binding has been observed to hepatic microsomes and neutrophils.[43,62] Metoclopramide is a substituted procainamide (see Fig. 3). It is not associated with the same type of idiosyncratic drug reactions, but its usual dose is about 10 mg/day compared with more than a gram a day for procainamide.

Nomifensine is an antidepressant drug that was withdrawn from the market because of a variety of allergic reactions, especially hemolytic anemia. The hemolytic anemia is associated with a variety of antibodies, some against drug metabolites derived from patient urine[63] and some against Rh blood-group specific antigens.[64] Nomifensine has a structure similar to that of procainamide (Fig. 3), and it is likely to be oxidized to hydroxylamine and nitroso metabolites. It is one of the few aryl amine drugs that is not substituted in the *para* position. It is also oxidized to a phenol[65] that has the potential to be further oxidized to a quinone methide (see Section 4.8).

Aminoglutethimide, an aromatase inhibitor, is used for the treatment of breast cancer. It is given at a dose of several hundred mg/day, and it is associated with a high incidence of a variety of adverse reactions including skin rashes, fever, agranulocytosis, thrombocytopenia and liver toxicity.[66] The *para* substituent is not electron withdrawing (Fig. 3), and it is metabolized to a hydroxylamine in humans.[67]

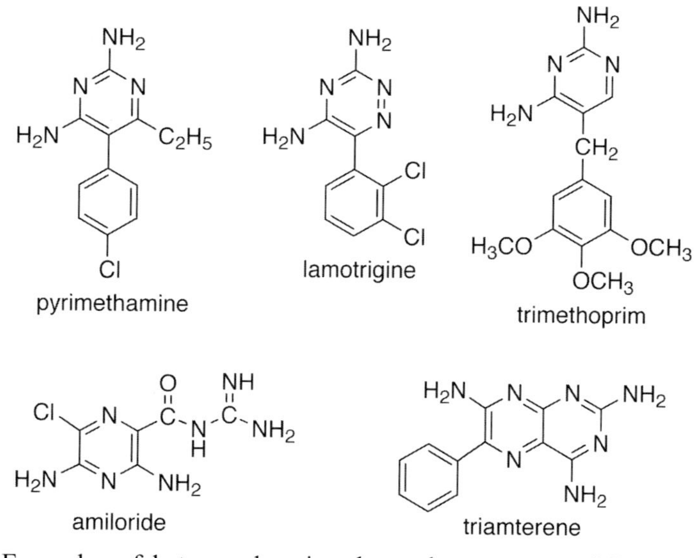

Figure 4. Examples of heteroaryl amine drugs that are more difficult to oxidize to hydroxylamines.

Aromatic amines containing one or more nitrogen atoms in the ring are electron deficient and are difficult to oxidize to hydroxylamines. Thus, drugs, such as pyrimethamine, lamotrigine, amiloride and triamterene (Fig. 4), do not appear to be oxidized to hydroxylamines, and idiosyncratic reactions associated with these drugs are likely due to other types of reactive metabolites. Specifically, trimethoprim forms a quinoneimine methide (see Section 4.8 and Fig. 19).

Sulfasalazine is a drug used to treat ulcerative colitis. It is reduced in the gut by anaerobic bacteria to 5-aminosalicylic acid and sulfapyridine. It is associated with a range of idiosyncratic reactions, most of which appear to be due to the sulfapyridine portion of the drug; however, 5-aminosalicylic acid is also associated with idiosyncratic drug reactions as described under quinone-type reactive metabolites (see Section 4.8 and Fig. 13).

There are several drugs that are amides of aromatic amines. In general, the hydrolysis of amides is slow; however, if the amine is aromatic, this significantly increases the rate of hydrolysis and releases an aromatic amine. Practolol and acebutolol are two amides that are hydrolyzed back to aromatic amines. Practolol is associated with an unusual oculomucocutaneous syndrome with lupus-like characteristics,[68,69] and was withdrawn from the market. Acebutolol is another ß-blocker that is associated with drug-induced lupus.[70] These are the only two ß-blockers that are associated with a significant incidence of autoantibodies[71] and the only two that are amides of aromatic amines. Some of the local anesthetics, such as procaine, are aryl amines; however, their dose and mode of administration leads to a different spectrum of idiosyncratic reactions. Several amide local anesthetics are hydrolyzed to aryl amines. Some people are allergic to local anesthetics; however, the incidence is lower than was previously believed.[72]

4.4   Aromatic nitro drugs

Aromatic nitro containing drugs are similar in many ways to aryl amines because the same hydroxylamine and nitroso metabolites are formed by the reduction of nitro groups as are formed by the oxidation of an aryl amine (Fig. 5). They also undergo one-electron reductions to form a radical cation. Although any of the steps involved in the reduction are potentially reversible, the conversions between the nitro and radical cation and the nitroso and hydroxylamine are especially facile. Drugs containing a nitro group are also usually associated with a high incidence of idiosyncratic drug reactions.

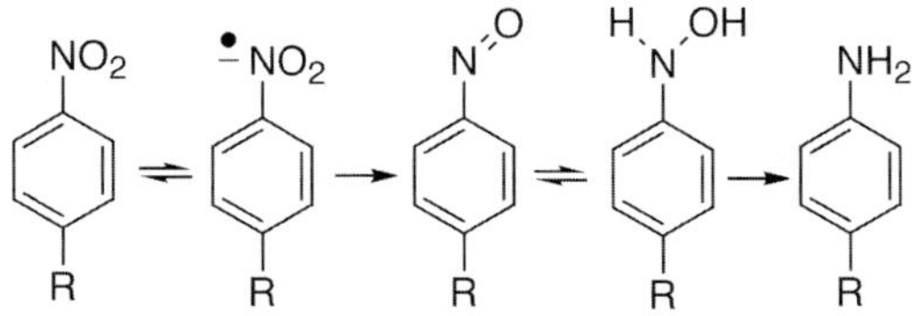

Figure 5. Reduction of aromatic nitro drugs.

Fig. 6 shows the structure of several drugs that have a nitro group. Tolcapone was the first of a new class of catechol-o-methyltransferase inhibitors approved for the treatment of Parkinson's disease. It is associated with a relatively high incidence of liver toxicity that led to its withdrawal from the market in several countries.[73] It has an aromatic nitro group, which is reduced to an amine, doubtless through nitroso and hydroxylamine intermediates.[74] The catechol also has the potential to be oxidized to an o-quinone as described below. Therefore, it is not surprising that it is associated with idiosyncratic drug reactions. However, entacapone has a very similar structure, but it does not appear to have the same liability with respect to liver toxicity as tolcapone. The basis for this difference is unclear although entacapone is a better substrate for glucuronyl transferase and so exposure to the parent drug and its reactive metabolites is likely to be less.[75] More recently there is evidence that the reduction of the nitro group of tolcapone leads to an quinone-imine and this does not appear to be a significant pathway for entacapone in humans.[211]

Chloramphenicol is an antibiotic that is associated with a relatively high incidence of aplastic anemia. It can also cause non-idiosyncratic bone marrow toxicity at high dose. The aplastic anemia is believed to be due to the nitroso metabolite.[76] This metabolite appears to be formed in both the gut and the bone marrow.[77,78] There have even been reports of aplastic anemia associated with chloramphenicol eye drops; however, it is unlikely that the chloramphenicol was causative in these cases. It must be remembered that, unlike agranulocytosis, which is usually drug-induced, most aplastic anemia is idiopathic and not caused by drugs.[79] Therefore, if a drug is given to a sufficient number of patients some will develop aplastic anemia by chance. Unlike oral chloramphenicol, there is no clear increase in the incidence of aplastic anemia associated with the use of chloramphenicol eye drops.[80] In addition to the nitroso and hydroxylamine metabolites that are formed, chloramphenicol is also a dichloroacetamide, which is oxidized to an acetyl chloride reactive metabolite analogous to that of halothane.[81] Thiamphenicol is an analog of chloramphenicol in which a methylsulfone group has replaced the nitro group. Thiamphenicol appeared to be safer than chloramphenicol;[82] however, there have been case reports of thiamphenicol-induced aplastic anemia.[83]

There are a few other drugs that contain an aryl nitro group. Although it is likely that the idiosyncratic reactions associated with these drugs are due to reactive species produced from this nitro group, there is no direct evidence for this. Dantrolene is a muscle relaxant and also used as an emergency treatment for malignant hyperthermia. It is not used extensively but it is associated with liver toxicity.[84] Dantrolene also contains a hydrazone and a furan ring that have the potential to form reactive metabolites (see Sections 4.5 and 4.11). Nimesulide was one of the first relatively selective COX 2 inhibitors. It is associated with a significant incidence of liver toxicity,[85] which is presumably due to the nitro group, and the nitro group is known to be metabolically reduced.[86] Although clonazepam also has a nitro group, it is not associated with a high incidence of idiosyncratic drug reactions. It is given in doses of only about 10 mg/day and this may explain its relative safety. Several of the calcium channel blockers, such as nifedipine, are nitroaromatics and yet are also not associated with a high incidence of the type of adverse reactions usually associated with this functional group. The dose is moderate (about 10 – 20 mg three times/day) and the half-life is short with little or no hepatic reduction of the nitro group,[87] which may explain the low incidence of idiosyncratic reactions. It is interesting to note that the nitro group on nifedipine is photo-reduced to a reactive nitroso derivative,[88] and nifedipine is associated with photodermatitis.[89]

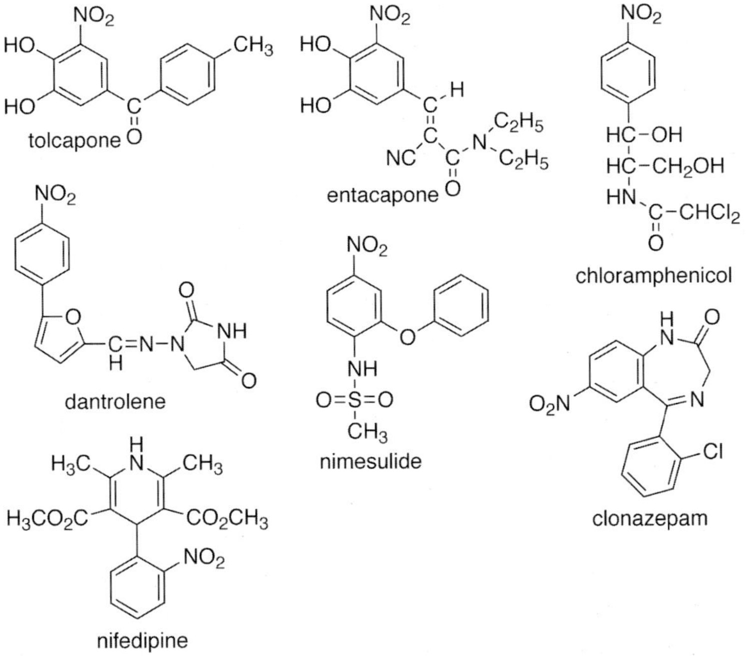

Figure 6. Structures of several drugs that have a nitro group.

There are also several drugs that are nitroheterocyclic compounds. Metronidazole and other nitroimidazoles are used as antibacterial and antiparasitic agents. They are more active against anaerobic organisms in which the nitro group is more readily reduced to form a reactive metabolite. Thus, the pharmacologic activity of these drugs depends on the formation of a reactive metabolite, but this can also lead to adverse drug reactions, and the selective toxicity is based on the greater degree of formation of reactive metabolites in the target organism. The reactive metabolite of metronidazole is a hydroxylamine as shown in Fig. 7.[90] Nitrofurantoin is another antibacterial agent that contains a nitro group, and it is associated with a pulmonary fibrosis and a few other types of idiosyncratic reactions.[91,92] Nitrofurantoin also contains a hydrazone moiety and a furan ring that could also form reactive metabolites; however, covalent binding was increased under reductive conditions thus suggesting that it was reduction of the nitro group that led to covalent binding.[93]

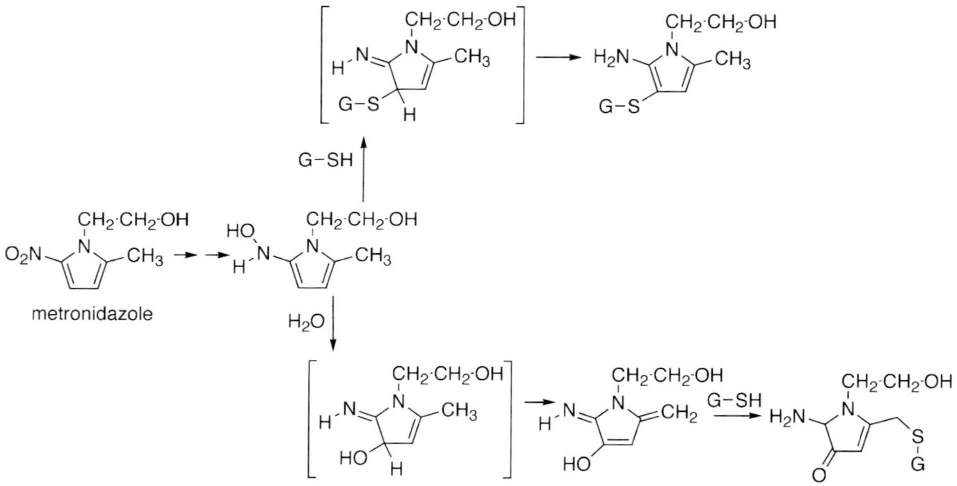

Figure 7. Bioactivation of metronidazole

4.5   *Hydrazines*

Hydrazines appear to be readily oxidized by a series of reactions to diazines and diazonium ions (Fig. 8). Diazonium ions lose molecular nitrogen to form very reactive carbocations. There is probably no better leaving group than molecular nitrogen, and it can lead to the formation of reactive species that would be difficult to form other ways. An even more prevalent pathway may involve oxidation to reactive free radicals as discussed in Section 4.17 (Fig. 45).

Hydralazine is an antihypertensive agent, although it is no longer commonly used. It is associated with the second highest incidence of drug-induced lupus of known drugs, especially when the daily dose is greater than 200 mg.[35] It is readily oxidized by hypochlorite to a reactive intermediate that can be trapped by *N*-acetyl cysteine (Fig. 8), and the same product was formed by activated neutrophils.[94] Although hydralazine is not associated with agranulocytosis there have been reports of pancytopenia. The most likely structure for the intermediate is a diazonium chloride, which either leads directly to a carbocation or undergoes an S_N2 reaction with the loss of molecular nitrogen. This reaction could occur in cells of the immune system, such as macrophages, and this could be involved in the mechanism of hydralazine-induced lupus; however, there is no direct evidence for this hypothesis. In contrast to hydralazine, dihydralazine is associated with hepatotoxicity rather than lupus. It is also metabolized to reactive metabolites as discussed in Section 2.2 on the hapten hypothesis and in the Section 4.17 on free radicals.

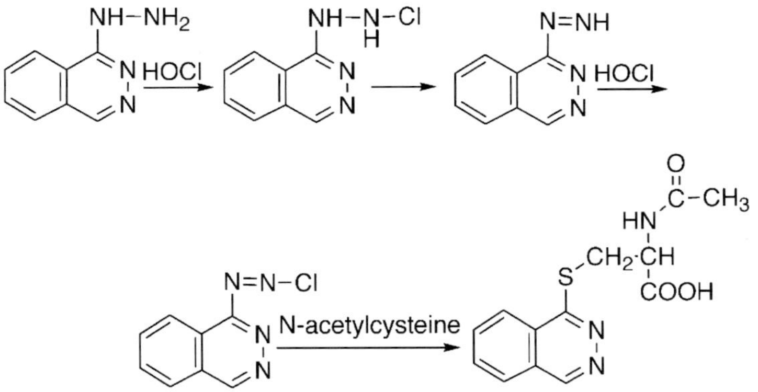

Figure 8. Proposed mechanism for the activation of hydralazine by HOCl and leukocytes.

Isoniazid is the principal drug used for the treatment of tuberculosis; however, it is associated with a relatively high incidence of severe hepatotoxicity, and it can also cause a lupus-like syndrome. In an animal model the primary pathway that appeared to be responsible for toxicity involved acetylation, hydrolysis of the hydrazide bond and oxidation of the acetyl hydrazine that is formed.[95] However, as mentioned above, one risk factor for hepatotoxicity in humans is the slow acetylator genotype. The idiosyncratic hepatotoxicity seen in humans could involve the direct oxidation of hydrazide group to a reactive metabolite, although even if the pathway leading to toxicity involves the formation of acetylhydrazine, acetylhydrazine is further acetylated to the relatively nontoxic diacetylhydrazine. The hydrazide group is oxidized by HOCl to give isonicotinic acid, and the pathway is likely similar to that seen with hydralazine.[96] The same pathway and/or a free radical pathway are likely to be catalyzed by P450.

4.6   Other nitrogen-containing aromatic compounds

Clozapine is a very effective antipsychotic agent, but its use is limited by a high incidence of agranulocytosis,[97] and it is also associated with liver toxicity and myocarditis.[98] It is metabolized to a reactive metabolite by the liver.[99] It is quite readily metabolized by activated neutrophils to a reactive intermediate, which covalently binds to neutrophils.[100] HOCl can mimic this oxidation, and a mass spectrum of the intermediate was obtained. The structure is formally a nitrenium ion, although the positive charge is highly delocalized, and when the reactive metabolite is trapped by glutathione, the major adducts are formed at the 6 and 9 positions (Fig. 9). Although one might expect the activation of neutrophils to be limited *in vivo* and this would limit covalent binding, *in vivo* covalent binding was readily detected to neutrophils that were taken from patients who were being treated with clozapine.[101] Most of the characteristics of clozapine-induced agranulocytosis are typical of an idiosyncratic reaction, i.e. the idiosyncratic nature and the usual delay of 6 weeks to 6 months between starting the drug and the onset of agranulocytosis. This suggests that the reaction is immune-mediated. On the other hand, if a patient who has a history of clozapine-induced agranulocytosis is rechallenged with clozapine, it requires just as long before the onset of agranulocytosis as it did on initial exposure.[102] This is not typical of an immune-mediated reaction in which there are memory T cells that lead to a rapid reaction on re-exposure. Clozapine-induced agranulocytosis could involve direct cytotoxicity, and the reactive metabolite of clozapine has been demonstrated to induce apoptosis *in vitro*.[103] However, it is not clear whether the concentrations of reactive metabolite generated *in vitro* would be formed *in vivo,* and the characteristics are also not typical of a reaction mediated by direct cytotoxicity.

Olanzapine is also an antipsychotic agent with a structure very similar to that of clozapine; however, it does not appear to cause agranulocytosis. It not only has a similar structure but it also is readily oxidized to a very similar reactive metabolite (Fig. 9), and the degree of covalent binding to activated neutrophils *in vitro* is not significantly different from that of clozapine (unpublished observation). One difference between the drugs is the therapeutic dose of clozapine is several hundred mg/day while the maximum recommended daily dose of olanzapine is just 10 mg/day. In contrast to clozapine, covalent binding of olanzapine to neutrophils *in vivo* could not be detected.[101] The reactive metabolite was also not as toxic to neutrophils as the reactive metabolite of clozapine.[104]

Aminopyrine and dipyrone (a sulfonate derivative of aminopyrine) are analgesics that are no longer available in many countries because they are associated with a high incidence of agranulocytosis. In fact they were one of the first drugs to be recognized to cause this syndrome.[105] Aminopyrine is known to form a relatively stable free radical.[29] Aminopyrine is also readily oxidized by HOCl, the major oxidant produced by activated neutrophils, to a very reactive dication (Fig. 10).[106] This dication is readily reduced by a one-electron reductant or simply co-proportionates with aminopyrine to form the free radical. It is likely that the dication and/or the free radical are responsible for aminopyrine-induced agranulocytosis.

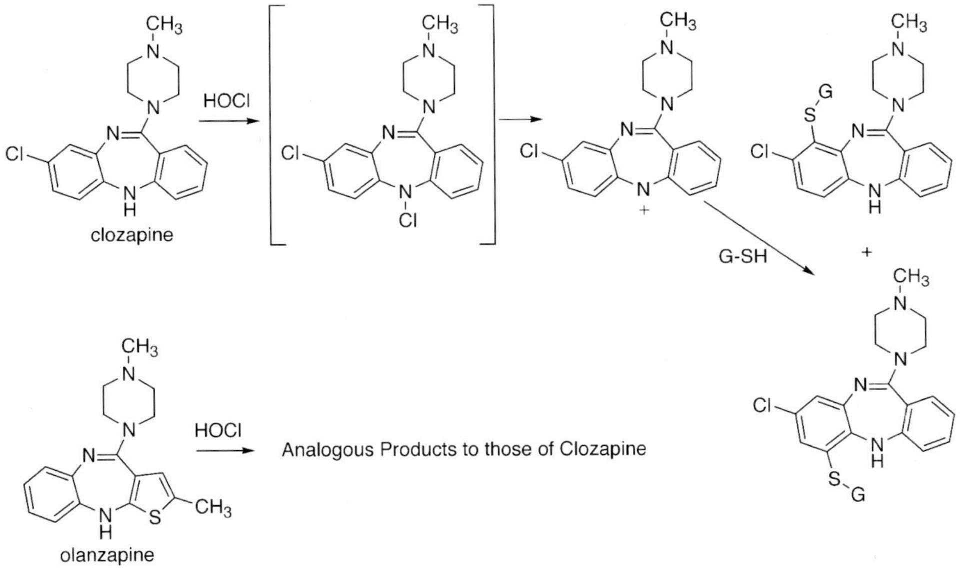

Figure 9. Bioactivation of clozapine and olanzapine.

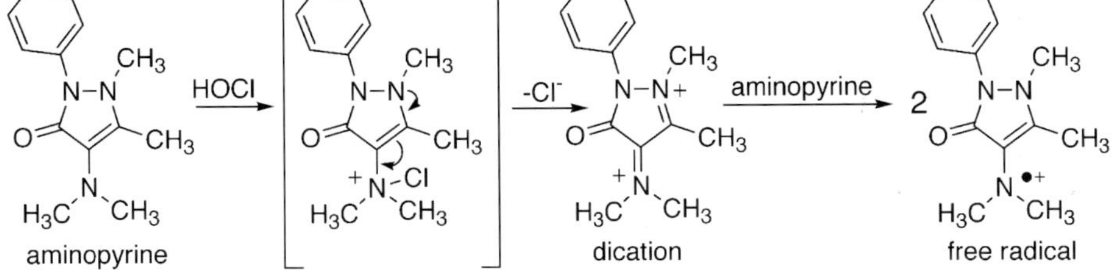

Figure 10. Oxidation of aminopyrine to a reactive dication and reduction to a radical cation.

4.7 Acyl Glucuronides and Co-A Esters

Most nonspecific nonsteroidal antiinflammatory agents (NSAIDs) are associated with a relatively high incidence of adverse drug reactions. The most common serious adverse reactions are gastrointestinal bleeding and pseudoallergic reactions, which appear to be related to shifting the metabolism of arachidonic acid from prostaglandins to leukotrienes.[107] Both of these effects are pharmacological and not dependent on a reactive metabolite. Therefore, changing the structure of the drug will not prevent these reactions except in so far as it makes the drug more COX-2 specific.[108] However, many NSAIDs are also associated with other types of idiosyncratic drug reactions, and of the drugs that have been removed from the market because of idiosyncratic drug reactions, a large proportion have been NSAIDs. Most nonspecific NSAIDs are carboxylic acids, and a major metabolic pathway for these drugs is glucuronidation. Glucuronides of carboxylic acids, known as acyl glucuronides, are weakly reactive, relatively hard electrophiles, which react best with primary amine nucleophiles (Fig. 11). The acyl glucuronides are made electrophilic by the carbonyl group and they are more reactive than the corresponding carboxylic acid because glucuronic acid is a better leaving group than hydroxide. This is in contrast to glucuronides of alcohols and phenols, known as ether glucuronides, which are not electrophilic. The reactivity of acyl glucuronides as well as the relative amount of *in vivo* covalent binding correlates with their ease of hydrolysis,[109] although this concept is controversial. Substitution on the carbon α to the carboxylic acid decreases reactivity because of steric hindrance, and most of the NSAIDs that had to be withdrawn have a low degree of α substitution. The glucuronide of valproic acid is highly substituted and its glucuronide has a low reactivity, yet it results in covalent binding comparable to more reactive acyl glucuronides.[110] This is presumably due to either an oxidative metabolite (see Section 4.9 and Fig. 25) or an acyl Co-A ester.

Carboxylic acids also form Co-A esters, which are reactive because the Co-A group is a relatively good leaving group, and these esters may also contribute significantly to the covalent binding associated with carboxylic acids (Fig. 11).[111] Furthermore, acyl glucuronides also rearrange by a mechanism known as the Amadori rearrangement, which also leads to covalent binding.[112] In contrast to the binding due to the direct reaction of an acyl glucuronide, the covalent binding involving an Amadori rearrangement retains the glucuronic acid moiety (Fig. 11). There is no question that these pathways activate carboxylic acids and lead to covalent binding. However, it has been noted that, although NSAIDs are associated with idiosyncratic drug reactions, the fibrates lead to comparable covalent binding and they are not associated with a significant incidence of idiosyncratic reactions.[113] It may be that, in addition to covalent binding, a danger signal is required for an idiosyncratic reaction, and other pathways, such as described below for indomethacin or diclofenac, may be required to supply this signal. We have also found that misoprostol, a prostaglandin E1 analog, inhibits penicillamine-induced lupus in an animal model,[114] and NSAIDs accelerate the reaction (unpublished observation). Therefore, the effects of NSAIDs on the immune system via alteration in prostaglandin levels may also play a role in the induction of idiosyncratic drug reactions.

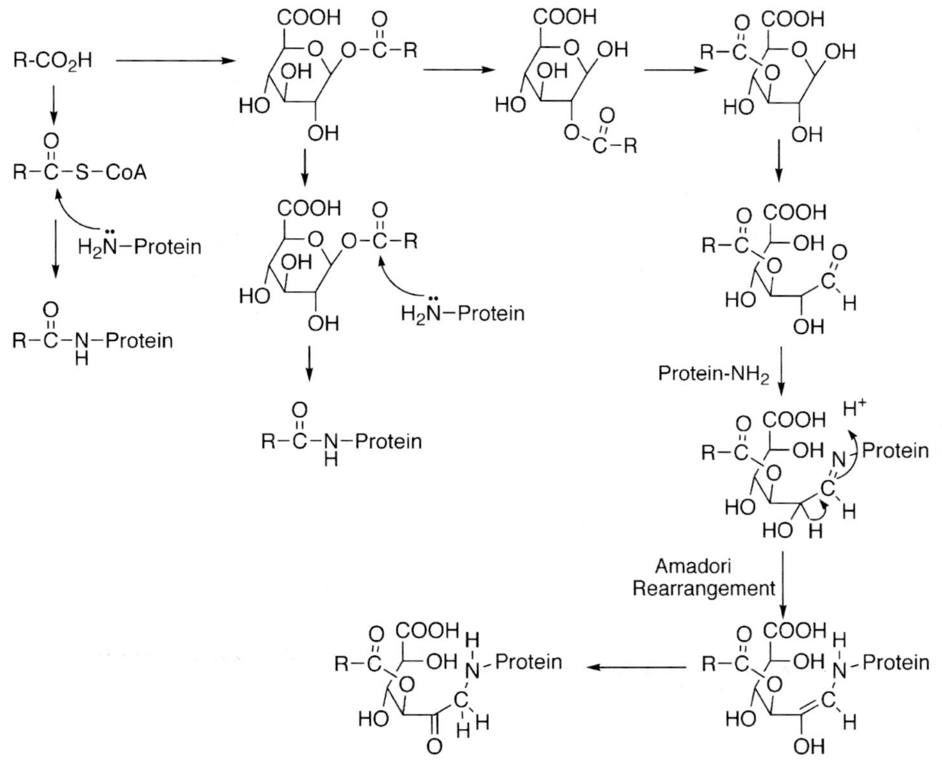

Figure 11. Mechanisms of covalent binding of carboxylic acids via Co-A esters, direct reaction of glucuronide conjugates with protein and after Amadori rearrangement of glucuronide conjugates.

Diclofenac is a popular NSAID, but is associated with liver toxicity and a lower incidence of agranulocytosis.[115,116] Covalent binding appears to be due to both the glucuronide and P450-generated reactive metabolites.[117] The glucuronide appears to be responsible for binding in the intestine, and it has been suggested that this may also be a factor in the induction of tolerance rather than just causing toxicity.[118] Diclofenac is also a substituted diphenylamine, and the binding mediated by P450 is likely due to the formation of quinoneimines as described in Section 4.8 and Fig. 17. These oxidative reactive metabolites may play an important role in the toxicity associated with diclofenac. The same is true for indomethacin (see Fig. 17), and this is likely responsible for the bone marrow toxicity that is associated with the drug.

Zomepirac was a popular NSAID that was often used instead of opiate analgesics. However, it was associated with a high incidence of what appeared to be IgE-mediated reactions, and it was eventually removed from the market. There is no evidence that zomepirac forms any reactive metabolite other than the acyl glucuronide or Co-A ester, and its glucuronide is one of the most reactive of the NSAIDs.[110] Tolmetin and ibufenac are other NSAIDs associated with idiosyncratic reactions that are most easily explained on the basis of an acyl glucuronide or Co-A ester, and this lead to the withdrawal of ibufenac. Bromfenac is another NSAID that was withdrawn, but it is also a primary aryl amine as well as being unsubstituted α to the carboxylate, so it is less clear what reactive metabolite might be responsible for the idiosyncratic reactions associated with its use.

Benoxaprofen was associated with an unusual combination of idiosyncratic reactions including photosensitivity and liver toxicity, which led to its withdrawal from the market.[119] However, it is less clear whether this was due to an acyl glucuronide or some other reactive metabolite. The glucuronide of benoxaprofen is less reactive than that of many other drugs,[120] and it is unlikely that the photosensitivity reactions were due to the glucuronide. It is possible that the heterocyclic ring plays a role in the high incidence of idiosyncratic reactions associated with benoxaprofen.

4.8   Quinone-type Reactive Metabolites

Probably the most common type of reactive metabolite is the quinone and related structures, such as quinoneimines, imidoquinones, quinone methides, etc. As mentioned earlier, the imidoquinone of acetaminophen was the first reactive metabolite that was associated with toxicity of a drug. Another simple example is benzoquinone itself, which is a significant metabolite of benzene, and benzene is associated with bone marrow toxicity.[121] Most drugs contain an aromatic ring, and it is very common to have two atoms *para* or *ortho* to one another that can form a quinone-like structure. In addition, even if there is only one such atom on a ring, i.e. oxygen or nitrogen, this usually activates the ring to further oxidation and directs that oxidation to the *para* or *ortho* position. Further oxidation leads to the quinone or quinone analog thus making this a very common type of reactive metabolite. Quinones are also very common in nature.[212]

Quinone-type metabolites act as soft electrophiles and usually react with soft nucleophiles, especially sulfhydryl groups, and glutathione conjugates are common. The glutathione conjugates are readily reoxidized to form another reactive electrophile.[122] The ability of protein conjugates of quinones to be reoxidized has the potential to lead to crosslinking of macromolecules. The major products observed from the reaction between quinones and nucleophiles involves a 1,4-type Michael addition, but more recently a labile 1,2-addition product was observed and this is referred to as an *ipso* adduct (Fig. 12).[123] Ipso adducts may be responsible for some of the toxicity associated with the some quinone-type metabolites because they can act as a transport form of the quinone. Quinones also readily undergo redox cycling and, therefore, have the potential to cause oxidative stress. The redox cycling can either involve one electron, with a semiquinone as the product, or it can involve two electrons with a hydroquinone as the product of reduction (Fig. 12). This redox cycling may play an important role in direct cytotoxicity as well as in producing a danger signal that increases the risk of an immune-medicated reaction. There are several good reviews of the chemistry of quinone-type reactive metabolites.[122,124]

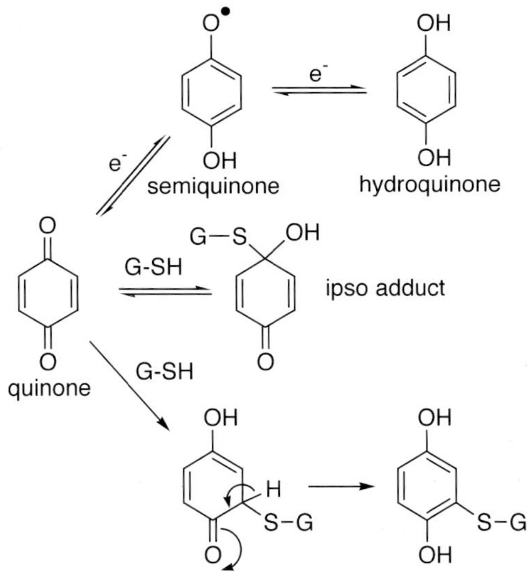

Figure 12. Reactions of quinones.

The following are specific examples but there are likely many more that have never been recognized. 5-Aminosalicylic acid is used for the treatment of inflammatory bowel disease. Its use is associated with a variety of idiosyncratic reactions including "allergic" reactions, pancreatitis, hepatitis, renal toxicity, a lupus-like syndrome and pulmonary hypersensitivity. This drug is readily oxidized to a quinoneimine by HOCl (Fig. 13)[125] or even by molecular oxygen.[126] This quinoneimine can react with glutathione to form a conjugate or with water to form the quinone. Oxidation products of the drug were found in the feces of patients taking the drug.[127] Thus, 5-aminosalicylic acid forms electrophilic metabolites and has both antioxidant[128] and prooxidant properties.[129] It is proposed that the ability of 5-aminosalicylic acid to inactivate HOCl produced by activated neutrophils may make a contribution to its therapeutic activity;[125] however, the reactive metabolites could also be responsible for idiosyncratic drug reactions.

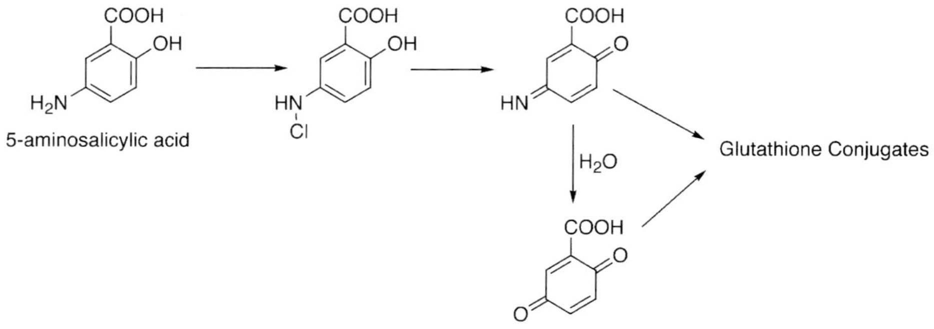

Figure 13. Bioactivation of 5-aminosalicylic acid.

Amodiaquine is a drug used to treat malaria. Its use is associated with a relatively high incidence of agranulocytosis and hepatotoxicity. It is easily oxidized to a reactive metabolite, presumably the quinoneimine (Fig. 14), by both hepatic microsomes[130] and activated neutrophils.[131] This quinoneimine is likely responsible for the idiosyncratic reactions associated with this drug, and anti-drug antibodies have been detected in patients with amodiaquine-induced adverse reactions.[132] Chemical modification of amodiaquine, especially with substitution of fluorine for the OH group, led to agents that were active against malaria but no longer formed a reactive metabolite, and this is an example of designing a reactive metabolite out of a drug candidate.[133]

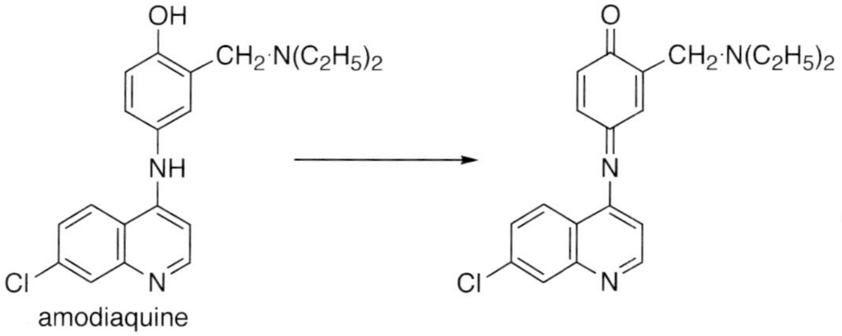

Figure 14. Bioactivation of amodiaquine

Phenytoin and carbamazepine are anticonvulsants that are associated with several idiosyncratic drug reactions, the major problem being a generalized hypersensitivity reaction (see Section 4.10). These adverse reactions have been blamed on arene oxide metabolites; however, both drugs have also been shown to form quinone-type reactive intermediates. In the case of phenytoin covalent binding has been linked to an *o*-quinone,[134] and in the case of carbamazepine, there is evidence for the formation of a quinoneimine (Fig. 15).[135] It also appears that 2-hydroxy and 3-hydroxycarbamazepine can be directly oxidized to reactive metabolites (unpublished observation) so there are many possible reactive metabolites for these drugs.

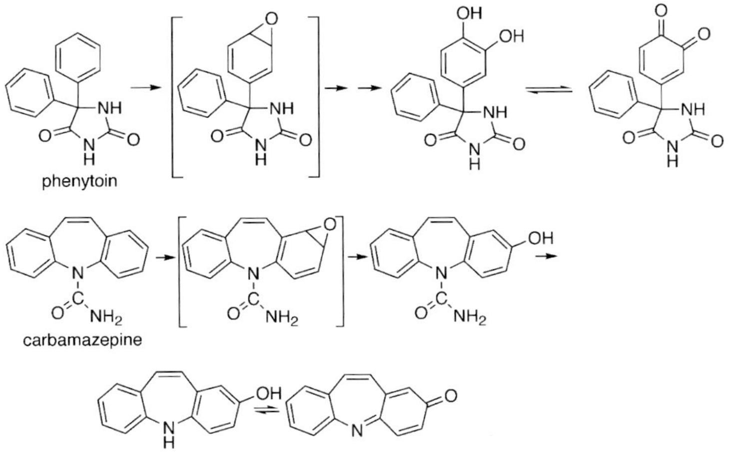

Figure 15. Bioactivation of phenytoin and carbamazepine

Vesnarinone is a phosphodiesterase inhibitor that was developed to treat congestive heart failure. It was found to be associated with a high incidence of agranulocytosis. It was also found to be oxidized by neutrophil-generated hypochlorite to a reactive imidoiminium ion (Fig. 16).[24] This pathway appears to be much less prevalent in the liver. This reactive metabolite is presumably responsible for the induction of agranulocytosis. Prinomide is a similar drug, which was being developed as an antiinflammatory drug. In clinical trials it was found to be associated with a 0.3% incidence of agranulocytosis and further development of the drug was halted. The major metabolite of the drug involves oxidation of the phenyl ring *para* to the amine group, and this is readily oxidized by myeloperoxidase and further hydrolyzed to benzoquinone.[136] Although benzoquinone is known to be toxic to bone marrow, based on the location of the radiolabel, much of the covalent binding is upstream of benzoquinone and probably involves an imidoquinone.

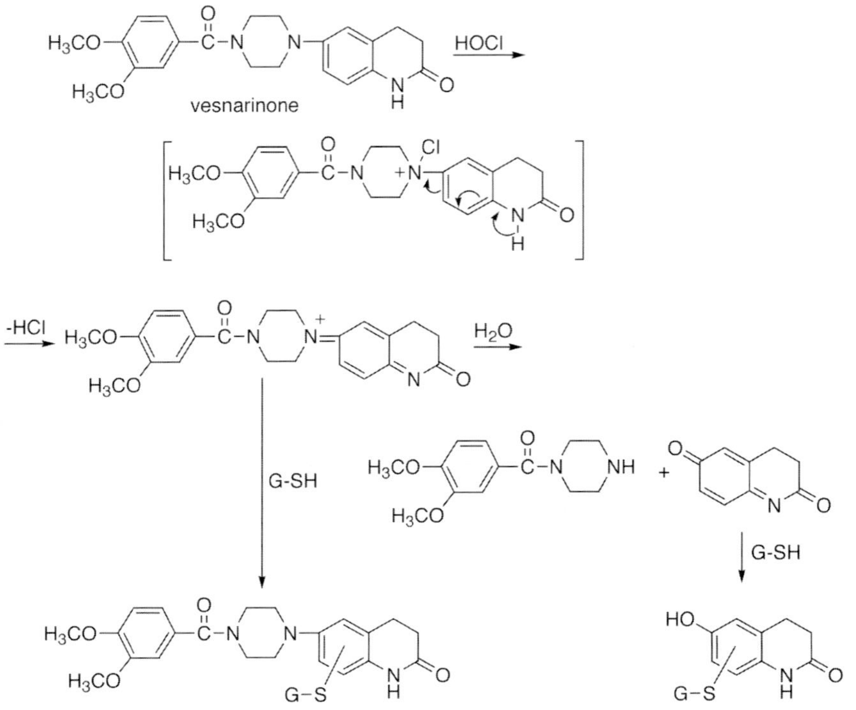

Figure 16. Bioactivation of vesnarinone

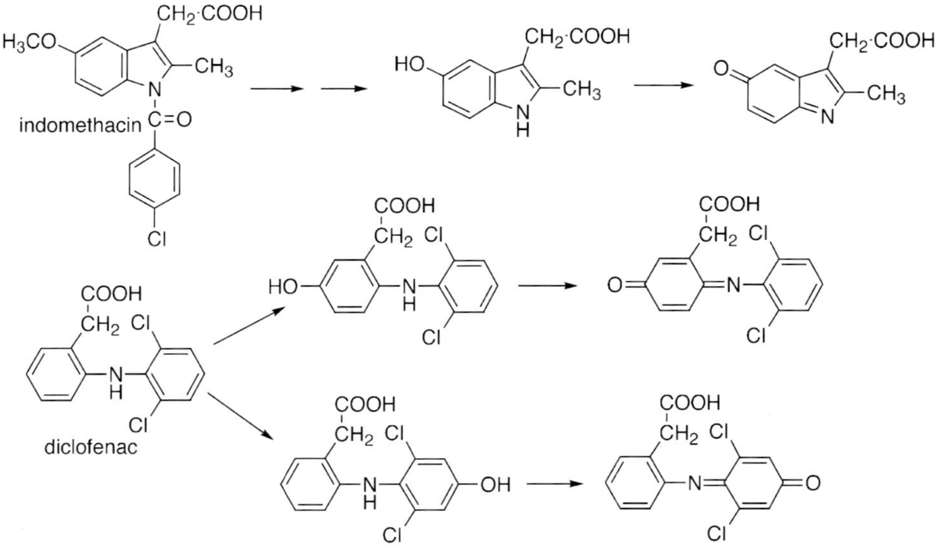

Figure 17. Bioactivation of indomethacin and diclofenac

Although all NSAIDs are associated with a relatively high incidence of adverse reactions, indomethacin is associated with a higher incidence than most, especially bone marrow toxicity. Two major metabolic pathways involve *o*-demethylation and hydrolysis, which lead to a metabolite with a phenol *para* to an indole nitrogen. This metabolite is readily oxidized to a reactive quinoneimine that is presumably responsible for some of the adverse reactions associated with this drug (Fig. 17).[137] It can also form an acyl glucuronide as discussed in Section 4.7. Diclofenac is another NSAID that is associated with a relatively high incidence of adverse reactions, especially hepatotoxicity. It is a diarylamine, which forms two different metabolites, each which a phenolic group *para* to the nitrogen, and both of these have the potential to be oxidized to reactive quinoneimines in the liver (Fig. 17).[138] Diclofenac is also oxidized to a quinoneimine by activated neutrophils and that may contribute to the agranulocytosis that is sometimes associated with the drug.[139]

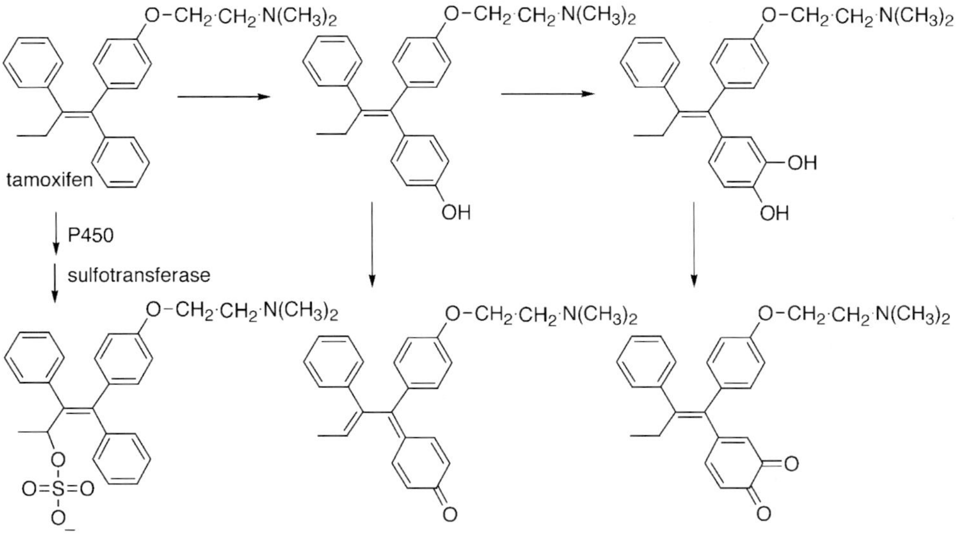

Figure 18. Bioactivation of tamoxifen

Estrogens are associated with an increased risk of uterine and breast cancer. Estrone and other estrogens are oxidized to reactive *o*-quinones.[140] Tamoxifen, a synthetic antiestrogen, can also form quinones and a quinone methide that can bind to DNA as well as an allylic sulfate as mentioned in Section 4.1 (Fig. 18).[141] It is unclear whether it is the effects of estrogens on DNA (either covalent binding or reactive oxygen species) or their proliferative effects that are responsible for the carcinogenic effects of estrogens in estrogen responsive tissue.[142] The dose of estrogens is much lower than most drugs, which should make the relevance of reactive metabolites less important. However, estrogens are presumably concentrated in the nucleus of estrogen-responsive tissue and this may make the local concentration of reactive metabolites higher than might be expected on the basis of dose. Another counter argument is that the carcinogenic effects do not parallel the estrogenic potency of the agents;[143] however, these results were obtained in animal models that may not reflect the same mechanism involved in cancer induction in humans.

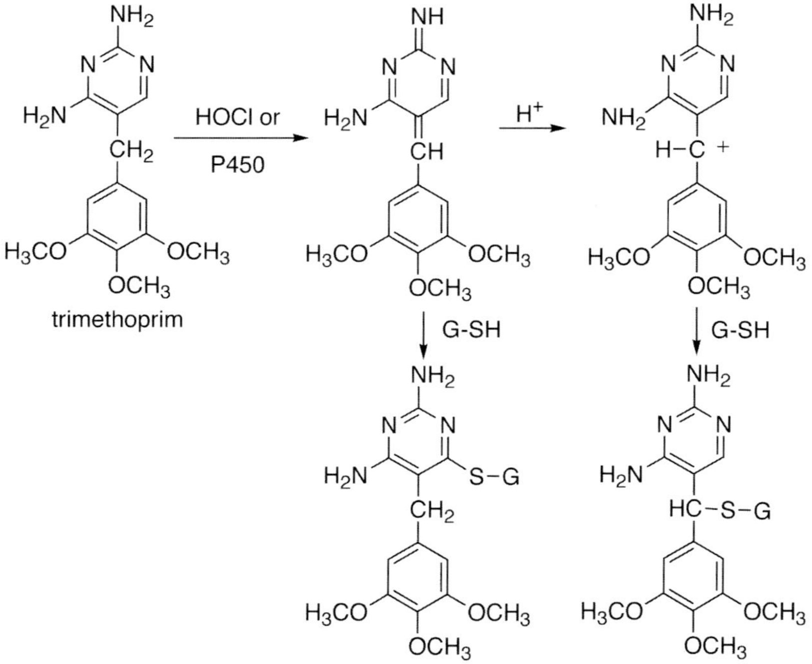

Figure 19. Bioactivation of trimethoprim

Trimethoprim is an antibacterial agent, which is usually given in combination with sulfamethoxazole. Although most of the adverse reactions associated with this combination are associated with the sulfa component, about 20% of the reactions appear to be due to trimethoprim.[144] Trimethoprim is oxidized to a reactive quinone methide[145] as shown in Fig. 19. Unlike most quinone methides in which attack is usually on the exocyclic carbon, probably because of the electron-withdrawing effect of the ring nitrogens, glutathione adds principally to the ring carbon. However, the addition is pH dependent because protonation of the imine leads to a carbocation. Trimethoprim also has the potential to form an *ortho* quinone by *o*-demethylation of the methoxy groups.

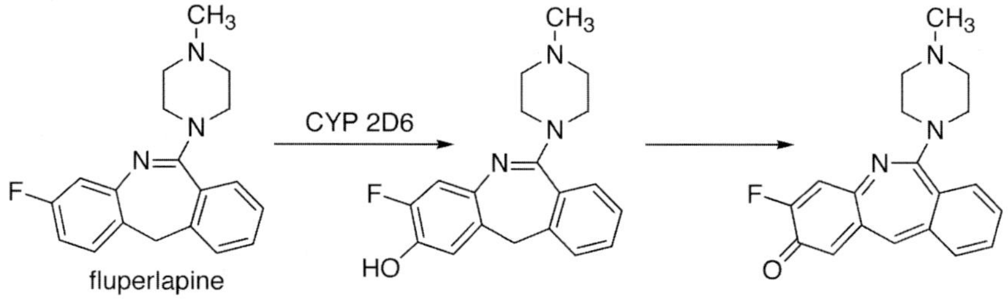

Figure 20. Bioactivation of fluperlapine

Fluperlapine is an analog of clozapine, but it does not have the nitrogen that is responsible for the formation of the reactive metabolite of clozapine so it should be safer. During early clinical trials there were cases that appeared to be drug-induced agranulocytosis,[146,147] although the details of the cases have not been published. The development of fluperlapine was halted so it is unclear what the risk of fluperlapine-induced agranulocytosis would have been. Although it does not form a nitrenium ion analogous to that of clozapine, the major metabolic pathway involves oxidation *para* to the nitrogen in the center ring mediated by CYP 2D6, and this makes further oxidation to a quinoneimine facile (Fig. 20). This is not a major pathway in clozapine. This reactive metabolite has a half-life and reactivity very similar to that of the nitrenium ion of clozapine.[148] Therefore, if fluperlapine does cause drug-induced agranulocytosis, it seems likely that it is due to this reactive metabolite.

Troglitazone is one of a new class of drugs used to treat type II diabetes. It was withdrawn from the market because of many cases of serious hepatotoxicity. One of the metabolites formed from this drug is a quinone methide.[149] This quinone methide may be responsible for the toxicity; however, troglitazone also forms an isocyanate and sulfenic acid as shown in Fig. 40.

Minocycline is the only one of the tetracyclines that is associated with a relatively high incidence of drug-induced liver toxicity and drug-induced lupus-like syndrome.[150] It is also unique among the tetracyclines in that it has a dimethylamino group in the 7 position. This group is *para* to a phenolic group and it appears that this is metabolized to a reactive quinone iminium ion (Fig. 21).[151,152] It seems likely that this pathway is responsible for the unique adverse reactions associated with minocycline.

Tacrine was one of the first agents shown to have some activity against Alzheimer's disease. Unfortunately it is also associated with a significant incidence of liver toxicity. Oxidation by hepatic microsomes leads to the formation of a reactive metabolite that can trapped with sulfhydryl containing agents. The pathway leading to the reactive metabolite appears to be 7-hydroxylation followed by further oxidation to the quinone methide as shown in Fig. 22.[153] It is likely that this metabolite is responsible for the hepatotoxicity of tacrine.

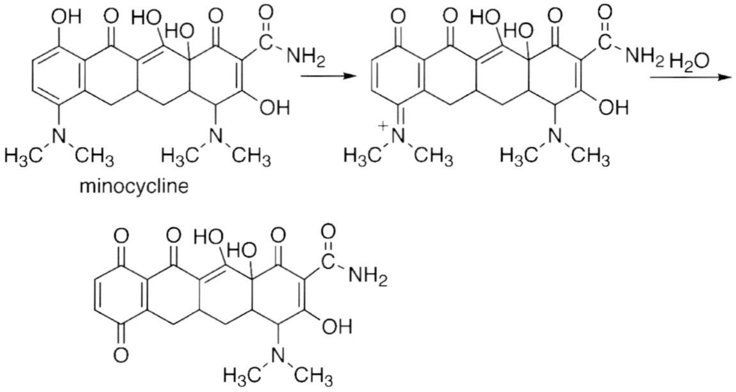

Figure 21. Proposed bioactivation of minocycline

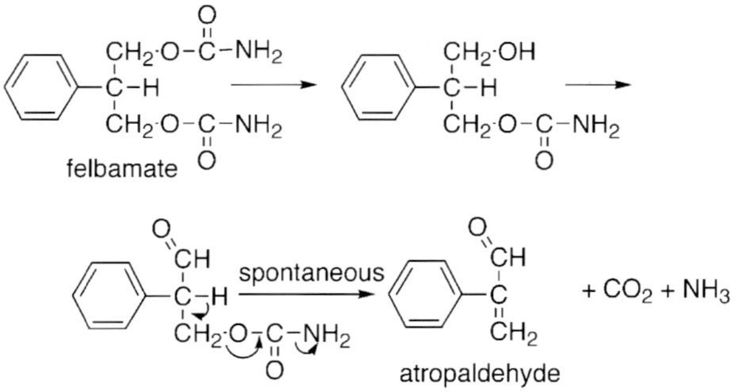

Figure 22. Bioactivation of tacrine

4.9   *Other Michael Acceptors*

Michael acceptors are compounds that have a double bond that has been activated by being next to a carbonyl or analogous group. Quinones are also Michael acceptors. As mentioned for quinones, Michael acceptors are usually soft electrophiles that react best with sulfhydryl groups.

Felbamate is an anticonvulsant, which appeared to be very safe in animal studies and clinical trials. However, after it was released on the market it was found to be associated with both aplastic anemia and liver toxicity. Macdonald postulated that this toxicity is due to the metabolite atropaldehyde (phenylacrolein).[154] The pathway leading to this metabolite is novel in that it involves hydrolysis of a carbamate followed by oxidation of the resultant alcohol to an aldehyde. This makes the *alpha* hydrogen more acidic and the aldehyde spontaneously loses carbon dioxide and ammonia to form atropaldehyde (Fig. 23). It is quite likely that atropaldehyde is responsible for both the bone marrow toxicity and liver toxicity associated with this drug. The amount of drug that is metabolized through this pathway is much less in rodents than in humans, and that may explain why it appeared to be so safe in animal tests.[155] This is a good example of how extrapolation of metabolism and toxicity data from animals to humans can be misleading.

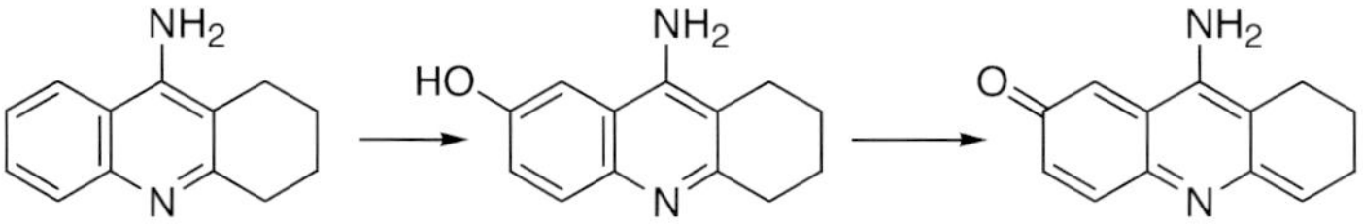

Figure 23. Bioactivation of felbamate

Terbinafine is a very effective antifungal agent, which has been associated with a variety of adverse reactions including bone marrow toxicity[156] and skin rashes, including erythema multiforme.[157] The most common idiosyncratic reaction is liver toxicity with a cholestatic or mixed pathology.[158,159] Terbinafine is metabolized by a variety of pathways, but one involves an *N*-dealkylation leading to an acrolein derivative as shown in Fig. 24. This metabolite was trapped with glutathione and the major product was found to involve a 1,6 addition.[160] This conjugate retains the acrolein structure and it was speculated that this conjugate is transported into bile where it reacts with structures in the biliary system. In some patients this could lead to the cholestatic-type injury that is associated with the drug through a mechanism similar to the classic biliary toxin, α-naphthylisot hiocyanate.[161]

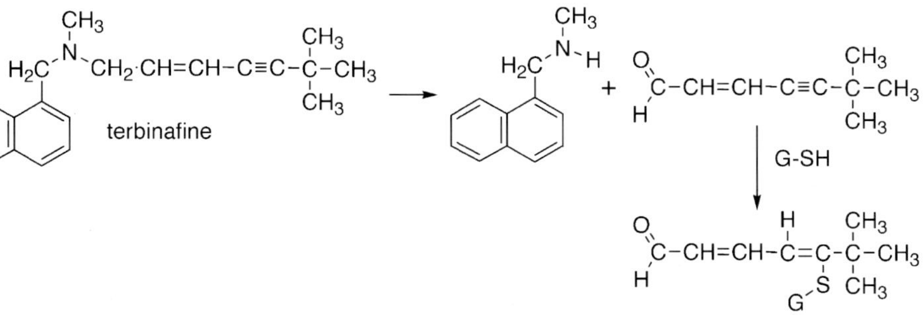

Figure 24. Bioactivation of terbinafine

Valproic acid is an anticonvulsant, which is associated with a different type of liver toxicity that resembles Rye's syndrome. This toxicity is believed to be due to oxidation of valproic acid to the electrophilic diene shown in Fig. 25. In order to block this pathway, a derivative was prepared with a fluorine in the 2 position. It was found not to cause toxicity comparable to valproic acid in an animal model, thus providing evidence to support this hypothesis.[162] However, the fluorinated derivative also did not form a Co-A conjugate; therefore, this is also compatible with the hypothesis that toxicity is due to a Co-A conjugate acting as an electrophile.[163]

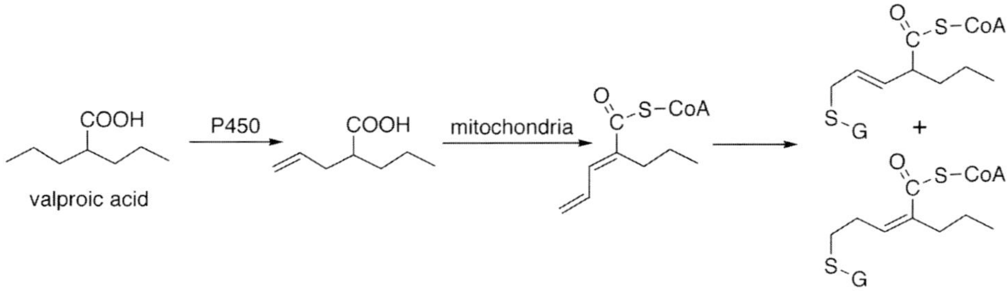

Figure 25. Bioactivation of valproic acid

Mianserin is an antidepressant that is associated with a significant incidence of agranulocytosis. It is oxidized by HOCl and activated neutrophils to a reactive iminium ion as shown in Fig. 26. Intermediate oxidation states were not observed.[164, 213]

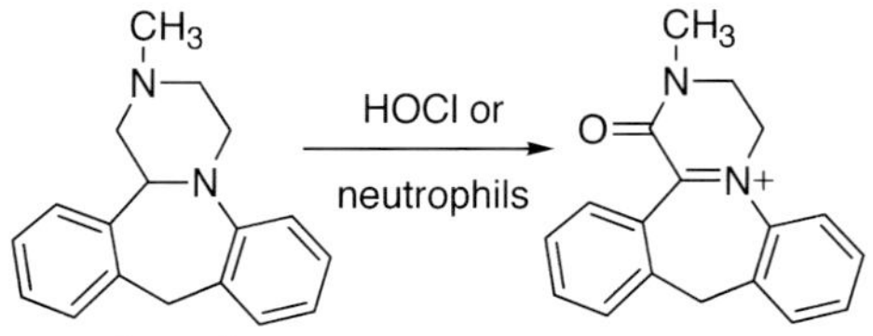

Figure 26. Bioactivation of mianserin

L-745,870 is a selective D4 antagonist that was being developed for the treatment of schizophrenia until it was found to form a reactive metabolite leading to a mercapturic acid conjugate. The presumed reactive metabolite is an imine methide similar to the reactive metabolite formed from 3-methylindole (Fig. 27), although the mechanism of formation is somewhat different.[165] Specifically the imine methide of 3-methylindole likely comes form oxidation of the methyl group followed by conjugation with sulfate and then loss of the sulfate, while it is speculated that the imine methide of L-745,870 arises from *N*-oxidation of the piperazine nitrogen followed by loss of the piperazine as a hydroxylamine.

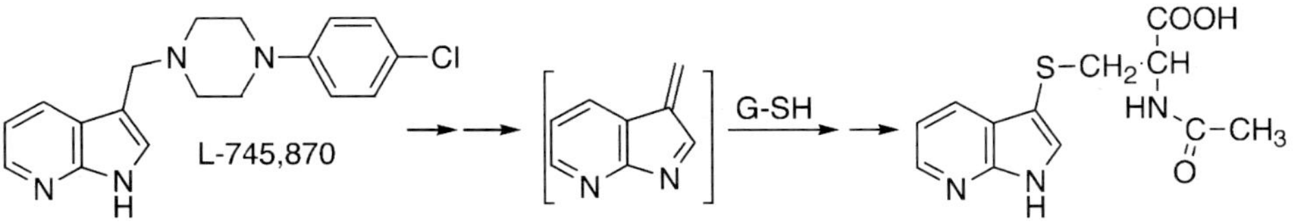

Figure 27. Bioactivation of L-745,870

A more recent issue is leflunomide. It is associated with a high incidence of liver toxicity as well as several other types of idiosyncratic drug reactions.[166] The major metabolite is a rearrangement product (Fig. 28)[167] that would be expected to be a good Michael acceptor; however, there does not appear to be any data to link the reactivity of this metabolite with the toxicity. It is also an amide of an aryl amine that could be released by hydrolysis.

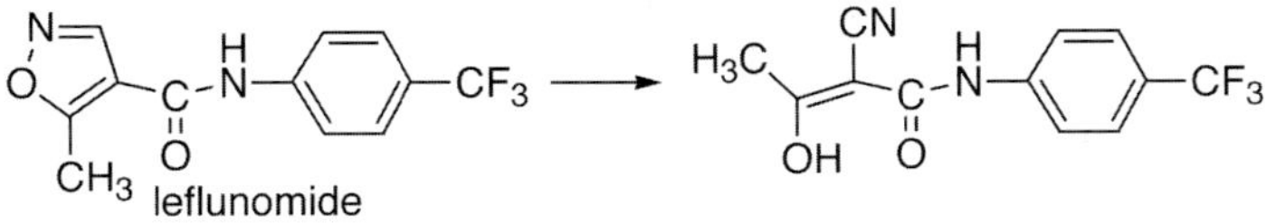

Figure 28. Bioactivation of leflunomide

4.10 *Metabolites, such as epoxides, that are reactive because of ring strain*

The most common metabolite that is reactive because of ring strain contains oxygen and is called an epoxide. The reactivity and significance of epoxides was first recognized when it was found that certain polycyclic aromatic hydrocarbons formed reactive epoxides that covalently bound to DNA. Furthermore, there was an association between this binding and the carcinogenic potential of the hydrocarbon. The reactivity of epoxides varies considerably and not all epoxides covalently bind to DNA. In fact simple epoxides are relatively unreactive. When the epoxide is formed from the oxidation of a benzene ring, there is a driving force for rearrangement to re-aromatize the ring, and these epoxides are often referred to as arene oxides. Most drugs contain a benzene ring and so this is a common pathway. The product of arene oxide rearrangement is a phenol, and it is usually assumed that the oxidation of a benzene ring to form a phenol involves an arene oxide intermediate (Fig 29). During the rearrangement, there is often migration of a substituent from one of the arene oxide carbons to the other, and this is referred to as the NIH shift.[168] At one time the presence of a NIH shift was considered proof of an arene oxide intermediate, but it is possible to write a mechanism for the shift in which a free arene oxide is never formed (Fig. 29).[169] Other reactions of arene oxides are also shown in Fig. 29. The reaction of an arene oxide with glutathione is shown as resulting in glutathione substitution in the *meta* position, but the reaction is not very selective and presumably does not involve the carbocation. In fact the reaction with glutathione may be largely mediated by glutathione transferase. In the case of an electron deficient aromatic ring, the oxidation to form a phenol appears to involve a "direct" oxidation not involving an arene oxide intermediate.[170]

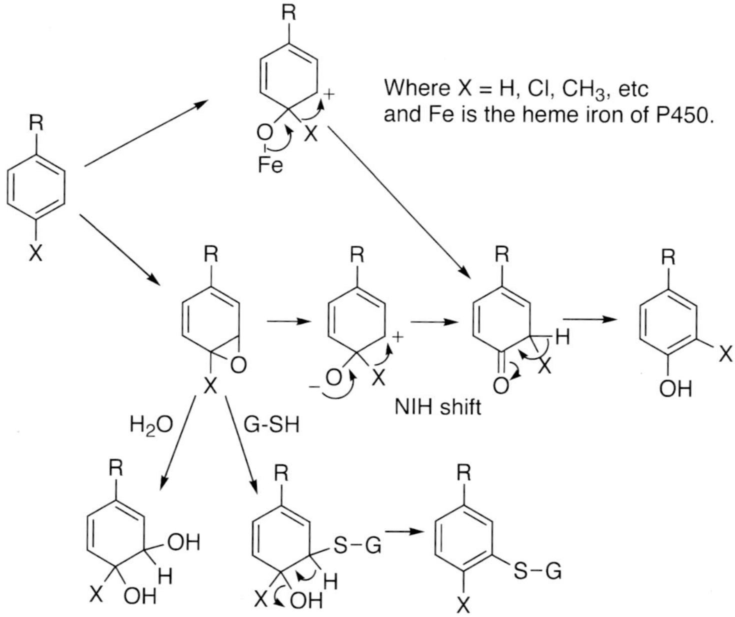

Where X = H, Cl, CH$_3$, etc
and Fe is the heme iron of P450.

H$_2$O   G-SH   NIH shift

Figure 29. Reactions of arene oxides

Alclofenac was withdrawn from the market because it is associated with a higher incidence of idiosyncratic reactions than most other NSAIDs.[171] It is a potent inhibitor of P450 and it was suggested that this is due to an epoxide metabolite

(Fig. 30).[172] Another presumed reactive metabolite of the drug is acrolein because a phenol is one of the known metabolites and this should come from an *o*-dealkylation that would lead to acrolein as the other product. Acrolein is a Michael acceptor (see Section 4.9) and a well-known toxic agent. These pathways could easily be anticipated and it would be wise to avoid this type of structure.

Phenytoin, carbamazepine and phenobarbital are aromatic anticonvulsants that are associated with a range of idiosyncratic drug reactions, but the most characteristic reaction is a syndrome called the anticonvulsant hypersensitivity syndrome.[173] This syndrome is characterized by rash, fever and lymphadenopathy. In addition, other organs are usually involved, such liver, bone marrow, thyroid and lungs, but the severity and pattern of organ involvement varies from one patient to another. If a patient has such a reaction to one of these anticonvulsants, they have a higher risk of having a similar reaction to one of the other two anticonvulsants, which suggests that there is some common feature to the reactive metabolite or mechanism involved.[173]

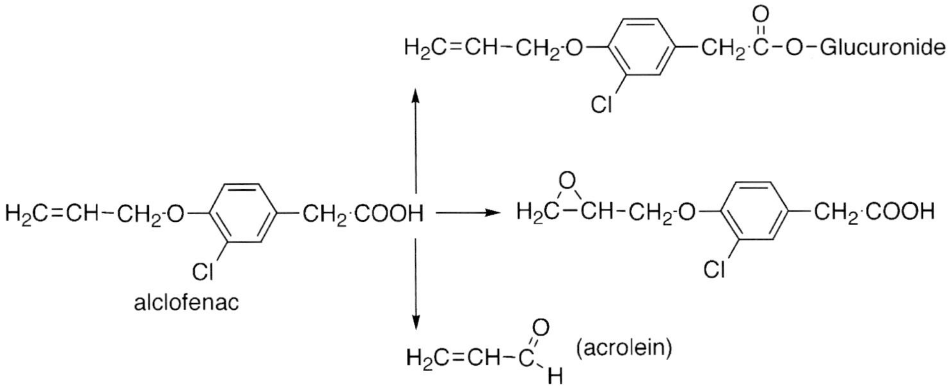

Figure 30. Bioactivation pathways of alclofenac

It was postulated that the aromatic anticonvulsant syndrome is due to arene oxide metabolites (see Fig 15). It was found that lymphocytes from patients with a history of such a reaction are more sensitive to the toxic effects of the drug *in vitro* in the presence of hepatic microsomes.[174] This suggests that the toxicity is due to a metabolite. Furthermore, the toxicity was increased by an inhibitor of epoxide hydrolase, which suggests that the toxic metabolite is an epoxide/arene oxide.[175] It was further postulated that it is a defect in detoxication, probably decreased epoxide hydrolase activity, which conferred susceptibility to the anticonvulsant hypersensitivity syndrome. However, when it was discovered that epoxide hydrolase activity is polymorphic, studies failed to find a correlation between the genotype associated with low activity and the risk that a patient would develop the syndrome.[37,176] Furthermore, investigators have failed to find evidence (other than the formation of phenolic metabolites) that arene oxides are formed, and specifically, no evidence for a glutathione conjugate of the arene oxide was found in humans. There is evidence for the formation of other reactive metabolites of these drugs as described in Sections 4.8 and 4.17.

As mentioned in the Introduction, penicillin and other β-lactams are classic drugs associated with idiosyncratic drug reactions, and more is known about the mechanisms of allergic reactions due to penicillin than any other type of idiosyncratic drug reaction. β-Lactams are reactive because of ring strain and they do not require metabolic activation in order to covalently bind to protein. They bind to amine and sulfhydryl nucleophiles on protein, and patients that have anaphylactic reactions have IgE antibodies against penicillin-modified proteins. Rearrangement products of penicillin can form reactive species (the so-called minor determinants) that bind to protein and, less commonly, lead to allergic reactions.[177]

Nitrogen and sulfur mustards are compounds that contain a halogen ß to a nitrogen or sulfur, respectively. These compounds rearrange by a nucleophilic attack of the nitrogen or sulfur on the carbon adjacent to the halogen leading to loss of the halogen and the formation of a highly reactive three-membered aziridinium or episulfonium ring, respectively (see Fig. 31). Sulfur mustards have been used as "war" gases and are cheap to manufacture. Nitrogen mustards are alkylating agents that are used for the treatment of cancer. Phenoxybenzamine (Fig 31) is also a nitrogen mustard that binds with some specificity to the α-adrenergic receptor, and it is used for the treatment of hypertension in patients with pheochromocytomas. In general these compounds are reactive without metabolic activation but a few, such as cyclophosphamide, do require metabolic activation. It is interesting to note that alkylating agents used for the treatment of cancer are not associated with an especially high incidence of idiosyncratic drug reactions. This could be due to the usual dosing schedule in which the drugs are only administered every three weeks. Although anticancer activity is believed to be due to covalent binding of the drug to DNA, most of the binding is to proteins.

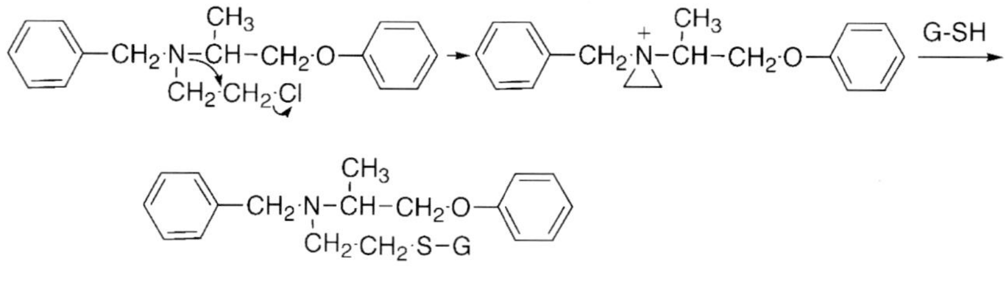

Figure 31. Reactivity of phenoxybenzamine

4.11 Furans and Thiophenes

Furosemide is a diuretic that is associated with a low but significant incidence of liver toxicity. Furans are oxidized, either through an epoxide or more directly, to a molecule that ring opens to a butene-1,4-dial (Fig. 32).[178] Therefore furans could be classed with the Michael acceptors because these dials are very good Michael acceptors.

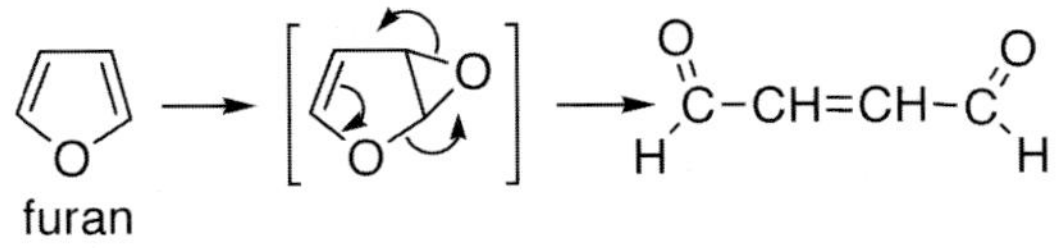

Figure 32. Oxidation of furan to butene-1,4-dial.

Thiophenes are believed to be oxidized to epoxides. However, at least in the case of tienilic acid (also known as ticrynafen), another diuretic that is associated with liver toxicity, it appears that the reactive metabolite is an S-oxide (Fig. 33).[179]

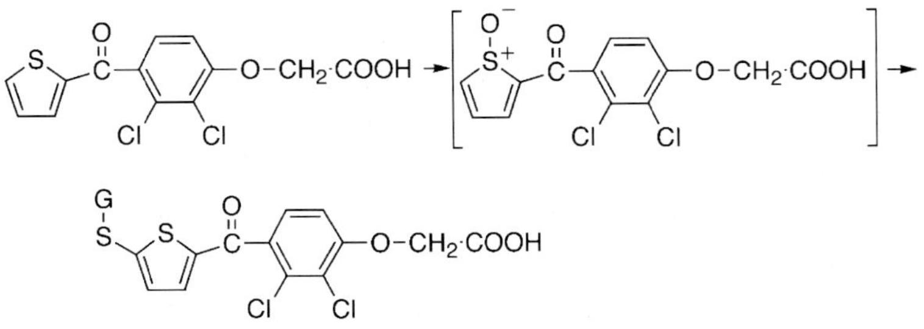

Figure 33. Bioactivation of tienilic acid

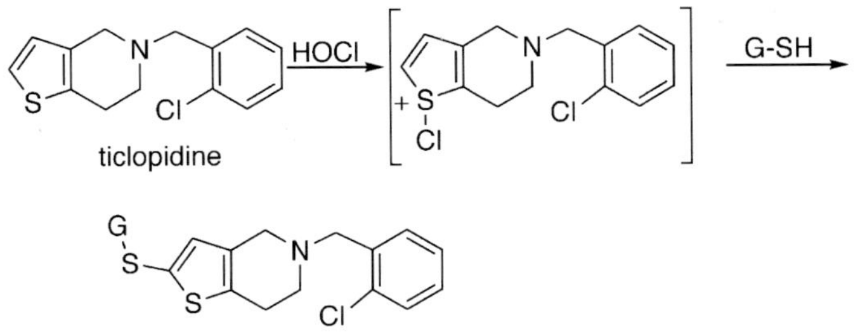

Figure 34. Activation of ticlopidine

Ticlopidine is a drug used to inhibit platelet aggregation, but its use is associated with a relatively high incidence of agranulocytosis, aplastic anemia and thrombocytopenia. It appears to be oxidized by activated neutrophils to a reactive S-chloro derivative that is similar to the S-oxide proposed for the toxic metabolite of ticrynafen (Fig. 34).[180]

4.12 Sulfhydryl-containing drugs

Drugs that contain a sulfhydryl group react with protein disulfides to form a new disulfide with the drug and releasing a free protein sulfhydryl group (Fig. 35). Therefore, no metabolic activation is required for the covalent binding of these drugs. In addition, sulfhydryl groups are easily oxidized to sulfenic acids, which react with protein sulfhydryl groups to form mixed disulfides. Thus, drugs that contain sulfhydryl groups react extensively with proteins. However, these disulfides usually undergo further exchange with other sulfhydryl-containing compounds, and the disulfide linkage is not as permanent as most covalent binding. None the less, such drugs are usually associated with a high incidence of allergic reactions so the binding probably survives antigen processing in antigen presenting cells.

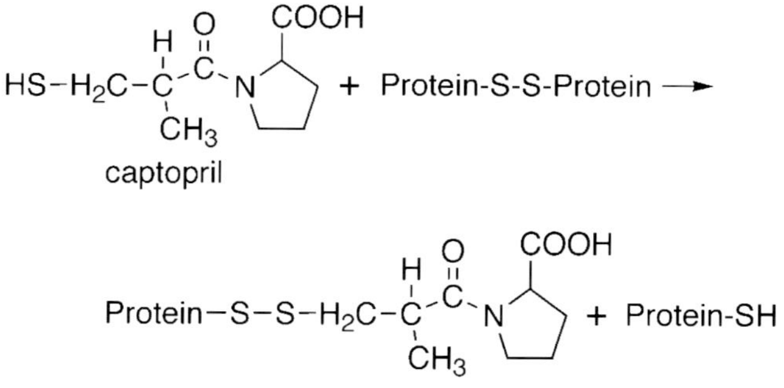

Figure 35. Reaction of captopril with proteins

Captopril was the first angiotensin converting enzyme inhibitor; however, it has been largely replaced by other agents because of a relatively high incidence of allergic reactions, especially skin rashes, as well as other types, such as bone marrow toxicity. It is known to form mixed disulfides with proteins, and this acts as a repository form of the drug because it is a reversible reaction.[181]

Penicillamine is a chelating agent that is used for the treatment of copper toxicity, and it is also a second line agent for the treatment of rheumatoid arthritis and scleroderma. Its use is associated with a wide range of unusual autoimmune reactions including a lupus-like syndrome, myasthenia gravis (a syndrome characterized by antibodies against the acetylcholine receptor that causes muscle weakness) as well as aplastic anemia.[182,183] It is also known to cause an autoimmune syndrome in Brown Norway rats.[184] Although a mixed disulfide with penicillamine can undergo exchange reactions with other sulfhydryl-containing molecules,[185] the steric effect of the methyl groups prevents ready exchange with penicillamine disulfide.

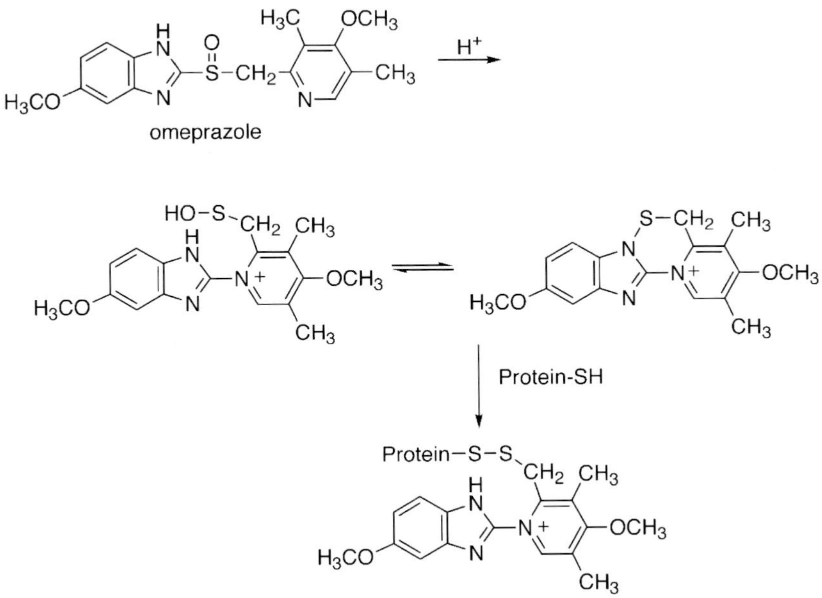

Figure 36. Activation of omeprazole

Omeprazole was the first marketed proton pump inhibitor used for the inhibition of stomach acid production. It undergoes an acid-catalyzed rearrangement leading to a reactive sulfenic acid (Fig. 36).[186] Omeprazole is sometimes associated with rash and itching; however, the incidence of idiosyncratic reactions is low. This is likely due to the low therapeutic dose and the fact that it is a weak base and preferentially distributes to parietal cells where it binds to the proton pump.

4.13 Thiono Sulfur Compounds

Thiono sulfur compounds, such as thiourea and thioacetamide, are well known to cause liver and bone marrow toxicity as well as several other types of adverse reactions.[187] In the case of thioacetamide, toxicity appears to be due to the 4-electron oxidation of the sulfur resulting in a sulfur dioxide leaving group and the formation of a very reactive electrophile. There are a few drugs that also contain a thiono sulfur group as one tautomer of the structure.

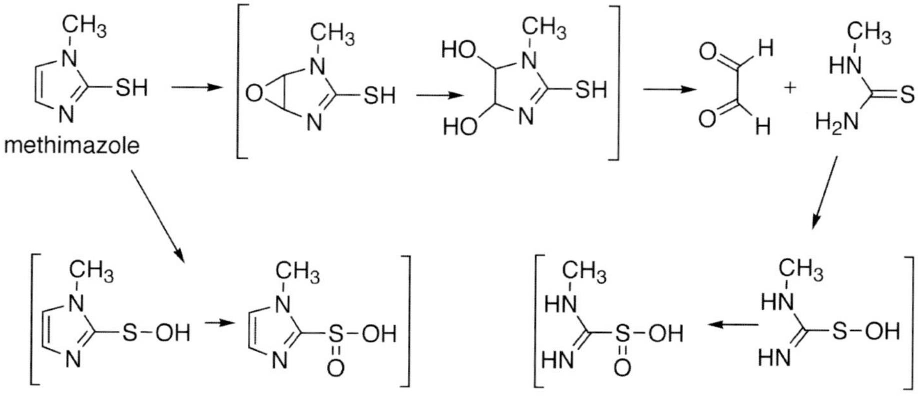

Figure 37. Bioactivation of methimazole

Methimazole is a thiono sulfur-containing drug that is used for the treatment of hyperthyroidism. It is associated with agranulocytosis and liver toxicity. It is metabolized first to *N*-methylthiourea by P450 and the flavin-containing monooxygenase. Saturation of the carbon-carbon double bond prevented liver toxicity in mice and *N*-methylthiourea caused similar toxicity, which indicates that the formation of *N*-methylthiourea is necessary for the toxicity, at least in the mouse model used in this study.[188] The *N*-methylthiourea required further oxidation, probably by the pathway shown in Fig. 37, in order to cause toxicity. It is likely that methimazole itself can undergo similar metabolic activation. This drug is associated with adverse reactions at a relatively low dose; for the initial treatment of hyperthyroidism the daily dose is only about 30 mg/day.

Propylthiouracil is another thiono sulfur-containing drug that is used to treat hyperthyroidism. It is also associated with a relatively high incidence of agranulocytosis and liver toxicity as well as a lupus-like syndrome. It also causes a lupus-like syndrome in cats.[189] It appears to be metabolized to a reactive metabolite by thyroid peroxidase, and this may play a role in its mechanism of action.[190] Furthermore, it appears to be metabolized to a reactive sulfenyl chloride and other reactive metabolites by activated neutrophils, and this likely explains why it is associated with a high incidence of agranulocytosis.[191] It is likely that it is also oxidized to reactive intermediates by P450/flavin-dependent monooxygenase, which could be responsible for liver toxicity.

N-(5-Chloro-2-methylphenyl)-*N*-(2-methylpropyl)thiourea is a drug that was being developed for the treatment of atherosclerosis; however, it was found to be metabolized to a reactive metabolite that formed a novel glutathione conjugate.[192] The proposed pathway for the bioactivation is shown in Fig. 38.

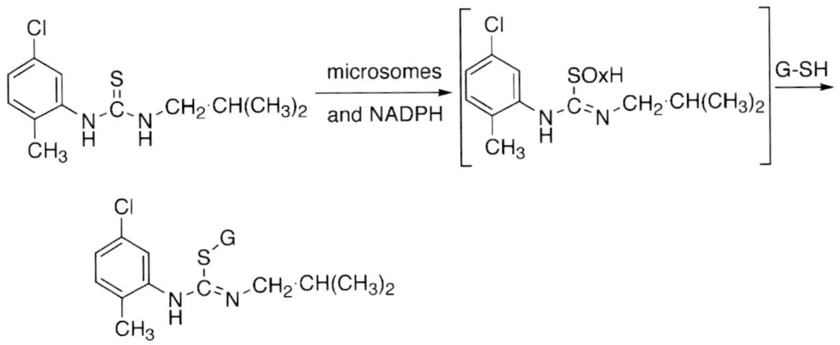

Figure 38. Bioactivation of *N*-(5-chloro-2-methylphenyl)-*N*-(2-methylpropyl) thiourea

4.14 Isocyanates

Isocyanates are electrophiles that react with primary amines and sulfhydryl groups. The reaction with sulfhydryl groups, such as glutathione, is reversible so that the glutathione conjugate can act to transfer this reactive species from one location to another. In general isocyanates are not formed by oxidation of drugs but rather from a rearrangement catalyzed by unknown enzymes. It has been known for some time that nitrosoureas rearrange to produce an isocyanate and a diazonium ion. More recently it was found that the antidiabetic drug, tolbutamide, also rearranges to form an isocyanate as shown in Fig. 39.[193]

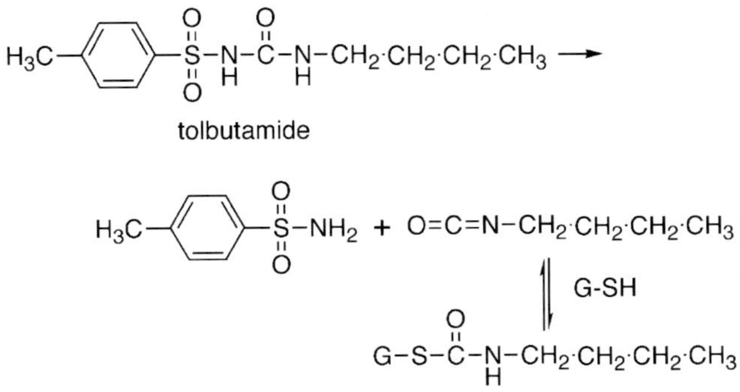

Figure 39. Bioactivation of tolbutamide to an isocyanate

Sulofenur is an anticancer agent that caused hemolytic anemia and methemoglobinemia. Its mechanism of action as an anticancer agent is unknown because it does not affect DNA, RNA or protein synthesis. The red cell toxicity was attributed to a chloroaniline metabolite, but it now appears that this metabolite is formed from a chlorophenylisocyanate intermediate by reaction with water.[194] An interesting finding was that the glutathione conjugate also shows anticancer activity.[195] As indicated above, the reaction of glutathione with an isocyanate is reversible and the glutathione conjugate may act as a transport system for the isocyanate. Therefore, the activity of the glutathione conjugate may be due to the isocyanate, and this may contribute to the anticancer activity of the drug.

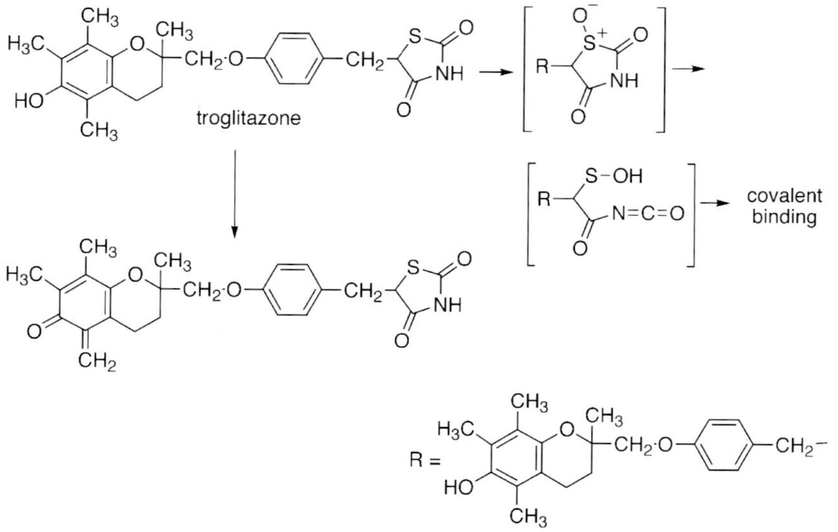

Figure 40. Bioactivation of troglitazone

Glitazones are used for the treatment of diabetes. Troglitazone, the first agent of the class to be marketed, was withdrawn from the market because of a relatively high incidence of severe hepatotoxicity. Troglitazone forms at least three reactive metabolites. The quinone methide was described above in Section 4.8. Oxidation of the thiazolidinedione ring leads to the formation of an isocyanate and a sulfenic acid as shown in Fig. 40.[149] It is unknown which metabolite or combination of metabolites is responsible for the liver toxicity of troglitazone. Another drug in the class is rosiglitazone, which is not associated with the same liver toxicity as troglitazone. Although two cases of liver failure have been described in association with rosiglitazone it is not clear that they were due to the drug.[196] One could speculate that because rosiglitazone does not form a quinone methide, this is the metabolite responsible for toxicity; however, the dose of rosiglitazone is also much lower than that of troglitazone (4 mg/day as opposed to 200 mg/day) so such a conclusion would be unwarranted.

4.15 Acetylenes

Terminal alkynes can be oxidized to very reactive ketene metabolites.[197] Examples include some of the synthetic progesterones and estrogens. The exact mechanism by which this metabolite is formed is unknown, but there is evidence that it involves an oxirene intermediate (Fig. 41).[198]

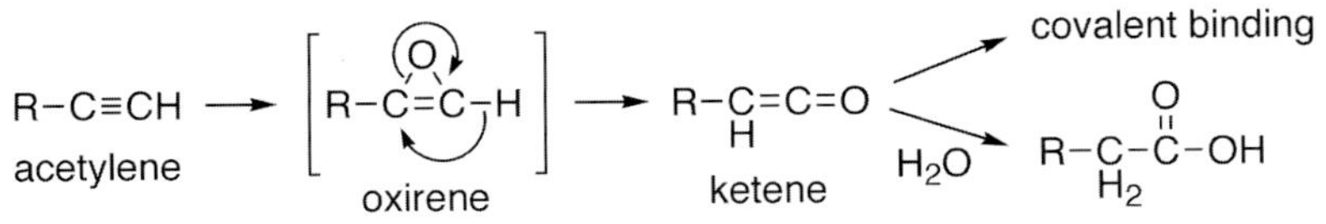

Figure 41. Proposed mechanism for the bioactivation of terminal acetylenes

4.16 Methylenedioxyphenyl compounds

The functional group methylenedioxyphenyl is oxidized to a reactive species, presumably a carbene (Fig. 42), which binds to the heme iron of P450 leading to mechanism-based inhibition of P450.[199,200] This property is used in the

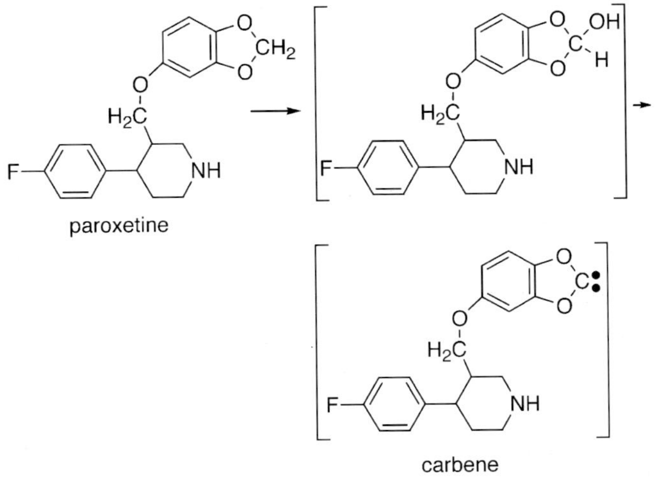

Figure 42. Bioactivation of methylenedioxyphenyl compounds

insecticide synergist piperonyl butoxide, which is combined with insecticides to inhibit their metabolism by insects. It is also present in several drugs, such as paroxetine and methylenedioxymetamphetamine (Ecstasy). This type of reactive metabolite is unlikely to directly lead to idiosyncratic drug reactions because it binds to heme iron not to proteins; however, it is likely to lead to drug interactions. In addition, this functional group is ultimately converted to a catechol, which can be further oxidized to a reactive *o*-quinone. This functional group is also present in many natural products. For example Kava contains methysticin, pepper contains piperine, nutmeg contains myristicin and sassafras and several essential oils contain safrole and/or isosafrole.

4.17 Free Radicals

Free radicals may be important for the mechanism of many idiosyncratic drug reactions.[41] However, their chemistry is more difficult to study than two electron chemistry, and there is less data on the association between the formation of free radicals and idiosyncratic drug reactions. Many drugs that form electrophilic reactive metabolites also form free radicals, and there is little data to indicate which is more important. The reactivity of free radicals is very different from that of electrophiles; they usually abstract a hydrogen from other molecules to generate a new free radical and seldom covalently bind to protein, although a molecule can be both a free radical and an electrophile. The most common site for hydrogen atom removal is from unsaturated lipids as shown in Fig. 43. Another important reaction of free radicals, especially carbon-centered free radicals, is with oxygen, which is itself a diradical. The resultant peroxy radical can lead to the generation of many new free radicals.

If hapten formation is necessary for the induction of an idiosyncratic reaction, free radicals should be less important than electrophilic reactive metabolites because they are less likely to bind to protein. On the other hand, the chain reaction initiated by a free radical can lead to reactive oxygen species and lipid peroxidation as well as other types of cell damage. Therefore, if a danger signal is important for the initiation of an idiosyncratic drug reaction then free radicals could be more important than electrophiles. If both covalent binding and cell stress are required for the induction of an idiosyncratic drug reaction, drugs that can form both types of reactive metabolites, e.g. aryl amines and quinones, may be the most likely to cause idiosyncratic reactions. This latter possibility appears to best fit the pattern of drugs that cause idiosyncratic drug reactions; however, in the absence of better quantitative data, it is difficult to be confident in this conclusion.

As indicated above, quinones readily undergo redox cycling involving either one or two electrons.[124] Aryl amines and nitro groups also readily undergo redox cycling often involving free radical intermediates.[201] Phenothiazines have been known for a long time to undergo a one-electron oxidation to form a nitrogen-centered free radical.[202] There is little data to indicate whether this pathway is important for the idiosyncratic liver or bone marrow toxicity associated with this class of drugs.

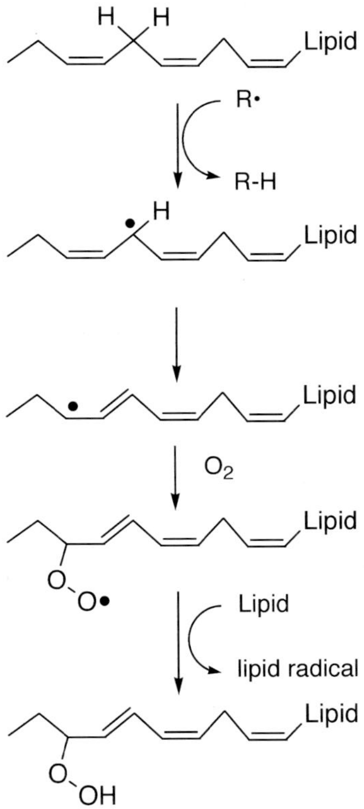

Figure 43. Reactions of free radicals with lipids (R• represents a free radical metabolite of a drug).

Aminopyrine is oxidized to a relatively stable free radical (Fig 10),[203] and again it is difficult to know the relative contribution of the dication and free radical to the mechanism of aminopyrine-induced agranulocytosis. Although clozapine is known to be oxidized by HOCl (and probably P450) to a nitrenium ion (Fig 9), it also appears to undergo a one-electron oxidation to a free radical.[204] This free radical was not observed directly but generation of a glutathione free radical was shown to be dependent on the presence of clozapine.

In addition to the other reactive metabolites, such as quinones and arene oxides, that have been proposed to be responsible for the adverse reactions associated with phenytoin (see Sections 4.8 and 4.10), it also appears to undergo a one electron oxidation of the nitrogen flanked by the carbonyl groups. By analogy with the one electron oxidation of succinimide, the resulting free radical would be expected to undergo a series of rearrangements, and products consistent with this pathway have been observed (Fig. 44).[205] In addition, incubation of phenytoin with prostaglandin synthase produced a carbon-centered free radical and the teratogenicity of phenytoin was blocked by a free radical trap,[206] which suggests that this type of toxicity is due to the free radical pathway. Unfortunately, there is no good animal model of the anticonvulsant hypersensitivity syndrome that could be used to determine which of the several pathways is most important for idiosyncratic reactions.

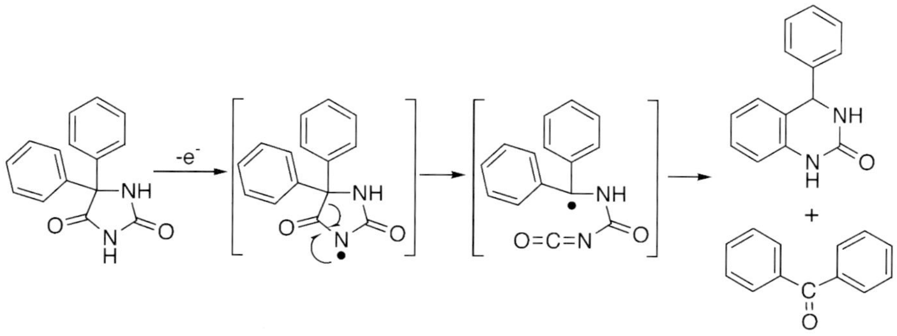

Figure 44. Proposed bioactivation of phenytoin through a free radical pathway

The metabolism of hydrazines usually involves the formation of free radicals, which often bind to the heme of P450, and the evidence for a free radical pathway is better than for a diazonium ion pathway (Section 4.5). It may be that binding to protein requires a 4 electron oxidation to a diazonium ion while binding to heme is a free radical process and the balance between these pathways may be different for different compounds.[207] Iproniazid and phenelzine are hydrazine-based monoamine oxidase inhibitors used for the treatment of depression. Iproniazid was removed from the market many years ago because of liver toxicity, and phenelzine also causes liver toxicity and is not commonly used. Their metabolism is complex but it appears that they are metabolized first to diazines and then to free radicals (Fig. 45).[208] Dihydralazine is oxidized to reactive metabolites, one of which is likely to be a free radical that binds to heme and leads to inactivation of P450.[209]

The oxidation of 3,5-(bis)carbethoxy-2,6-dimethyl-4-ethyl-1,4-dihydro-pyridine, an analog of some of the calcium channel blockers, is oxidized by

P450 to an ethyl free radical that also binds to the heme portion of the enzyme as shown in Fig 46.[210]

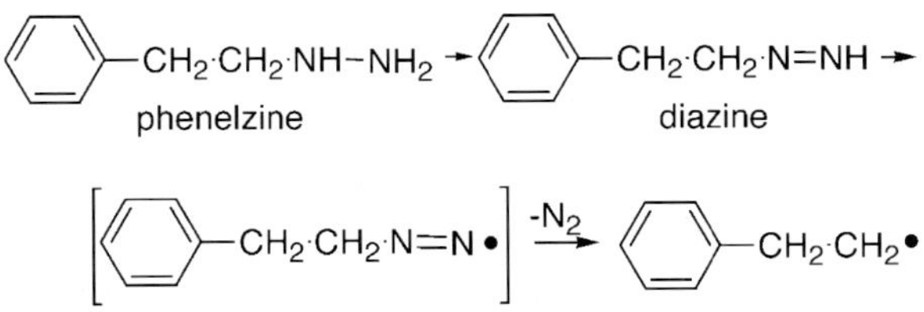

Figure 45. Free radical bioactivation of hydrazines

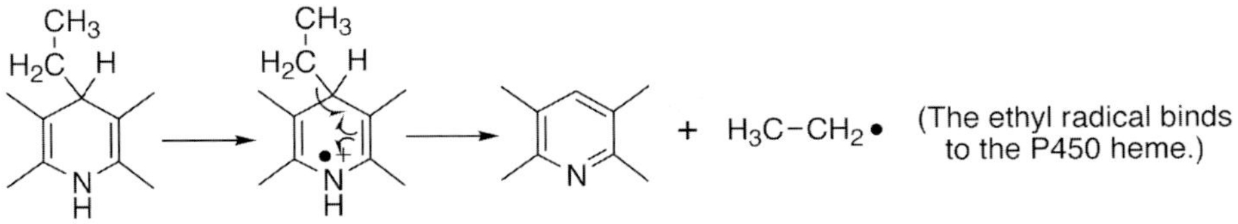

Figure 46. Oxidation of a dihydropyridine to alkyl free radicals

5   THE USE OF BIOACTIVATION SCREENS FOR LEAD DRUG SELECTION

The process of avoiding reactive metabolites starts with avoidance of functional groups that are associated with a high incidence of idiosyncratic drug reactions unless that functional group is absolutely essential for the desired pharmacological activity. Within this group I would include primary aromatic amines (unless it is on a highly electron deficient ring), nitro groups, hydrazines, sulfhydryl groups, thiono sulfur, thiophenes, furans and structures that readily form quinones or related metabolites or other Michael acceptors. I would not include carboxylic acids in this list, although if the expected dose of the drug were high, it may have the potential to lead to allergic reactions. The next step is to guess how the drug will be metabolized and to try to predict if that would likely result in a reactive species.

Initial data on how the drug is metabolized can provide further clues as to possible reactive metabolite formation. Obviously if a glutathione conjugate or mercapturic acid is a significant metabolite, this is a signal that there may be a problem. However, other pathways may also provide clues. For example, although it is possible to guess that a Michael acceptor would be formed by the metabolism of terbinafine, the detection of the secondary amine shown in

Fig. 24 implies the formation of the reactive aldehyde. Terbinafine is also an example of the need for balance. I suspect, without detailed knowledge, that the allylamine is required for pharmacological activity, which could make it difficult to design formation of a reactive metabolite out of the molecule. Furthermore, the *N*-dealkylation leading to the Michael acceptor appears to be a relatively minor pathway. The result is a drug with very good efficacy, and although it is associated with significant idiosyncratic drug reactions, it has a reasonable safety profile. Another example is vesnarinone. The first clue we had to the formation of a reactive metabolite was the deamination of the aromatic amine as shown in Fig. 16. In general, deamination of an aromatic amine does not occur because there is no hydrogen on the α carbon. The only reasonable mechanism for this deamination involves a reactive imidoiminium ion, and this led to successful attempts to trap this reactive species (see Section 4.8). We were also able to directly observe this reactive metabolite by mass spectrometry.

Although these two steps require a significant deal of thought and it is necessary to balance the possibility of reactive metabolites against other considerations, they do not require additional experiments and I consider them to be essential for prudent drug development. Unfortunately it is not always easy to predict the formation of reactive metabolites, and it is even more difficult to predict quantitative aspects of reactive metabolite formation. Therefore, more active screens can provide very important additional information. Unfortunately, no one screening method will detect all reactive metabolites, and there are many mechanisms by which false negative and false positive results may be obtained.

Probably the easiest screen for reactive metabolites is to use LC/MS/MS for the analysis of metabolites with a search for fragments involving the neutral loss of 129 mass units, which is a characteristic fragment of glutathione conjugates. This is best done in an *in vitro* metabolic screen because, *in vivo,* glutathione conjugates are often converted to mercapturic acids or transported into bile and extensively degraded by bacteria in the gut and not observed. However, no single *in vitro* system can duplicate all of the pathways that occur *in vivo.* For example, the oxidation of vesnarinone by P450 appears very slow, and little reactive metabolite would likely be formed in a microsomal or hepatocyte culture; however, with activated neutrophils the oxidation is very rapid and extensive. In the case of felbamate, as shown in Fig. 23, the first step in the formation of the reactive metabolite involves hydrolysis of the carbamate. This does not occur at a significant rate in hepatocytes and probably involves a plasma esterase. Furthermore, not all reactive metabolites form observable glutathione conjugates, either because they are very reactive and react with the first molecule they come in contact with, or because they are too hard and react better with nitrogen nucleophiles, or because the glutathione conjugate is unstable. The reaction of glutathione with some electrophiles, such as acrolein and isocyanates, is reversible and, depending on sample preparation, the conjugate could be lost. Another example is a nitrenium ion where, if the glutathione reacts with the nitrogen instead of a carbon, the product will be unstable. Such electrophiles are more likely to react with a carbon on the imidazole ring of histidine or the C-8 of guanosine.

Quantification of reactive metabolites requires the use of radiolabeled drug and, in some circumstances, this can be an impediment. The simplest method is to incubate radiolabeled drug with rat hepatic microsomes. A rule of thumb proposed by Dr. L. Pohl is if the binding is less than 50 pmole/mg of protein, the binding is unlikely to be associated with a significant incidence of idiosyncratic drug reactions, and this is discussed in detail in the next chapter by Dr. Baillie. Although there is not sufficient data to know how accurate this rule of thumb is, it is very useful to have such a rule of thumb to emphasize that some degree of covalent binding is acceptable and probably unavoidable. There are also many ways in which false negative and false positive results can be obtained. If the major enzyme responsible for bioactivation is not P450 then this system will not reflect *in vivo* bioactivation. For the same reasons given above, this system would likely fail to reveal potential problems with vesnarinone and felbamate. It also may overestimate the degree of covalent binding that would occur *in vivo* because the normal scavenging systems are absent. The use of hepatocytes instead of hepatic microsomes has the advantages that human hepatocytes are available and they provide a better reflection of the pattern of metabolites formed in humans. They also contain many systems to scavenge reactive metabolites that microsomes lack. However, they still only reflect reactive metabolites formed in the liver and, therefore, would be unlikely to accurately reflect metabolic activation of vesnarinone or felbamate. The addition of other metabolic systems, such as activated neutrophils or simply chemical oxidation by HOCl, would provide further information, but it would be impractical to screen with every possible system that could lead to metabolic activation. However, it is often possible to guess how a drug might be bioactivated, and the appropriate metabolic systems could be chosen on this basis.

Another possible problem is that if several steps are involved in the formation of a reactive metabolite and the intermediate metabolites have long half-lives, the intermediate metabolites are unlikely to reach concentrations *in vitro* that accurately reflect the pathways that occur *in vivo*. Furthermore, some idiosyncratic drug reactions may reflect specific patterns of metabolism that are unique to certain individuals, and unless hepatocytes were obtained from such individuals, the results would not detect possible problems that might occur in these individuals. However, genetic polymorphism of drug metabolism does not appear to be the major basis for the idiosyncratic nature of these adverse reactions.

As indicated earlier, covalent binding may not be the only mechanism involved in the association between bioactivation and the induction of idiosyncratic drug reaction. In fact, it may not even be a major factor. It is possible that the induction of cell stress is just as important. However, we do not yet know anything about the possible association between certain types of cell stress and the induction of idiosyncratic drug reactions. Therefore, no suggestions can be made as to what pattern (or more likely patterns) of cell stress will predict a drug candidate's potential to cause idiosyncratic drug reactions. Many *in vitro* methods are available to measure drug toxicity; however, I suspect that the mechanisms of idiosyncratic reactions is too complex to be modeled in such *in vitro* assays and they have little predictive value. However, if we had a better understanding of the mechanisms of idiosyncratic reactions, I suspect that *in vivo* patterns of cell stress might provide a much better assessment of the risk that a potential drug would cause idiosyncratic drug reactions.

As indicated above, no one screen for bioactivation will work for all drugs and types of metabolic activation. Furthermore, the significance of the data obtained is often hard to determine. Despite all of the unknowns and possible caveats involved in the screening of drug candidates for bioactivation, it is likely that the rational use of a combination of such studies at an early stage in drug development would greatly improve the selection of the best candidate for further development. One of the most important factors is that the people involved in the process of drug development be cognizant of the issue of bioactivation and make the best use of the studies available as well as the best use of the data obtained from such studies. In the next chapter of this book, Dr. Baillie has provided an excellent snapshot of how such studies are done at Merck.

6   SUMMARY AND CONCLUSIONS

There is a large amount of circumstantial evidence to suggest that most idiosyncratic drug reactions are due to reactive metabolites rather than due to the parent drug itself; however, it must be stated that this hypothesis has not been proven. The exact basis for this association is unknown but there are two major hypotheses: 1. reactive metabolites can act as haptens, which can lead to an immunological reaction, and 2. reactive metabolites can cause tissue damage or stress, which leads to a danger signal and stimulation of an immune response. These two hypotheses are not mutually exclusive and there is evidence that both may be important; however, the evidence is far from unambiguous for either hypothesis. In addition, in some cases the association between bioactivation and idiosyncratic drug reactions may be due to direct cytotoxicity, and the basis for the idiosyncratic nature of the reaction may not involve the immune system.

Without a better understanding of the mechanisms involved in idiosyncratic drug reactions it is difficult to judge the significance of reactive metabolite formation observed in drug candidates. An added complication is that many drugs form several reactive metabolites, and without valid animal models, it is impossible to be certain of the relative contribution of each. Despite these uncertainties, as well as the difficulties in screening for reactive metabolites as outlined above, it is highly likely that such screening at an early stage of drug development followed, where possible, by designing the metabolic pathway out of the molecule, would significantly improve the safety of new drugs. As always, dose is important, and other things being equal, more potent drugs will be safer drugs. However, it is also true that if the criteria are made too stringent it would result in failure to develop many very important drugs. Many considerations must be balanced; however, in looking at the structures of many drugs that have been marketed it does not appear that much consideration has been given to the avoidance of reactive metabolites. The price of failure is too high, and even a modest improvement in the rate at which drugs fail because of adverse reactions would have a major impact on drug development.

REFERENCES

1. Lasser, K. E., *et al. J. Am. Med. Assoc.* **2002**, *287*, 2215.
2. Miller, J. A. *Cancer Res.* **1970**, *30*, 559.
3. Mitchell, J. R., *et al. J. Pharmacol. Exp. Ther.* **1973**, *187*, 185.
4. Davies, D. M. *Textbook of Adverse Drug Reactions* **1985** (Oxford University Press, Oxford).
5. Landsteiner, K.; Jacobs, J. *J. Exp. Med.* **1935**, *61*, 643.
6. Vergani, D., *et al. N. Engl. J. Med.* **1980**, *303*, 66.
7. Njoku, D., *et al. Anesth. Analg.* **1997**, *84*, 173.
8. Bourdi, M., *et al. Chem. Res. Toxicol.* **1996**, *9*, 1159.
9. Lecoeur, S., *et al. Chem. Res. Toxicol.* **1994**, *7*, 434.
10. Bourdi, M., *et al. J. Clin. Invest.* **1990**, *85*, 1967.
11. Masubuchi, Y.; Horie, T. *Chem. Res. Toxicol.* **1999**, *12*, 1028.
12. Uetrecht, J. P. *CRC Crit. Rev. Toxicol.* **1990**, *20*, 213.
13. Cambridge, G.; Wallace, H.; Bernstein, R. M.; Leaker, B. *Br. J. Rheumatol.* **1994**, *33*, 109.
14. Pumford, N. R.; Halmes, N. C. *Annu. Rev. Pharmacol. Toxicol.* **1997**, *37*, 91.
15. Uetrecht, J. P. *Chem. Res. Toxicol.* **1999**, *12*, 387.
16. Fearon, D. T. *Nature* **1997**, *388*, 323.
17. Matzinger, P. *Annu. Rev. Immunol.* **1994**, *12*, 991.
18. Matzinger, P. *Semin. Immunol.* **1998**, *10*, 399.
19. Gallucci, S.; Matzinger, P. *Curr. Opin. Immunol.* **2001**, *13*, 114.
20. Fischl, M. A.; Dickinson, G. M.; La Voie, L. *J. Am. Med. Assoc.* **1988**, *259*, 1185.
21. Levy, M. *Drug Saf.* **1997**, *16*, 1.

22. Ellrodt, A. G., *et al. Ann. Intern. Med.* **1984**, *100*, 197.
23. Meyers, D. G., *et al. Am. Heart J.* **1985**, *109*, 1393.
24. Uetrecht, J. P.; Zahid, N.; Whitfield, D. *J. Pharmacol. Exp. Ther.* **1994**, *270*, 865.
25. Bonierbale, E., *et al. Chem. Res. Toxicol.* **1999**, *12*, 286.
26. Zanni, M. P., *et al. J. Clin. Invest.* **1998**, *102*, 1591.
27. von Greyerz, S.; Zanni, M.; Schnyder, B.; Pichler, W. J. *Clin. Exp. Allergy* **1998**, *28 Suppl 4*, 7.
28. Uetrecht, J. P. *Drug Metab. Rev.* **1992**, *24*, 299.
29. Lasker, J. M., *et al. J. Biol. Chem.* **1981**, *256*, 7764.
30. Purohit, V.; Basu, A. K. *Chem. Res. Toxicol.* **2000**, *13*, 673.
31. Ueda, O., *et al. Xenobiotica* **2001**, *31*, 33.
32. Monks, T. J., *et al. Toxicol. Appl. Pharmacol.* **1990**, *106*, 1.
33. Meyer, U. A. *Lancet* **2000**, *356*, 1667.
34. Huang, Y. S., *et al. Hepatology* **2002**, *35*, 883.
35. Cameron, H. A.; Ramsay, L. E. *Br. Med. J.* **1984**, *289*, 410.
36. Woosley, R. L., *et al. N. Engl. J. Med.* **1978**, *298*, 1157.
37. Gaedigk, A.; Spielberg, S. P.; Grant, D. M. *Pharmacogenetics* **1994**, *4*, 142.
38. Green, V. J.; Pirmohamed, M.; Kitteringham, N. R.; Knapp, M. J.; Park, B. K. *Br. J. Clin. Pharmacol.* **1995**, *39*, 411.
39. Leeder, J. S. *Epilepsia* **1998**, *39*, S8.
40. Ramakrishna, K. V.; Fan, P. W.; Boyer, C. S.; Dalvie, D.; Bolton, J. L. *Chem. Res. Toxicol.* **1997**, *10*, 887.
41. Mason, R. P.; Chignell, C. F. *Pharmacol. Rev.* **1982**, *33*, 189.
42. Smith, C. V.; Hughes, H.; Mitchell, J. R. *Mol. Pharmacol.* **1984**, *26*, 112.
43. Uetrecht, J. P. *J. Pharmacol. Exp. Ther.* **1985**, *232*, 420.
44. Skipper, P. L.; Stillwell, W. G. *Methods Enzymol.* **1994**, *231*, 643.
45. Mahmud, R., *et al. Toxicology* **1997**, *117*, 1.
46. Tingle, M. D.; Park, B. K. *Br. J. Clin. Pharmacol.* **1993**, *36*, 31.
47. Grant, D. M.; Goodfellow, G. H.; Sugamori, K.; Durette, K. *Pharmacology* **2000**, *61*, 204.
48. Roujeau, J. C.; Stern, R. S. *N. Engl. J. Med.* **1994**, *331*, 1272.
49. Cribb, A. E.; Miller, M.; Tesoro, A.; Spielberg, S. P. *Mol. Pharmacol.* **1990**, *38*, 744.
50. Cribb, A. E.; Spielberg, S. P. *Clin. Pharmacol. Ther.* **1992**, *51*, 522.
51. Cribb, A. E., *et al. Chem. Res. Toxicol.* **1996**, *9*, 500.
52. Cribb, A. E.; Pohl, L. R.; Spielberg, S. P.; Leeder, J. S. *J. Pharmacol. Exp. Ther.* **1997**, *282*, 1064.
53. Knowles, S.; Shapiro, L.; Shear, N. H. *Drug Saf.* **2001**, *24*, 239.
54. Hornsten, P.; Keisu, M.; Wiholm, B. *Arch. Dermatol.* **1990**, *126*, 919.
55. Kromann, N. P.; Vilhelmsen, R.; Stahl, D. *Arch. Dermatol.* **1982**, *118*, 531.
56. Grossman, S. J.; Jollow, D. J. *J. Pharmacol. Exp. Ther.* **1988**, *244*, 118.
57. Israili, Z. H., *et al. J. Pharmacol. Exp. Ther.* **1973**, *187*, 138.
58. Uetrecht, J.; Zahid, N.; Shear, N. H.; Biggar, W. D. *J. Pharmacol. Exp. Ther.* **1988**, *245*, 274.
59. Lawson, D. H.; Jick, H. *Br. J. Clin. Pharmacol.* **1977**, *4*, 507.

60. Uetrecht, J. P.; Sweetman, B. J.; Woosley, R. L.; Oates, J. A. *Drug Metab. Disp.* **1984**, *12*, 77.

61. Uetrecht, J.; Zahid, N.; Rubin, R. *Chem. Res. Toxicol.* **1988**, *1*, 74.

62. Uetrecht, J.; Sokoluk, B. *Drug Metab. Dispos.* **1992**, *20*, 120.

63. Salama, A.; Mueller-Eckhardt, C. *N. Engl. J. Med.* **1985**, *313*, 469.

64. Salama, A.; Mueller-Eckhardt, C. *Blood* **1986**, *68*, 1285.

65. Lindberg, R. L.; Syvalahti, E. K. *Clin. Pharmacol. Ther.* **1986**, *39*, 378.

66. Stuart-Harris, R. C.; Smith, I. E. *Cancer Treat. Rep.* **1984**, *11*, 189.

67. Dalrymple, P. D.; Nicholls, P. J. *Xenobiotica* **1988**, *18*, 75.

68. Raftery, E. B.; Denman, A. M. *Br. Med. J.* **1973**, *2*, 452.

69. Skegg, D. C.; Doll, R. *Lancet* **1977**, *2*, 475.

70. Record, N. B. *Ann. Int. Med.* **1981**, *95*, 326.

71. Wilson, J. D.; Bullock, J. Y.; Sutherland, D. C.; Main, C.; O'Brien, K. P. *Br. Med. J.* **1978**, *1*, 14.

72. Rood, J. P. *Br. Dent. J.* **2000**, *189*, 380.

73. Watkins, P. *Neurology* **2000**, *55*, S51.

74. Jorga, K.; Fotteler, B.; Heizmann, P.; Gasser, R. *Br. J. Clin. Pharmacol.* **1999**, *48*, 513.

75. Lautala, P.; Ethell, B. T.; Taskinen, J.; Burchell, B. *Drug Metab. Dispos.* **2000**, *28*, 1385.

76. Yunis, A. A.; Miller, A. M.; Salem, Z.; Corbett, M. D.; Arimura, G. K. *J. Lab. Clin. Med.* **1980**, *96*, 36.

77. Scheline, R. R. *Pharmacol. Rev.* **1973**, *25*, 451.

78. Isildar, M.; Abou-Khalil, W. H.; Jimenez, J. J.; Abou-Khalil, A.; Yunis, A. A. *Toxicol. Appl. Pharmacol.* **1988**, *94*, 305.

79. Young, N.; Alter, B. P. *Aplastic Anemia: Acquired and Inherited* **1994** (W. B. Saunders Company, London).

80. Wiholm, B. E., *et al. Brit. Med. J.* **1998**, *316*, 666.

81. Pohl, L. R.; Nelson, S. D.; Krishna, G. *Biochem. Pharmacol.* **1978**, *27*, 491.

82. Ferrari, V. *Sexually Transmit. Dis.* **1984**, *11*, 336.

83. De Renzo, A.; Formisano, S.; Rotoli, B. *Haematologica* **1981**, *66*, 98.

84. Utili, R.; Boitnott, J. K.; Zimmerman, H. J. *Gastroenterology* **1977**, *72*, 610.

85. Merlani, G., *et al. Eur. J. Clin. Pharmacol.* **2001**, *57*, 321.

86. Mingatto, F. E., *et al. Br. J. Pharmacol.* **2000**, *131*, 1154.

87. Rosen, G. M.; Demos, H. A.; Rauckman, E. J. *Toxicol. Lett.* **1984**, *22*, 145.

88. de Vries, H.; Beijersbergen van Henegouwen, G. *J. Photochem. Photobiol. B: Biology* **1998**, *43*, 217.

89. Zenarola, P.; Gatti, S.; Lomuto, M. *Dermatologica* **1991**, *182*, 196.

90. Kedderis, G. L.; Miwa, G. T. *Drug Metab. Rev.* **1988**, *19*, 33.

91. Back, O.; Lundgren, R.; Wiman, L. *Lancet* **1974**, *1*, 930.

92. Sharp, J. R.; Ishak, K. G.; Zimmerman, H. J. *Ann. Intern. Med.* **1980**, *92*, 14.

93. Boyd, M. R.; Stiko, A. W.; Sasame, H. A. *Biochem. Pharmacol.* **1979**, *28*, 601.

94. Hofstra, A.; Uetrecht, J. P. *Chemico-Biol. Interact.* **1993**, *89*, 183.

95. Timbrell, J. A.; Mitchell, J. R.; Snodgrass, W. R.; Nelson, S. D. *J. Pharmacol. Exp. Ther.* **1980**, *213*, 364.

96. Hofstra, A. H.; Angela Li-Muller, S. M.; Uetrecht, J. P. *Drug Metab. Dispos.* **1992**, *20*, 205.

97. Idanpaan-Heikkila, J.; Alhava, E.; Olkinuora, M.; Palva, I. P. *Eur. J. Clin. Pharmacol.* **1977**, *11*, 193.

98. Killian, J. G.; Kerr, K.; Lawrence, C.; Celermajer, D. S. *Lancet* **1999**, *354*, 1841.

99. Pirmohamed, M.; Williams, D.; Madden, S.; Templeton, E.; Park, B. K. *J. Pharmacol. Exp. Ther.* **1995**, *272*, 984.

100. Liu, Z. C.; Uetrecht, J. P. *J. Pharmacol. Exp. Ther.* **1995**, *275*, 1476.

101. Gardner, I.; Leeder, J. S.; Chin, T.; Zahid, N.; Uetrecht, J. P. *Mol. Pharmacol.* **1998**, *53*, 999.

102. Guest, I.; Sokoluk, B.; MacCrimmon, J.; Uetrecht, J. *Toxicology* **1998**, *131*, 53.

103. Williams, D. P.; Pirmohamed, M.; Naisbitt, D. J.; Uetrecht, J. P.; Park, B. K. *Mol. Pharmacol.* **2000**, *58*, 207.

104. Gardner, I.; Zahid, N.; MacCrimmon, D.; Uetrecht, J. P. *Mol. Pharmacol.* **1998**, *53*, 991.

105. Hoffman, A. M.; Butt, E. M.; Hickey, N. G. *J. Am. Med. Assoc.* **1934**, *102*, 1213.

106. Uetrecht, J. P.; Ma, H. M.; MacKnight, E.; McClelland, R. *Chem. Res. Toxicol.* **1995**, *8*, 226.

107. Szczeklik, A.; Nizankowska, E.; Sanak, M.; Swierczynska, M. *Curr. Opin. Allergy Clin. Immunol.* **2001**, *1*, 27.

108. Szczeklik, A., *et al. Clin. Exp. Allergy* **2001**, *31*, 219.

109. Benet, L. Z., *et al. Life. Sci.* **1993**, *53*, L141.

110. Bailey, M. J.; Dickinson, R. G. *Chem. Res. Toxicol.* **1996**, *9*, 659.

111. Benet, L. Z.; Grillo, M. P.; M. P. *Chem. Res. Toxicol.* **2003**, *15*, 1309.

112. Smith, P. C.; Benet, L. Z.; McDonagh, A. F. *Drug Metab. Dispos.* **1990**, *18*, 639.

113. Boelsterli, U. A. *Cur. Drug Metab.* **(in press)**,

114. Sayeh, E.; Uetrecht, J. P. *Toxicology* **2001**, *163*, 195.

115. Dunk, A. A.; Walt, R. P.; Jenkins, W. J.; Sherlock, S. S. *Br. Med. J. Clin. Res. Ed.* **1982**, *284*, 1605.

116. Colomina, P.; Garcia, S. *DICP* **1989**, *23*, 507.

117. Hargus, S. J., *et al. Chem. Res. Toxicol.* **1994**, *7*, 575.

118. Ware, J. A.; Graf, M. L.; Martin, B. M.; Lustberg, L. R.; Pohl, L. R. *Chem. Res. Toxicol.* **1998**, *11*, 164.

119. Halsey, J. P.; Cardoe, N. *Br. Med. J. (Clin Res Ed)* **1982**, *284*, 1365.

120. Qiu, Y.; Burlingame, A. L.; Benet, L. Z. *Drug Metab. Dispos.* **1998**, *26*, 246.

121. Ross, D.; Siegel, D.; Schattenberg, D. G.; Sun, X. M.; Moran, J. L. *Environ. Health Perspect.* **1996**, *104 Suppl 6*, 1177.

122. Monks, T. J.; Hanzlik, R. P.; Cohen, G. M.; Ross, D.; Graham, D. G. *Toxicol. Appl. Pharmacol.* **1992**, *112*, 2.

123. Chen, W., *et al. Biochemistry* **1999**, *38*, 8159.

124. Bolton, J. L.; Trush, M. A.; Penning, T. M.; Dryhurst, G.; Monks, T. J. *Chem. Res. Toxicol.* **2000**, *13*, 135.

125. Liu, Z. C.; McClelland, R. A.; Uetrecht, J. P. *Drug Metab. Dispos.* **1995**, *23*, 246.

126. Palsmeier, R. K.; Radzik, D. M.; Lunte, C. E. *Pharm. Res.* **1992**, *9*, 933.

127. Jensen, J., *et al. Biochem. Pharmacol.* **1993**, *45*, 1201.

128. Pearson, D. C.; Jourd'heuil, D.; Meddings, J. B. *Free Radic. Biol. Med.* **1996**, *21*, 367.

129. Grisham, M. B.; Ware, K.; Marshall, S.; Yamada, T.; Sandhu, I. S. *Dig. Dis. Sci.* **1992**, *37*, 1383.

130. Maggs, J. L.; Kitteringham, N. R.; Breckenridge, A. M.; Park, B. K. *Biochem. Pharmacol.* **1987**, *36*, 2061.

131. Clarke, J. B.; Maggs, J. L.; Kitteringham, N. R.; Park, B. K. *Int. Arch. Allergy Appl. Immunol.* **1990**, *91*, 335.

132. Clarke, J. B.; Neftel, K.; Kitteringham, N. R.; Park, B. K. *Int. Arch. Allergy Appl. Immunol.* **1991**, *95*, 369.

133. Ruscoe, J. E., *et al. J. Pharmacol. Exp. Ther.* **1995**, *273*, 393.

134. Munns, A. J.; De Voss, J. J.; Hooper, W. D.; Dickinson, R. G.; Gillam, E. M. *Chem. Res. Toxicol.* **1997**, *10*, 1049.

135. Ju, C.; Uetrecht, J. P. *J. Pharmacol. Exp. Ther.* **1999**, *288*, 51.

136. Parrish, D. D.; Schlosser, M. J.; Kapeghian, J. C.; Traina, V. M. *Fundam. Appl. Toxicol.* **1997**, *35*, 197.

137. Ju, C.; Uetrecht, J. P. *Drug Metab. Dispos.* **1998**, *26*, 676.

138. Shen, S.; Marchick, M. R.; Davis, M. R.; Doss, G. A.; Pohl, L. R. *Chem. Res. Toxicol.* **1999**, *12*, 214.

139. Miyamoto, G.; Zahid, N.; Uetrecht, J. P. *Chem. Res. Toxicol.* **1997**, *10*, 414.

140. Iverson, S. L.; Shen, L.; Anlar, N.; Bolton, J. L. *Chem. Res. Toxicol.* **1996**, *9*, 492.

141. Fan, P. W.; Bolton, J. L. *Drug Metab. Dispos.* **2001**, *29*, 891.

142. Yager, J. D.; Liehr, J. G. *Annu. Rev. Pharmacol. Toxicol.* **1996**, *36*, 203.

143. Liehr, J. G. *Mol. Pharmacol.* **1983**, *23*, 278.

144. Carr, A.; Penny, R.; Cooper, D. A. *AIDS* **1993**, *7*, 65.

145. Lai, W. G.; Zahid, N.; Uetrecht, J. P. *J. Pharmacol. Exp. Ther.* **1999**, *291*, 292.

146. Muller-Oerlinghausen, B. *Arzneimittelforschung* **1984**, *34*, 131.

147. Mann, K.; Bartels, M.; Gartner, H. J.; Wagner, H. W.; Heimann, H. *Pharmacopsychiat.* **1987**, *20*, 155.

148. Lai, W. G.; Gardner, I.; Zahid, N.; Uetrecht, J. P. *Drug Metab. Dispos.* **2000**, *28*, 255.

149. Kassahun, K., *et al. Chem. Res. Toxicol.* **2001**, *14*, 62.

150. Shapiro, L. E.; Knowles, S. R.; Shear, N. H. *Arch. Dermatol.* **1997**, *133*, 1224.

151. Doerge, D. R.; Divi, R. L.; Deck, J.; Taurog, A. *Chem. Res. Toxicol.* **1997**, *10*, 49.

152. Shapiro, L. E.; Uetrecht, J.; Shear, N. H. *J. Am. Acad. Dermatol.* **2001**, *45*, 787.

153. Madden, S.; Woolf, T. F.; Pool, W. F.; Park, B. K. *Biochem. Pharmacol.* **1993**, *46*, 13.

154. Thompson, C. D.; Kinter, M. T.; Macdonald, T. L. *Chem. Res. Toxicol.* **1996**, *9*, 1225.

155. Dieckhaus, C.; Miller, T.; Sofia, R. D.; Macdonald, T. *Chem. Res. Toxicol.* **2000**, *28*, 814.

156. Gupta, A. K.; Soori, G. S.; Del Rosso, J. Q.; Bartos, P. B.; Shear, N. H. *J. Am. Acad. Dermatol.* **1998**, *38*, 765.

157. Gupta, A. K., *et al. Br. J. Dermatol.* **1998**, *138*, 529.

158. Mallat, A.; Zafrani, E. S.; Metreau, J. M.; Dhumeaux, D. *Dig. Dis. Sci.* **1997**, *42*, 1486.

159. Gupta, A. K.; del Rosso, J. Q.; Lynde, C. W.; Brown, G. H.; Shear, N. H. *Clin. Exp. Dermatol.* **1998**, *23*, 64.

160. Iverson, S. L.; Uetrecht, J. P. *Chem. Res. Toxicol.* **2001**, *14*, 175.

161. Jean, P. A.; Bailie, M. B.; Roth, R. A. *Biochem. Pharmacol.* **1995**, *49*, 197.

162. Tang, W.; Borel, A. G.; Fujimiya, T.; Abbott, F. S. *Chem. Res. Toxicol.* **1995**, *8*, 671.

163. Grillo, M. P.; Chiellini, G.; Tonelli, M.; Benet, L. Z. *Drug Metab. Dispos.* **2001**, *29*, 1210.

164. Iverson, S.; Zahid, N.; Uetrecht, J. P. *Chemico-Biol. Interact.* **(in Press)**,

165. Zhang, K. E., *et al. Drug Metab. Dispos.* **2000**, *28*, 633.

166. Charatan, F. *Brit. Med. J.* **2002**, *324*, 869.

167. Davis, J. P.; Cain, G. A.; Pitts, W. J.; Magolda, R. L.; Copeland, R. A. *Biochemistry* **1996**, *35*, 1270.

168. Jerina, D. M.; Daly, J. W. *Science* **1974**, *185*, 573.

169. Guengerich, F. P. *Chem. Res. Toxicol.* **2001**, *14*, 611.

170. Hanzlik, R. P.; Hogberg, K.; Judson, C. M. *Biochemistry* **1984**, *23*, 3048.

171. Hort, J. F. *Curr. Med. Res. Opin.* **1975**, *3*, 333.

172. Brown, L. M.; Ford-Hutchinson, A. W. *Biochem. Pharmacol.* **1982**, *31*, 195.

173. Shear, N. H.; Spielberg, S. P. *J. Clin. Invest.* **1988**, *82*, 1826.

174. Spielberg, S. P.; Gordon, G. B.; Blake, D. A.; Goldstein, D. A.; Herlong, H. F. *N. Engl. J. Med.* **1981**, *305*, 722.

175. Spielberg, S. P.; Gordon, G. B.; Blake, D. A.; Mellits, E. D.; Bross, D. S. *J. Pharmacol. Exp. Ther.* **1981**, *217*, 386.

176. Green, V. J., *et al. Biochem. Pharmacol.* **1995**, *50*, 1353.

177. Saxon, A.; Beall, G. N.; Rohr, A. S.; Adelman, D. C. *Ann. Intern. Med.* **1987**, *107*, 204.

178. Chen, L. J.; Hecht, S. S.; Peterson, L. A. *Chem. Res. Toxicol.* **1995**, *8*, 903.

179. Valadon, P.; Dansette, P. M.; Girault, J. P.; Amar, C.; Mansuy, D. *Chem. Res. Toxicol.* **1996**, *9*, 1403.

180. Liu, Z. C.; Uetrecht, J. P. *Drug. Metab. Dispos.* **2000**, *28*, 726.

181. Migdalof, B. H., *et al. Drug Metab. Rev.* **1984**, *15*, 841.

182. Jaffe, I. A. *Springer Semin. Immunopathol.* **1981**, *4*, 193.

183. Weiss, A. S.; Markenson, J. A.; Weiss, M. S.; Kammerer, W. H. *Am. J. Med.* **1978**, *64*, 114.

184. Donker, A. J.; Venuto, R. C.; Vladutiu, A. O.; Brentjens, J. R.; Andres, G. A. *Clin. Immunol. Immunopath.* **1984**, *30*, 142.

185. Coleman, J. W.; Foster, A. L.; Yeung, J. H. K.; Park, B. K. *Biochem. Pharmacol.* **1988**, *37*, 737.

186. Wallmark, B. *Methods Find. Exp. Clin. Pharmacol.* **1989**, *11*, 101.

187. Neal, R. A.; Halpert, J. *Annu. Rev. Pharmacol. Toxicol.* **1982**, *22*, 321.

188. Mizutani, T.; Yoshida, K.; Murakami, M.; Shirai, M.; Kawazoe, S. *Chem. Res. Toxicol.* **2000**, *13*, 170.

189. Aucoin, D. P., *et al. J. Pharmacol. Exp. Ther.* **1985**, *234*, 13.

190. Engler, H.; Taurog, A.; Nakashima, T. *Biochem. Pharmacol.* **1982**, *31*, 3801.

191. Waldhauser, L.; Uetrecht, J. *Drug Metab. Dispos.* **1991**, *19*, 354.

192. Stevens, G. J., *et al. Chem. Res. Toxicol.* **1997**, *10*, 733.

193. Guan, X., *et al. Chem. Res. Toxicol.* **1999**, *12*, 1138.

194. Jochheim, C. M.; Davis, M. R.; Baillie, K. M.; Ehlhardt, W. J.; Baillie, T. A. *Chem. Res. Toxicol.* **2002**, *15*, 240.

195. Guan, X., *et al. Drug Metab. Dispos.* **2002**, *30*, 331.

196. Scheen, A. J. *Drug Saf.* **2001**, *24*, 873.

197. Ortiz de Montellano, P. *Cytochrome P450: Structure, Mechanism, and Biochemistry* **1995** (Plenum Press, New York).

198. Mutlib, A., *et al. Chem. Res. Toxicol.* **2000**, *13*, 775.

199. Dickins, M.; Elcombe, C. R.; Moloney, S. J.; Netter, K. J.; Bridges, J. W. *Biochem. Pharmacol.* **1979**, *28*, 231.

200. Murray, M. *Curr. Drug Metab.* **2000**, *1*, 67.

201. Kiese, M. *Pharmacol. Rev.* **1966**, *18*, 1091.

202. Sackett, P. H.; McCreery, R. L. *J. Med. Chem.* **1979**, *22*, 1447.

203. Eling, T. E.; Mason, R. P.; Sivarajah, K. *J. Biol. Chem.* **1985**, *260*, 1601.

204. Fischer, V.; Haar, J. A.; Greiner, L.; Loyd, R. V.; Mason, R. P. *Mol. Pharmacol.* **1991**, *40*, 846.

205. Parman, T.; Chen, G.; Wells, P. G. *J. Biol. Chem.* **1998**, *273*, 25079.

206. Wells, P. G.; Zubovits, J. T.; Wong, S. T.; Molinari, L. M.; Ali, S. *Toxicol. Appl. Pharmacol.* **1989**, *97*, 192.

207. Goodwin, D. C.; Aust, S. D.; Grover, T. A. *Chem. Res. Toxicol.* **1996**, *9*, 1333.

208. Ortiz de Montellano, P. R.; Watanabe, M. D. *Mol. Pharmacol.* **1987**, *31*, 213.

209. Bourdi, M.; Tinel, M.; Beaune, P. H.; Pessayre, D. *Mol. Pharmacol.* **1994**, *45*, 1287.

210. Augusto, O.; Beilan, H. S.; Ortiz de Montellano, P. R. *J. Biol. Chem.* **1982**, *257*, 11288.

211. Smith, K. S.; Smith P. L.; Heady, T. N.; Trugman, J. M.; Harman, W. D.; Macdonald, T. L. *Chem. Res. Toxicol.* **2003**, *16*, 123.

212. Monks, T. J; Jones, D. C. *Curr. Drug Metab.* **2002**, *3*, 425.

213. Lambert, C.; Park, B. K.; Kitteringham, N. R. *Biochem. Pharmacol.* **1989**, *38*, 2853.

Chapter 4

Chemically Reactive Metabolites in Drug Discovery and Development

Thomas A. Baillie, PhD, DSc

*Merck Research Laboratories, Department of Drug Metabolism, WP75A-303
West Point, PA 19486, USA*

From the standpoint of drug discovery in the pharmaceutical industry, the formation of chemically reactive intermediates raises a number of complex issues, such as: 'How best to detect reactive metabolites, and at what stage should a search be performed for such species?' 'How to interpret the generation of reactive electrophiles in the context of overall risk assessment, and how much significance should one place on covalent binding data?' 'Given the lack of correlation between covalent modification of proteins and drug-mediated toxicities, should we even care about metabolic activation in the selection of lead candidates for development?' The answers to these questions are far from straightforward, and the approach adopted may vary widely depending on the pharmaceutical company involved and, in some instances, on the viewpoint of the individual investigator. The discussion which follows reflects the approach adopted by Merck & Co., and is founded on the premise that exposure of biological systems to chemically reactive electrophiles generated during the metabolism of a drug candidate is an undesirable feature of any drug development program.

Following the pioneering work of Brodie, Gillette, Mitchell and their colleagues at the National Institutes of Health during the 1970s, the "Covalent Binding Theory" of foreign compound-induced liver injury emerged as a plausible basis for the hepatotoxic properties of a structurally diverse group of xenobiotics.[1,2] According to this theory, the metabolic activation of foreign compounds (typically catalyzed by liver enzymes) afforded products that covalently modified hepatocellular proteins in a process that, by poorly defined mechanisms, ultimately led to necrosis or other forms of toxic insult to the liver.[3-5] In support of this view, it was demonstrated using animal models, that the zonal pattern of covalent binding in liver tissue usually correlated well with the pattern of tissue damage associated with a given hepatotoxin, that inducers of drug metabolizing enzyme systems (notably cytochromes P-450) increased the level of covalent binding and exacerbated toxicity, whereas pretreatment with enzyme inhibitors had the opposite effect, and that depletion (or supplementation) of hepatic glutathione pools modulated the hepatotoxic response to the agent in question.[6] Further elaboration of this theory over the intervening years led to the development of a conceptual framework upon which to rationalize the relationship between drug metabolism and various types of drug toxicity,[7] and the current status of this field is reviewed in depth in the accompanying chapter by Dr. Uetrecht. However, the Covalent Binding Theory was challenged following observations that some compounds bind covalently to proteins yet fail to cause appreciable toxicity. A classic example of this phenomenon is found with 3'-hydroxyacetanilide, a positional isomer of acetaminophen (4'-hydroxyacetanilide); both isomers undergo hepatic cytochrome P-450-mediated oxidation to electrophilic quinone-type intermediates which react with glutathione and form covalent adducts with proteins, yet only acetaminophen causes hepatic necrosis under normal conditions.[8-11] The reasons for such differences in behavior remain obscure, although it may be speculated that different reactive metabolites target different proteins, some of which may be more critical than others to cell viability, or that different protein adducts may differ in their ability to 'signal' the electrophilic insult and thereby recruit appropriate cellular defenses.[12-15] It is also possible that some covalent binding may, in fact, represent a 'detoxification' event, whereby circulating proteins such as serum albumin serve as macromolecular 'trapping agents' for potentially harmful electrophiles.[15]

Regardless of the underlying basis for the apparent differences in behavior of reactive metabolites, these observations have significant implications for the development of drug candidates that are subject to metabolic activation. First, it should be acknowledged that, in light of our limited understanding of the molecular mechanisms of cell toxicity, it is not possible to predict, *a priori,* whether a given drug candidate which undergoes metabolic activation is likely to cause toxicity when dosed to animals or humans. Second, even if such a drug candidate were to fail to cause liver damage (or injury to other organs) in preclinical safety assessment studies, there remains the concern that it may still elicit immune-mediated adverse events in humans (idiosyncratic toxicity) through haptenization of proteins.[16-18] This is a significant concern since it is generally agreed that there is no good animal model for the human immune system, and therefore conventional preclinical safety studies may fail to identify those agents which give rise to drug-protein covalent adducts and thereby have the potential to induce hypersensitivity reactions in man.

Based upon the above uncertainties over the toxicological consequences of reactive metabolite formation, the general strategy adopted by Merck Research Laboratories is to evaluate the formation of electrophilic intermediates as early as reasonably possible in the selection of drug candidates for clinical development, with the goal of minimizing the generation of such species by informed structural modification of lead compounds. Typically, this entails a sequence of studies, the first being an *in vitro* "trapping" experiment conducted with liver microsomal incubations fortified with NADPH and an appropriate nucleophile (usually GSH, but in some cases cyanide); analysis of the resulting products by LC-MS/MS techniques is employed to reveal the formation of conjugates with the added nucleophile, the structures of which allow one to deduce the identities of the reactive electrophiles from which they derived.[19] In cases where appreciable levels of such adducts are detected, attempts are made to modify the structure of the compound in question in such a way as to obviate the offending metabolic pathway. For those compounds which fail to form abundant conjugates with GSH or cyanide, and which possess the requisite characteristics of a development candidate (high degree of efficacy and potency in animal models, acceptable pharmacokinetics in preclinical species, appropriate physico-chemical properties, etc), a radiolabeled analog of the compound (preferably ^{14}C-labeled) is prepared and *in vitro* covalent binding experiments are performed in liver preparations from animals and humans. Hepatic microsomes are employed to assess metabolic activation catalyzed primarily by oxidative enzymes, while the use of freshly isolated hepatocytes serves to reveal metabolic activation processes that depend upon the full complement of cellular enzyme systems. Covalent binding studies with hepatocytes also are of value in obtaining information on the 'balance' between metabolic activation and detoxification in an intact cellular environment, which cannot be obtained readily from experiments with subcellular fractions alone. Finally, *in vivo* covalent binding studies are carried out in intact rats, where the level of adducts to both liver and plasma proteins is determined at appropriate intervals after administration of a standard oral dose of the test compound.

In view of the relatively high levels of specific radioactivity of ^{14}C and ^{3}H in common synthetic intermediates now available commercially, radiolabeled drug candidates for binding studies can be prepared on a routine basis labeled with ^{14}C at 50 mCi/mol and with ^{3}H at 5 Ci/mol. Although dilution of such tracers with 'cold' drug may be performed prior to the conduct of covalent binding studies, the specific activities involved nevertheless are such that some measurable level of irreversible binding to proteins almost always is detectable by liquid scintillation counting following incubations *in vitro* or dosing *in vivo* and digestion of protein with strong alkali. The question, then, becomes, "How much apparent covalent binding is acceptable in deciding whether to advance a drug candidate into development?" Merck's approach to this question was to take the levels of covalent adducts typically found in the livers of animals given a dose of a prototypic hepatotoxin (such as acetaminophen) sufficient to cause hepatic necrosis (approximately 1 nmol drug equivalent / mg protein), and to reduce this figure by 20-fold to give a conservative target 'threshold' value for *in vivo* covalent binding of 50 pmol drug equivalents / mg total liver protein. This value also corresponds to a level of radioactivity that is approximately 20-times background under normal conditions, and thus provides a suitable dynamic range for measurement of covalently bound drug-protein adducts. It should be emphasized that the figure of 50 pmol / mg protein is viewed as a *target* upper limit, and not as an absolute threshold above which a compound would not be advanced into development, since it is recognized that other factors must be taken into consideration in arriving at this decision. Such factors would include the intended therapeutic indication (disabling or life-threatening disease?), the availability of existing treatments, the target patient population, the projected clinical dose and duration of therapy, novelty of the drug target, and the chemical tractability of the structural series. However, levels of adducts to liver proteins *in vivo* in excess of 200 pmol / mg protein normally would be viewed as unacceptable, and efforts would be made to modify the structure to reduce the metabolic liability in question. A flow chart depicting the decision tree for candidate selection based upon covalent binding data is depicted in Figure 1.

From the foregoing discussion, it will be evident that the covalent binding studies performed at the three levels of experimental complexity, namely liver microsomes, hepatocytes and whole animals, coupled with metabolite identification studies in the same preparations, fulfill different functions in evaluating the extent to which drug candidates give rise to reactive, potentially toxic metabolites. Thus, the two *in vitro* systems afford a mechanistic insight into the processes that lead to metabolic activation of the compound-of-interest (and upon which rational structural modification may be based), while the *in vivo* experiment plays a key role in the decision whether to advance the candidate into development. By way of illustrating the application of this strategy to candidate selection in a contemporary drug discovery program, a series of metabolism and covalent binding experiments were performed on the substituted pyrazinone derivative (Compound 'A', Figure 2) which emerged as a potential development candidate in a program aimed at discovering an orally-active inhibitor of the enzyme thrombin. When incubations of Compound 'A' were performed in liver microsomal preparations from rats and humans fortified with GSH, two isomeric GSH conjugates ('GSH-1' and 'GSH-

2', Figure 2) were detected by LC-MS/MS techniques.[20] While the structure of the former adduct was deduced readily by collision-induced dissociation (CID) MS/MS analysis and ^{1}H NMR spectroscopy as the product of addition of GSH at the methyl group attached to carbon-6 of the pyrazinone ring, a number of NMR experiments (gHMBC, gHSQC, 2D NOESY) were necessary to elucidate the structure of the rearranged adduct GSH-2. A third metabolite, the benzylic alcohol resulting from hydroxylation of compound 'A' at the 6-methyl substituent, also was identified in these studies. The proposed origins of these metabolites are shown in Figure 2, which points to the operation of two distinct, cytochrome P-450-mediated metabolic activation processes. Thus, Compound 'A' undergoes a two-electron oxidation, either directly or *via* dehydration of the hydroxymethyl metabolite referred to above, to generate an electrophilic imine-methide. Capture of this intermediate by GSH affords adduct GSH-1. In the second activation process, it is proposed that oxidative attack of the pyrazinone ring generates an unstable epoxide intermediate, which is attacked regiospecifically by GSH to afford a cyclic carbinolamide. Tautomeric ring-opening of the latter species to the acyclic methyl ketone, followed by *syn/anti* isomerization of this substituted amidine, leads to the formation of a new ring system which eliminates the elements of water to generate the stable imidazole derivative, GSH-2.

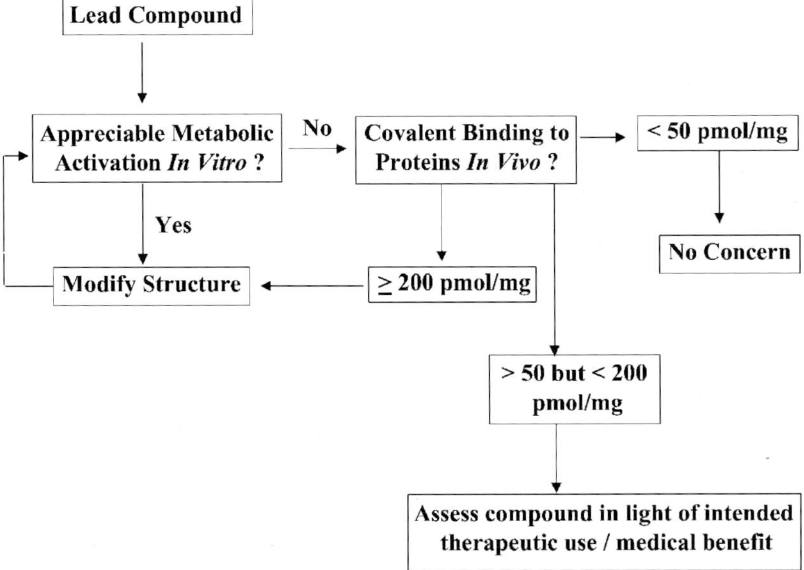

Figure 1. Decision tree for assessing the suitability of lead compounds for development based on metabolic activation considerations. (As indicated in the text, a liability with regard to metabolic activation is viewed as only one of a number of criteria in the selection process).

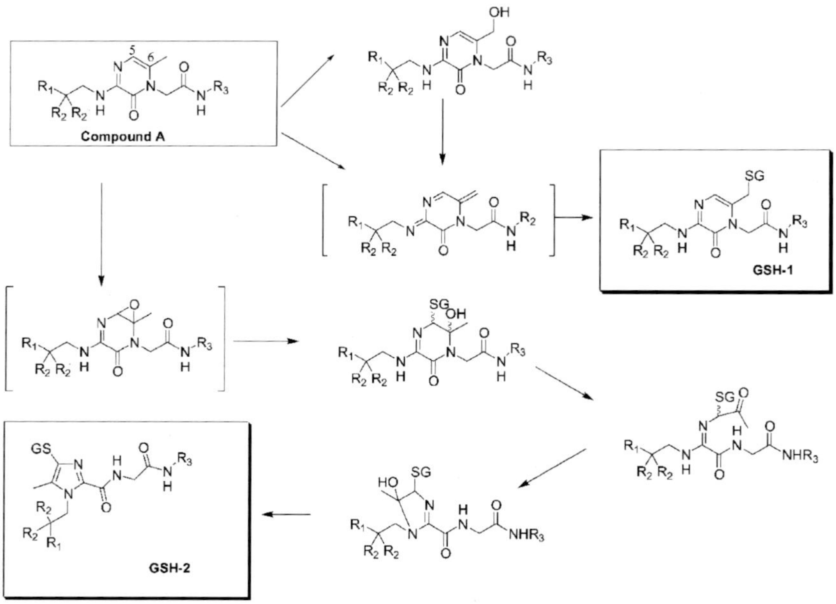

Figure 2. Proposed scheme for the metabolism of Compound 'A' to reactive intermediates and subsequent formation of glutathione conjugates GSH-1 and GSH-2. Substituents R_1 and R_3 represent simple aromatic moieties, while R_2 = F.

In light of concerns raised by the detection of products of metabolic activation of Compound 'A,' a radiolabeled analog of this drug candidate was prepared with a single atom of ^{14}C incorporated at the methine position (carbon-5) of the pyrazinone ring, and a series of covalent binding studies were performed with this tracer, both *in vitro* and *in vivo*. In liver microsomal preparations from both rats and humans, levels of irreversibly bound radioactivity reached values of approximately 85 pmol/mg protein after a 60-minute incubation with Compound 'A' at 2 μM. When [^{14}C]Compound 'A' was dosed orally to rats *in vivo* (10 mg/kg), the level of binding to plasma and liver proteins at the 6-hour time point was 76 and 105 pmol/mg protein, respectively. Based upon these findings, it was clear that Compound 'A' was subject to metabolism *in vivo* to reactive intermediates that covalently modified tissue proteins, and that the likely underlying mechanisms involved oxidation of the pyrazinone ring and its 6-methyl substituent. Given this information, it was reasoned that simple modification of the core structure of Compound 'A' by replacement of the 6-methyl group by chlorine (to afford Compound 'B') should serve to suppress metabolism to reactive intermediates. Thus, formation of the imine methide intermediate could not occur from Compound 'B,' while placement of the electron-withdrawing halogen on the pyrazinone ring should render this heterocycle less susceptible to epoxidation. These predictions were borne out in a similar series of parallel metabolism and covalent binding experiments with Compound 'B,' which not only retained many of the desirable pharmacological and pharmacokinetic properties of its predecessor, but gave rise to barely detectable amounts of GSH adducts in liver microsomal preparations and to 5- to 10-fold lower levels of covalent adducts to plasma and liver proteins in rats *in vivo* than was observed with comparable doses of Compound 'A.' Moreover, when the covalent binding of Compound 'B' was examined in liver microsomes from rat, dog, monkey and human, it was found that the level of adducts to preparations from human liver (24 pmol/mg protein at 60 minutes) was lower than observed with any of the animal species. Based on the outcome of these comparative metabolism and covalent binding studies, Compound 'B' was selected for full development and currently is undergoing clinical trials as an orally active anticoagulant for the treatment of deep vein thrombosis, pulmonary embolism and thromboembolic stroke.

In conclusion, it should be emphasized that measurements of covalent binding as currently performed merely reflect exposure of biological systems to chemically-reactive metabolites, and cannot be translated directly to predictions of drug-induced toxicity, either in animals or in human subjects. However, the value of such studies in a pharmaceutical research environment lies at the *discovery phase,* when choices typically have to be made between a number of compounds in the selection of a candidate for preclinical and clinical development and where decisions need to be made as to where the sponsor wishes to "place their bets." It is imperative, therefore, that there be close collaboration between scientists in Drug Metabolism and Medicinal Chemistry at this critical phase of drug research, and that the joint efforts of these two groups are supported adequately by a radiosynthesis function that can prepare ^{14}C- or ^{3}H-labeled tracers of selected compounds in a timely fashion. While future research in the fields of biochemical and molecular toxicology may shed light on those characteristics of chemically-reactive metabolites that dictate their toxic potential, the experimental strategy outlined above represents a practical approach to the issue of metabolic activation, the use of which hopefully will contribute to the development of novel therapeutic agents that exhibit a high degree of both safety and efficacy.

REFERENCES

1. Brodie, B. B., Reid, W. D., Cho, A. K., Sipes, G., Krishna, G., *et al., Proc. Natl. Acad. Sci. (USA)* **1971,** *68,* 160.
2. Gillette, J. R., Mitchell, J. R., and Brodie, B. B., *Annu. Rev. Pharmacol.,* **1974,** *14,* 271.
3. Nelson, S. D., and Pearson, P. G., *Annu. Rev. Pharmacol. Toxicol.,* **1990,** *30,* 169.
4. Boelsterli, U. A., *Drug Metab. Rev.,* **1993,** *25,* 395.
5. Hinson, J. A., Pumford, N. R., and Nelson, S. D., *Drug Metab. Rev.,* **1994,** *26,* 395.
6. Reed, D. J., In, Bioactivation of Foreign Compounds, Anders M. W., ed., Academic Press, New York, **1985,** pp. 71-107.
7. Park, B. K., Naisbitt, D. J., Gordon, S. F., Kitteringham, N. R., and Pirmohamed, M., *Toxicology,* **2001,** *158,* 11.
8. Streeter, A. J., Bjorge, S. M., Axworthy, D. B., Nelson S. D., and Baillie, T. A., *Drug Metab. Dispos.,* **1984,** *12,* 565.
9. Tirmenstein, M. A., and Nelson, S. D., *J. Biol. Chem.,* **1989,** *264,* 9814.
10. Myers, T. G., Dietz, E. C., Anderson, N. L., Khairallah, E. A., Cohen, S. D., and Nelson, S. D., *Chem. Res. Toxicol.,* **1995,** *8,* 403.
11. Bessems, J. G. M. and Vermeulen, N. P. E., *Crit. Rev. Toxicol.,* **2001,** *31,* 55.
12. Cohen, S. D., Pumford, N. R., Khairallah, E. D., Boekelheide, K., Pohl, L. R., Amouzadeh, H. R., and Hinson, J. A., *Toxicol. Appl. Pharmacol.,* **1997,** *143,* 1.
13. Pumford, N. R., and Halmes, N. C., *Annu. Rev. Pharmacol. Toxicol.,* **1997,** *37,* 91.
14. Qiu, Y., Benet, L. Z., and Burlingame, A. L., *J. Biol. Chem.,* **1998,** *273,* 17940.
15. Koen, Y. M., and Hanzlik, R. P., *Chem. Res. Toxicol.,* **2002,** *15,* 699.
16. Park, B. K., Pirmohamed, M., and Kitteringham, N. R., *Chem. Res. Toxicol.,* **1998,** *11,* 969.
17. Uetrecht, J. P., *Chem. Res. Toxicol.,* **1999,** *12,* 387.
18. Naisbitt, D. J., Gordon, S. F., Pirmohamed, M., and Park, B. K., *Drug Safety,* **2000,** *23,* 483.
19. Baillie, T. A., and Davis, M. R., *Biol. Mass Spectrom.,* **1993,** *22,* 319.
20. Singh, R., Silva-Elipe, M. V., Pearson, P. G., Arison, B. H., Wong, B. K., White, R., Yu, X., Burgey, C. S., Linn, J. H., and Baillie, T. A., Chem. Res. Toxicol., **2003,** *16,* 198.

Chapter 5

Cytochrome P450 and its Place in Drug Discovery and Development

Dennis Smith

*Pharmacokinetic and Metabolism, Pfizer Global Research and Development
Sandwich, Kent, United Kingdom, CT13 9NJ*

INTRODUCTION

The field of cytochrome P-450 (P450) research and its impact on drug discovery follows a path of novel innovative findings, often underestimated in potential impact at the time, leading to substantial change in the drug research process. By far the biggest catalyst for change was the finding that P450 existed in multiple families, forms or isoenzymes. It was this finding and a chain of events and research that moved P450 from a largely academic pursuit to one with a critical bearing on the pharmaceutical industry and the discovery and development of drugs. The series of findings was not a linear path. The purpose of this chapter is to allow the reader to experience the journey of P450 discovery and understand the background to its pivotal role in research within the pharmaceutical industry.

1   FROM SINGLE ENZYME TO SUPERFAMILY OF HUMAN ISOFORMS

Initially P450 was viewed as a single entity (enzyme, isoform etc.) with little or no selectivity. This view began to be changed by results showing that phenobarbitone (PB) and 3-methylcholanthrene (3-MC) induced a second form P448 (differentiated by spectrophotometry) in rodents. In 1973[1,2] classic experiments performed by amongst others Anthony Lu began a new understanding of the P450 system. Studies of the substrate specificities of the liver microsomal cytochrome P450 fractions from control and PB treated rats revealed that the cytochrome P450 from control rats was catalytically different from the P450 from the induced rats. The cytochrome P450 fraction from control rats had slightly greater activity than the P450 fraction from PB induced rats for 3,4-benzpyrene hydroxylation. Both fractions had similar activity for ethylmorphine N-demethylation and for testosterone hydroxylation at positions 6 beta and 7 alpha, while the cytochrome P450 fraction from PB induced rats was much more active than the P450 fraction from control rats for benzphetamine and chlorcyclizine N-demethylation, pentobarbital oxidation, and testosterone hydroxylation at position 16 alpha. The substrate specificity of the cytochrome P448 fraction from 3-MC treated rats differed from that of the cytochrome P450 fractions from both control and PB treated rats. In further experiments the CO-binding cytochromes P-450 and P-448 were isolated from microsomes of 3-MC treated rats. When rats were sacrificed 3 and 48 hr after administration of a single dose of 3MC, the maximum absorption peak of the reduced CO difference spectrum was different, i.e., 450 nm for the 3-hr preparation and 448 nm for the 48-hr preparation. Enzymatically, the 48-hr preparation was very active for the hydroxylation of 3,4-benzpyrene, whereas the 3-hr preparation was much less active for 3,4-benzyprene hydroxylation. These data, along with other studies, indicated that inducing agents resulted in the formation of a new P450 like hemoproteins active in the metabolism of xenobiotics, and with substrate selectivity different to the native protein. Work in rodents led to the concept of multiple forms, but a clinical curiosity was to lead to major findings in humans with a huge impact on first clinical pharmacology, then drug safety and ultimately drug design.

Pioneering work in humans with debrisoquine led to the concept of extensive (EM) and poor metabolisers (PM) (c.f., Chapter 11) and the link between genetics and drug hydroxylation and ultimately pharmacokinetic variation. In an otherwise normal human volunteer experiment Robert Smith received debrisoquine, an early hypotensive agent, and unlike his fellow subjects experienced marked postural hypotension. When the urines were analysed the primary metabolite of debrisoquine, 4-hydroxydebrisoquine metabolite was present in all the volunteers with the exception of Smith. In a series of pivotal experiments performed by Smith and others, debrisoquine and its primary metabolite were measured in the urine of volunteers after a single oral dose. The ratio between excreted debrisoquine and its metabolite (metabolic ratio) was calculated and when plotted as a frequency plot was shown to be bimorphically distributed in the study population.[3] Subsequent family studies supported the view that alicyclic 4-hydroxylation of debrisoquine was controlled by a single autosomal gene and that a defect in this metabolic step was caused by a recessive allele.[4] The phenotyping by debrisoquine allowed humans to be grouped into EM and PM and comparison of this status with the pharmacokinetics of other administered drugs. The defect in the 4-hydroxylation of debrisoquine appeared to be due to the absence or deficiency of a specific form of P450. The search for an animal model discovered that, whilst Sprague-Dawley (SD) rats possessed the ability to readily hydroxylate debrisoquine, females of the Dark Agouti (DA) strain did not, although total cytochrome P450 levels were similar. A minor P450 was purified,[5] by Guengerich and his group, from male SD rats and designated as P450UT-H. P450UT-H differed from 8 other purified rat liver P450s as judged by peptide mapping and immunochemical analysis and thus appeared to be isoenzymic with these other P450s. P450UT-H exhibited considerably more debrisoquine activity than any of the other purified P 450s and, on a total P450 basis, more than total microsomal P450. Antibodies raised against P450UT-H specifically recognized P450UT-H and inhibited >90% of the debrisoquine hydroxylase activity present in SD rat liver microsomes. The level of P450UT-H in SD rat liver microsomes accounted for <10% of the total P450, as judged by immunoquantitation. These assays also indicated that the level of P450UT-H in female DA rat liver microsomes is only 5% of that in male or female SD rat liver microsomes, consonant with the view that deficiency of this form of P450 is responsible for the defective debrisoquine hydroxylation activity in the former animals. The results clearly demonstrated the role of a specific P450 in a model for genetic polymorphism of oxidative drug metabolism.

These results allowed the unravelling of the mystery in human as debrisoquine hydroxylase could be characterised in man and led to the single isoenzyme we now term CYP2D6. Antibodies prepared to a cytochrome P450UT-H shown to be responsible for debrisoquine hydroxylation in rats inhibited the oxidation[6] of debrisoquine, sparteine, encainide, and propranolol by human liver microsomes. These drugs were associated with the CYP2D6 phenotype in clinical studies and in human liver microsomes. The antibodies recognized a single polypeptide of Mr 51,000, the intensity of this band was correlated with debrisoquine 4-hydroxylase activity when liver microsomes from 44 organ donors were examined. The antibodies did not inhibit the oxidation of other cytochrome P450 substrates. From a chance clinical observation the trail had led to a single protein. The protein (CYP2D6) could be expressed differently in individuals leading to variation in pharmacokinetics. Moreover this P450 metabolised a discrete, structurally related subset of compounds.

Subsequent work with the purified CYP2D6 and later combinant derived forms together with clinical data from phenotyped and ultimately genotyped individuals eventually provided clear substrate structure activity relationships for the isozyme and a clear rationale as to why certain chemotypes and pharmacophores are associated with high pharmacokinetic variability in humans due to metabolism by CYP2D6. Substrates listed above include, β-blockers, and Na⁺channel blocker class I anti-arrhythmics. Another important drug class are tricyclic antidepressants. In brief, the structural similarities of many of the substrates and inhibitors in terms of position of hydroxylation, overall structure (aryl-alkylamine) and physiochemistry (ionised nitrogen at physiological pH) have allowed template models such as that illustrated as Figure 1 to be constructed.

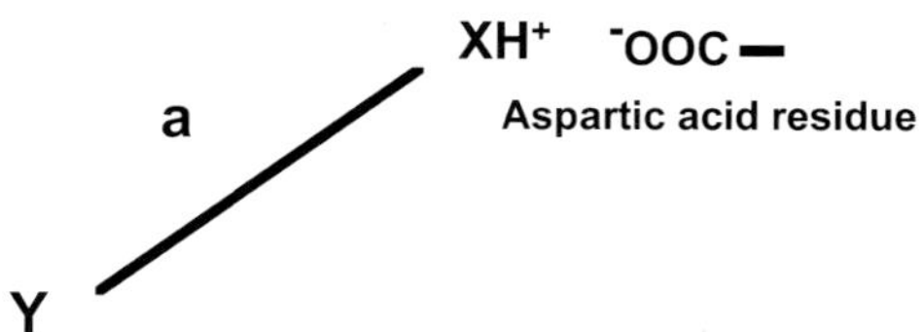

Figure 1. Template model for CYP2D6 with Y the site of oxidation, a is the distance from Y to a heteroatom which is positively charged (normally 5-7 angstroms).

All the template models produced have the same common features of a basic nitrogen atom at a distance of 5-7Å from the site of metabolism, which is in general on or near a planar aromatic system. It is currently believed that aspartic acid residue 301 provides the carboxylate residue which binds the basic nitrogen of the substrates.

With CYP2D6 therefore the catalytic selectivity relies heavily on a substrate-protein interaction. The relative strength of the proposed ion pair between the basic nitrogen and the active site aspartic acid means that the affinity for substrates will be high. This is borne out by the enzyme having lower K_m and K_i values than other P450s. Thus CYP2D6 is often a major enzyme in drug oxidation for those compounds fitting the structural requirements despite its low abundance in human liver.

Similar strategies involving animal and human tissue sources, enzyme purification, antibody purification to the early work on CYP2D6 were used to probe the wide variation in the calcium channel blocker nifedipine's clinical pharmacokinetic variation. Studies with rat liver indicated that the cytochrome P450 forms designated P450UT-A and P450PCN-E were the major contributors to microsomal nifedipine oxidation.[7]   The P450 that oxidized nifedipine

(P450NF) was purified and now designated as CYP3A4. Antibodies raised to P-450NF inhibited the nifedipine oxidase activity in human liver microsomes. A monoclonal antibody raised to the human P450 reacted with both human P450NF and rat P450PCN-E. Analysis of a bank (n=39) of human liver microsomal samples with anti-P450NF antibodies revealed the same 52,000-dalton polypeptide, corresponding to P450NF. The level of nifedipine oxidation was highly correlated with the amount of P-450NF, suggesting that the amount of the P450NF may be a major factor in influencing the level of catalytic activity in humans as well as rats. P450NF also was responsible for the major proportion of human liver microsomal aldrin epoxidation, d-benzphetamine N-demethylation, 17 α-estradiol 2- and 4-hydroxylations, and testosterone 6 β-hydroxylation.

Work with the purified CYP3A4 and subsequent recombinant derived forms provided remarkably different substrate structure activity relationships. The isoenzyme had a remarkably diverse series of substrates with no apparent structural similarities other than certain labile positions were the site of metabolism. The enzyme has been shown to be the principal participant in N-demethylation reactions where the substrate is a tertiary amine. The list of substrates includes erythromycin, ethylmorphine, lidocaine, diltiazem, tamoxifen, toremifene, verapamil, cocaine, amiodarone, alfentanil and terfenadine. Allylic and benzylic position carbons, such as those present in quinidine, steroids and cyclosporin, are also particularly prone to oxidation by CYP3A4.

The major human CYPs can be characterised in terms of their substrate selectivites as:

CYP1A2: Neutral or basic lipophilic planar molecules with at least one putative H-bond donating site. Principal substrate is theophylline.

CYP2D6: Aryl-alkyl-amines (basic) with site of oxidation a discrete distance from a protonated nitrogen. Substrates are lipophilic, particularly when measured or calculated for the neutral form. Principle substrates are β-adrenoceptor blockers, Class I anti-arrythmics and tricyclic anti-depressants. Often hydroxylation occurs in an aromatic ring or an accompanying short alkyl side chain.

CYP2C9: Neutral or acidic molecules with site of oxidation a discrete distance from H-bond donor or possibly anionic heteroatom. Molecules tend to be amphipathic with a region of lipophilicity at the site of hydroxylation and an area of hydrophobicity around the H-bond forming region. Principal substrates are non-steroidal anti-inflammatory agents. Oxidation often occurs in an aromatic ring or an accompanying short alkyl side chain.

CYP3A4: Lipophilic, neutral, or basic molecules with site of oxidation often nitrogen (N-dealkylation) or allylic positions. Wide range of substrates covering all fields of pharmaceuticals.

CYP2E1: Small (molecular weight of 200 daltons or less) normally lipophilic linear and cyclic molecules. Principal pharmaceutical compounds are volatile anaesthetics.

The above experimental trail began to answer many questions on human pharmacokinetic variability, the effect of co-medication (as detailed below), and species differences in metabolism. The concept of a superfamily with variable expression, related but not identical forms across species and isoform substrate specificity was radically different to the previous single protein view. The academic interest moved rapidly to one of huge industrial importance.

2   INHIBITION OF HUMAN ISOFORMS

All substrates of P450 have the ability to act as competitive inhibitors of P450 enzymes (reference to DDI chapter). Some compounds are activated to meta-stable or stable complexes during metabolism to irreversible or slowly reversible inhibitors. These compounds are known as "mechanism based inhibitors". Of particular importance, because of there widespread useage, are several macrolide antibiotics, including troleandomycin, erythromycin, and roxithromycin. They have been shown to inhibit CYP3A4 catalytic activities by forming inactive $P450{\cdot}Fe^{2+}$-metabolite complexes after metabolic activation.[8] The complexes prevent metabolism of other substrates and this inhibition of the enzyme has been implicated as being the principal mechanism causing drug-drug interactions when macrolide antibiotics are administered simultaneously with other drugs to human patients. Very potent inhibitors of P450 often include a nitrogen-containing heterocycle (pyridine, imidazole or triazole) capable of forming a lone pair ligand interaction with the heme of P450. In a study of the inhibition of the activity of recombinant CYP3A4 with 30 diverse chemicals there was a general, strong correlation between the IC_{50} value and lipophilicity (LogD 7.4). This relationship was strengthened further by subdividing the structures studied into two distinct subpopulations: either the absence or presence of a sterically uninhindered N-containing heterocycle, more specifically a pyridine, imidazole, or triazole function. The presence of these structural motifs increased the potency of CYP3A4 inhibition by approximately 10-fold for a given lipophilicity (LogD 7.4 value). More detailed analyses of a larger data set by the authors demonstrated that the inhibitory potency of the pyridine structure can be attenuated through direct steric effects or electronic substitution resulting in a modulation of the pKa of the pyridine nitrogen, thereby influencing its ability to interact with the P450 heme.[9] In other studies directly comparing very closely related structures the ligand interaction contributed some 6 Kcals of binding energy to the interaction (3 orders of magnitude increase in potency).

Although such heterocyles are essential for the activity of certain drugs (azole antifungals: 14 alpha demethylase inhibitors, see below), their incorporation into molecules is commonplace to increase solubility (for instance, by replacing a phenyl group). Incorporation in a form in which the nitrogen is sterically unhindered will potentially render the compound a potent inhibitor of P450s. Ketoconazole is an example of this type of inhibitor, a widely prescribed oral and topical antifungal agent that is an extremely potent CYP3A4 ligand and inhibitor. Drug interactions have been described for a considerable period in the literature, but became the centre of attention due to a number of deaths occurring by fatal arrhythmia (extensive QT prolongation) with the drug terfenadine, the most widely prescribed antihistamine at the time. Terfenadine is in fact a prodrug, well absorbed but converted to a zwitterionic metabolite by first pass metabolism in the liver. This metabolite is formed by oxidation of one of the methyls of the tertiary-butyl group to a primary alcohol, which is subsequently oxidized to a carboxylic acid (Figure 2). Another major route is oxidative N-dealkylation to give 4-(hydroxydiphenylmethyl) piperidine, a metabolite that is largely inactive.

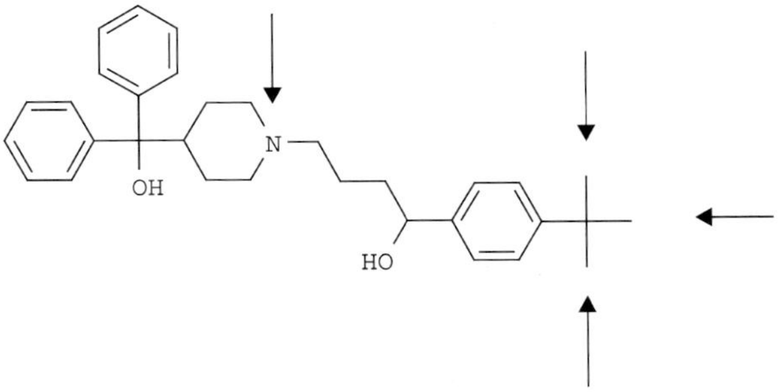

Figure 2. Principal sites of metabolism of terfenadine. The tertiary butyl group has 3 equivalent methyls although only one is oxidized to the hydroxy and subsequent carboxylic acid metabolites.

The physicochemical properties of this metabolite give terfenadine its favourable non-sedating properties. It was noted that deaths occurred most frequently when co-prescribed with ketoconazole or erythromycin. Controlled clinical studies[10] examined the effects of ketoconazole on the pharmacokinetics and electrocardiograph repolarization pharmacodynamics (QT intervals) of terfenadine in men and women. After achieving a steady state while taking terfenadine, daily concomitant oral ketoconazole was added to the subjects' regimen. The subjects had detectable levels of unmetabolized terfenadine after treatment with ketoconazole, which was associated with QT prolongation. In the period without treatment with ketoconazole, terfenadine was not detectable, although high concentrations of the zwitterionic metabolite were detected and there were no changes of QT interval. In a similar study[11] subjects were given terfenadine to steady state before initiation of oral erythromycin. Some of the subjects accumulated unmetabolized terfenadine after administration of erythromycin. Accompanying changes were noted in QT intervals and ST-U complexes in a subset of the subjects who accumulated terfenadine. These studies conclusively showed that ketoconazole and erythromycin inhibited the metabolism of terfenadine, leading to accumulation of terfenadine. In certain individuals this accumulation of the parent drug was associated with altered cardiac repolarization.

In further studies[12] the zwitterionic metabolite formed by oxidation of the tertiary butyl group and the metabolite formed by oxidative N-dealkylation showed similar rates of formation across a microsomal liver bank and correlated with rates of nifedipine aromatisation (a marker of cytochrome CYP3A4) but not with selective substrates for other human P450s. The reaction was inhibited by gestodene, a selective mechanism-based inactivator of CYP3A4 and antibodies raised against CY3A4. The oxidation of terfenadine was also catalysed by purified human liver microsomal CYP3A4 and by yeast recombinant CYP3A4. These results provide evidence that CYP3A4 is the pivotal enzyme in the metabolism of terfenadine. Addition of ketoconazole[12] reduced the rate of formation of both metabolites in a manner consistent with competitive inhibition. Inhibition constants (K_i) were 0.024 µM for the de-alkylation pathway, and 0.237 µM for the hydroxylation pathway. Using the in vitro K_i values and the clinical plasma ketoconazole concentrations, a simulation indicated that plasma terfenadine levels during coadministration of ketoconazole would increase by a factor ranging from 13- to 59-fold relative to without ketoconazole.

The experimental approaches referred to above allowed, for the first time, rationalization of a very serious drug interaction down to the protein and molecular level and was a landmark series of investigations. By defining which P450 isozyme metabolised a drug, it was possible to immediately predict the medicants that when co-prescribed could lead to drug interactions. When conducted early in the programme of discovery or development the involvement of polymorphically expressed P450s (such as CYP2D6) can be identified and the involvement of these P450 enzymes be attenuated by targeted medicinal chemistry or factored into the development to investigate. The approaches adopted for terfenadine in defining the P450 isozyme metabolising the drug in 1993 form the basic protocol used today in academic and pharmaceutical company laboratories in characterizing the P450 isoenzymes involved in the clearance of existing and new chemical entities. This work indicates that CYP3A4 is the predominant enzyme when all drug substrates are considered (Figure 3).

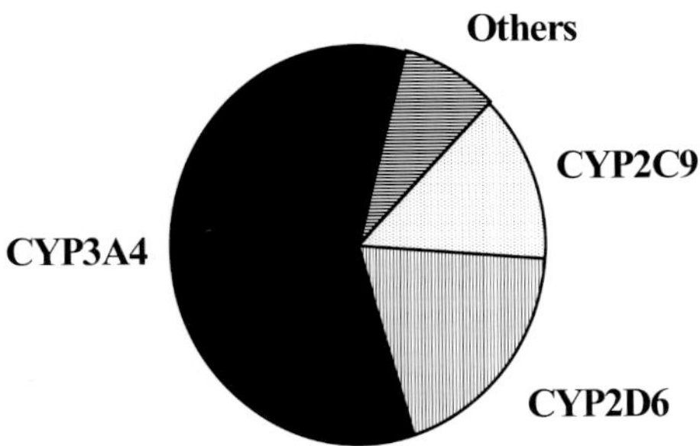

Figure 3. Relative number of pharmaceuticals metabolised by the major human P450s

In addition the battery of selective probe substrates can be used to test the effect of existing and new chemical entities as inhibitors of the major isoenzymes. The work forms a key part of the labeling of a new drug and underpins the strategy for clinical pharmacology studies that investigate co-medication. Such is the impact that drug discovery programmes try to avoid potent P450 inhibitors and screen for them early in the discovery cycle. To this end assays have been developed that use high-throughput multiwell plate-based, direct, fluorometric assays for the activities of the principal drug-metabolizing P450s. These assays rely on fluorescent substrates that may or may not be selective for the isoforms, and individually expressed recombinant human P450s. In less than a decade the impact of terfenadine has redefined the screening sequences used to discover new drugs and stimulated technological advancement really not foreseeable from the beginning of the 1990's.

3   INDUCTION OF HUMAN ISOFORMS

The evolution of the human isoforms into a superfamily, capable of metabolising most molecules, has also developed a system of control whereby the continuing "challenge" of compound may lead to an elevation in this defence system (induction). For the most promiscuous isoform, CYP3A4, the mechanism for this is the occupancy of the orphan nuclear receptor PXR.[13] PXR activation also triggers the induction of CYP2C9 and p-glycoprotein. The PXR receptor has evolved to permit it to function as the switch for a defensive reaction to a wide range of chemical structures. Like other nuclear receptors, PXR contains both a DNA-binding domain and a ligand-binding domain. PXR binds to the xenobiotic DNA response elements in the regulatory regions of CYP3A genes as a heterodimer with the 9-cis retinoic acid receptor, also known as the retinoid X receptor (RXR). A small number of polar residues are spaced throughout the smooth, hydrophobic ligand-binding pocket. The character of the ligand pocket mirrors the character of most of the known PXR ligands, which are generally hydrophobic and contain a small number of polar groups capable of hydrogen bonding.[14] The unique composition of the ligand pocket not only allows PXR to bind a diverse set of chemicals, but also permits a single ligand to dock in multiple orientations. This binding mode stands in sharp contrast to other nuclear receptor-ligand interactions, which have evolved to be highly specific. Although it binds a diverse range of substrates, PXR also exhibits some specificity, as evidenced by the differences in the activation profile of PXR across species. For instance, human PXR is activated by rifampicin and the cholesterol-lowering drug SR12813, whereas mouse PXR is not. Mouse PXR is activated by the synthetic steroid 5-pregnen-3-ol-20-one-16-carbonitrile, whereas the human receptor is not. Overall, although showing some selectivity, the PXR receptor is promiscuous, like the P450 isoform it triggers the induction of. The "non-specific " nature of the binding is reflected in many of the ligands showing relatively weak (micromolar) affinity. The drugs, which induce clinically, are usually a subset of these.[15] In most cases inducers are not potent ligands for the receptor, binding constants being in the μmolar range rather than the nmolar range seen with traditional pharmacological potency. Some drugs do need to be at μmolar systemic concentrations, however. Most clinical inducers are drugs used at high doses to achieve high systemic concentrations. Table 1 lists most of the better-known clinical inducers. The high clinical doses reflect the weak potency of some of the drugs. For instance the Na^+ channel blockers have affinities for the channel of 3,9 and 25μM (moricizine, phenytoin and carbemazepine, respectively) and have to therefore achieve these concentrations to be active pharmacologically. With anti-infectives there is the need to dose to IC_{95} or greater. For instance, although efavirenz is a potent inhibitor of wild type RT HIV (K_i = 3nM) there is a need to go to higher concentrations to reach the IC_{95} for the virus. Moreover with resistant mutants, the drug is considerably less potent and a large increase in concentration and dose is required.

Table 1 Clinical inducers of CYP3A4 and the dose administered and associated with induction

	Dose (mg/day)
Carbamazepine	400-1200
Phenytoin	350-1000
Rifampicin	450-600
Phenobarbitone	70-400
Troglitazone	200-600
Efavirenz	600
Nevirapine	400
Moricizine	100-400
Probenicid	1000-2000

Induction may lower the concentration of the inducer, auto-induction, or lower the concentration of co-adminstered drugs. Whilst not normally a safety concern (as with inhibition, see above) the effect can be dramatic in decreasing the effectiveness of drugs. A simple case with obvious serious potential outcomes would be lowering the concentrations of an oral contaceptive and rendering it ineffective.

Apart from anti-infectives particularly in the anti-viral area (for reasons discussed above) few new drug introductions are inducing agents. Partly this reflects the selection of highly potent and selective agents. These often can replace the existing drug therapy from clinical usage. An example of this is the anti-diabetic compound troglitazone (Figure 4 A). This was used at a high clinical dose (Table 1) and associated with enzyme induction: troglitazone lowered the plasma concentrations of known CYP3A4 substrates such as cyclosporine, terfenadine, atorvastatin, and ethinylestradiol. In contrast the structurally related rosiglitazone (Figure 4 B) is used at 2-12 mg and shows no evidence of enzyme induction. Concomitant administration of rosiglitazone (8mg) has been shown not to affect the pharmacokinetics of the CYP 3A4 substrates. The clinical dose used closely relates to the receptor potency of these agents. For instance the EC_{50} values of troglitazone and rosiglitazone, for affinity against the peroxisome proliferator-activated receptor γ ligand binding domain are 322 nM and 36 nM, respectively. Corresponding figures for elevation of P2 mRNA levels as a result of peroxisome proliferator-activated receptor γ agonism are 690 and 80 nM, respectively. This increase in potency is even more marked in intact human adipocytes with affinities for PPAR-γ of 1050 and 40 nM respectively. Thus in all measures of pharmacological potency troglitazone is close to or in the micromolar range whereas those for rosigliazone are in the nanomolar region. Even without knowing the PXR affinities it is predictable which compound is likely to be associated with P450 induction clinically.

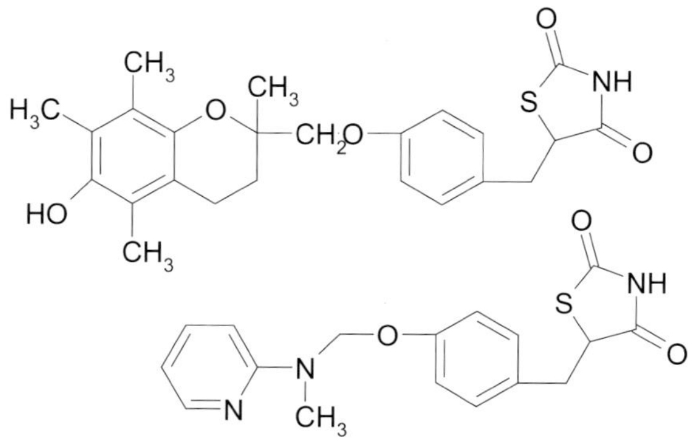

Figure 4 . Structures of Troglitazone (A) and rosiglitzone (B)

Medicinal chemistry now looks upon drug metabolism and pharmacokinetic data as essential in the design of new molecules. Traditionally, potency was the driving factor in the decision on which compounds to make (rational design, structure-activity-led drug design) but now absorption, clearance etc. are equally important. Selectivity is always another key area for focus. The promiscuous nature of receptors like PXR, where many compounds of diverse structure exhibit micromolar affinity, leads to the conclusion that potency against the target is still of great importance to provide margins against undesirable effects such as P450 induction.

4   DIVERSITY OF HUMAN ISOFORMS

Evolution of the P450 system has led to a super-family of isoforms capable of metabolising virtually any lipophilic molecule. This is not just due to a family of relatively selective forms metabolising diverse chemotypes. Most of the isoforms accept compounds in different orientations to the normal prime-binding mode. For instance, CYP2D6 is responsible for the N-dealkylation of a number of compounds. Some of the isoforms show much greater diversity of structures: CYP3A4, the most abundant isoform, shows the greatest promiscuity. This promiscuity has been ascribed to a large lipophilic binding site capable of binding substrates in multiple orientations. At least three different binding modes have been described for three difference structural classes of compound (steroids, benzodiazepines, dihydropyridine Ca^{++} channel blockers). Many substrates ranging from small rigid chemotypes, such as benzodiazepines (midazolam), to large flexible structures, such as cyclosporin, can be metabolised at more than one site by the one isoform (CYP3A4). Moreover the same site of catalysis can involve multipe isoforms. The involvement of multiple P450 isoforms is often indicated by biphasic Eadie-Hofstee enzyme kinetic plots when using human microsomes, and suggests a high affinity component and a lower affinity component. Which is the clinically important enzyme depends of course on what concentrations the drug achieves in man. Detailed work on imipramine has shown that CYP2C19 and 1A2 are involved in the N-demethylation whilst the 2-hydroxylation is mediated largely by CYP2D6 with a contribution from CYP2C19.[16] The combination of multiple isoforms and promiscuity leads to a formidable challenge to the medicinal chemist trying to attenuate metabolism. Manipulation of the structure, to remove metabolically vulnerable functionalities or positions, may simply result in an increase in the rate of metabolism in an alternative position in the molecule. This process of metabolic switching was elegantly demonstrated experimentally by Ling and Hanzlik.[17] Liver microsomes from phenobarbital induced rats were shown to oxidize toluene to benzyl alcohol and o-, m- and p-cresol (ratio of 69 : 31). Stepwise deuteration of the Me group caused progressive decreases in the yield of benzyl alcohol relative to the cresols (ratio 24 : 76 for toluene-d3). This is in effect blocking a major site of metabolism with the smallest of structural perturbation. However, stepwise deuteration also had the effect of inducing stepwise increase in total oxidation, giving rise to an inverse isotope effect overall. Throughout the series the ratios of cresol isomers remain constant. The results were interpreted in terms of product release for the benzyl alcohol being slower than the release of cresols), the rate being slow enough to be partially rate-limiting in turnover. The overall finding of "blocking a site of metabolism", triggering metabolic switching, and causing a net acceleration of turnover is obviously extremely frustrating. Switching between isoforms with structural changes also can occur.

Lipophilicity is a key property involved in metabolism rate. This reflects the essentially hydrophobic nature of P450 binding sites. Indeed, even CYP2D6 which is heavily dependent on a salt-bridge (ion-pair) interactive between the basic nitrogen of its substrates and aspartate in the active site of the receptor also requires substrates to bind to a phenylalanine residue close to the heme via a hydrophobic or π-π interaction. It may also reflect the greater opportunity for productive metabolism of a molecule if lipophilicity is regarded essentially as the amount of carbon (and sulphur) in the molecule.

5   RESPONSE TO P450 METABOLISM IN DRUG DESIGN

Medicinal chemists have responded to metabolism by CYPs in a variety of ways. Overall reduction of lipophilicity is one such tactic. In some cases the P450 involved is partially selective and metabolises compounds in a discrete region. Stable functionality can be incorporated in this region to attenuate or block metabolism. For instance cardioselectivity for β-adrenoceptor agents is conferred by substitution in the para position of the phenoxy-propanolamine skeleton. Oxidation of the para position or the substituent (methoxyethyl in the case of the cardioselective drug metoprolol) at this position are the major biotransformations for these compounds. The P450 responsible for the metabolism is the "regioselective" CYP2D6. Drugs such as metoprolol with vulnerable para substituents show high clearance ($CL_{i,\,free}$ 33ml/min/kg), resultant low bioavailability (38%), and also a short systemic half-life (3 hours). Manoury *et al* synthesized[18] the series of compounds leading to betaxolol with the belief that the bulky stable substituents in the para position (Figure 5) would be cardioselective and resistant to metabolism.

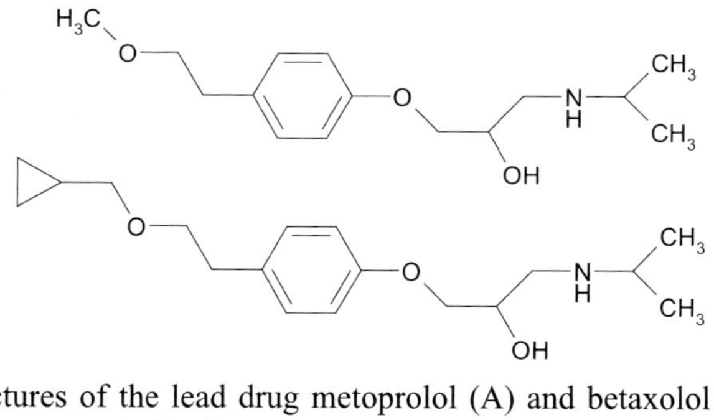

Figure 5. Structures of the lead drug metoprolol (A) and betaxolol (B), an analogue designed to be more metabolically stable.

Beside the steric bulk of the substituent, cyclopropyl is much more stable to hydrogen abstraction than other alkyl functions and represents an ideal terminal group.   These changes make betaxolol a compound with much improved pharmacokinetics ($CL_{i,\,free}$ of 13ml/min/kg, 90% bioavailability and an 18-hour half life) compared to other lipophilic analogues (cf metoprolol).

Halogen groups are often used to block metabolism or potentially deactivate ring systems. For instance Tolbutamide (Figure 6) is an oral hypoglaecemic compound of the sufonylurea class. The compound is metabolised in the benzylic methyl group by the "regioselective" CYP2C9, as the major clearance mechanism. Chlorpropamide is a related compound, but with a chlorine function in this position. These changes greatly increase metabolic stability and give chlorpropamide a lower clearance and a longer half-life than tolbutamide (approximately 35 hours compared to 5 hours), resulting in a substantial increase in duration of action.

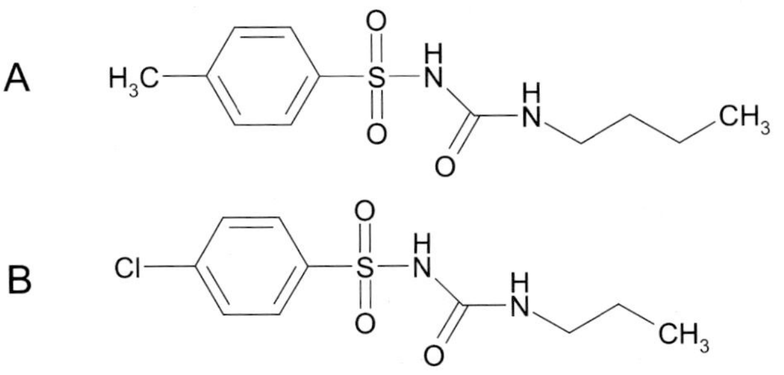

Figure 6. Structures of tolbutamide (A) and chlorpropamide (B), a more metabolically stable analogue

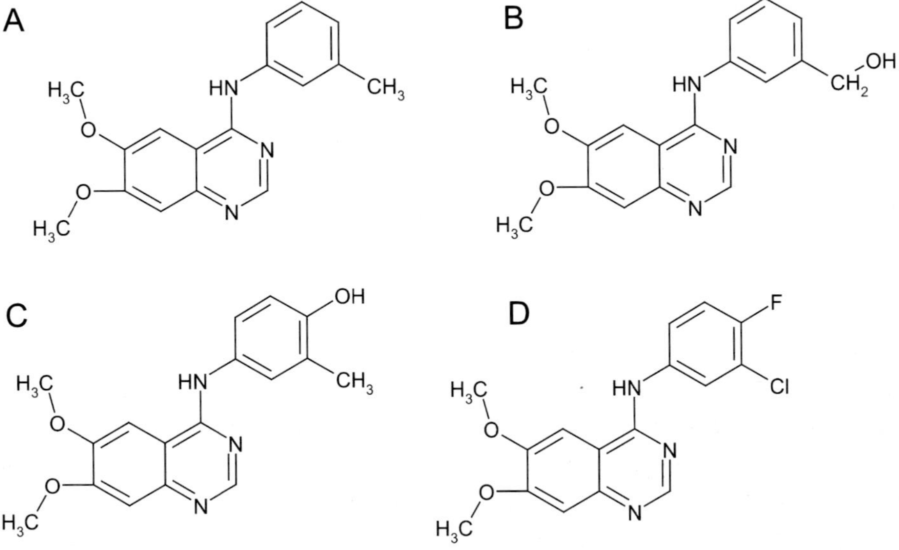

Figure 7. Structures of the lead anti-cancer molecule (A) and its two metabolites (B,C) and the metabolically stable derivative (D)

In the search for new anti-cancer agents the lead compound (Figure 7 A) is a potent inhibitor of EGFR-TK and of EGF-stimulated human tumour cell growth.[19] On oral dosing to mice, the compound showed evidence of antitumour activity but was rapidly metabolised with a systemic half-life of approximately 1 hour. Two major metabolites were identified, one formed by oxidation of the methyl group producing the corresponding benzyl alcohol (B), and the other from oxidation at the *para*-position of the aniline moiety producing the corresponding phenol metabolite (C). The SAR for activity allowed the scope to block or deactivate these routes: substitution with chlorine in place of the methyl group and introduction of a fluorine at the *para*-position of the aniline gave a metabolically stable compound (D). This compound did show a slight loss of potency *in vitro*, relative to the lead compound growth but had much-improved efficacy on oral dosing due to a reduced clearance and 3-fold longer half life.

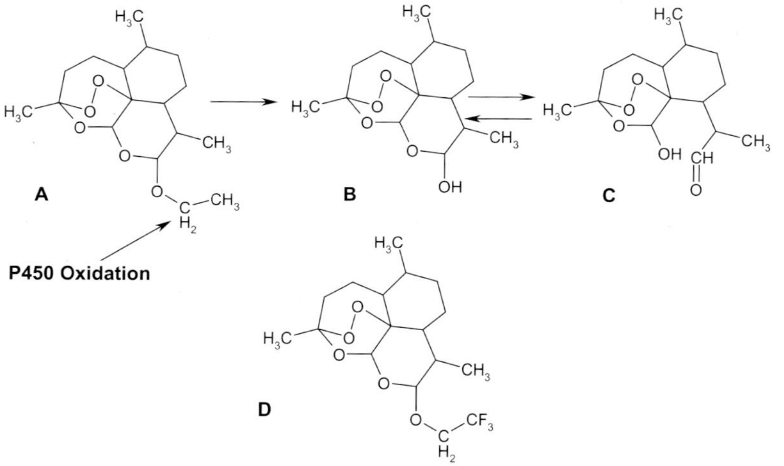

Figure 8. Metabolism of dihydroartemisinin (DHA) ether prodrugs (A), by cytochrome P450 to DHA (B) and its corresponding aldehyde (C) and a more stable fluorinated ether analogue (D).

Stabilisation of drug molecules to oxidation extends from aromatic rings to alkyl chains. Artemisinin, a sesquiterpene bearing an endoperoxide linkage, isolated from the Chinese medicinal herb *Artemisia annua L.*, is an antimalarial drug limited in use by poor solubility. More soluble ether derivatives (Figure 8 A) have been produced of artemisinin and the related dihydroartemisinin (DHA). These act as prodrugs for DHA which itself has a very short plasma half-life. The main pathway of metabolism of the ether prodrugs is rapid hydroxylation by cytochrome P450 enzymes to generate a hemiketal intermediate and subsequent loss of the ether, producing the active compound DHA (Figure 7 B) in equilibrium with its corresponding aldehyde (Figure 7 C). The rapid conversion to DHA and rapid clearance of these products still renders them unsuitable drugs. One approach to prolong the half-life of DHA is to design new ether pro-drugs that are more slowly converted by cytochrome P450 (or new hemiacetals more stable than DHA). The introduction of a fluorinated substituent at the terminal methyl carbon of DHA ethers (Figure 7 D) leads[20] to a slower rate of oxidative dealkylation (protection toward oxidative processes, provided by a fluoroalkyl group, extends to adjacent CH or CH_2 groups) which appear considerably more active *in vivo*.

Unfortunately due to the promiscuity of P450s and the problems alluded to above, often several strategies have to be employed in combination. Allylic and benzylic positions are extremely vulnerable to the hydrogen abstraction mechanism of P450s and their replacement in a drug candidate is often fruitful in lowering metabolic lability. SCH48461 (Figure 9 A) is a cholesterol absorption inhibitor metabolised in a number of positions including benzylic hydroxylation. Dugar and co-workers[21] substituted oxygen for the C-3' carbon to remove this site of metabolism. In itself, this change produces an electron rich phenoxy moiety which is more amenable to aromatic hydroxylation than the original phenyl group. This is often the situation faced where the change is beneficial in stability to the original metabolic step, but produces vulnerability elsewhere in the molecule. Subsequently fluorine was introduced in the para-position to produce the eventual more stable derivative (Figure 9 B).

Figure 9.   Synthetic strategies to overcome benzylic hydroxylation in SCH48461 (A) a cholesterol absorption inhibitor.   Positions of metabolism in particular benzylic hydroxylation (1) and to a lesser extent para-hydroxylation (2) are marked.

Tertiary amines are particularly labile positions due to the ease of abstraction by the P450 for one of the nitrogen electron lone pair. Secondary amines are more resistant to P450 metabolism due to the decreased substitution on the nitrogen

(2° *vs* 3°) stabilising the nitrogen to electron abstraction (decreased radical stability). Floyd *et al*[22] capitalised on this whilst synthesizing benzozapepinone analogues of the benzothiazepinone calcium channel blocking drug diltiazem. Metabolism of these types of compounds were by (1) N-demethylation, (2) conversion to an aldehyde (precursor of an acid), (3) deacetylation and (4) *O*-demethylation. For the tertiary amine-containing compounds the electron abstraction from the nitrogen is the first step to both the N-desmethyl and aldehyde products (a total of 84% of the total metabolism). N-1 pyrrolidinyl derivatives were designed to achieve metabolic stability via the decreased radical stability of secondary compounds and steric hindrance afforded by β-substitution (Figure 10).

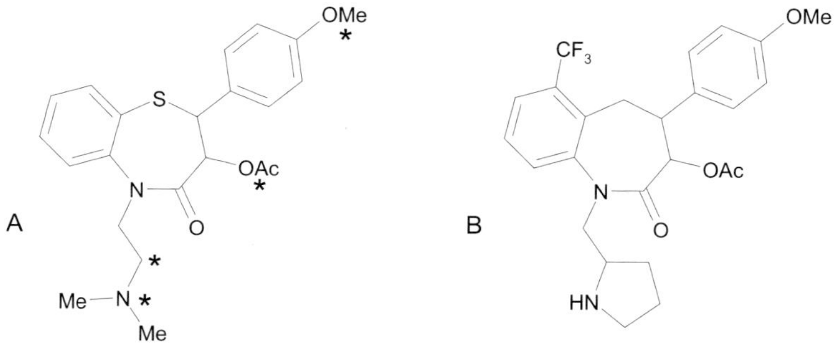

Figure 10.   Structures of diltiazem and a benzozapepinone analogue resistant to metabolism.

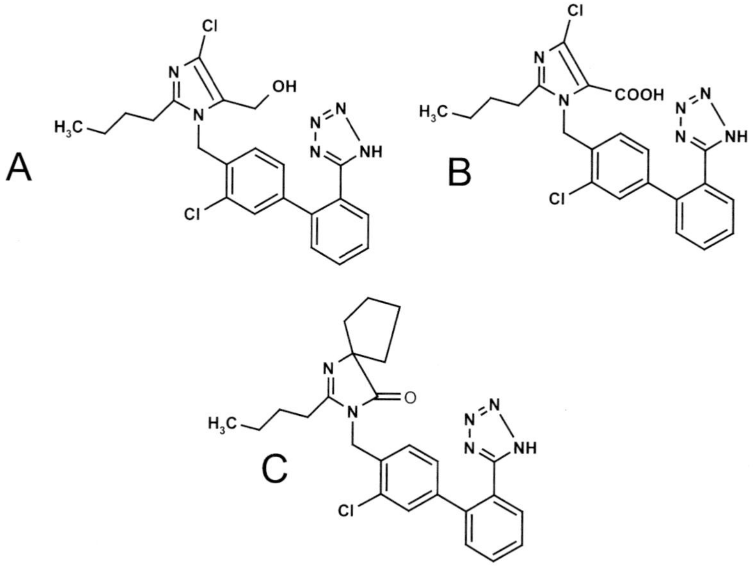

Figure 11. Structures of losartan (A), its active metablite EXP3174 (B) and irbesartan (C)

An unusual position for CYP lability is the hydroxyl group. Normally drug disposition scientists would expect alcohol dehydrogenase and aldehyde oxidase enzyme systems to be involved or the enzymes of drug conjugation. Losartan (Figure 11) is the first launched and the market leading angiotensin receptor antagonist. The drug is 33% bioavailable with a short half-life (1-3h) but is 14% converted to a potent, longer half life (5-10 hours) carboxylic acid metabolite[23] mainly by CYP2C19 and CYP3A4. Whilst in itself this is advantageous, variability is naturally introduced by this step. Irbesartan is a derivative in which losartan's metabolically labile 5-hydroxymethyl group has been replaced by 4-hydroxyimidazole, which exists preferentially in the dihydrimidazol-4 -one form. (Figure 12).

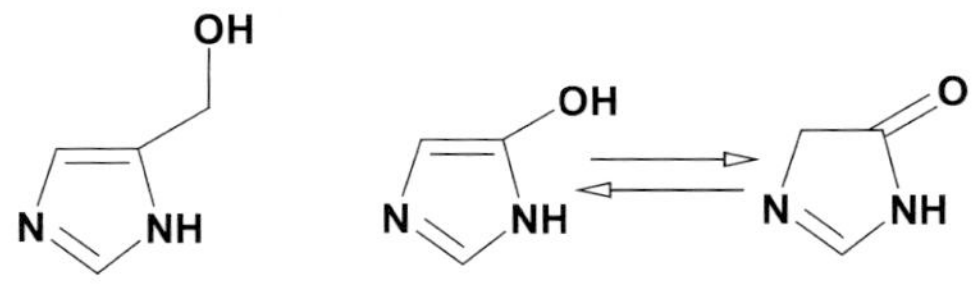

Figure 12. 5-(hydroxymethyl)imidazole and its replacement pharmacophore by 4-hydroxyimidazole, which exists preferentially in the dihydrimidazol-4 -one form.

Because the position 4 of the imidazole requires a lipophilic substitute (chlorine in losartan) and simple 5-substituents are unstable, 5,5-disubstitution resulted in the spiropentane compound being the selected development candidate[24] (Figure 11).

6   P450 AS A DRUG TARGET

Cytochrome P450 14-sterol demethylases (CYP51s) catalyze the oxidative removal of the 14-methyl group of lanosterol to form ergosterol, an important constituent of the cell membrane in fungi, by causing the elimination of 14R methyl group of lanosterol to give the C14-C15 unsaturated sterol. During the catalytic cycle, a substrate undergoes three successive monooxygenation reactions resulting in formation of 14-hydroxymethyl, 14-carboxaldehyde, and 14-formyl derivatives followed by elimination of formic acid with introduction of a 14,15 double bond. The accumulation of 14R-methylated sterols in azole-treated fungal cells affects membrane structure and functions, resulting in an inhibition of the growth of fungi. Differential inhibition of this enzyme between pathogenic fungi and man is the basis for the clinically important activity of these azole antifungal agents. The specificity of the inhibitors is determined by the differential complementarity between the structure of the agent and the active sites of the fungal and host enzymes.[25] The azole antifungal agents in clinical use contain either two or three nitrogens in the azole ring and are thereby classified as imidazoles (e.g., ketoconazole and miconazole, clotrimazole) or triazoles (e.g., itraconazole, fluconazole and voriconazole), respectively (Figure 12).

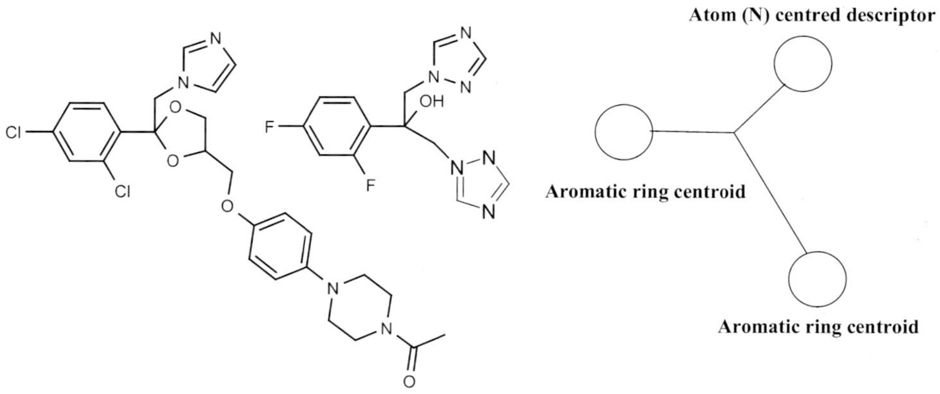

Figure 12.　Structures of ketoconazole and　fluconazole shown together with a pharmacophore model for azole antifungal agents[26.]

With the exception of ketoconazole, use of the imidazoles is limited to the treatment of superficial mycoses, whereas the triazoles have a broad range of applications in the treatment of both superficial and systemic fungal infections. Another advantage of the triazoles is their greater affinity for fungal rather than mammalian cytochrome P450 enzymes, which contributes to an improved safety profile. X-ray crystal structures of CYP51 show the inhibitors bound in the active site such that the imidazole or triazole ring is positioned perpendicular to the porphyrin plane with a ring nitrogen atom coordinated to the heme iron (see above).[27] Whilst exhibiting some selectivity towards CYP51 the azole antifungals are inhibitors of mammalian P450s metabolizing exogenous compounds.[28] This is not too surprising due to their mode of action (ligand interaction with the P450 heme), the binding energy (affinity) it imparts (see above), and the promiscuity of many of the mammalian P450s metabolizing endogenous compounds. For instance ketoconazole and itraconazole are potent inhibitors, both *in vitro* and *in vivo,* of CYP3A4. Not surprisingly the range of interactions is large and potentially serious (see above). In addition to the case of terfenadine outlined above, coadministration of ketoconazole or itraconazole with drugs metabolized by CYP3A4 such as immunomodulators (cyclosporin, tacrolimus), hypnotics (alprazolam, triazolam, midazolam), calcium channel blockers (nifedipine, felodipine), lipid modulators (simvastatin, lovastatin, vincristine), or antihistamines (astemizole) results in reduction in metabolic clearance and elevated concentrations in the circulation.

Aromatase is the cytochrome P450 enzyme responsible for the last step of estrogen biosynthesis. The estrogen biosynthesis in postmenopausal women occurs principally at extragonadal sites, among which the adipose tissue plays a major role. The peripheral production of estrogens is of particular relevance for the growth and development of breast tumors, and it provides an explanation of the occurrence of this kind of cancer in postmenopausal women. The aromatization step is generally considered the rate-limiting step in the synthesis of estrogens. Aromatase (CYP 19) carries out the conversion of androgen to estrogen (i.e., the aromatization reaction). The first non-steroidal inhibitor was an existing drug: aminoglutethimide (Figure 13).[29] The hypothesis that the AG mode of binding could involve interaction of a nitrogen atom with the heme iron of CYP19 led to the design of azole containing aromatase inhibitors, and from this work emerged drugs such as fadrozole, letrozole, vorozole and anastrozole (Figure 13).

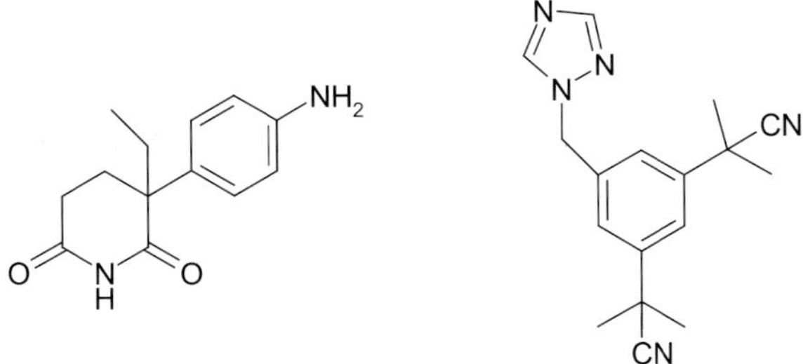

Figure 13. Structures of aminoglutethimide and anastrozole

Drug interactions with the azole non steroidal aromatase inhibitors seem less problematic than with the azole antifungal agents.[30] The results of *in vitro* studies show that although anastrozole, for instance, can inhibit CYP1A2, 2C9, and 3A4 mediated metabolism, the relatively weak potency of the compound (around 10 micromolar) would not be expected to cause significant interactions in the *in vivo* clinical setting (clinical dose is around 1mg): the IC_{50} for inhibition of human placental aromatase activity (15 nM) gives a margin of selectivity of at least 500-fold over inhibition of the drug metabolising CYPs.

7 SUMMARY

Proteins play a central role in the drug discovery and development process. They represent the target that the industry attempts to agonise or antagonise by designing and synthesising ligands with affinity for them. Each drug target protein has a period in the spotlight followed by a minor supporting role as a selectivity screen (broad ligand binding). But few proteins have taken centre stage in such a dramatic fashion. In the last decade the discovery of multiple forms of P450, the variation in expression of those forms, and their inhibition in selective or semi-selective manner by concomitant drugs has reshaped whole departments and changed the way we conduct drug discovery and development. No-one could have seen how *in vitro* enzymology data would be seen as vital to clinical research planning, No-one either would have envisaged from the depths of the drug metabolism laboratories that CYP2D6 and CYP3A4 would be terms commonly used in daily conversation by the heads of some the largest pharmaceutical research laboratories around the world.

REFERENCES

1. Lu, A. Y. H.; West, S. B.; Ryan, D.; Levin, W. *Drug Metab. Dispos.* **1973**, *1*, 29.
2. Lu, Anthony Y. H.; Levin, Wayne; West, Susan B.; Jacobson, Martin; Ryan, Dene; Kuntzman, R.; Conney, A. H. D. *J. Biol. Chem.* **1973**, *248*, 456.
3   Mahgoub, A.; Idle, J. R.; Dring, L. G.; Lancaster, R.; Smith, R. L. *Lancet* **1977**, *8038* 584.
4. Kahn, G. C.; Boobis, A. R.; Murray, S.; Plummer, S.; Brodie, M. J.; Davies, D. S. *Br. J. Clin. Pharmacol.* **1982**, *13*, 594.
5. Larrey, D.; Distlerath, L. M.; Dannan, G. A.; Wilkinson, G. R.; Guengerich, F. P. *Biochemistry* **1984**, *23*, 2787.
6. Distlerath, L. M.; Guengerich, F. P. *Proc. Natl. Acad. Sci. U. S. A.* **1984**, *81*, 7348.
7. Guengerich, F. P.; Martin, M. V.; Beaune, P. H.; Kremers, P.; Wolff, T.; Waxman, D. J. *J. Biol. Chem.* **1986**, *261*, 5051.
8. Yamazaki, H; Shimada, T. *Drug Metabolism and Disposition* **1998**, *26*, 1053.
9. Riley, R. J.; Parker, A. J.; Trigg, S.; Manners, C. N. *Pharmaceutical Research* **2001**, *18*, 652.
10. Honig, P. K.; Wortham, D. C.; Zamani, K.; Conner, D. P.; Mullin, J. C.; Cantilena, L. R. *J. Am. Med. Assoc.* **1993**, *269*, 1513.
11. Honig, P. K.; Woosley, R. L.; Zamani, K.; Conner, D. P.; Cantilena, L. R., Jr. *Clin. Pharmacol. Ther.* **1992**, *52*, 231.
12. Yun, C. H.; Okerholm, R. A., Guengerich, F. P. *Drug Metab. Dispos.* **1993**, *21*, 403.
13. Moltke, L. L. Von; Greenblatt, D. J.; Duan, S. X.; Harmatz, J. S.; Shader, R. I. *J. Clin. Pharmacology* **1994**, *34*, 1222.

14. Watkins, R. E.; Wisely, G. B.; Moore, L. B.; Collins, J. L.; Lambert, M. H.; Williams, Shawn P.; Willson, Timothy M.; Kliewer, Steven A.; Redinbo, Matthew R. *Science* **2001**, *292*, 2329.

15. Smith, D. A. *European Journal of Pharmaceutical Sciences* **2000**, *11*, 185.

16. Koyama, E.; Chiba, Kan; T., Masanao; I., Takashi. *Journal of Pharmacology and Experimental Therapeutics* **1997**, *281*, 1199.

17. Ling, K. H. J.; Hanzlik, R. P. *Biochem. Biophys. Res. Commun.* **1989**, *160*, 844.

18. Manoury P M; Binet J. L.; Rousseau J.; Lefevre-Borg F.; Cavero I. G. *Journal of Medicinal Chemistry* **1987**, *30*, 1003.

19. Barker, A. J.; Gibson, K. H.; Grundy, W.; Godfrey, A. A.; Barlow, J. J.; Healy, M. P.; Woodburn, J. R.; Ashton, S. E.; Curry, B. J.; Scarlett, L.; Henthorn, L.; Richards, L. *Bioorganic & Medicinal Chemistry Letters* **2001**, *11*, 1911.

20 Nga, T. T. T.; Menage, C.; Begue, J.-P.; Bonnet-Delpon, D.; Gantier, J.-C.; Pradines, B.; Doury, J.-C.; Thac, T. D. *Journal of Medicinal Chemistry* **1998**, *41*, 4101.

21 Dugar, S.; Yumibe, N.; Clade, J.W.; Vizziano, M.; Huie, K.; Heek, M. V.; Compton, D.S.; Davis, H. R.; *Biorganic Medicinal Chemistry Letters*, **1996**, *6*, 1271.

22. Floyd D. M.; Kimball S. D.; Krapcho J.; Das J.; Turk C. F.; Moquin R. V.; Lago M. W.; Duff K. J.; Lee V. G.; White R. E., *Journal of Medicinal Chemistry* **1992**, *35*, 756.

23. Israili, Z. H. *Journal of Human Hypertension* **2000**, *14*, S73.

24. Bernhart, C. A.; Perreaut, P. M.; Ferrari, B. P.; Muneaux, Y. A.; Assens, J. L. A.; Clement, J.; Haudricourt, F.; Muneaux, C. F.; Taillades, J. E.; et al. *J. Medicinal Chemistry* **1993**, *36*, 3371.

25. Sheehan D.J., Hitchcock C.A., Sibley C.M., *Clinical Microbiology Reviews*, **1999**, *12*, 40.

26. Talele, T. T.; Kulkarni, V. M. *Journal of Chemical Information and Computer Sciences* **1999**, *39*, 204.

27. Podust, L. M.; Poulos, T. L.; Waterman, M. R. *Proceedings of the National Academy of Sciences of the United States of America* **2001**, *98*, 3068.

28. Venkatakrishnan, K.; von Moltke, L. L; Greenblatt, D. *J. Clinical Pharmacokinetics* **2000**, *38*, 111.

29. Recanatini, M.; Cavalli, A.; Valenti, P. *Medicinal Research Reviews* **2002**, *22*, 282.

30. Grimm, S. W.; Dyroff, M. C. *Drug Metabolism and Disposition* **1997**, *25*, 598.

Chapter 6

Cytochrome P450 in Laboratory Animal Species

Margit Spatzenegger[1], Stephanie L. Born[2], James R. Halpert[1]

[1]*Department of Pharmacology and Toxicology,*
 University of Texas Medical Branch at Galveston, TX 77555

[2]*Inflammatory Biology, Bristol-Myers Squibb Company, Wilmington, DE 19880*

1   INTRODUCTION

Cytochromes P450 exhibit a high degree of structural and functional similarity across species lines. Data from experimental animal species, such as rodents, dogs, and non-human primates, have been key to our understanding of xenobiotic biotransformation processes. Today we continue to utilize animal models to understand complex multi-organ biotransformation mechanisms, to predict toxicity, and to define pharmacokinetic and pharmacodynamic relationships. *In vitro* tools play an ever larger role in predicting metabolites, estimating clearance, and in answering mechanistic questions. As the study of xenobiotic metabolism has matured, it has also become clear that direct extrapolation between experimental animal species and humans is limited by differences in enzyme activity, abundance, specificity and regulation, among other reasons. Our challenge is to select the appropriate tools to answer complex questions. This chapter provides an introduction to many issues that impact metabolism studies and the application of data obtained in experimental animal species to humans.

2   LABORATORY ANIMAL MODELS AS PREDICTORS OF HUMAN METABOLISM

Animal models are tools. They do not, in general, provide information that can be directly extrapolated to humans. However, drug metabolism, therapeutic efficacy and toxicology studies will all benefit from the selection of an animal model species that closely approximates human metabolism. For perspective, Table 1 contains data on the qualitative value of different animal species as predictors of human drug metabolism. Good or fair predictions were found 92% of the time in the case of the rhesus monkey.[1] However, despite the fact that non-human primates are oftentimes more predictive of human metabolism, the bulk of investigative and regulatory studies are conducted in rats and dogs. Thus, gaining an understanding of species differences in metabolism is essential.

Table 1. Laboratory Animal Drug Metabolism Pathways as Qualitative Models for Humans[1]

	Percent of occasions		
Rating: Pattern As Model for Human	Rat	Other Nonprimate (Dog, Rabbit, Guinea Pig)	Rhesus Monkey
Good	29	32	73
Fair	12	27	19
Poor	20	9	4
Invalid	42	32	4

2.1   Basic approaches

The complexities of xenobiotic metabolism are best approached using a combination of *in vitro* and *in vivo* studies. Early in the life of a potential new drug, *in vitro* metabolism studies are conducted that compare the properties of the compound in hepatic microsomes or S9 fractions from several species of interest. The rate of compound biotransformation (sometimes referred to as metabolic stability) in liver homogenate fractions, and a determination of the major P450 forms involved in the metabolism of the compound, provide valuable data for the selection of animal species for *in vivo* studies. Recombinant human and rat P450s are now commercially available, and can be used to investigate species differences in enzyme specificity. Optimally, the compound will not be extensively metabolized by polymorphically expressed human P450 forms (such as CYP2D6). Further *in vitro* studies that examine intestinal permeability and transport aid in screening and selecting compounds that are most likely to have appropriate pharmacokinetic properties. A more comprehensive pattern of metabolism, including phase I and phase II metabolites, can be observed by using isolated hepatocytes in culture. However, metabolite identification, whether it be from media, from cultured hepatocytes or from blood samples, is normally performed only on a relatively "mature" compound. Together, a panel of *in vitro* assays should reflect *in vivo* metabolism, detect species differences, and predict compound clearance, thereby aiding in the selection of the best experimental animal species.[2] Following the initial *in vitro* screens, selected compounds are advanced for pharmacokinetic studies. Optimally, the compound will have a reasonable half-life in at least one rodent and one non-rodent species.

In addition to these pharmacokinetic considerations, animal models for drug discovery and development studies are also selected based on the disease models available for study. For example, rheumatoid arthritis disease models are available principally in mice and rats. In general, dogs and non-human primates are less desirable as disease models due to size and compound dosing requirements, and difficulty of handling. In many cases, the disease model available and the optimal species with regard to drug metabolism will not be the same. In these instances, efficacious dosages of a compound in a disease model should be paired with data on species differences in drug metabolism prior to dose extrapolation to man.

In vitro toxicology studies can aid in predicting toxic metabolite formation, but intact animal models continue to be the gold standard in toxicology testing. Dosages are determined based upon efficacy studies in disease models and an appropriate safety margin. Prior to the first dose in man, various toxicology studies must be performed. The studies required are dependent upon intended duration of administration in the human test subjects. For example, repeated dosing studies of one month duration in two species are required prior to the conduct of 1 month phase I human studies. Most often, qualitatively similar toxicity patterns are observed regardless of interspecies pharmacokinetic differences. The assumption is made that if high multiple doses are safe in animals, then low single doses will probably be safe in humans. This assumption generally holds true, as mammals are a monophyletic group and homogenous in physiology.[3,4] Further, cellular transport and biochemical mechanisms seem to be similar across species.[5]

One important consideration in comparing laboratory animal models to humans is that, in general, rodents have a higher metabolic rate than humans. This difference may affect the types of metabolites produced and the pharmacokinetics of the drug. For extrapolation purposes it is important to understand basic species differences in cytochrome P450 enzymes. For example, CYP2A-mediated coumarin 7-hydroxylase activities in man have been shown to be up to 200-fold higher than in rat.[6,7] The CYP2C9-related protein in monkey catalyzes diclofenac with a V_{max} value similar to man. However, a 25-35-fold higher K_m value was found, indicating a much higher affinity of human CYP2C9 towards diclofenac than monkey CYP2C9-related protein.[8,9] CYP3A enzymes exhibit sex-related differences in rat, with the male rat having 5-10-fold higher activities than the female rat.[10] Similar CYP3A activities were observed in the male rat and in humans.[11] However, total cytochrome P450 content across species is fairly consistent when expressed on a per milligram microsomal protein, or per gram of liver basis.[12] Total hepatic cytochrome P450 content also seems to correlate with body weight.[13] Therefore, drugs that are metabolized by multiple cytochrome P450 enzymes seem to have a good correlation of intrinsic clearance with total cytochrome P450.

For further information on the relationships between the pharmacokinetics of P450s and physiology the reader is referred to some of the recent reviews of this subject.[14-19]

2.2   Interspecies scaling

The most important approach used to combine pharmacokinetics with physiology, and to extrapolate information obtained from animal studies to humans is *interspecies scaling*. Over the years *allometry*, which is based on the assumption that the relationships between anatomy and physiological functions are similar among mammalian species, has become a useful tool to correlate pharmacokinetic parameters with body weight from different animal species.[20] Huxley[21] showed a linear relationship between organ weight and body weight with the following equation:

$$\log Y = \log a + b \log W \tag{2.2.1}$$

Y is organ size, W is body weight, log a is the value of log Y for 1-kg body weight and b shows the size effects (slope).
The antilog of Eq. 2.2.1. is termed the power function.

$$Y = aW^b \tag{2.2.2}$$

Clearance (Cl) is the most important parameter for characterizing drug removal from the body.[12] *Intrinsic clearance* (Cl_{int}) is the rate of drug elimination divided by the unbound drug concentration at the elimination site. Organ clearance in the well-stirred model is described by Eq. 2.2.3.

$$Cl_{organ} = Q_{organ} \left[\frac{f_{u,b} \cdot Cl_{int}}{Q_{organ} + \left(f_{u,b} \cdot Cl_{int} \right)} \right] \qquad (2.2.3)$$

Q_{organ} is the organ blood flow, $f_{u,b}$ is the fraction unbound in the blood and Cl_{int} is the maximal ability of the organ to remove a drug by all pathways in the absence of protein binding or blood flow limitations.

Eq. 2.2.2. can be applied to the estimation of Cl resulting in the simple allometric Eq. 2.2.4., where the Cl of a compound is plotted against the *body weight* on a log-log scale.

$$Cl = a \ W^b \qquad (2.2.4.)$$

W is the body weight and *a* and *b* are the coefficient and exponent of the allometric equation, respectively.

Several studies have shown that Cl is well described by Eq. 2.2.4. for all of the animal species studied.[22] Examples of compounds whose Cl correlates with body weight include phenytoin, diazepam, phenobarbital, and antipyrine, drugs that are metabolized by multiple forms of hepatic cytochrome P450.[23] The clear exception is man, where predicted values of the intrinsic clearance of these drugs are consistently larger than those observed.

The problem of scaling laboratory animal metabolism and clearance to that of humans may be overcome in part by accounting for the life span and brain weight of the laboratory animal species. Boxenbaum[19] concludes that short-living species metabolize compounds quickly, and long-living species catalyze slowly. At the end of the maximum life span, each species will have cleared the same volume of xenobiotic per kg body weight. Therefore, observed clearance values in the different animal species are multiplied by their respective maximum life-span potential (MLP). From the allometric equation, clearance x MLP was estimated in man, the result divided by the MLP of man (8.18×10^5 h) to predict the clearance in man (Eq. 2.2.5.).

$$Cl = a \ (MLP \times Clearance)^b \ / \ 8.18 \times 10^5 \qquad (2.2.5.)$$

MLP in years was calculated by Sacher[24]:

$$MLP \ (years) = 185.4 \ (BW)^{0.636} (W)^{-0.225} \qquad (2.2.6.)$$

Both brain weight (BW) and body weight (W) are in kg.

The allometric relationship between intrinsic clearance of unbound compounds, MLP, and body weight provides a significantly better correlation of human clearance data with those predicted from other animals compared with Eq. 2.2.4.

Besides the rather theoretical importance of cytochrome P450 to explain certain differences between animal and man, the big practical question remains: Can allometry successfully predict the clearance of drugs metabolized by cytochrome P450 and is it possible to link the knowledge of a particular P450 enzyme responsible for the metabolism of a drug and interspecies scaling? Several recent studies combined *in vivo* and *in vitro* metabolism data, including microsome binding values and the animal scaling factor, in efforts to more effectively compare metabolism across species.[25, 26] The 'rule of exponents', proposed by Mahmood and Balian, is one way to improve the predictive performance of allometry for the clearance of drugs catalyzed by cytochrome P450.[27, 28] It uses three different approaches such as: (1) simple allometry,[19] (2) the product of body weight and maximum life-span potential[19] and 3) the product of clearance and brain weight vs body weight.[28] The choice of one of the three methods depends on the value of the exponent b calculated by the simple allometric equation. The exponent indicates which equation should be used for the calculation of the clearance of a specific compound. Table 2 shows which method is most predictive for humans based on the exponent of simple allometry.

Table 2. Proposed 'rule of exponents' by Mahmood and Balian[28]

Exponent of simple allometry	Method of choice
0.55 to 0.70	Simple allometry
0.71 to 1.0	CL x MLP
≥ 1.0	CL x brain weight

The clearance data on a wide variety of drugs that are metabolized mainly by CYP3A4, CYP2D6, or by more than one P450 enzyme, were studied in at least three species (mouse, rat, rabbit, guinea-pig, monkey, or dog), and were analyzed by the 'rule of exponents'.[29] The results indicate that a good correlation exists between observed and predicted clearance values for most drugs using the 'rule of exponents'. However, the knowledge of a P450 enzyme responsible for metabolism of a particular drug does not provide guidance for the failure or success of allometry in predicting clearance in man.

3   FROM ANIMAL TO MAN: COMPARISON OF CATALYTIC SELECTIVITY BY ONE SUBFAMILY

Cytochrome P450 substrate specificity is often similar within a subfamily across species.[30] However, important differences exist that limit direct extrapolation between species. Before beginning projects that involve reaction phenotyping, the selection of marker substrates, or the identification of a predictive toxicology species, one should be familiar with basic species differences in cytochrome P450 substrate specificity. The following section provides an introduction to some of these differences in laboratory animal species.

3.1   CYP1A

CYP1A1/1A2 enzymes have been extensively studied because of their importance in the activation of carcinogens. CYP1A1 and its associated activities are well conserved in mammals.[31] As CYP1A1 is expressed in very low levels in human liver and is essentially an extrahepatic enzyme in man, many studies have focused on the more highly expressed rodent CYP1A1 forms.[32] Most experimental animal species efficiently catalyze the CYP1A1-mediated epoxidation of polycyclic aromatic hydrocarbons.[33, 34] Shou *et al.*[35] compared the role of cDNA-expressed rat, mouse, rabbit and human CYP1A1 activities towards benzo[*a*]pyrene, a potent carcinogen, and its metabolite benzo[*a*]pyrene *trans*-7,8-dihydrodiol, which must be activated by biotransformation to bind to DNA and initiate tumors.[36] CYP1A1 of all tested species catalyzed the biotransformation of these two compounds. The metabolism by rat and rabbit was similar to that by the human form (Table 3). However, mouse CYP1a1 was 40 times more active than any human, rabbit, or rodent P450 towards benzo[*a*]pyrene, and 10 times more active towards benzo[*a*]pyrene *trans*-7,8-dihydrodiol.

In contrast to polycyclic aromatic hydrocarbons, mutagenic and carcinogenic heterocyclic amines are metabolized predominately by CYP1A2.[37] The contribution of CYP1A1 and CYP1A2 towards N-hydroxylation, the step of mutagenic activation, is species-dependent, and is different for each heterocyclic amine substrate.[38] CYP1A2 has been purified from mouse, rat, rabbit, and man, and is well conserved in function across species.[39] CYP1A2 genes were also identified in the dog and monkey.[40, 41, 42] The example of the food-borne carcinogen 2-amino-3,8-dimethylimidazol[4,5-f]quinoxaline (MeIQx) demonstrates how defining the metabolite of interest, as well as recognizing species differences in P450 expression, are crucial in the selection of an animal model. In rodents, MeIQx is both C- and N-oxidized,[43] whereas humans efficiently N-hydroxylate the amine but do not form the C-oxidation product.[44] Activation of MeIQx to its genotoxic derivative involves CYP1A2-mediated N-hydroxylation of the exocyclic amine function.[45] Therefore, because both rodents and humans constitutively express hepatic CYP1A2,[46] rodents are a valuable model to assess human risk of genotoxicity from MeIQx (Table 3). In contrast, the cynomologus monkey *Macaca fascicularis* is resistant to MeIQx-related carcinogenicity because it does not express CYP1A2 and is unable to convert MeIQx to its N-hydroxylamine.[47, 48]

Table 3. Prediction of species most comparable with man with respect to cytochrome P450 activity

Human P450	Substrate	Species most similar to man
CYP1A1	benzo[*a*]pyrene	rat, rabbit[35]
CYP1A2	MeIQx[(a)]	rat, mouse[46]
CYP2A6	coumarin	monkey, mouse, hamster, rabbit[55]
CYP2B6	7-EFC[(b)]	mouse[63]
CYP2C19	mephenytoin	monkey[63, 78]
CYP2D6	bufuralol	dog[63, 82]
CYP2E1	chlorzoxazone	no large species differences[63]
CYP3A4	testosterone	mouse, male rat[63]
CYP4A11	lauric acid	no large species differences[63]

(a) MeIQx, 2-amino-3,8-dimethylimidazol[4,5-f]quinoxaline
(b) 7-EFC, 7-ethoxy-4-trifluoromethylcoumarin.

3.2   *CYP2A*

CYP2A enzymes have been identified in mouse, rat, rabbit, monkey, hamster, and humans.[49] Despite their sequence similarity (e.g. mouse CYP2a4 and CYP2a5 differ in only 11 amino acids) CYP2A enzymes differ significantly in catalytic specificity. Some CYP2A forms have been classified by Honkakoski and Negishi[49] with regard to steroid hydroxylation. One branch includes rat CYP2A1, human CYP2A9, and mouse Cyp2a12, which catalyze steroid 7α-hydroxylation. Another branch includes rat CYP2A2 and mouse Cyp2a4, steroid 15α-hydroxylases. Rat CYP2A3, mouse CYP2a5, hamster CYP2A8, rabbit CYP2A10, the CYP2A isoforms of the baboon and cynomologus monkey, and human CYP2A6 all have coumarin 7-hydroxylase activity (Table 3).[39] Coumarin 7-hydroxylation is used as marker substrate for 2A forms across species. However, *in vitro* and *in vivo* studies have shown that coumarin metabolism is complex, and differs significantly between species.[50] Coumarin may be metabolized by hydroxylation at all six possible positions to yield 3-, 4-, 5, 6-, 7- and 8-hydroxycoumarins, and by opening the lactone ring to yield various products. The first step in coumarin metabolism by the latter pathway is the formation of a toxic coumarin 3,4-epoxide intermediate, which rearranges to form *o*-hydroxy phenylacetaldehyde.[51] In human subjects, coumarin is extensively metabolized to 7-hydroxycoumarin. In contrast, lactone ring hydrolysis predominates in the rat.[52] Indeed, coumarin is hepatotoxic in the rat, where only 0.7% of the parent compound is metabolized to 7-hydroxycoumarin.[53] Strain differences in CYP2A activity have also been identified. In the mouse, DBA/1J and DBA/2J have high coumarin 7-hydroxylase activity, whereas strains such as C57BL/6J have low activity.[54] The CYP isoform responsible for coumarin 7-hydroxylation in mouse liver is Cyp2a5, [55] which is predominantly expressed in female mouse liver.[49]

3.3   CYP2B

Cytochrome P450 2B enzymes have been characterized in rats, rabbits, dogs, monkeys and humans.[39] In addition, the mouse expresses multiple Cyp2b enzymes (Cyp2b9, 2b10, and 2b13).[41, 42] CYP2B forms in rat and rabbit are highly inducible by barbiturates, were among the first forms purified, and played a major role in the development of P450 structural models.[56] Rat CYP2B1 and 2B2, rabbit CYP2B4 and 2B5, and dog CYP2B11 all oxidize steroids; however each of these enzymes shows a unique steroid hydroxylase profile.[57] Although rat CYP2B1 and 2B2 differ only in 14 amino acid residues, the two expressed CYP2B2 variants (CYP2B2 and CYP2B2$_{FF}$) showed very different androstenedione and testosterone hydroxylase activities and profiles compared with CYP2B1 and only 1-5% of the total androstenedione hydroxylase activity compared with CYP2B1.[58] CYP2B2$_{FF}$ produced mainly 7α-OH and 15α–OH metabolites, whereas CYP2B2 produced 15α–OH and 6β-OH as major metabolites. However, similar hydroxylation rates for CYP2B1 and CYP2B2 variants were found with progesterone.[58] Rat and rabbit steroid hydroxylase profiles differ widely. Rabbit CYP2B4 hydroxylates androstenedione nearly exclusively at the 16β-position, while CYP2B5 preferentially hydroxylates at the 16α- and 15α-positions. The high 15α- hydroxylation activity seems to be unique among the CYP2B enzymes.[59, 60] Dog CYP2B11 produces 16α- and 16β-hydroxyandrostenedione in equal proportions, whereas testosterone is preferentially hydroxylated at the 16α-position.[61] In contrast to the important role of the CYP2B enzymes in rodents, rabbit and dog, steroid metabolism by the human ortholog CYP2B6 is negligible.[39, 62]

Other marker substrates of CYP2B forms, including human CYP2B6, are 7-ethoxy-4-trifluoromethylcoumarin and 7-pentoxyresorufin. A comparison of the enzyme kinetics of these substrates in rat, dog and rabbit liver microsomes did not reveal large differences in K_m and V_{max} values among species. In contrast, monkey liver microsomes showed the highest V_{max} value of all species investigated. Mouse liver microsomes exhibited 7-ethoxy-4-trifluoromethylcoumarin kinetics most similar to humans (Table 3).[63]

Finally, it is interesting to note that rodents, especially the rat, are good predictive models for metabolism of the anticancer prodrug cyclophosphamide. This drug is activated in the liver by 4-hydroxylation, and is deactivated by N-dechloroethylation.[64, 65] Both pathways are observed in rat and man, with 4-hydroxylation catalyzed mainly by CYP2B1 in rat and by CYP2B6 in man.[66]

3.4 CYP2C

The CYP2C subfamily contains forms that vary widely in sequence identity (ranging from 68% to 95%) and substrate specificity. CYP2C enzymes have been identified in mouse, rat, rabbit, dog, monkey, and man.[40, 41] In the rabbit alone, seven different CYP2C forms have been characterized (CYP2C1, CYP2C2, CYP2C3, CYP2C4, CYP2C5, CYP2C14, CYP2C16).[67] Although both CYP2C4 and CYP2C5 have considerable progesterone 21-hydroxylase activity and exhibit 95% sequence identity, the K_m for CYP2C4 is more than 10-fold higher than that of CYP2C5.[68] In rat liver, steroid 21-hydroxylase activity is catalyzed mainly by CYP2C6.[69] Although Richardson *et al.*[70] reported that human CYP2C19 catalyzes the 21-hydroxylation of progesterone, in other species this activity is catalyzed essentially only by adrenal CYP21A.[71, 72] CYP2C3 is the main progesterone 6β-hydroxylase in rabbit liver microsomes, but in rat and man this reaction is typical of CYP3A enzymes.[73] Furthermore, CYP2C3, CYP2C2, CYP2C14, and CYP2C16 have 16α-hydroxylase activity.[74]

The metabolism of mephenytoin, a marker CYP2C substrate, illustrates how P450s may also differ in their stereoselective metabolism of compounds.[75,76,77] Mephenytoin exists as both an *R*- and *S*-enantiomer, with both forms being 4'-hydroxylated by CYP2C enzymes. In rabbit, dog and rat, the *R*-enantiomer is preferentially 4'-hydroxylated at a rate 2-6 times higher than that of the *S*-enantiomer. In female mice, both enantiomers are equally good substrates, and in the rat, CYP2C11 preferentially metabolizes the *S*-enantiomer.[78] Both human CYP2C19 and monkey CYP2C forms preferentially 4'-hydroxylate *S*-mephenytoin and exhibit similar K_m values, indicating that monkey may be useful as a model of human CYP2C metabolism (Table 3).[63, 78,]

3.5   CYP2D

Human CYP2D6, a debrisoquine 4-hydroxylase, is polymorphically expressed in the human population,[79] making it difficult to predict the metabolic fate of CYP2D6 substrates in patients. Four CYP2D genes (CYP2D1, 2D2, 2D3, and 2D4) have been identified in the rat.[80] Interestingly, rat CYP2D activities are inhibited by quinine, whereas human CYP2D6 is sensitive to inhibition by quinidine.[81]

A comparison of the CYP2D marker activity, bufuralol 1'-hydroxylation, in liver microsomes of different species, showed that rat, mouse, rabbit, and monkey exhibit a K_m value about 10 times lower than CYP2D6, whereas the V_{max} value is about 6 to 10 times higher than that of the human enzyme.[63] However, the kinetic behavior of two canine CYP2D15 variants and human CYP2D6 were similar toward bufuralol, metoprolol, and dextromethorphan (Table 3).[82] Furthermore, quinidine inhibition obtained in dog liver microsomes, as well as with recombinant CYP2D15, was similar to that obtained in human liver microsomes and recombinant CYP2D6.[63, 82]

3.6   CYP2E

The CYP2E subfamily consists of a single form in most species. CYP2E1 has been purified from rat, rabbit and humans,[39] and has been detected in mouse, dog and monkey tissues.[40, 41] CYP2E2 is expressed only in neonatal rabbits,[83] and exhibits substrate specificities similar to CYP2E1.[39]

CYP2E1 is responsible for the metabolism of small molecules, such as ethanol and benzene,[84] as well as fatty acids, where the major metabolite is a ω-1 alcohol.[85, 86] Chlorzoxazone is a marker substrate for CYP2E1 in mouse, rat, rabbit, dog, micropig, monkey and man.[63] With the exception of the micropig, the K_m and V_{max} values were observed in a very narrow range (Table 3).[63] However, other P450 forms, such as 1A2, may contribute to chlorzoxazone metabolism in rodents.[87]

CYP2E1 is known for its involvement in the bioactivation of substances to carcinogenic and hepatotoxic products,[88] and its inducibility by ethanol, acetone, and isoniazid. Bernauer *et al.*[89] investigated the interspecies variability in CYP2E1 expression and chlorzoxazone hydroxylation in the bone marrow, which is an important target organ for the toxic effects of several compounds, including benzene. Chlorzoxazone hydroxylation activities of Wistar rats were similar to that observed in mice. However, rabbit showed significantly higher chlorzoxazone hydroxylation activity than rat and rabbit.

3.7   CYP3A

CYP3A genes have been identified in the mouse, rat, rabbit, dog, minipig, monkey and human.[40, 41] Five CYP3A enzymes (CYP3A1, 3A2, 3A9, 3A18, and 3A23) have been detected in rats,[90, 91] and six CYP3a enzymes (Cyp3a11, 3a13, 3a16, 3a25, 3a41, and 3a44) have been reported in mouse.[42] Only two forms (CYP3A12 and 3A26) have been found in dog and a single form (CYP3A6) is expressed in the rabbit.[42] Approximately one-half of the pharmaceuticals on the market today are metabolized by human CYP3A4.[31] However, comparison of the catalytic activities of CYP3A enzymes from different species has been limited in part because of the initial difficulties encountered during CYP3A heterologous expression and enzyme reconstitution.[92]

Testosterone 6β-hydroxylation is a commonly used marker activity for CYP3A enzymes.[91] Rat CYP3A enzymes exhibit high rates of testosterone 6β-hydroxylation, but also catalyze the formation of 2β- and 16β-hydroxytestosterone. 6β-hydroxylation of testosterone showed no significant differences among various species with the exception of the dog (higher K_m value) [93] and the female rat (higher K_m and lower V_{max} values).[63]   Rat has a sex-related difference in hepatic CYP3A activity.[94] For example, testosterone 6β-hydroxylation in female rats is much lower than in male rats with K_m and V_{max} values 7-fold higher and 9-fold lower, respectively, in female rats. Interestingly, CYP2C forms are thought to contribute to the formation of 6β-hydroxytestosterone in the mouse, rat, rabbit, dog, micropig and monkey,[63] and rabbit CYP2C3 is known to be the predominant steroid 6β-hydroxylase in that species.[39] CYP2C forms do not contribute to testosterone 6β-hydroxylation in humans, suggesting that CYP2C and CYP3A enzymes share important similarities in laboratory animal species that are not present in man. With respect to observed kinetic parameters in liver microsomes for testosterone 6β-hydroxylation, mouse and male rat appear to be the most similar species to man (Table 3).[63]

However, Born *et al.*[95] and Fraser *et al.*[96] demonstrated that canine CYP3A12 catalyzes the hydroxylation of steroids including progesterone, testosterone, and androstenedione at rates comparable with human CYP3A4. In contrast, canine CYP3A26 has much lower 6β-hydroxylase activity, although CYP3A12 and 3A26 exhibit 96% amino acid sequence identity.[96] For example CYP3A26 displays only 2% of the activity of CYP3A12 for the formation of 6β-hydroxyprogesterone, but 22% of the CYP3A12 activity for 2β-hydroxytestosterone.[96] Interestingly, rabbit CYP3A6 metabolizes steroids and macrolides at rates comparable to those of canine CYP3A12.[95]

Information regarding the metabolic activities of each mouse Cyp3a enzyme is limited at present. However, it is known that microsomal testosterone 6β-hydroxylation is higher in females than in males, perhaps due to Cyp3a41 activity, which is a female- specific form.[97]

Midazolam, another marker substrate of CYP3A4, showed significant interspecies variability in metabolism when compared in dogs, cynomolgus monkeys, rhesus monkeys, and humans.[98] Higher rates of midazolam 1'-hydroxylation were observed in dogs and monkeys in comparison with humans, with the clearance in dog being 11-fold higher than in man, and 7-fold higher than in both monkey species. No significant differences were observed between cynomolgus and rhesus monkey.

3.8   CYP4A

The CYP4A subfamily includes twenty isoforms in nine different mammalian species.[99] CYP4A enzymes are mainly involved in the ω-hydroxylation of fatty acids, prostaglandins, and eicosanoids, and they are inducible by hypolipidemic agents such as clofibrate and lovastatin.[100, 101] The induction of CYP4A-mediated lauric acid 12(ω)-hydroxylation is associated with peroxisome proliferation in rodents.[102, 103, 104, 105]

Lauric acid 12(ω)-hydroxylase activity appears to be well conserved in mouse, rat, rabbit, dog, micropig, monkey and humans (Table 3).[63] In addition to catalyzing the hydroxylation of the terminal ω-carbon, CYP4A enzymes hydroxylate the (ω-1)-position of saturated and unsaturated fatty acids.[106, 107, 108] However, other forms of cytochrome P450 enzymes as CYP2A1, 2C6, 2C11, and 2E1 also hydroxylate fatty acids such as lauric acid at the (ω-1)-position.[109]

In rats, the CYP4A subfamily consists of CYP4A1, 4A2, 4A3, and 4A8.[110] CYP4A1 hydroxylates palmitate and laurate at the ω and ω-1 positions, with the ω/ω-1 ratio being 9-fold greater for laurate hydroxylation.[108] For CYP4A3, the ω/ω-1 ratio for laurate and palmitate hydroxylation shows only a three-fold difference.

The rabbit expresses four CYP4A enzymes (CYP4A4, 4A5, 4A6, and 4A7).[111, 112] These enzymes exhibit 85% sequence identity and have overlapping substrate specificities.[113] The substrate spectrum of CYP4A4 is unique, as it metabolizes long chain fatty acids and prostaglandins, whereas it oxidizes lauric acid only at low rates or not at all.[114] CYP4A5, 4A6, and 4A7 preferentially hydroxylate lauric acid at the ω-position, but only CYP4A6 and 4A7 catalyze arachidonic acid hydroxylation at the ω-position. CYP4A5, which is the most abundant of the CYP4A forms in rat liver and kidney, and is constitutively expressed, resembles the substrate specificity patterns of human CYP4A11.[113, 115] Like CYP4A5, CYP4A11 catalyzes primarily the ω-hydroxylation of medium-chain-length fatty acids, e.g. lauric acid.[116]

4   REGULATION OF LABORATORY ANIMAL CYTOCHROME P450

Cytochrome P450 expression is modulated by endogenous hormones, cytokines, and xenobiotics. In this section, species differences in the regulation and induction of P450s involved in drug metabolism will be discussed, focusing on differences in receptor ligand-binding domains and induction mechanisms.

4.1   Regulation of CYP1 genes

CYP1A1, 1A2 and 1B1 are regulated by the aryl hydrocarbon receptor (Ah receptor/ AHR). The binding of polycyclic or halogenated aromatic hydrocarbons (PAH and HAH, respectively) to the AHR results in the dissociation of the AHR/ substrate unit from a complex comprised of two molecules of a 90-kD heat-shock protein (hsp90) and other proteins. The AHR associates with the AHR nuclear translocator protein (ARNT) to generate the heterodimeric aryl hydrocarbon receptor complex (AHRC).[117, 118] Activation of the AHRC occurs through interaction with dioxin-responsive enhancer (DRE) sequences located within the promoter regions of CYP1 genes. This interaction disrupts the chromatin structure and allows the association of other factors involved in transcriptional initiation.[119, 120] This regulatory process is well conserved across species and is responsible for the induction of CYP1 genes by PAH.[118, 121]

The AHR has been identified, cloned and characterized in several mammalian species (Table 4).[122]

Species differences have been observed relative to the abundance of AHR, and the binding affinity of AHRs towards TCDD, 3-methylcholanthrene (3-MC) and benzo[a]pyrene in rat, hamster, rabbit, mouse, and guinea pig.[123, 124] Although no clear relationship exists between CYP1 inducibility and AHR expression, affinity or molecular size,[125] a relationship between the facility with which the transformation of the AHR occurs and TCDD toxicity has been observed. For example, AHRs from guinea pigs (TCDD sensitive) are more easily transformed than those of hamsters (TCDD resistant).[126, 127]

Much of our current understanding of AHR-mediated CYP1A1 regulation is due to studies conducted in mice.[128] Mouse and human AHR and ARNT proteins are approximately 20% identical in amino acid sequence and are similar in structure.[118] Mice exhibit a strain-dependent genetic polymorphism in Cyp1a inducibility, [129, 130] with the DBA/2 strain being resistant to Cyp1a induction by 3-MC, and the C57BL/6 strain being highly responsive to 3-MC.[131] Although some reports suggest that the lack of Cyp1a induction by PAHs in DBA/2 mouse liver is due to the rapid metabolism and elimination of these compounds,[118] other data suggest that strain differences are due to differences in ligand binding affinity.[132]

Table 4. Ah receptor in mammalian species[122]

Species	Cloning/ sequencing	Size (kDa)[a]
Human (*Homo sapiens*)	+	106-110
Monkey (*Macaca mulatta*)		113
Mouse (*Mus musculus*, several strains)	+ (2 forms)	95-104
Deer mouse (*Peromyscus maniculatus*)		130
Rat (*Rattus norvegicus*)	+	101-106
Rabbit (*Oryctolagus cuniculus*)	+	104
Hamster (*Mesocricetus auratus*)		124
Guinea pig (*Cavia porrcellus*)		103

(a) Molecular mass of protein as estimated on denaturing polyacrylamide gels. This mass may differ somewhat as compared to mass predicted from the deduced amino acid sequence.

4.2 Regulation of CYP2B genes

Although the induction of CYP2B enzymes by phenobarbital (PB) has been studied for more than 35 years, key induction pathway elements were only recently discovered. Using mouse primary hepatocytes in culture, Honkakoski and Negishi[133] examined the transcriptional inducibility of Cyp2b genes by PB-like chemicals, and found that the the phenobarbital-responsive enhancer (PBRE) of Cyp2b10 is highly similar to that of rat CYP2B2, suggesting that PBRE sequences are conserved across species. Further, the resistance of Cyp2b9 to BP-mediated induction was attributed to mutations within the PBRE.[133] Another major breakthrough was the discovery of the constitutive active receptor (CAR)[134, 135] as a key factor that binds PBRE and induces PB-associated genes in transfected mouse hepatocyte cell lines.[136] Based on these and other studies, it is currently thought that PB-induced transcription of CYP2B genes results from the translocation of cytoplasmatic CAR into the hepatocyte nucleus, where CAR forms a heterodimer with retinoid X receptor and activates the NR1 enhancer within the PBRE.[136, 137]

A similar regulatory paradigm for CYP2B genes has been identified in humans. The 16-bp NR1 site of human PBRE differs by only one base from mouse NR1, making NR1 the most conserved sequence between the PBRE elements of mouse Cyp2b10, rat CYP2B1 and 2B2, and human CYP2B6.[138, 139] As in rodents, human CAR translocates into the nucleus in response to PB, and binds to PBRE sequences. However, human and rodent CAR share only 70% amino acid identity in their ligand binding domains,[134, 135] and this might result in marked pharmacological differences between rodent and human orthologs. Studies showed that treatment of rat with PB increases CYP2B1 up to 50-100-fold, whereas human CYP2B6 is induced to a much lesser extent than the rodent counterparts.[140] Species differences also exist in the regulation of CAR by steroid hormones and other CAR repressors. In the mouse and rat, CAR is activated or repressed in response to endogenous steroid hormones. Estrogen activates CAR, leading to the induction of CYP2B genes. In contrast, androstenol, progesterone, and androgens repress CAR activation.[117, 141, 142] Since estrogens and androgens are metabolized by CYP2B enzymes, CYP2B induction leads to increased estrogen metabolism, resulting in lowered estrogen levels. It is interesting that human CAR is not inhibited by androgens and other rodent CAR repressors, and does not respond well to estrogens under experimental conditions.[141, 142]

Significant sex differences in hepatocellular CAR levels have been identified. CAR expression is extremely low in hepatocyte nuclei of PB-induced female rats, as compared to male rats.[143] In ovariectomized female mice, treatment with estradiol at pharmacological levels resulted in the accumulation of CAR in the nucleus. In contrast, accumulation of CAR in the nucleus was prevented by endogenous levels of androgens in male mice.[142] These sex differences might also occur in humans, as significant differences in plasma androstane levels have been reported in men and women.[144]

Compounds such as clotrimazole and 1,4-bis[2-(3,5-dichloropyridyl-oxy-)] benzene differ in their pharmacological effects on mouse and human CAR, and also exert differential effects on the pregnane X receptor (PXR) in these species. These data suggest that certain compounds may regulate both CYP2B expression (via CAR) and CYP3A expression (via PXR) through common nuclear receptor signaling pathways.[145]

4.3   Regulation of CYP3A genes

CYP3A forms metabolize a wide range of structurally diverse drug compounds, and are induced by steroids, antibiotics and other substrates.[146] Historically, concomitant clinical use of multiple CYP3A substrates has resulted in serious drug interactions, a fact that emphasizes the importance of understanding CYP3A specificity and regulation.

As with other P450s, CYP3A regulation differs between species. For example, the thiazolidinedione antidiabetic drug troglitazone was developed using rodent models of insulin resistance, and did not display toxicity in this species.[147, 148] However, troglitazone administration to humans resulted in heightened CYP3A activity[149] and the increased clearance of other CYP3A substrates, such as oral contraceptives. The failure of the rodent model to predict this response in humans is due to the fact that troglitazone activates the human pregnane X receptor (PXR), which regulates CYP3A genes, whereas this compound does not bind to the mouse or rat PXR. Indeed, many species differences have been observed in the effects of xenobiotics on CYP3A expression.[150, 151] Rat, rabbit, and human CYP3A enzymes are each induced by dexamethasone, but the anti-gucocorticoid pregnenolone 16α-carbonitrile (PCN) is an efficient CYP3A inducer only in the rat (Table 5).[152]

Species differences in CYP3A regulation appear to result from differences in dexamethasone-responsive sequence motifs and PXR amino acid sequences. The rat CYP3A2 and CYP3A23 genes contain a DR3 motif (direct repeat, separated by 3 bp; AGTTCA-N_3–AGTTCA) whereas human CYP3A4, CYP3A5 and rabbit CYP3A6 contain an ER6 motif (everted repeat, separated by 6 bp; TGAACT-N_6–AGGTCA) (Table 5).[151] With regard to PXR, rat and mouse PXR share 97% amino acid sequence identity and are similarly regulated. Although the DNA-binding domains of rabbit, rodent, and human PXR are 95% identical, the LBD regions share only about 80% amino acid identity. Species differences in PXR activation have been examined via CYP3A gene induction in primary hepatocytes.[153] Rifampicin and troglitazone are efficient activators of human and rabbit PXR, but the compounds have little effect on rat and mouse PXR.[153] In addition, PCN activates rat and mouse PXR much more efficiently than the human and rabbit receptor (Table 5). Although these results suggest that rabbit PXR is closely related to the human ortholog,[154] rabbit and human PXR differ markedly in their response to synthetic steroids such as dexamethasone and cyproterone acetate. As already observed for the activation of CAR, there might be cross-talk between the various receptor signaling pathways. For example, there is evidence that mouse Cyp3a induction is not mediated exclusively by PXR, and that the PB-mediated increase in Cyp3a11 expression might be regulated by CAR.[136, 137, 145]

Table 5. Induction of CYP3A in different species

Species	CYP3A	AGGCA-based DNA response element[a]	CYP3A inducing agents
Man	3A4, 3A5, 3A7	ER6	dexamethasone, rifampicin, troglitazone
Rabbit	3A6	ER6	dexamethasone, rifampicin, troglitazone
Rat	3A1, 3A, 3A23	DR3	dexamethasone, PCN
Mouse	3a11	DR3	dexamethasone, PCN

(a) Shown are the hexameric repeat motifs, including the indicated number of base pair spacing between repeats.

4.4   *Regulation of CYP4A genes*

CYP4A gene expression is induced in both liver and kidney by acidic xenobiotics, hypolipidemic fibrate drugs and phthalate ester plasticizers.[155] The peroxisome proliferator-activated receptor-α (PPARα) is responsible for CYP4A induction, and is named for the fact that receptor binding also induces peroxisomal enzymes and hepatic peroxisome proliferation.[156] Although long-term administration of peroxisome proliferators causes liver cancer in animals, chronic use of hypolipidemic drugs such as gemfibrozil and fenofibrate is not associated with increased cancer risk in humans. Several studies report large species differences in sensitivity to peroxisome proliferators ,[157, 158, 159] with the mouse and rat being most sensitive > rabbits and hamsters > Rhesus monkeys and humans.[160, 161] Guinea pigs and marmosets do not respond to peroxisome proliferating agents.[162]

Despite these differences in response to peroxisome proliferating agents, PPARα appears to be highly conserved,[163] and activates transcription of responsive genes by binding to peroxisome proliferator response elements (PPREs) as a heterodimer with RXR in all species.[164, 165, 166] In comparison to the highly-responsive mouse liver, human liver contains 10-fold less PPARα mRNA,[167, 168] and a fraction of this mRNA lacks exon 6 and does not encode a functional receptor protein.[169] This reduced expression of functional human PPARα could be responsible for the low sensitivity of humans to the hepatocarcinogenic actions of peroxisome proliferator chemicals.

CYP4A mRNA expression may also be induced in rodents in response to inflammatory processes and fasting. Not only are these effects PPARα dependent,[159] but they are also tissue- and species-specific. In mouse, treatment with bacterial endotoxin lipopolysaccharide induces Cyp4a10 and 4a14 expression in kidney, but not in liver. In the F344 rat, CYP4A mRNAs are induced in both tissues.[170]

Clearly, understanding species differences in cytochrome P450 regulation can be key to predicting or elucidating mechanisms involved in drug interactions and toxicity.

The following section describes examples of how animal models may be genetically altered, via transgenesis or knockout techniques, and used to answer specific questions about enzyme specificity or regulation.

5   NON-NATIVE ANIMAL MODELS

Characterization of an individual cytochrome P450 form in an intact animal is difficult due to overlapping enzyme substrate specificities. Genetically engineered animal models, such as the gene knockout mouse and the transgenic mouse, are powerful tools in the elucidation of metabolic and toxic pathways *in vivo*. "Humanized" mice, created by knocking out rodent genes and replacing them with their human counterparts, are currently being examined as models that may provide metabolism data directly relevant to humans.

5.1   The gene knockout mouse model

The involvement of cytochrome P450 enzymes in the bioactivation of xenobiotics can be studied through the knockout of cytochrome P450 genes, the knockout of receptors responsible for cytochrome P450 regulation, or through the targeted interruption of other genes that might influence cytochrome P450 expression and activity. Gene knockout animal models are suitable for large scale *in vivo* screening of potentially toxic drugs, and can be used as bioassay models for drug metabolism and carcinogenicity.

5.1.1 Knockout of cytochrome P450 genes

Cytochrome P450 knockout mouse models are invaluable tools in the field of toxicology, providing definitive data on the involvement of an enzyme in a particular metabolic pathway. Therefore, the interruption of cytochrome P450 genes involved in the bioactivation of non-toxic molecules into toxic compounds is of major interest.

CYP1A1 is involved in the epoxidation of polycyclic aromatic hydrocarbons.[33, 35] The knockout of Cyp1a1 in mice suggested a paradoxical involvement of this enzyme in benzo[*a*]pyrene-induced toxicity.[171] The absence of Cyp1a1 appears to protect the animal from benzo[*a*]pyrene-mediated liver toxicity and death by decreasing the formation of large amounts of toxic metabolites. However, the slower metabolic clearance of benzo[*a*]pyrene in Cyp1a1 (-/-) mice resulted in a larger number of benzo[*a*]pyrene-DNA adducts.

The Cyp1a2-deficient mouse line[172] has been used to define marker substrates of this important enzyme. In animals deficient in Cyp1a2, the half-life of caffeine was 7-times longer than in wild-type strains,[173] and production of the 3-demethylated xanthine metabolite was substantially decreased compared with normal mice, implicating Cyp1a2 as the main catalyst of this reaction. As the Cyp1a1 gene remains fully active in the Cyp1a2 knockout animal, it represents a tool to discriminate the *in vivo* substrate specificities of CYP1A1 and CYP1A2.

A knockout mouse line has also been used to determine whether Cyp1a1 or Cyp 1b1 catalyzes the bioactivation of 7,12-dimethylbenz[*a*]anthracene (DMBA) and might be responsible for DMBA-mediated tumorgenesis.[174, 175] Although Cyp1a1 was induced to the same extent in both wild-type and Cyp1b1-null mice and would have been expected to contribute to carcinogenesis of DMBA,[176] Cyp1b1-null mice were completely resistant to DMBA-induced tumors. The results established that Cyp1b1 alone is responsible for the carcinogenicity of DMBA.[177]

In a similar manner, knockout mice have been used to identify the relative importance of CYP1A and CYP2E forms in acetaminophen (APAP) bioactivation to the hepatotoxic reactive metabolite, *N*-acetyl-*p*-benzoquinone imine (NAPQI).[178] Assays with native and recombinant enzymes suggested that both CYP1A2 and CYP2E1 were involved in NAPQI formation.[179] Although Cyp2e1-null mice showed lower sensitivity to acetaminophen toxicity than Cyp1a2-null mice, protection against acetaminophen toxicity was highest when both enzymes were deleted, which indicated that both enzymes play a significant role in NAPQI formation.[178]

Interestingly, although cytochrome P450 expression is critical to the clearance of xenobiotics, most P450 knockout mice do not exhibit significantly abnormal phenotypes, suggesting that these enzymes have no critical role in mammalian development and physiological homeostasis.[177]

5.1.2 *Knockout of receptor genes responsible for cytochrome P450 regulation*

In contrast to P450 knockout mice, the disruption of genes involved in cytochrome P450 regulation has resulted in phenotypic differences from wild-type mice. Null mouse lines have been generated for receptor genes important in the regulation of CYP1A1/2, CYP2B, CYP3A and CYP4A.

Knockout of the AHR gene was achieved independently by two different groups.[180, 181] Despite some differences, both AHR-deficient mouse lines have remarkable phenotypes, such as impaired development of the liver and a very low constitutive hepatic expression of AHR-regulated genes, including CYP1A2. Administration of dioxin to AHR (-/-) mice fails to induce Cyp1a1 or Cyp1a2.[182] Thus, these knockout mice represent a novel model system to characterize AHR-dependent and independent effects of environmental toxicants. However, the hepatic defects observed in AHR knockout mice limit their application in toxicological studies because of the lack of control mice with normal AHR expression but similar liver deficits.

Deletion of the CAR gene has provided a mechanism to examine the impact of CYP2B enzymes on xenobiotic metabolism and toxicity. Indeed, studies in CAR-null mice suggested that modulation of CAR activity in humans might significantly affect the metabolism of drugs and other xenobiotics.[183] In mice lacking the CAR gene, Cyp2b10 gene expression was not induced by phenobarbital , and phenobarbital-like inducers did not increase the hepatotoxic effects of cocaine.[183] Furthermore, knockout animals showed decreased biotransformation of zoxazolamine, a muscle relaxant, resulting in a prolonged duration of zoxazolamine-mediated paralysis. Interestingly, although CAR and PXR have some overlapping regulatory roles,[145] no compensatory effects of PXR was observed in the CAR null mouse.

PXR, which regulates CYP3A genes, can be activated by pregnane-16α-carbonitrile (PCN), leading to induction of the hepatic Cyp3a11 gene in mice.[184] These effects are totally absent in PXR knockout mice,[185] which confirmed the crucial role of PXR in the induction of CYP3A. Interestingly, the deletion of PXR increased basal Cyp3a11 transcription and CYP3A activity in one of the knockout mouse lines indicating a possible compensatory action of CAR.[185] However, this cross-talk between PXR and CAR is controversial.[186, 187]

PPARα-deficient mouse models have played an important role in the understanding of lipid metabolism. These mice are heavier than their wild-type counterparts and have an increased abundance of fat droplets in liver and kidney.[188] After treatment with clofibrate, Cyp4a induction was not observed, confirming the important role of PPARα in the induction of these fatty acid metabolizing enzymes. PPARα-deficient mice also exhibited significantly higher basal levels of serum cholesterol, in particular high density lipoprotein cholesterol, compared with wild-type controls.[189]

5.1.3   Knockout of other genes that influence cytochrome P450 expression

The knockout of genes that encode transporter proteins and inflammatory molecules has demonstrated the interdependence between apparently unrelated pathways. For example, the P-glycoprotein (Pgp) efflux pump is known to modulate the hepatocellular concentration of CYP3A substrates and inducers, including glucocorticoids and macrolide antibiotics.[190, 191] Studies in mice lacking the *mdr1a* gene, which encodes the Pgp protein, showed that Pgp is a major determinant of rifampicin-inducible expression of Cyp3a in mice.[192] Upon oral treatment with rifampicin the drug concentration in the livers of *mdr1a* null mice was higher than in the wild-type controls. Consistent with enhanced levels of rifampicin in the *mdr1a* null mice, lower doses of rifampicin were able to induce Cyp3a protein expression to a greater extent in the knockout mice compared with control mice.

A knockout mouse deficient in tumor necrosis factor-alpha (TNFα) signaling aided in elucidating the relationship between the inflammatory mitogenic effect of phenobarbital and the induction of hepatic CYP2B and CYP3A.[193] Phenobarbital-mediated Cyp2b and CYP3a induction was significantly higher in TNFα knockout mice than in wild-type mice. In addition, the nuclear accumulation of CAR was elevated, suggesting that endogenous TNFα signaling influences phenobarbital induction of CYP2B by inhibiting the nuclear accumulation of CAR.

Another interesting unexpected interaction among nuclear hormone receptors, transporters, and cytochrome P450 was discovered by using bile acid receptor (BAR) knockout mice.[194] Bile acids regulate gene expression through BAR in hepatic nuclei.[195] In mice lacking BAR, fecal excretion of bile acids is impaired due to decreased expression of the canalicular bile salt transporter, sister of P-glycoprotein (SPGP).[196, 197] Schuetz *et al.*[194] showed that mouse SPGP promoters are regulated by BAR and that loss of SPGP-mediated bile acid efflux leads to compensatory up-regulation of ABC transporters and CYP3A and CYP2B by the activation of PXR.

Although knockout models have provided valuable mechanistic information, this technique fails to address specific issues surrounding human P450 enzymes and their roles in metabolism and toxicity. The humanized mouse model is a novel approach to circumventing certain issues surrounding species differences, and is discussed in the following section.

5.2   The humanized mouse model

Humanized mouse lines offer the possibility to perform experiments in the context of an entire animal, but with the human genes and proteins of interest.

These rodent models overcome many of the limitations of *in vitro* studies, are continuously renewable and completely standardized. By knocking out genes for P450s or receptors and replacing them with their human counterparts, mouse models can be created that display a human drug-response profile and provide a unique tool to examine drug metabolism and predict drug-drug interactions in pre-clinical studies.

The CYP2D6 humanized mouse was one of the earliest humanized P450 models.[198] Debrisoquine metabolism in the humanized CYP2D6 mouse model mimicked the pharmacokinetic and metabolic properties of CYP2D6-mediated biotransformation. Although debrisoquine 4-hydroxylase activity is nearly absent in wild-type mice,[199] this activity is readily detected in the humanized mouse.[198] Debrisoquine has been used as a clinical probe for the CYP2D6 polymorphism. However, many nongenetic factors such as age, sex, smoking, and pathological status can influence drug metabolism and interfere with an experimentally determined phenotype or genotype.[200] The CYP2D6 humanized mouse model offers the possibility to investigate phenotype-genotype correlation under controlled conditions.

As mentioned earlier, the cytochrome P450 genes are not the only factors that affect drug metabolism. To mimic human biotransformation, the receptors that regulate cytochrome P450 induction must also be considered. One such experiment involved disrupting the gene for mouse PXR and replacing it with the human gene.[201] The replacement of a single transcriptional regulator was successful in altering CYP3A expression to fit the human phenotype. In the first step, mice were generated that expressed both mPXR and hPXR/ SXR in their livers, and these animals responded to both the rodent-specific inducer PCN and the human-specific inducer rifampicin. By knocking out the mPXR, the humanized mouse responded only to rifampicin, therefore gaining a complete human profile of CYP3A inducibility.[202]

The humanized mouse represents a major step in generating animal models that can be directly extrapolated to humans. Future generations of humanized P450 models containing more than one human P450 gene will advance the understanding of the complexity of the human drug response.

6   CONCLUSIONS

Over the past decade, considerable advances have been made in the study of cytochrome P450 in laboratory animals. Species differences in catalytic selectivity and enzyme regulation have been investigated, and genetic engineering techniques now make it possible to examine human cytochrome P450 enzymes in intact laboratory animals. These humanized animal models may one day overcome certain species differences, making it possible to study the effects of discrete changes in animal physiology on drug metabolism. However, much remains to be learned about the structure, regulation and function of cytochrome P450 enzymes. With the help of proteomic and other novel experimental tools, protein interactions may be mapped and characterized in animals and humans, increasing our ability to effectively use animal models to predict human drug metabolism.

REFERENCES

1. Caldwell, J. *Drug Metab. Rev.* **1981,** *12,* 221.
2. Wrighton, S.A.; Ring, B.J.; VandenBranden, M. *Toxicol. Pathol.* **1995,** *23,* 199.
3. Boxenbaum, H.; Dilea, C. *J. Clin. Pharmacol.* **1995,** *35,* 957.
4. Sacher, G.A. in *Handbook of the Biology of Aging* (Finch, C. E.; Hayflick, L. eds.), Van Nostrand Reinhold, New York **1977,** 582.
5. Greenberg, M. J. *Am. Zool.* **1985,** *25,* 737.
6. Pearce, R.; Greenway, D; Parkinson, A. *Arch. Biochem. Biophys.* **1992,** *298,* 211.
7. Yamazaki, H.; Mimura, M.; Sugahara, C.; Shimada, T. *Biochem. Pharmacol.* **1994,** *48,* 524.
8. Sharer, J. E.; Shipley, L. A.; Vanderbranden, M. R.; Binkley, S. N.; Wrighton, S. A. *Drug Metab. Dispos.* **1995,** *23,* 1231.
9. Bogaards, J. J. P.; Bertrand, M.; Jackson, P.; Oudshoorn, M. J.; Weavers, R. J.; van Bladeren, P. J.; Walther, B. *Xenobiotica* **2000,** *30,* 1131.
10. Werner, M.; Birner, G.; Dekant, W. *Drug Metab. Dispos.* **1995,** *23,* 861.
11. Tomlinson, E. S.; Maggs, J. J.; Park, B. K.; Back, D. J. *J. Steroid Biochem. Mol. Biol.* **1997,** *62,* 345.
12. McNamara, P. J. in *Pharmaceutical Bioequivalence* (Welling P. G.; Tse F. L. S.; Dighe S. V. eds.), Marcel Dekker, New York, Basel, Hong Kong **1991,** 267.
13. Kato, R. *International Encyclopedia Pharmacology and Therapy.* Oxford: Pergamon Press **1982,** 99.
14. Ritschel, W. A. *S. T. P. Pharm.* **1987,** *3,* 125.
15. Mordenti, J.; Chappel, W. in *Toxicokinetics and New Drug Development* (Yacobi, A.; Skelly, J. P.; Batra, V. K. eds.), Pergamon Press, New York **1989,** 42.
16. Wilkinson, G. R. *Pharmacol. Rev.* **1987,** *39,* 1.
17. Sawada, Y.; Hanano, M.; Sugiyama, Y.; Iga, T. *J. Pharmacokinet. Biopharm.* **1985,** *13,* 477.
18. Boxenbaum, H. *J. Pharmacokinet. Biopharm.* **1982,** *10,* 201.
19. Boxenbaum, H. *Drug Metab. Rev.* **1984,** *15,* 1071.
20. Dedrick, R. L. *J. Pharmacokinet. Biopharm.* **1973,** *1,* 435.
21. Huxley, J. S. *Problems of Relative Growth.* Methuen, London, **1932.**
22. Benet, L. Z.; Øie, S.; Schwartz J. B. in *Goodman and Gilman's The Pharmacological Basis of Therapeutics* (Hardman, J. G.; Goddman Gilman, A.; Limbird, L. E. eds.) McGraw-Hill, Inc., New York **1996,** 1707.
23. Boxenbaum, H. *J. Pharmacokinet. Biopharm.* **1980,** *8,*165.

24. Sacher, G. in *CIBA Foundation Colloquia on Aging* (Wolstenholme, G. E. W.; O'Connor, M. eds.), Churchill, London **1959**, 115.

25. Naritomi, Y.; Terashita, S.; Kimura, S.; Suzuki, A.; Kagayama, A.; Sugiyama, Y. *Drug Metab. Dispos.* **2001**, *29*, 1316.

26. Lave, T.; Dupin, S.; Schmitt, C.; Chou, R. C., Jaeck, D.; Coassolo, P. *J. Pharm. Sci.* **1997**, *86*, 584.

27. Mahmood, I.; Balian, J. D. *J. Pharm. Sci.* **1996**, *85*, 411.

28. Mahmood, I.; Balian, J. D. *Xenobiotica* **1996**, *26*, 887.

29. Mahmood, I. *Drug Metabol. Drug Interact.* **2001**, *18*, 135.

30. Gonzalez, F. J. *Pharmacol. Ther.* **1990**, *45*, 38.

31. Guengerich, F. P. in *Cytochrome P450. Structure, Mechanism, and Biochemistry* (Ortiz de Montellano, P.R. ed.), Plenum Press, New York and London **1995**, 473.

32. Funae, Y.; Imaoka, S. in *Cytochrome P450. Handbook of Experimental Pharmacology,* Vol. 105 (Schenkman, J. B.; Greim, H. eds.), Springer-Verlag, Berlin and Heidelberg **1993**, 221.

33. Gelboin, H. V. *Physiol. Rev.* **1980**, *60*, 1107.

34. Yang, S. K.; Roller, P. P.; Gelboin, H.V. in *Carcinogenesis-A Comprehensive Survey. Polynuclear Aromatic Hydrocarbons: Chemistry, Metabolism and Carcinogenesis,* Vol. 3 (Jones, P.W.; Freudenthal, R. I. eds.), Raven Press, New York **1978**, 285.

35. Shou, M.; Korzekwa, K. R.; Crespi, C. L.; Gonzalez, F. J.; Gelboin, H. V. *Mol. Carcinog.* **1994**, *10*, 159.

36. Jerina, D. M. *Drug Metab. Dispos.* **1983**, *11,*1.

37. McManus, M.E.; Burgess,W. M.; Veronese, M. E.; Huggett, A.; Quattrochi, L. C.; Tukey, R. H. *Cancer Res.* **1990**, *50*, 3367.

38. Snyderwine, E.G.; Battula, *N. J Natl. Cancer Inst.* **1989**, *81*, 223.

39. Guengerich, F. P. *Chemico-Biol. Interact.* **1997**, *106,*161.

40. Nelson, D. R.; Kamataki, T.; Waxman, D. J.; Guengerich, F. P., Estabrook, R. W.; Feyereisen, R.; Gonzalez, F. J.; Coon, M. J.; Gunsalus, I. C.; Gotoh, O.; Okuda, K.; Nebert, D. W. *DNA Cell Biol.* **1993**, *12*, 1.

41. Nelson, D. R.; Koymans, L.; Kamataki, T.; Stegeman, J. J.; Feyereisen, R.; Waxman, D. J.; Waterman, M. R.; Gotoh, O.; Coon, M. J.; Estabrook, R. W.; Gunsalus, I. C.; Nebert, D. W. *Pharmacogenetics* **1996**, *6*, 1.

42. http://drnelson.utmem.edu/Nomenclature.html

43. Watkins, B. E.; Susuki, M.; Wallin, H.; Wakabayashi, K.; Alexander, J.; Vanderlaan, M.; Sugimura, T.; Esumi, H. *Carcinogenesis* **1991**, *12*, 2291.

44. Zhao, K.; Murray, S.; Davies, D. S.; Boobis A. R.; *Gooderham*, N. J. *Carcinogenesis* **1994**, *15*, 1285.

45. Murray, B. P.; Edwards, R. J.; Murray, S.; Singleton, A. M.; Davies, D. S.; Boobis, A. R. *Carcinogenesis* **1993**, *14*, 585.

46. Sesardic, D.; Boobis, A. R.; Murray, B. P., Murray, S.; Segura, J.; De La Torre, R.; Davies, D. S. *Br. J. Clin. Pharmacol.* **1990**, *29*, 651.

47. Adamson, R. H.; Snyderwine, E. G.; Thorgeirsson, U. P.; Schut, H. A.; Turesky, R. J.; Thorgeirsson, S. S.; Takayama, S.; Sugimura, T. **1995**, *Int. Symp. Princess Takamatsu Cancer Research Fund* **1995**, *21*, 289.

48. Edwards, R. J.; Murray, B. P.; Murray, S.; Schultz, T.; Neubert, D.; Gant, T. W.; Thorgeirsson, S. S.; Boobis, A. R.; Davies, D. S. *Carcinogenesis* **1994**, *15*, 829.

49. Honkakoshi, P.; Negishi, M. *Drug Metab. Rev.* **1997**, *29*, 977.

50. Lake, B. G. *Food Chem. Toxicol.* **1999**, *37*, 423.

51. Born, S. L.; Rodriguez, P. E.; Eddy, C. L.; Lehman-McKeeman, L. D. *Drug Metab. Dispos.* **1997**, *25*, 1318.

52. Booth, A. N.; Masri, M. S.; Robbins, D. J.; Emerson, O. H.; Jones, F. T.; DeEds, F. *J. Biol. Chem.* **1959**, *234*, 946.

53. Van Sumere, C. F.; Teuchy, H. *Arch. Int. Physiol. Biochim.* **1971**, *79*, 665.

54. Wood, A. W. *J. Biol. Chem.* **1979**, *254*, 5641.

55. Chang, T. K. H.; Waxman, D. J. in *Cytochrome P450: Metabolic and Toxicological Aspects* (Ioannides, C. ed.), CRC Press, Boca Raton, FL: **1996**, 99.

56. Haugen, D.A.; van der Hoeven, T.A.; Coon, M.J. *J. Biol. Chem.* **1975**, *250, 3567.

57. Kedzie, K.M.; Grimm, S.W.; Chen, F.; Halpert, J. R. *Biochim. Biophys. Acta* **1993**, *1164*, 124.

58. Strobel, S. M.; Halpert, J. R. *Biochemistry* **1997**, *36*, 11697.

59. Szklarz, G. D.; He,Y. Q.; Kedzie, K. M.; Halpert, J. R.; Burnett, V. L. *Arch. Biochem. Biophys.* **1996**, *327*, 308.

60. Ryan, R.; Grimm, S.W.; Kedzie, K.M.; Halpert, J.R.; Philpot, R.M. *Arch. Biochem. Biophys.* **1993**, *304*, 454.

61. Hasler, J. A.; Harlow, G. R.; Szklarz, G. D.; John, G. H.; Kedzie, K. M.; Burnett, V. L.; He Y.A.; Kaminsky, L. S.; Halpert, J. R. *Mol. Pharmacol.* **1994**, *46*, 338.

62. Hanna, I. H.; Reed, J. R.; Guengerich, F. P.; Hollenberg, P. F. *Arch. Biochem. Biophys.* **2000**, *376*, 206.

63. Bogaards, J. J. P.; Bertrand, M.; Jackson, P.; Oudshoorn, M. J.; Weaver, R. J.; van Bladeren, P. J.; Walther, B. *Xenobiotica* **2000**, *30*, 1131.

64. Clarke, L.; Waxman, D. J. *Cancer Res.* **1989**, *49*, 2344.

65. Yu, L.; Waxman, D. J. *Drug Metab. Dispos.* **1996**, *24*,1254. 66. Waxman, D. J.; Azaroff, L. *Biochem. J.* **1992**, *281*, 577.

67. Johnson, E.F.; Kronbach, T.; Hsu, MH *FASEB J.* **1992**, *6*, 700.

68. Kronbach, T.; Larabee, T.M.; Johnson, E.F. *Proc.Natl. Acad. Sci. U.S.A.* **1989**, *86*, 8262 .

69. Endoh, A.; Natsume, H.; Igarashi, Y. *J. Steroid Biochem. Mol. Biol.* **1995**, *54*, 163.

70. Richardson, T. H.; Jung, F.; Griffin, K. J.; Wester, M.; Raucy, J. L.; Kemper, B.; Bornheim, L. M.; Hassett, C.; Omiecinski, C. J.; Johnson, E. F. *Arch. Biochem. Biophys.* **1995**, *323*, 87.

71. Dieter, H. H.; Muller-Eberhard, U.; Johnson, E. F. *Biochem. Biophys. Res. Commun.* **1982**, *105, 515.

72. Miller, W.L.; Morel, Y. *Ann. Rev. Genet.* **1989**, *23*, 371.

73. Waxman, D. J.; Attisano, C.; Guengerich, F. P.; Lapenson, D. P. *Arch. Biochem. Biophys.* **1988**, *263*, 424.

74. Johnson, E. F.; Kronbach, T.; Hsu, M. H. *FASEB J.* **1992**, *6*, 700.

75. Yasumori, T.; Chen, L.; Nagata, K.; Yamazoe, Y.; Kato, R. *J. Pharmacol. Exp. Ther.* **1993**, *264*, 89.

76. Chauret, N.; Gauthier, A.; Martin, J.; Nicoll-Griffith, D. A. *Drug Metab. Dispos.* **1997**, *25*, 1130.

77. Akrawi, S. H.; Wedlund, P. J. *Eur. J. Drug Metab. Pharmacokin.* **1989**, *14,*195.

78. Yasumori, T.; Chen, L.; Nagata, K.; Yamazoe, Y.; Kato, R. *J. Pharmacol. Exp. Ther.* **1993,** *264,*89.

79. Gonzalez, F. J.; Skoda, R. C.; Kimura, S.; Umeno, M.; Zanger, U. M.; Nebert, D.W.; Gelboin, H.V.; Hardwick, J. P.; Meyer, U. A. *Nature* **1988,** *331,* 442.

80. Matsunaga, E.; Zanger, U. M.; Hardwick, J. P.; Gelboin, H. V.; Meyer, U. A.; Gonzalez, F. J. *Biochemistry* **1989,** *28,* 7349.

81. Strobl, G. R.; von Kruedener, S.; Stoeckigt, J.; Guengerich, F. P.; Wolff, T. *J. Med. Chem.* **1993,** *36,* 1136.

82. Roussel, F.; Duignan, D. B.; Lawton, M. P.; Obach, R. S.; Strick, C. A.; Tweedie, D. J. *Arch. Biochem. Biophys.* **1998,** *357,* 36.

83. Ding, X.; Pernecky, S. J.; Coon, M. J. *Arch. Biochem. Biophys.* **1991,** *291,* 270.

84. Gonzalez, F. J.; Gelboin, H. V. *Drug Metab. Rev.* **1994,** *26,* 165.

85. Clarke, S. E.; Baldwin, S. J.; Bloomer, J. C.; Ayrton, A. D.; Sozio, R. S.; Chenery, R. J. *Chem. Res. Toxicol.* **1994,** *7,* 836.

86. Amet, Y.; Berthou, F.; Baird, S.; Dreano, Y.; Bail, J. P.; Menez, J. F. *Biochem. Pharmacol.* **1995,** *50,* 1775.

87. Jayyosi, Z.; Knoble, D.; Muc, M.; Erick, J.; Thomas, P. E.; Kelley, M. *J. Pharmacol. Exp. Ther.* **1995,** *273,* 1156.

88. Koop, D.R. *FASEB J.* **1992,** *6,* 724.

89. Bernauer, U.; Vieth, B.; Ellrich, R.; Heinrich-Hirsch, B.; Jänig, G. R.; Gundert-Remy, U. *Arch. Toxicol.* **2000,** *73,* 618.

90. Nagata, K.; Gonzalez, F. J.; Yamazoe, Y.; Kato, R. *J. Biochem.* **1990,** *107,* 718.

91. Mahnke, A.; Stronkamp, D.; Roos, P. H.; Hanstein, W. G.; Chabot, G. G.; Nef, P. *Arch. Biochem. Biophys.* **1997,** *337,* 62.

92. Imaoka, S.;Imai, Y.; Shimada, T.; Funae, Y. *Biochemistry* **1992,** *31,* 6063.

93. Ciaccio, P. J.; Halpert, J. R. *Arch. Biochem. Biophys.* **1989,** *271,* 284.

94. Tomlinson, E. S.; Maggs, J. J.; Park, B. K.; Back, D. J. *J. Steroid Biochem. Mol. Biol.* **1997,** *62,* 345.

95. Born, S. L.; Fraser, D. J.; Harlow, G.R.; Halpert; J. R. *J. Pharmacol. Exp. Ther.* **1996,** *278,* 957.

96. Fraser, D. J.; Feyereisen, R.; Harlow, G. R.; Halpert, J. R. *J. Pharmacol. Exp. Ther.* **1997,** *283,* 1425.

97. Sakuma, T.; Takai, M.; Endo, Y.; Kuroiwa, M.; Ohara, A.; Jarukamjorn, K.; Honma, R.; Nemoto, N. *Arch. Biochem. Biophys.* **2000,** *377,*153.

98. Sharer, J. E.; Shipley, L. A.; Vandenbranden, M. R.; Binkley, S. N.; Wrighton, S. A. *Drug Metab. Dispos.* **1995,** *23,* 1231.

99. Okita, R. T.; Okita, J. R. *Curr. Drug. Metab.* **2001,** *2,* 265.

100. Amacher, D. E.; Beck, R.; Schomaker, S. J.; Kenny, C. V. *Toxicol. Appl. Pharmacol.* **1997,** *142,* 143.

101. Kocarek, T. A.; Reddy, A. B. *Drug Metab. Dispos.* **1996,** *24,* 1197.

102. Sharma, R.; Lake, B. G.; Foster, J.; Gibson, G. G. *Biochem. Pharmacol.* **1988,** *37,* 1193.

103. Dirven, H. A. A. M.; de Bruijn, A. A. G. M.; Sessink, P. J. M.; Jongeneelen, F. J. *J. Chromatogr.* **1991,** *564,* 266.

104. Makowska, J. M.; Gibson, G. G. *J. Biochem. Toxicol.* **1992,** *7,* 183.

105. Amet, Y.; Adas, F.; Nanji, A. A. *Clin. Exp. Ther.* **1998,** *22*, 1493.

106. Bains, S. K.; Gardiner, S. M.; Mannweiler, K.; Gillett, D.; Gibson, G. G. *Biochem. Pharmacol.* **1985,** *34*, 3221.

107. Sharma, R. K.; Doig, M. V.; Lewis, D. F. V.; Gibson, G. G. *Biochem. Pharmacol.* **1989,** *38*, 3621.

108. Aoyama, T.; Hardwick, J. P.; Imaoka, S.; Funae, Y.; Gelboin, H. V.; Gonzalez, F. J. *J. Lipid Res.* **1990,** *31*, 1477.

109. Tanaka, S.; Imaoka, S.; Kusunose, E.; Kusunose, M.; Maekawa, M.; Funae, Y. *Biochim. Biophys. Acta* **1990,** *1043*, 177.

110. Bleicher, K. B.; Pippert, T. R.; Glaab, W. E.; Skopek, T. R., Sina, J. F.; Umbenhauer, D. R. *J. Biochem. Mol. Toxicol.* **2001,** *15*, 133.

111. Johnson, E. F.; Walker, D. L.; Griffin, K. J.; Clark, J. E.; Okita, R. T.; Muerhoff, A. S.; Masters, B. S. S. *Biochemistry* **1990,** *29*, 873.

112. Loughran, P. A.; Roman, L. J.; Aitken, A. E.; Miller, R. T., Masters, B. S. S. *Biochemistry* **2000,** *39*, 15110.

113. Roman, L. J.; Palmer, C. N. A.; Clark, J. E.; Muerhoff, A. S.; Griffin, K. J.; Johnson, E. F.; Masters, B. S. S. *Arch. Biochem. Biophys.* **1993,** *307*, 57.

114. Nishimoto, M.; Clark, J. E.; Masters, B. S. S. *Biochemistry* **1993,** *32*, 8863.

115. Powell, P. K., Wolf, I.; Lasker, J. M. *Arch. Biochem. Biophys.* **1996,** *335*, 219.

116. Hosny, G.; Roman, L. J.; Mostafa, M. H.; Masters, B. S. S. *Arch. Biochem. Biophys.* **1999,** *366*, 199.

117. Waxman, D. J. *Arch. Biochem. Biophys.* **1999,** *369*, 11.

118. Hankinson, O. *Annu. Rev. Pharmacol. Toxicol.* **1995,** *35*, 307.

119. Whitlock, J. P., Jr. *Chem. Res. Toxicol.* **1993,** *6*, 754.

120. Whitlock, J .P., Jr.; Okino, S. T.; Dong, L.; Ko, H. P.; Clarke-Katzenberg, R.; Ma, Q.; Li, H. *FASEB J.,* **1996,** *10*, 809.

121. Sogawa, K.; Fujii-Kuriyama, Y. *J. Biochem.* **1997,** *122*, 1075.

122. Hahn M. E. *Comp. Biochem. Phys. Part C* **1998,** *121*, 23.

123. Denison, M. S.; Wilkinson, C. F. *Eur. J. Biochem.,* **1985,** *147*, 429.

124. Gasiewicz, T. A.; Rucci, G. *Mol. Pharmacol.,* **1984,** *26*, 90.

125. Poland, A.; Glover, E. *Biochem. Biophys. Res. Commun,* **1987,** *146,* 1439.

126. Harper, P. A.; Giannone, J. V.; Okey, A.B. *Mol. Pharmacol,* **1992,** *42*, 603.

127. Henry, E. C.; Rucci, G.; Gasiewicz, T. A. *Biochemistry,* **1989,** *28*, 6430.

128. Ma, Q. *Curr. Drug Metabol.,* **2001,** *2*, 149.

129. Thomas, P. E.; Kouri, R. E.; Hutton, J. J. *Biochem. Genet.,* **1972,** *6*, 157.

130. Nebert, D. W.; Goujon, K. M.; Gielen, J. E. *Nat. New Biol.,* **1972,** *236*, 107.

131. Poland, A.; Glover, E.; Robinson, J. R.; Nebert, D. W. *J. Biol. Chem.,* **1974,** *249*, 5599.

132. Chang, C-Y.; Smith, D.R. ; Prasad, V. S.; Sidman, C. L.; Nebert, D. W.; Puga, A. *Pharmacogenetics,* **1993,** *3*, 312.

133. Honkakoski, P; Negshi, M; *J. Biol. Chem.,* **1997,** *272*, 14943.

134. Baes, M.; Gulick, T.; Choi,H. S.; Martinoli, M. G.; Simha, D.; Moore, D. D. *Mol. Cell. Biol.,* **1994,** *14*, 1544.

135. Choi, H. S.; Chung, M. ; Tzameli, I. ; Simha, D. ; Lee, Y. K. ; Seol, W. ; Moore, D. D. *J. Biol. Chem.,* **1997,** *272*, 23565.

136. Honkakoski, P.; Moore, R.; Washburn, K. A.; Negishi, M. *Mol. Pharmacol.*, **1998**, *5*, 597.

137. Kawamoto, T.; Sueyoshi, T.; Zelko, I.; Moore, R.; Washburn, K.; Negishi, M. *Mol. Cell. Biol.*, **1999**, *19*, 6318.

138. Corcos, L.; Lagadic-Gossmann, D. *Pharmacol. Toxicol.*, **2001**, *89*, 113

139. Sueyoshi, T.; Kawamoto, T.; Zelko, I.; Honkakoski, P.; Negishi, M. *J. Biol. Chem.*, **1999**, *274*, 6043.

140. Chang, T. K.; Yu, L.; Maurel, P. ; Waxman, D. J. *Cancer Res.*, *57*, 1946.

141. Kawamoto, T. ; Kakizaki S. ; Yoshinari, K.; Negishi, M. *Mol. Endocrinol.*, **2000**, *14*, 1897.

142. Zelko, I.; Negishi, M. *Biochem. Biophys. Res. Commun.*, **2000**, *277*, 1.

143. Yoshinari, K.; Sueyoshi, T.; Negishi, M. *FASEB J. Suppl.*, **2000**, *14*, A1334.

144. Bicknell, D. C.; Gower, D. B.; *J. Steroid Biochem.*, **1976**, *7*, 451.

145. Moore, L. B.; Parks, D. J.; Jones, S. A.; Bledsoe, R. K.; Consler, T. G.; Stimmel, J. B.; Goodwin, B.; Liddle, C.; Blanchard, S. G.; Willson, T. M.; Collins, J. L.; Kliewer, S A. *J. Biol. Chem.*, **2000**, *275*, 15122.

146. Thummel, K. E.; Wilkinson, G. R. *Annu. Rev. Pharmacol.Toxicol.*, **1998**, *38*, 389.

147. Lehmann, J. M., Moore, L. B., Smith-Oliver, T. A., Wilkison, W. O., Willson, T. M.; Kliewer, S. A.; *J. Biol.Chem.*, **1995**, *270*, 12953.

148. Willson, T. M.; Cobb, J. E.; Cowan, D. J.; Wiethe R.W.; Correa, I. D.; Prakash, S. R.; Beck, K. D.; Moore, L. B.; Kliewer, S. A.; Lehmann, J. M. *J. Med. Chem.*, **1996**, *39*, 665.

149. Michalets, E. L. *Pharmacotherapy*, **1998**, *18*, 84.

150. Kocarek, T. A. ; Schuetz, E G.; Strom, S. C.; Fisher, R. A.; Guzelian, P. S. *Drug Metabol. Dispos.*, **1995**, *23*, 415.

151. Barwick, J. L.; Quattochi, L. C.; Mills, A. S.; Potenza, C.; Tukey, R. H.; Guzelian, P. S. *Mol. Pharmacol.*, **1996**, *50*, 10.

152. Wrighton, S. A.; Schuetz, E. G.; Watkins, P. B.; Maurel, P.; Barwick, J.; Bailey, B. S.; Hartle, H. T.; Young, B.; Guzelian, P. *Mol.Pharmacol.*, **1985**, *28,* 312.

153. Jones, S. A.; Moore, L. B. ; Shenk, J. L.; Wisely, G. B.; Hamilton, G. A.; McKee, D. D.; Tomkinson, N. C. O.; LeCluyse, E. L.; Lambert M. H.; Willson, T. M.; Kliewer, S. A., Moore, J. T. *Mol. Endocrinol.*, **2000**, *14*, 27.

154. Blumberg, B.; Sabbagh, W.; Juguilon, H.; Bolado, J.; van Meter, C. M.; Ong, E. S.; Evans, R. M. *Genes Dev.*, **1998**, *12*, 3195.

155. Rao, M. S.; Reddy, J. K. *Carcinogenesis*, **1987**, *8*, 631.

156. Issemann, I.; Green, S. *Nature*, **1990**, *347*, 645.

157. Gonzalez, F. J.; Peters, J. M.; Cattley, R. C. *J. Natl. Cancer Inst.*, **1998**, *90*, 1702.

158. Palmer, C. N. A.; Hsu, M-H.; Griffin, K. J. ; Raucy, J. L. ; Johnson, E. F. *Mol. Pharmacol.*, **1998**, *53*, 14.

159. Barclay, T. B.; Peters, J. M.; Sewer, M. B.; Ferrari, L.; Gonzalez, F. J.; Morgan, E. T. *J. Pharmcol. Exp. Ther.*, **1999**, *290*, 1250.

160. Reddy, J. K.; Lalwani, N. D., Qureshi, S. A.; Reddy, M. K.; Moehle, C. M. *Am. J. Pathol.*, **1984**, *114*, 171.

161. Lalwani, N. D.; Reddy, M. K.; Ghosh, S.; Barnard, S. D.; Molello, J. A.; Reddy, J. K. *Biochem. Pharmacol.*, **1985**, *34*, 3473.

162. Lake, B. G.; Evans, J. G.; Gray, T. J.; Korosi, S. A.; North C. J. *Toxicol. Appl. Pharmacol.*, **1989**, *99*, 148.

163. Dreyer, C.; Krey, G.; Keller, H.; Givel, F.; Heftenbein, G.; Wahli, W. *Cell*, **1992**, *68*, 879.

164. Bardot, O.; Aldridge, T. C.; Latruffe, N.; Green, S. *Biochem. Biophys. Res. Commun.*, **1993**, *192*, 37.

165. Gearing, K. L., Göttlicher , M.; Teboul, M; Widmark, E.; Gustafsson, J-A. *Proc. Natl. Acad. Sci. USA*, **1993**, *90*, 1440.

166. Keller, H.; Dreyer, C.; Medin, J.; Mahfoudi, A; Ozato, K.; Wahli, W. *Proc. Natl. Acad. Sci. USA*, **1993**, *90*, 2160.

167. Mukherjee, R.; Jow, L.; Noonan, D.; McDonnell, D. P. *J. Steroid Biochem. Mol. Biol.*, **1994**, *51*, 157.

168. Sher, T.; Yi, H-F.; McBride, O. W.; Gonzalez, F. J. *Biochemistry*, **1993**, *32*, 5598.

169. Tugwood, J. D.; Aldridge, T. C.; Lambe, K. G.; Macdonald, N.; Woodyatt, N. J. *Ann. N. Y. Acad. Sci.*, **1996**, *804*, 252.

170. Sewer, M. B.; Koop, D. R.; Morgan, E. T. *J. Pharmacol. Exp. Ther.*, **1996**, *280*, 1445.

171. Uno, S.; Dalton, T. P.; Shertzer, H. G.; Genter, M. B.; Warshawsky, D.; Talaska, G.; Nebert, D. W. *Biochem. Biophys. Res. Commun.*, **2001**, *289*, 1049.

172. Pineau, T.; Fernandez-Salguero, P.; Lee, S. S. T.; McPhail, T.; Ward, J. M.; Gonzalez, F. J. *Proc. Natl. Acad. Sci. USA*, **1995**, *92*, 5134.

173. Buters, J. T. M.; Tang, B. K.; Pineau, T.; Gelboin, H.; Kimura, S.; Gonzalez, F. *Pharmacogenetics*, **1996**, *6*, 291.

174. Otto, S.; Marcus, C.; Pidgeon, C.; Jefcoate, C. *Endocrinology*, **1991**, *129*, 970.

175. Crespi, C. L.; Penman, B. W.; Steimel, D. T.; Smith, T.; Yang, C. S.; Sutter, T. R. *Mutagenesis*, **1997**, *12*, 83.

176. Buters, J. T. M.; Sakai, S.; Richter, T.; Pineau, T.; Alexander, D. L; Savas, U.; Doehmer, J.; Ward, J. M.; Jefcoate, C. R.; Gonzalez, F. J. *Proc. Natl. Acad. Sci. USA*, **1999**, *96*, 1977.

177. Gonzalez, F. J.; Kimura, S. *Cancer Lett.*, **1999**, *143*, 199.

178. Zaher, H.; Buters, J. T. M.; Ward, J. M.; Bruno, M. K.; Lucas, A. M.; Stern, S. T.; Cohen, S. D.; Gonzalez, F. J. *Toxicol. Appl. Pharmacol.*, **1998**, *152*, 193.

179. Raucy, J. L.; Lasker, J. M.; Lieber, C. S.; Black, M. *Arch. Biochem. Biophys.*, **1989**, *271*, 270.

180. Fernandez-Salguero, P.; Pineau, T.; Hilbert, D.; McPhail, T.; Lee, S. S. T.; Kimura, S.; Nebert, D. W., Rudikoff, S.; Ward, J. M.; Gonzalez, F. J. *Science*, **1995**, *268*, 722.

181. Schmidt, J. V.; Su, G. H.; Reddy, J. K.; Simon, M. C.; Bradfield, C. A. *Proc. Natl. Acad. Sci. USA*, **1996**, *93*, 6731.

182. McKinnon, R.; Nebert, D. W. *Clin. Exp. Pharmacol. Physiol.*, **1998**, *25*, 783.

183. Wel, P.; Zhang, J.; Egan-hafley, M.; Liang, S.; Moore, D. D. *Nature*, **2000**, *407*, 920.

184. Bertilsson, G.; Heidrich, J.; Svensson, K.; Asman, M.; Jendeberg, L.; Sydow-Backman, M.; Ohlsson, R.; Postlind, H.; Blomquist, P.; Berkenstam, A. *Proc. Natl. Acad. Sci. USA*, **1998**, *95*, 12208.

185. Staudinger, J. L.; Goodwin, B.; Jones, S. A.; Hawkins-Brown, D.; MacKenzie K. I., LaTour, A.; Liu, Y.; Klaassen, C. D.; Brown, K. K.; Reinhard, J. *Proc. Natl. Acad. Sci. USA*, **2001**, *98*, 3369.

186. Xie, W.; Barwick, J. L.; Downes, M.; Blumberg, B.; Simon, C. M.; Nelson, M. C.; Neuschwander-Tetri, B. A.; Brunt, E. M.; Guzelian, P. S.; Evans, R. M. *Nature*, **2000**, *406*, 435.

187. Xie, W.; Barwick, J. L.; Simon, C. M.; Pierce, A. M.; Safe, S.; Blumberg, B.; Guzelian, P. S.; Evans, R. M. *Genes Dev.*, **2000**, *14*, 3014.

188. Lee, S. S.; Pineau ,T.; Drago, J.; Lee, E. J.; Owens, J.W.; Kroetz, D. L.; Fernandez-Salguero, P. M.; Westphal, H.; Gonzalez, F.J. *Mol. Cell. Biol.*, **1995**, *15*, 3012.

189. Peters, J. M.; Hennuyer, N.; Staels, B.; Fruchart, J.C.; Fievet, C.; Gonzalez, F. J.; Auwerx, J. *J. Biol. Chem.*, **1997**, *272*, 27307.

190. Ueda, K.; Okamura, N.; Hirai, M.; Tanigawara, Y.; Saeki, T.; Kioka, N.; Komano, T.; Hori, R. *J. Biol. Chem.*, **1992**, *267*, 24248.

191. Ford, J. M.; Hait, W. N. *Cytotechnology*, **1993**, *12*, 171 .

192. Schuetz, E. G.; Schinkel, A. H.; Relling, M. V.; Schuetz, J. D. *Proc. Natl. Acad.Sci. USA*, **1996**, *93*, 4001.

193. Van Ess, P. J.; Mattson, M. P.; Blouin, R. A. *J. Pharmacol. Exp. Ther.*, **2002**, *300*, 824.

194. Schuetz, E. G.; Strom, S.; K. Yasuda, Lecureur, V.; Assem, M.; Brimer, C.; Lamba, J.; Kim, R. B.; Ramachandran, V.; Komoroski, B. J.; Venkataramanan, R.; Cai, H.; Sinal, C. J.; Gonzalez, F. J.; Schuetz, J. D. *J. Biol. Chem.*, **2001**, *276*, 39411.

195. Makishima, M.; Okamoto, A. Y.; Repa, J. J.; Tu, H.; Learned, R. M.; Luk, A.; Hull, M. V.; Lustig, K. D.; Mangelsdorf, D. J.; Shan, B. *Science*, **1999**, *284*,1362.

196. Sinal, C. J.; Tohkin, M.; Miyata, M.; Ward, J.M.; Lambert, G.; Gonzalez, F. J. *Cell*, **2000**, *102*, 731.

197. Childs, S.; Yeh, R. L.; Georges, E.; Ling, V. *Cancer Res.*, **1995**, *55*, 2029.

198. Corchero, J.; Granvil, C. P.; Akiyama, T. E.; Hayhurst, G. P.; Pimprale, S.; Feigenbaum, L.; Idle, J. R.; Gonzalez, F. J. *Mol. Pharmacol.*, **2001**, *60*, 1260.

199. Masubuchi, Y.; Iwasa, T.; Hosokawa, S.; Suzuki, T.; Horie, T.; Imaoka, S.; Funae, Y.; Narimatsu, S. *J. Pharmacol. Exp. Ther.*, **1997**, *282*, 1435.

200. Wolf, C. R.; Smith, G. *IARC Sci. Publ.*, **1999**, *128*, 209.

201. Xie, W; Barwick, J. L.; Downes, M.; Blumberg, B.; Simon, C. M.; Nelson, M. C.; Neuschwander-Tetri, B. A.; Brunt, E. M.; Guzelian, P. S.; Evans, R. M. *Nature*, **2000**, *406*, 435.

202. Xie, W.; Evans, R. M. *DDT*, **2002**, *7*, 509.

Chapter 7

Typical and Atypical Enzyme Kinetics

J. Brian Houston[1], Kathryn E. Kenworthy[2] and Aleksandra Galetin[1]

[1]*School of Pharmacy and Pharmaceutical Sciences, University of Manchester, Manchester, UK*
[2]*GlaxoSmithKline, The Frythe, Welwyn, Herts, UK*

ABSTRACT

The importance of *in vitro* systems in drug metabolism has increased significantly in recent years. Quantitative kinetic studies can provide predictions of *in vivo* metabolic stability and inhibitory potential of drugs. The choice of *in vitro* system (e.g., hepatocytes, microsomes or heterologous expression systems) and the conditions adopted (e.g., metabolite formation vs. substrate depletion approach, initial rate conditions vs. generic conditions, incubation matrix composition, substrate/enzyme concentration ratio, presence of organic solvents) allow optimisation of the methodology for the desired end-point.

A major assumption for much of the analysis of both clearance and inhibition data is the existence of a single active site and the appropriateness of Michaelis-Menten kinetics. For example, the calculation of intrinsic clearance for a particular metabolic pathway can be calculated from the ratio of the V_{max} and K_m. Similarly for inhibition studies, although IC_{50} values are generated to provide information on inhibitory potential of many compounds, a more accurate and supposedly non-substrate-dependent inhibitor constant (K_i) is more unequivocal but requires adoption of Michaelis-Menten principles for its determination.

This classical approach is valid for many drug metabolizing enzymes but not all. Atypical kinetic features are observed for some CYP substrates (most notably CYP3A4), e.g., autoactivation, heteroactivation, partial inhibition, substrate inhibition and lack of inhibitory reciprocity; these phenomena cannot be readily interpreted using the single active site approach. Evidence of multiple binding sites for CYP3A4 has been indicated by mechanistic studies, site directed mutagenesis and binding studies with non-metabolised peptides. Multi-site kinetic models can be derived and used for the analysis of more complex kinetic and inhibition profiles, using the same rapid equilibrium/steady-state assumptions as the single-site model. Such models suggest that 2 or 3 binding sites may exist for a given CYP3A4 substrate and/or effector. Although relatively complex these models share common features, depending on the substrate kinetics and the drug-drug interactions observed *in vitro*, that allow a generic model to be proposed.

This chapter will discuss both the classical Michaelis-Menten and the multisite approaches used for analysis of the kinetics associated with most drug metabolizing enzymes. Additionally, a number of atypical non-Michaelis-Menten kinetic phenomena are discussed and their interpretation in an *in vivo* context is reviewed. It will also focus on the utility and relevance of typical and atypical enzyme kinetic data in the prediction of *in vivo* clearance and drug-drug interaction potential from *in vitro* data.

1   INTRODUCTION

Over the last decade the use and expectations from *in vitro* systems in the drug metabolism areas has grown exponentially. At the beginning of the 1990s *in vitro* systems were used essentially for either mechanistic studies or as metabolite 'factories'. At the start of the 21st century *in vitro* approaches to drug metabolism are being employed far more extensively and for a wide variety of purposes. Arguably their most valuable use is to provide insight into the cellular processing of drug molecules to form metabolites that allows extrapolation beyond the simple experimental *milieu* to the complex, multifaceted *in vivo* situation. The combination of *in vitro* systems together with *in silico* techniques, either working in parallel as a concerted strategy or offers a predictive approach that is unlikely to be rivalled for several years. *In vitro* systems in the context of this chapter include isolated and recombinant enzymes, cellular and subcellular preparations and precision-cut tissue slices, but not isolated perfused organs.

Three major reasons have contributed to the growth and acceptance of *in vitro* systems by drug metabolism scientists:

(i)   The rapid increase in our understanding of the drug metabolizing enzymes, notably CYP, that has resulted from advances in molecular biology. This has provided a sufficiently detailed background to the drug metabolism processes to build a framework for extrapolation of *in vitro* information.

(ii)   The developments in analytical chemistry that have provided a substantial increase in sensitivity (LC-MS/MS) and allowed rapid throughput of biological samples.

(iii)   The increased availability of human tissue allowing the *in vitro* approaches to be routinely applied to human systems. This has received further impetus from ethical and social pressures, particularly in Europe, for the reduction, refinement and replacement of animals.

The increased use of *in vitro* approaches has played a key role in the 'coming of age' of drug metabolism; the discipline is no longer regarded as a qualitative science but can provide quantitative information that in many cases offers predictive capability.

1.1 General considerations of in vitro experimentation

The selection of the most appropriate *in vitro* systems is often dependent upon available expertise and tissue source rather than scientific reasoning. It is common knowledge that hepatic microsomes do not contain all the drug metabolizing enzymes. Also, even if it is only microsomal enzymes that are of interest, it is difficult to optimize for both CYPs and UGT enzymes at one time. Therefore, the case is often made for hepatocytes as the ideal drug metabolizing system. While this cannot be refuted, this criterion is applicable only to hepatocytes from freshly supplied tissue. There are many outstanding questions over the utility of cultured and cryopreserved cells and few laboratories have local access to fresh human tissue. Human microsomes remain the most used and well understood system. The potential of recombinant enzymes has yet to be fully realized as a routine enzyme source for quantitative studies. However, they offer many attractions as they avoid issues associated with the involvement of multiple enzymes and hence are preferred for mechanistic modelling. They also eliminate dilemmas in the selection of appropriate donor tissue that arises from the large inter-individual expression of the various CYPs in the population.

Perhaps more important than which *in vitro* system to use is the experimental approach to take. There are two distinct approaches commonly used for *in vitro* kinetic experiments: (i) metabolite formation, and (ii) drug substrate depletion. The former is the traditional one where a full characterization of the kinetic properties of one drug substrate(s) is carried out to provide well-defined parameter estimations, for example V_{max}, K_m, CL_{int}, K_i. For this approach metabolite formation kinetics must be obtained under initial rate conditions and this is addressed in detail later in this chapter. However when it is the total CL_{int} for the drug that is required as a measure of metabolic stability, the use of the second approach is an alternative. Both half-lives and CL terms can be derived from a depletion time profile.[1,2]

The obvious and major advantage of the depletion approach is that it requires no prior knowledge of the particular metabolic pathways important for the drug under study. Normally only a single substrate concentration is used, usually a low one in order to reduce the chances of saturating the enzymes. This approach has found favour within the Pharmaceutical Industry for use with new chemical entities synthesized as part of a discovery program. However, for many established drugs the full metabolic profile is also unknown and a measure of total metabolic stability cannot be achieved by any other method. When faced with a need for kinetic information on a large number of drugs in a short period of time, the depletion method is often carried out in a 'high throughput' mode with minimal experimentation to allow many substrates to be studied simultaneously. This is achieved by adopting a generic method with no optimization of conditions. While generating rapid data, such approaches can only screen and are best used to answer specific questions rather than for attempting to obtain a quick 'guestimate' of the 'true' kinetic parameters. Of concern is the practise of using arbitrary experimental rather than elected conditions for initial rates studies that are key to enzyme kinetic studies. In particular the length of the incubation is often quite long to ensure at least 20% loss of drug substrate. Issues such as enzyme inactivation during the incubation and end-product inhibition from build up of metabolites remain to be systematically addressed. Few investigations have directly compared the two approaches[3, 4] and mixed success is reported.

Ideally an *in vitro* incubation system should mimic the physiological situation and any major difference may be criticised and the validity of *in vitro* data may be questionable.[5] Due to the lipophilic nature of many drug substrates, organic solvents are often required for their solubilization, and are, therefore, present in *in vitro* incubation systems. The commonly observed inhibitory effects of organic solvents observed in human liver microsomes, recombinant enzymes[6,7] and human hepatocytes[8] are dependent on the enzyme under investigation, the solvent selected and its concentration. For example, methanol and acetonitrile represent the most suitable organic solvents for *in vitro* metabolic studies for CYP3A4,[9] their concentration should not exceed 1% v/v in order to avoid any possible inhibitory effect. The appropriateness of microsomal supplements (e.g. serum albumin) remains of concern.[5]

The catalytic activity of the CYP3A enzymes, CYP3A4 in particular, is also affected by various biochemical factors. These include the ratio of the accessory electron-transfer proteins NADPH-cytochrome P450 reductase and cytochrome b_5, membrane lipid composition and ionic strength of the *in vitro* matrix.[10, 11, 12] Mäenpää *et al*[13] have reported in detail the substrate-dependent effect of the ionic strength of buffer solutions on CYP3A4-mediated reactions. Concentrations of accessory proteins can differ considerably between human liver microsomes and cDNA-expressed CYPs[12] and their addition may increase the turnover of some substrates with certain CYPs.[14] Divalent metal ions, particularly Mg^{2+}, may enhance catalysis in some CYP3A4 systems[15] and even cause topological alterations of the CYP3A4 active site at high concentrations.[16]

For inhibition studies, high enzyme concentrations can result in inhibitor depletion and some drugs may bind extensively to microsomal protein, reducing the free inhibitor concentration.[17] Use of recombinant enzymes in inhibition screening reduces the opportunity of non-specific binding to protein/lipid, as high protein concentrations can be avoided due to the high level of enzyme expression.[18] Successful incorporation of microsomal and plasma protein binding in the *in vitro* clearance and K_i data, is highly dependent on the character of the drug (basic or acidic) and the extent of binding to the protein.[19, 20]

1.2 Scope of this chapter

Any quantitative *in vitro* methodology in drug metabolism requires consideration of enzyme kinetic principles and this is the focus of this chapter. Enzyme kinetics are presented in the context of defining drug metabolic clearance and drug inhibition potential, first from the classic view of Michaelis-Menten for a single site drug-enzyme interaction model and then exploring the multisite approaches that are necessary for CYP3A4, and possibly other human enzymes. The former has found wide and successful use in drug metabolism and advantage has been taken of the comprehensive experimental design possible for both broad ranging and tightly controlled experimental conditions that have been used by enzymologists for decades. In contrast the use of multisite enzyme kinetics has yet to be widely embraced. While there is still much to learn in establishing the general utility of relatively complex enzyme models, the alternative of using simple yet inappropriate methods of analysis has lead to much confusion and inaccurate information in the literature. We hope this chapter will not only alert new investigators to some of the complexities of *in vitro* kinetics but also encourage more development in this challenging and fascinating aspect of quantitative prediction in drug metabolism.

2   THE MICHAELIS-MENTEN APPROACH FOR ANALYSIS OF ENZYME KINETIC DATA

Enzyme kinetic studies are used to understand the nature of the active site of an enzyme and the likely nature of an enzyme reaction in a physiological environment. Enzyme catalysed reactions are influenced by a number of factors including enzyme concentration, ligand concentrations (substrates, products, inhibitors and activators), pH, ionic strength and temperature. The Michaelis-Menten equation is most commonly used to determine kinetic constants (V_{max} and K_m) in the analysis of data from *in vitro* characterizations of CYP metabolism. Reasons for performing enzyme kinetic studies and the assumptions relating to the use of this approach with respect to substrate and enzyme concentrations in CYP studies needs to be considered.

2.1   Enzyme kinetic studies in drug metabolism

Determination of kinetic parameters is useful for a number of reasons:

(i)   The K_m is an indicator of the affinity of an enzyme for a given substrate and shows how readily the drug metabolising enzyme can be saturated and at what concentration this may occur *in vivo*. In drug metabolism the substrate concentration is related to the dose of the drug or pharmaceutical agent, and will vary over time. In some rare circumstances, an enzyme that is readily saturated *in vivo* may lead to nonlinear kinetics and this may cause unexpected adverse events and makes prediction of drug response difficult.

(ii)   Kinetic parameters provide information on the mechanism of action of the enzyme. They may also highlight the presence of activators or other physiological modulators of activity. Drug metabolism scientists are generally interested in the properties of the substrate or drug agent, rather than the enzyme itself. However enzyme kinetics can be a useful diagnostic tool for investigating multiple enzymes in an *in vitro* system or allosteric phenomena.

(iii) Enzymes from different sources can be compared, since the K_m is constant for a given enzyme. This allows validation of cDNA expressed enzymes as sources of drug metabolising enzymes.

(iv)   When the K_m is known the assay conditions can be adjusted in further experiments to investigate multiple components in an enzyme reaction. For example, the enzyme can be saturated to determine the V_{max} and a substrate concentration well below the K_m can be used to investigate high affinity components in mixed enzyme sources, e.g., human liver microsomes.

(v)   High V_{max}/K_m ratios or intrinsic clearance (CL_{int}) highlights the best substrates for an enzyme. In drug metabolism this is usually an undesirable property since the drug will be metabolised very rapidly and may not be therapeutically useful *in vivo*. High affinity substrates may also be significant inhibitors of the enzyme resulting in undesirable drug-drug interactions.

2.2   Assumptions in enzyme kinetics

The almost universally applied approach for investigation of enzyme kinetics in *in vitro* drug metabolism is the single substrate steady-state approach. Two key assumptions are made with this approach; the first is that the drug is treated as if it is the *single substrate*, however, CYP mediated reactions also have NADPH and oxygen as substrates in addition to the test drug. NADPH is added to the reaction at a saturating concentration and the oxygen concentration is assumed to be constant within the limitations of the experimental design. The second assumption is that the enzyme is present at low concentrations relative to the substrate and thus at steady-state the formation of the enzyme-substrate complex, ES, is rapid. This low enzyme: high substrate ratio should be maintained for the duration of the experiment and there should be no significant impact on the substrate concentration as a consequence of enzyme binding or the substrate being metabolised. Under pre-steady-state kinetics or at high enzyme concentrations the mathematics is far more complex, hence the *steady-state* approach is used.[21]

2.3   Methodological considerations

The co-substrates, NADPH or UDPGA, should not be rate-limiting and are normally present at saturating concentrations in the incubation (e.g., concentrations of NADPH of 0.5 to 1 mM and UDPGA of 3 to 5 mM are typically used). The ionic strength, pH and temperature should be physiologically relevant (e.g., the use of phosphate or tris buffers at concentrations of 50 to 100 mM at pH 7.4 and 37°C) and should remain constant throughout the reaction.

The enzyme concentration should be very much less than the substrate concentration and low enough such that the amount of product formed is negligible compared to the substrate concentration. Up to 5 % conversion of substrate to product over the duration of the assay is generally considered acceptable, i.e. only 1/20 of the substrate is exhausted. If too much substrate is exhausted the rate of reaction could decline either due to the reduction in substrate concentration or the significant build up of product causing product inhibition. In order to select a suitable enzyme concentration the linearity of the reaction with respect to the enzyme/protein concentration should be determined. If the reaction is linear at a low substrate concentration it is likely it will also be suitable for high substrate concentrations. The incubation time should also be tailored such that the amount of product formed is negligible compared to the substrate concentration. Very short incubation times are in reality impractical and may introduce further errors. Long incubation times unnecessarily prolong the experiment and may result in degradation of the enzyme and reagents/ substrates used. Linear conditions should be established using a higher enzyme concentration than required, which would then also be suitable for a lower enzyme concentration if used. In practice, for many drug metabolising enzymes a protein concentration of 0.1 - 0.5 mg/ml and an incubation time of 5 - 10 min will produce a linear reaction with readily detectable rates of reaction.

The substrate concentration range needs to be considered with respect to the same considerations given to the enzyme concentration and incubation time. In an incubation, the substrate will be present either free or as part of an enzyme-substrate complex. Steady-state kinetic analysis assumes that $[S]_{total} = [S]_{free}$, therefore $[S]_{free} \gg [ES]$ and the formation of the enzyme substrate complex does not alter the substrate concentration. This status can be achieved if the $[S] \gg [E]$ or if $K_m \gg [E]$. This is generally the case in enzyme kinetic studies using microsomes. For example a typical human liver microsomal incubation of 0.5 mg protein/ml may contain as much as a 0.1 µM concentration of CYP3A4 (assuming 0.6 nmol CYP/mg protein, and that 1/3 of the CYP is CYP3A4). In this case the assumptions above would only be valid at substrate concentrations of at least 2 µM (preferably 10 µM) and higher. Alternatively, if $K_m \gg [E]$ (i.e. $K_m > 10$ µM), the [S] would only need to be as a low as 2 µM for an accurate determination of K_m (i.e., the lowest concentration 1/5 of K_m). Somewhat lower substrate concentrations could be justified at lower protein concentrations or higher substrate concentrations could be used with enzymes that are less abundant in human liver microsomes. As new technologies for the sensitive analysis of drug and metabolites emerge and cDNA expression systems allow production of relatively high concentrations of individual enzymes, the low end of the substrate concentration range remains set at 1 –2 µM, in order to remain within the assumptions of steady-state kinetics. The upper limit of the substrate concentration range used *in vitro* is often limited by the physicochemical properties of the substrate. Solubility is often an issue at physiological pH. In reality, few drugs achieve *in vivo* concentrations greater than 100 µM, thus little relevant information is gained at *in vitro* concentrations higher than this value. Very high kinetic constants are unlikely to have any physiological relevance to the disposition or toxicity of the drug. With the substrate concentration range routinely defined as 1 to 100 µM, a suitable number of points must be included for accurate kinetic analysis which adequately span the K_m. A range from one fifth to five-fold the K_m would be suitable (one tenth to ten-fold ideal) if solubility and analytical methodology allows. Literature surveys suggest that 8 – 10 data points are acceptable for most *in vitro* drug metabolism studies, however up to 30 points may be evaluated if more complex kinetics is encountered, e.g., presence of multiple enzymes or allosteric kinetics.

2.4 Analysis of substrate kinetic data

A velocity equation for a typical enzyme reaction can be described by the following equation; where E represents enzyme, S the substrate, ES the enzyme-substrate complex and P the product.

$$E + S \Leftrightarrow ES \Rightarrow E + E + P \tag{7.1}$$

The initial reaction velocity for this reaction can be described by the Michaelis-

Menten equation; where v represents velocity, [S] the substrate concentration, V_{max} the maximal velocity and K_m the concentration giving half the maximal velocity.

$$v \;=\; \frac{V_{max} \times [S]}{K_m + [S]} \tag{7.2}$$

In order to analyse enzyme kinetic data with the Michaelis-Menten equation the data should first be plotted and examined. The most appropriate method of analysis currently is by nonlinear regression of the untransformed data using a suitable data-fitting software package. Transformed plots can be useful for interpolation of the parameters directly from the graph and are also useful

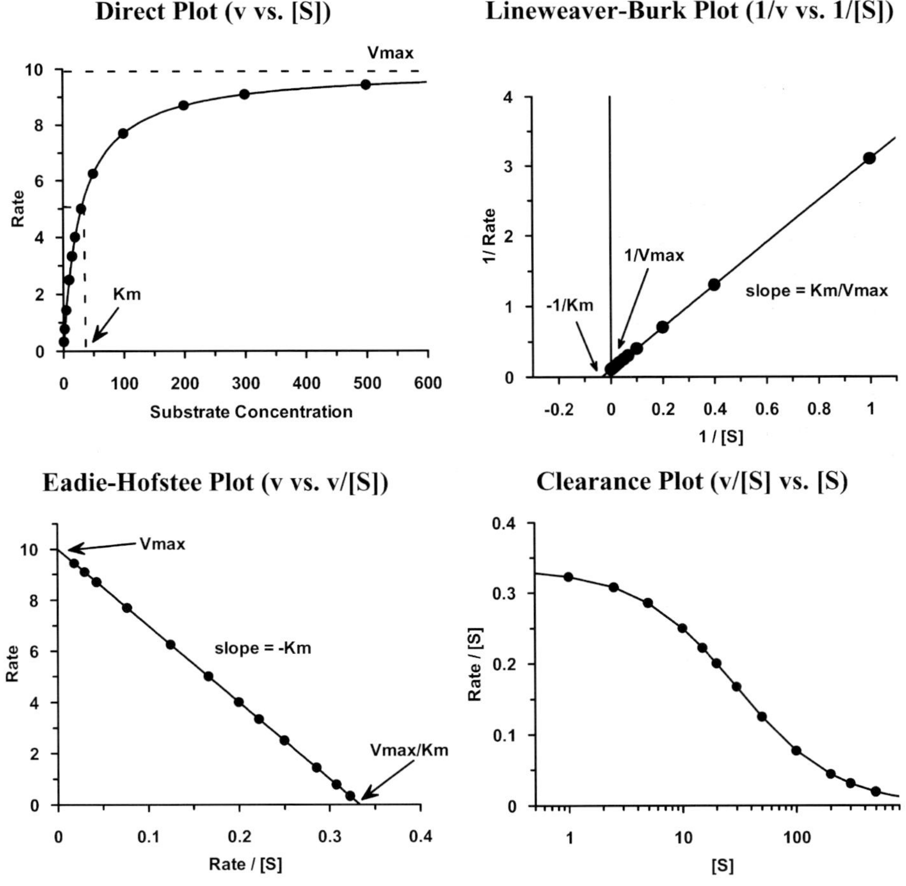

Figure 1. Analysis of enzyme kinetic data

diagnostic tools for determining deviations from simple Michaelis-Menten kinetics. Data may be plotted directly, or may be transformed to an Eadie-Hofstee plot or Lineweaver-Burk plot (Figure 1). The advantages and disadvantages of these linear transforms are discussed below. An additional plot valuable for identifying the quality of the data and the appropriateness of the model adopted is the clearance plot (v/S vs. [S]). The plateau at low [S] represents the intrinsic clearance (CL_{int}).

$$\frac{v}{[S]} = \frac{V_{max}}{K_m + [S]} \tag{7.3}$$

(i) Direct plot

The direct plot is the preferred method of data analysis, followed by nonlinear regression of the data. When using a wide substrate concentration range, the low concentration data appear compressed, although the high concentration data are easily seen. It is also difficult to visualise V_{max} and see any deviations from the perfect rectangular hyperbola as described by the Michaelis-Menten equation. This plot is useful for presentation, but may hide the low concentration data and any deviations from simple enzyme kinetics, such as autoactivation. The parameters K_m and V_{max} can be approximated from this graphical plot, however linear transforms give more accurate determinations in the absence of computer aided fitting.

(ii) Lineweaver-Burk plot

The reciprocal of the rate data plotted against the reciprocal of the substrate concentration. When using a wide substrate concentration range this plot tends to compress the high concentration data, although the low concentration data are easily visualised. The V_{max} and K_m can be estimated from this plot graphically and any major deviations from the Michaelis-Menten straight line are more easily seen. However, the data are not uniformly spread on this transform and the fit may be weighted to the lowest concentration data which are more likely to be the least accurately determined. This type of plot is not recommended for analysing enzyme kinetic data.

(iii) Eadie-Hofstee plot

The rate data plotted against the rate divided by the substrate concentration (v vs. v/[S]). This plot evenly distributes both the low and high concentration data, making it easy to visualise the Michaelis-Menten parameters and it is useful for detecting any deviations from simple enzyme kinetics.

2.4.1 Identifying deviations from simple Michaelis-Menten kinetics

Deviations from simple Michaelis-Menten kinetics may occur for a number of reasons and it is essential to eliminate the effect of any non-enzymatic processes which may impinge on the shape of the rate-substrate concentration profile.[22] Table 1 lists some of the processes that would lower the concentration of substrate available to the enzyme relative to the concentration calculated, following the addition of a particular quantity of substrate to the incubation. Three of these processes involve saturable events and hence the impact would be concentration dependent - maximal at the low concentrations and tapering off to no effect at high concentrations. Ideally, turnover of substrate should be <10% in order to comply with initial rate conditions yet analytical limitations may prevent this and correction for loss of substrate during the incubation is required in order to avoid artefactual conclusions. Alterations due to experimental conditions, e.g., solubility limited data, may be observed more clearly on an Eadie-Hofstee plot, where a deviation from the straight line is observed in the high concentration range. Deviations related to the characteristics of particular enzymes and their substrates, together with how to analyse these data more comprehensively, are dealt with fully in the next section of this chapter.

Table 1. Examples of artefactual sources for sigmoidicity and convexity in rate substrate concentration profiles[23]

Substantial substrate depletion during the incubation period
Saturable futile binding within the incubation matrix
Saturable cellular efflux processes
Poorly defined limit of analytical quantitation
Aqueous solubility limitations

When a mixture of enzymes is present in the incubation, as is the case in liver microsome preparations, what appears to be a 'normal' Michaelis-Menten curve may actually be a combination of two (or more) contributing enzymes (Figure 2). This can be observed more clearly on the Eadie-Hofstee plot when the two components are separated by a clear difference in their affinities for each enzyme. These data can be simply described as a sum of two Michaelis-Menten equations:

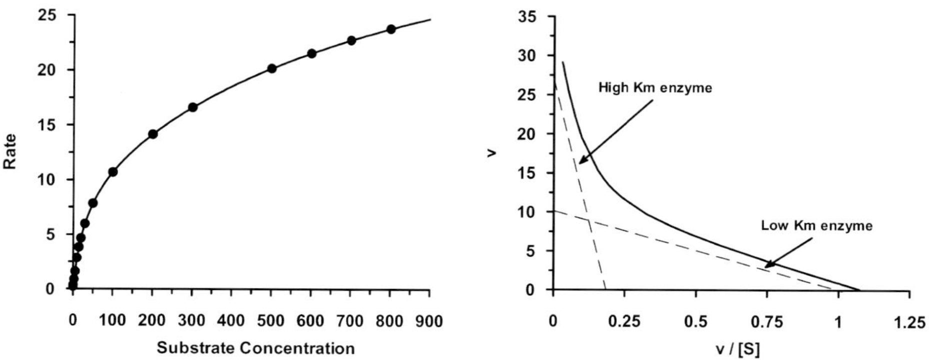

Figure 2. Direct and Eadie-Hofstee plots for a two enzyme system

$$v = \frac{V_{max\,1} \times [S]}{K_{m\,1} + [S]} + \frac{V_{max\,2} \times [S]}{K_{m\,2} + [S]} \qquad (7.4)$$

Note that the individual K_{ms} for each enzyme are not at a tangent to the data. If the K_{ms} for the two enzymes are not sufficiently different the data will appear as if only one enzyme is present and the two enzymes cannot be distinguished. In Figure 2, the K_{ms} differ by *ca* 30-fold. Equally, if the velocity of one component is much lower than the other it is likely that the low velocity component could easily be overlooked. Often the second site requires a high concentration to saturate and only a linear CL term can be identified (e.g., ethoxycoumarin O-deethylase[3]).

2.4.2   Determination of intrinsic clearance

In vitro affinity measurements of a drug for an enzyme need to be related to the behaviour of the enzyme *in vivo* in order to determine its likely influence on the disposition and clearance of a drug. The cornerstone parameter used for *in vitro-in vivo* extrapolation is the term intrinsic clearance, CL_{int}.[24] This term was coined in the 1970s to describe the rate of metabolism of a drug under *in vivo* conditions. Intrinsic clearance is a pure measure of enzyme activity towards a drug and is not influenced by other physiological determinants of liver clearance such as hepatic blood flow or drug binding within the blood matrix. As with all clearance terms, it has units of volume rate (e.g. ml/min) and acts as a proportionality constant to describe the relationship between rate of metabolism of a drug and its concentration at the enzyme site (C_E).

$$\text{Rate of metabolism} = CL_{int} \bullet C_E \tag{7.5}$$

The free concentration of drug within the liver is assumed to equate with C_E. From a biochemical viewpoint CL_{int} can be considered in terms of the enzyme parameters of the Michaelis-Menten relationship. Under linear conditions when C_E is 10% or less of the K_m, the Michaelis Menten equation reduces to:

$$\text{Rate of metabolism} = \frac{V_{max} \times C_E}{K_m} \tag{7.6}$$

which is analogous to the equation above, i.e.

$$CL_{int} = V_{max}/K_m = \frac{\text{rate of metabolism}}{C_E} \tag{7.7}$$

In this manner the Michaelis-Menten parameters used in *in vitro* determinations can be used to obtain an *in vitro* intrinsic clearance that can be used to scale to the *in vivo* situation. This is the subject of further discussion in Chapter 13.

2.5   Analysis of enzyme inhibition data

Evaluation of enzyme inhibition *in vitro* has become a routine method in drug metabolism for the prediction and evaluation of drug-drug interactions, particularly in the field of cytochrome P450. Such information is now expected from regulatory authorities. Early evaluation of drug-drug interactions can also prevent costly mistakes later in the drug development process. *In vitro* systems, such as liver microsomes, are useful as they can be used to compare preclinical species with man.

2.5.1   Types of reversible inhibition

Inhibition may be reversible or irreversible, this section will consider reversible inhibition processes only, irreversible inhibition is dealt with elsewhere in this book. The velocity equations describing each type of inhibition and the resulting changes to V_{max} and K_m are shown in Table 2. The main types of inhibition[21] are as follows:

(i)   Competitive inhibition: A competitive inhibitor is a substance that combines with an enzyme in such a manner that it prevents substrate binding. It may bind to exactly the same site, or partly share or mask the substrate-binding site. In competitive inhibition substrate and inhibitor binding is mutually exclusive and there is reciprocity of inhibition.

(ii) Non-competitive inhibition: A simple non-competitive inhibitor has no effect on substrate binding and vice versa. S and I bind reversibly and independently, but bound inhibitor does not inactivate the enzyme, but has an effect of apparently decreasing the amount of enzyme present.

(iii) Uncompetitive inhibition: Pure uncompetitive inhibition is rare in unireactant systems and even rarer in *in vitro* drug metabolism studies. In uncompetitive inhibition the inhibitor only binds to enzyme that already has substrate bound. This favours the equilibrium to ES complex and therefore the K_m is decreased.

(iv) Linear Mixed-type Inhibition: This is actually a form of non-competitive inhibition and is sometimes referred to as full non-competitive inhibition. In this case both the substrate and inhibitor can bind independently to the enzyme. However, the binding of the inhibitor to the enzyme affects the binding of the substrate and binding of the substrate affects binding of the inhibitor. The degree of this effect is the same for inhibitor on substrate as substrate on inhibitor in order to maintain steady-state, as defined by parameters K_i' and K_m'.

2.5.1.1 Methodological Considerations

The steady-state kinetic approach also applies to enzyme inhibition and the assumptions are similar to those described previously. In addition to the substrate concentration, the inhibitor concentration should also be much higher than the enzyme concentration, such that inhibitor binding to the enzyme does not alter the free concentration of the inhibitor. The inhibitor concentrations used should span the K_i or IC_{50}, and an ideal experimental design would include inhibitor concentrations of $0K_i$ (control), $1/3\ K_i$, K_i, $3K_i$ and $10\ K_i$. For IC_{50} determinations the range of inhibition should be from virtually no inhibition to virtually complete inhibition in concentrations evenly spread on a log scale (e.g. 0, 0.1, 0.3, 1, 3, 10, 30 and 100 μM). The substrate concentration range should be in a similar range as that for determination of V_{max} and K_m, e.g. $1/3K_m$, K_m, $3K_m$ and $10K_m$.

2.5.1.2 K_i Determinations

Determination of the K_i, the inhibition constant, is used when complete *in vitro* characterisation of an inhibitor is required. Velocity equations for K_i determinations generate a 3D plot, as there are two variable or x-axis parameters ([S] and [I]). This may be difficult to directly visualise, hence data are generally plotted and evaluated as a series of 2D plots, with one line for each substrate or inhibitor concentration. Data may be linearised in the form of an Eadie-Hofstee plot (v vs. v/[S]) or as a Dixon plot (1/v vs. [I]). The K_i is not readily determined from the Eadie-Hofstee plot, but this may be a useful way of presenting the data and determining the mechanism of inhibition involved. In the Dixon plot the K_i is the point where all the lines intercept. However, using this method to determine K_i can be very inaccurate as there may not be a discreet intersection of the lines. K_i should always be determined by simultaneous nonlinear regression of the entire untransformed data set using a suitable software package.

2.5.1.3 *IC$_{50}$ Determinations*

IC_{50} determinations are frequently used in *in vitro* inhibition experiments, as they are quicker and easier to conduct than full K_i determinations. However, no information is obtained on the inhibition mechanism and the IC_{50} can be substantially different from the K_i depending on the substrate concentration used. At a substrate concentration at or near the K_m, the IC_{50} is equivalent or within 2-fold of the K_i irrespective of the inhibition mechanism (see Table 2). IC_{50} plots are conventionally plotted as rate vs. log[I], and the curve can be described by the following equation:

$$v = \frac{v_0}{1 + \left(\dfrac{[I]}{IC_{50}}\right)^S} + b \tag{7.8}$$

Where v_0 is the control rate, b is the background or uninhibitable activity, s is the slope factor and IC_{50} is the inhibitor concentration that reduces the control rate by 50%. For selective assays there should be no background activity and the slope should equal 1 irrespective of the mechanism of inhibition. As for K_i determinations, the IC_{50} should always be calculated by nonlinear regression of the untransformed data.

Table 2. Velocity equations for reversible inhibition and the relationship between IC_{50} and K_i.

Inhibition Type	Velocity Equation	Effect of increasing [I]		$IC_{50} - K_i$ relationship	
		Change to V_{max}	Change to K_m	General equation	when $[S] = K_m$
Competitive	$$v = \frac{V_{max} \times [S]}{K_m \left(1 + \frac{[I]}{K_i}\right) + [S]}$$	$\leftrightarrow$	$\uparrow$	$$IC_{50} = K_i \left(1 + \frac{[S]}{K_m}\right)$$	$IC_{50} = 2K_i$
Non-competitive	$$v = \frac{V_{max} \times [S]}{K_m \left(1 + \frac{[I]}{K_i}\right) + [S]\left(1 + \frac{[I]}{K_i}\right)}$$	$\downarrow$	$\leftrightarrow$	$$IC_{50} = K_i$$	$IC_{50} = K_i$
Uncompetitive	$$v = \frac{V_{max} \times [S]}{K_m + [S]\left(1 + \frac{[I]}{K_i}\right)}$$	$\downarrow$	$\downarrow$	$$IC_{50} = K_i \left(1 + \frac{K_m}{[S]}\right)$$	$IC_{50} = 2K_i$
Linear Mixed $(K_i' > K_i)$	$$v = \frac{V_{max} \times [S]}{K_m \left(1 + \frac{[I]}{K_i}\right) + [S]\left(1 + \frac{[I]}{K_i'}\right)}$$	$\downarrow$	$\uparrow$	$$IC_{50} = K_i' \left(\frac{K_m + [S]}{K_m' + [S]}\right)$$	$IC_{50} = K_i$
Linear Mixed $(K_i' < K_i)$		$\downarrow$	$\downarrow$		

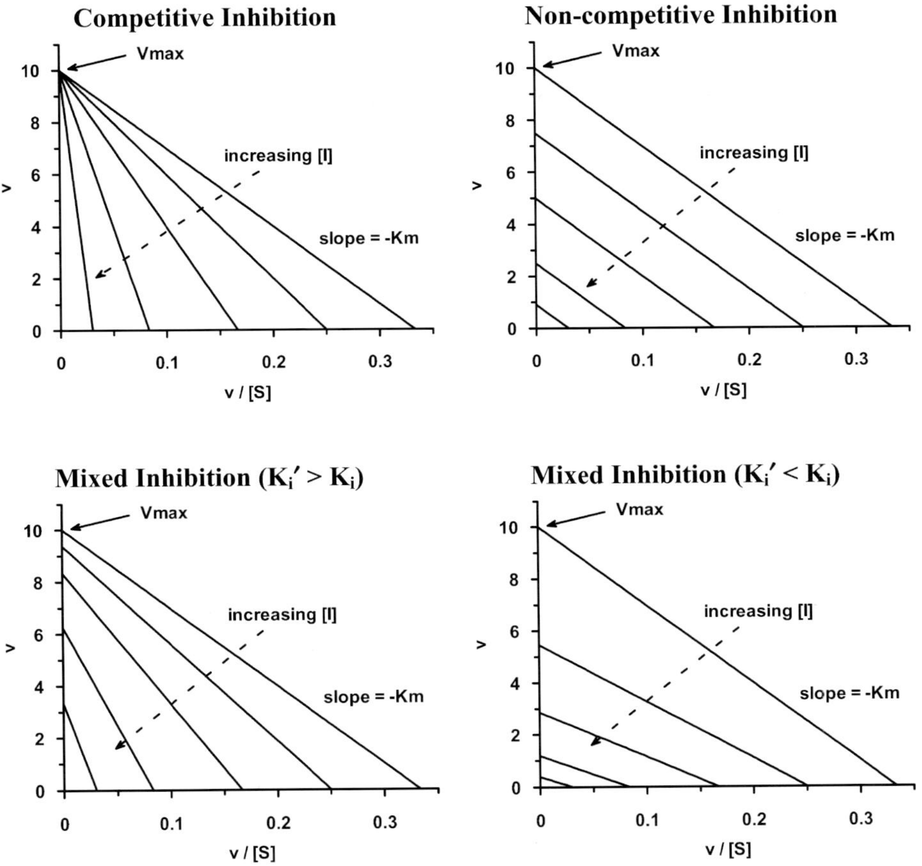

Figure 3. Eadie-Hofstee plots of enzyme inhibition

3   MULTISITE APPROACH FOR ANALYSIS OF ATYPICAL KINETIC DATA

The single-site Michaelis-Menten kinetic approach does not accommodate all kinetic features observed for certain CYP reactions, for example auto- and heteroactivation, substrate inhibition, substrate-dependent effects, partial/ cooperative inhibition and pathway differential effects. Most of these non-Michaelis-Menten properties are associated with CYP3A4 *in vitro*, but recent studies indicate atypical kinetics for some other human enzymes, particularly CYP2C9[25, 26, 27] and UGT1A1[28] and UGT2B7.[29]

The finding that a number of CYP3A4 substrates do not conform to the expected competitive type of interaction indicates the existence of and interaction between several binding sites on the enzyme. The large CYP3A4 active site may allow the simultaneous presence of multiple molecules (at least two) and the exact binding conformations depend on the substrates involved, their relative concentration and affinity for the enzyme. The following section summarises the atypical effects mentioned above, including those observed when more than one molecule of the same substrate is present at the active site (homotropic) and interactions involving two different substrates (heterotropic).

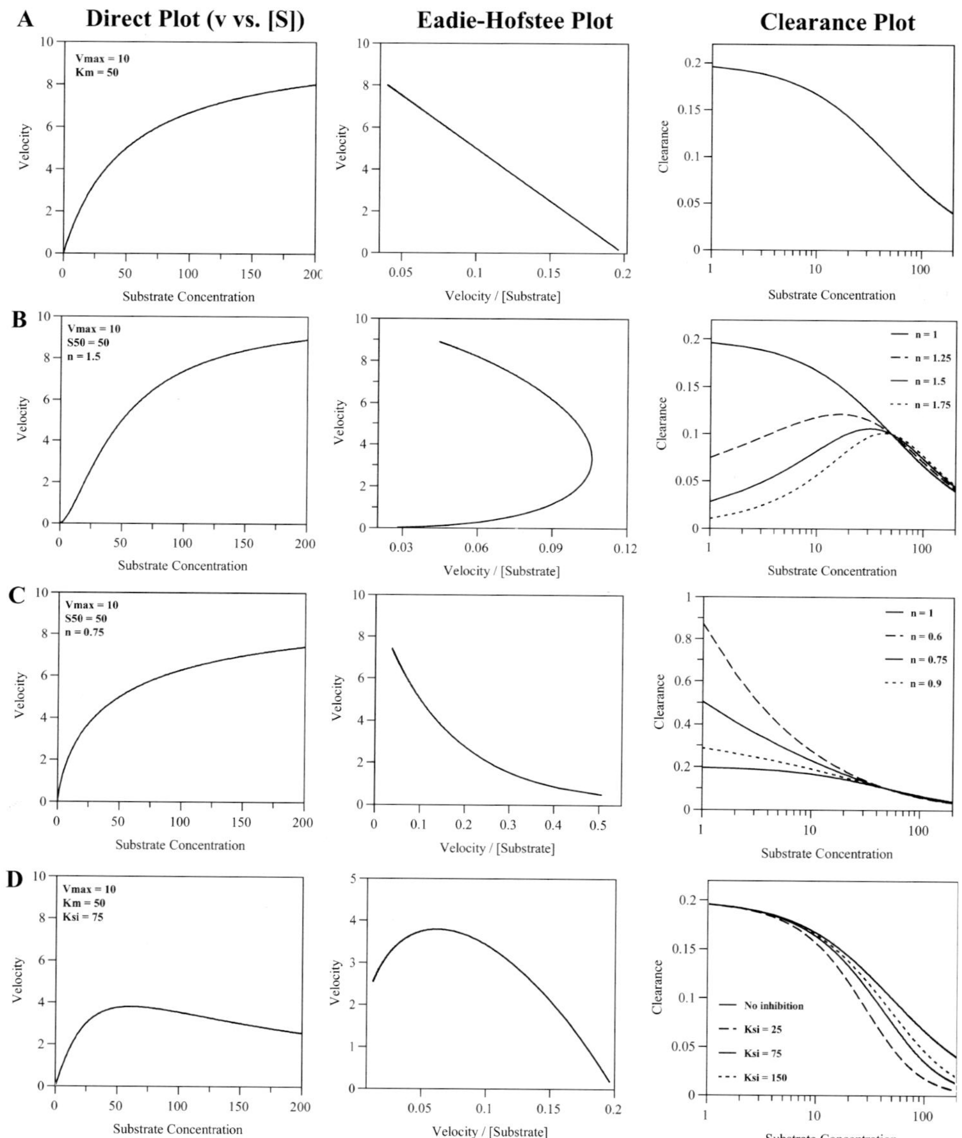

Figure 4. Kinetic profile simulations with the resulting Eadie-Hofstee and CL_{int} plots for Michaelis-Menten kinetics (A), autoactivation (B), negative cooperativity (C) and substrate inhibition (D)

3.1 Homotropic Effects

Homotropic effects are associated with the alterations in either binding affinity or the rate of product formation after the binding of a second molecule of the same substrate to the enzyme active site. The net effects, resulting in either a substrate concentration-dependent increase (Figure 4B) or decrease (Figure 4C and D) in CYP activity, are discussed in further detail in the following section. For reference, Figure 4A represents standard Michaelis-Menten kinetics.

3.1.1 Substrate Activation

Since the first demonstration of autoactivation for progesterone 6β-hydroxylation reported by Schwab *et al*,[30] numerous drugs have shown deviations from standard hyperbolic kinetics. Positive cooperativity, characterised by a sigmoidal saturation profile and deviations from linearity in Eadie-Hofstee plots, have been observed for various CYP3A4 substrates, and in various *in vitro* systems, as shown in Table 3.

The deviation from Michaelis-Menten kinetics occurs at low substrate concentrations and it might not be readily apparent from a direct plot. The Eadie-Hofstee plot can be used as a diagnostic tool as the marked curvature or 'boomerang' shape is more clearly observed (Figure 4B). This type of data is indicative of cooperativity and suggests the presence of multiple binding sites. In order to explain and obtain the appropriate fit for the homotropic data, several models have been proposed, from the empirical Hill equation (Eq. 7.9) to more complex mechanistic ones.

$$v = \frac{V_{max} \times [S]^n}{S_{50}{}^n + [S]^n} \tag{7.9}$$

Where the substrate concentration resulting in 50% of v_{max} (S_{50}) is analogous to the Km parameter in the Eq. 72. When n (Hill coefficient) is equal to 1 the equation simplifies to the Michaelis-Menten equation. Values of n greater than 1 (positive cooperativity) give sigmoidal direct plots and convex Eadie-Hofstee plots as shown in Figure 4B. If n is less than 1 this is known as negative cooperativity (Figure 4C), which is kinetically indistinguishable from two-enzyme kinetics as described earlier. The Hill equation is useful for describing the data for a single substrate but does not provide any mechanistic explanation of the interactions between multiple substrates. Autoactivation is characterised by the dependence of clearance on the substrate concentration and requires an alternative to CL_{int} for scaling of *in vitro* data, CL_{max}. Eq. 7.10 is suggested as an alternative parameter and provides an estimate for the maximum clearance when the enzyme is fully activated before the saturation occurs.[22,23]

Table 3. CYP3A4 substrates displaying sigmoidal kinetics *in vitro* (autoactivation)

Substrate	Metabolic Pathway	In Vitro System	Reference
Aflatoxin B1	3-hydroxylation 8,9-epoxidation	HLM[a], E.Coli[b] Fusion protein[c]	Ueng et al, 1997[31]
Amitriptyline	N-demethylation	HLM, Fusion protein E.Coli HLM	Shaw et al, 1997[10] Ueng et al, 1997[31] Schmider et al, 1995[32]
7-benzyloxy-4-trifluoromethyl-coumarin	O-dealkylation	Baculovirus	Lu et al, 2001[33]
Carbamazepine	10,11-epoxidation	HLM, Hep G2 E.Coli Hep G2[d]	Kerr et al, 1994[34] Ueng et al, 1997[31] Korzekwa et al, 1998[25]
Diazepam	3-hydroxylation N-demethylation	HLM	Andersson et al, 1994[35]
		HLM, Baculovirus	Shou et al, 1999[36]
		B-lymphoblastoid	Kenworthy et al, 2001[37]
	3-hydroxylation	HLM, SUPERSOMES™ SUPERMIX™, HL Pool	Bicknell et al, 1999[38]
Dextromethorphan	N-demethylation	RLM[e], Hepatocytes	Witherow and Houston, 1999[22]
17ß-Estradiol	2-hydroxylation	E.Coli	Ueng et al, 1997[31]
Nifedipine	Oxidation	Baculovirus, Fusion protein	Shaw et al, 1997[10]
Nordiazepam	3-hydroxylation	HLM, Baculovirus	Shou et al, 1999[36]
Temazepam	N-demethylation		
Testosterone	6ß-hydroxylation	HLM, Baculovirus E.Coli, Fusion protein E.Coli B-lymphoblastoid Duodenal S9 fraction	Lee et al, 1995[39] Ueng et al, 1997[31] Harlow & Halpert, 1998[40] Kenworthy et al, 2001[37] Johnson et al, 2001[41]
Progesterone	6ß-hydroxylation	E.Coli	Ueng et al, 1997[31]

[a] HLM- Human liver microsomes; [b] Reconstituted system; [c] CYP3A4 – NADPH P450 Reductase Fusion Protein; [d] *Vaccinia* virus expressed CYP3A4 in Hep G2 cells, [e] Rat liver microsomes

$$CL_{max} = \frac{V_{max}}{S_{50}} \times \frac{(n-1)}{n\,(n-1)^{1/n}} \qquad (7.10)$$

Multisite kinetic models can be employed for the analysis of atypical interactions[37, 42] and are adopted from Segel.[21] These kinetic models are based on rapid equilibrium and steady-state approach (outlined in section 2.2 and 2.3), allowing the simultaneous fit of multiple sets of data to a single equation. The rapid equilibrium assumption, i.e., the rate at which ES complex dissociates is much faster than the rate of product formation, is also one of the basic assumptions for standard Michaelis-Menten kinetics. One must be aware of the possibility that the enzyme-product complex (EP) could also reduce the enzyme availability for the substrate interaction, causing a decreased rate of the reaction.[43] In order to keep the modelling relatively simple, EP complexes have not been included in the total sum of the product forming complexes in the model derivation.

The simplest multisite model accommodating atypical kinetic properties, when 2 molecules of the same substrate bind to the active site, is presented in the Figure 5. It is assumed that the two binding sites (SE and ES) are equivalent and no orientation difference in binding of S to E is defined. Alterations in either binding affinity or catalytic efficiency upon binding of a second substrate molecule to a vacant site can describe data from substrates showing positive cooperativity and substrate inhibition. The interaction of the substrate molecules is quantified by the velocity equation shown below:

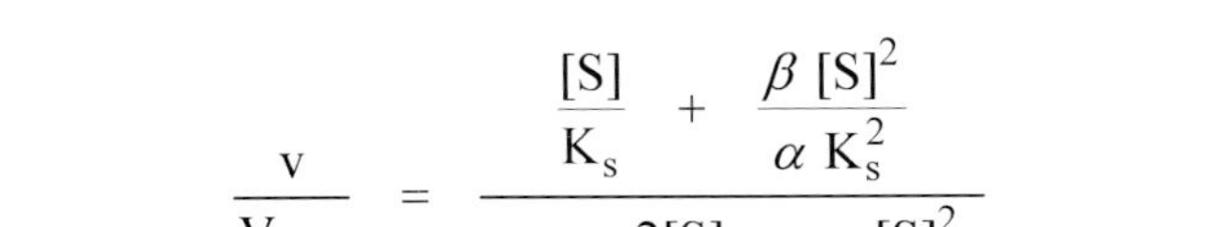

$$E + P \xleftarrow{K_p} SE \xrightarrow{\alpha K_s} SES \xrightarrow{\beta K_p} \begin{array}{l} SE + P \text{ or} \\ ES + P \end{array}$$

$$\Big\downarrow K_s \qquad\qquad \Big\downarrow \alpha K_s$$

$$E \xrightarrow{K_s} ES \xrightarrow{K_p} E + P$$

$$\frac{v}{V_{max}} = \frac{\dfrac{[S]}{K_s} + \dfrac{\beta\,[S]^2}{\alpha\,K_s^2}}{1 + \dfrac{2[S]}{K_s} + \dfrac{[S]^2}{\alpha\,K_s^2}} \qquad (7.11)$$

Figure 5. Kinetic model for an enzyme with two-substrate binding sites, the second substrate (S) molecule binds cooperatively

Autoactivation may result in either increased binding affinity for a second substrate molecule (dissociation constant K_s changes by the factor $\alpha < 1$), or changes in the effective catalytic rate constant (K_p) by the factor β from SES complex ($\beta > 1$). Changes in α or β in the opposite direction can yield negative cooperativity ($\alpha > 1$, resulting in biphasic kinetic profile, $\beta < 1$ resulting in substrate inhibition). Negative cooperativity results in a concave Eadie-Hofstee plot (Figure 4C) identical in form to the apparent biphasic kinetics involving a high-affinity, low-capacity and a low-affinity, high-capacity enzyme.[23] These biphasic-type of curves are less common for CYP3A4 than autoactivation (Table 3) or substrate inhibition (Table 4). The proposed model does not distinguish between the simultaneous binding of multiple molecules within a single active site and the binding of two molecules to two distinct binding sites. This two-site model has been successfully applied for the analysis of diazepam and testosterone kinetic properties in human lymphoblast-expressed CYP3A4;[37] the link with the parameters obtained by Hill equation is shown in Table 5.

Different approaches to the above have been used by other investigators; the similarities and differences are summarised below. Korzekwa *et al* [25] have proposed that changes in the binding affinity and the catalytic efficiency for the single- (SE/ES) and two-substrate bound form (SES) are characterised by two K_m and two V_{max} values. The ratio of these two K_m or V_{max} values corresponds to the α and β interaction factors, respectively (two-site model, Figure 5); e.g., $\alpha < 1$ (two-site model) and $K_{m2} < K_{m1}$[25] for positive cooperativity.

Additionally, Shou *et al* [36] have suggested that two binding sites within the CYP3A4 active site differ in their physicochemical properties (steric hindrance, hydrophobic interaction and electron characteristics); i.e., simultaneous binding of two substrates in a cooperative manner is assumed, but with the orientation difference in binding of S to E (SE$\neq$ES). The fit obtained from this kinetic model for diazepam, temazepam and nordiazepam metabolism in human liver microsomes and recombinant CYP3A4 was in good agreement with the simpler equations,[23, 25] as they both describe the reduced binding affinity and lower catalytic capacity of ES complex compared to ESS ($K_{m1} > K_{m2}$, $V_{max1} < V_{max2}$). The main advantage of the model with two distinct sites is the possibility to fully characterise, from the mechanistic point of view, the nature of cooperative binding of two substrate molecules in the inner portions of CYP3A4 active site.

3.1.2 Substrate Inhibition

The phenomenon of substrate inhibition is characterised by the typical decline in the rate of product formation with the increasing substrate concentration.

The clearance plot initially follows the Michaelis-Menten type, but due to the inhibition effect the decrease in the saturation portion of the curve is more rapid (Figure 4D). Although the full mechanism is unknown, ignoring the substrate inhibition effect at higher concentration points can lead to erroneous estimates of the kinetic parameters.[23, 44] Recently, several authors have proposed the two-site model approach for the substrate inhibition phenomenon,[45,46] observed with different CYP enzymes (Table 4).

Table 4. CYP3A4 substrates displaying substrate inhibition *in vitro*

Substrate	CYP Isoform	Metabolic Pathway	In Vitro System	Reference
Benzyl-oxyresorufin		Oxidation	Baculovirus	Lin et al, 2001[46]
Midazolam	CYP3A4	1'-hydroxylation	HLM	Kronbach et al, 1989[47]
			B-lymphoblastoid	Ghosal et al, 1996[48]
			HLM	von Moltke et al, 1996[49] Khan et al, 2002[50]
Nifedipine		Oxidation	B-lymphoblastoid SUPERSOMES™ SUPERMIX™	Galetin et al, 2001a[42] Bicknell, unpublished
Terfenadine		C-hydroxylation N-dealkylation	B-lymphoblastoid	Kenworthy et al, 1999[51]
Triazolam		1'-hydroxylation	HLM	Schrag and Wienkers, 2001a[45]
Testosterone		6ß-hydroxylation	Baculovirus	Lin et al, 2001[46]
Progesterone		6ß-hydroxylation		
Celecoxib	CYP2C9	C-hydroxylation	HLM Baculovirus	Tang et al, 2000[52] Lin et al, 2001[46]
Dextromethor-phan	CYP2D6	O-demethylation	Baculovirus	Lin et al, 2001[46]
Ethoxyresorufin	CYP1A2	O-deethylation		
Diazepam	CYP2C19	3-hydroxylation N-demethylation	SUPERSOMES™	Bicknell et al, 1999[38]

The two-site model described for autoactivation (Figure 5) can also be applied in substrate inhibition circumstances, assuming the decrease in the rate of product formation from a two-substrate bound complex, described by the interaction factor $\beta<1$. Alteration of this kinetic model incorporates sequential binding of substrate molecules, i.e., the substrate inhibition site cannot be occupied until the active site is filled and the second site may be independent from the active site. The earlier onset of substrate inhibition produces a higher K_s/K_i ratio, allowing the application of the model suggested by Houston and Kenworthy[23] and justifies the assumption of only one catalytically active site (V_{max} is equivalent to $K_p[E]_t$, where $[E]_t$ is the total enzyme concentration). This two-site model approach accommodates well the substrate inhibition kinetics observed for terfenadine[51] and nifedipine.[42] A comparison between this analysis and the commonly used uncompetitive substrate inhibition model (Eq. 7.12) is shown in Table 5.

Table 5. Kinetic parameters for various CYP3A4 substrates in human lymphoblast-expressed CYP3A4 (mean ± SE) using one or two-site model approach

CYP3A4 Substrate	Michaelis-Menten / Hill Equation				Two-site Model		
	$K_m/S50$	V_{max}	CL_{int}/CL_{max}	K_s	V_{max}	α	β
	μM	pmol/min/ pmolCYP	$\mu L/min/$ pmolCYP	μM	pmol/min/ pmolCYP		
Testosterone 6β-hydro-xylation[a]	42.7 ± 0.8	16.5 ± 0.2	0.22	93.5 ± 4.8	16.5 ± 0.2	0.15 ± 0.02	2
Diazepam 3-hydro-xylation	140 ± 8	30 ± 1	0.12	421 ± 33	28 ± 1	0.09 ± 0.02	2
Diazepam N-demethy-lation	152 ± 56	2.7 ± 0.5	0.011	252 ± 40	2.5 ± 0.3	0.28 ± 0.18	2
Nifedipine oxidation[b]	20.9 ± 2.2	9.7 ± 0.7	0.46	33.4 ± 4.9	12.8 ± 1.5	/	0.36 ± 0.08
Terfenadine oxidation[b]	5.1 ± 0.7	1.30 ± 0.09	0.25	9.2 ± 1.8	3.4 ± 0.4	/	0.25 ± 0.06

[a] Sigmoidal kinetics, Hill coefficient $=1.34 \pm 0.02$, 1.42 ± 0.04 and 1.20 ± 0.14 for testosterone, diazepam 3-hydroxylation and N-demethylation, respectively. CL_{max} calculated as defined by Eq. 7. 10. [b] Substrate inhibition, $K_{si} = 306 \pm 55$ and 53 ± 9 μM for nifedipine and terfenadine, respectively; CL_{int} calculated as V_{max}/K_m ratio

$$v = \frac{V_{max}}{(1 + (K_m/S) + (S/K_{si}))} \tag{7.12}$$

where K_{si} represents the inhibition constant for the substrate.

3.2 Heterotropic Effects

Heterotropic effects (either activation or inhibition) require a more elaborate approach than homotropic effects as they involve the simultaneous binding of two different substrates to the active site and there is an increased number of enzyme species involved. Due to the fast release of products (but slower compared to the binding equilibrium) each compound is considered independently and only the metabolism of the substrate is considered. Although the velocity equation does not accommodate the metabolism of the modifier, it takes into account differences in binding affinity and catalytic efficiency resulting from modifier binding.

A summary of various inhibition effects that are accommodated by two-site kinetic models and the corresponding interaction factors associated with changes in either binding affinity (α, δ) or rate of product formation (β, γ) are presented in Figure 6. An identical model can be applied for activation; the only difference is the substitution of the inhibition terms (I, K_i) with the ones for activation (A, K_a). Table 6 provides several examples of each of these scenarios. The advantage of the generic two-site model shown in Figure 6 is the possibility to include/exclude in the data modelling the interaction factors α, β, δ and γ, depending on the substrate kinetics and the changes observed in the presence of another substrate in order to simplify the model for data fitting.

Interaction factors associated with the changes in binding affinity (K_s/K_i) are: α - homotropic, δ - heterotropic cooperativity and α_I – cooperative binding of an I/A. Interaction factors associated with the changes in catalytic rate constant (K_p) include β (from SES) and γ (MES). Changes in the binding affinity for a second molecule of the same substrate/modifier are defined by the interaction factor α. Therefore, it can either be associated with substrate binding constant (K_s), for substrates showing sigmoidal kinetics (e.g. testosterone), or with inhibitor binding constant (K_i), in the case of cooperative inhibition. The interaction factor δ describes a heterotropic effect and is required for binding affinity alterations when a complex with two different substrate molecules (MES) is formed. The net effect of the increased affinity ($\delta < 1$) can vary from either heteroactivation or inhibition depending on the corresponding changes in the rate of product formation (γ) in the presence of a modifier as illustrated by the cases in Table 6.

The alterations in the K_p are associated with the interaction factors β and γ. These changes are either a result of binding of a second substrate molecule (SES, e.g substrate inhibition, $\beta K_p < K_p$) or a modifier molecule (MES, heteroactivation $\gamma > 1$, inhibition $\gamma < 1$). When there is no alteration in the rate of product formation due to the binding of a second substrate molecule, β is 2, assuming the equivalence of the two substrate-binding sites. In many cases a modifier can act in a concentration-dependent manner, causing an activation at low substrate concentrations and inhibition at higher concentrations. This has been observed for the effects of terfenadine[56] and quinidine[42] on midazolam and testosterone on nordiazepam formation.[37] It can be rationalised by a relatively low value of the interaction factor γ (<2) and at low S concentrations the rank order of metabolite formation is MES/SEM > SE/ES/SES. However, at higher concentrations the competition of the modifier with the S at both sites causes inhibition (Figure 7A). With large values of γ (>2), the activation is more pronounced and occurs over a wider range of substrate concentrations, as illustrated in the Figure 7B.

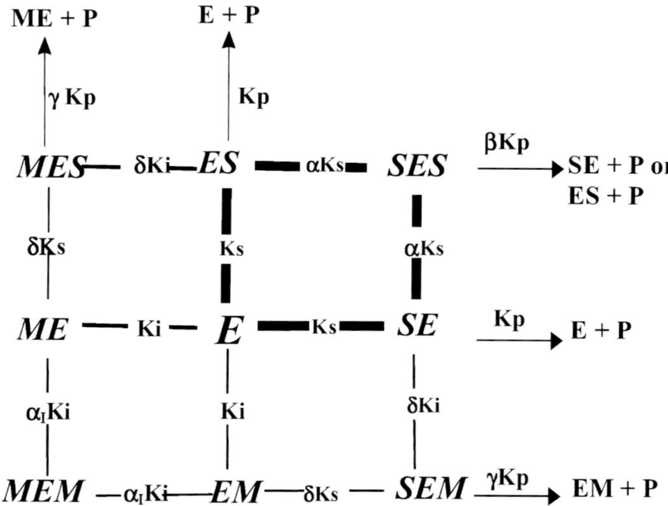

$$\frac{v}{V_{max}} = \frac{\dfrac{[S]}{K_S} + \dfrac{\beta[S]^2}{\alpha K_S^2} + \dfrac{\gamma\,[S][I]}{\delta\,K_S K_i}}{1 + \dfrac{2[S]}{K_S} + \dfrac{[S]^2}{K_S^2} + \dfrac{2[S][I]}{\delta\,K_S K_i} + \dfrac{2[I]}{K_i} + \dfrac{[I]^2}{\alpha_1 K_i^2}} \qquad (7.13)$$

Figure 6. Generic two-site model for CYP3A4 interactions and the corresponding equation[42]

Table 6. Correlation of multi-site kinetic model interaction factors with various effects on CYP3A4 activity

Substrate Characteristics	Effect on CYP3A4	Interaction Factor			Example
		α	γ*	δ	
	Heterotropic activation		>1	<1	QUI effect on FEL, SV[42], DIC[53], WAR[54]
Hyperbolic kinetics	Heterotropic inhibition	1	<1	<1	HAL and MDZ effect on FEL[55] HAL, NIF and FEL effect on QUI[a] NIF and FEL effect on SV[a]
			≤1	>1	NIF and FEL effect on MDZ[55,b] DZ, TST, and TRF effect on ER[b,c]
	Partial inhibition		1	>1	NIF effect on FEL[55] TST effect on MDZ[55,56], ER[57] and TRF[56] TFR effect on MDZ[56]
Substrate inhibition	Heterotropic inhibition	< 1 (β <1, could = γ)		≤1	QUI, HAL, MDZ and FEL effect on NIF[42,55] DZ and TST effect on TRF[51]
Sigmoidal kinetics	Heterotropic activation[d]	<1	>1	= α	TST effect on DZ[37], PRO effect on TST[a]
	Heterotropic inhibition[d]		1	≤1	QUI effect on TST[42], DZ effect on TST[37]

QUI – quinidine; FEL – felodipine; SV – simvastatin; DIC – diclofenac; WAR – warfarin; HAL – haloperidol; NIF – nifedipine; MDZ – midazolam; DZ – diazepam; TST – testosterone; TRF – terfenadine; ER – erythromycin; PRO – progesterone; *β =2 (equivalent binding sites). [a] Galetin, unpublished data, [b] Cooperative binding of an I (α_I <1), [c]Kenworthy, unpublished data, [d] Three-site kinetic model

Even though the values of α, β, γ and δ are often substrate-pair dependent, certain trends can be rationalised from the kinetic properties of both the substrate and modifier. These will be outlined throughout the remainder of this chapter in order to demonstrate the selection of an appropriate multisite kinetic model for the analysis of complex interactions.

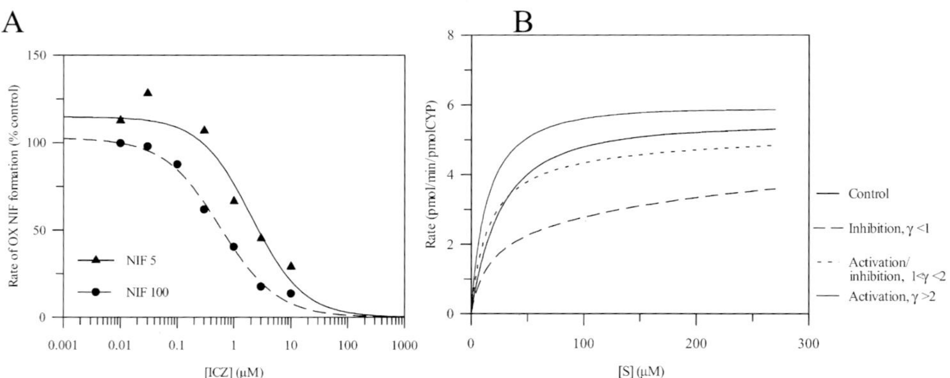

Figure 7. Concentration-dependent differential effect IC^{50} plots of itraconazole effect on nifedipine metabolism at two substrate concentrations (Galetin, unpublished data) (A) and velocity curves corresponding to different values of the interaction factor γ (B).

3.2.1   Heteroactivation

Heteroactivation of metabolite formation for some CYP3A4 substrates (Table 7) results from either changes in binding affinities or product formation in the presence of a modifier or a combination of both effects. The first attempt to explain mechanistically the heteroactivation phenomenon came from Shou *et al.*[58] Their study using recombinant CYP3A4 expressed in Hep G2 cells demonstrated a stimulation of benzo[a]pyrene and phenanthrene metabolism by α-naphthoflavone, with no alteration in K_m values, as would be the case if the interaction between the substrates was of competitive nature. These results together with the fact that α-naphthoflavone was found to be a CYP3A4 substrate itself, led to the hypothesis that both the substrate and activator could bind simultaneously in the active site. The main deficiency of the model is the inability to explain the cooperativity observed for some of the CYP3A4 substrates, as it was assumed that the substrate binding constants were unaffected by the presence of the modifier.

In order to describe fully the activator-enzyme interaction and the conformational changes observed, various analyses have been reported.[25, 31, 36, 37, 40, 59, 60, 61] The complexity and differences between these various methods depend on whether there is simultaneous binding of two molecules[25, 36, 60, 62] or a

Table 7. Heteroactivation of various CYP3A4 and CYP2C9 substrates *in vitro*

Activator	Substrate	CYP 450 Isoform	Metabolic pathway	*In vitro* system	Reference
Testosterone	Diazepam	CYP3A4	3-hydroxylation N-demethylation	B-lymphoblastoid	Kenworthy et al, 2001[37]
	Midazolam		4-hydroxylation	HLM	Mäenpää et al, 1998[13]
				HLM, B-lymphoblastoid	Wang et al, 2000[56]
	Quinidine		N-oxidation	HLM	Nielsen et al, 1999[64]
	17ß-Estradiol		2-hydroxylation	HLM	Kerlan et al, 1992[65]
Quinidine	Diclofenac	CYP3A4	5-hydroxylation	HLM	Tang et al, 1999[66]
				Baculovirus Human hepatocytes	Ngui et al, 2000[53]
	Fentanyl		N-dealkylation	HLM	Feierman & Lasker, 1996[67]
	Meloxicam		5'-hydroxylation	HLM	Ludwig et al, 1999[68]
	S-Warfarin		4'-hydroxylation	HLM	Ngui et al, 2001[54]
	R-Warfarin		10-hydroxylation	HLM	Ngui et al, 2001[54]
	Testosterone		6ß-hydroxylation	3A4/HR/MDR[+b]	Baron et al, 2001[69]
α-Naphtho-flavone	Benzo[a]pyrene	CYP3A4		Hep G2	
	Phenathrene			Hep G2	Shou et al, 1994[58] Korzekwa et al, 1998[25]
	Testosterone		6ß-hydroxylation	E.Coli	Harlow & Halpert, 1998[40]
	Progesterone		6ß-hydroxylation		
	Carbamazepine		10,11-epoxidation	E.Coli	Ueng et al, 1997[31]
				HLM, Hep G2	Kerr et al, 1994[34]
	Aflatoxin B1		8,9-epoxidation	E.Coli	Ueng et al, 1997[31]
	17ß-Estradiol		2-hydroxylation	HLM	Kerlan et al, 1992[65]
				E.Coli	Ueng et al, 1997[31]
	Midazolam		1'-hydroxylation	HLM	Mäenpää et al, 1998[13]
	Diazepam		3-hydroxylation N-demethylation	HLM	Andersson et al, 1994[35]
Progesterone	Carbamazepine	CYP3A4	10,11-epoxidation	HLM, Hep G2	Kerr et al, 1994[34]
	17ß-Estradiol		2-hydroxylation	HLM	Kerlan et al, 1992[65]
Dapsone	S-Flurbiprofen	CYP2C9	4'-hydroxylation	HLM, Baculovirus	Korzekwa et al, 1998[25]
	S-Naproxen		Demethylation	Baculovirus	
	Piroxicam		5'-hydroxylation		Hutzler et al, 2001[26]

separate effector-binding site[31, 37] is assumed. These two approaches /models are not mutually exclusive; the only question is the proximity of two binding sites.

Initial suggestions by Korzekwa *et al*[25] that modifier will bind to the same active site as the substrate, was based on the similar K_a value for α-naphthoflavone activation of phenanthrene metabolism and its K_m value, but were ruled out by site-directed mutagenesis studies on residue Phe-304.[63] In addition, lack of mutual competitive inhibition in the case of cooperative substrates, as observed in the aflatoxin B1-α-naphthoflavone,[31] testosterone-diazepam[37] and testosterone-7-benzyloxyquinoline interactions[33] favours the possibility that the site for metabolism and activation by a modifier can be dissociated and related to different binding locations. Existence of a distinct effector-binding site is the principal characteristic of three-site models; an approach sometimes required to describe the heteroactivation of substrates also showing cooperativity in their binding to the active site.

3.2.2 *Inhibition at Multiple Sites*

Interactions with multiple binding sites may result in inhibition at only one of the binding sites or a differential effect at each site, thus confounding the straightforward prediction of potential *in vivo* drug-drug interactions. A stepwise guide to the analysis of such data is shown in Scheme 1. IC_{50} plots, together with the observations in changes to the kinetic parameters (V_{max} and K_m) in the presence of the modifier provide a valuable initial step in the modelling process. This is followed by a more precise description of the molecular events, incorporating the binding of multiple substrate/inhibitor molecules, using the steady state and rapid equilibrium approach for the analysis of the interactions. Various kinetic models and their corresponding equations can be derived, based on the existence of substrate and modifier binding domains within the active site and incorporating the kinetic interaction factors that occur when substrate and/or modifier bind to the enzyme. Simulated curves can be generated from such models which are then compared with the observed data to aid selection of the most appropriate model.

The inhibition at multiple sites can be characterised by using the following interaction factors when applying multisite kinetic models:

1. Alterations in the binding affinity and an effect on K_s and K_i e.g. (a) increased binding affinity for the formation of ISE/ESI/ISES complexes, $\delta < 1$ or (b) $\delta > 1$ formation of ISE/SEI/ISES is non-favourable.

2. Alterations in the rate of metabolite formation and an effect on K_p; e.g complexes containing I are less productive ($\gamma < 1$) or non-productive, $\gamma = 0$ (examples in Table 6).

Some of the distinct types of inhibition and pathway differential phenomena will be discussed below in more detail.

IC$_{50}$ plots

⬇

Kinetic profiles (S + modifier)

⬇

Preliminary kinetic parameters (V$_{max}$, K$_m$)

⬇

Multisite kinetic models (K$_s$, K$_i$, α, β, δ, γ)

⬇

Simultaneous fit of multiple sets of data (GraFit, WinNonlin)

⬇

Selection of the best fit based on statistical parameters

⬇

Simulated velocity curves (rate vs [S])

Scheme 1. Data analysis of multisite inhibition data

3.2.2.1 *Positive and negative cooperativity in inhibition*

Cooperative binding of a second inhibitor molecule to the enzyme results in a characteristic inhibition profile. The slope of the IC$_{50}$ plots differs from unity as seen for the standard one-site Michaelis-Menten type of inhibition (Figure 8A).

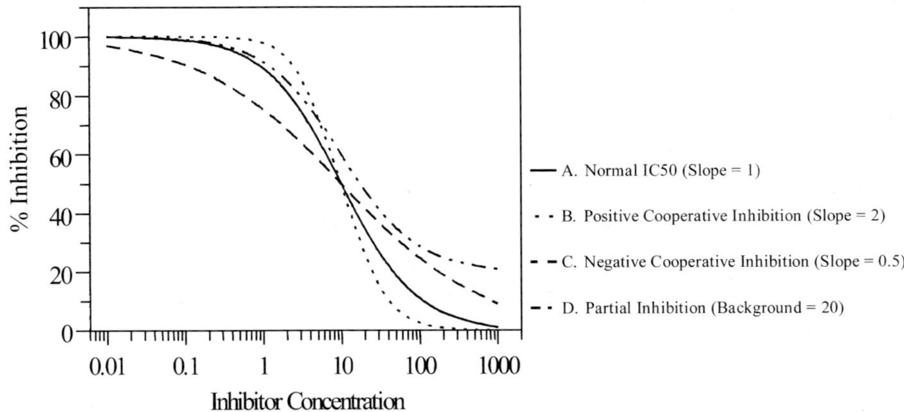

Figure 8. The effects of cooperative and partial inhibition on the IC$_{50}$ plot

Positive cooperative inhibition is analogous to substrate autoactivation, where the binding affinity of the second inhibitor molecule is increased in the presence of the first. This type of cooperativity is characterised by changes to the K_i value by the factor α_I (< 1), an enhanced extent of inhibition and steeper IC_{50} plots with increasing concentrations of the inhibitor (Curve B, Figure 8). Together with the alterations in the binding affinity, the cooperativity observed in the IC_{50} plots can also be attributed to changes in the product formation (K_p) by the factor γ, causing a decrease in the overall rate of the reaction when $\gamma < 1$. In the case where no changes to the effective catalytic rate constant are observed ($\gamma = 1$) the corresponding equation can be simplified by eliminating γ.

The opposite effect (i.e., slope is <1), decreased binding of the second inhibitor molecule is observed in the presence of the first (negative cooperativity), and full inhibition is not seen even at relatively high inhibitor concentrations (Curve C).

3.2.2.2 Partial inhibition

Partial inhibition is characterized by incomplete inhibition, even at saturating inhibitor concentrations. The shallow slope of the IC_{50} plots (Figure 9A) represents the case of partial inhibition, where inhibition is achieved at only one of the two catalytically active sites. Unlike one-site inhibition, the velocity of the reaction cannot be minimized to zero, as substrate-inhibitor-enzyme complex is still productive (S and I simultaneously bound and have access to the active oxygen). Both the ES and ESI complexes are equally able to form the product; therefore the K_p and V_{max} will be unchanged.

Partial inhibition is typically characterised by a value of $\delta > 1$ (Table 6), as the affinity of a second inhibitor molecule for a binding site is decreased compared to the first (e.g. 1'-OH midazolam inhibition by testosterone, Figure 9B). This type of inhibition may be a result of either partial interaction at both sites (e.g. testosterone effect on erythromycin[57]) or via competition at only one binding site; allowing the application of a simpler kinetic model for the data analysis in the latter case. An example of the latter type is the interaction between nifedipine and its analogue felodipine, where the presence of nifedipine in the active site partially shields only one of felodipine sites from the active oxygen.[55]

Another way of distinguishing partial from pure competitive inhibition is by plotting the rates of metabolism at fixed substrate concentrations in the presence of increasing inhibitor concentrations, as shown in Figure 9C. Unlike competitive inhibition (where the velocity of the reaction is minimized at high concentrations of inhibitor), the plot for partial or negative cooperative inhibition plateaus (ES and ESI are still productive) at a level dictated by the value of the interaction factor δ.

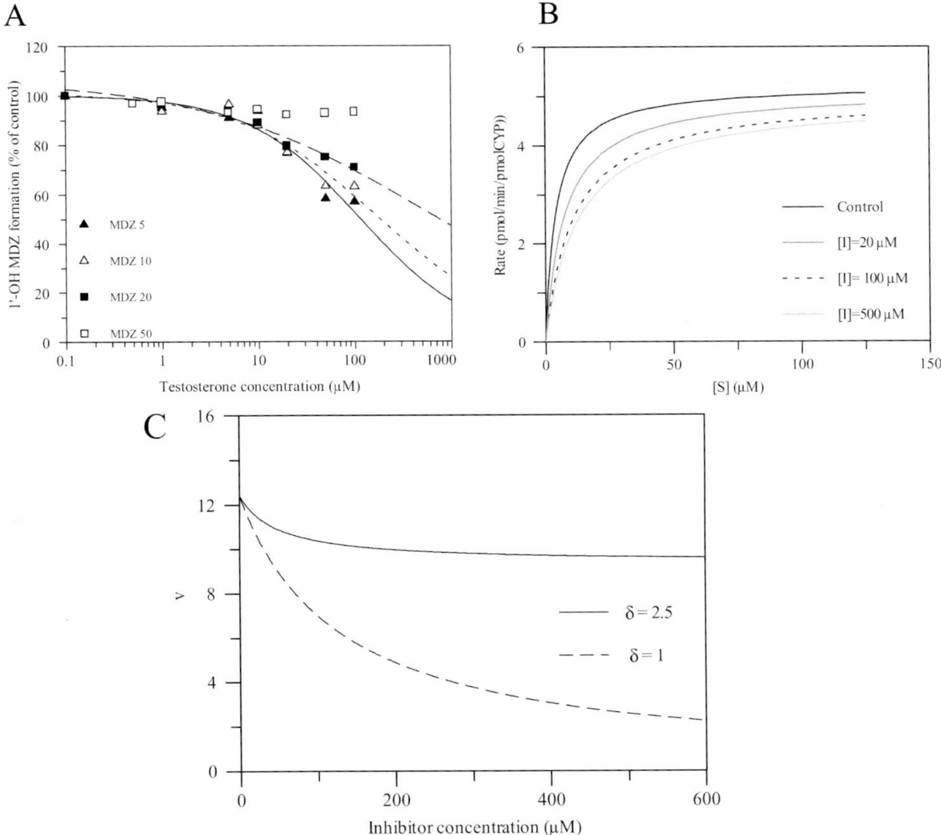

Figure 9. Partial inhibition with the increasing I concentrations example of the IC_{50} plots (A) and rate vs. [S]. Comparison between partial and pure competitive inhibition model at fixed S concentration (C).[55] IC50 plots for the effect of testosterone on midazolam 1'-hydroxylation, slope < 1 and high background (uninhibited activity) are associated with partial inhibition (A); nifedipine effect on felodipine cannot be explained by simple competitive inhibition at one site (C).

3.2.2.3 Inhibition of substrates displaying atypical kinetics

Kinetic models for interactions involving substrates demonstrating atypical kinetics can be derived from the simple two-site model (Figure 5) assuming cooperative substrate-binding sites within the active site ($\alpha < 1$). To simplify the data analysis, substrate-binding sites are assumed to be equivalent (therefore β equals 2 in the corresponding equations, and cancels out as V_{max} is equal to $2K_p[E]_t$). Interestingly, Shou *et al*[36] could show no superiority for a model when different ES and SE conformations are assumed.

Loss of sigmoidicity in the presence of a modifier is a common observation (e.g. midazolam, nifedipine, felodipine effect on testosterone[55]). One possible scenario occurs if an inhibitor causes a conformational change that prevents the interaction between two substrate-binding sites and the substrate-induced increase in the affinity of the vacant substrate site. High concentrations of these modifiers cause the loss of the substrate cooperativity, giving linear Eadie-Hofstee plots, normally seen for hyperbolic type of kinetics (Figure 10A).

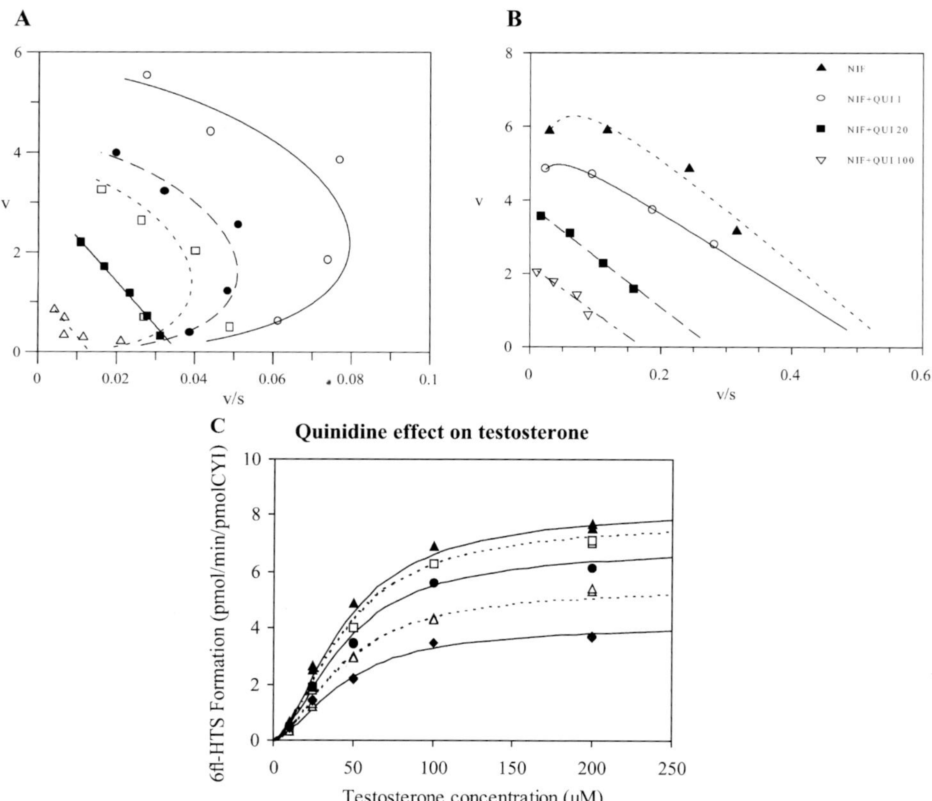

Figure 10. Loss of cooperativity (A) and substrate inhibition kinetic properties (B) in the presence of a modifier and an example of maintained sigmoidicity (C); Eadie-Hofstee plots for 6β-hydroxytestosterone and oxidised nifedipine formation in the presence of nifedipine and quinidine, respectively (A and B). Kinetic profiles for 6β-HTS formation at QUI concentrations of 0 µM (▲), 5 µM (□), 20 µM (●), 50 µM (△) and 100 µM (■); the solid and dashed lines represent the simultaneous fit to the Eq. 7.14 at 10, 25, 50, 100 and 200 µM TST concentration (C).

An analogous situation is observed for substrates with substrate inhibition kinetic properties. The reduced product formation at higher S concentrations associated with this phenomenon is defined by $\beta<1$ (K_p from SES). When γ is comparable to β, the rate of metabolite formation from IES complex (γK_p) is analogous to SES (βK_p) and the substrate inhibition trend remains. However, at high S and I concentrations a non-productive inhibitor complex dominates, changing the profile to a hyperbolic type. The example of this is shown in quinidine effect on nifedipine (Figure 10B).

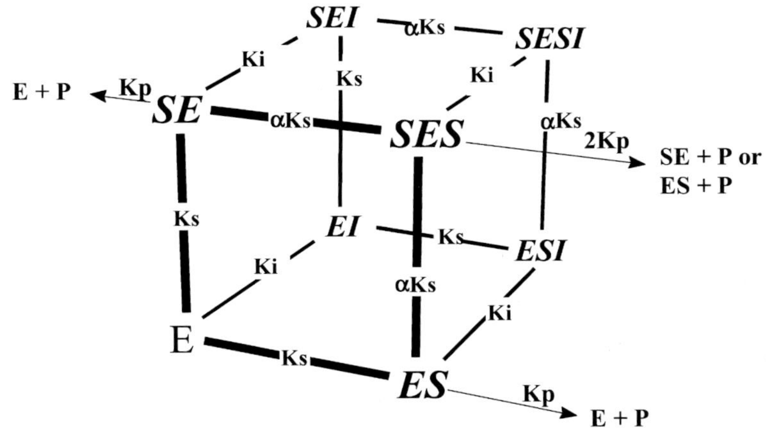

$$\frac{v}{V_{max}} = \frac{\dfrac{[S]}{K_S} + \dfrac{[S]^2}{\alpha\,K_S^2}}{1 + \dfrac{2[S]}{K_S} + \dfrac{[S]^2}{\alpha\,K_S^2} + \dfrac{[I]}{K_i} + \dfrac{2[S][I]}{K_S K_i} + \dfrac{[S]^2[I]}{\alpha\,K_S^2\,K_i}} \qquad (7.14)$$

Figure 11. Three-site kinetic model for an enzyme where S binds cooperatively in the presence or absence of I at a distinct site.

If, however, the substrate cooperativity remains apparent even at very high modifier concentrations, a more complex 3-site approach is required. These complex kinetic models assume that both substrate and effector can bind to two sites, and that one site is unique to either molecule.[37] The kinetic model shown in Figure 11 and the corresponding equation (Eq. 7.14) represents a case where a substrate binds cooperatively in the presence or absence of any concentration of inhibitor. The interaction between two substrate molecules and the sigmoidal properties of the substrate are unaffected by increasing inhibitor concentration (Figure 10C), suggesting that the inhibitor acts at a distinct effector site. Inhibition is not consistent with a competitive type, as the modifier changes V_{max}, rather than the substrate binding constant K_s. This three-site model has been applied to the effects of diazepam[37] and quinidine[42] on testosterone.

In the cases where binding of an inhibitor to the separate effector site causes an alteration in K_i ($\delta K_i > K_i$), consistent with a negative cooperative effect, partial inhibition is observed with the increasing inhibitor concentration (haloperidol effect on testosterone[42]).

3.2.3   Pathway Differential Effects

Another atypical characteristic observed for CYP3A4 is an alteration in regioselectivity, i.e. activation of the substrate metabolism at one position, while inhibiting the metabolite formation at the other. This phenomenon has been reported in various *in vitro* systems, with α-naphthoflavone[31, 44] and testosterone[56, 70] as modifiers. Existence of a separate effector site or two distinct and independent binding domains were suggested in order to describe this phenomenon.

A three-site kinetic model, with 2 distinct substrate-binding sites (each preferable for one midazolam pathway and defined by respective K_s and K_p values) best accommodated the pathway differential effects of quinidine[42] and testosterone[55] on midazolam. Competition for the mutual binding site between midazolam and testosterone/quinidine caused the inhibition of 1'-OH pathway. In contrast, binding of the modifier molecule (either quinidine or testosterone) to the distinct effector site causes an allosteric effect on the substrate site preferential for 4-OH midazolam, stimulating this pathway ($\lambda K_{p2} > K_{p2}$). The successful application of this model for the effects of testosterone and quinidine is encouraging as it illustrates that modifier effects are not always unique to one particular modifier.

3.3   Criteria for Selection of an Appropriate Multisite Kinetic Model in Prediction of Drug Interactions

Differential effects (observed particularly with α-naphthoflavone and quinidine as modifiers),[42, 63, 70] poor correlation among the CYP3A4 substrates and the range of variable and substrate-dependent effects[33, 56, 71, 72] indicate that each substrate may have a preferential position within the active site. The exact binding conformations depend on the substrates involved, their relative concentration and affinity for the enzyme. The complexity of a potential kinetic interaction with a modifier of CYP3A4 (inhibitor or activator) is highly dependent on the number of binding sites for both substrate and modifier, and their possible overlap.

The initial selection of a model and interaction factors is highly correlated with both the S and I kinetics (hyperbolic, substrate inhibition or sigmoidal), as illustrated by α and β values in Scheme 2. The generic two-site model (Figure 6) can accommodate a variety of inhibition/activation profiles observed for substrates with hyperbolic or substrate inhibition kinetic properties. Models with three binding sites (distinct effector-binding domain) are sometimes more appropriate to describe the interactions of substrates which show cooperative binding to the active site (e.g. testosterone).

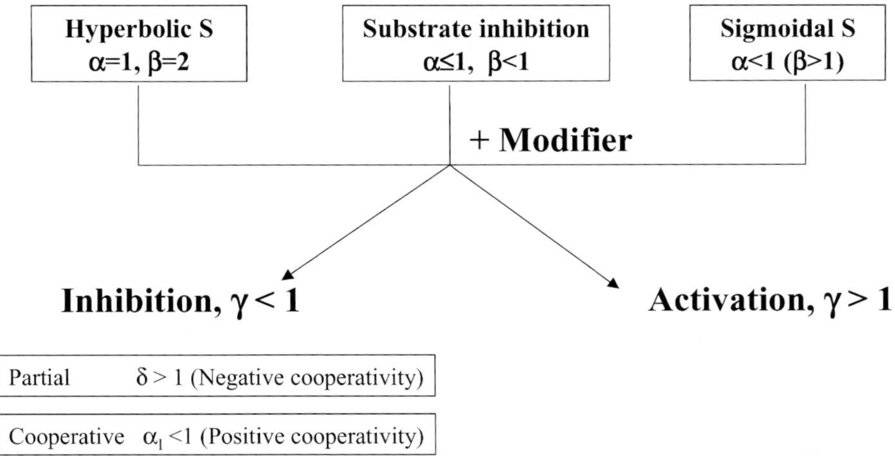

Scheme 2. Summary of interaction factors in multisite models

Whenever possible the simplest kinetic model should be employed to obtain the kinetic parameters and describe the observed interaction, and in some cases the inclusion of all possible enzyme species in the data analysis may not be necessary (e.g., steric restrictions resulting in the interaction at only one site). However, one has to be aware of certain limitations of the one-site model in the analysis of enzymes known to exhibit atypical interactions. For example, in the case of cooperative inhibition, a competitive inhibition model can generate a satisfactory fit for the range of low I concentrations, but the enhancement of the inhibition observed with increasing concentrations of inhibitor cannot be predicted. This was observed in the effect of felodipine on 1'-OH midazolam formation, shown in Table 6. Additionally, certain data sets may show some of the features of competitive inhibition, but not consistently. The example of this is the substrate-concentration dependence of IC_{50} values with the unchanged S_{50} values observed in quinidine effect on testosterone.[42] Application of one-site models for interactions involving substrates with sigmoidal (testosterone, diazepam) or substrate-inhibition kinetic properties (terfenadine, nifedipine) is particularly problematic and misuse of a one-site model may lead to erroneous estimation of kinetic parameters which are important in *in vivo* prediction (CL_{int}, CL_{max}, K_i, K_a).

In some instances more than one kinetic model can be used to describe the inhibition profiles observed. The least number of parameters, lowest standard errors of the parameter estimates, and consistency with kinetic properties of both the substrate and modifier represent the principal criteria for the selection of a particular model. Goodness of fit is determined by comparison of statistical parameters (F test, χ^2 and Akaike information criterion values) between the models and a reduction in the standard errors of the parameter estimates. In some cases the estimates of kinetic parameters (K_i/K_a) from different but similar models are comparable, indicating the robust nature of multisite kinetic models.

The use of multiple substrates, at various substrate concentrations, is recommended for characterisation of CYP3A4 interactions due to the phenomena of multisite kinetics. Selection of substrates for these drug interaction studies should be based on three distinct CYP3A4 substrate groups first identified by Kenworthy *et al* [71], including (1) benzodiazepine group (2) large molecular weight group (erythromycin, cyclosporine, testosterone and more recently quinidine) and (3) nifedipine/felodipine, each showing distinctly different behaviour towards a range of different modifiers. Multisite kinetic analysis of mutual interactions of prototypical substrates (e.g midazolam, testosterone, nifedipine[55]) supported the hypothesis of distinct and preferential binding domains for each substrate subclass. However, these interaction studies also indicated the existence of one common site for all subgroups of CYP3A4 substrates. Therefore, in addition to different subclasses, application of a range of substrates with differential kinetic properties is relevant for the accurate evaluation of a potential inhibitor. The applicability of parameters determined by multisite approach (K_i or δK_i) for prediction of *in vivo* drug-drug interactions remains to be evaluated.

4   *IN VIVO* RELEVANCE OF *IN VITRO* ATYPICAL KINETICS AND CURRENT PERSPECTIVE

The *in vivo* implications of *in vitro* atypical kinetic phenomenon for CYP3A4 remains uncertain. Yet it is important to address the practical problem of dealing with data that cannot be described by the Michaelis-Menten model in order to make *in vivo* predictions, particularly as it is the major human cytochrome, CYP3A4, for which these complications are most commonly noted. Considerations of how we can accommodate these characteristics into strategies for *in vitro-in vivo* scaling have been discussed here and elsewhere.[23] The kinetics in the *in vitro* subsystem will always need to be characterized in a fully comprehensive manner with limits that may extend beyond those observed *in vivo*. Only then can we abstract useful parameters from the *in vitro* kinetic analysis for *in vivo* extrapolation.

Within the drug metabolism literature there is extensive use of Michaelis-Menten kinetics and that is entirely appropriate in many cases when there is clear involvement of, for example CYP2D6, 1A2 or 2C19. There are also literature examples of the use of standard Michaelis-Menten hyperbolic curves forced through data that clearly show atypical kinetic features apparently not appreciated by the investigators. In some of these cases the paucity of data points precludes any meaningful selection of an alternative model. Many authors seem uncertain of how to deal with atypical kinetics and there is a common held belief that it is purely a microsomal phenomenon. It is better to limit discussion of such data to a purely qualitative analysis rather than to adopt unsuitable models.

Autoactivation is not a phenomenon limited only to microsomal incubations. Recent studies have shown sigmoidicity for CYP3A4 reactions in freshly isolated rat hepatocytes using dextromethorphan,[22] as well as in cryopreserved human hepatocytes using alprazolam, flunitrazepam and triazolam.[73] It has been proposed that any endogenous activator(s) would be washed out of both *in vitro* preparation during either the isolation procedures, in the case of hepatocytes, or the homogenisation/centrifugation steps, in the microsomal case. It is of interest that the extent of sigmoidicity for these reactions are at least as pronounced in isolated hepatocyte systems as in corresponding microsomal preparations from the same species.

In addition, heteroactivation of midazolam by α-naphthoflavone,[13] diclofenac and warfarin metabolism by quinidine[53, 54] and of alprazolam, flunitrazepam and triazolam by testosterone[73] has been demonstrated in human hepatocytes. It would therefore appear prudent to assume that *in vivo* the CYP3A system is activated as endogenous steroid hormones (e.g. testosterone and progesterone) and dietary flavones are established as the prototypic activators. Thus the phenomenon of activation should be incorporated into the treatment of *in vitro* data when prediction of *in vivo* events is the aim of the study. This will not be a trivial issue, and heteroactivation is likely to be an important source of variability between individuals due to different dietary intakes and hormonal changes that will compound further the issue of variability in the expression of these enzymes.

The clinical significance of *in vitro* activation is still largely unexplored.

However, carbamazepine when coadministered with felbamate provides a good example of heteroactivation as a mechanism for a clinical drug-drug interaction.[74] Meta analysis of 8 separate studies shows that coadministration of the heteroactivator felbamate lowers the steady-state plasma concentration of carbamazepine, as a result of increased clearance. In this particular case the degree of effect is not large (25%), but it is likely that further examples of possibly clinically relevant CYP3A4 cooperativity will be uncovered soon.

In addition, there is *in vivo* confirmation of *in vitro* activation effects reported in animal studies. It is unfortunate that species differences in CYP3A have hampered identification of a good animal model to allow a detailed exploration of activation in systems like perfused organs as well as *in vitro*.[75] The most comprehensive work has been carried out on the diclofenac-quinidine interaction. A 50% decrease in diclofenac steady-state plasma concentrations was demonstrated upon the coadministration of quinidine in rhesus monkeys that was consistent with interaction seen *in vitro*.[52, 53] Quinidine has also been shown to stimulate meloxicam metabolism by increasing the contribution of CYP3A4 over CYP2C9 to the overall metabolism, as a result of heteroactivation.[68] In certain circumstances the metabolic pathway activated (e.g. 10-hydroxylation of R-warfarin by quinidine), although confirmed *in vivo*, is not a substantial enough pathway to significantly alter the overall elimination, and hence pharmacokinetics, of the drug. This is probably a very common effect as many drugs have a partial dependence on CYP3A4 and for others activation is only apparent in one pathway. A case in point is triazolam which has 2 equally important metabolites, one activated (1'-hydroxylation) and the other inhibited (4-hydroxylation) by testosterone, thus there is no net effect in the *in vitro* depletion half-life, and presumably the *in vivo* clearance, in the presence of this modifier.[76]

An additional important consideration is the impact of activation on the estimation of metabolic stability of drugs by the use of drug substrate depletion. It is common practice within the Pharmaceutical Industry to use a single, low substrate concentration (often 1μM, based on the rationale that this concentration should be well below the unknown K_m value). If consideration is not given to the phenomenon of activation then clearance is likely to be underestimated. Taking the example of diazepam, the microsomal half-life at 1μM is five times longer than at 100μM (close to the substrate concentration for CL_{max}) reflecting the nonactivated and fully activated clearance. There is also an analogous decrease in microsomal half-life of diazepam at the low substrate concentration in the presence of testosterone.[76]

The Pharmaceutical Industry, and to a lesser degree academia, live in the 'age of prediction', where the major objective of an *in vitro* system is to provide valuable quantitative information on metabolic stability and inhibition potential that complements other pre-clinical information to characterize the pharmacokinetic properties of a new drug. It is the expectation of the Regulatory Authorities[5, 77, 78] that this information should be submitted as part of the regulatory package. The importance placed on this *in vitro* information is substantial and expectations are high. Some might regard this enthusiasm for *in vitro* methodology as a little premature! Definitely there are still major issues to be addressed in the interpretation of much *in vitro* information, especially in the area of drug-drug interactions and this is limiting progress within the pharmaceutical industry. The high level of activity surrounding the CYP and other drug metabolising enzymes continues to provide much needed molecular information and the further technological developments that will follow will require adoption of kinetic approaches as discussed here. Therefore, continued refinement is necessary as our understanding increases to ensure optimal use of *in vitro* kinetic data.

REFERENCES

1. Obach, S.; Baxter, J,G.; Liston, T.E.; Silber, B.M.; Jones, B.C.; MacIntyre, F.; Rance, D.J.; Wastall, P. *J Pharmacol. Exp.Ther.* **1997** *283*, 46.
2. Obach, S.R. *Drug Metab. Dispos.***1999**, *27*, 1350.
3. Carlile, D.J.; Stevens, A.J.; Ashforth, E.I.L.;Waghela, D.; Houston, J.B. *Drug Metab. Dispos.* **1998**, *26*, 216.
4. Obach, R.S.; Reed-Hagen, A.E. *Drug Metab. Dispos.* **2002**, *27*, 1350.
5. Tucker, G.T.; Houston, J.B.; Huang, S-M. *Clin. Pharmacol. Ther.* **2001**, *70*, 103.
6. Hickman, D.; Wang, J-P.; Wang, Y.; Unadkat, J.D. *Drug Metab. Dispos.* **1998**, 26, 207.
7. Chauret, N.; Gauthier, A.; Nicoll-Griffith, D.A. *Drug Metab. Dispos.***1998**, *26, 1*.
8. Easterbrook, J.; Lu, C.; Sakai, Y.; Li, A.P. *Drug Metab. Dispos.* **2001**, *29, 141*.
9. Busby Jr., W.F.; Ackermann, J.M.; Crespi,C.L. . *Drug Metab. Dispos.* **1999, 27**, 266.
10. Shaw, P.M.; Hosea, N.A.; Thomspon, D.V.; Lenius, J.M.; Guengerich, F.P. *Biochem. Biophys.***1997**, *348*, 107.
11. Yamazaki, H.; Nakajima, M.; Nakamura, M.; Asahi, S.; Shimada, N.; Gillam, E.M.J.; Guengerich, F.P.; Shimada. T.; Yokoi, T. *Drug Metab* Dispos **1999**, *27*, 999.
12. Venkatakrishnan, K.; von Moltke, L.L.; Court, M.H.; Harmatz, J.S.; Crespi, C.L.; Greenblatt, D.J. *Drug Metab. Dispos.* **2000,** *28*, 1493-1504.
13. Mäenpää, J.; Hall, S.D.; Ring, B.J.; Strom, S.C.; Wrighton, S.A. *Pharmacogenetics* **1998, *8*,** 137.
14. Voice, M.W., Zhang, Y.; Wolf, C.R.; Burchell, B.; Friedberg, T. *Arch. Biochem. Biophys.* **1999**, *366*, 116.
15. Guengerich, F.P. *Annu. Rev. Pharmacol. Toxicol.* **1999**, *39*, 1.

16. Schrag, M.L.; Wienkers, L.C. *Drug Metab. Dispos.* **2000**, *28*, 1198.

17. Tran, T.H.; von Moltke, L.L.; Venkatakrishnan, K.; Granda, B.W.; Gibbs, M.A.; Obach, R.S.; Harmatz, J.S.; Greenblatt, D.J.; *Drug Metab. Dispos.* **2002**, *30*, 1441.

18. Crespi, C.L.; Stresser, D.M. *J Pharm. Toxicol. Meth.* **2000**, 44, 325.

19. Obach, S. *Drug Metab.Dispos.* **1997**, *25*, 1359.

20. McLure, J.A; Miners, J.O.; Birkett, D.J. *Br. J. Clin. Pharmacol.* **2000**, *49*, 453.

21. Segel, I.H. In *Enzyme Kinetics: Behaviour and Analysis of Rapid Equilibrium and Steady State enzyme Systems*; Wiley & Sons Inc: New York, NY, **1975**.

22. Witherow, L.Y.; Houston, J.B. *J. Pharmacol. Exp. Ther.* **1999**, *58*, 290

23. Houston, J.B.; Kenworthy, K.E. *Drug Metab. Dispos.* **2000**, *28*, 246.

24. Houston, J.B. *Biochem Pharmacol* **1994,** *47*, 1469.

25. Korzekwa, K.R.; Krishnamachary, N.; Shou, M.; Ogai, A.; Parise, R.A.; Rettie, A.E.; Gonzalez, F.J.; Tracy, T.S. *Biochemistry* **1998**, *37*, 4137.

26. Hutzler, J.M.; Hauer, M.J.; Tracy, T.S. *Drug Metab. Dispos.* **2001a,** *29*, 1029.

27. Hutzler, J.M.; Frye, R.F.; Korzekwa, K.R.; Branch, R.A.; Huang, S-M.; Tracy, T.S. *Eur. J. Pharm. Sci.* **2001b,** *14*, 47.

28. Williams, J.A.; Ring, B.J.; Cantrell, V.E.; Campanelle, K.; Jones, D. R.; Hall, S. D.; Wrighton, S.A. *Drug Metab Dispos* **2002**, *30*, 1266.

29. Stone, A.N.; Mackenzie, P.I.; Galetin, A.; Houston, J.B.; Birkett, D.J.; Miners, J.O. (unpublished data)

30. Schwab, G.E.; Raucy, J.L.; Johnson, E.F. *Mol. Pharmacol.* **1988**, *33*, 493.

31. Ueng, Y-F.; Kuwabara, T.; Chun, Y-J.; Guengerich, F.P. *Biochemistry* **1997**, *36*, 370.

32. Schmider, J.; Greenblatt, D.J.; von Moltke, L.L.; Harmatz, J.S.; Shader, R.I. *J. Pharmacol. Exp. Ther.* **1995,** *275,* 592.

33. Lu, P.; Lin, Y.; Rodrigues, A.D.; Rushmore, T.H.; Baillie, T.A.; Shou, M. *Drug Metab. Dispos.* **2001**, *29*, 1473.

34. Kerr, B.M.; Thummel, K.E.; Wurden, C.J.; Klein, S.M.; Kroetz, D.L.; Gonzalez, F.J.; Levy, R.H. *Biochem. Pharmacol.* **1994**, *47*, 1969.

35. Andersson, T.; Miners, J.O.; Veronese, M.E.; Birkett, D.J. *Br. J. Clin. Pharmacol.* **1994**, *38*, 131.

36. Shou, M.; Mei, Q.; Ettore, W. Jr.; Dai, R.; Baillie, T.; Rushmore, T.H. *Biochem. J.* **1999**, *340*, 845.

37. Kenworthy, K.E.; Clarke, S.E.; Andrews, J.; Houston, J.B. *Drug Metab. Dispos.* **2001**, *29*, 1.

38. Bicknell, H.A.; Worboys, P.D.; Evans, D.; Houston, J.B. *Proceeding of 7th Stowe School Symposium on Drug Metabolism.* **1999**, 1.

39. Lee, C.A.; Kadwell, S.H.; Kost, T.A.; Serabjit-Singh, C.J. *Arch. Biochem. Biophys.* **1995**, *319*, 157.

40. Harlow, G.R.; Halpert, J.R. *Proc. Natl. Acad. Sci.* **1998**, *95*, 6636.

41. Johnson, T.N.; Taneer, M.S.; Taylor, C.J.; Tucker, G.T. *Br. J. Clin. Pharmacol.* **2001**, *51*, 451.

42. Galetin, A.; Clarke, S.E.; Houston J.B.; *Drug Metab. Dispos.* **2002a,** 30, 1512.

43. Narasimhulu, S.; Havran, L.M.; Axelsen, P.H.; Winkler, J.D. *Arch. Biochem. Biophys.* **1998**, *353*, 228.

44. Shou, M.; Dai, R.; Cui, D.; Korzekwa, K.R.; Baillie, T.A.; Rushmore, T.H. J. *Biol. Chem.* **2001a**, *276*, 2256.
45. Schrag, M.L.; Wienkers, L.C. *Drug Metab. Dispos.* **2001a**, *29*, 70.
46. Lin, Y.; Lu, P.; Tang, C.; Mei, Q.; Sandig, G.; Rodrigues, A.D.; Rushmore, T.H.; Shou, M. *Drug Metab. Dispos.* **2001**, *29*, 368.
47. Kronbach, T.; Mathys, D.; Umeno, M.; Gonzalez, F.J.; Meyer, U.A. *Mol. Pharmacol.* **1989**, *36*, 89.
48. Ghosal, A.; Satoh, H.; Thomas, P.E.; Bush, E.; Moore, D. *Drug Metab. Dispos.*
49. von Moltke, L.L.; Greenblatt, D.J.; Schmider, J.; Duan, S.X.; Wright, C.E.; Harmatz, J.S.; Shader, R.I. *J. Clin. Pharmacol.* **1996**, *36, 783.*
50. Khan, K.K.; He, Y.Q.; Domanski, T.L.; Halpert, J.R. *Mol Pharmacol.* **2002**, 495.
51. Kenworthy , K.E.; Andrews, J.; Clarke, S.E.; Houston, J.B. *Proceeding of 7ᵗʰ Stowe School Symposium on Drug Metabolism.* **1999**, 44.
52. Tang, C.; Shou, M.; Mei, Q.; Rushmore, T.H.; Rodrigues, A.D. *J. Pharmacol. Exp. Ther.* **2000**, *293*, 453.
53. Ngui, J.S.; Tang, W.; Stearns, R.A.; Shou, M.; Miller, R.R.; Zhang, Y.; Lin, J.H.; Baillie, T.A. *Drug Met. Dispos.* **2000**, *28*, 1043.
54. Ngui, J.S.; Chen, Q.; Shou, M.; Wang, R.W.; Stearns, R.A.; Baillie, T.A.; Tang, W. *Drug Metab. Dispos.* **2001**, *29*, 877.
55. Galetin, A.; Clarke, S.E.; Houston J.B. Drug Metab. Dispos. **2003**, *31*, in press.
56. Wang, R.W.; Newton, D.J.; Liu, N.; Atkins, W.M.; Lu, A.Y.H. *Drug Metab. Dispos.* **2000**, *28*, 360.
57. Wang, R.W.; Newton, D.J.; Scheri, T.D.; Lu, A.Y.H. *Drug Metab. Dispos.* **1997**, *25,* 502.
58. Shou, M.; Grogan, J.; Mancewicz, J.A.; Krausz, K.W.; Gonzalez, F.J.; Gelboin, H.V.; Korzekwa, K.R. *Biochemistry* **1994**, *33*, 6450.
59. Domanski, T.L.; HE, Y-A.; Khan, K.K.; Roussel, F.; Wang, Q.; Halpert, J.R. *Bichemistry* **2001**, *40*, 10150.
60. Hosea, N.A.; Miller, G.P.; Guengerich, F.P. *Biochemistry* **2000**, *39*, 5929.
61. Tang, W.; Stearns, R.A. *Curr. Drug. Metab.* **2001**, *2*, 185.
62. Shou, M.; Lin, Y.; Lu, P.; Tang, C.; Mei, Q.; Cui, D.; Tang, W.; Ngui, J.S.; Lin, C.C.; Singh, R.; Wong, B.K.; Yergey, J.A.; Lin, J.H.; Pearson, P.G.; Baillie, T.A.; Rodrigues, A.D.; Rushmore, T.H. *Curr. Drug Metab.* **2001b**, *2*, 17.
63. Domanski,, T.L.; HE, Y-A.; Harlow, G.R.; Halpert, J.R. *J. Pharmacol. Exp. Ther.* **2000**, *293*, 585.
64. Nielsen, T.L.; Rasmussen, B.B.; Flinois, J-P.; Beaune, P.; Brøsen, K. *J Pharmacol. Exp.Ther.***1999** *289*, 31.
65. Kerlan, V.; Dreano, Y.; Bercovici, J.P.; Beaune, P.H.; Floch, H.H.; Berthou, F. *Biochem. Pharmacol.* **1992**, *44*, 1745.
66. Tang, W.; Stearns, R.A.; Kwei, G.Y.; Iliff, S.A.; Miller, R.R.; Egan, M.A.; Yu, N.X.; Dean, D.C.; Kumar, S.; Shou, M.; Lin, J.H.; Baillie, T.A. *J. Pharmacol. Exp. Ther.* **1999**, *291*, 1068.
67. Feierman, D.E.; Lasker, J.M. (1996) *Drug Metab. Dispos.* **1996**, 24, 932.
68. Ludwig, E.; Jochen, S.; Beschke, K.; Ebner, T. *J. Pharmacol. Exp. Ther.* **1999**, *290,* 1.

69. Baron, J.M.; Goh, L.B.; Yao, D.; Wolf, R.C., Friedberg, T. *J Pharmacol. Exp.The.* **2001**, *296*, 351.
70. Schrag, M.L.; Wienkers, L.C. *Arch. Biochem. Biophys.* **2001b**, *391*, 49.
71. Kenworthy, K.; Bloomer, J.C.; Clarke, S.E.; Houston, J.B. *Br. J. Clin. Pharmacol.* **1999**, *48*, 716.
72. Stresser, D.M.; Blanchard, A.P.; Turner, S.D.; Erve, J.C.L.; Dandeneau, A.A.; Miller, V.P.; Crespi, C.L. *Drug Metab. Dispos.* **2000**, *28*, 1440.
73. Rawden, C.H.; Hallifax, D.; Houston, J.B. (submitted for publication).
74. Egnell, A-C.; Houston J.B.; Boyer, S. J. Pharmacol. Exp. Ther. **2003**, *305*, 1251.
75. Chan, Q.; Strauss, J.; Ngui, J.; Fenyk-Melody, J.; Kerrick, G.; Stearns, R.; Evans, D.; Tang, W. *Drug Met.Rev.* **2002**, *34*, 64.
76. Rawden, C.H.; Tindall, A.; Hallifax, D.; Houston, J.B. (submitted for publication).
77. FDA Guidance for Industry: *Drug Metabolism/Drug Interaction Studies in the Drug Development Process: Studies in Vitro.* http://www.fda.gov/cder/guidance.htm
78. Yuan, R.; Madani, S.; Wei, X-X.; Reynolds, K.; Huang, S-M. *Drug Metab. Dispos.* **2002**, *30*, 1311.

Chapter 8

Cytochrome P450 Reaction Phenotyping

Larry C. Wienkers, Ph.D. and Jeffrey C. Stevens, Ph.D.

Global Drug Metabolism, Pharmacia Corporation,
301 Henrietta St., 7265-300-306, Kalamazoo, MI 49007

ABSTRACT

Clearance of a drug from the body occurs by excretion of the unchanged drug or by biotransformation leading to one or more metabolites. For drugs whose route of elimination occurs by metabolism, changes in the particular enzymatic pathway can have marked effects on the safety and/or efficacy of the drug. This effect is particularly pronounced if clearance is mediated via a single enzymatic pathway, where inter-individual genetic variability in enzyme expression may lead to large differences in systemic drug exposure. Furthermore, elimination pathways that are governed via metabolism are often susceptible to inhibition or induction by the co-administration of a second drug. For many drugs, elimination is often times mediated *via* enzymatic oxidation by a member of the cytochrome P450 superfamily of enzymes. Therefore, an understanding of the P450 enzymes responsible for the elimination of a drug is important to avoid and/or predict untoward effects associated with multiple drug therapy. This chapter describes some of the history and objectives of P450 reaction phenotyping studies. First, fundamental considerations of experimental design, including the conduct and interpretation of enzyme kinetic determinations are discussed, followed by an evaluation of the various in vitro experiments which make-up what is collectively described as P450 reaction phenotyping, including the use of correlation analysis of enzyme activities, chemical and antibody inhibitors, and expressed P450s. Particular emphasis is placed on describing the strengths and weaknesses associated with the design and interpretation of P450 reaction phenotyping studies.

1   INTRODUCTION

Drug metabolism is an important component of the various physiological processes that contribute to the pharmacokinetic profile of therapeutic agents. In general terms, most drug metabolism reactions may be characterized as the biotransformation of lipophilic, nonpolar pharmacologically active molecules to polar, pharmacologically inactive products (e.g., metabolites). Traditionally, drug metabolism reactions have been categorized into two groups; phase-I reactions which involve the addition or removal of a functional group to increase the polarity of a compound (e.g., phenol formation or O-dealkylation) and phase-II reactions which involve the conjugation of the drug with hydrophilic molecules such as sulfates and glucuronides. While both reactions are important determinants in the disposition of a drug, greater than 60% of therapeutic agents prescribed today are cleared via phase-I oxidative metabolism by cytochrome P450 (CYP) enzymes, and as a consequence, a significant amount of attention has centered on this family of enzymes.[1] The importance of the cytochromes P450 in drug metabolism is well recognized by global regulatory agencies in that the individual P450-mediated pathways associated with the metabolism of new chemical entities (NCE) must be characterized as part of a new drug application (NDA). CYP reaction phenotyping may generally be described as a set of in vitro experiments with the collective goal of identifying the human P450 enzymes responsible for the biotransformation of a given drug. Such information may help to retrospectively address or even predict potential variability in patient pharmacokinetics associated with a given drug therapy. For instance, if a drug is principally metabolized by CYP2D6, one in ten patients may be a poor metabolizer (PM) of this drug since 5-10% of Caucasians are devoid of this enzyme.[2] Additionally, for drugs that are predominantly metabolized by CYP2C19, between 12-23% of Asians may be categorized as PMs[3] (An overview of polymorphic expression and pharmacogenetics is provided in Chapter 11). Given that exposure is central to the magnitude and duration of pharmacological activity for most drugs, P450-mediated clearance of a drug may represent an important determinant in efficacy. Conversely, undesired effects of some drugs whose pharmacological action is governed through CYP2D6 or CYP2C19 may be exaggerated in individuals who are genetically deficient in one of these enzymes. In addition to polymorphically expressed enzymes, identification of the specific P450(s) responsible for the metabolism of a drug is also useful in understanding or anticipating many clinically important drug-drug interactions.[4]

During the course of drug therapy, patients may receive two or more drugs simultaneously. If these drugs are metabolized or somehow interact with the same drug-metabolizing enzyme, a condition exists which may manifest itself in a drug-drug interaction.[5] The clinical significance of a pharmacokinetically-based drug interaction is primarily dependent upon how well the change in drug concentration is tolerated by the system. For compounds with a narrow therapeutic index, an increase in drug exposure may precipitate harmful side effects. The magnitude of change in the concentration of a drug is dependent upon the number of elimination pathways associated with clearance.[6] If drug clearance is largely dependent upon a single enzyme, inhibition and/or induction of the enzyme will have a profound effect on drug clearance in comparison with a drug that has several metabolic routes of elimination. For example, for a drug primarily metabolized by CYP2C9 (e.g., (S)-warfarin), one would predict that the rate of metabolism of the drug would be *increased* by rifampin and other drugs known to induce this P450.[7] Alternatively, the rate of metabolism of the drug would be predicted to be *decreased* by co-administration of fluconazole and other drugs that inhibit CYP2C9 activity.[8] In this light, CYP2C9 inducers may attenuate the therapeutic effect of a drug that is metabolized by this enzyme [9] while CYP2C9 inhibitors exaggerate the pharmacologic and toxic effects of the drug.[10]

Interestingly, while oxidation reactions of xenobiotics have been recognized for over 50 years,[11-13] the concept of P450 reaction phenotyping as a course of study in drug development is a relatively new scientific endeavor. The origin of P450-dependent metabolism and ultimately reaction phenotyping stems from the finding of a unique carbon monoxide binding pigment in rat liver microsomes first reported in 1958.[14] This observation was soon followed by the work of Sato and Omura, who coined the name 'P450' for the hepatic pigment which possessed a maximal absorbance at 450 nm when bound to carbon monoxide in the reduced state.[15] The physiological role of the P450 as an oxygen-activating terminal oxidase was established for microsome-catalyzed reactions of a wide variety of substrates through a series of photochemical action spectrum experiments.[16,17] These studies suggested that a single P450 form having broad substrate specificity was participating in all reactions since each produced the same spectrum. However, in a relatively short period of time, evidence for the presence of multiple forms of P450 in liver microsomes was demonstrated. Welch et al. studied the 6β–, 7α–, and 16α–hydroxylation of testosterone by rat liver microsomes, and found that one hydroxylation pathway could be selectively stimulated or inhibited without influencing the other pathways.[18] In addition, Sladek and Mannering observed that the cytochrome P450 in liver microsomes of methylcholanthrene-treated rats could be distinguished spectrally from the P450 of untreated rats.[19] However, it was not clear whether these two spectral patterns represented distinct molecular species of cytochrome P450 or rather two interconvertible physical states of a single enzyme. The first successful purification and reconstitution of a cytochrome P450 system provided an enzymatic approach for separation of P450s into distinct proteins.[20] In subsequent experiments, two isolated and reconstituted P450 forms were found to exhibit different specificities for benzphetamine and benzopyrene metabolism.[21,22] These observations led to an aggressive effort to catalog information regarding individual P450s (for both experimental animals and humans) and their catalytic activities both in vitro and in vivo.[23,24] Over the course of the following decade, the sequencing, cloning and expression of cytochrome P450s in various systems was established,[25-27] with the first commercial source of recombinant P450 enzymes made available in the early1990s.[28] As the number and diversity of the P450 enzymes expanded, a systematic classification that grouped P450s into families, subfamilies and single enzymes based upon amino acid sequence was developed.[29] In humans, at least 53 different CYP genes have been categorized into eighteen families.[30] Of the eighteen human P450 families known, enzymes from the CYP1, CYP2 and CYP3 families carry out most xenobiotic biotransformation processes. In humans, the most important CYPs from the point of view of drug metabolism are CYP1A2, CYP2B6, CYP2C8, CYP2C9, CYP2C19, CYP2D6, CYP2E1 and CYP3A4/5.

Relatively few substrates have been described for CYP1A2, -2B6 and –2E1. Human CYP1A2 is primarily a hepatic form and accounts for ~15% of the total P450 expressed in the liver.[31,32] Substrates for CYP1A2 include: caffeine,[33] bropirimine,[34] tacrine,[35] mexiletine,[36] and theophylline.[37] In addition, CYP1A2 also participates in the metabolic activation of a variety of environmental toxicants.[38] The amount of CYP2B6 expressed in the liver appears to be low and highly variable.[39] Substrates of CYP2B6 include the HIV-1 reverse transcriptase inhibitor nevirapine,[40] bupropion,[41,42] 7-ethoxy-4-trifluoromethylcoumarin (7-EFC), (S)-mephenytoin (N-demethylation reaction),[43] RP 73401[44] and selegiline.[45] In addition, CYP2B6 expression appears to be co-regulated with CYP3A4 or at least to be increased by the same inducers.[46] CYP2E1 accounts for about 6% of total hepatic P450 content, with expression regulated by many physiological and environmental factors.[47,48] Among the relatively few drugs that are substrates for CYP2E1 are dapsone,[49] disulfiram,[50] acetaminophen,[51] chlorzoxazone[52] and enflurane.[53] However, CYP2E1 does play an important role in the metabolism of a large number organic solvents and anesthetics.[54,55]

The CYP2C subfamily consists of three members in human liver, namely CYP2C8, CYP2C9, and CYP2C19. It is the second most abundant P450 subfamily, representing about 20% of the total hepatic P450.[32] CYP2C8 is expressed at very low levels in liver, and to date has not been demonstrated to play an important role in drug metabolism. However, the oncology drug, taxol, is partly metabolized by CYP2C8[56] as is the antiepileptic drug carbamazepine[57] and amodiaquine.[58] CYP2C8 is also involved in the metabolism of retinol and retinoic acid.[59] CYP2C9 is the major 2C enzyme expressed in human liver,[60] and has been shown to be genetically polymorphic with different alleles that produce protein variants.[61] CYP2C9 does have a major role in the metabolism of several clinically important drugs, including many weakly acidic drugs such as the non-steroidal anti-inflammatory compounds,[62,63] in addition to (S)-warfarin,[64] lornoxicam,[65] tolbutamide[66] and phenytoin.[67] As described previously, CYP2C19 is polymorphically expressed, with 5-30% of the population exhibiting the PM phenotype and subsequently expressing less active or completely inactive enzyme. CYP2C19 is involved in the metabolism of a variety of drugs including some barbiturates,[68,69] (S)-mephenytoin,[70] (R)-warfarin,[71] omeprazole[72] and proguanil.[73]

Polymorphic expression is also characteristic of the one functional gene in the CYP2D subfamily, namely CYP2D6.[74] CYP2D6 represents between 1 and 5% of total P450,[32,75] with 5-10% of the Caucasian population deficient (PM) with respect to this enzyme.[76] While CYP2D6 represents a relatively small portion of the total hepatic P450, this enzyme is responsible for metabolizing a significant number of medications.[1] For example, CYP2D6 is responsible for the metabolism of certain opioids,[77] neuroleptics,[78] antidepressants[79] and cardiac medications.[80] Because these medications are often used in combination therapy and may have narrow therapeutic indices, CYP2D6 is often associated with clinically significant pharmacokinetic interactions.[81,82] In light of the numerous drugs on the market that are metabolized by CYP2D6 and the polymorphic nature of the expression of this protein, many pharmaceutical companies seek to avoid CYP2D6-mediated metabolic pathways associated with NCEs in early drug discovery.[83]

For the CYP3A subfamily, CYP3A4, CYP3A5 and CYP3A7 are considered important for human drug metabolism. Together, these enzymes comprise ~30% of hepatic P450 content.[32] Within this subfamily, CYP3A4 is the most abundant form in human liver and is expressed in several tissues, with expression in the liver and small intestine of primary interest due to their importance in presystemic or 'first-pass' clearance of orally administered drugs.[84,85] CYP3A5 is a relatively minor, polymorphic form, with 25-40% of individuals expressing appreciable levels in the liver.[86] The members of the CYP3A subfamily have overlapping substrate specificities and participate in about half of the P450 drug oxidation reactions reported to date.[1,87] Substrates of CYP3A4 vary in size from small molecules, such as acetaminophen, to extremely large compounds such as cyclosporin A (molecular weights of 151 and 1201, respectively).[88] In addition, CYP3A4 substrates span most therapeutic classes, including CNS (midazolam),[89] cardiovascular (diltiazem and nifedipine),[90,91] and antiviral (delavirdine and ritonavir).[92,93] Moreover, CYP3A4 is inducible by several drugs,[94] thus producing inter-individual variation both in terms of bioavailability and drug-drug interactions.[95]

In light of the wide variety of structurally diverse substrates, the active site of CYP3A4 is generally considered to be spacious. In addition, for many reactions CYP3A4 demonstrates a variety of atypical kinetic profiles including positive cooperativity[96] and substrate inhibition.[97,98] Several factors may contribute to the atypical kinetics associated with CYP3A4 reactions. Molecular modeling[99] and mechanistic studies[100,101] suggest that the complex effects observed with select CYP3A4 substrates may be attributable to the binding of multiple substrates within the active site of the enzyme[102,103] (see also the chapter on Enzyme Kinetics). Consistent with the hypothesis of multiple CYP3A4 binding sites, recent in vitro interaction profiles with various CYP3A4 substrates have been shown to be highly substrate-dependent.[104,105]

Despite the broad and overlapping substrate specificities associated with P450s, the ability of multiple enzymes to metabolize a given substrate is often dependent on concentration. Restated, while various enzymes may metabolize a particular substrate they do so with different substrate affinities and capacities. When substrate concentrations are reflective of in vivo conditions,[106] the enzyme kinetics that govern the system frequently dictate that a single P450 enzyme is the primary catalyst. Thus, the objective of P450 reaction phenotyping is not the identification of all the P450 enzymes able to oxidize a particular substrate but rather to characterize the enzyme(s) that are associated with the key metabolic pathways and ultimately the clearance of the drug.

Four *in vitro* experimental approaches provide the basis for reaction phenotyping. As individual experiments, the information that may be gleaned is limited and therefore a combination of approaches is necessary to identify the human P450 enzyme(s) responsible for metabolizing a drug. The principal approaches to P450 reaction phenotyping are briefly described below:

I.   *Enzyme Kinetics.* The first experiments in reaction phenotyping involve the determination of the Michaelis-Menten constants for the various metabolic pathways associated with the drug in human liver microsomes. The importance of the accurate characterization of the kinetic parameters underwrites the success of the reaction phenotyping study as this information provides a rationale for the drug concentrations used in the later experiments.

II.   *Correlation Analysis.* Typically in this experiment the rate of reaction for a particular biotransformation pathway is measured across a panel of human liver microsomes and correlated against the reaction rates for an established P450 marker substrate in the same microsomal samples.

III. *Recombinant Enzymes.* For these studies, the rate of metabolite formation is assessed across a battery of individual cDNA expressed human P450 enzymes.

IV. *Chemical and Antibody Inhibition.* These experiments examine the effect of co-incubation of known chemical inhibitors or inhibitory antibodies against selected P450 enzymes on the formation of a specific drug metabolite by human liver microsomes.

An obvious prerequisite to conducting the series of experiments to support a P450 reaction phenotyping effort is having a thorough understanding of the overall metabolic profile of the drug and whether or not the drug actually undergoes P450-mediated oxidative metabolism. Although the majority of documented drug oxidations are P450-mediated, other enzymes may be involved in carrying out similar types of reactions. Table 1 represents a short list of known oxidative reactions in which the metabolite formed is a common endpoint for P450 or another enzyme. A more in depth overview of drug oxidation reactions is found in Chapter 14 (Non-P450 Oxidations).

Table 1. Common enzymatic reactions shared with microsomal P450 pathways.

Reaction	Example of Reaction	Enzyme
S-Oxidation	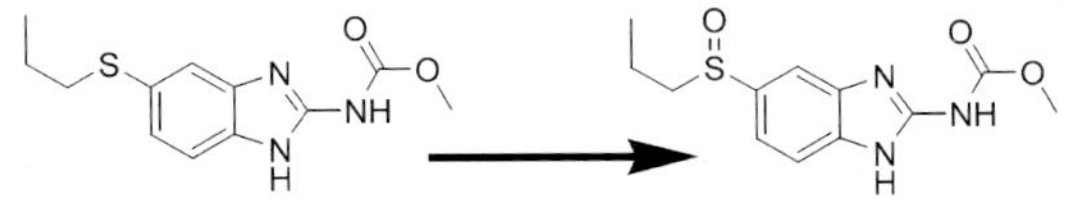	FMO / P450
N-Oxidation	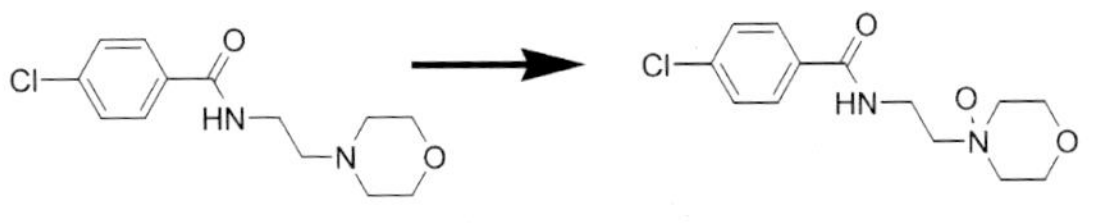	FMO / P450
N-Dealkylation	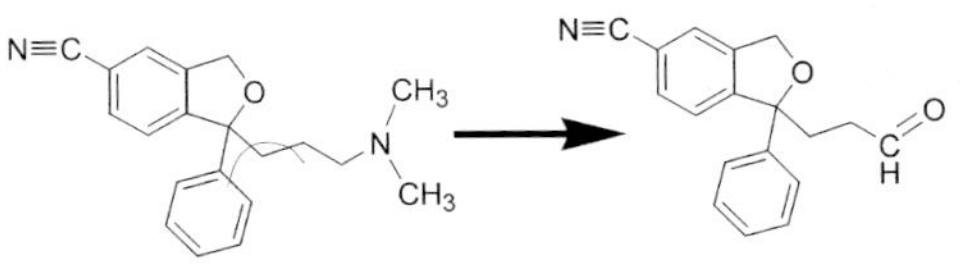	MAO / P450
Hydrolysis	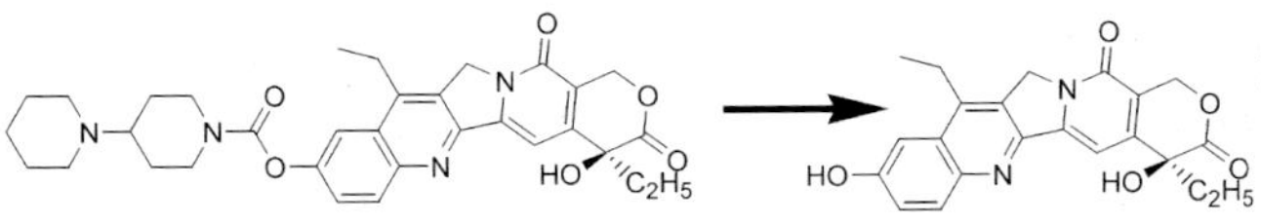	Esterase / P450

2　EFFECT OF INCUBATION CONDITIONS ON REACTION PHENOTYPING

Several factors should be considered when establishing reagents and conditions for reaction phenotyping experiments. These include; (a) the generation, characterization and maintenance of a set or 'bank' of human liver microsome samples, (b) the effect of organic solvents on enzyme activity, (c) the non-specific binding of substrate to microsomal protein, and (d) the linearity of product formation with microsomal protein concentration, incubation time, and substrate concentration.

One of the most important determinants in the accuracy and utility of reaction phenotyping work is the establishment of a characterized human liver microsome bank. This bank should contain samples from a range of individual subjects (n=10-20), with particular attention to the donor medication profiles due to the role of previous drug administration in CYP induction. Ideally, the bank would contain samples that are distinctive for polymorphic P450 forms, such as CYP2D6 and 2C19. Also, genotyping of the tissue DNA for the major human P450 forms is recommended in order to identify samples that contain allelic variants of particular P450 forms. Regarding tissue source, cadaver tissue is essentially useless for microsome preparation due to the rapid degradation of P450 and associated activities.[107] In contrast, organ donor tissue is preferred due to the effectiveness of current procurement techniques that allow for processing and/or storage of liver tissue for up to 72 hours post-procurement. Although liver microsomes prepared from fresh tissue have been reported to have higher P450 content and activity in comparison to preparations from frozen tissue,[108] practical considerations often dictate that liver tissue be stored at −80°C for future processing (as ~0.5 inch pieces frozen in liquid nitrogen to preserve enzyme activity). Data on the effects of freezer storage of human liver microsomes have demonstrated the stability of P450-dependent enzyme activities for at least five years.[109,110] In addition, microsome samples subjected for up to 10 freeze/thaw cycles did not show alteration of enzyme activities.[109]

Two of the most basic, yet critical steps in the characterization of human liver microsome samples are the determinations of spectral P450 content and protein concentration. The spectral P450 analysis should show that the majority of P450 is able to bind carbon monoxide in the reduced state, giving a maximal absorbance at ~450 nm for the complex.[111] Samples where the peak absorbance occurs at 420 nm contain inactive protein ("P420") and are likely to produce erroneous results for reaction phenotyping studies, especially those involving comparisons of enzyme activity and immunodetectable levels of P450 forms. The protein determination must be accurate since enzyme activities are normally expressed on a per milligram microsomal protein basis. The microsome bank should then be characterized using CYP-selective enzyme activities. Determinations for the 'major' P450 forms, such as 3A4, 1A2, 2C9, 2C19, and 2D6 are required. The addition of characterization data for other P450 forms such as 2E1, 2C8, 2B6, 3A5, and 4F1 is suggested as this provides a high degree of certainty that any future correlation analyses are accurate. Finally, a pooled human liver microsome sample, normally consisting of three to ten individual human liver microsome samples, is suggested. This sample is useful for the determination of reaction conditions (linearity of enzyme activity with time and protein) and enzyme kinetics without introducing interindividual variability.

Due to the high lipophilicity of many NCEs, the substrate must often be added to the microsomal incubation in an organic solvent. Several recent reports have investigated the effects of various solvents on P450-specific activities.[112-116] In general, dimethylsulfoxide (DMSO) has been shown to have significant inhibitory effects on CYP3A4, -2E1, and -2C19 enzyme activity, even at concentrations as low as 0.1% of the total reaction volume.[113,116] Also, ethanol is a potent inhibitor of CYP2B6, -2C19, and -2D6 while having no effect on CYP2C8, -2C9, or -3A4 at 1% final concentration.[113] Methanol and acetonitrile appear to have the best overall compatibility with enzyme activity at final incubation concentrations of $\leq 1\%$. However, the report by Tang et al.[112] showed that the effect of acetonitrile on CYP2C9 enzyme activity is substrate-dependent, and may potentially lead to the erroneous estimation of kinetic parameters and scaling of drug metabolic clearance.

An assumption for microsomal incubations is that the P450 has access to the total concentration of substrate. However, several investigators have shown that non-specific binding of substrate to microsomal protein can produce altered kinetic estimates. Figure 1 shows that the K_m of the highly protein bound compound *A* increased proportionally to the 5-fold increase in microsomal protein concentration.[117] In contrast, a compound with low microsomal protein binding did not show a proportional change in K_m with protein concentration. Given that the metabolic efficiency of a P450 can be expressed as V_{max}/K_m or 'intrinsic clearance', the increased K_m for the highly protein bound compound would result in a lower intrinsic clearance. Since intrinsic clearance is often used to scale *in vitro* results to *in vivo* clearance, an underestimation of drug clearance would result. Some investigators have shown that the accuracy of *in vitro-in vivo* scaling can be improved by incorporating a 'microsomal binding factor' into a prediction of *in vivo* clearance.[118] Finally, the high level of CYP expression (0.2-1.0 nmol/mg) for commercially available expressed P450 products combined with the lower limits of analytical detection afforded by LC/MS technology now allow for microsomal incubations and kinetic analysis to be performed with 10 to 100-fold less protein per incubation. As discussed in the section on expressed P450s, this phenomenon of substrate binding to microsomes may account for discrepancies in K_m values measured in various in vitro systems (expressed P450s versus liver microsomes) by different laboratories.

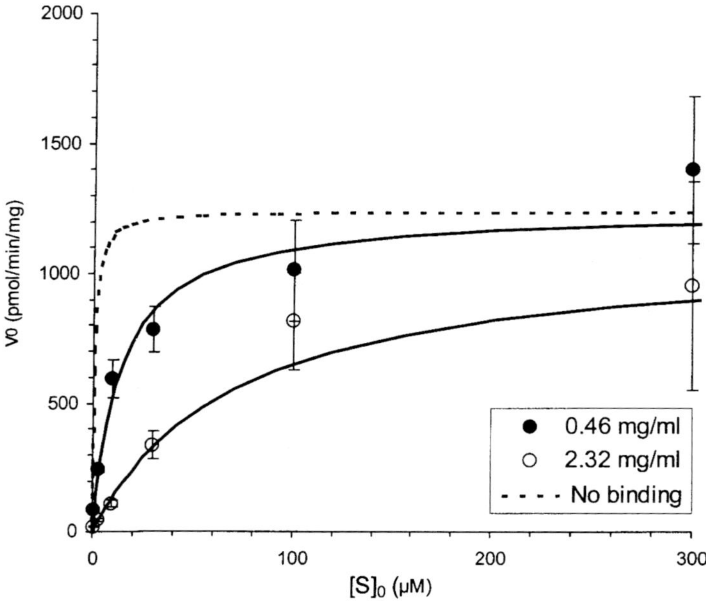

Figure 1. Saturation plot for compound A at two different microsomal protein (0.46 and 2.3 mg/ml) and enzyme (0.25 and 1.0 µM CYP) concentrations. Solid lines represent the best fit to the data from individual runs using the Michaelis-Menten equation. The dashed line represents a simulation using the estimated parameters, following correction for nonspecific microsomal binding. Error bars represent standard deviation.

Conditions for the linear production of metabolite with microsomal protein concentration and incubation time must also be established. A basis tenant of enzyme kinetics (see following Section) is that the rate of metabolism is first-order with respect to the amount of enzyme. With the depletion of >10-20% of substrate, reaction kinetics should be analyzed by a modified or 'integrated' form of the Michaelis-Menten equation that accounts for the reduced substrate concentration with time.[119] Preliminary experiments should be performed using a fixed substrate and protein concentration (liver microsomes or expressed enzyme), with metabolite production monitored over time. One of the potential problems in defining reaction linearity was illustrated recently[101]

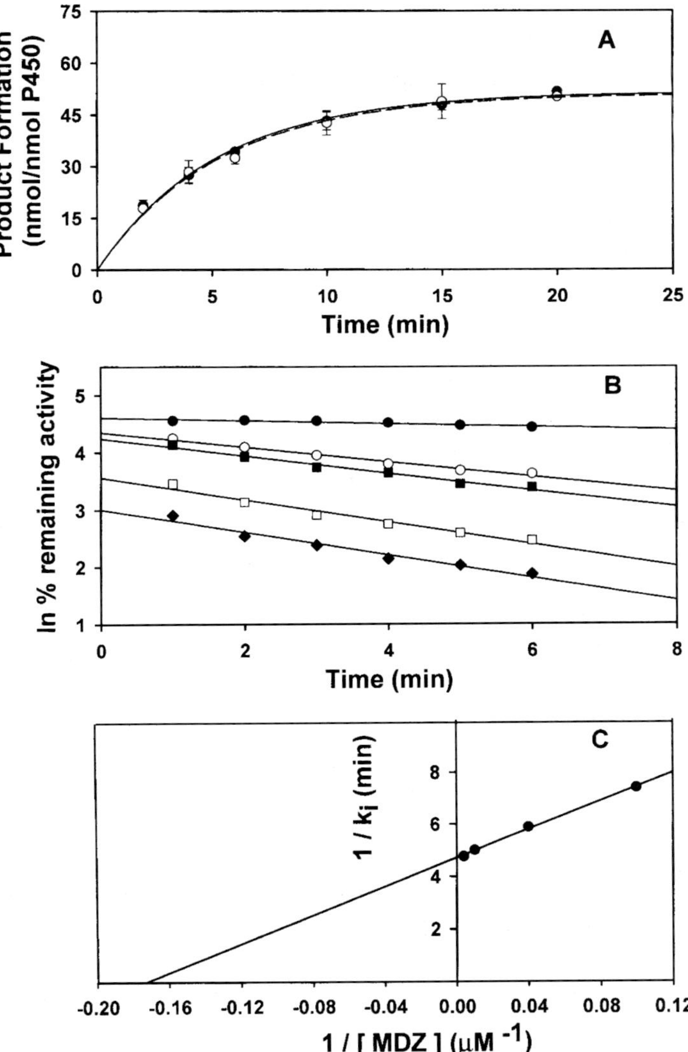

Figure 2. A, the time dependence of MDZ (250 μM) hydroxylation by CYP3A4; ● and O represent 4-OH and 1'-OH MDZ, respectively. The solid lines through the experimental data points show the fit to the first-order exponential rate equation, r = R_{max} × exp (-k_{obs} × t). B, time- and concentration-dependent inactivation of the wild-type enzyme by MDZ as measured by the percentage decrease in progesterone 6β-hydroxylation. The concentrations of MDZ were 0 (●), 10 (O), 25 (■), 100 (□), and 250 (◆) μM. The lines shown through experimental points were generated by linear regression analysis of the natural logarithm of the residual activity as a function of time. Rate constants of inactivation (k_i) are derived from the negative slope of the lines, whereas the extent of reversible inhibition is reflected by a decrease in the extrapolated activity at zero preincubation time compared with the methanol control. C, double-reciprocal plot of the rate constants of inactivation as a function of MDZ concentration.

for the hydroxylation of midazolam by CYP3A4. These authors observed that metabolite formation was linear for <5 minutes despite minimal substrate consumption. This was attributed to mechanism-based inactivation of CYP3A4 by midazolam (Figure 2), thus nullifying the first-order nature of the enzyme concentration relative to substrate hydroxylation. In summary, it is important that incubation conditions for microsomal experiments are established prior to, and then consistently applied to, all future reaction phenotyping experiments.

3   ENZYME KINETICS AND REACTION PHENOTYPING

As mentioned earlier, the goal of CYP reaction phenotyping is to extrapolate information gleaned from *in vitro* approaches to predict the *in vivo* condition.[120] Thus successful *in vitro/in vivo* correlations rely upon the accuracy of the calculated *in vitro* intrinsic clearance (V_{max}/K_m) of a given enzymatic pathway.[121,122] These experiments are typically conducted with individual human liver microsome samples, and analysis of metabolite formation is accomplished using a sensitive, robust analytical method. The kinetic parameters, K_m and V_{max}, for P450-catalyzed reactions are determined by analysis of initial velocities from a series of substrate concentrations, using a fixed incubation time and enzyme concentration. The Michaelis-Menten constant (K_m) is a measure of the affinity of an enzyme for a given substrate, or the substrate concentration at which half of the enzyme molecules are bound with substrate.[119] Thus, an enzyme with a high affinity for a substrate will have a low K_m resulting in 50% of the enzyme molecules with bound substrate at a relatively low concentration of substrate. In contrast, an enzyme with a low affinity for a substrate has a high K_m value, thus requiring a higher concentration of substrate to drive 50% of the enzyme molecules into complexes with the substrate. As a consequence, kinetic analysis requires that the concentration of substrate must span values above and below the apparent K_m to produce a hyperbola and generate reliable values for the kinetic parameters (Figure 3).

Unfortunately, an issue that is particularly germane to P450 enzyme kinetics is the relatively high K_m values for most P450 substrates compared to low affinity constants usually observed for physiologically based enzyme reactions. A familiar consequence of this lower affinity is a reduction in the number of data points that may be achieved at substrate concentrations in excess of the K_m value due to poor aqueous solubility of the drug. As a result, many P450 reactions never achieve zero order or saturated kinetics. This condition is observed in the kinetic evaluation of the rate of formation of m-chlorophenylpiperazine, the major CYP3A4-mediated metabolite of trazodone.[124] As evidenced from the V vs S plot presented in Figure 4, the concentration of trazodone never

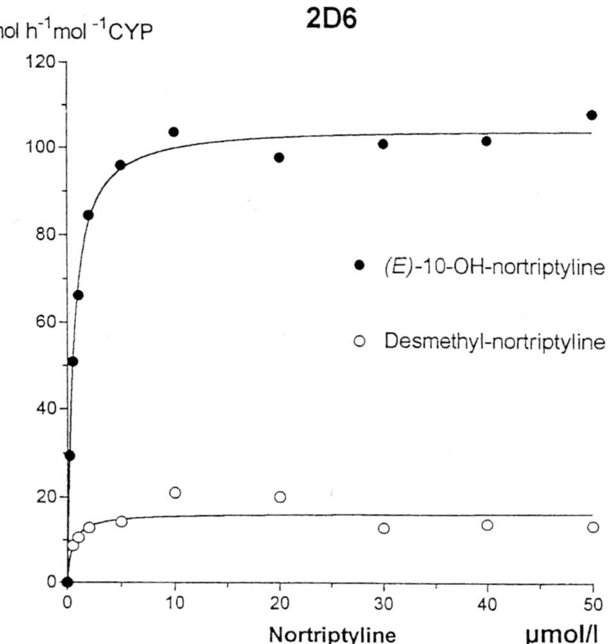

Figure 3. Reaction rates for CYP2D6 catalyzed hydroxylation and demethylation of nortriptyline. [123]

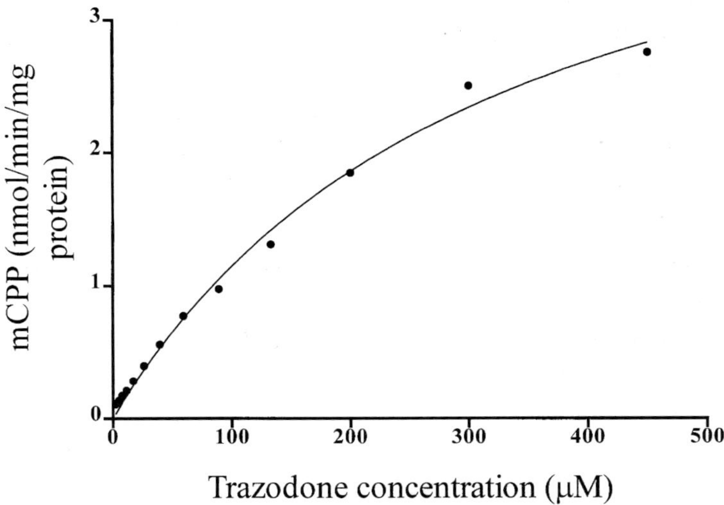

Figure 4. Formation of mCPP (y axis, nmol/min/mg protein) at various concentrations of the substrate trazodone (x axis, µM). The solid line was determined by nonlinear least squares regression analysis.[124]

achieves saturating conditions and therefore the kinetics values are calculated based upon mixed-order kinetics. Conversely, if the K_m value is very low, the analytical methodologies may not be sufficiently sensitive to detect metabolite formation without significant consumption of substrate (e.g.,> 10%). In contrast to the situation where multiple P450s contribute to the formation of a single metabolite, there are several examples where a single P450 metabolizes a substrate to multiple products.[125] Under these circumstances, the rates (i.e. V_{max}) of formation may vary for each metabolite (Figure 3); however each reaction should be defined by a distinct K_m value.[126]

The rate of an enzymatic reaction depends on the concentrations of both the substrate and the enzyme. When the amount of enzyme is much less than the amount of substrate, the reaction is considered pseudo first order, or in other words, the more enzyme present the faster the reaction. This phenomenon is particularly apparent when examining enzyme kinetics for a given drug across a number of different human liver microsomal preparations. For example, levels of some P450 enzymes may vary more than 30-fold across a human liver bank. In this case the calculated V_{max} for the substrate would also be expected to vary 30-fold. This variability in V_{max} values across a series of human liver samples is illustrated in Figure 5, where the rates of hydroxybupropion formation are compared for four different human livers.[127] This example also illustrates the concept that despite observed variability in V_{max} among human liver microsomal samples, the apparent K_m value for a substrate that is metabolized by a single P450 enzyme should remain relatively constant across all the livers tested. A primary source of variability in K_m determinations can be attributed to the amount of non-specific protein that occurs in the microsomal incubation. In this circumstance, high levels of protein binding can effectively reduce the amount of unbound drug available to the enzyme for metabolism (see Section 2.2 on Incubation Conditions).

The preferred method for analyzing enzyme kinetic data is to fit the data directly to the Michaelis-Menten equation using nonlinear regression. However, prior to nonlinear regression analysis it is often useful to perform a linear transformation of the original kinetic data to visually analyze the results.[128] This is usually accomplished by one of two graphical methods; Lineweaver-Burk (1/V against 1/[S] plot) and Eadie-Hofstee (V against V/[S] plot).[129,130] However, while both methods have great utility for the empirical interpretation of enzyme kinetic data, they have limited accuracy and therefore are not recommended to calculate final K_m or V_{max} constants. For instance, Lineweaver-Burk plots generate straight lines when fitted by eye, and this may give a misleading impression of the relative experimental error (Figure 6). Because of this limitation, Lineweaver-Burk plots are most often used in conjunction with inhibition studies, as the

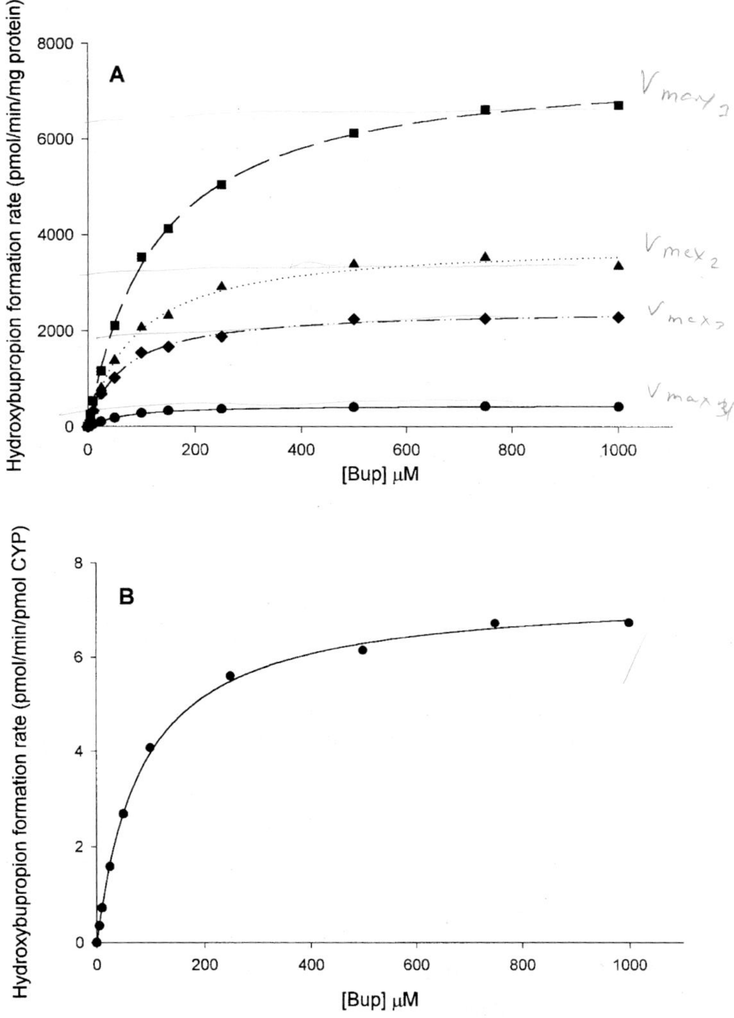

Figure 5. The velocity of formation of hydroxybupropion versus the concentration of bupropion (μM) for four different human livers (A) (velocity units = pmol of hydroxybupropion formed/min/mg of protein) and microsomes from lymphoblastoid cells expressing CYP2B6 (B) (velocity units = pmol of hydroxybupropion formed/min/pmol of CYP2B6).[127]

intercept of the lines is diagnostic of the type of inhibition. Likewise, Eadie-Hofstee analysis does not fulfill all the conditions for a least-squares method since the error-containing variable (velocity) is incorporated within both axes and thus may provoke an angular distortion of the error. Unlike the V vs [S] and

Lineweaver-Burk plots, the Eadie-Hofstee plot is very sensitive to aberrations of simple Michaelis-Menten single enzyme kinetics. Thus an Eadie-Hofstee plot will yield a straight line with a slope equal to $- K_m$ for single enzyme kinetics. However, if more than one enzyme contributes to metabolite formation, the plot will be curved, thus reflecting the different K_m values for each enzyme (Figure 7).

Many investigators therefore routinely employ the Eadie-Hofstee graphical analysis of kinetic data as a diagnostic tool to detect multiple enzyme kinetics. In addition, the Eadie-Hofstee plot is very sensitive to atypical kinetics associated with P450-mediated drug oxidations such as substrate inhibition (Figure 8) and allosterism (Figure 9). A detailed overview of atypical kinetics associated with P450 drug reactions is presented in Chapter 7.

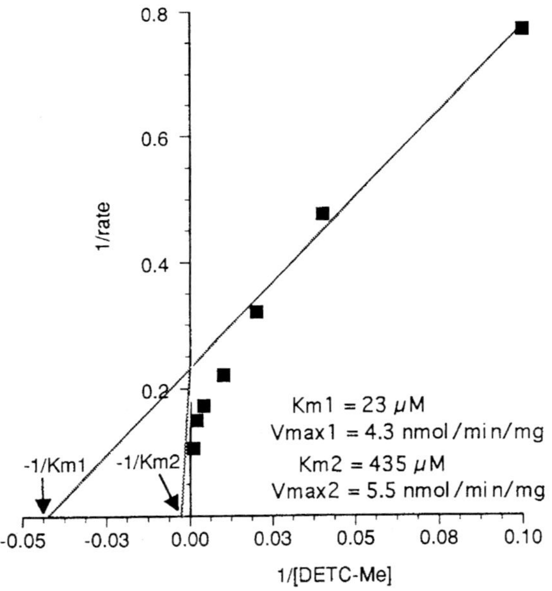

Figure 6. Lineweaver-Burk plot for DETC-Me sulfoxidation by human liver microsomes.[131]

It should be noted that a straight line on an Eadie-Hofstee plot does not always indicate a single enzyme reaction, as it is possible for two enzymes with similar K_m values to contribute to the formation of a single metabolite. In this case, other reaction phenotyping experiments must be relied upon to fully distinguish

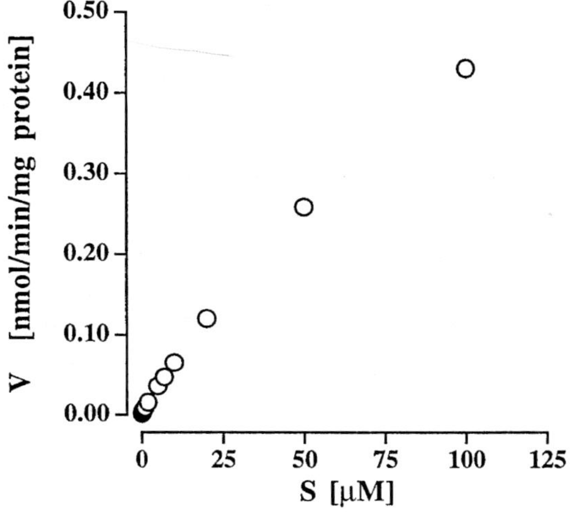

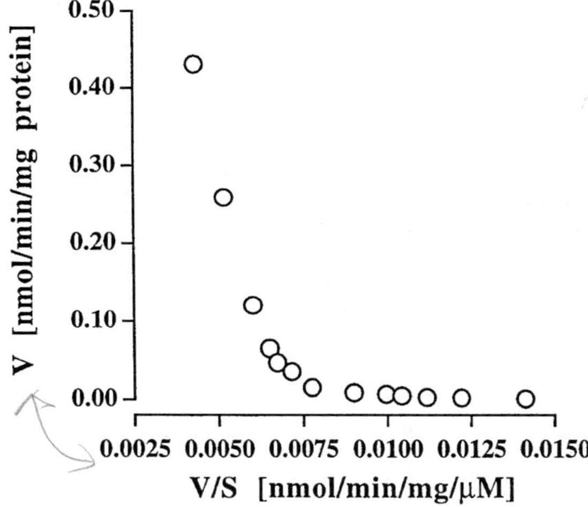

Figure 7. Upper panel, velocity of pentoxifylline formation vs. lisofylline concentration. Lower panel, Eadie-Hofstee plot of upper panel data. Each data point represents the mean value of duplicate incubation.[132]

between single and multiple enzyme kinetics. A mechanism to probe the relative contribution of the individual enzymes towards metabolite formation is through the addition of a selective P450 enzyme inhibitor in kinetic experiments. The choice of inhibitor is detailed in a later section of this chapter, with specific attention to the concept that the potency and selectivity of inhibition are paramount to successful mechanistic evaluation of multiple enzyme kinetic studies. An

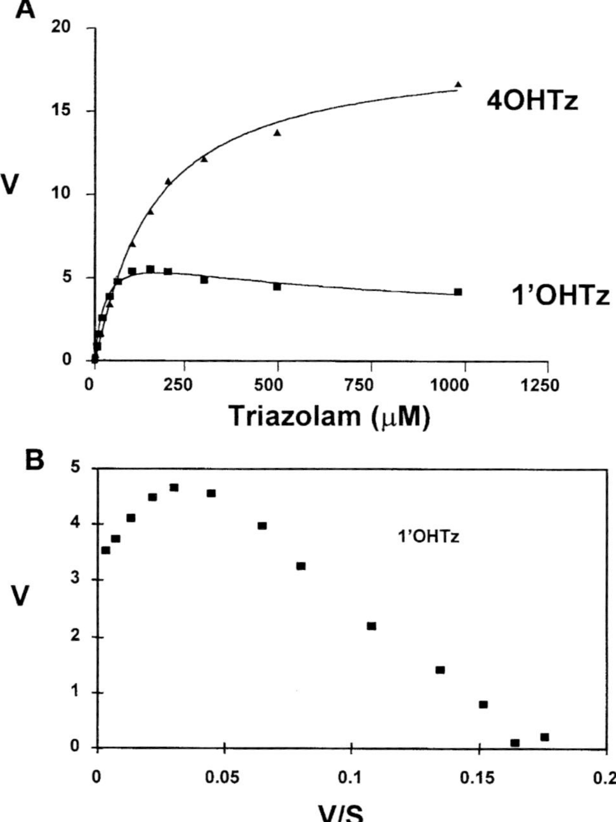

Figure 8. A, substrate-velocity curves for 4OHTz (▲) and 1'OHTz (■); B, Eadie-Hofstee plot of 1'OHTz data. *The curves in A yielded the following kinetic constants for 1'OHTz formation: 1'OHTz (r^2 = 0.99) V_{max}, 6.7 ± 0.4; K_{m1}, 40 ± 3; K_{m2}, 239 ± 95; , 0.38 ±0.05. Velocity is expressed in nmol/min/mg of protein, and assays were performed in duplicate.* [97]

example of the use of a specific P450 inhibitor to evaluate multiple enzyme reactions is observed in the kinetic characterization of (R)-8-hydroxywarfarin formation in human liver microsomes. Initial kinetic experiments revealed a curved Eadie-Hofstee plot which suggested that formation of (R)-8-hydroxywarfarin was mediated by at least two enzymes, CYP1A2 and a second unknown enzyme (Figure 10). In a subsequent experiment, human liver microsomes were pretreated with the potent CYP1A2 mechanism-based inhibitor furafylline and the resulting kinetic analysis revealed a straight line Eadie-Hofstee plot that represented the kinetic characteristics.[71] Moreover, it is possible to observe enzyme kinetics that suggest two enzymes contributing to metabolite formation only to determine that several P450 enzymes contribute to the oxidation of the molecule (e.g., amitriptyline).[134]

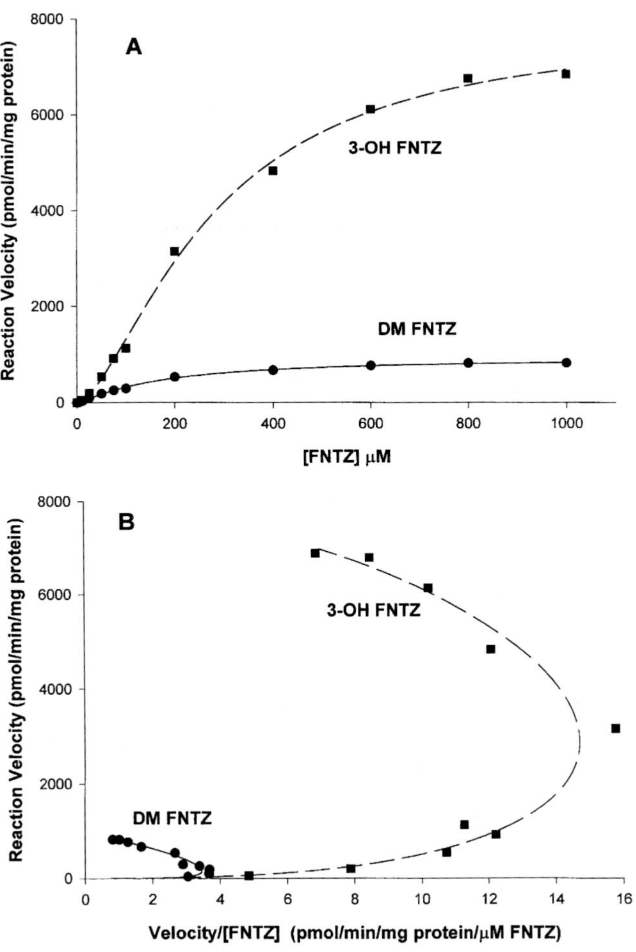

Figure 9. Flunitrazepam (FNTZ) biotransformation by a representative human liver sample. A, velocity of formation of desmethyl-FNTZ and 3-OH FNTZ versus the concentration of FNTZ for one representative liver (HLM4). (Velocity units = pmol of metabolite formed/min/mg of protein). Symbols are experimental data points, representing an average of duplicate incubations. Lines are fitted functions described by a Hill equation. B, Eadie-Hofstee plots for velocity (formation of DM FNTZ or 3-OH FNTZ) versus the velocity/concentration of FNTZ for the same liver sample (HLM4). Symbols are experimental data points, representing an average of duplicate incubations. HLM, human liver microsomes.[133]

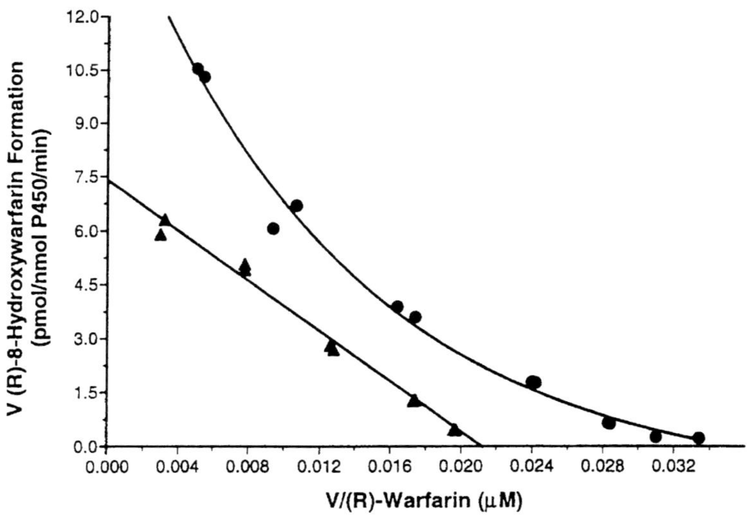

Figure 10. Eadie-Hofstee plots for the rates of formation of (R)-8-hydroxywarfarin in control (circles) and furafylline-pretreated (triangles) human liver microsomes (HL122).[71]

4   CORRELATION ANALYSIS

Correlation analysis is a procedure that exploits the wide inter-individual variability in the expression of various P450 enzymes.[32] Briefly, correlation analysis involves the comparison of the inter-liver variability in the rate of formation of a specific drug metabolite with the measured activities towards CYP marker substrates (Table 2). The statistical significance of the linear correlation coefficient (e.g., coefficient of determination) is calculated for each P450 tested. Typically, if one P450 enzyme is responsible for the formation of a metabolite, a linear relationship will exist between some measure of a given P450 and metabolite formation across a bank of human liver microsomal preparations. As there exists a relationship between relative P450 enzyme expression and catalytic activity for a particular enzyme,[135-137] metabolite formation can be compared against relative P450 content or catalytic activity. For example, the correlation plot presented in Figure 11 depicts a comparison between the rates of bupropion hydroxylation and (S)-mephenytoin N-demethylation, a marker reaction for CYP2B6 catalytic activity,[43] across 16 human liver microsomal preparations.[42] Alternatively, the data depicted in Figure 12 demonstrates a linear relationship between the rate of 10-hydroxy (R)-warfarin formation and CYP3A4 content estimated from Western blots.[8] In addition to correlating the rate of metabolite formation against a given P450 enzyme content or catalytic activity, it has been recently demonstrated that P450 enzyme activity also reasonably correlates with mRNA levels as determined by RT-PCR.[138]

Table 2. Structures of P450-marker substrates commonly used in correlation studies.

P450 Enzyme	P450 Marker Substrate	P450 Enzyme	P450 Marker Substrate
CYP1A2	ethoxyresorufin	CYP2D6	dextromethorphan
CYP2A6	coumarin	CYP2E1	chlorzoxazone
CYP2C9	diclofenac	CYP3A4	midazolam
CYP2C19	(S)-mephenytoin	CYP3A4	testosterone

While the concept of in vitro correlation analysis is rather straight forward, there are a number of technical factors that must be considered prior to data generation: (I) Enzyme kinetic parameters for the molecule must be clearly established. If the drug appears to be metabolized primarily by a single enzyme, then the correlation analysis should be conducted at a substrate concentration that reflects

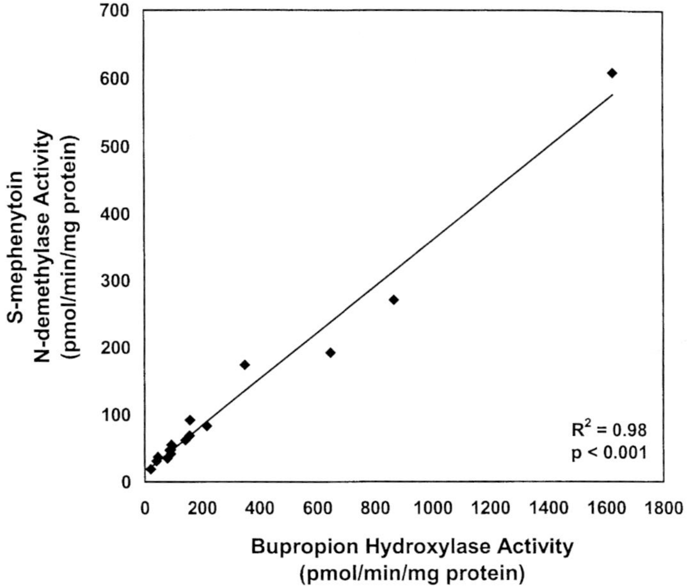

Figure 11. Correlations between P450 isoform-selective catalytic activities and BUP hydroxylase activity were examined in the 16 HLMs assayed at a substrate concentration of 500 µM BUP. BUP hydroxylation significantly correlated with S-mephenytoin N-demethylation (19-609 pmol/min/mg of protein; $r^2 = 0.98$, P <0.001) [42].

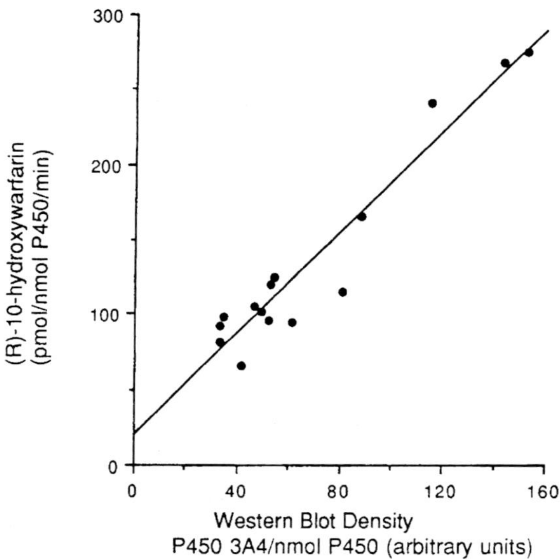

Figure 12. Correlation of (R) -10-hydroxywarfarin formation rates from 1 mM (R) -warfarin with P4503A4 content estimated from Western blots ($r^2 = 0.92$, y-intercept 18 pmol•nmolP450^{-1}•min^{-1}).[8]

V_{max} conditions (solubility permitting). Under these conditions, the differences in rates observed across a bank of human liver microsome samples will reflect enzyme variability and will not be adversely affected by consumption of the substrate. However, if more than one enzyme appears to contribute to metabolite formation, measurement of enzyme activity at reduced or "physiologically relevant" concentrations followed by correlation analysis may aid in establishing the primary P450 involved in the metabolism of a given drug. In some instances it may therefore be advisable to conduct correlation analysis experiments at two different substrate concentrations (one low and the other relatively high) as a mechanistic tool for determining relative enzyme contribution to metabolite formation.[139]

(II) The proposed human liver microsome samples should be evaluated for any potential advantageous correlations between the different P450 enzymes. For example, it is possible for CYP2C8 and CYP3A4 activities to significantly correlate across a limited number of human liver samples, possibly due to similar mechanisms of enzyme regulation. In this case, while the drug may be selectively metabolized by CYP2C8, the correlation results will suggest the involvement of both enzymes. To guard against the potential of generating a false correlation, the human liver bank should be prescreened and livers removed from testing in order to insure a lack of correlation ($r^2 \leq 0.3$) across all P450 marker activities or protein levels.[140,141]

(III) The number of liver microsome samples used in the initial screen and relative variability in the different P450 enzyme activities (or protein levels) is a critical factor in conducting meaningful correlation analysis experiments. It appears that at least ten, but preferably closer to twenty samples, is optimum for conducting correlation studies. If the number of livers is less than ten, the statistical power of the correlation is markedly weakened and data interpretation compromised (i.e., false correlation). The potential to be misled by a low number of liver samples incorporated into a correlation study is illustrated in the preliminary identification of the P450 responsible for taxol 6α-hydroxylation.[142] In these experiments the investigators observed a significant correlation ($r = 0.954$) between the rates of 6α-hydroxytaxol formation and CYP3A4 activity as measured by midazolam 1-hydroxylation (Figure 13). However, in subsequent studies, the biotransformation of taxol to 6α-hydroxytaxol was determined to be principally mediated by CYP2C8.[143] With respect to the relative P450 catalytic activities for each human liver, it is important to avoid clustering of activities for specific P450s (Figure 14), rather the activities for a given P450 should be uniformly spread across the liver samples used in the correlation study.

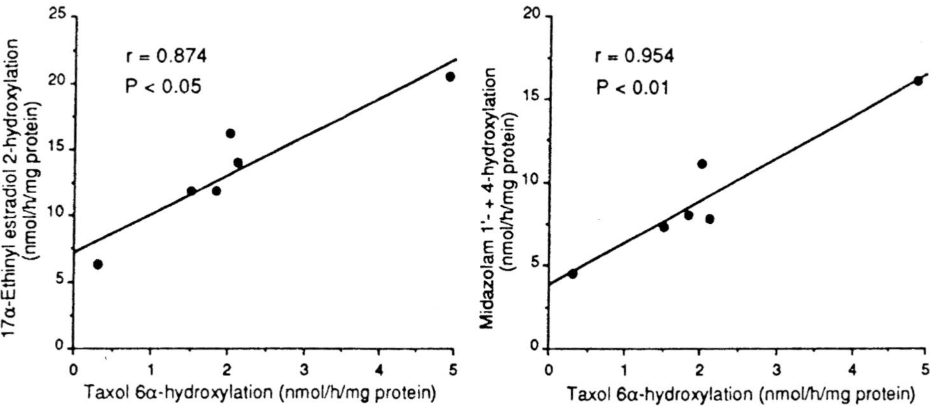

Figure 13. Correlation of taxol (30 μM) metabolism with the metabolism of 17α-ethinyl estradiol (50 μM) (left) and midazolam (60 μM) (right).[142]

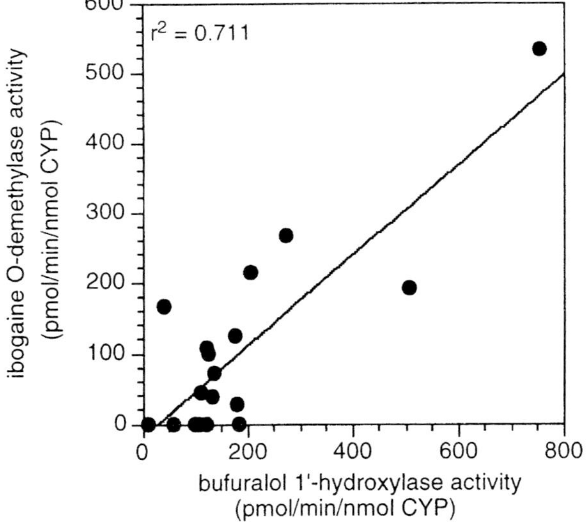

Figure 14. Correlation between bufuralol 1'-hydroxylase activity and ibogaine O-demethylase activity in a panel of human liver microsomes. Ibogaine (1.0 μM) was incubated with human liver microsomes from 19 individual donors (0.5 mg/ml), and 12-hydroxyibogamine was determined. Each point represents the average of duplicate determinations.[144]

Even with the above technical precautions in place, multiple P450 enzymes contributing to the formation of a single metabolite may confound correlation studies. For some analyses, a reasonable correlation is achieved; however, the

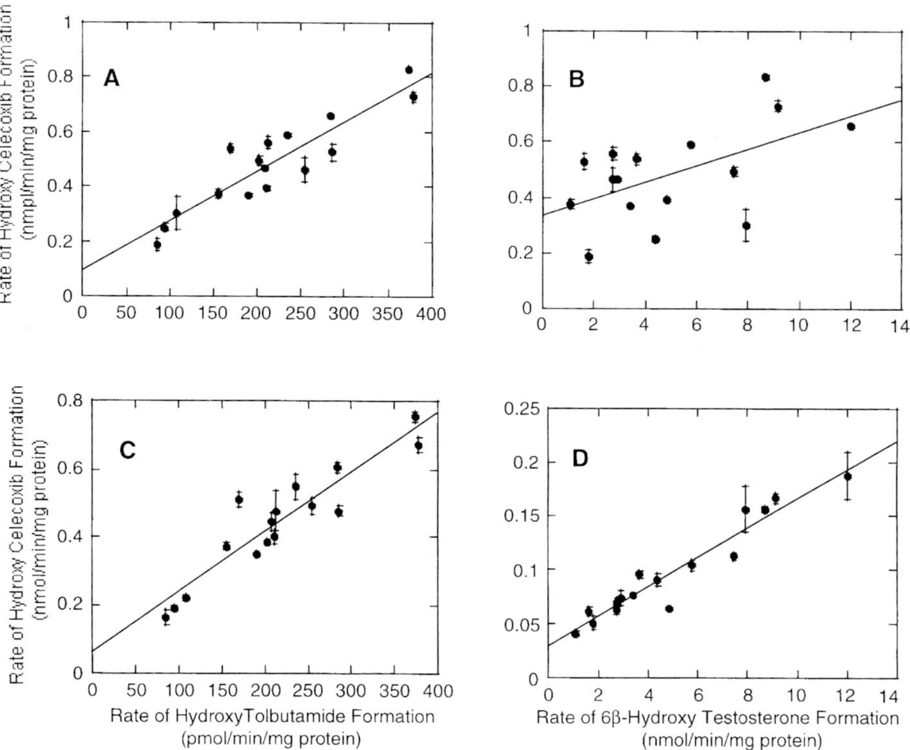

Figure 15. Correlation between hydroxycelecoxib formation and tolbutamide methyl-hydroxylase activity (A and C) and testosterone 6β-hydroxylase activity (B and D) in human liver microsomal samples (n = 16 subjects). A, hydroxycelecoxib formation in the absence of mAbs versus tolbutamide methyl-hydroxylase activity. B, hydroxycelecoxib formation in the absence of mAbs versus testosterone 6β-hydroxylase activity. C, hydroxycelecoxib formation in the presence of anti-CYP3A4 mAb versus tolbutamide methyl-hydroxylase activity. D, hydroxycelecoxib formation in the presence of anti-CYP2C9 versus testosterone 6β-hydroxylase activity.[145]

observation of a high correlation coefficient with a positive y-axis intercept is typically indicative of a second enzyme contributing to metabolite formation. For example, in studies investigating the metabolism of the substance P receptor antagonist CJ-12458, a strong correlation between metabolite formation and CYP2D6 activity was observed. However, the correlation plot with CYP2D6 included a positive y-intercept and further investigations revealed that in addition to a major contribution by CYP2D6, CYP3A4 also contributed to metabolite formation, albeit at a reduced level.[118] If the correlation analysis or existing phenotyping data (e.g., expressed P450s) suggests a catalytic role for a second P450, the identity and overall magnitude of the contribution to metabolite formation may be investigated using either the antibody or chemical inhibition techniques described in this Chapter. In these studies, an antibody or chemical inhibitor is used to selectively block the catalytic activity of a specific P450, which

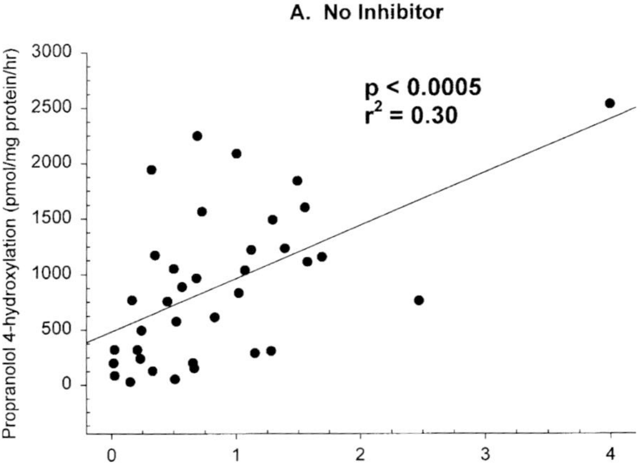

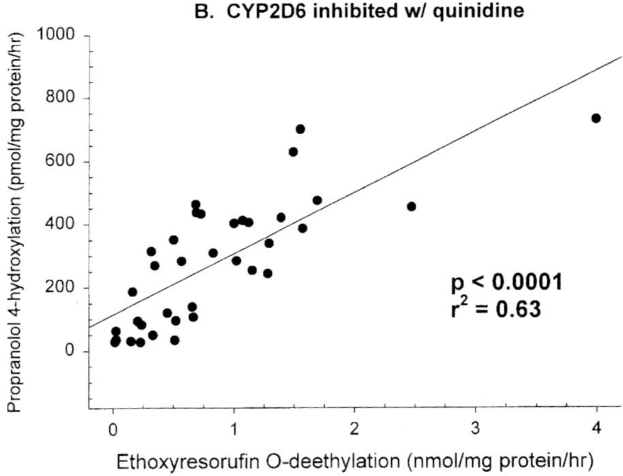

Figure 16. Confirmation of propranolol 4-hydroxylase activity of CYP1A2. Correlation and inhibition studies were conducted to confirm positive findings of propranolol 4-hydroxylase activity from the recombinant CYP studies. Correlation between propranolol 4-hydroxylation and ethoxyresorufin O-deethylation (EROD) (model CYP1A2-mediated reaction) was determined for 36 HLM samples. Propranolol 4-hydroxylation was determined with a substrate concentration of 100 μM propranolol in the absence and presence of 2.5 μM quinidine (potent, selective inhibitor of CYP2D6).[147]

allows for the residual microsomal catalytic activity to represent the contribution of the second P450 to metabolite formation. The use of inhibitory antibodies in multiple P450 correlation analyes is illustrated in Figure 15 which describes a series of mechanistic studies that established a role for CYPs 2C9 and 3A4 in hydroxycelecoxib formation.[145] In a second example, the correlation between propranolol 4-hydroxylation and ethoxyresorufin O-deethylation (EROD) (CYP1A2 marker activity) was determined for 36 HLM samples to be weak ($r^2 = 0.3$) (Figure 16). However, when the same microsomal incubations were conducted in the presence of the selective CYP2D6 inhibitor[146] quinidine (2.5 µM), the correlation was dramatically improved ($r^2 = 0.63$).[147] When correlation studies are conducted in the presence and absence of a selective chemical inhibitor or antibody for a given P450, the participation of that particular P450 becomes readily apparent. Thus, correlation analysis represents a valuable tool in characterizing the metabolism of a substrate by multiple P450s.

5   EXPRESSED P450s AND REACTION PHENOTYPING

The characterization and commercialization of expressed P450s by the mid-1990s made available human enzymes that had the practical advantages of a renewable resource and reduced biohazard liability over human liver microsomes. From a scientific perspective, the utility of cDNA expressed P450s (also called 'recombinant P450s') was recognized as part of a standard approach for assessing the enzymology of drug metabolism in a guidance from the Food and Drug Administration.[4,148] Thus, the ability of a panel of expressed P450s to metabolize a specific NCE in large part reduces reaction phenotyping to the simplest components of enzyme and substrate. However, the 'user-friendly' experimental approach offered by these systems often belies difficulties in the interpretation of results and use in context with other methods for understanding the overall metabolic fate of a NCE. This Section will focus primarily on the use of expressed CYPs for reaction phenotyping experiments and directly assessing NCE metabolic stability. Other applications, such as P450 inhibition screening, enzyme bioreactor systems for metabolite production, or as a tool for studying bioactivation-dependent mutagenesis or genotoxicity[149] are beyond the scope of this Chapter.

In a recent review on the current status of expressed P450s for drug metabolism studies, the authors state that the current needs lie in the application of these systems to drug discovery and development rather than for further refinements in enzyme expression technology.[150] While an in-depth consideration of systems and techniques for P450 expression is the subject of another Chapter, some general comments on these topics is relevant to reaction phenotyping. Numerous expression systems have been developed for the stable or transient expression of catalytically active P450 forms, including bacteria (*E. coli*), yeast, insect (Sf9 or *T. ni*), and mammalian (i.e., lymphoblastoid) models (reviewed elsewhere [151,152]). *E. coli* is an attractive system due to the combination of high levels of enzyme expression, the capacity of the cells to grow to high density, low cost, and ease of maintenance,[153,154] however, the mammalian P450 cDNA usually requires modification in order to obtain protein expression. In addition, *E. coli* does not express a protein homologous to mammalian NADPH-P450 reductase, thus requiring the co-expression of reductase.[155,156] The ability of bacterial systems to produce large amounts of protein is evidenced by the expression of a modified rabbit P450 2C5 in *E. coli* for use in the generation of the first crystal structure of a mammalian P450.[157] Yeast systems are relatively inexpensive and simple to set up, and modification of the P450 cDNA coding region does not appear to be a requirement for the expression of protein.[158,159] The baculovirus transient expression system is restricted to insect cells and following optimization of expression conditions, allows P450s to be expressed at very high levels (>1 nmol/mg of cell microsomal protein). Commercial sources of enzyme contain co-expressed reductase and often cytochrome b_5, thus ensuring that the supply of reducing equivalents to the P450 catalytic cycle is not rate-limiting to overall substrate turnover. Finally, P450 stable expression in B lymphoblastoid cells has been used in toxicology applications due to their readily detectable genetic mutation endpoints.[160,161]

Regardless of the source/expression system of the P450, certain catalytic parameters should be solely a function of the enzyme and thus consistent across systems. Specifically, the affinity of the enzyme for a particular substrate (K_m) or inhibitor (K_i or IC_{50}) for a single enzyme process should be similar for determinations using the expressed enzyme or human liver microsomes. This criterion is therefore used as a point of evidence in the identification of the P450 form responsible for a particular reaction in human liver microsomes. Friedberg et al.[152] expressed various P450s in the systems described above and compared the kinetic parameters for model substrate oxidations. For CYP2D6 and 3A4 dependent activities, the K_m values were consistent across three models (*E. coli*, yeast, CHO cells). In addition, these K_m values compared well with those obtained with CYP3A4 or 2D6 expressed in human lymphoblastoid cells or baculovirus expression systems.[162,163] Inter-laboratory variability of K_m values for (S)-warfarin hydroxylation by CYP2C9 has been shown to be less than 2-fold, regardless of the expression system.[164,165] It should be noted that apparent K_m values determined using expressed P450 forms tend to be lower than the corresponding K_m values determined in human liver microsomes, particularly with the high levels (>1 nmol/mg) of expressed P450 generated by baculovirus systems. For example, a 2.5-fold increase in the apparent K_m value for CYP2C9-

catalyzed diclofenac hydroxylation was observed with a 500-fold increase in microsomal protein.[150]   Assuming that the expressed enzyme experiment accurately depicts the role of a single enzyme reaction in human liver microsomes, this change in K_m is likely due to the reduced level of non-specific protein binding that accompanies the necessary reduction in the total amount of protein necessary as the investigator transitions from human liver microsomes to expressed enzyme while still maintaining linear substrate turnover. This problem can be largely avoided by standardization at the lowest reasonable protein content for expressed enzyme experiments.

The effect of cytochrome b_5 co-expression on enzyme activity has been shown to be dependent on the particular P450 enzyme. For example, cytochrome b_5 has little if any effect on CYP2C9 activity, but can stimulate CYP3A4-dependent activities by as much as 10-15 fold.[166,167] This effect of b_5 likely extends to the entire CYP3A (human) subfamily, and to complicate matters, has shown substrate-dependency for CYP3A forms and CYP2B6.[166,168] In addition, several laboratories have observed lower substrate turnover for CYP3A4 expressed in yeast that may not be directly attributable to levels of reductase and cytochrome b_5.[169]

There are several examples where the use of cDNA expressed P450s has aided or complicated the interpretation of reaction phenotyping studies. Many of the problems from early reports (pre-1994) largely reflect limitations of expression technology and commercial availability of the majority of P450s. As described in an earlier Section on Correlation Analysis, there is an example where two separate P450-reaction phenotyping reports implicated CYP3A[142] and CYP2C8[143] as the primary catalysts of taxol 6α-hydroxylation. Kumar et al. found that although certain CYP3A substrates produced inhibition of taxol 6α-hydroxylation in human liver microsomes, the CYP3A4 transfected cell line produced only negligible activity.[142] Rahman et al. used the advantage of expressed CYP2C8 to clearly demonstrate taxol 6α-hydroxylase activity and comparable K_m values for expressed enzyme vs. human liver microsomes, and as a standard for the accurate quantitation of human liver microsomal 2C8 levels for correlation analysis of enzyme level with activity.[143] Also within the CYP2C subfamily, the lack of expressed P450 data and likely confusion on the state of P450 nomenclature led to the erroneous initial assignment of CYP2C9 rather than 2C19 as the P450 responsible for the hydroxylation of omeprazole.[170] Finally, using the available technology but again lacking a panel of expressed P450 forms, chlorzoxazone 6-hydroxylation was initially proposed as a CYP2E1-specific reaction.[171] Several years later however, other investigators demonstrated that CYP1A2 and 3A4 contribute to this activity and can confound the interpretation of reaction phenotyping studies.[172,173]

Despite the current advanced state of the art and the commercialization of numerous expressed P450 forms, examples where only a limited number of expressed enzymes are evaluated during reaction phenotyping studies continue to appear. The rationale of examining only the 'major' P450 forms (typically 1A2, 2C9, 2C19, 2D6 and 3A4) based on older reports[32] is dangerous, if not flawed and requires careful re-examination. First, the categorization of human P450 forms as 'major' or 'minor' must be placed into context relative to a stated parameter. The most common index is the average percentage of a particular P450 relative to the total liver microsomal P450 content. It is important to note that the frequently cited data of Shimada et al. (1994) on the relative expression of human hepatic P450 forms, while substantiated for certain P450 forms such as 3A4 by numerous other laboratories, likely underestimated levels of CYP2B6 [39,174] and was unable to distinguish P450 forms within certain subfamilies (CYP2Cs). Therefore, contexting the role of particular P450s to the overall metabolism of a NCE is an exercise that should follow, rather than proceed, expressed enzyme studies. In the same theme, the role of a particular P450 in overall human drug metabolism, as measured by the number of literature references of CYP-specific metabolism determined by reaction phenotyping, is simply a general metric that has been erroneously applied to experimental design.

Not withstanding these few exceptions, current discrepancies in the literature on the interpretation of expressed enzyme data within the realm of P450 reaction phenotyping are generally qualitative, as investigators attempt to context these results to overall in vitro (human liver microsome studies) and in vivo metabolism (scaling of metabolic clearance to predict clinical pharmacokinetics) for the NCE (reviewed elsewhere [150,175]). In order to develop a continuum for studies with expression systems to human liver microsomes and ultimately to in vivo metabolic clearance, two basic approaches have been proposed. One approach is the determination of a "relative activity factor", obtained by dividing the maximal activity for an isoform-specific reaction in human liver microsomes by the maximal activity for the same reaction for the expressed enzyme.[176] Thus, differences in substrate turnover by the expressed protein versus liver microsomes can be factored. However, the non-physiological concentration of accessory proteins (P450 reductase and cytochrome b_5) for expression systems has been shown to be a source of error for this approach in some instances.[177] By comparison, numerous other studies had scaled the relative involvement of individual P450 forms by using the absolute levels of expression of a particular CYP in human liver microsomes determined by immunoquantitation. Currently, a hybrid of these approaches involving the incorporation of the affinity (K_m), capacity (V_{max}), and relative expression of the individual CYP in human liver microsomes into a normalized RSF or "relative substrate-activity factor" appears best suited.[168] These authors also emphasize important assumptions of the RSF method; linearity of activity with enzyme amount, fidelity of the P450 forms for the diagnostic substrate, and equal influence of non-P450 factors (i.e., cytochrome b_5 content, buffer composition, etc.) on the diagnostic and test substrates. Since relative expression of P450 forms translates into significant interindividual variability, further work to model the variability for the RSF and thus predict clinical variability for a given biotransformation process in vivo are necessary.

6   CHEMICAL INHIBITORS AND REACTION PHENOTYPING

The study of chemical inhibition of oxidative drug metabolism has shown tremendous progression from the early toxicology and pharmacology studies where enzyme inhibition was commonly measured by the effect of various xenobiotics on the hexobarbital-induced sleeping time of rats.[178] An excellent case study on the emergence of enzyme inhibition and clinical drug interactions in the late 1970s and 80s involves the H_2-receptor blocker cimetidine.[179,180] Despite having a large therapeutic safety index and a reasonably low affinity for human cytochromes P450, the second generation drug ranitidine was found to not inhibit human P450s,[181] and was able to quickly capture the predominant share of the anti-ulcer therapeutic market based largely on the improvement in clinical safety. Today, specialized applications within the category of enzyme inhibition include the use in examining the mechanism of substrate oxidation,[182-184] as probes for defining the enzyme active site,[185-187] as modulators of in vivo drug efficacy or toxicity,[1,188] and finally, as biochemical tools for P450 reaction phenotyping.[189] There is necessarily considerable overlap among these classifications, with improved understanding of the structural determinants of enzyme inhibition resulting in the design of inhibitors with increased affinity and selectivity. As a result, improved modeling and chemical template development is possible at the drug discovery stage and the potential for drug interactions can be more accurately assessed (see Chapter 9 on Drug-Drug Interactions). Thus this section will include some general discussion of the properties of inhibitors and enzymes that comprise the structural basis for P450 inhibition; however, the primary focus will be on the applications and limitations of chemical inhibitors in human P450 reaction phenotyping studies.

As with immunoinhibition studies, the limiting factor in the use of chemical inhibitors for in vitro studies has historically been the lack of adequate selectivity among P450 forms. Selectivity, in turn, is largely defined by two factors, the mechanism of inhibition, and the relative affinity of both the inhibitor and the test substrate for the enzyme (see also Chapters 7 and 9 on Enzyme Kinetics and Drug-Drug Interactions for related discussions). An understanding of processes of enzyme inhibition is necessary since this can influence selectivity, experimental conduct and the interpretation of results. The most common mechanism of P450 inhibition is the reversible binding of a compound to the enzyme in a competitive manner. This is simply the interaction of two substrates at the same binding site of the enzyme. An example of competitive inhibition is shown in Figure 17 as a Dixon plot of the inhibition of CYP2C19-catalyzed (S)-mephenytoin 4'-hydroxylation by (R)-warfarin.[71] The relative affinity of the inhibitor is defined as the concentration required for half-maximal inhibition, or K_i. Competitive inhibition that involves the interaction of a substrate nitrogen atom with the P450 heme is generally dramatic. Possibly the most studied P450 inhibitor, ketoconazole, exemplifies these properties. Ketoconazole and many imidazole agents were developed as therapeutic agents based on the ability to inhibit a critical P450 involved in fungal biosynthesis. This compound interacts with several human P450 forms as well, specifically by the coordination of the imidazole function of the molecule with the ferrous (reduced) form of the P450 heme. An additional inhibitory characteristic of ketoconazole and imidazole agents in general is the strong interaction of the hydrophobic portion of the molecule with the lipophilic binding site that is characteristic of P450s.[190]

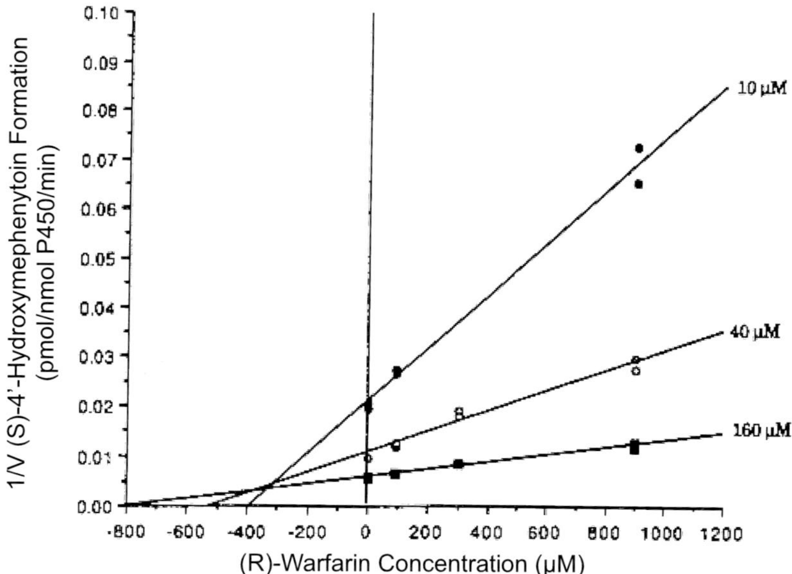

Figure 17. Dixon plot for the inhibition of (S)-mephenytoin (10, 40, and 160 uM) 4'-hydroxylation by (R)-warfarin (0, 90, 300, and 900 uM) catalyzed by human liver microsomes (HL122).

A competitive inhibitor may also act as a substrate, with binding followed by a catalytic step that results in non-competitive enzyme inhibition. Examples of compounds that undergo metabolism to form an intermediate that complexes with the P450 include the calcium channel blocker diltiazem and many macrolide antibiotics such as erythromycin and troleandomycin.[191,192,192] Figure 18 shows the ability of diltiazem to inhibit CYP3A4-dependent midazolam hydroxylation while forming a characteristic complex that can be measured spectrophotometrically.[193] This complex formation is clearly reversible in vitro (by oxidation of the heme iron), but is essentially irreversible in vivo and can result in severe impairment of drug metabolism capacity.[194,195] Alternatively,

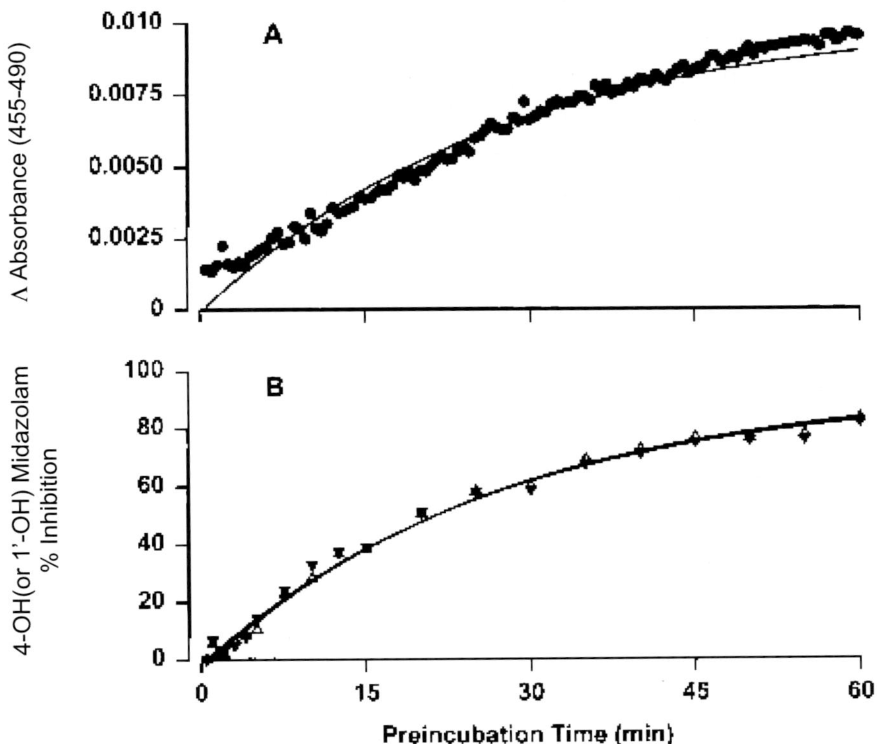

Figure 18. The simultaneous measurement of metabolite intermediate complex formation by diltiazem and the effect of diltiazem preincubation on midazolam hydroxylation in human liver microsomes. A, metabolite intermediate complex formation by diltiazem. B, effect of diltiazem preincubation on midazolam hydroxylation in human liver microsomes. DTZ (100μM) was incubated with human liver microsomes and NADPH for various times, and then added to a tube containing midazolam and NADPH. The midazolam reaction was terminated after a 5-min incubation. Each line is the line of best fit for the data. λ=0.037min^{-1}; ▼, 1'-hydoxymidazolam; Δ, 4-hydroxymidazolam.[193]

metabolism-dependent enzyme inhibition or suicide inactivation results in a covalent modification of the enzyme protein or the heme and is irreversible. This process contains a rate component (k_{inact}) that reflects the pseudo-first order loss of activity, and an affinity component (K_i), to describe the interaction of the substrate (inhibitor) and enzyme. Figure 2 illustrates the mechanism-based inactivation of CYP3A4 by midazolam. Additional experimental approaches to demonstrate inactivation include the incubation of the enzyme (human liver microsomes or expressed P450) with the inhibitor and NADPH followed by extensive dialysis of the sample to remove any unbound inhibitor. The determination of the residual enzyme activity, assayed with a panel of P450 marker substrates, yields an evaluation of the selectivity of the inactivator. Furthermore, parallel determinations of the amount of spectrally detectable P450 remaining should provide strong evidence of the mechanism of inactivation. Specifically, inactivation via protein modification will decrease enzyme activity but have no effect on spectrally determined P450 levels. In contrast, gestodene and furafylline are selective mechanism-based inactivators of CYP3A4 and 1A2, respectively, that produce inactivation through heme modification and subsequent destruction. Thus the NADPH-dependent loss of enzyme activity will be accompanied by a similar loss of spectrally detectable P450. Mechanism-based inactivators have been shown to have exquisite selectivity in some cases, with a single amino acid change in a rat CYP2B1 variant dictating inactivation by a derivative of chloramphenicol.[196] These studies have evolved to site-directed mutagenesis of P450 active site 'hot spots' coupled with in silico modeling of the substrate-enzyme interaction.[197,198] One future application of this research would be the study of human P450 variants and the ability to predict how these enzymes will respond to inhibitors when compared with the wild type human P450 form.

Prior to the availability of P450 marker substrates and commercially available expressed P450s, studies to examine the selectivity of chemical inhibitors were at a significant disadvantage. For example, cimetidine and SKF-525a were used as general P450 inhibitors. However, these compounds are no longer used due to the lack of selectivity.[199,200] Similarly, diethyldithiocarbamate and disulfiram were reported to be CYP2E1-selective,[201] however, the effect on CYP2A6 was not discernable until several years later when 2A6 was expressed and fully characterized.[202] The use of orphenadrine as an inhibitor of CYP2B6 was perpetuated in the literature until a comprehensive study showed that this compound is actually more selective for CYP2D6 (Figure 19).[203]

The assumption that a P450 marker substrate can be used as a selective inhibitor can also be flawed. An example is the use of nifedipine to measure CYP3A4 inhibition of the metabolism of the anti-hypertensive drug irbesartan, where these chemical inhibition studies produced the only reaction phenotyping results to implicate CYP3A4.[204] The explanation that nifedipine also inhibits CYP2C9 ($K_i \approx 20$ μM) clarified the interpretation. More recent studies have attempted to define the chemical inhibition of two highly related P450 forms, CYP3A4 and 3A5, with some selectivity observed for ketoconazole and fluconazole.[205]

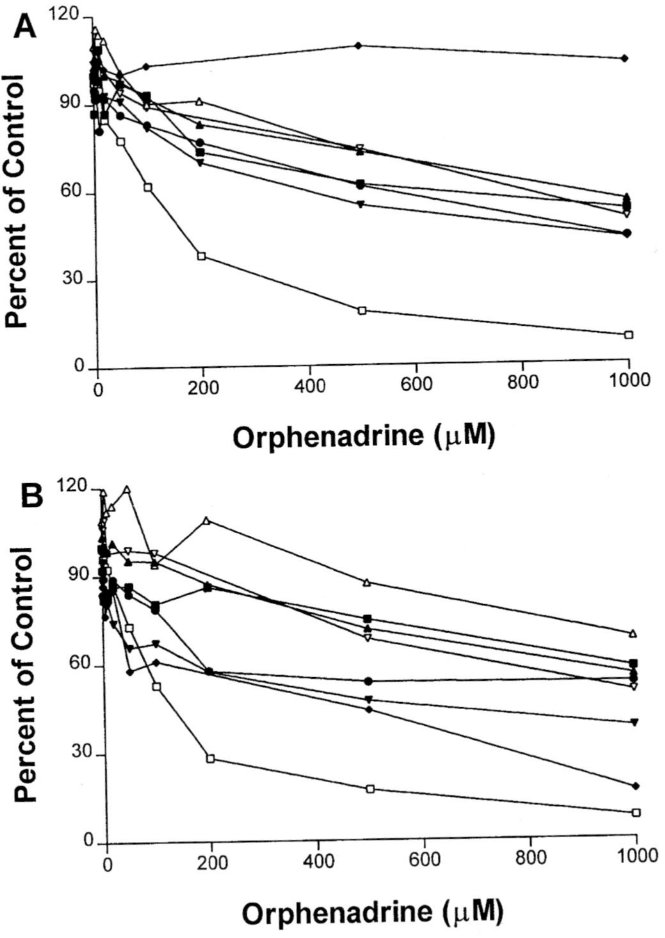

Figure 19. Effect of orphenadrine on human P450 enzyme-catalyzed reactions in human liver microsomes.[203]　　(A) Orphenadrine in the competitive inhibition protocol and (B) orphenadrine in the preincubation (mechanism-based inactivation) protocol. Enzyme reactions assayed are: phenacetin O-deethylation for CYP1A2 (■); coumarin 7-hydroxylation for CYP2A6 (▲); 7-ethoxytrifluoromethylcoumarin O-deethylation for CYP2B6 (▼); tolbutamide 4-hydroxylation for CYP2C9 (♦); (S)-mephenytoin 4-hydroxylation for CYP2C19 (●); bufuralol 1-hydroxylation for CYP2D6 (□); chlorzoxazone 6-hydroxylation for CYP2E1 (Δ); and nifedipine oxidation for CYP3A4 (∇).

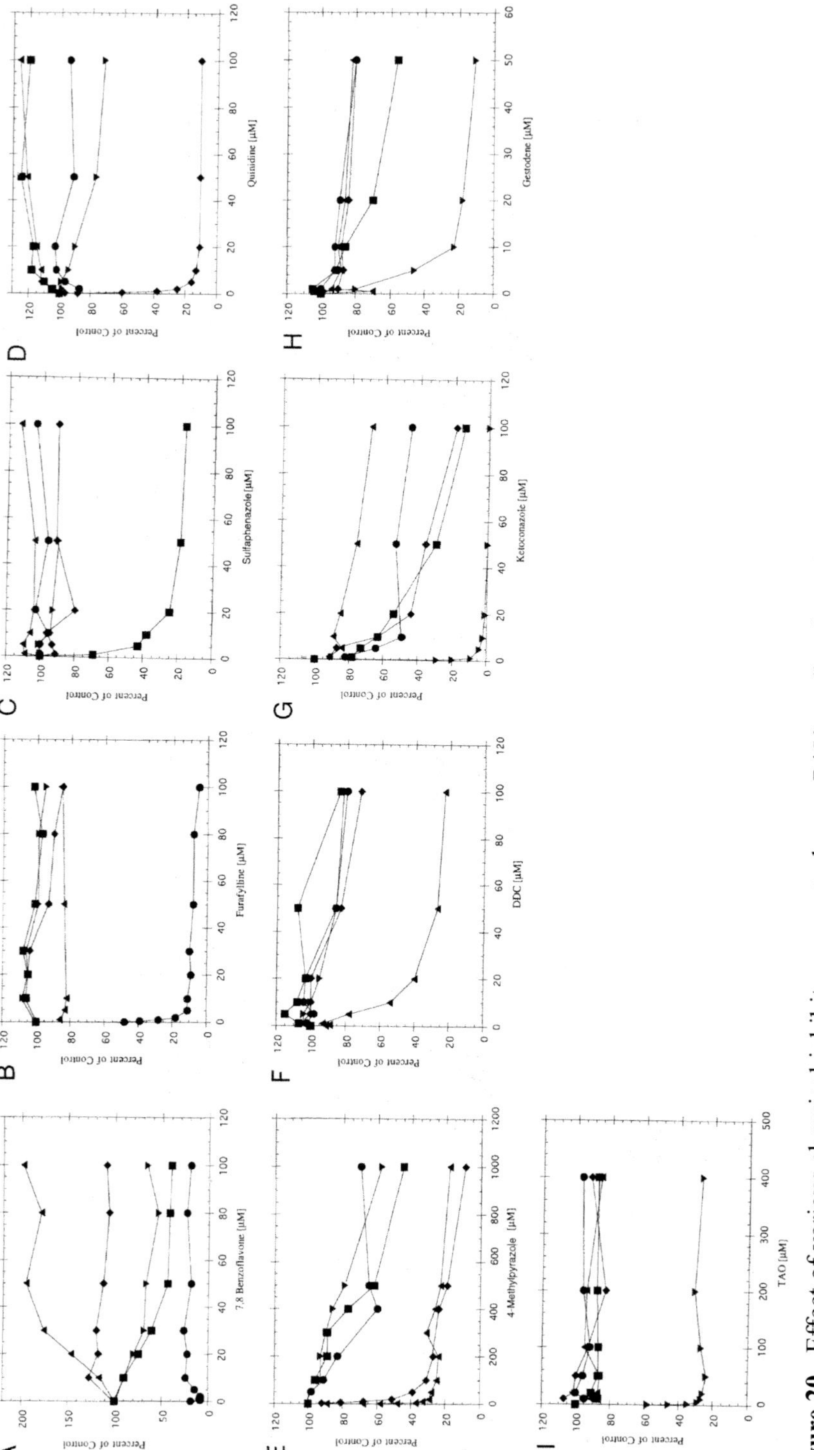

Figure 20. Effect of various chemical inhibitors on cytochrome P450-mediated reactions in human liver microsomes. Effects of (A) 7,8,-benzoflavone, (B) furafylline, (C) sulfaphenazole, (D) quinidine, (E) 4-methylpyrazole, (F) DDC, (G) ketoconazole, (H) gestodene and (I) TAO were determined on phenacetin O-deethylation (●), tolbutamide methyl hydroxylation (■), bufuralol 1'-hydroxylation (◆), chlorzoxazone 6-hydroxylation (▲), and testosterone 6-hydroxylation (▼). Activities are expressed as a percentage of control activity in the presence of 1% methanol.

With the advancement of P450-marker substrates and commercially available recombinant P450s, several laboratories undertook more comprehensive studies on the selectivity of the most commonly used P450 inhibitors.[206-209] Figure 20 shows the concentration-dependent inhibition of CYP1A2, -2C9, -2D6, -2E1, and -3A4 by nine common inhibitors.[206] These compounds represent all of the previously discussed mechanisms of inhibition. Therefore, the experimental

Table 3. The chemical structure of commonly used P450 specific inhibitors.

P450 Enzyme	P450 selective inhibitor	P450 Enzyme	P450 selective inhibitor
CYP1A2	furafylline	CYP2C19	(+)-N-3-benzyl-nirvanol
CYP2A6	methoxsalen	CYP2D6	quinidine
CYP2B6	No selective inhibitor available, recommend CYP2B6 antibody	CYP2E1	No selective inhibitor available, recommend CYP2E1 antibody
CYP2C9	sulfaphenazole	CYP3A4	ketoconazole

protocol for the incubations containing the mechanism-based inactivators troleandomycin, gestodene, furafylline, and diethyldithiocarbamate included a preincubation of the test inhibitor with NADPH and human liver microsomes prior to the evaluation of residual P450-marker enzyme activity. The results for ketoconazole and 7,8-benzoflavone (also known as α-naphthoflavone) highlight the role of inhibitor concentration in selectivity, as both CYP3A4 and 1A2 are selectively inhibited at ~1 μM inhibitor, although all P450 enzymes studied are effected at higher concentrations. Of course, certain P450s, such as 2C19, 2C8, and 2B6 were not evaluated in each of these studies so selectivity is always relative to the number of P450 forms evaluated.

Some advantages are evident when multiple inhibitors are used within a single microsomal incubation. First, chemical inhibition can be used to define and eliminate the contribution of a suspected P450 to a reaction pathway. For example, quinidine has been used to eliminate the contribution of CYP2D6, and thus strengthen the correlation of the remaining enzyme activity in a panel of human liver microsomes (see Figure 7). Also worth noting are limitations of multiple inhibitor experiments. For example, simultaneous stimulation and inhibition of CYP3A4-dependent 7-benzyloxy-4-trifluromethyl coumarin (BFC) dealkylase by the CYP1A2 inhibitor 7,8-benzoflavone has been observed (Figure 21). These results may confound the

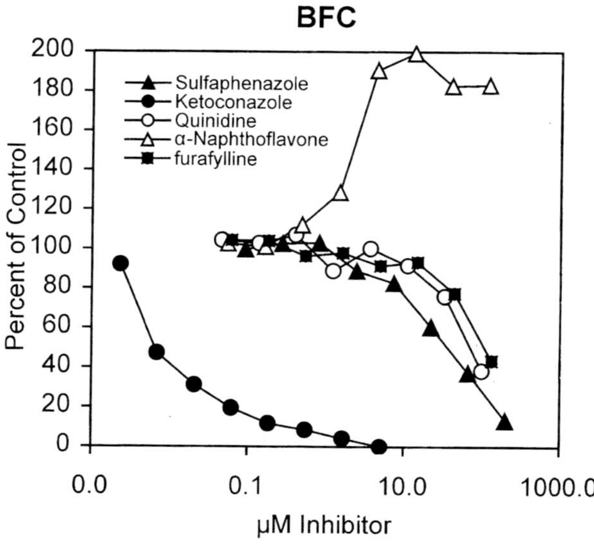

Figure 21. Effect of P450 isoform-selective chemical inhibitors on BFC dealkylase activity in HLM.[211]

interpretation of more complex chemical inhibition studies where two distinct P450s (one form being CYP3A4) are involved in common reaction pathways. This stimulation phenomenon has been reproduced in several laboratories for CYP3A4[210] [97,103] and is discussed in more detail in the chapter on Enzyme Kinetics.

In addition to the concept of concentration-dependency of enzyme inhibition, several other experimental parameters should be considered. First, studies to determine the effect of a panel of selective P450 inhibitors on a specific reaction pathway should use a pooled sample (N=3 or greater) of human liver microsomes to avoid potential errors caused by polymorphic P450 expression in individual human liver microsome samples. In contrast, a panel of individual microsome samples is required if the objective is to attempt to quantitate the role of specific P450s to a particular reaction.[175] Alternatively, the use of a panel of expressed P450s provides direct experimental evidence of the selectivity of chemical inhibitors. As might be predicted for mechanism-based inactivation, the level of enzyme expression and the presence/level of accessory proteins (cytochrome b_5 and reductase) have been shown to effect inhibition. Figure 22 shows the effect

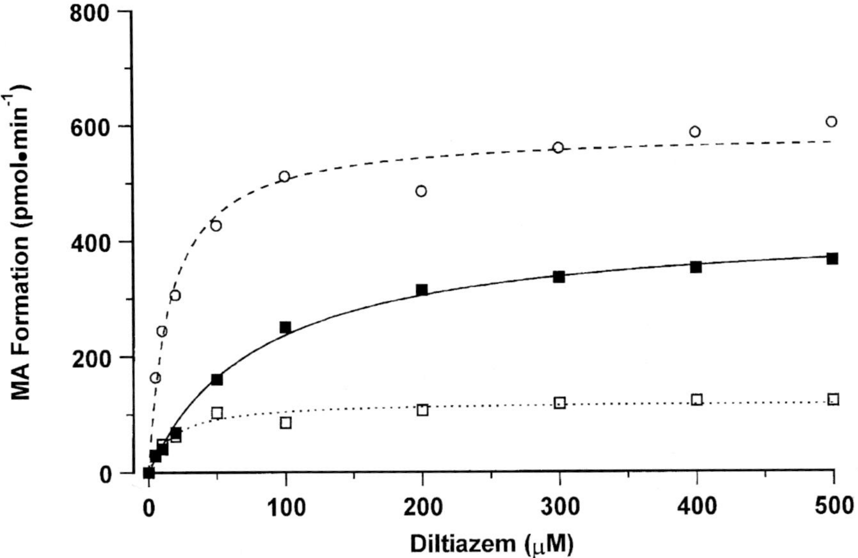

Figure 22. Diltiazem N-demethylation in microsomes of cDNA-expressed CYP3A. Diltiazem (5-500 μM) was incubated with 40 pmol of cDNA-expressed CYP3A and NADPH for 5 min. Each symbol represents the average of two points. Each line represents the line of best fit with a single-enzyme Michaelis-Menten equation. (O) CYP3A4 (+b_5); (■) CYP3A5; (□) CYP3A4.

of b_5 co-expression on the mechanism-based inactivation of CYP3A4 and 3A5 by diltiazem.[193] Another factor that can influence mechanism-based inactivation data is NADPH-dependent enzyme degradation (inactivation) in the absence of inhibitor. CYP2E1 has been shown to be particularly vulnerable, with rapid degradation of enzyme activity and protein following a five minute incubation with NADPH.[211,212] This highlights the importance of control samples for chemical inhibition studies. In examining mechanism-based inactivation, the effect of NADPH is the correct control to compare with the effect of inhibitor and NADPH. If the inhibitor is added to the incubation in an organic solvent, the total percentage of solvent (inhibitor and substrate) should be kept to $\leq 1\%$ of the total volume (see also section 2.2 on Incubation Conditions). Finally, as with enzyme kinetic determinations, the effect of inhibitor binding non-specifically to microsomal protein can substantially reduce the free concentration of inhibitor and artificially increase the K_i value. For this reason, the inhibition constant determined for a defined enzyme-inhibitor interaction is likely to be lower when using expressed P450s (lower protein concentration) compared with human liver microsomes (higher protein concentration).

7   ANTIBODIES

Several applications for specific human P450 antibodies to reaction phenotyping analysis have been developed. The most studied and relevant approach is as a biological inhibitor of enzyme activity, thus allowing the direct assessment of the role of specific P450s to the metabolism of the NCE in an enzyme mixture such as human liver microsomes.[213-215] More specifically, the use of antibodies for quantitatively predicting the role of individual P450s to a particular reaction in context with other complementary reaction phenotyping approaches has shown both advantages and limitation. As described earlier in this chapter, antibodies can also be used for the immunoquantitation of specific P450s within microsome samples, data that can then be used for correlation analysis to enzyme activity (CYP marker substrate or NCE metabolism) in a bank of human liver microsome samples. Immunodetectable levels of P450 forms can also be incorporated into extrapolations of in vitro results to in vivo relevance. This section will review the role of antibodies in reaction phenotyping including elements of the experimental approaches, applications, and limitations.

Prior to a discussion on experimental procedures, some clarification of terminology for antibodies is required. First, monoclonal antibodies, by definition, recognize a single antigenic determinant or epitope and are strongly preferred to the corresponding polyclonal reagent due to the general increased specificity. Also, three categories can be used to describe the nature of the interaction of the antibody with the antigen (P450 enzyme); inhibitory of enzyme activity, immunoreactive or able to recognize the protein using ELISA or Western blot analysis, or both inhibitory and immunoreactive. Recent detailed work on the mapping of specific P450 amino acid regions as epitopes has greatly advanced the ability to predict the ultimate effect of antibody recognition.[214,216-219] However, this science still contains a sizeable element of art as evidenced by problems due to common epitopes among similar P450 forms but also of an antibody that can differentiate between allelic forms of CYP2C9 that differ by only a single amino acid.[220]

The conduct of experiments to determine the effect of a P450 antibody on a particular metabolic reaction in human liver microsomes is relatively simple, and shares many characteristics of chemical inhibition. As with other reaction phenotyping approaches, it is absolutely critical that the experimental system be validated *a priori*, that is, the antibody must be thoroughly characterized with regards to the selectivity of inhibition to determine cross-reactivity with other P450s. As illustrated below, numerous errors in the application and interpretation of antibody inhibition studies can be attributed to the lack of characterization of antibody specificity. Given the much-improved quality of current commercially available antibody preparations, full characterization is unnecessary. However, for inhibition experiments, controls to measure (a) the ability of the antibody to inhibit an established marker P450 activity, and (b) the effect of control IgG from the same manufacturer/process on the metabolism of the NCE, are essential. Similarly, immunoquantitation experiments should contain a standard curve of the cDNA-expressed P450 form under investigation (antigen) *and* highly related P450 forms (negative controls). Inhibition of enzyme activity is typically determined using a pooled human liver microsome sample, a fixed concentration of the NCE, and increasing concentrations of antibody (represented as mg IgG/ mg protein or nmol P450) to the particular P450 form. This concept is illustrated in Figure 23, which shows the effect of an anti-CYP3A4 antibody on CYP3A4-marker testosterone 6β–hydroxylation for human liver microsomes and cDNA-expressed CYP3A4. It should be noted that many investigators have found the recombinant enzyme system to be more sensitive to antibody inhibition, both in the amount of antibody required and maximal inhibition observed, when compared with human liver microsomes. Following the determination of the optimal ratio of antibody to microsomal protein, this ratio can then be fixed and used to examine the range of inhibition of enzyme activity in a set of individual human liver microsome samples, thereby providing a measure of the range of interindividual variability in enzyme level and biotransformation via that particular P450. For example, the interindividual variability (N=19 human liver microsome samples) in a CYP2C9-dependent hydroxylation reaction is shown in Figure 24.

This example illustrates that the addition of an anti-2C9 monoclonal antibody to the human liver microsome incubations was able to inhibit approximately 80% of drug X hydroxylation, thus suggesting a dominant role for CYP2C9 in this reaction. This approach was applied to chlorzoxazone 6-hydroxylation using an anti-2E1 antibody, resulting in an average of 50% inhibition and thus supporting the involvement of P450 forms other than 2E1 to this activity.[173,221] Two other experimental approaches for antibody inhibition have been used to resolve

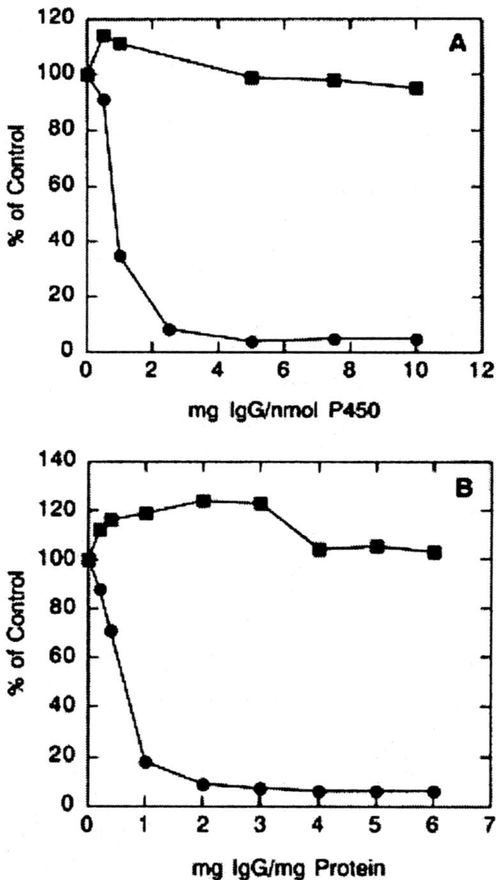

Figure 23.   Inhibition of testosterone 6β-hydroxylation by anti-peptide IgG. Human liver microsomes containing 0.1 nmol cytochrome P450 (A) or microsomes (0.25 mg) prepared from human B-lymphoblastoid cells expressing human CYP3A4 and P450 reductase (B) were preincubated with 0.05-1.5 mg of preimmune IgG (■) or anti-peptide IgG (●). The control activities of testosterone 6β-hydroxylation were 3.15 nmol/min/ nmol P450 for human liver microsome and 1.08 nmol/min/mg protein for microsome from cells expressing human CYP3A4 and P450 reductase.

difficult reaction phenotyping problems. First, two or more antibodies to distinct P450s can be added to human liver microsome incubations to estimate the total contribution of multiple P450s to a particular reaction. For example, Lemoine et al. (1993) and Mei et al. (1999) used multiple antibodies to inhibit human liver microsomal imipramine N-demethylation and diazepam N-demethylation, respectively. This concept is also illustrated in Figure 6 where the addition of antibodies to CYP2C9 and CYP3A4 followed by correlation analysis was used to differentiate the involvement of enzymes in the metabolism of a COX-2 inhibitor. Alternatively, the contribution of a potentially competing P450

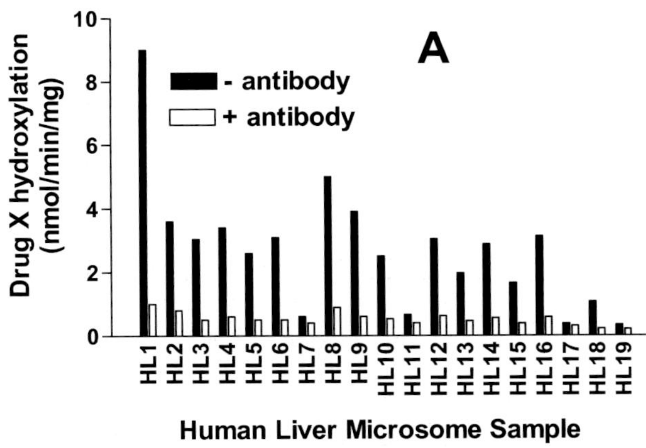

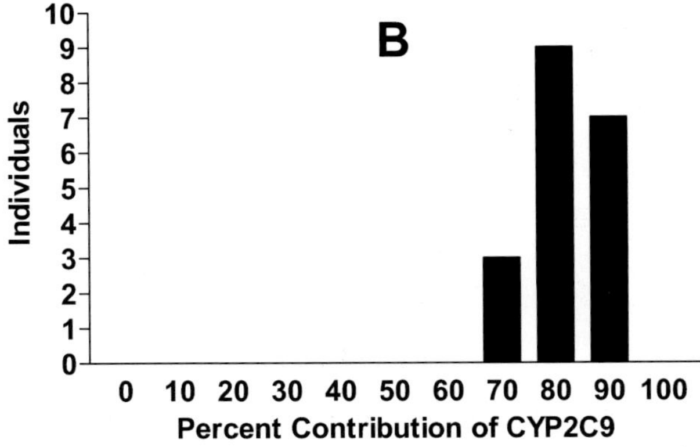

Figure 24. Panel A, Inhibition of drug X hydroxylase activity by antibody to CYP2C9 for 19 individual human liver microsome samples. Activity in the presence of antibody is indicated by the open bars, with the solid bars representing activity in the presence of control antibody. Panel B shows the contribution of CYP2C9 to enzyme activity as determined from the percent inhibition calculated for each individual.

to the rate of NCE metabolism in a panel of human liver microsome samples can be measured by suppression of one of the P450s by the corresponding antibody, thus strengthening the correlation for the rate of NCE metabolism with the marker enzyme activity for the other P450.[145] Due to the confounding factors often observed in enzyme activity correlation analysis (see earlier section), this approach was often used to decipher the involvement of 2D6 in biotransformation reactions prior to the commercial availability of expressed P450s.

A primary limitation of anti-P450 antibodies for reaction phenotyping has been the cross-reactivity with related P450s, producing errors in both enzyme inhibition studies and correlation analyses produced from P450 immunoquantitation. The technologies involved in the generation and/or isolation of the antigen needed for antibody production parallels the improvements in antibody specificity over the past 20 years. The pioneering work in the 1970s on the purification and characterization of rat P450s produced the first antigens. With further advances in P450 purification however, the vast majority of the resultant antibodies were later found to cross react with related rat P450s. Many of these antibodies were also subsequently applied to in vitro studies using human liver microsomes, with inconsistent results. Ultimately, the clear demonstration of cross-reactivity of antibodies to rat 2B1 with human 2E1 highlighted the problems associated with the cross-species use of antibodies.[222] In fact, these authors were able to use this anti-rat 2B1 antibody to identify four human liver P450s, none of which corresponded to the homologous human form (CYP2B6).[70] Later, the development of P450 expression systems as a source of purified antigen coupled with monoclonal technology resulted in substantial improvements in antibody specificity.[213,223] Most recently, antibodies developed to CYP-specific peptides have shown specificity to allelic forms.[220]

An additional limitation of the use of P450 antibodies is that the inhibition of reactions known to be highly specific to a particular P450 form is rarely 100%, suggesting that some small fraction of the enzyme retains function. Of course, this phenomenon can confound the interpretation that partial antibody inhibition of the metabolism of a NCE is instead due to the involvement of a distinct P450 form.[217,224,225] For example, CYP2C9-marker diclofenac 4-hydroxylation was inhibited by 85-90% with a 2C9-specific monoclonal antibody.[226] This likely represents maximal inhibition for human liver microsomes, and the fact that an additional ~5% inhibition was observed with other 2C antibodies does not necessarily implicate other 2C enzymes in this reaction. Numerous other examples of this limitation exist, and proposed explanations include the possibility that antibody inhibition of some P450s such as 2E1 may be less efficient for smaller substrates.[201] This could lead to the phenomenon of substrate-dependent inhibition as has been described for the interaction of chemical inhibitors with CYP3A4, and further highlights the need for careful experimental controls.

The use of antibodies for P450 immunoquantitation is also prone to the same errors caused by limitations of antibody specificity or by over-interpretation of the experimental results. Examples within the 3A and 2B families illustrate the advancements of the past ten years in this area and challenges for the future. Until recently, antibodies to human 3A forms (whether developed toward the human or rat 3A forms) were assumed to recognize all highly-related 3A forms (e.g., CYP3A4, -3A5, and -3A7).[227] Antibody specific to 3A5 (anti-peptide monoclonal) is now available, and immunochemical differentiation of 3A4 and 3A7 in fetal/neonatal liver microsome samples has also been demonstrated.[217] In addition, numerous variants of CYP3A4 [228,229] have been identified, and the effect of amino acid changes on enzyme activity and reactivity with antibodies is largely uninvestigated. For CYP2B6, protein immunoquantitation has improved with antibody specificity, as the antigen has evolved from that of a monkey 2B protein [32] to the rat 2B1 form[39] and finally to an·anti-2B6 specific peptide.[230] However, as with the 3A forms, variants of 2B6 have been identified,[231] with the effects on drug metabolism yet to be determined.

8   CONCLUDING REMARKS

Given the correct experimental design and data interpretation, cytochrome P450 reaction phenotyping can provide a biochemical explanation for, and/or prediction of, interindividual variability in drug metabolism. Over the past 25 years, tremendous progress has been made in the generation and characterization of a now commercially available 'tool kit' for reaction phenotyping studies. However, this evolution continues as new findings and technologies often require the reevaluation of reagents (i.e., marker substrate, chemical inhibitors, antibodies) and occasionally changes in the interpretation of existing data. A common theme for this chapter has been that a single reaction phenotyping approach cannot support a conclusion of the P450(s) responsible for a particular biotransformation step. In addition, the data must ultimately be interpreted in context with the overall biotransformation of the drug, with consideration for extrahepatic metabolism, enzyme levels, and routes of drug administration, distribution, and elimination. In addition to the generally accepted four approaches of correlation analysis, chemical and antibody inhibition, and metabolism by expressed P450s, a strong argument can be made that pharmacogenetic considerations are a logical fifth element for future P450 reaction phenotyping work. Although the influence of pharmacogenetics on drug metabolism currently draws considerable attention in drug discovery and development, the recent completion of the human genome project promises more detailed explanations of the basis for pharmacokinetic and pharmacodynamic response. Further understanding of the role of factors such as single nucleotide polymorphisms, allelic variants, and nuclear xenobiotic receptors in the expression and activity of drug metabolizing enzymes will allow for more definitive statements on the role of human P450s in the metabolic clearance of NCEs and will ultimately contribute to the development of more personalized drug therapy to improve safety and efficacy.

9   ACKNOWLEDGEMENTS

Drs. Stevens and Wienkers would like to acknowledge the support and patience of their families during the writing of this chapter. The authors acknowledge the comments and guidance of Drs. Peter Fan and J. Matthew Hutzler in reviewing the material.

REFERENCES

1. Bertz, R. J. and Granneman, G. R. *Clinical Pharmacokinetics* **1997,** *32*, 210.
2. Bertilsson, L. *Clinical Pharmacokinetics* **1995,** *29*, 192.
3. Goldstein, J. A., Ishizaki, T., Chiba, K., de Morais, S. M., Bell, D., Krahn, P. M., and Evans, D. A. *Pharmacogenetics* **1997,** *7*, 59.
4. Peck, C. C., Temple, R., and Collins, J. M. *JAMA* **1993,** *269*, 1550.
5. Chong, P. H., Seeger, J. D., and Franklin, C. *American Journal of Medicine* **2001,** *111*, 390.
6. Kunze, K. L. and Trager, W. F. *Drug Metabolism and Disposition: The Biological Fate of Chemicals* **1996,** *24*, 429 .
7. Heimark, L. D., Gibaldi, M., Trager, W. F., O'Reilly, R. A., and Goulart, D. A. *Clinical Pharmacology and Therapeutics* **1987,** *42*, 388.
8. Kunze, K. L., Wienkers, L. C., Thummel, K. E., and Trager, W. F. *Drug Metabolism and Disposition: The Biological Fate of Chemicals* **1996,** *24*, 414.
9. Lee, C. R. and Thrasher, K. A. *Pharmacotherapy* **2001,** *21*, 1240.
10. Crussell-Porter, L. L., Rindone, J. P., Ford, M. A., and Jaskar, D. W. *Archives of Internal Medicine* **1993** , *153*, 102.
11. Brodie, B. B., Gillette, J. R., and LaDu, B. N. *Annual Review of Biochemistry* **1958,** 427.
12. Axelrod J *Journal of Biological Chemistry* **1955,** *214*, 753.
13. Boyland, E. and Levi A. *Biochemical Journal* **1935,** *29*, 2679.
14. Klingenberg, M. *Archives of Biochemistry and Biophysics* **1958,** *75*, 376.
15. Omura, T. and Sato, R. *Journal of Biological Chemistry* **1962,** 1375.
16. Estabrook, R. W., Cooper, D. Y., and Rosenthal, O. *Biochemische Zeitschrift* **1963,** *338*, 741.

17. Cooper, D. Y., Levine, S., Narasimhule, S., Rosenthal, O., and and Estabrook, R. W. *Science* **1965**, *147*, 400.
18. Welch, R. M., Harrison, Y. E., and Burns, J. J. *Toxicology and Applied Pharmacology* **1967**, *10*, 340.
19. Sladek, N. E. and Mannering, G. J. *Molecular Pharmacology* **1969**, *5*, 186.
20. Lu, A. Y. and Coon, M. J. *Journal of Biological Chemistry* **1968**, *243*, 1331.
21. Ryan, D., Lu, A. Y., West, S., and Levin, W. *Journal of Biological Chemistry* **1975**, *250*, 2157.
22. Thomas, P. E., Lu, A. Y., Ryan, D., West, S. B., Kawalek, J., and Levin, W. *Molecular Pharmacology* **1976**, *12*, 746.
23. Gander, J. E. and Mannering, G. J. *Pharmacology and Therapeutics* **1980**, *10*, 191.
24. Guengerich, F. P., Dannan, G. A., Wright, S. T., Martin, M. V., and Kaminsky, L. S. *Biochemistry* **1982**, *21*, 6019.
25. Estabrook, R. W., Mason, J. I., Simpson, E. R., Peterson, J. A., and Waterman, M. R. *Advances in Enzyme Regulation* **1991**, *31*, 365.
26. Aoyama, T., Yamano, S., Guzelian, P. S., Gelboin, H. V., and Gonzalez, F. J. *Proceedings of the National Academy of Sciences of the United States of America* **1990**, *87*, 4790.
27. Asseffa, A., Smith, S. J., Nagata, K., Gillette, J., Gelboin, H. V., and Gonzalez, F. J. *Archives of Biochemistry and Biophysics* **1989**, *274*, 481.
28. Crespi, C. L., Langenbach, R., and Penman, B. W. *Progress in Clinical and Biological Research* **1990**, *340B*, 97.
29. Nebert, D. W., Adesnik, M., Coon, M. J., Estabrook, R. W., Gonzalez, F. J., Guengerich, F. P., Gunsalus, I. C., Johnson, E. F., Kemper, B., Levin, W., et al. *DNA* **1987**, *6*, 1.
30. Nelson, D. R. *Archives of Biochemistry and Biophysics* **1999**, *369*, 1.
31. Hakkola, J., Pasanen, M., Purkunen, R., Saarikoski, S., Pelkonen, O., Maenpaa, J., Rane, A., and Raunio, H. *Biochemical Pharmacology* **1994**, *48*, 59.
32. Shimada, T., Yamazaki, H., Mimura, M., Inui, Y., and Guengerich, F. P. *Journal of Pharmacology and Experimental Therapeutics* **1994**, *270*, 414.
33. Bloomer, J. C., Clarke, S. E., and Chenery, R. J. *Xenobiotica* **1995**, *25*, 917.
34. Wynalda, M. A., Hauer, M. J., and Wienkers, L. C. *Drug Metabolism and Disposition: The Biological Fate of Chemicals* **1998**, *26*, 1048.
35. Spaldin, V., Madden, S., Adams, D. A., Edwards, R. J., Davies, D. S., and Park, B. K. *Drug Metabolism and Disposition: The Biological Fate of Chemicals* **1995**, *23*, 929.
36. Nakajima, M., Kobayashi, K., Shimada, N., Tokudome, S., Yamamoto, T., and Kuroiwa, Y. *British Journal of Clinical Pharmacology* **1998**, *46*, 55.
37. Ha, H. R., Chen, J., Freiburghaus, A. U., and Follath, F. *British Journal of Clinical Pharmacology* **1995**, *39*, 321.
38. Guengerich, F. P., Parikh, A., Turesky, R. J., and Josephy, P. D. *Mutation Research* **1999**, *428*, 115.
39. Code, E. L., Crespi, C. L., Penman, B. W., Gonzalez, F. J., Chang, T. K., and Waxman, D. J. *Drug Metabolism and Disposition: The Biological Fate of Chemicals* **1997**, *25*, 985.

40. Erickson, D. A., Mather, G., Trager, W. F., Levy, R. H., and Keirns, J. J. *Drug Metabolism and Disposition: The Biological Fate of Chemicals* **1999**, *27*, 1488.
41. Lindley, C., Hamilton, G., McCune, J. S., Faucette, S., Shord, S. S., Hawke, R. L., Wang, H., Gilbert, D., Jolley, S., Yan, B., and LeCluyse, E. L. *Drug Metabolism and Disposition: The Biological Fate of Chemicals* **2002**, *30*, 814.
42. Faucette, S. R., Hawke, R. L., Lecluyse, E. L., Shord, S. S., Yan, B., Laethem, R. M., and Lindley, C. M. *Drug Metabolism and Disposition: The Biological Fate of Chemicals* **2000**, *28*, 1222.
43. Heyn, H., White, R. B., and Stevens, J. C. *Drug Metabolism and Disposition: The Biological Fate of Chemicals* **1996**, *24*, 948.
44. Stevens, J. C., White, R. B., Hsu, S. H., and Martinet, M. *Journal of Pharmacology and Experimental Therapeutics* **1997**, *282*, 1389.
45. Hidestrand, M., Oscarson, M., Salonen, J. S., Nyman, L., Pelkonen, O., Turpeinen, M., and Ingelman-Sundberg, M. *Drug Metabolism and Disposition: The Biological Fate of Chemicals* **2001**, *29*, 1480.
46. Goodwin, B., Moore, L. B., Stoltz, C. M., McKee, D. D., and Kliewer, S. A. *Molecular Pharmacology* **2001**, *60*, 427.
47. Novak, R. F. and Woodcroft, K. J. *Archives of Pharmacal Research* **2000**, *23*, 267.
48. Carriere, V., Berthou, F., Baird, S., Belloc, C., Beaune, P., and de Waziers, I. *Pharmacogenetics* **1996**, *6*, 203.
49. Mitra, A. K. , Thummel, K. E., Kalhorn, T. F., Kharasch, E. D., Unadkat, J. D., and Slattery, J. T. *Clinical Pharmacology and Therapeutics* **1995**, *58*, 556.
50. Kharasch, E. D., Hankins, D. C., Jubert, C., Thummel, K. E., and Taraday, J. K. *Drug Metabolism and Disposition: The Biological Fate of Chemicals* **1999**, *27*, 717.
51. Manyike, P. T., Kharasch, E. D., Kalhorn, T. F., and Slattery, J. T. *Clinical Pharmacology and Therapeutics* **2000**, *67*, 275.
52. Lucas, D., Ferrara, R., Gonzalez, E., Bodenez, P., Albores, A., Manno, M., and Berthou, F. *Pharmacogenetics* **1999**, *9*, 377.
53. Garton, K. J., Yuen, P., Meinwald, J., Thummel, K. E., and Kharasch, E. D. *Drug Metabolism and Disposition: The Biological Fate of Chemicals* **1995**, *23*, 1426.
54. Tanaka, E., Terada, M., and Misawa, S. *Journal of Clinical Pharmacy and Therapeutics* **2000**, *25*, 165.
55. Lucas, D., Ferrara, R., Gonzales, E., Albores, A., Manno, M., and Berthou, F. *Toxicology Letters* **2001**, *124*, 71.
56. Harris, J. W., Rahman, A., Kim, B. R., Guengerich, F. P., and Collins, J. M. *Cancer Research* **1994**, *54*, 4026.
57. Kerr, B. M., Thummel, K. E., Wurden, C. J., Klein, S. M., Kroetz, D. L., Gonzalez, F. J., and Levy, R. H. *Biochemical Pharmacology* **1994**, *47*, 1969.
58. Li, X. Q., Bjorkman, A., Andersson, T. B., Ridderstrom, M., and Masimirembwa, C. M. *Journal of Pharmacology and Experimental Therapeutics* **2002**, *300*, 399.
59. McSorley, L. C. and Daly, A. K. *Biochemical Pharmacology* **2000**, *60*, 517.
60. Goldstein, J. A. and de Morais, S. M. *Pharmacogenetics* **1994**, *4*, 285.
61. Rettie, A. E., Wienkers, L. C., Gonzalez, F. J., Trager, W. F., and Korzekwa, K. R. *Pharmacogenetics* **1994** , *4*, 39.

62. Leemann, T., Transon, C., and Dayer, P. *Life Sciences* **1993**, *52*, 29.
63. Tracy, T. S., Marra, C., Wrighton, S. A., Gonzalez, F. J., and Korzekwa, K. R. *Biochemical Pharmacology* **1996**, *52*, 1305.
64. Rettie, A. E., Korzekwa, K. R., Kunze, K. L., Lawrence, R. F., Eddy, A. C., Aoyama, T., Gelboin, H. V., Gonzalez, F. J., and Trager, W. F. *Chemical Research in Toxicology* **1992**, *5*, 54.
65. Bonnabry, P., Leemann, T., and Dayer, P. *European Journal of Clinical Pharmacology* **1996**, *49*, 305.
66. Miners, J. O. and Birkett, D. J. *British Journal of Clinical Pharmacology* **1998**, *45*, 525.
67. Hall, S. D., Hamman, M. A., Rettie, A. E., Wienkers, L. C. , Trager, W. F., Vandenbranden, M., and Wrighton, S. A. *Drug Metabolism and Disposition: The Biological Fate of Chemicals* **1994**, *22*, 975.
68. Kobayashi, K., Kogo, M., Tani, M., Shimada, N., Ishizaki, T., Numazawa, S., Yoshida, T., Yamamoto, T., Kuroiwa, Y., and Chiba, K. *Drug Metabolism and Disposition: The Biological Fate of Chemicals* **2001**, *29*, 36.
69. Adedoyin, A., Prakash, C., O'Shea, D., Blair, I. A., and Wilkinson, G. R. *Pharmacogenetics* **1994**, *4*, 27.
70. Wrighton, S. A., Stevens, J. C., Becker, G. W., and VandenBranden, M. *Archives of Biochemistry and Biophysics* **1993**, *306*, 240.
71. Wienkers, L. C., Wurden, C. J., Storch, E., Kunze, K. L. , Rettie, A. E., and Trager, W. F. *Drug Metabolism and Disposition: The Biological Fate of Chemicals* **1996**, *24*, 610.
72. Abelo, A., Andersson, T. B., Antonsson, M., Naudot, A. K., Skanberg, I., and Weidolf, L. *Drug Metabolism and Disposition: The Biological Fate of Chemicals* **2000**, *28*, 966.
73. Helsby, N. A., Ward, S. A., Edwards, G., Howells, R. E., and Breckenridge, A. M. *British Journal of Clinical Pharmacology* **1990**, *30*, 593.
74. Nelson, D. R., Koymans, L., Kamataki, T., Stegeman, J. J., Feyereisen, R., Waxman, D. J., Waterman, M. R., Gotoh, O., Coon, M. J., Estabrook, R. W., Gunsalus, I. C., and Nebert, D. W. *Pharmacogenetics* **1996**, *6*, 1.
75. Pelkonen, O., Maenpaa, J., Taavitsainen, P., Rautio, A., and Raunio, H. *Xenobiotica* **1998**, *28*, 1203.
76. Shimada, T., Tsumura, F., Yamazaki, H., Guengerich, F. P., and Inoue, K. *Pharmacogenetics* **2001**, *11*, 143.
77. Wilcox, R. A. and Owen, H. *Anaesthesia and Intensive Care* **2000**, *28*, 611.
78. Shin, J. G., Soukhova, N., and Flockhart, D. A. *Drug Metabolism and Disposition: The Biological Fate of Chemicals* **1999**, *27*, 1078.
79. Brachtendorf, L., Jetter, A., Beckurts, K. T., Hoscher, A. H., and Fuhr, U. *Pharmacology and Toxicology* **2002**, *90*, 144.
80. Botsch, S., Gautier, J. C., Beaune, P., Eichelbaum, M., and Kroemer, H. K. *Molecular Pharmacology* **1993**, *43*, 120.
81. Naranjo, C. A., Sproule, B. A., and Knoke, D. M. *International Clinical Psychopharmacology* **1999**, *14* Suppl 2, S35.
82. Tanaka, E. and Hisawa, S. *Journal of Clinical Pharmacy and Therapeutics* **1999**, *24*, 7.

83. Kariv, I., Fereshteh, M. P., and Oldenburg, K. R. *J Biomol Screen* **2001**, *6*, 91.

84. Shen, D. D., Kunze, K. L., and Thummel, K. E. *Adv Drug Deliv Rev* **1997**, *27*, 99.

85. Hall, S. D., Thummel, K. E., Watkins, P. B., Lown, K. S., Benet, L. Z., Paine, M. F., Mayo, R. R., Turgeon, D. K., Bailey, D. G., Fontana, R. J., and Wrighton, S. A. *Drug Metabolism and Disposition: The Biological Fate of Chemicals* **1999**, *27*, 161.

86. Gibson, G. G., Plant, N. J., Swales, K. E., Ayrton, A., and El-Sankary, W. *Xenobiotica* **2002**, *32*, 165.

87. Thummel, K. E. and Wilkinson, G. R. *Annual Review of Pharmacology and Toxicology* **1998**, *38*, 389.

88. Guengerich, F. P. *Annual Review of Pharmacology and Toxicology* **1999**, *39*, 1.

89. Thummel, K. E., Shen, D. D., Podoll, T. D., Kunze, K. L., Trager, W. F., Hartwell, P. S., Raisys, V. A., Marsh, C. L., McVicar, J. P., Barr, D. M., and et, a. l. *Journal of Pharmacology and Experimental Therapeutics* **1994**, *271*, 549.

90. Sutton, D., Butler, A. M., Nadin, L., and Murray, M. *Journal of Pharmacology and Experimental Therapeutics* **1997**, *282*, 294.

91. Beaune, P. H., Umbenhauer, D. R., Bork, R. W., Lloyd, R. S., and Guengerich, F. P. *Proceedings of the National Academy of Sciences of the United States of America* **1986**, *83*, 8064.

92. Voorman, R. L., Payne, N. A., Wienkers, L. C., Hauer, M. J., and Sanders, P. E. *Drug Metabolism and Disposition: The Biological Fate of Chemicals* **2001**, *29*, 41.

93. Kumar, G. N., Dykstra, J., Roberts, E. M., Jayanti, V. K., Hickman, D., Uchic, J., Yao, Y., Surber, B., Thomas, S., and Granneman, G. R. *Drug Metabolism and Disposition: The Biological Fate of Chemicals* **1999**, *27*, 902.

94. Luo, G., Cunningham, M., Kim, S., Burn, T., Lin, J., Sinz, M., Hamilton, G., Rizzo, C., Jolley, S., Gilbert, D., Downey, A., Mudra, D., Graham, R., Carroll, K., Xie, J., Madan, A., Parkinson, A., Christ, D., Selling, B., LeCluyse, E., and Gan, L. S. *Drug Metabolism and Disposition: The Biological Fate of Chemicals* **2002**, *30*, 795.

95. Lehmann, J. M., McKee, D. D., Watson, M. A., Willson, T. M., Moore, J. T., and Kliewer, S. A. *Journal of Clinical Investigation* **1998**, *102*, 1016.

96. Ueng, Y. F., Kuwabara, T., Chun, Y. J., and Guengerich, F. P. *Biochemistry* **1997**, *36*, 370.

97. Schrag, M. L. and Wienkers, L. C. *Drug Metabolism and Disposition: The Biological Fate of Chemicals* **2001**, *29*, 70.

98. Lin, Y., Lu, P., Tang, C., Mei, Q., Sandig, G., Rodrigues, A. D., Rushmore, T. H., and Shou, M. *Drug Metabolism and Disposition: The Biological Fate of Chemicals* **2001**, *29*, 368.

99. Harlow, G. R. and Halpert, J. R. *Proceedings of the National Academy of Sciences of the United States of America* **1998**, *95*, 6636.

100. Schrag, M. L. and Wienkers, L. C. *Archives of Biochemistry and Biophysics* **2001**, *391*, 49.

101. Khan, K. K., He, Y. Q., Domanski, T. L., and Halpert, J. R. *Molecular Pharmacology* **2002**, *61*, 495.

102. Atkins, W. M., Wang, R. W., and Lu, A. Y. *Chemical Research in Toxicology* **2001**, *14*, 338.

103. Shou, M., Grogan, J., Mancewicz, J. A., Krausz, K. W., Gonzalez, F. J., Gelboin, H. V., and Korzekwa, K. R. *Biochemistry* **1994**, *33*, 6450.

104. Wang, R. W., Newton, D. J., Liu, N., Atkins, W. M., and Lu, A. Y. *Drug Metabolism and Disposition: The Biological Fate of Chemicals* **2000**, *28*, 360.

105. Kenworthy, K. E., Bloomer, J. C., Clarke, S. E., and Houston, J. B. *British Journal of Clinical Pharmacology* **1999**, *48*, 716.

106. Rane, A., Wilkinson, G. R., and Shand, D. G. *Journal of Pharmacology and Experimental Therapeutics* **1977**, *200*, 420.

107. Powis, G., Jardine, I., Van Dyke, R., Weinshilboum, R., Moore, D., Wilke, T., Rhodes, W., Nelson, R., Benson, L., and Szumlanski, C. *Drug Metabolism and Disposition: The Biological Fate of Chemicals* **1988**, *16*, 582.

108. Tredger, J. M. and Chhabra, R. S. *Drug Metabolism and Disposition: The Biological Fate of Chemicals* **1976**, *4*, 451.

109. Pearce, R. E., McIntyre, C. J., Madan, A., Sanzgiri, U., Draper, A. J., Bullock, P. L., Cook, D. C., Burton, L. A., Latham, J., Nevins, C., and Parkinson, A. *Archives of Biochemistry and Biophysics* **1996**, *331*, 145.

110. Yamazaki, H., Inoue, K., Turvy, C. G., Guengerich, F. P., and Shimada, T. *Drug Metabolism and Disposition: The Biological Fate of Chemicals* **1997**, *25*, 168.

111. Omura, T., Sato, R., Cooper, D. Y., Rosenthal, O., and Estabrook, R. W. *Federation Proceedings* **1965**, *24*, 1181.

112. Tang, C., Shou, M., and Rodrigues, A. D. *Drug Metabolism and Disposition: The Biological Fate of Chemicals* **2000**, *28*, 567.

113. Busby, W. F. Jr, Ackermann, J. M., and Crespi, C. L. *Drug Metabolism and Disposition: The Biological Fate of Chemicals* **1999**, *27*, 246.

114. Easterbrook, J., Lu, C., Sakai, Y., and Li, A. P. *Drug Metabolism and Disposition: The Biological Fate of Chemicals* **2001**, *29*, 141.

115. Hickman, D., Wang, J. P., Wang, Y., and Unadkat, J. D. *Drug Metabolism and Disposition: The Biological Fate of Chemicals* **1998**, *26*, 207.

116. Chauret, N., Gauthier, A., and Nicoll-Griffith, D. A. *Drug Metabolism and Disposition: The Biological Fate of Chemicals* **1998**, *26*, 1.

117. Kalvass, J. C., Tess, D. A., Giragossian, C., Linhares, M. C., and Maurer, T. S. *Drug Metabolism and Disposition: The Biological Fate of Chemicals* **2001**, *29*, 1332.

118. Obach, R. S. *Drug Metabolism and Disposition: The Biological Fate of Chemicals* **2001**, *29*, 1057.

119. Segel, I. H. Enzyme Kinetics, John Wiley and Sons, Inc., New York, NY, **1972**.

120. Houston, J. B. *Biochemical Pharmacology* **1994**, *47*, 1469.

121. Rane, A., Oelz, O., Frolich, J. C., Seyberth, H. W., Sweetman, B. J., Watson, J. T., Wilkinson, G. R., and Oates, J. A. *Clinical Pharmacology and Therapeutics* **1978**, *23*, 658.

122. Obach, R. S. *Drug Metabolism and Disposition: The Biological Fate of Chemicals* **1999**, *27*, 1350.

123. Olesen, O. V. and Linnet, K. *Pharmacology* **1997**, *55*, 235.

124. Rotzinger, S., Fang, J., and Baker, G. B. *Drug Metabolism and Disposition: The Biological Fate of Chemicals* **1998**, *26*, 572.

125. Guengerich, F. P. *Chemical Research in Toxicology* **2001,** *14,* 611.

126. Porter, W. R., Branchflower, R. V., and Trager, W. F. *Biochemical Pharmacology* **1977,** *26,* 549.

127. Hesse, L. M., Venkatakrishnan, K., Court, M. H., von Moltke, L. L., Duan, S. X., Shader, R. I., and Greenblatt, D. J. *Drug Metabolism and Disposition: The Biological Fate of Chemicals* **2000,** *28,* 1176.

128. Dowd, J. E. and Riggs, D. S. *Journal of Biological Chemistry* **1965,** *240,* 863.

129. Lineweaver, H. and Burk, D. *Journal of the American Chemical Society* **1934,** *56,* 658.

130. Hofstee BHJ *Science* **1952,** *116,* 329.

131. Madan, A., Parkinson, A., and Faiman, M. D. *Drug Metabolism and Disposition: The Biological Fate of Chemicals* **1995,** *23,* 1153.

132. Lee, S. H. and Slattery, J. T. *Drug Metabolism and Disposition: The Biological Fate of Chemicals* **1997,** *25,* 1354.

133. Hesse, L. M., Venkatakrishnan, K., von Moltke, L. L., Shader, R. I., and Greenblatt, D. J. *Drug Metabolism and Disposition: The Biological Fate of Chemicals* **2001,** *29,* 133.

134. Venkatakrishnan, K., Greenblatt, D. J., von Moltke, L. L., Schmider, J., Harmatz, J. S., and Shader, R. I. *Journal of Clinical Pharmacology* **1998,** *38,* 112.

135. Forrester, L. M., Henderson, C. J., Glancey, M. J., Back, D. J., Park, B. K., Ball, S. E., Kitteringham, N. R., McLaren, A. W., Miles, J. S., Skett, P., et al. *Biochemical Journal* **1992,** *281 (Pt 2),* 359.

136. Lucas, D., Berthou, F., Dreano, Y., Lozac'h, P., Volant, A., and Menez, J. F. *Alcoholism, Clinical and Experimental Research* **1993,** *17,* 900.

137. Snawder, J. E. and Lipscomb, J. C. *Regulatory Toxicology and Pharmacology* **2000,** *32,* 200.

138. Rodriguez-Antona, C., Donato, M. T., Pareja, E., Gomez-Lechon, M. J., and Castell, J. V. *Archives of Biochemistry and Biophysics* **2001,** *393,* 308.

139. Kassahun, K., McIntosh, I. S., Shou, M., Walsh, D. J., Rodeheffer, C., Slaughter, D. E., Geer, L. A., Halpin, R. A., Agrawal, N., and Rodrigues, A. D. *Drug Metabolism and Disposition: The Biological Fate of Chemicals* **2001,** *29,* 813.

140. Kumar, G. N., Dubberke, E., Rodrigues, A. D., Roberts, E., and Dennisen, J. F. *Drug Metabolism and Disposition: The Biological Fate of Chemicals* **1997,** *25,* 110.

141. Wienkers, L. C., Steenwyk, R. C., Sanders, P. E., and Pearson, P. G. *Journal of Pharmacology and Experimental Therapeutics* **1996,** *277,* 982.

142. Kumar, G. N., Walle, U. K., and Walle, T. *Journal of Pharmacology and Experimental Therapeutics* **1994,** *268,* 1160.

143. Rahman, A., Korzekwa, K. R., Grogan, J., Gonzalez, F. J., and Harris, J. W. *Cancer Research* **1994,** *54,* 5543.

144. Obach, R. S., Pablo, J., and Mash, D. C. *Drug Metabolism and Disposition: The Biological Fate of Chemicals* **1998,** *26,* 764.

145. Tang, C., Shou, M., Mei, Q., Rushmore, T. H., and Rodrigues, A. D. *Journal of Pharmacology and Experimental Therapeutics* **2000,** *293,* 453.

146. Guengerich, F. P., Muller-Enoch, D., and Blair, I. A. *Molecular Pharmacology* **1986,** *30,* 287.

147. Johnson, J. A., Herring, V. L., Wolfe, M. S., and Relling, M. V. *Journal of Pharmacology and Experimental Therapeutics* **2000**, *294*, 1099.

148. Davit, B., Reynolds, K., Yuan, R., Ajayi, F., Conner, D., Fadiran, E., Gillespie, B., Sahajwalla, C., Huang, S. M., and Lesko, L. J. *Journal of Clinical Pharmacology* **1999**, *39*, 899.

149. Langenbach, R., Smith, P. B., and Crespi, C. *Mutation Research* **1992**, *277*, 251.

150. Crespi, C. L. and Miller, V. P. *Pharmacology and Therapeutics* **1999**, *84*, 121.

151. Guengerich, F. P., Parikh, A., Johnson, E. F., Richardson, T. H., von Wachenfeldt, C., Cosme, J., Jung, F., Strassburg, C. P., Manns, M. P., Tukey, R. H., Pritchard, M., Fournel-Gigleux, S., and Burchell, B. *Drug Metabolism and Disposition: The Biological Fate of Chemicals* **1997**, *25*, 1234.

152. Friedberg, T., Pritchard, M. P., Bandera, M., Hanlon, S. P., Yao, D., McLaughlin, L. A., Ding, S., Burchell, B., and Wolf, C. R. *Drug Metabolism Reviews* **1999**, *31*, 523.

153. Gold, L. *Methods in Enzymology* **1990**, *185*, 11.

154. Iwata, H., Fujita, K., Kushida, H., Suzuki, A., Konno, Y., Nakamura, K., Fujino, A., and Kamataki, T. *Biochemical Pharmacology* **1998**, *55*, 1315.

155. Pritchard, M. P., Glancey, M. J., Blake, J. A., Gilham, D. E., Burchell, B., Wolf, C. R., and Friedberg, T. *Pharmacogenetics* **1998**, *8*, 33.

156. Dong, J. and Porter, T. D. *Archives of Biochemistry and Biophysics* **1996**, *327*, 254.

157. Cosme, J. and Johnson, E. F. *Journal of Biological Chemistry* **2000**, *275*, 2545.

158. Imaoka, S., Enomoto, K., Oda, Y., Asada, A., Fujimori, M., Shimada, T., Fujita, S., Guengerich, F. P., and Funae, Y. *Journal of Pharmacology and Experimental Therapeutics* **1990**, *255*, 1385.

159. Masimirembwa, C. M., Otter, C., Berg, M., Jonsson, M., Leidvik, B., Jonsson, E., Johansson, T., Backman, A., Edlund, A., and Andersson, T. B. *Drug Metabolism and Disposition: The Biological Fate of Chemicals* **1999**, *27*, 1117.

160. Crespi, C. L., Penman, B. W., Gonzalez, F. J., Gelboin, H. V., Galvin, M., and Langenbach, R. *Biochemical Society Transactions* **1993**, *21*, 1023.

161. Penman, B. W., Chen, L., Gelboin, H. V., Gonzalez, F. J., and Crespi, C. L. *Carcinogenesis* **1994**, *15*, 1931.

162. Yamazaki, H., Guo, Z., Persmark, M., Mimura, M., Inoue, K., Guengerich, F. P., and Shimada, T. *Molecular Pharmacology* **1994**, *46*, 568.

163. Paine, M. J., Gilham, D., Roberts, G. C., and Wolf, C. R. *Archives of Biochemistry and Biophysics* **1996**, *328*, 143.

164. Haining, R. L., Hunter, A. P., Veronese, M. E., Trager, W. F., and Rettie, A. E. *Archives of Biochemistry and Biophysics* **1996**, *333*, 447.

165. Crespi, C. L. and Miller, V. P. *Pharmacogenetics* **1997**, *7*, 203.

166. Yamazaki, H., Nakano, M., Imai, Y., Ueng, Y. F., Guengerich, F. P., and Shimada, T. *Archives of Biochemistry and Biophysics* **1996**, *325*, 174.

167. Shaw, P. M., Hosea, N. A., Thompson, D. V., Lenius, J. M., and Guengerich, F. P. *Archives of Biochemistry and Biophysics* **1997**, *348*, 107.

168. Roy, P., Yu, L. J., Crespi, C. L., and Waxman, D. J. *Drug Metabolism and Disposition: The Biological Fate of Chemicals* **1999**, *27*, 655.

169. Brian, W. R., Sari, M. A., Iwasaki, M., Shimada, T., Kaminsky, L. S., and Guengerich, F. P. *Biochemistry* **1990**, *29*, 11280.
170. Chiba, K., Kobayashi, K., Manabe, K., Tani, M., Kamataki, T., and Ishizaki, T. *Journal of Pharmacology and Experimental Therapeutics* **1993**, *266*, 52.
171. Peter, R., Bocker, R., Beaune, P. H., Iwasaki, M., Guengerich, F. P., and Yang, C. S. *Chemical Research in Toxicology* **1990**, *3*, 566.
172. Gorski, J. C., Jones, D. R., Wrighton, S. A., and Hall, S. D. *Xenobiotica* **1997**, *27*, 243.
173. Ono, S., Hatanaka, T., Hotta, H., Tsutsui, M., Satoh, T., and Gonzalez, F. J. *Pharmacogenetics* **1995**, *5*, 143.
174. Yang, T. J., Krausz, K. W., Shou, M., Yang, S. K., Buters, J. T., Gonzalez, F. J., and Gelboin, H. V. *Biochemical Pharmacology* **1998**, *55*, 1633.
175. Rodrigues, A. D. *Biochemical Pharmacology* **1999**, *57*, 465.
176. Crespi, C. L., Steimel, D. T., Penman, B. W., Korzekwa, K. R., Fernandez-Salguero, P., Buters, J. T., Gelboin, H. V., Gonzalez, F. J., Idle, J. R., and Daly, A. K. *Pharmacogenetics* **1995**, *5*, 234.
177. Venkatakrishnan, K., von Moltke, L. L., Court, M. H., Harmatz, J. S., Crespi, C. L., and Greenblatt, D. J. *Drug Metabolism and Disposition: The Biological Fate of Chemicals* **2000**, *28*, 1493.
178. Burke, M. D. *Biochemical Pharmacology* **1981**, *30*, 181.
179. Gerber, M. C., Tejwani, G. A., Gerber, N., and Bianchine, J. R. *Pharmacology and Therapeutics* **1985**, *27*, 353.
180. Puurunen, J., Sotaniemi, E., and Pelkonen, O. *European Journal of Clinical Pharmacology* **1980**, *18*, 185.
181. Knodell, R. G., Browne, D. G., Gwozdz, G. P., Brian, W. R., and Guengerich, F. P. *Gastroenterology* **1991**, *101*, 1680.
182. Halpert, J. R. *Annual Review of Pharmacology and Toxicology* **1995**, *35*, 29.
183. Guengerich, F. P. *Toxicology Letters* **1994**, *70*, 133.
184. Ortiz de Montellano, P.R. Cytochrome P450: Structure, Function, and Biochemistry **1995**, Plenum, New York.
185. Szklarz, G. D. and Halpert, J. R. *Drug Metabolism and Disposition: The Biological Fate of Chemicals* **1998**, *26*, 1179.
186. Chan, W. K., Sui, Z., and Ortiz de Montellano, P. R. *Chemical Research in Toxicology* **1993**, *6*, 38.
187. CaJacob, C. A., Chan, W. K., Shephard, E., and Ortiz de Montellano, P. R. *Journal of Biological Chemistry* **1988**, *263*, 18640.
188. Ito, K., Iwatsubo, T., Kanamitsu, S., Nakajima, Y., and Sugiyama, Y. *Annual Review of Pharmacology and Toxicology* **1998**, *38*, 461.
189. von Moltke, L. L., Greenblatt, D. J., Schmider, J., Wright, C. E., Harmatz, J. S., and Shader, R. I. *Biochemical Pharmacology* **1998**, *55*, 113.
190. Baldwin, S. J., Bloomer, J. C., Smith, G. J., Ayrton, A. D., Clarke, S. E., and Chenery, R. J. *Xenobiotica* **1995**, *25*, 261.
191. Pessayre, D., Larrey, D., Funck-Brentano, C., and Benhamou, J. P. *Journal of Antimicrobial Chemotherapy* **1985**, *16 Suppl A*, 181.
192. Pershing, L. K. and Franklin, M. R. *Xenobiotica* **1982**, *12*, 687.

193. Jones, D. R., Gorski, J. C., Hamman, M. A., Mayhew, B. S., Rider, S., and Hall, S. D. *Journal of Pharmacology and Experimental Therapeutics* **1999**, *290*, 1116.

194. May, D. C., Jarboe, C. H., Ellenburg, D. T., Roe, E. J., and Karibo, J. *Journal of Clinical Pharmacology* **1982**, *22*, 125.

195. von Rosensteil, N. A. and Adam, D. *Drug Safety* **1995**, *13*, 105.

196. Kedzie, K. M., Balfour, C. A., Escobar, G. Y., Grimm, S. W., He, Y. A., Pepperl, D. J., Regan, J. W., Stevens, J. C., and Halpert, J. R. *Journal of Biological Chemistry* **1991**, *266*, 22515.

197. Domanski, T. L. and Halpert, J. R. *Curr Drug Metab* **2001**, *2*, 117.

198. Kent, U. M., Juschyshyn, M. I., and Hollenberg, P. F. *Curr Drug Metab* **2001**, *2*, 215.

199. Buening, M. K. and Franklin, M. R. *Drug Metabolism and Disposition: The Biological Fate of Chemicals* **1974**, *2*, 386.

200. Jenner, S. and Netter, K. J. *Biochemical Pharmacology* **1972**, *21*, 1921.

201. Guengerich, F. P., Kim, D. H., and Iwasaki, M. *Chemical Research in Toxicology* **1991**, *4*, 168.

202. Yun, C. H., Shimada, T., and Guengerich, F. P. *Molecular Pharmacology* **1991**, *40*, 679.

203. Guo, Z., Raeissi, S., White, R. B., and Stevens, J. C. *Drug Metabolism and Disposition: The Biological Fate of Chemicals* **1997**, *25*, 390.

204. Bourrie, M., Meunier, V., Berger, Y., and Fabre, G. *Drug Metabolism and Disposition: The Biological Fate of Chemicals* **1999**, *27*, 288.

205. Gibbs, M. A., Thummel, K. E., Shen, D. D., and Kunze, K. L. *Drug Metabolism and Disposition: The Biological Fate of Chemicals* **1999**, *27*, 180.

206. Newton, D. J., Wang, R. W., and Lu, A. Y. *Drug Metabolism and Disposition: The Biological Fate of Chemicals* **1995**, *23*, 154.

207. McNamee, J. P. and Marks, G. S. *Drug Metabolism and Disposition: The Biological Fate of Chemicals* **1996**, *24*, 872.

208. Chang, T. K., Gonzalez, F. J., and Waxman, D. J. *Archives of Biochemistry and Biophysics* **1994**, *311*, 437.

209. Bourrie, M., Meunier, V., Berger, Y., and Fabre, G. *Journal of Pharmacology and Experimental Therapeutics* **1996**, *277*, 321.

210. Maenpaa, J., Hall, S. D., Ring, B. J., Strom, S. C., and Wrighton, S. A. *Pharmacogenetics* **1998**, *8*, 137.

211. Stresser, D. M., Turner, S. D., Blanchard, A. P., Miller, V. P., and Crespi, C. L. *Drug Metabolism and Disposition: The Biological Fate of Chemicals* **2002**, *30*, 845.

212. Goasduff, T. and Cederbaum, A. I. *Archives of Biochemistry and Biophysics* **1999**, *370*, 258.

213. Mei, Q., Tang, C., Assang, C., Lin, Y., Slaughter, D., Rodrigues, A. D., Baillie, T. A., Rushmore, T. H., and Shou, M. *Journal of Pharmacology and Experimental Therapeutics* **1999**, *291*, 749.

214. Wang, R. W. and Lu, A. Y. *Drug Metabolism and Disposition: The Biological Fate of Chemicals* **1997**, *25*, 762 .

215. Shou, M., Lu, T., Krausz, K. W., Sai, Y., Yang, T., Korzekwa, K. R., Gonzalez, F. J., and Gelboin, H. V. *European Journal of Pharmacology* **2000**, *394*, 199.

216. Leeder, J. S., Gaedigk, A., Lu, X., and Cook, V. A. *Molecular Pharmacology* **1996,** *49,* 234.
217. Wang, R. W., Newton, D. J., Liu, N. Y., Shou, M., Rushmore, T., and Lu, A. Y. *Drug Metabolism and Disposition: The Biological Fate of Chemicals* **1999,** *27,* 167.
218. Parimoo, B. and Thomas, P. E. *Archives of Biochemistry and Biophysics* **2000,** *380,* 117.
219. Adams, D. A., Edwards, R. J., Davies, D. S., and Boobis, A. R. *Biochemical Pharmacology* **1997,** *54,* 189.
220. Krausz, K. W., Goldfarb, I., Yang, T. J., Gonzalez, F. J., and Gelboin, H. V. *Xenobiotica* **2000,** *30,* 619.
221. Yamazaki, H., Guo, Z., and Guengerich, F. P. *Drug Metabolism and Disposition: The Biological Fate of Chemicals* **1995,** *23,* 438.
222. Wrighton, S. A., Vandenbranden, M., Becker, G. W., Black, S. D., and Thomas, P. E. *Molecular Pharmacology* **1992,** *41,* 76.
223. Belloc, C., Baird, S., Cosme, J., Lecoeur, S., Gautier, J. C., Challine, D., de Waziers, I., Flinois, J. P., and Beaune, P. H. *Toxicology* **1996,** *106,* 207.
224. Thummel, K. E., Kharasch, E. D., Podoll, T., and Kunze, K. *Drug Metabolism and Disposition: The Biological Fate of Chemicals* **1993,** *21,* 350.
225. Lemoine, A., Gautier, J. C., Azoulay, D., Kiffel, L., Belloc, C., Guengerich, F. P., Maurel, P., Beaune, P., and Leroux, J. P. *Molecular Pharmacology* **1993,** *43,* 827.
226. Krausz, K. W., Goldfarb, I., Buters, J. T., Yang, T. J., Gonzalez, F. J., and Gelboin, H. V. *Drug Metabolism and Disposition: The Biological Fate of Chemicals* **2001,** *29,* 1410.
227. Wheeler, C. W., Wrighton, S. A., and Guenthner, T. M. *Biochemical Pharmacology* **1992,** *44,* 183.
228. Eiselt, R., Domanski, T. L., Zibat, A., Mueller, R., Presecan-Siedel, E., Hustert, E., Zanger, U. M., Brockmoller, J., Klenk, H. P., Meyer, U. A., Khan, K. K., He, Y. A., Halpert, J. R., and Wojnowski, L. *Pharmacogenetics* **2001,** *11,* 447.
229. Sata, F., Sapone, A., Elizondo, G., Stocker, P., Miller, V. P., Zheng, W., Raunio, H., Crespi, C. L., and Gonzalez, F. J. *Clinical Pharmacology and Therapeutics* **2000,** *67,* 48.
230. Stresser, D. M. and Kupfer, D. *Drug Metabolism and Disposition: The Biological Fate of Chemicals* **1999,** *27,* 517.
231. Lang, T., Klein, K., Fischer, J., Nussler, A. K., Neuhaus, P., Hofmann, U., Eichelbaum, M., Schwab, M., and Zanger, U. M. *Pharmacogenetics* **2001,** *11,* 399.

Chapter 9

Drug-Drug Interactions and the Cytochromes P450

Kenneth A. Bachmann[1], Barbara J. Ring[2] and Steven A. Wrighton[2]

[1] *The University of Toledo, College of Pharmacy, 2801 Bancroft St., Toledo, OH 43606*

[2] *Department of Drug Disposition, Lilly Research Laboratories, Eli Lilly and Company, Indianapolis, IN 46285*

1   INTRODUCTION

1.1   The Numerology of Drug-Drug Interactions

Clearly the potential for drugs to interact with one another has been and continues to be a significant health concern. Adverse drug reactions (ADRs) may be responsible for up to 10% of hospital admissions.[1] Drug-drug interactions have been implicated in approximately 20% of ADRs.[2] It is generally accepted that the probability for ADRs associated with drug-drug interactions will increase as the number of concomitantly consumed drugs increases. The incidence of ADRs attributed to drug-drug interactions may range from under 5% when few drugs are simultaneously used to more than 80% when several drugs are used.[3]

Estimates of the *per capita* number of different prescriptions consumed concomitantly vary. However, data from the 1999 National Ambulatory Medical Care Survey (NAMCS) database[4] reveal that 4.3% of outpatient visits in the U.S. resulted in at least six prescriptions issued per visit, 17% of all outpatient visits resulted in two prescriptions issued per visit, and 9% of all outpatient visits resulted in three prescriptions issued. An analysis of the 1997 NAMCS database focusing on prescription use among the elderly found that four or more prescriptions were issued in 17% of the outpatient visits made by the elderly.[5] If one were to estimate the potential number of two-drug drug-interactions omitting consideration of more than two drugs at a time interacting with one another, then those outpatient visits resulting in at least six prescriptions could each potentially result in fifteen or more different drug-drug interactions (The number, N, of possible two-drug combinations when n drugs are consumed is given by: N = n!/2!(n-2)!).[6] Likewise, for each of the outpatient visits by the elderly resulting in at least four prescriptions, a total of at least six different drug-drug interactions would be theoretically possible.

Of course, not every potential pair of interacting drugs is expected to, in fact, interact adversely or interact at all. Moreover, there is evidence that the actual incidence of hospital admissions or mortality attributable to drug-drug interactions is relatively low.[7,8] Nevertheless, the potential for such interactions coupled with a pattern of multiple prescription use by many patients certainly raises the issue of drug-drug interactions and morbidity associated with the resulting ADRs to a level of public health concern that also extends into the realm of drug development. Indeed, during the past fifteen years the pharmaceutical industry has dramatically increased its reporting to the FDA of drug-drug interactions discovered during the development process.[9] In response, the FDA has issued two guidances for the industry regarding *in vitro*[10] and *in vivo*[11] studies of drug-drug interactions. In addition, the European Agency for the Evaluation of Medicinal Products has issued a combined *in vitro* and *in vivo* drug interaction guidance.[11]

Some drug-drug interactions arise from the inadvertent use of two drugs producing similar pharmacological outcomes resulting in additive or even synergistic effects. Likewise, the inadvertent use of two drugs producing opposing pharmacological outcomes can lead to therapeutic failure. Paying careful attention to the known pharmacodynamic activities of drugs can usually avert these pharmacodynamic interactions. On the other hand, many drug-drug interactions are more insidious and less obvious to predict since they arise from pharmacokinetic interactions between agents. These interactions occur, for example, if one drug accelerates the disposition kinetics of another. In this case the outcome may be therapeutic failure. If one drug slows the disposition kinetics of another, the outcome may be an unexpected increase in its pharmacodynamic effects leading to an ADR. The most frequently reported drug-drug interactions in clinical practice tend to be manifestations of the latter.

1.2 *The Centrality of the Cytochromes P450 in Drug-Drug Interactions*

It has become increasingly clear that the alteration of the catalytic activity of the key hepatic drug metabolizing enzymes, the cytochromes P450 (CYPs), is often at the heart of drug-drug interactions leading to ADRs. In this scenario one drug functions to inhibit the hepatic oxidation of another that is mediated by a CYP.

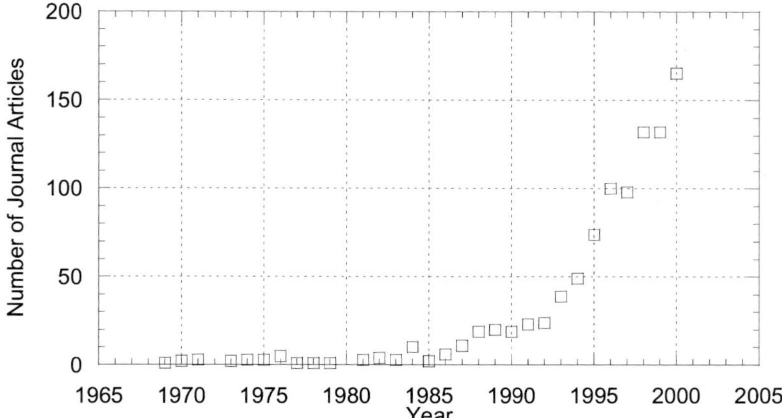

Figure 1. Exponential increase in annual publication drug-interaction papers associated with modification of CYP activity. Data from Chemical Abstracts and Medline searches using SciFinder Scholar™.

Collectively the CYP enzymes are responsible for the Phase I metabolism of approximately three-fourths of all drugs.[12] Among the hepatic CYP enzymes CYP3A4 accounts for 30-60% of their hepatic content[13,14] and is responsible for the metabolism of approximately half of all drugs.[12] CYP2C9 and CYP2C19 account for about 20%, of the hepatic CYP content, CYP1A2 accounts for approximately 13%, and CYP2D6 accounts for only about 2-3%,[3,13] even though it is responsible for the metabolism of 20-25% of all drugs.[12,15] While it is apparent that many drugs may be substrates for the same CYP enzyme, it is also the case that many drugs are metabolized by multiple CYP enzymes. It is no wonder then, that the CYP enzymes are implicated in many drug-drug interactions, and that the processing of new chemical entities (NCEs) by CYP enzymes and the effects of NCEs on CYP enzymes is of concern to clinicians, patients, the pharmaceutical industry, and regulatory agencies. In fact this area is of such interest that complete annual meetings have been devoted to drug metabolism with a significant focus on drug-drug interactions.[16] Additionally, the current level of interest on the rather central role of CYP enzymes in accounting for and predicting drug-drug interactions can be seen in the number of published papers relating these two topics (i.e. CYP enzymes and drug-drug interactions). Figure 1 depicts the number of annually published papers pertaining to the cytochromes P450 and drug-drug interactions for the period 1969-2000.

Within the past two years alone two prestigious books on drug-drug interactions have emerged. Significant space in both of these books is devoted to drug-drug interactions associated with CYP activity,[17,18] including a plurality of chapters in the most recently published book.[18]

1.3   Clinical Ramifications of CYP-Based Drug-Drug Interactions

Recent examples of the clinical, regulatory, and marketing consequences of CYP-based drug-drug interactions amplify the need to evaluate the likelihood for such interactions as early as possible in the process of drug development. Cisapride, a gastrointestinal prokinetic agent that is processed largely by CYP3A, can elicit serious ventricular arrhythmias including torsades de pointes upon its accumulation. Accordingly, since cisapride is processed by CYP3A, its use with CYP3A inhibitors is contraindicated.[19] Examples of such inhibitors include ketoconazole, indinavir, ritonavir, clarithromycin, and erythromycin. When, in the face of warnings, 30% of prescribers continued to issue prescriptions for cisapride to patients taking regimens that included drugs likely to cause ADRs with cisapride, the FDA restricted the sale of cisapride in the U.S.A.[20] Both terfenadine and subsequently astemizole were also withdrawn from the market owing to an unacceptable potential to prolong the QT interval and cause torsade de pointes.[21] One of the risk factors for these ADRs was the inhibition of the oxidation of terfenadine and astemizole to their less cardiotoxic metabolites, fexofenadine and norastemizole, respectively, by CYP3A inhibitors such as itraconazole,[22] ketoconazole, and erythromycin.[23]   Another recent example is mibefradil which was a calcium channel blocker with the novel feature of "vascular-specificity". Mibefradil was recommended for heart failure, since it lowered both heart rate and blood pressure without producing a negative inotropic effect, and was approved by the FDA in 1997. However, it became clinically apparent that mibefradil inhibited the metabolism of a large number of drugs processed by CYP3A and CYP2D6. This led to the voluntary withdrawal of mibefradil from the market on June 8, 1998, and the subsequent circulation of a "Dear Doctor" letter by Roche that listed twenty-six different drugs with which mibefradil might interact.[24]

The examples of clinically significant drug-drug interactions arising from the interaction of drugs with the cytochromes P450 are truly legion, and in addition to the two books cited above, there have been numerous review articles devoted either largely or exclusively to delineating those types of drug-drug interactions.[8,14,23,25,26] In addition, CYP-based interactions have been reviewed in the context of therapeutic category. For example, the CYP-based interactions between anticancer agents and other drugs,[27] antidepressants as perpetrators of drug-drug interactions,[28] antiepileptic drugs,[29,30,31] and antiulcer drugs[32,33] have been reviewed. The HIV protease inhibitors and non-nucleoside reverse-transcriptase inhibitors are CYP3A4 inhibitors, and their interactions with one another and with other drugs have also been reviewed.[34] However, even within well-intended reviews of CYP-based drug-drug interactions, interactions are included with outcomes that can only be qualitatively predicted, e.g. one drug might be expected to slow the clearance of another by inhibiting its CYP-related metabolic biotranformation.

Despite all the reporting of potential drug-drug interactions there are some stunning examples of very dramatic effects of CYP-based drug-drug interactions. Neuvonen's group at the University of Helsinki studied the quantitative effects of erythromycin and itraconazole as inhibitors of gastrointestinal epithelial CYP3A on the bioavailability of buspirone in a randomized, double-blind, double-dummy crossover study.[35] They found that four-day regimens of itraconazole or erythromycin increased buspirone AUC by 19-fold and 6-fold respectively, and increased buspirone Cmax by 13-fold and 5-fold, respectively. However, the plasma half-life of buspirone was not extended by either erythromycin or itraconazole. Moreover, they found that the regimens of both erythromycin and itraconazole elicited significant impairment of psychomotor performance and increased the side effects of buspirone, the former having been assessed by both the Digit Symbol Substitution test and the Critical Flicker Fusion test.[35] Similarly, when the effects of a two-week regimen of phenytoin on the pharmacokinetics of itraconazole were evaluated in a randomized, parallel group design, the findings were pronounced.[36] Phenytoin, as an inducer of CYP3A4, reduced itraconazole's AUC to less than 10% of its value in the absence of phenytoin treatment, and simultaneously shortened itraconazole's half-life by more than 80%.

1.4   *Electronic Sources of Information on CYP-Based Drug-Drug Interactions*

Two well-maintained websites that allow users to explore hypotheses about drug-drug interactions that are specifically dependent on CYP activities and alterations thereof have emerged in recent years. The Cytochrome P450 Drug Interaction Table was created by Dr. David Flockhart. This is a searchable database maintained by the Indiana University School of Medicine listing many of the most prominent substrates for each CYP enzyme, the major CYP inhibitors, and the major CYP inducers.[37] Some K_i values for the inhibitors are provided in the database, however links to abstracts of published papers in which K_i values have been measured are also given. A second database is available at the Gentest website, http://www.gentest.com. The searchable Human P450 Metabolism Database can be queried by enzyme, e.g. CYP1A2; therapeutic drug category, e.g. antiepileptic; reaction type, eg., demethylation; chemical or drug name, e.g.

mephenytoin; or type of CYP interaction, e.g., substrate, inhibitor, or inducer. While no specific K_i data are included in the tabular output, each line of each table ends with a list of literature citations. If one were interested, for example, in evaluating the potential interactions of the antiepileptic drug, phenytoin, a search on phenytoin would result in an output with thirty-one separate listings (including those for S-mephenytoin) identifying phenytoin or S-mephenytoin as substrates, inhibitors, or inducers of either CYP2B6, CYP2C18, CYP2C19, CYP2C8, CYP2C9, CYP3A4, CYP3A5, or CYP3A7. Separate citations are accessible by a "view detail" link at the end of each line of the presented table. An additional search for a presumed interacting drug would likewise reveal those CYP enzymes for which the second drug is a known substrate, inhibitor, or inducer. Apart from a small number of K_i values or pointers to the published literature that may contain K_i or IC_{50} values, the current CYP-focused drug-interaction databases provide principally qualitative insights into drug-drug interactions.

The types of quantitative data that would enhance the assessment of the clinical significance of drug-drug interactions would include the fractions of absorbed dose of CYP substrates that are metabolized by relevant processing CYPs (f_m), the K_i values for drugs that are CYP inhibitors, and the concentration of inhibitor achieved at the enzyme active site, typically the liver ($[I]_{hepatic}$).[38] The ratio of $[I]/K_i$ acquired from *in vitro* experiments (see below) should enable the characterization of a putative inhibitory reaction as low risk, $0.01 < [I]/K_i < 0.1$; medium risk, $0.1 < [I]/K_i < 1.0$; or high risk, $[I]/K_i > 1.0$.[39] A surrogate measure such as the average steady-state plasma concentration of the drug may be substituted for $[I]_{hepatic}$, with the caveat that compartmental partitioning and non-specific tissue-sequestration might give rise to significant misspecification of $[I]_{hepatic}$. According to Yamano et al[40] the hepatic concentrations of an inhibitor *in vivo* ($[I]_{hepatic}$), can be estimated from concentrations of inhibitor within hepatocytes *in vitro* ($[I]_{in\ vitro}$), in the medium ($[I]_{med}$), and in blood ($[I]_b$) from:

$$[I]_{hepatic} = ([I]_{in\ vitro}/[I]_{med})\ [I]_b * f_u$$

where f_u is the fraction unbound in blood.

In any case, Rodrigues et al have made a plea for industrial, academic, and regulatory scientists to redouble efforts to create an on-line quantitative database for predicting drug-drug interactions stemming from the alteration of CYP activity.[38]

The input for quantitative databases of CYP-based drug-drug interactions will be derived from published, ongoing, and future drug metabolism studies conducted both *in vivo* and *in vitro*. The pharmaceutical industry is moving rapidly to establish high throughput strategies for characterizing the stability of NCEs among each of the CYP enzymes, as well as for evaluating the inductive and inhibitory effects of NCEs on the CYP enzymes. Some high throughput technology for predicting CYP-based drug-drug interactions has been commercialized by Gentest, In Vitro Technologies, and others, and can be implemented by contract research organizations enabling small biotechnology and pharmaceutical companies to outsource the evaluation of CYP-interaction potential of their NCEs in relatively early stages of pre-clinical development. Details of *in vitro* enzyme profiling, prediction of CYP inhibition and CYP induction are described elsewhere in this chapter.

2 *IN VITRO* DRUG-DRUG INTERACTIONS: DETERMINING THE POTENTIAL OF AN NCE TO INHIBIT THE CYPS

Over the past 10 to 15 years the identification of substrates that are selectively metabolized *in vitro* by specific CYPs (Table 1) allows for the investigation of the inhibition by an NCE of metabolism mediated by this enzyme system.[39,41,42] The results from the *in vitro* inhibition studies, yielding K_i or IC_{50} values allows for the prediction of the potential for an NCE to inhibit specific CYPs *in vivo*. This information is used by the clinician to design *in vivo* interaction studies that are relevant for the NCE. In addition, the results of the *in vitro* studies may be reported to the regulatory agencies, included in the product insert, and in some cases, substitute for clinical interaction studies.

In order to obtain valid results from in vitro drug-drug interaction studies, an understanding of the enzyme kinetics of the biotransformation being inhibited is required.[43] First, the initial rate conditions (e.g. reaction rate linear with time and enzyme concentration) for the formation of the metabolite of interest by the specific CYP must be determined. Using these initial rate conditions, the metabolism of the substrate by the specific CYP is determined over a wide range of substrate concentrations in order to determine the enzyme kinetic parameters of the biotransformation (see Chapter 7 on enzyme kinetics). The first kinetic parameter of interest is the K_m value, which is defined as the substrate concentration at half maximal velocity and is a reflection of the affinity of the substrate for the enzyme of interest. The second kinetic parameter determined is the maximal velocity or V_{max} of the biotransformation. Next, the ability of the NCE to inhibit the enzyme of interest is determined. For these inhibition studies, several concentrations of the NCE are added to the incubations at several substrate concentrations that encompass the K_m value (3-5x above and below the K_m value) and the formation rate of the metabolite of interest is determined. These results are fit to the various models of inhibition using nonlinear regression analyses to determine the type of inhibition exhibited by the NCE and an apparent K_i value (the dissociation constant for the enzyme-inhibitor complex). It is important to note that to determine an accurate K_i value, the results must be fit to the appropriate model of enzyme inhibition as determined by defined statistical criteria. The models of inhibition to be examined for appropriate fit of the data are competitive, non-competitive, uncompetitive, and mixed competitive/non-competitive.[44]

Table 1. Form-Selective Catalytic Activities of the Human
Cytochromes P450[39,41,42]

Cytochrome P450	Form-Selective Catalytic Activity
CYP1A2	Phenacetin O-deethylase
	Caffeine N3-demethylase
	Theophylline 8-hydroxylase
	Ethoxyresorufin O-deethylase
CYP2A6	Coumarin 7-hydroxylase
CYP2B6	S-mephenytoin N-demethylase
CYP2C8	Taxol 6-hydroxylase
CYP2C9	Tolbutamide 4-hydroxylase
	Diclofenac 4'-hydroxylase
	S-Warfarin 4' hydroxylase
CYP2C19	S-Mephenytoin 4'-hydroxylase
CYP2D6	Bufuralol 1'-hydroxylase
	Debrisoquine 4-hydroxylase
	Dextromethorphan O-demethylase
CYP2E1	N-Nitrosodimethylamine N-demethylase
	Chlorzoxazone 6-hydroxylase
CYP3A	Midazolam 1'-hydroxylase
	Testosterone 6β-hydroxylase
	Erythromycin N-demethylase
	Nifedipine oxidation

A less labor intensive method to predict the potential of an NCE to cause a drug-drug interaction is to determine the concentration of inhibitor which inhibits 50% (IC_{50}) of the biotransformation of a CYP-selective substrate at a single substrate concentration. If the IC_{50} value is much greater than the plasma concentration for that inhibitor, it can be predicted that the NCE is not a clinically significant inhibitor of the CYP. However, there are several significant concerns with this less rigorous prediction method. The first major issue is that the relationship between the IC_{50} value and K_i value for an inhibitor-CYP complex depends on the type or mechanism of inhibition, which is not known. Specifically, only in the case of non-competitive inhibition are the IC_{50} and K_i values actually equal. For competitive or mixed competitive/non-competitive inhibition relationships,[44] IC_{50} and K_i values are not always equal. With competitive inhibition, the IC_{50} is equal to $K_i(1+(S/K_m))$ where S is substrate concentration. Thus, as the concentration of the substrate approaches the K_m value, the relationship simplifies to IC_{50} is equal to 2-times the K_i value. In those situations where the single concentration of substrate used in the determination of IC_{50} exceeds the K_m value, the IC_{50} value determined is substantially greater than the K_i value. Since the type of inhibition is not known and cannot be determined from studies yielding only IC_{50} values, the conservative assumption should be made that the IC_{50} value is larger than the actual K_i value. Therefore, in the prediction of the clinical significance of the inhibition of a CYP by an NCE, if only an IC_{50} value is used, an underestimation of the potential inhibition is likely. However, if the concentration of substrate used in the IC_{50} determinations is at or below the K_m value for the biotransformation, the problem of comparing the IC_{50} value with the K_i value is minimized.

As a special case, in vitro interaction studies may also be performed to address a question concerning the interaction of an NCE and a specific drug (versus a specific enzyme). An important aspect for the interpretation of such studies is to understand the overall metabolic clearance of the substrate being inhibited. Specifically, if a minor route of metabolism is being inhibited by the NCE, the overall significance of the inhibition of that route to the clearance of the drug is minimal and therefore not likely to be clinically relevant.[45] However, it is possible to use such studies in lieu of *in vivo* interaction studies, thus saving time and money.

2.1 *In Vitro to In Vivo Extrapolations*

In theory, the K_i value determined for an enzyme–inhibitor complex is a constant and is independent of the substrate used in the K_i determination.[44] Thus, the K_i value may be utilized in predictions of the potential of the NCE to inhibit all biotransformations mediated by the particular CYP investigated. As this area of research has evolved these studies are now performed early in development of the drug and thus the concentration of the NCE at which the desired pharmacology will be achieved is unknown. As a general rule, however, if the anticipated circulating levels of the NCE approach the K_i value for the interaction, there is a potential for an interaction to occur. As the development of the NCE continues, maximal concentrations (C_{max}) of the NCE are determined *in vivo*. Once the C_{max} values are known, predictions of the inhibitory potency of the NCE for the CYPs in the patient population can be performed. There are several caveats concerning these predictions that must be considered by the investigators. First, there must be knowledge of the dependence of the clearance of the drug on the particular route of metabolism that is mediated by the inhibited CYP. Recent consensus of opinion suggests that the inhibited route of metabolism should account for 30% of the clearance of the compound for inhibition of that pathway to have a significant impact on total clearance of the drug.[46] Further, an understanding of the inhibitor concentration at the active site of the enzyme is also important. However, it is usually not possible to determine the concentration of inhibitor within the cell at the active site of the enzyme and thus the circulating concentration of inhibitor is used in these predictions (see earlier discussion of hepatic concentration). Another factor that impacts the reliability of these predictions is the level of protein binding of the inhibitor, which affects the drug concentration available to interact with the CYP. Pharmacokinetic theory suggests that only unbound drug should be available to interact with the active site of the enzyme. However, the real relationships between NCE plasma concentration, liver concentration and concentration at the CYP active site are not known. Therefore, since there are a number of assumptions and caveats around these predictions, they should currently be considered qualitative in nature.

Keeping in mind the caveats already discussed, prediction of the inhibition by an NCE of metabolism mediated by a CYP is possible if the therapeutic concentration of an NCE and the K_i value for a particular drug-CYP interaction are known. The predicted percent inhibition of the *in vivo* metabolism is calculated as: percent inhibition = (1-(inhibited velocity of the biotransformation / uninhibited velocity)) *100.[44] For the uninhibited rate of reaction or velocity, the Michaelis-Menten formula is substituted into the above relationship and for the inhibited velocity the formula for the type of inhibition observed for the enzyme-inhibitor complex is utilized.[43] For the majority of cases, that is for competitive, noncompetitive or mixed types of inhibition, the additional assumption is made that the substrate concentration *in vivo* is substantially smaller than the K_m value of the biotransformation. Thus the above relationship for the 4 models of inhibition reduces to: percent inhibition = (I) / (I+K_i)) *100 where I is defined as the inhibitor concentration.

The process of extrapolating the in vitro inhibition results to the patient as discussed above, is a relatively qualitative approach. In fact, regulatory agencies, academic scientists and the pharmaceutical industry have generally accepted a rank ordering of the relative potential of a NCE to cause interactions as the result of the inhibition of the various CYPs examined *in vitro*.[39,46] These rankings defines the inhibitory potential of a NCE broadly into three categories; likely, possible, and unlikely[46] or alternatively, high risk, medium risk and low risk.[39] These categories are related to the *in vitro* K_i values or when plasma concentrations of the NCE are known, the I/K_i ratio where I is the maximum plasma concentration of the NCE. The likely (high risk) category represents those compounds with a K_i value of <1 μM or an I/K_i ratio of >1, the possible (medium risk) category is defined as K_i values between 1 and 50 μM or I/K_i of between 1 and 0.1, and finally the unlikely (low risk) category is defined as K_i value > 50 μM or an I/K_i ratio of <0.1. With such information, the potential of drug-drug interactions with a NCE and drugs metabolized by the CYPs may be put into perspective before the clinical development of the NCE. For those compounds in clinical development, this information helps define inclusion and exclusion criteria for various protocols. In addition, more rigorous physiologically-based methods for the prediction of *in vivo* drug–drug interaction from *in vitro* data have been examined. However, as a general rule most of these predictions have not been deemed acceptable except when correction factors of usually an empirical nature are included in the calculations. Recently a number of reviews concerning these more quantitative approaches have been published.[47, 48, 49, 50, 51]

2.2 Irreversible or Quasi-Irreversible Inhibition

The common feature of irreversible and quasi-irreversible inhibitors is that the compounds are catalytically activated by the CYP to an intermediate that binds the CYP thus inactivating it. Structural features of a NCE that tend to result in metabolic activation include alkyl and aromatic amines, terminal olefinic or acetylenic groups, and methylenedioxy compounds.[52] In the case of irreversible inhibitors (also referred to as suicide inhibitors), the metabolic intermediate formed is of sufficient reactivity to covalently bind the protein and/or the heme components of the CYP. For quasi-irreversible inhibitors, the metabolic intermediate coordinates with the heme of the CYP so tightly that the intermediate is essentially irreversibly bound. Both of these types of inhibitors are more broadly known as mechanism-based inhibitors. The resultant inactivation of the CYP by the mechanism-based inhibitor is only reversed by the synthesis of new enzyme. The recovery of catalytic activity of the enzyme pool is therefore not dependent on the dissociation of the enzyme–inhibitor complex but by the rate of synthesis of new CYP.

The potential of a NCE to be a mechanism-based inhibitor can be determined by *in vitro* techniques.[53,54] The hallmark of such inhibitors is that their inhibition is time- and concentration-dependent since catalytic turnover is required for inhibition. To determine the *in vitro* potential of a NCE to cause mechanism-based inhibition, the NCE at several concentrations is first preincubated with the CYP (in human liver microsomes or recombinant enzyme) in the presence or absence of NADPH. After various times of preincubation, an aliquot of the mixture is diluted 10- to 20-fold into a second incubation mixture for the determination of a CYP specific biotransformation. The remaining catalytic activity of the CYP is determined and kinetic parameters of inactivation determined.[54,55] The parameter of k_{inact} or the maximal inactivation rate of the CYP by the NCE and the K_I value or the dissociation constant for the inactivator–CYP complex have utility in the extrapolation of the *in vitro* results to the *in vivo* situation.[54,55] Specifically, from a ratio of these two parameters (i.e. k_{inact}/K_I) the mechanism-based inhibitor potency, termed inactivation clearance (Cl_{inact}), is determined. By comparing the Cl_{inact} of a NCE with those of known clinically significant mechanism-based inhibitors like erythromycin and diltiazem, the clinical significance of the mechanism-based inhibition of the NCE may be put into perspective.

3   IN VIVO DRUG-DRUG INTERACTION STUDIES

Identifying drug-drug interactions and quantifying them for new drugs allows decision-makers both in drug development and in drug regulation to develop strategies for avoiding interactions or adjusting dosages, thereby enabling the marketing of such drugs even in the face of predicted drug-drug interactions. Drug-interaction data gathered in early preclinical investigations can be used to eliminate the need for clinical investigations in instances where no evidence of any drug-drug interactions can be ascertained. However, more positive findings can also be used to devise the subsequent clinical studies. In fact, even for those drugs that successfully pass *in vitro* drug-interaction screens it would be difficult to avoid an analysis of their drug-interaction potential in humans, since such studies are often mandated by regulatory agencies. For example, the Committee for Proprietary Medicinal Products (CPMP) of The European Agency for the Evaluation of Medicinal Products stipulates in its "Note for Guidance on the Investigation of Drug Interactions" that *"in vitro* data should mainly be used qualitatively due to the uncertainty regarding *in vitro/in vivo* correlation. Thus, based on *in vitro* data, applicants should design and perform relevant *in vivo* studies".[11] The regulatory position therefore, is that *in vitro* studies will inform investigators as to the types of *in vivo* drug-drug interaction studies that will be necessary for approval and labeling purposes, however they will not necessarily supplant *in vivo* studies. The November, 2000 conference of the European Federation of Pharmaceutical Sciences (EUFEPS) held in Basel, Switzerland arrived at a perspective compatible with regulatory agencies. At this meeting a consensus was reached recommending deferral from using *in vitro* interaction data in decisions to undertake *in vivo* drug-drug interaction studies except when the *in vitro* studies afforded a high level of confidence that no inhibition of drug metabolism occurred.[39]

Table 2. The effects of new drugs on the pharmacokinetics of these drugs is suggested in characterizing the drug interaction potential of new drugs as CYP inhibitors.

Enzyme	FDA recommended substrates[10]	EUFEPS recommended Substrates[39]
CYP1A2	theophylline	caffeine
CYP2C9	S-warfarin	tolbutamide
CYP2C19		mephenytoin omeprazole
CYP2D6	Desipramine	Debrisoquine
CYP2E1		Chlorzoxazone
CYP3A4	midazolam buspirone felodipine simvastatin†	Midazolam (oral or oral and i.v.) Midazolam (oral) plus erythromycin (i.v.) simvastatin atorvastatin

†lovastatin could be used instead of simvastatin.

Clinical drug interaction screening occurs largely during Phase I and II of clinical trials and in postmarketing surveillance studies. As with *in vitro* studies, clinical study design should try to establish whether or not a new drug is likely to affect the disposition kinetics of other drugs, and/or whether existing drugs, foodstuffs, or herbals/nutraceuticals are likely to affect the disposition kinetics of the new drug. In the former case the FDA Guidance for Industry refers to the new drug as the "interacting drug" and the target drugs as "substrates". In the latter case, the new drug would be viewed as the substrate. Another way to characterize the role of a new drug in a drug-interaction would be as either the "perpetrator" or the "victim". These terms have been used by Greenblatt and von Moltke, and add clear and simple meaning to the putative roles of interacting drugs.[55,56]

For the purposes of satisfying regulatory requirements *in vivo* drug-interaction studies should conform to the designs specified in the guidance documents from the FDA and CPMP.[11] Several different study designs are recommended depending upon the usual use patterns of the interacting agents (chronic consumption versus single dose), the breadth of the therapeutic range of the interactants (narrow vs. broad), the pharmacokinetic and pharmacodynamic characteristics of the interactants, and whether or not the expected effect of the perpetrator is likely to be inhibition of the metabolism of the victim. If the investigational drug is to be studied as a perpetrator in the drug-interaction, and if the hypothesis to be tested is that it is a CYP inhibitor, then the victim drugs specified in Table 2 are recommended.

Table 3. Probes of CYP enzymes that can be used to predict clinical
drug-interaction outcomes.

Enzyme	Probe	Reaction
CYP1A2	caffeine	caffeine 3-demethylation; paraxanthine 7-demethylation
	theophylline	theophylline-1-demethylation
	tolbutamide	tolbutamide hydroxylation
CYP2C9	phenytoin	phenytoin 4'-hydroxylation
	warfarin	(S)-warfarin 6- and 7-hydroxylation
	losartan	losartan oxidation
CYP2C19	mephenytoin	(S)-mephenytoin 4'-hydroxylation
	omeprazole	omeprazole 5'-hydroxylation
	proguanil	proguanil oxidation
CYP2D6	dextromethorphan	dextromethorphan O-demethylation
	debrisoquin	debrisoquin 4-hydroxylation
	sparteine	sparteine N1-oxidation
	metoprolol	(R)-metoprolol O-demethylation; metoprolol hydroxylation
CYP2E1	chlorzoxazone	chlorzoxazone 6'-hydroxylation
CYP3A4	cortisol	cortisol 6β hydroxylation
	^{14}C-erythromycin	erythromycin N-demethylation
	midazolam	midazolam 1'- and 4-hydroxylation
	dapsone	dapsone N-hydroxylation
	lidocaine	lidocaine N-deethylation
	alfentanil	piperidine N-dealkylation
	nifedipine	nifedipine dehydrogenation

If the new drug is demonstrated to be an effective inhibitor of one or more CYPs, then interactions with other victim drugs may be explored based on the likelihood of the co-administration of the combination in clinical practice. Negative findings in the initial set of clinical experiments suggests that less sensitive victim drugs will not be subject to interactions with the new drug.[11]

If the new drug is itself a substrate for one or more CYPs (determined as described elsewhere in this text), then its role as a victim drug must be explored. Moreover, the effects of both inhibitory and inductive perpetrators must be explored. The FDA Guidance recommends the use of ketoconazole and rifampin as perpetrators of inhibition and induction, respectively, of new drugs that are known to be CYP3A substrates. It further recommends against the use of broad-spectrum CYP inhibitors such as cimetidine that can inhibit multiple metabolic pathways and the activity of drug transporters. The Guidance does not restrict the evaluation of new drugs as potential perpetrators of drug-drug interactions to the use of the substrates listed in Table 2, and a comparison of the FDA and EUFEPS recommended victim drugs is provided in the table. Indeed, a healthy variety of CYP-selective probes have emerged in recent years. Many of them are listed in Table 3.

A minimally invasive probe of CYP3A is ^{14}C-erythromycin, since its metabolic clearance by N-demethylation can be quantitated as expired $^{14}CO_2$.[57] One might surmise that characterizations of enzyme phenotype using any or all probes for a specific enzyme ought to give comparable results. Unfortunately, experience indicates otherwise. In fact, the correlations made between any two probe-based measures of CYP3A activity have been dismal to non-existent. Poor correlations exist between the erythromycin breath test (ERBT) and urinary excretion of 6β-hydroxycortisol,[58] the ERBT and alfentanil clearance,[59] urinary excretion of 6β-hydroxycortisol or the ERBT with urinary dapsone recovery ratios,[60] or the ERBT with midazolam clearance.[61] Urinary ratios of dextromethorphan to 3-methoxymorphinan correlated poorly with plasma ratios of hydroxymidazolam/midazolam, midazolam clearance, and urinary ratios of 6β-hydroxycortisol/cortisol.[62] These disappointing correlations have been attributed to the following phenomena:

- dietary factors that modify CYP3A activity
- overlapping substrate specifity of probes (i.e. processing by multiple CYPs)
- extrahepatic processing of probes in GI epithelium and kidney
- metabolism by CYP3A5 or CYP3A7 enzymes in addition to CYP3A4
- co-processing of a probe by transport proteins such as p-glycoprotein

Ideally, the use of any drug as a selective CYP probe *in vivo* will have been validated. However, there is not universal agreement on the criteria for

Table 4. Criteria used in the validation of probe-based methods for phenotyping CYP activity. (modified from Zaigler et al.) [1]

In vivo measure must correlate with activity of the target enzyme derived from human liver

In vivo measure must correlate with content of target enzyme derived from human liver

In vivo measure must correlate with the fractional clearance of probe drug mediated by target enzyme

In vivo measure must correlate with other validated phenotyping procedures

Subjects treated with known inhibitors of target enzyme must exhibit a reduction in the *in vivo* measure of enzyme activity

Subjects treated with known inducers of target enzyme must exhibit an increase in the *in vivo* measure of enzyme activity

Patients with severe liver disease must exhibit a reduction in the *in vivo* measure of enzyme activity

Patients undergoing liver transplant surgery must exhibit a reduction in the *in vivo* measure of enzyme activity during the anhepatic phase of surgery

There must be demonstrated *in vitro* specificity for the enzyme and the metabolite being formed and measured *in vivo*

The procedure must give reproducible results

The procedure must correlate with genetic polymorphism if it exists

The procedure should give results independent of confounding influences such as urinary flow, creatinine clearance, activity of other enzymes

Analytical methods should adhere to general validation criteria

Invasiveness should be low

Probe should be well tolerated

No blood sampling should be required

No radioactivity should be used

The procedure should be benign to subjects or patients

The procedure should be technically easy

The procedure should be inexpensive

[1] Zaigler, M., Tantcheva-Poor, I., and Fuhr, U. Int. J. Clin. Pharmacol. Ther. **2000**,37,1-9.d

validation, and if all of the criteria listed below (Table 4) were required, then no drug could be accepted as a selective CYP probe.

The combined administration of multiple probes, i.e. "cocktails" to simultaneously phenotype several CYP activities was originally proposed by Breimer.[63] Criticisms of the cocktail approach focus primarily on probe-toxicity.[64] Nevertheless, more recent cocktail strategies have been proposed. The combined use of caffeine, chlorzoxazone, dapsone, debrisoquin, and mephenytoin (termed the Pittsburgh cocktail) has been proposed to phenotype CYP1A2, CYP2E1, CYP3A4, CYP2D6, and CYP2D6 activities.[65] The combined administration of all five drugs yielded quantitative urinary recovery ratios that were not significantly different from ratios established by the administration of each individual probe by itself, suggesting that these probes could, in fact, be used in a "cocktail." The administration of the another cocktail termed the Cooperstown cocktail containing caffeine, dextromethorphan, omeprazole, and midazolam as probes of CYP1A2, CYP2D6, CYP2C19, and CYP3A4, respectively, has also been shown to yield urinary recovery ratios or plasma clearance (for midazolam) values that were not significantly different than those established by the independent administration of each separate probe.[66] Both the "Pittsburgh" cocktail[65] and the "Cooperstown" cocktail[66] were reported to cause no adverse effects apart from mild sedation or drowsiness.

In the EUFEPS conference an uneasy agreement was reached regarding the use of mixtures of CYP substrates (as victim drugs) in "cocktails" in order to confirm *in vitro* predictions of enzyme-specific drug-drug interactions. One basis for arguing for proceeding with "cocktail" studies was the observation that one compound, shown to be a potent *in vitro* inhibitor of several common CYP enzymes, nevertheless failed to exhibit significant inhibitory effects in subsequent "cocktail" studies.[39] This type of predictive error might be averted if hepatocytes, rather than microsomes or heterologous expression systems are used, since phase II reactions ought to proceed in hepatocytes.[40] Conversely, in our own experience with rofecoxib as a perpetrator, we found significant and dose-dependent CYP1A2 inhibition in normal healthy subjects using theophylline as a probe,[67] even though the precedent *in vitro* findings predicted no interaction. The absence of an *in vitro* inhibitory effect of rofecoxib on CYP1A2 may have been methodologically grounded, with a short pre-incubation period failing to generate sufficient reactive metabolic intermediates to inhibit the enzyme. Accordingly, the EUFEPS conference recommended a 30-minute pre-incubation of the perpetrator with the addition of EDTA to scavenge metabolites arising from peroxidation.[39] The EUFEPS conference agreement about "cocktail" strategies for investigating drug-drug interactions was that the issue requires additional study, and in particular attention must be paid to the statistical power or sensitivity of this approach.

3.1   *CYP-Based Interactions with Nutraceuticals and Food*

The remarkable popularity of phytotherapy as part of "alternative" medical treatments is evidenced by the multi billion-dollar/year market for these substances in the United States.[68] Since some of the phytochemicals in herbal remedies do have well established biochemical/physiological effects and have, in fact, been demonstrated to provide beneficial therapeutic effects in animal models (e.g. St. Johns wort)[69,70] or in humans (e.g. St. Johns wort[68] and Saw Palmetto) it is not unrealistic to expect that herbals might also be capable of causing adverse effects. Indeed, several adverse drug interactions have already been with reported with herbal medicines[71-75] notably with St. John's wort[68] which may increase hepatic or gastrointestinal CYP3A activity,[76] or increase the expression of intestinal p-glycoprotein. In addition, the multiple phytochemicals in an herbal preparation, such as hypericin and hyperforin in St. John's wort, may exhibit complex effects on xenobiotic processing systems. Hypericin is a relatively potent inhibitor of CYP2C9, CYP2D6, and CYP3A4, a modest inhibitor of CYP2C19, and a poor inhibitor of CYP1A2.[77] Hyperforin is a potent inhibitory of CYP3A4, though a less potent inhibitor of CYP2C9 and CYP2D6.[77] St. John's wort also contains I3,II8-biapigenin which is a potent inhibitor of CYP2C9 and a very potent inhibitor of CYP3A4. Yet, sustained ingestion of St. John's wort increases CYP3A4 activity as measured by the urinary 6β-hydroxycortisol/cortisol ratio,[76] and also increases p-glycoprotein activity. Recently hyperforin, a constituent of St. John's wort, has been shown to be a potent inducer of human hepatic CYP3A4, and that the inductive effect is mediated through high affinity binding to the pregnane X receptor (PXR).[78]

The use of herbals is anticipated to increase dramatically and herbals are statutorily regarded as not being drugs. Therefore, the potential for their interactions with prescription or even OTC drugs and for those interactions to thwart systematic identification via the FDA's spontaneous reporting system is quite substantial. This concern about the potential for herbal-drug interactions was specifically mentioned in the President's FY 2000 NIH budget proposal. The American Society for Pharmacology and Experimental Therapeutics endorsed the following language in the Senate report on the portion of the NIH bill that dealt with the National Center for Complementary and Alternative Medicine (NCCAM): "Increasing numbers of Americans are using herbal products, and sound and rigorous research will help determine the interactive effects of herbal remedies with prescription drugs".[79]

Constituents of food have been shown capable of causing profound effects on the metabolism of drugs. Consumption of grapefruit juice is now known to significantly decrease the activity of CYP3A thereby increasing the bioavailability of some CYP3A substrates such as midazolam and cyclosporine, or decreasing the total clearance of others such as dihydropyridine calcium channel blockers and terfenadine.[80] More recently it was shown that frequent grapefruit juice consumption (three times daily for two days) increased peak plasma concentrations of cisapride, increased its AUC, and prolonged its half-life.[81] These alterations in dispositional kinetics of cisapride were not only statistically significant; they were clinically meaningful. Grapefruit juice ingestion elicited a four-fold increase in buspirone C_{max} and a ten-fold increase in buspirone AUC.[82] It has been shown to completely inhibit the oxidation of amiodarone to desethylamiodarone.[83] The time-course for the grapefruit effect has been studied in the context of its interaction with simvastatin. In that study, the pharmacokinetics of simvastatin in healthy volunteers was evaluated immediately after the last dose of a three-day regimen (three times per day) of "double strength" grapefruit juice, and again 24 hours later, 3 days later, and 7 days after the final "dose" of grapefruit juice. When simvastatin was taken at the time of the last "dose" of grapefruit juice, the C_{max} and AUC were over 12-times higher than when simvastatin was taken absent any grapefruit juice.[84] Increases in these parameters fell to about 2-2.5 fold higher than control values when simvastatin was withheld until 24 hours after the last grapefruit "dose", then to 1.5 fold higher than control values when simvastin was taken three days after the last grapefruit "dose." Normal simvastatin pharmacokinetic parameters emerged at one week after the last "dose" of grapefruit juice. It should be noted that if grapefruit juice were being studied as a perpetrator and simvastatin as a victim in accordance with the FDA Guidance, that even after three days of abstinence from grapefruit juice, the impact of prior grapefruit juice ingestion on simvastatin would likely be regarded as clinically significant.

It is now generally accepted that the bioflavonoid-derived naringin, once suspected as the grapefruit juice constituent responsible for CYP inhibition, is not the inhibiting species. More recent attention has shifted to furanocoumarins present in grapefruit juice such as 6',7'-dihydroxybergamottin, however even the findings pertaining to the role of this substance in inhibiting CYP3A activity are equivocal.[85-87] The most compelling evidence suggests that furanocoumarin dimers GF-I-1 and GF-I-4 are the most potent inhibitors of CYP3A4 with K_i values < 0.5 µM compared to 40 µM for bergamottin.[88]

Brassica vegetables have been known for many years to be capable of inducing some of the CYP enzymes,[89] probably CYP1A2. More recently red wine has been shown to be capable of inhibiting human CYP3A via mechanism-based inhibition.[90] Since the inhibitory effect on CYP3A was demonstrated with red wine rather than white wine, and since the phytoalexin, resveratrol, is found in red wine but not in white wine, its effects on CYP3A activity were studied.[91] Resveratrol was found to inhibit CYP3A, but to be a less potent mechanism-based inhibitor than erythromycin or TAO. Indeed, when resveratrol was fractionally removed from red wine, significant competitive inhibition of CYP3A still occurred, suggesting that the ability of red wine to inhibit CYP3A activity is due only in part to resveratrol, and that other, as yet unidentified, red wine constituents are largely responsible for the effect.[93] Others have similarly reported that red wine constituents in addition to resveratrol are responsible for mediating the inhibition of CYP1A, CYP2E1, and CYP3A.[92]

3.2 Transporter-Mediated Drug-Drug Interactions Masquerading as CYP-Based Interactions

St. John's wort has recently been shown to significantly lower blood levels of digoxin, a substrate of p-glycoprotein in normal subjects.[93] These changes were on the order of 25-33%, and were theorized to arise from the induction of p-glycoprotein by hypericin, and/or pseudohypericin, and/or hyperforin. The induction of CYP3A and p-glycoprotein appear to represent pleiotropic responses to the same challenge.[95] Whether such pleiotropism exists, it has nevertheless been shown that the active phytochemicals in St. John's wort are capable of altering the disposition kinetics of digoxin in humans. Transporters such as p-glycoprotein are looming as an increasingly important targets for putative drug-drug interactions.[94] Tangeretin, a flavonoid in orange and tangerine juice, has been demonstrated to increase the activity of CYP3A *in vitro*, although tangerine juice appears to be without any effect on CYP3A in humans.[95] However, tangeretin and other polymethoxylated flavones from orange juice were moderately potent inhibitors of p-glycoprotein function in Caco-2 cells.[96]

The components of grapefruit juice, orange juice and apple juice were investigated as perpetrators for their effects on fexofenadine transport *in vitro* and fexofenadine disposition kinetics in humans. While generally weak inhibitory activity was demonstrated on p-glycoprotein mediated efflux of digoxin or vinblastine in Caco-2 cells, all three substances appeared to be potent inhibitors of fexofenadine uptake mediated by OATp in HeLa cells.[97] This was also reflected in a 63%-73% decrease in the oral bioavailability of fexofenadine mediated by these fruit juices in humans, presumably due to decreased OATp-mediated uptake.[97] Others have reported an approximate 30% decrease in the bioavailability of nizatidine when ingested with apple sauce, cranberry juice, or vegetable juice, but these findings pre-dated the current understanding of the role of transporters in drug disposition.[98]

That multiple xenobiotic processing systems such as CYP3A, p-glycoprotein, and OATp might be simultaneously affected by one or multiple components of natural substances or even by a single new chemical entity underscores the necessity of human *in vivo* interaction experiments. In these *in vivo* studies, the combined effects of multiple components within the test substance can be evaluated directly on several pharmacokinetic outcome measures.

3.3   *Pharmacogenetics and CYP-Based Drug-Drug Interactions*

For nearly a half-century it has been recognized that the metabolic disposition of some drugs is genetically controlled. The first published suggestion of the genetic basis for the polymorphic metabolism of isoniazid occurred in 1958.[99] Evans et al, in a classic investigation somewhat reminiscent in design of the current application of probe drugs for phenotyping drug metabolizing activities *in vivo*, demonstrated that slow metabolism of isoniazid was inherited as an autosomal recessive trait.[100] The CYP enzymes have been extensively evaluated for their roles in polymorphic drug metabolism.[101-103] As noted by Vesell, "…that while genetic polymorphisms occur in most, if not all, human cytochrome P450 (CYP) isozymes, the great majority of functional genetic polymorphisms reside in only four: CYP2A6, CYP2C9, CYP2C19, and CYP2D6".[104] He also noted that the mutations at those genetic loci and the alleles associated with them are being discovered at so rapid a pace that a web site, http://www.imm.ki.se/ cypalleles, has been established to track them. The current state of the art, and the rapidly emerging state of the art have significant clinical ramifications. Among them are/will be the ability to tailor dosages of drugs that are substrates for one of those enzymes specifically to an individual's genotype.[105] In the context of drug development, some companies have made strategic decisions to drop candidate drugs that are primary substrates for one of the CYPs responsible for functional genetic polymorphisms. Likewise, in the case of victim drugs that are processed by a single isoform, and perpetrator drugs that induce or inhibit a single isoform, drug-drug interaction outcomes are fairly easy to predict. For example, while quinidine significantly slows the O-demethylation of codeine to morphine by CYP2D6 in extensive metabolizers of CYP2D6 substrates (EMs), neither the pharmacokinetics nor the pharmacodynamics of codeine are altered by quinidine in poor metabolizers of CYP2D6 substrates (PMs).[106] Nevertheless, it should be pointed out again that many drugs are substrates for more than one CYP enzyme, and many inhibitors and inducers alter the activity of more than just one CYP enzyme. Therefore, drug-drug interaction outcomes involving perpetrator and/or victim drugs that affect or are processed by multiple enzymes, respectively, can become relatively difficult to predict, even in individuals who are PMs with regard to one CYP, but who are normal or EMs with regard to other CYPs. Two interesting studies illustrate this point. Rifampin is a relatively non-selective inducer of CYPs and it also induces UDP glucuronsyl transferase. Codeine is partially oxidized by CYP2D6 to morphine, and PMs lacking functional CYP2D6 experience a reduced analgesic effect from codeine.[107] A regimen of rifampin administered to panels of EMs and PMs alike increased the partial metabolic clearances of codeine by glucuronidation, N-demethylation, and O-demethylation in EMs but not by O-demethylation in PMs. Nevertheless, the total oral clearance of codeine was significantly increased in both EMs and PMs after rifampin treatment, and rifampin treatment attenuated both respiratory and psychomotor effects of codeine in EMs but not in PMs.[108] Propafenone is also processed by multiple CYPs including CYP2D6. Accordingly one might be tempted to predict that an enzyme inducer such as rifampin might not significantly alter the disposition kinetics of propafenone in PMs. However, when the effects of a regimen of rifampin on the disposition kinetics of propafenone were investigated in a panel of six EMs and another panel of six PMs the following occurred: (1) Intravenous clearance of propafenone was unchanged in EMs and doubled in PMs; (2) Oral clearance of propafenone was unchanged by rifampin treatment in both EMs and PMs; (3) The bioavailability of propafenone decreased from 30% to 10% in EMs, and from 81% to 48% in PMs; (4) The N-dealkylation, and glucuronidation of oral propafenone were increased by rifampin treatment in both EMs and PMs, (5) the 5-hydroxylation of propafenone was inceased by rifampin in EMs, but not in PMs; (6) Propafenone-induced QRS prolongation was significantly reduced by rifampin in both EMs and PMs.[109] It was projected that rifampin treatment is capable of decreasing the pharmacologically active concentration of propafenone by about 70% in both EMs and PMs with the potential to cause a loss of dysrhythmia control regardless of CYP2D6 phenotype.

4   CONCLUSION

With the advent of a high level of polypharmacy, the potential for drug-drug interactions has become a significant health care concern. However, through the use of *in vitro* methods followed by rational *in vivo* studies, our understanding of the potential of drug-drug interactions between existing drugs and drug candidates in development has greatly increased. The evolution of these combined studies and the information they provide will hopefully improve the outcome of polypharmacy for the patient.

REFERENCES

1. Kohler, G.; Bode-Boger, S.; Busse, R.; Hoopmann, M.; Welete, T.; Boger, R. *Int. J. Clin. Pharmacol. Ther.* **2000**, *38*, 504-513.
2. Levy, M.; Kewitz, H.; Altwein, W.; Hillebrand, J.; Eliakim, M.; *Eur. J. Clin. Pharmacol.* **1980**, *17*, 25-31.
3. Weaver, R. *Xenobiotica* **2001**, *31*, 499-538.
4. ftp://ftp.cdc.gov/pub/Health_Statistics/NCHS/Datasets/NAMCS/
5. Huang, B.; Bachmann, K.; He, X.; Chen, R.; McAllister, J.; Wang, T. *Pharmacoepidemiology and Drug Safety* **2002**, *11*, 127-134.

6. Blaschke T.; Cohen S.; Tatro D.; Rubin P.L.; Jarvik F. et al., (Eds) Raven Press, New York, 1981, pp 11-26.

7. Jankel, C.; Fitterman, L.; *Drug Safety* **1993**, *9*, 51-59.

8. Seymour, R.; Routledge, P.; *Drugs and Aging* **1998**, *12*, 485-494.

9. Marroum, P.; Uppoor, R.; Parmelee, T.; Ajayi, F.; Burnett, A.; Yuan, R.; Svadjian, R.; Lesko, L.; Balian, J.; *Clin. Pharmacol. Ther.* **2000**, *68*, 280-285.

10. Guidance for Industry. Center for drug Evaluation and Research, Department of Health and Human Services, U.S. Food and Drug Administration, Rockville, MD, **1997**, http://www.fda.gov/cder/guidance/clin3.pdf.

11. Note for Guidance on the Investigation of Drug Interactions. Committee for Proprietary Medicinal Products (CPMP), The European Agency for the Evaluation of Medicinal Products, **1997**.

12. Wilkinson G. "Pharmacokinetics." *In Goodman & Gilman's The Pharmacological Basis of Therapeutics* 10th edition, J. Hardman; L. Limbird; A. Goodman Gilman, Eds., McGraw Hill, New York, pp 3-29, **2001**.

13. Ortiz de Montellano, P. In *Handbook of Drug Metabolism*, T. Woolf (Ed.) Marcel Dekker, New York, pp 109-130, **1999**.

14. Michalets, E. Pharmacotherapy **1998**, *18*, 84-112.

15. Smith, D.; Abel, S.; Hyland, R.; Jones, B. Xenobiotica **1998**, *28*, 1095-1128.

16. Cooperman, L.; and Batcheller, J.; *CBIs 2nd Annual Predicting Drug Metabolism:In Vitro and in Vivo Screening Techniques to Minimize Attrition and Potential Drug-Drug interactions*. CBI, Woburn, MA, **2001**.

17. Levy, R.; Thummel, K.; Trager, W.; Hansten, P.; Eichelbaum, M. (Eds.) *Metabolic Drug Interactions*, Lippincott Willims & Wilkins, Philadelphia, PA, **2000**.

18. Rodrigues, D. (Ed.):*Drug-Drug Interactions*. Marcel Dekker, Inc., New York, NY, **2001**.

19. Michalets, E. L.; and Williams, C. R. *Clin. Pharmacokinet.* **2000**, *39*, 49-75.

20. Henney, J. *JAMA* **2000**, *283*, 2228.

21. Taglialatela, M.; Timmerman, H.; Annunziato, L. *Trends in Pharmacology* **2000**, *21*, 52-56.

22. Crane, J. K.; Shih, H. T. *Am. J. Med.* **1993**, *95*, 445-446.

23. Flockhart, D. A.; Tanus-Santos, J. E. *Arch. Intern. Med.* **2002**, *162*, 405-412.

24. http://www.fda.gov/medwatch/safety/1998/posico2.htm

25. Thummel, K.; Wilkinson, G. *Ann. Rev. Pharmacol. Toxicol.* **1998**, *38*, 389-430.

26. Roos, T.; Merk, H. F., *Drugs* **2000**, *59*,181-192.

27. Kivisto, K.; Kroemer, H.; and Eichelbaum, M. *Br. J. Clin. Pharmacol.* **1995**, *40*, 523-530.

28. Richelson, E. *Mayo Clin. Proc.* **1997**, *72*, 835, 847.

29. Levy, R. H.. *Epilepsia* **1995**, *36 (Suppl 5),* S8-S13.

30. Benedett, M. S. *Fund. Clin. Pharmacol.* **2000**, *14*, 301-319.

31. French, J.; Gidal, B. *Epilepsia* **2000**, *41(Suppl 8)*, S30-S36.

32. Dal Negro, R. *Clin Pharmacokinet.* **1998**, *35*, 135-150.

33. Unge, P.; Andersson, T. *Drug Safety* **1997**, *16*, 171-179.

34. Piscitelli, S.; Gallicano, K. *New Engl. J. Med.* **2001**, *344*, 984-996.

35. Kivisto, K.; Lamberg, T.; Kantola, T.; and Neuvonen, P. *Clin. Pharmacol. Ther.* **1998**, *62*, 348-354.

36. Ducharme, M. P.; Slaughter, R. L.; Warbasse, L. H.; Chandrasekar, P. H.; Van de Velde, V.; Mannens, G.; Edwards, D. J. *Clin. Pharmacol. Ther.* **1996**, *58*, 617-624.

37. http://www.drug-interactions.com, 2002. This website is also accessible at: http://medicine.iupui.edu/flockhart

38. Rodrigues, A.; Winchell, G.; and Dobrinska, M. *J. Clin. Pharmacol.* **2001**, *41*, 368-373.

39. Tucker, G.; Houston, B.; and Huang, S.-M. *Clin. Pharmacol. Ther.* **2001**, *70*, 103-114.

40. Yamano, K.; Yamamoto, K.; Kotaki, H.; Takedomi, S.; Matsuo, H.; Sawada, Y.; Iga, T. *Drug Metab. Dispo.* **1999**, *27*, 1225-1231.

41. Wrighton, S. A.; VandenBranden, M.; Stevens, J. C.; Shipley, L. A.; Ring, B. J. *Drug Metab Rev* **1993**, *25*, 453-484.

42. Parkinson, A. *Toxicol. Path.* **1996**, *24*, 45-57.

43. Segel, I. H. *Enzyme Kinetics*, John Wiley and Sons, Inc., New York, NY, **1975**.

44. Cheng, Y.-C.; Prusoff, W. H.. *Biochem. Pharmacol.* **1973**, *22*, 3099-3108.

45. Tucker, G. T.; *Int. J. Clin. Pharmacol. Ther. Toxicol.* **1992**, *30*, 550-553.

46. Wrighton, S. A.; Schuetz, E. G.; Thummel, K. E.; Shen, D. D.; Korzekwa, K. R.; Watkins, P. B. *Drug Metabol. Rev.* **2000**, *32*, 339-361.

47. Thummel, K. E.; Kunze, K. L.; Shen, D. D. *Metabolic Drug Interactions.* Levy, R. H.;, Thummel, K. E.; Trager, W. F.; Hansten, P. D.; Eichelbaum, M., (Eds.). Lippincott, Williams and Wilkins, Philadelphia, PA, pp 3-19, **2000**.

48. Bachmann, K. A.; Ghosh, R. *Curr. Drug Metabol.* **2001**, *2*, 299-314.

49. Ito, K.; Iwatsubo, T.; Kanamitsu, S.; Nadajima, Y.; Sugiyama, Y. *Ann. Rev. Pharmacol. Toxicol.* **1998**, *38*, 461-499.

50. Janis-Dow, C. A.; Pearl, M. L.; Watkins, P. B.; Blake, D. S.; Klecker, R. W.; Collins, J. M. *Am. J. Clin. Oncol.* **1997**, *20*, 595-599.

51. Venkatakrishnan, K.; von Moltke, L. L.; Greenblatt, D. J. *J. Clin. Pharmacol.* **2001**, *41*, 1149-1179.

52. Ortiz de Montellano, P. R.; Correia, M. A. *Cytochrome P450: Structure, Mechanism and Biochemisty* (second edition) Ortiz de Montellano, P. R. (Ed.) Plenum Press, New York, NY, pp 305-366, **1995**.

53. Mayhew, B. S.; Jones, D. R.; Hall, S. D. *Drug Metab. Dispo.* **2000**, *28*, 1031-1037.

54. Jones, D. R.; Gorski, J. C.; Hamman, M. A.; Mayhew, B. S.; Rider, S.; Hall, S. D. *J. Pharmacol. Exp. Therap.* **1999**, *290*, 1116-1125

55. Greenblatt, D. J.; von Moltke, L. L. *Metabolic Drug Interactions*, Levy, R. H.; K. E. Thummel, et. al. (Eds). Lippincott, Williams and Wilkins, Philadelphia, USA, pp 259-270, **2000**.

56. Greenblatt, D. J.; von Moltke, L. L. *Drug-Drug Interactions*, Rodrigues, A. D. (Ed.) Marcel Dekker, New York, USA, pp 565-584, **2002**.

57. Watkins, P. B.; Murray, S. A.; Winkelman, L. G.; Heuman, D. M.; Wrighton, S. A.; Guzelian; P. S. *J. Clin. Invest.* **1989**, *83*, 688-697.

58. Watkins, P. B.; Turgeon, D.; Saenger, P.; Lown, K.; Kolars, J.; Hamilton, T.; Fishman, K.; Guzelian, P.; Voorhees, J. *Clin. Pharmacol. Ther.* **1992**, *52*, 265-273.

59. Krivoruk Y.; Kinirons, M.; Wood, A.; Wood, M. *Clin. Pharmacol. Ther.* **1994**, *56*, 608-614.

60. Kinirons, M.; O'Shea, D.; Downint T.; Fitzwilliams, A.; Joellenbeck, L.; Groopman, J.; Wilkinson, G.; Wood, A. *Clin. Pharmacol. Ther.* **1993**, *54*, 621-629.

61. Kinirons, M.; O'Shea, D.; Kim R.; Groopman, J.; Thummel, K.; Wood, A. J.; Wilkinson, G. *Clin. Pharmacol. Ther.* **1999**, *66*, 224-231.

62. Streetman, D. S.; Bertino Jr., J. S.; Nafziger, A. N. *Pharmacogenetics* **2000**, *10*, 187-216.

63. Breimer, D. D.; Schellens, J. H. M. *Trends in Pharmacol. Sci.* **1990**, *11*, 223-225.

64. Paolin M.; Biagi, G. L.; Bauer, C.; and Cantelli-Forti, G. *J. Clin. Pharmacol.* **1993**, *33*, 1011-1012.

65. Frye, R. G.; Matzke, G. R.; Adedoyin, A.; Porter, J. A.; Branch, R. A. *Clin. Pharmacol. Ther.* **1997**, *62*, 365-376.

66. Streetman, D. S.; Bleakley, F. F.; Kim., J. S.; Nafziger, A. N.; Leeder, J. S.; Gaedigk, A.; Gotschall, R.; Kearns, G. L.; and Bertino Jr., J. S. *Clin. Pharmacol. Ther.* **2000**, *68*, 375-383.

67. Bachmann, K.; White, D.; Jauregui, L.; Schwartz, J.; Agrawal, N.; Mazenko, R.; Porras, A. (abstract) *Exp. Biol. 2002*, Annual Meeting of ASPET, New Orleans, USA **2002**.

68. Wilt, M.; Ishani, A.; Start, G.; MacDonald, R.; Lau, J.; Mulrow, C. *JAMA*, **1998**, *11*, 1604-1609.

69. Muller, W.; Singer, A.; Wonnemann, M.; Hafner, U.; Rolli, M.; Schafer, C. *Pharmacopsychiat.* **1998**, *31*, 16-21.

70. Bhattacharya, S.; Chakrabarti, A.; Chatterjee S. *Pharmacopsychiat.* **1998**, *31*, 22-29.

71. D'Arcy, P. F. *Drug React. Toxicol. Rev.* **1993**, *12*, 147-162

72. Fugh-Berman, A. *Lancet* **2000**, *355*, 134-138

73. Ernst, E. *Lancet* **1999**, *354*, 2014-2015.

74. Piscitelli, S.; Burstein, A.; Chaitt, D.; Alfaro, R.; Falloon, J. *Lancet* **2000**, *355*, 547-548

75. Ruschitzka, F.; Meier, P.; Turina, M.; Luscher, T.; Noll, G. *Lancet* **2000**, *355*, 548-549

76. Roby, C.; Anderson, G.; Kantor, E.; Dryer, D.; Burstein, A. H. *Clin. Pharmacol. Ther.* **2000**, *67*, 451-457

77. Obach, S. J. *Pharmacol. Exp. Ther.* **2000**, *294*, 88-95.

78. Moore, L.; Goodwin, B.; Jones, S.; Wisely, G.; Serabjit-Singh, C.; Willson, T.; Collins, J.; Kliewer, S. *Proc. Natl. Acad. Sci.* **2000**, *97*, 7500-7502

79. Anonomous, "Public Affairs Update" *The Pharmacologist* **1999**, *41*, 198.

80. Kupferschmidt, H.; Ha, H.; Ziegler, W.; Meier, P.; Krahenbuhl, S. *Clin. Pharmacol. Therap.* **1995**, *58*, 20-28

81. Kivisto, K.; Lilja, J.; Backman, J.; Neuvonen, P. *Clin Pharmacol. Ther.* **1999**, *66*, 448-453

82. Lilja, J.; Kivisto, K.; Backman, J.; Lamberg, T.; Neuvonene, P. *Clin. Pharmacol. Therap.* **1998**, *58*, 20-28

83. Libersa, C.; Brique, S.; Motte, K.; Caron, J.; Guedon-Moreau, L.; Humbert, L.; Vincent, A.; Devos, P.; Lhermitte, M. A.. *Br. J. Clin. Pharmacol.* **2000**, *49*, 373-378

84. Lilja, J.; Kivisto, K.; Neuvonen, P. *Clin. Pharmacol. Ther.* **2000**, *68*, 384-390.

85. Schmiedlin-Ren, P.; Edwards, D.; Fitzsimmons, M.; He, K.; Lown, K.; Woster, P.; Rahman, A.; Thumel, K.; Fisher, J.; Hollenberg, P.; Watkins P. *Drug Metab. Dispo.* **1998**, *25*, 1228-1233

86. Bailey, D.; Kreeft, J.; Munoz, C.; Freeman, D.; Bend J. *Clin. Pharmacol. Therap.* **1998**, *64*, 248-256

87. Bailey, D.; Malcolm, J.; Arnold, O.; Spence, J. *British J. Clin. Pharmacol.* **1998**, *46*, 101-110.

88. Wichittra, T.; Guo, L.-Q.; Fukuda, K.; Ohta, T.; Yamazoe, Y. *Arch. Biochem. Biophys.* **2000**, *378*, 356-363

89. Anderson, K.; Conney, A.; Kappas, A. *Prog. Clin Biol. Res.* **1986**, *214*, 39-54

90. Chan, W. K.; Nguyen, L. T.; Miller, V. P.; Harris, R. Z. *Life Sci.* **1998**, *62*, L135-142.

91. Chan, W.; Delucchi, A. *Life Sci.* **2000**, *67*, 3103-3112.

92. Piver, B.; Berthou, F.; Dreano, Y.; Lucas, D. *Toxicol Letters* **2001**, *125*, 83-91.

93. Johne, A.; Brockmoller, J.; Bauer, S.; Maurer, A.; Langhenrich, M.; Roots, I. *Clin. Pharmacol. Ther.* **1999**, *66*, 338-345

94. Yu, D. *J. Clin. Pharmacol.* **1999**, *39*, 1203-1201

95. Backman, J.; Maenpaa, J.; Belle, D.; Wrighton, S.; Kivisto, K.; Neuvonen, P. *Clin Pharmacol. Ther.* **2000**, *67*, 382-390.

96. Takanaga, H.; Ohnishi, A.; Yamada, S.; Matsuo, H.; Morimoto, S.; Shoyama, Y.; Ohtani, H.; Sawada, Y. *J. Pharmacol. Exp. Ther.* **2000**, *293*, 230-236.

97. Dresser, G.; Bailey, D.; Leake, B.; Schwarz, U.; Dawson, P.; Freeman, D.; Kim, R. *Clin. Pharmacol. Ther.* **2002**, *71*, 11-20.

98. Sullivan, T. J.; Reese, J.; Bachmann, K.; Jauregui, L.; Miller, K.; Scott, M.; Levine, L. *Am. J. Therapeut.* **1995**, *2*, 275-278.

99. Mitchell, R.; Riemensnider, D.; Harsch, J.; Bell, J., in *Transactions of the 17th Conference on Chemotherapy of Tuberculosis,* Veterans Admin. Army Navy, Washington, D.C., USA, pp 77-85, **1958**

100. Evans, D.; Manley, K.; McKusick, V. *Brit. Med. J.* **1960**, *2*, 485-491.

101. Evans, D. A. P. *Genetic Factors in Drug Therapy*, Cambridge University Press, Cambridge, UK, **1993**

102. Kalow, W. *Eur. J. Clin. Pharmacol.* **1987**, *31*, 633-641.

103. Ensom, M. H.; Chang, T. K.; Patel, P. *Clin. Pharmacokinet.* **2001**, *40*, 783-802.

104. Vesell, E. S. *Pharmacology* **2000**, *61*, 118-123.

105. Pillans, P. I. *Int. Med. J.* **2001**, *31*, 476-478.

106. Caraco, Y. *Therapeutic Drug Monitoring* **1998**, *20*, 517-524.

107. Desmeules, J.; Gascon, M.; Dayer, P.; Magistris, M. *Eur. J. Clin. Pharmacol.* **1991**, *41*, 23-26.

108. Caraco, Y.; Sheller, J.; Wood, A. J. J. *J. Pharmacol. Exp. Ther.* **1997**, *281*, 330-336.

109. Dilger, K.; Greiner, B.; Fromm, M.; Hofman. U.; Kroemer, H.; Eichelbaum, M. *Pharmacogenetics* **1999**, *9*, 551-559.

Chapter 10

CYP Gene Induction by Xenobiotics and Drugs

Academic authors:
Jean-Marc Pascussi, Sabine Gerbal-Chaloin, Martine Daujat, Lionel Drocourt,
Lydiane Pichard-Garcia, Marie-José Vilarem and Patrick Maurel

INSERM U128, 1919 Route de Mende, 34293 Montpellier, France

Drug Industry authors:
Sylvie Klieber, François Torreilles, Martine Bourrié, François Guillou and
Gérard Fabre

Sanofi-Synthélabo Recherche, 314 Avenue du Professeur Blayac, 34295 Montpellier, France

1   INTRODUCTION

Induction of cytochromes P450 (CYPs) and other xenobiotic metabolizing and transporting systems (XMTS) is known to be responsible for drug-drug interactions and other deleterious effects of drugs in man.[1,2] See also www.drug-interactions.com. During the early development of a new drug it is therefore desirable to predict its capacity to induce CYPs and XMTS. For this purpose it is necessary, first to understand the mechanism of induction at the molecular level and second to have access to adequate and reliable methods of inducer screening.

Although the implication of the aryl hydrocarbon receptor (AhR) in the induction of CYP1s and other systems has been known for more than 20 years,[3,4] the mechanism of induction of the CYP2 and CYP3A families has been clarified only recently. Two xenoreceptors, the pregnane X receptor (PXR) and the constitutive androstane receptor (CAR), members of the nuclear receptor superfamily, have been characterized and shown to play a major role in the inducibility of these CYP genes and other XMTS in response to xenobiotics.[5-11]

The aim of this chapter is to review the clinical and pharmacological consequences of CYP1-3 gene induction, the recent developments in the understanding of the mechanisms of CYP gene induction, including the role of nuclear receptors such as AhR, PXR and CAR, the possible interference between xenobiotic-signalling pathways and other endogenous signalling pathways, and the methods and tools for screening xenobiotic inducers in man both *in vitro* and *in vivo*.

2   CLINICAL AND PHARMACOLOGICAL CONSEQUENCES OF CYP1-3 GENE INDUCTION IN MAN

Approximately 70% of human liver CYP is accounted for by CYP1A2, -2A6, -2B6, -2C (-2C8, -2C9, 2C18 and -2C19), -2D6, -2E1 and -3A (-3A4 and -3A5) enzymes. Among these, the CYP2C and CYP3A family members are the most abundant, accounting for 20 and 30 % of the total CYP, respectively.[12,13] In addition, CYP3A plays a major role in the oxidative metabolism of more than 50% of marketed drugs, but also contributes to the metabolism of many endogenous compounds such as hormones (estrogens, testosterone, progesterone) and bile salts. One of the intriguing aspects of the CYPs is that most of them are inducible. From a biological point of view, induction is an adaptative response which, in contrast to CYP inhibition that is associated with an almost immediate response, is a slow regulatory process.

In drug therapy, major concerns are related to CYP induction.[1] See also the review by Niemi et al. (*Clin. Pharmacokinet.* in press).

• Induction produces a reduction of the pharmacological effects because of increased drug metabolism. This mainly concerns drugs for which pharmacological activity is associated with the unchanged drug. Hence, administration of rifampicin in long-term cyclosporin-treated patients, results in low blood concentrations of the immunosuppressive agent,[14] associated with frequent acute allograft rejection.[15,16] For CYP-activated prodrugs, enzyme induction produces an increase in the pharmacological effects.

• Induction may create an undesirable imbalance between "toxication" and "detoxication". Induction of drug metabolizing enzymes may led to a decrease in toxicity through acceleration of detoxification, or to an increase in toxicity caused by increased formation of reactive metabolites. A good example of an adverse consequence due to enzyme induction is the increased toxicity of paracetamol (acetaminophen). Long-term consumption of alcohol (a CYP2E1 inducer) is associated with liver damage from therapeutic doses (< 4 g/day) of paracetamol.[17] Indeed CYP2E1 is mainly involved in converting paracetamol to its reactive intermediate, i.e. *N*-acetyl-*p*-amino-benzoquinone.

• Another clinically important drug interaction with CYP3A4-type inducers involves oral contraceptives. This results in menstrual disturbance and unplanned pregnancies as a consequence of the increased metabolism of both estrogenic and progesterogenic components of oral contraceptives.[18,19]

The clinical significance of a metabolic drug interaction will depend mainly on the magnitude of the change in the concentration of the active species (parent drug and/or active metabolite and/or toxic metabolite) at the site of pharmacological action and on the therapeutic index of the drug. The smaller the difference between toxic and effective concentration, the greater the likelihood that a drug interaction will have serious clinical consequences.

3   MECHANISMS OF CYP GENE INDUCTION

Studies on enzyme induction have played a prominent role in the discovery and characterization of cytochromes P450 in mammals. Compounds such as polycyclic aromatic hydrocarbons and phenobarbitone were defined as the first two classes of prototypical inducers of CYP1As and CYP2Bs more than 30 years ago.[20] Later on, other classes of enzyme inducers were pregnenolone, 16α-carbonitrile and rifampicin (as prototypical inducers of the CYP3A family);[21-25] peroxysome proliferators (CYP4A family);[24,25] and alcohols and imidazole derivatives (CYP2E1).[26,27] Except for the last case, these inducers generally and primarily act at the transcriptional level through specific nuclear receptors. In this chapter, we shall restrict our review to the induction of CYP1A, CYP2B, CYP2C and CYP3A families.

3.1   Induction of the CYP1A family

Induction of CYP1A1 and CYP1A2 in response to polycyclic and halogenated aromatic hydrocarbons has been known for more than 30 years and shown to be mediated through the activation of the aryl hydrocarbon receptor (AhR).[3,4,28] More recently, the list of XMTS inducible by these compounds has increased with the inclusion of CYP1B1, UDP-glucuronosyltransferase, glutathione S-transferase, aldehyde dehydrogenase, NADPH oxidoreductase, and other genes involved in the regulation of cell cycle and differentiation.[29,30] A number of genes are also negatively regulated by AhR including estrogen receptor, tissue plasminogen activator and others.[31] On the other hand, the list of possible inducers of CYP1As has been extended to other molecules the chemical structure of which differs widely from that of the classical aryl hydrocarbon receptor (AhR) ligands.[32-37] These compounds generally bind poorly or not at all to the AhR and are believed to induce CYP1As either through the classical pathway, or by being biotransformed into "true" AhR ligands, or by interfering with various processes involved in AhR activation.[38]

Induction of CYP1As is mediated through a heterodimeric transcription factor consisting of AhR and the aryl hydrocarbon receptor nuclear translocator (ARNT).[3,4] Both AhR and ARNT belong to the basic-helix-loop-helix (HLH) Per/ARNT/Sim (PAS) family of transcription factors.[39] In its inactive form, AhR is located in the cytosol as a multiprotein complex including two heat shock proteins HSP90,[40] the co-chaperone p23,[41] and the hepatitis-B-virus X-associated protein 2.[42] These proteins play important roles in AhR signalling by controlling various steps such as ligand binding, cellular localization, translocation to the nucleus, heterodimerisation with ARNT, ubiquitination and phosphorylation of this receptor. Upon ligand binding, AhR translocates to the nucleus where it associates with ARNT to form the functional transcription factor complex AhRC. This complex binds to conserved sequences of DNA termed xenobiotic responsive elements (XRE, canonic sequence: 5'-CACGCNA/T-3') and, after appropriate recruitment of various co-activators, enhances the transcription of target genes.[31]

The AhR is rapidly degraded by the ubiquitin-cytoplasmic proteasome pathway after activation by its ligands. Inhibition of degradation results in increased target gene induction.[43] Thus, the AhR protein concentration is an important parameter of the cell response to the ligand (in terms of induction of CYP1A1 by dioxin for example) and cells might affect AhR signalling in part by controlling the rate of degradation of this receptor by the 26S proteasome.[44,45] In the nucleus, the AhRC associates with several co-activators including CBP/P300 and members of the NcoA/SRC-1/pCIP family, in a ligand-dependent manner.[31,46] These proteins enhance AhRC-mediated gene transcription, first by bridging the AhRC with the core transcription initiation complex, and second by facilitating the opening of chromatin through their intrinsic histone acetylase activity.

3.2   Cross-talks between the AhR pathway and other transcriptional factors and signal transduction pathways

3.2.1 Hypoxia-inducible factor 1α

Hypoxia-inducible factor 1α (HIF-1α) is another member of the basic-helix-loop-helix (HLH) Per/ARNT/Sim (PAS) family of transcription factors which is regulated through oxygen concentration changes.[47] Transcriptional activity of HIF-1α is controlled by O_2-dependent hydroxylations of key aminoacid residues which affect first the stability of the protein (*via* the N-terminal transactivation domain, TAD) and second its interaction with coactivators such as CBP/P300 and SRC-1 (on the C-terminal transactivation domain).[48] Briefly, under normoxia conditions, HIF-1α is rapidly degraded and unable to recruit coactivators. Under hypoxia conditions, hydroxylations of the critical residues in both C- and N-

TAD are inhibited so that the HIF-1α protein is stabilized, exhibits increased accumulation, translocates into the nucleus, forms heterodimers with ARNT, and recruits coactivators. This eventually results in the activation of target genes encoding notably erythropoietin (EPO), vascular endothelial growth factor and glycolytic enzymes through the transactivation of hypoxia responsive elements (HRE). The finding that both AhR and HIF-1α form heterodimers with a common partner ARNT suggested that a cross-talk between both signalling pathways might occur upon activation (Figure 1). Two questions may therefore be asked: (i) is hypoxia able to interfere with AhR-mediated activation of target genes? (ii) are AhR activators like dioxin able to interfere with the expression of genes controlled by HIF-1α? Poellinger and coworkers[49] demonstrated that the affinity of HIF-1α for ARNT is greater than the affinity of AhR for this common partner so that that once HIF-1α is activated it interferes with the dioxin signalling pathway. Similar conclusion was reached by Bradfield and coworkers.[50] However, and this might be consistent with the relative affinities of HIF-1α and AhR for ARNT, the reverse was not true, that is, dioxin does not inhibit the expression of HIF-1α-responsive genes. Interestingly, Bradfield and coworkers demonstrated that EPO promoter harbors several functional XRE which make this gene a target of AhR.

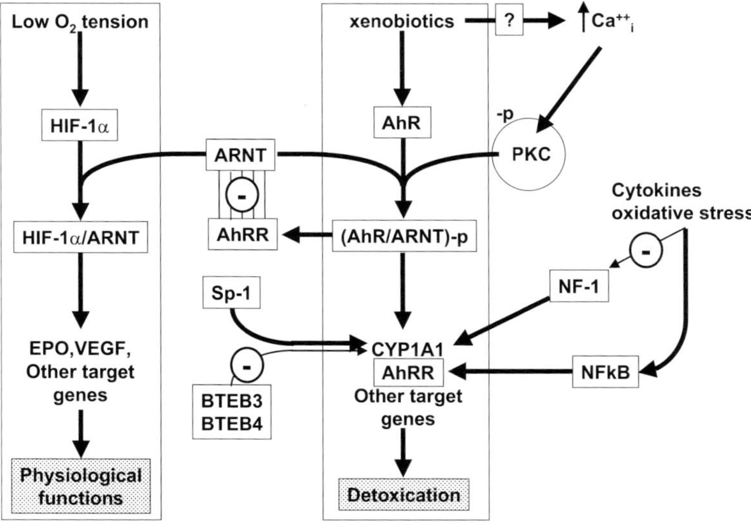

Figure 1. Schematic representation of cross-talk between the xenoreceptor AhR and various nuclear receptors and transcription factors. HIF: hypoxia inducible factor ; ARNT : AhR nuclear translocator ; AhRR : AhR repressor ; BTEB3/4 : basic transcription element binding proteins 3 and 4; NF1 : nuclear factor 1; PKC : protein kinase C ; Ca^{2+}_{+i} : intracellular calcium concentration. Broad arrows : stimulatory pathways; thin arrows with (-): inhibitory pathways. See text for detail (Section 3.2.)

3.2.2 AhR repressor

In addition to the numerous genes regulated through AhR, the AhR repressor (AhRR) has been recently identified as a target of AhR.[51,52] This factor is another member of the PAS family of transcription factors, closely related to AhR. AhRR is located in the nucleus and, in contrast to AhR, binds to ARNT in a ligand-independent manner and does not interact with HSP90. Heterodimer AhRR-ARNT binds XRE sequences but has no transactivational activity. Thus AhRR might exert an inhibitory action on AhR function by two mechanisms : (i) sequestration of ARNT through formation of transcriptionally inactive AhRR-ARNT complexes, and (ii) competition of these complexes with the AhR-ARNT heterodimer for DNA target sequences (Figure 1). Thus the fact that AhRR is a target of AhR suggests the existence of feedback inhibition of AhR function. This is reminiscent of previous observations on the superinduction of CYP1A1 in response to AhR ligands in the presence of cycloheximide, a protein synthesis inhibitor, in various cellular systems.[53] Superinduction was interpreted as reflecting the inhibition of synthesis of a short half-life repressor protein which is likely to be AhRR.

AhRR is regulated not only through XRE but also through GC boxes (also called BTE, basic transcription elements), most likely *via* a cooperative interaction between AhR and Sp-1 as previously shown for CYP1A1 regulation.[54] The regulatory region of AhRR also contains a NFkB site that functions as an inducible enhancer in response to the phorbol-ester TPA.[51] It has been shown that NFkB and AhR-ARNT interact with each other and, in the context of the AhRR promoter (where both responsive elements are correctly juxtaposed), these factors cooperate to enhance the gene expression. Interestingly, NFkB is activated in response to various cytokines such as Il-1, TNF and IFNγ,[55] which have been shown to inhibit *CYP1A* expression.[56,57] AhRR could therefore play a role in the negative control of *CYP1A* genes either through cytokines or through other NFkB-activating stimuli (Figure 1).

3.2.3 Sp-1 and NF1 transcription factors

The Sp-1-like family of transcription factors is characterized by a highly conserved DNA-binding domain, consisting of three zinc fingers.[58] This family has currently 18 members. Sp-1-like transcription factors have been shown to regulate either positively or negatively (depending on the promoter context) many genes exhibiting the so-called GC boxes or basic transcription elements (BTE) in their promoter. For example Sp-1, the first identified member of this family, has been previously shown to participate in the expression of CYP1A1 at both basal and induction levels.[59] On the other hand, other Sp-1-like proteins including BTEB3 and BTEB4 (basic transcription element binding proteins) have recently been characterized and shown to down-regulate CYP1A1 expression.[60] BTEB4 is constitutively expressed at high level in the hepatocyte nucleus and its subcellular location is not regulated by AhR ligands. It has been proposed that BTEB3/4 proteins bind the BTE motif of the CYP1A1 promoter and compete with Sp-1 in repressing the gene expression by recruiting the mSin3A co-repressor which possesses a histone deacetylase activity (Figure 1). The regulation of these Sp-1-like proteins is currently unknown but it is likely that changes affecting their relative distribution will impact on the expression of target genes including CYP1As. Whether BTEB4 is responsible for the very low level of expression of CYP1A1 in the normal liver, in the absence of an AhR ligand, is currently unknown.

Although the implication of AhR and Sp-1-like proteins in CYP1As regulation is well established,[59] other factors have been suspected to contribute as well. This is the case for NF1 (nuclear factor 1 or CCAAT Box transcription factor). A NF1 site was first reported to be critical for mouse *cyp1A1*[61] and more recently several investigators identified a functional NF1 element in the CYP1A1 [62,63] and 1A2 promoter.[64] Notably, Morel and Barouki showed that mutation of the NF1 site in human CYP1A1 promoter strongly decreases the activity of the promoter. A problem with the actual role of NF1 *versus* Sp-1 in CYP1A1 expression is that, both NF1 and Sp-1 response elements overlap in this gene. Whether these two factors compete or cooperate on the CYP1A1 promoter is currently unknown. It has been suggested that NF1 is involved in the down-regulation of human CYP1A1 expression by oxidative stress.[62,63] Indeed, under various stress stimuli including inflammatory cytokines, glutathione depletion, heat and osmotic shocks and chemical stress, the transcriptional activity of NF1 is inhibited presumably through oxidation of Cys427 present in the trans-activating domain, and this leads to decreased activity of the CYP1A1 promoter (Figure 1). Interestingly, Sp-1 also has been shown to be a possible target of oxidative stress, through the oxidation of the cysteine sulfhydryl groups which are involved in maintaining the zinc finger structure of these transcription factors.[65]

3.2.4 Intracellular calcium

Another cellular factor recently suspected to be involved in AhR-mediated gene activation is intracellular calcium $(Ca^{2+})_i$. Calcium is a well known second messenger of transcriptional modulation.[66] In fact, changes in Ca^{2+}_i are rapidly reflected in concomitant changes in the activity and cellular location of a number of protein kinases including PKA (protein kinase A), PKC (protein kinase C), MAPKs (mitogen-activated proetin kinases), and Ca^{2+}/calmodulin-dependent protein kinases including CaMK and CaMKK. A number of previous reports consistently established that phosphorylation of both AhR and ARNT is required for full transcriptional activity of AhRC and that PKC is involved in this process.[67,68] Investigating the induction of CYP1A1 in the Caco-2 cell line by oltipraz, a cruciferous vegetable constituent exhibiting chemoprotective capacity, Le Ferrec et al. suspected that Ca^{2+}_i could be implicated in one of the early steps controlling AhR-mediated gene activation.[69] These authors observed that oltipraz induces a rapid and transient increase in Ca^{2+}_i. Using a chelator of intracellular calcium and an inhibitor of store-operated calcium channels, they unraveled a functional link between the oltipraz-mediated Ca^{2+}_i and the induction of CYP1A1 through the AhR pathway. Whether the increase in Ca^{2+}_i is the initial step leading to the activation of PKC which, in turn, leads to the phosphorylation and full activation of AhRC components and eventually to the transactivation of target genes remains to be established (Figure 1). The nature of the form of PKC specifically involved in AhRC phosphorylation is currently unknown. However, differential expression patterns and responses to activator stimuli of PKC isoforms is likely to impact strongly on the cell- and tissue-specific and developmental modulations of AhRC activity.

3.3 Induction of CYP2 and CYP3 families

In contrast to the situation of the CYP1As, the mechanisms of induction of CYP2B and CYP3A genes remained an enigma until recently. In 1998, while investigating the hormonal activation of orphan receptors, S Kliewer and coworkers identified a new member of the nuclear hormone receptor family which they designated PXR (NR1I3) for pregnane X receptor, because it was primarily activated by natural pregnanes.[5] However and surprisingly, this receptor also appeared to be activated by both synthetic glucocorticoids and antiglucocorticoids, including dexamethasone and pregnenolone 16α-carbonitrile, two well-known inducers of CYP3As.[5,6] Simultaneously, two other groups isolated and characterized the same receptor designated by them SXR (steroid and xenobiotic receptor) and PAR (pregnane activated receptor), respectively.[70,71] In the meantime, Negishi and coworkers demonstrated that CAR (NR1I2), the previously characterized constitutive androstane receptor, was able to transactivate the phenobarbital-responsive element of CYP2B genes.[11,72,73] Collectively, these observations established, for the first time, consistent mechanistic links between CYP3A and CYP2B induction and the xenobiotic-mediated activation of nuclear receptors. Various orthologous forms of PXR have now been isolated from different organisms including mammals (humans, monkey, pig, dog, rabbit, rodents), fish and chicken indicating that the presence of signalling pathways mediated by this xenobiotic-sensing receptor has been preserved during species evolution.[74] The evolutionary mechanisms by which the members of the PXR subfamily of receptors diverged, presumably from the fish PXR common ancestor, is however currently unknown.

3.3.1 Nuclear hormone receptors

Nuclear hormone receptors (NHRs) constitute a superfamily of structurally related, ligand-activated transcriptional regulators that include more than 50 distinct proteins.[75-77] They are categorized into three subclasses. Class I contains receptors for steroid hormones such as progesterone, androgens, mineralocorticoids, estrogens and glucocorticoids. In the resting state, these receptors are sequestered in the cytosol as a multiprotein-complex in which they are associated with molecular chaperones, heat shock proteins (HSPs) and other proteins. After binding the hormone, these receptors undergo a series of structural and functional changes leading to dissociation from HSPs, homodimerisation, phosphorylation, migration into the nucleus, binding to DNA at specific hormone response elements located in the promoter of the target genes and recruitment of and interaction with coactivators, leading to the activation of the basal transcription machinery which eventually results in the initiation of transcription.[78] The hormone response elements of Class I receptors consist of two half-sites organized as a palindrome in which they are separated by three nucleotides. Class II contains receptors for thyroid hormone, vitamin D3, 9-cis retinoic acid and all-trans retinoid receptors. In the absence of ligand, these receptors are present in the nucleus as a multiprotein-complex including their common heterodimer partner RXR and other proteins such as corepressors. This complex is bound to its hormone response element in the promoter of target genes. Class II response elements consist of two half-sites organized generally as direct repeats (DRn) in which both halves are separated by a spacer whose length (n nucleotides) determines the receptor selectivity. Upon ligand binding, the receptor undergoes a conformational change that results in the release of corepressors, recruitment of coactivators and activation of target gene transcription. NHRs of Class III contain orphan receptors, such as PXR and CAR, whose natural ligands have not yet been identified and for which only minimal information is generally available concerning the mechanism of transactivation.[79]

Although each nuclear receptor binds preferentially to a specific DNA sequence,[77,80] there have recently been indications however that a given receptor (whatever family it belongs to) may bind to and transactivate different responsive elements. Thus for example, the steroid hormone receptors (NR3C subfamily) bind classically and almost exclusively as homodimers to palindromic sequences separated by 3 nucleotides. However, the glucocorticoid or the estrogen receptors have been shown to bind also to direct repeats with different spacings between half sites (including DR2, DR5, DR6, DR9) as well as to an ER9 (everted repeat), although binding to these motifs is weaker than to the palindrome.[81] Similarly, the androgen receptor may bind to a DR1 motif in addition to binding to the classical palindrome.[82] The apparent ability of a given nuclear receptor to target similar but distinct DNA sequences is believed to result from the flexibility of either the ligand and/or DNA binding domains, the intervening linker region, or the DNA template itself. From a functional point of view, this apparent versatility is expected, and was actually shown, to lead to functional cross-talk between nuclear receptors and several examples of this will be given later on PXR/CAR and other nuclear receptors.

3.3.2 The Pregnane X Receptor (PXR)

The pregnane X receptor (PXR, NR1I3) is a class III nuclear receptor which forms functional heterodimers with RXRα. It is expressed mainly in the liver, small intestine, colon as well as in the kidney.[5,6] This receptor binds generally with low affinity (EC_{50} in the micromolar range) to a wide variety of structurally diverse exogenous and endogenous compounds including drugs such as rifampicin, phenobarbital, nifedipine and other calcium channel blockers; clotrimazole, mifepristone, metyrapone, steroid hormones and metabolites such as progesterone, estrogens, corticosterone, 5β-pregnanes; and androstanol, and dietary compounds such as coumestrol and hyperforin.[6,70,71,83-87] The last compound appears to be the most potent PXR activator with an EC_{50} of 23nM. A well-known peculiarity of CYP3A inducibility is its species-specificity. For example, rifampicin is a good inducer in man but not in the rat, while PCN is a good inducer in the rat but is not in man. Similarly, dexamethasone is a potent inducer in the rat but a moderate inducer of human CYP3A.[88] Interestingly, the same species-specificity is observed when comparing the activation of human and rat PXR by these compounds, thus providing a major argument in favor of the main role for this receptor in CYP3A regulation.[89] Crystal structure analyses reveal that the ligand-binding domain of human PXR is highly hydrophobic and flexible, allowing lipophilic molecules of different size to bind in multiple orientations.[90] Recently, 38 single nucleotide polymorphisms (SNPs) have been identified in human PXR gene.[91] However, none of them was located in the ligand-binding site and their implication remains unclear.

PXR appears to be responsible for the xenobiotic-mediated induction of a battery of genes including CYP3A4,[6,70,71] CYP3A7,[92,93] CYP2B6,[94] CYP2C8 and CYP2C9,[95-97] MDR1,[9] BSEP [99] and MRP2.[100] Kliewer et al. demonstrated that ligand binding induces the interaction between PXR and the coactivator protein SRC-1.[5] Two distinct PXR-responsive elements have been identified in the 5'-flanking region of CYP3A4. The first of these is the proximal PXR responsive element located at -160. It consists of an everted repeat of the nuclear receptor half-site AGGTCA separated by 6 nucleotides (ER6); this element is necessary but not sufficient for full transactivation of the CYP3A4 promoter.[6] A second distal PXR response element has been identified and located between -7800 and -7200.[101] This element is composite and consists of two direct repeats separated by 3 nucleotides (DR3), encompassing an ER6 motif. Apparently, a cooperative interaction between both distal and proximal PXR response elements is necessary for full PXR-mediated induction of CYP3A4.

Studies in PXR-null mice suggest that PXR is not only a xenobiotic sensor, but also plays a role in the protective mechanism against secondary bile acid-induced liver injury, in part through its role as an inducer of CYP3A.[102-104] Indeed, bile acids such as lithocholic acid and 6-ketolithocholic acid are PXR ligands.[105] Wild mice pretreated with pregnenolone 16α-carbonitrile (a potent PXR activator in rodents) are protected against lithocholic acid-induced liver injury, while PXR-null mice are not. In addition, experiments in PXR-null mice demonstrated that PXR is a negative regulator of the CYP7A1,[103] the enzyme controlling the rate-limiting step in bile acid biosynthesis.

3.3.3 The Constitutive Androstane Receptor (CAR)

The Constitutive Androstane Receptor (CAR, NR1I2) is a class III orphan receptor originally characterized as a transactivator of retinoid acid response elements (RARE), even in the absence of ligand, hence its name.[106] Like PXR, CAR is predominantly expressed in the liver and intestine.

In contrast to PXR, however, CAR is located in the cytosol under normal conditions, i.e. in mice hepatocytes *in vivo* and in primary human hepatocytes.[107-109] In response to activators such as phenobarbital, CAR translocates to the nucleus where it forms a functional heterodimer with RXRα which eventually activates the transcription of target genes after binding to appropriate response elements (see below). The molecular mechanism responsible for this translocation is still unclear, but it seems to involve a specific phosphorylation-sensitive step, as suggested by the interfering effect of okadeic acid an inhibitor of phosphatase 1 and 2A.[109,110] In this respect, it is interesting that previous reports have emphasized the importance of phosphorylation-dependent steps in CYP2B induction.[107,111]

CAR behaves in a manner somewhat opposed to the classical hormone nuclear receptors. In classical receptors, ligand binding mediates conformational changes in the LBD-AF-2 domain which result, notably, in enhanced recruitment and binding of coactivator(s) such as SRC-1 required for full transcriptional activity. In CAR, the active conformation seems to be natural and the LBD-AF2 domain of the protein interacts with coactivators in a ligand-independent manner. However, some ligands upon binding might affect the protein in such a way that an inactive conformation is induced. These ligands are called inverse agonists.[112] Like PXR, CAR exhibits a pronounced species-specificity of ligands, activators and inverse agonists.[86] Androstanol is an inverse agonist of both human and mouse CAR, while clotrimazole is an inverse agonist of human CAR only. Progesterone (a PXR activator) and androgens have been shown to inhibit significantly the transcriptional activity of mouse CAR only.[113] 5β-pregnane-3,20 dione is an activator and ligand of human CAR only. In contrast, 1,4-bis(2-(3,5-dichloropyridyloxy))benzene (TCPOBOP), estradiol and estrone are ligands and activators for mouse CAR only. Some of the activators like TCPOBOP and estrogens are able to reverse the inhibition of mouse CAR induced by the presence of inverse agonists. However, some of the CAR activators are not necessarily ligands of this receptor *in vitro*. This is notably the case for phenobarbital, which does not influence CAR-SRC-1 binding in both man and mouse. Finally, the physiological significance of the activation and deactivation of mouse CAR by estrogens and androgens, respectively, is unclear. Whether this is at the origin of the sex dimorphism observed in the phenobarbital induction of CYP2B genes in some rodent strains is currently unknown.[114,115]

In man, recent data show that CAR mediates the phenobarbital induction of UDP-glucuronosyltransferase (UGT1A1),[116] CYP2B6,[7,117] CYP3A4 [7,118] and CYP2C9 [96] and enoyl-CoA hydratase /3-hydroxy acyl CoA-dehydrogenase gene.[119] In mice, increased liver weight and rate of DNA synthesis in response to phenobarbital are mediated by CAR.[120] Several groups have identified a complex phenobarbital-response element module (PBREM) which consists of two nuclear receptor binding sites (termed NR1 and NR2) and one NF1 binding site.[7,109] NR1 and NR2 are both imperfect DR4 motifs essential for phenobarbital induction of CYP2B genes. In human CYP2B6, PBREM is located between -1684 and -1733, and has been shown to bind to and be transactivated by the CAR/RXR heterodimer.[9,94] HepG2 or HuH7 cell lines that were genetically modified in such a way as to constitutively express CAR were shown to re-express CYP2B6 [7] and CYP3A4.[88] Finally, *in vivo* disruption of CAR in mice completely abrogates phenobarbital- and TCPOBOP-mediated induction of cyp2b10 and cyp3a genes, demonstrating the pivotal role of this receptor in the xenobiotic induction of some P450s.[121]

3.4   Cross-talk between PXR and CAR

It has long been known that a given CYP gene (i.e. CYP3A4) may be differentially induced by a number of xenobiotics (including rifampicin, phenobarbital, clotrimazole, etc.) and that a given xenobiotic (i.e. rifampicin) may differentially induce distinct CYP genes (including CYP3A, CYP2B, CYP2C). Some molecular mechanisms may now be proposed to explain, at least in part, these intriguing observations. Shortly after the role of PXR and CAR in the induction of CYP3A and CYP2B genes, respectively, had been established, it became evident that xenobiotic inducers could not be classified merely as being PXR- or CAR-specific activators. Indeed, many compounds are able to interact with both receptors as ligands, activators and/or inverse agonists.[86,122] For example, phenobarbital and 5β-pregnane-3,20 dione are activators of both CAR and PXR. On the other hand, PXR activators such as clotrimazole, androstanol and progesterone appear to be ligands/deactivators of CAR. Other notable examples include dexamethasone an activator of both GR and PXR, and RU486 an antagonist of GR but an activator of PXR. Similarly, it rapidly became evident that responsive genes could not be classified merely as being specifically activated by CAR or PXR. For example, the CYP3A4 PXR-response elements are also recognized and transcriptionally activated by CAR;[7,121] CAR/RXR heterodimer binds the rat CYP2B1 DR4 more efficiently than does the PXR/RXR, and the PXR/RXR binds more efficiently the rat CYP3A1 proximal DR3 PXRE than does the CAR/RXR.[123] The CAR response element of CYP2C9 is also targeted by PXR.[96] Another example is that of the PPAR-responsive element of enoyl-CoA hydratase /3-hydroxy acyl CoA-dehydrogenase gene that may be activated by CAR.[119] In fact, these receptors are able to bind to and transactivate different DNA response elements. The nature of these closely related but distinct elements may influence both the binding affinity of the receptor heterodimer and its ability to interact with the transcriptional machinery, thus providing a basis for the differential induction. In addition, differences in the response to CYP inducers may be due, in part, to the relative abundance of these receptors, as they compete for cofactors such as SCR-1 and RXRα, and ligands. The extent of such cross-talk would be gene- and inducer (ligand)-specific and would depend on the expression levels of these receptors in the cell.

In the 1980s, P450 scientists were fascinated by the finding that CYP proteins exhibited a broad and overlapping substrate specificity. Twenty years later, it has become clear that the nuclear receptors involved in the control of some of these CYP genes similarly exhibit a broad and overlapping ligand/activator specificity and response element specificity. This is likely to increase the organism's ability to detect and respond to a wide variety of potentially harmful xenobiotics.

3.5   Cross-talk between PXR and CAR and other nuclear receptors

During the last few years, cross-talk between PXR/CAR and other nuclear receptors including the glucocorticoid receptor (GR),[87,88,96,108,124-126] the vitamin D receptor (VDR),[127,128] and some of the receptors controlling bile acids homeostasis has been reported.[102-104,129,130] This aspect is extremely important as it leads first to an understanding of the consequences of various pathologies on drug metabolism and secondly to an explanation of other deleterious effects of drugs.

3.5.1 The Glucocorticoid Receptor

The early finding that glucocorticoids were potent inducers of CYP3A in rodents, suggested that GR was involved. However, in 1984 Guzelian and coworkers [131,132] clearly demonstrated that GR was not directly responsible for this induction and the actual role of this receptor remained the object of numerous controversies; for a more extensive review see ref. 126. Basically, GR might affect gene expression either directly through the transcriptional activation of GRE(s) in the promoter of "primary" glucocorticoid-responsive genes (such as tyrosine aminotransferase for example), or indirectly by up- (or down-) regulating one (or several) protein effector(s) controlling the transcription of "secondary" glucocorticoid-responsive genes. Literature survey indicates that only a few CYP genes may be considered as being primary GR-responsive. Recently, we demonstrated that *CYP2C9* enters this category.[95,96] We showed that this gene responds to submicromolar concentrations of dexamethasone in human hepatocytes and subsequently identified and characterized a functional GRE at position −1648/−1684 in the promoter. Other functional GREs have been identified in other *CYP* genes, including the human *CYP3A5* gene with two half-sites (TGTTCT) separated by 160 nucleotides at position −891/−1109[133] and rat *CYP3A1* at position −1960 (GGCACAnnnTGTTAT).[134] In contrast, many investigations suggest that several CYPs could be "secondary" glucocorticoid-responsive genes.

We and others have observed that induction of CYP2B, CYP2C and CYP3A by phenobarbital and rifampicin in cultured rodent or human hepatocytes is potentiated by pre-treatment of cells with dexamethasone or glucocorticoids.[107,108,125,135-140] Using human hepatocyte cultures, we showed that this potentiation was: (i) obtained with submicromolar concentrations of dexamethasone, (ii) time-dependent (at least 48 hours to reach a maximum) and iii) abolished if cells were treated with cycloheximide, an inhibitor of protein synthesis. This suggested that the potentiation was due to an essential protein factor, the expression of which was GR-dependent. We hypothesized and subsequently demonstrated that this potentiation was related to the induction of PXR and CAR by glucocorticoids.[88,108,125] We, and others[141] also reported the dexamethasone-mediated induction of RXR as well. Using a number of different tests, we showed that this effect was mediated through the direct transcriptional activation of these genes by GR. This has recently been confirmed by the identification of a functional GRE in the human CAR gene promoter.[124] Thus, CAR appears to be a primary glucocorticoid-responsive gene. PXR is suspected to be a primary glucocorticoid-responsive gene as well but the presence of a functional GRE in its promoter remains to be demonstrated. These studies lead us to propose the existence of a cascade of signal transmission GR-CAR/PXR-CYP2/3-monooxygenase activities (Figure 2). According to this model, the response to glucocorticoids may be divided into two steps (two different receptors) depending on the glucocorticoid concentration: (i) GR for physiological doses (submicromolar) and (ii) PXR for stress induced or pharmacological (bolus) doses (supramicromolar). Physiological levels of glucocorticoids activate GR which controls the basal levels of PXR, CAR and RXR. These receptors in turn can further function as rate limiting positive transcription regulators for glucocorticoid- and xenobiotic-mediated responses.

What might be the clinical consequences of such a cascade? Indeed, it is likely that any process that affects the expression and/or transcriptional activity of GR will lead to a down-regulation of PXR/CAR and eventually of CYP and other XMTS regulated by these xenoreceptors. A process which could possibly affect the cascade is inflammation through the action of cytokines. It has been known for many years that the ability of the liver to metabolize drugs and xenobiotics is reduced during inflammation and infection and this has been shown to be related to cytokines.[56,57,142-147] Several scenarios have been proposed for this effect, including induction of heme oxygenase and an increased rate of CYP degradation through a NO signalling pathway.[148,149] However, data obtained on primary human hepatocytes do not support these suggestions.[57] Thus, the molecular mechanisms responsible for this decrease remain to be identified. Our hypothesis is that cytokines exert their action by interfering with the GR-PXR/CAR-CYP2/3 cascade. This is based on the finding that: (i) pretreatment of primary hepatocytes by IL-6 reduces the expression of both PXR and CAR (in addition to CYPs) [147] and (ii) that transcription factors such as AP-1 and NFkB have been shown to bind to and inactivate GR.[150,151] It is well established that activation of these factors is mediated by a number of possible stimuli including cytokines (this point has recently been confirmed in our laboratory), oxidative stress, chemicals, etc. We therefore propose that pro-inflammatory cytokines activate NFkB (and/or AP-1) and that this leads to the inhibition of GR transcriptional activity which, in turn, results in the down-regulation of PXR and CAR and by cascade of CYP2/3A and other XMTS genes (Figure 2).

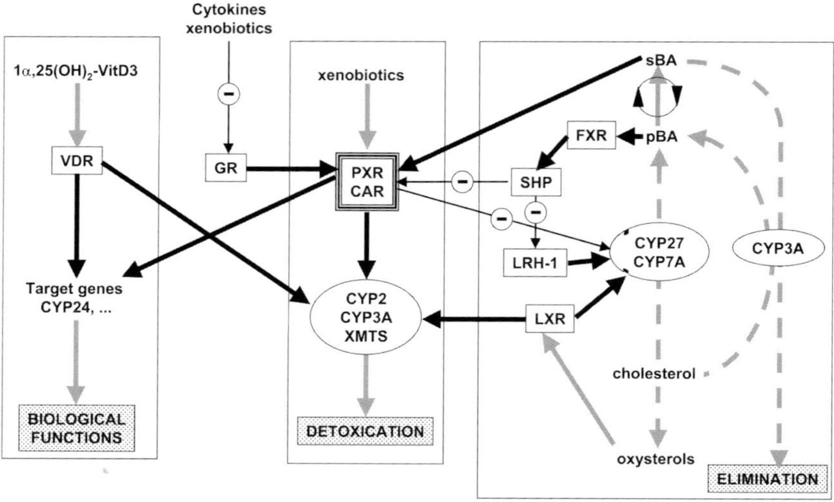

Figure 2. Schematic representation of cross-talk between xenoreceptors PXR and CAR and various nuclear receptors controlling endogenous metabolic pathways. VDR : vitamin D receptor ; GR : glucocorticoid receptor ; FXR : farnoid X receptor ; SHP : small heterodimer partner ; LRH-1 : liver receptor homolog-1; LXR : liver X receptor ; pBA and sBA : primary and secondary bile acids. Broad arrows : stimulatory pathways; thin arrows with (-): inhibitory pathways ; intermitent arrows : metabolic pathways. See text for detail (Section 3.4.)

3.5.2 The Vitamin D Receptor (VDR)

Schmiedlin-Ren et al.[152] were the first to suggest the possible implication of the vitamin D receptor (VDR, NR1I1) in CYP3A4 gene expression. They showed that 25-hydroxy vitamin D3 behaves as a transcriptional inducer of CYP3A4 in the colic carcinoma cell line Caco-2 as well as in the human intestinal LS180 cell line.[127] Moreover, 1α,25-dihydroxy vitamin D3 has been shown to induce the expression of CYP2B6, CYP2C9 and CYP3A4 mRNAs in human hepatocytes.[128] Although PXR is closely related to VDR with 63% homology of the amino acid sequences in their DNA binding domains, their ligand binding domains exhibit only 37% identity.[5] Consistent with this is the finding that human PXR is not activated by vitamin D or its hydroxy derivatives, and that VDR is not activated by PXR agonists, such as rifampicin or phenobarbital.[128] This raises the question of how vitamin D3 derivatives induce CYP3A4 and other CYP2B/2C genes?

Upon vitamin D hydroxy derivative binding, VDR forms a heterodimer with RXRα which then transactivates vitamin D response elements (VDRE) present in the regulatory region of the target genes.[153,154] The consensus VDRE is an imperfect direct repeat of a core hexanucleotide sequence, (G/A)GGT(G/C)A, with a three-nucleotide spacer (DR3). We and others have shown that the heterodimer VDR/RXRα is able to bind to and transactivate the PXR/CAR responsive elements previously identified in the 5' promoter of CYP2B6, CYP2C9, and CYP3A4 genes including ER6, DR3 and DR4 motifs.[127,128] This is consistent with the notion introduced in a previous section that nuclear receptors may bind to and transactivate responsive elements other than their consensus element, and clearly offers another example of nuclear receptor versatility. Interestingly, previous investigations identified other VDRE motifs including a DR4 (for which VDR exhibited a higher affinity than for DR3), a DR6 and an inverted palindrome IP9.[155-157]

From a functional point of view, this is expected to generate interferences between the respective signalling pathways controlled by these receptors, that is : (i) control of CYP2/3A gene expression by vitamin-D hydroxy-derivatives through VDR and (ii) control of VDR-responsive genes by xenobiotics through PXR/CAR. Although the first alternative is not likely to produce significant effects due to the very low level of vitamin-D hydroxy-derivatives, the second one is likely to occur and might well be of real concern. Indeed, prolonged therapy with rifampicin has been shown to cause vitamin-D deficiency. Moreover, rifampicin and phenobarbital are two of the drugs most frequently administered to patients suffering from osteomalacia, a disease characterized by a defect in bone mineralization due to a decreased blood level of vitamin D.[158-161] Since neither CYP2B6, CYP2Cs nor CYP3A4 are involved in the metabolism of vitamin D, induction of these CYPs cannot explain these observations. In fact, the major route of degradation of vitamin D is the oxidation of the side chain catalyzed by vitamin D-24 hydroxylase (CYP24), an enzyme which is highly induced by 1α,25 dihydroxy vitamin D through VDR.[162] It is therefore possible that by activating PXR and/or CAR, xenobiotics and drugs stimulate the expression of VDR target genes thus producing the observed deficiency in Vitamin D leading to osteomalacia. This is important as several such drugs are currently in clinical use, including notably barbiturates, antibiotics, HIV protease inhibitors, isophosphamide, and carbamazepine.

3.5.3 SHP and FXR

PXR activates the CYP2 and CYP3A families not only in response to xenobiotics but also in response to endogenous molecules such as catatoxic steroids and secondary bile acid derivatives including lithocholic acid and ketolithocholic acid derivatives.[163] In addition, PXR regulates the expression of genes involved in the biosynthesis, transport, and metabolism of bile acids including CYP7α,[163] Oatp2,[102,103] and CYP3A.[130] Thus, it has been proposed that PXR serves as a physiological sensor not only for xenobiotics but also for secondary bile acids, by coordinately regulating the expression of genes involved in the reduction of the concentration of these toxic compounds.

Recently we observed that the short heterodimer partner SHP,[164] originally shown to bind to and inhibit CAR[165] and other nuclear receptors[164,166-173] is a negative regulator of PXR transcriptional activity as well (Pascussi et al. to be published). SHP lacks a conventional DBD and is mostly expressed in the liver [165] where its expression is under the control of FXR (farnesoid X receptor). FXR is activated by primary bile acids such as cholic acid and chenodeoxycholic acid.[174] CYP7α the rate-limiting enzyme in bile acid-biosynthesis, is known to be subject to primary bile acid feedback regulation *via* SHP.[166] Using *in vitro* tests and primary hepatocytes we have shown that SHP interacts with PXR in a ligand-dependent manner and inhibits its transcriptional activity. This inhibition is reversed by SRC-1 a transcriptional coactivator of PXR. Pretreatment of hepatocytes with chenodeoxycholic acid results in the induction of SHP, a concomitant down-regulation of CYP7α, and a decrease in CYP3A4, CYP2B6 and CYP2C9 gene induction in response to the PXR activator rifampicin. Two complementary mechanisms can thus be proposed: (i) SHP interacts directly with PXR and weakens its binding to DNA as proposed previously for RAR; (ii) SHP blocks the AF-2 activation domain as proposed previously for ER.[168] Previous observations by others support these conclusions. Handschin et al. reported that cotreatment of a chicken hepatoma cell line LMH with cholic acid or chenodeoxycholic acid and phenobarbital or clotrimazole (activators of CAR and PXR respectively) reduced CYP3A induction.[175] Paolini et al. reported that taurochenodeoxycholic acid, a FXR ligand,[174] reduced CYP3A-associated monooxygenase activities *in vivo* in rats.[176] Finally, Schuetz et al. observed a strong induction of CYP3A11 and CYP2B10 in FXR -/- mice.[99] Collectively, these data suggest that SHP interacts with and inhibits the transcriptional activity of PXR, in addition to CAR. As both receptors control the inducible expression of CYPs and other genes involved not only in the metabolism and elimination of xenobiotics but also in the biosynthesis and catabolism of bile acids, it is likely that a functional interaction occurs between bile acid homeostasis and the xenobiotic-mediated CYP induction through SHP.

4   *IN VITRO* SYSTEMS TO SCREEN CYP INDUCERS

In contrast to the screening of enzyme inhibitors which may be made through the use of subcellular fractions such as liver microsomes or cytosol, recombinant or purified enzymes, the screening of inducers requires more sophisticated systems/approaches including: (i) *in vivo* studies with specific drug probes, (ii) fully functional cellular systems in which gene expression is maintained, or iii) specific *in vitro* methods for evaluating the capacity of the given drug to activate specific xenoreceptors. Data obtained with animal models (rat, mouse) either *in vivo* or *in vitro* may not be relevant to the human situation because the xenobiotic-mediated activation of xenoreceptors appears to be species specific.[89] For example, rifampicin and omeprazole, two inducers of *CYP3A* and *CYP1A*, respectively, in man, are not inducers in the rat or the mouse; likewise pregnenolone 16α-carbonitrile, a potent CYP3A inducer in the rat, is not an inducer in man.[32,177] Thus, the prediction of enzyme induction in man necessitates the use of experimental systems of human origin.

The mechanistic considerations presented above aid the planning and design of different experimental approaches for the screening of inducers, using more practical *in vitro* and cell bioassays. These will be briefly described in the next sections, the reader being referred to published papers for more detailed experimental protocols. Since CYP genes are in general more strongly responsive to xenobiotics than other XMTS, the demonstration that a particular molecule is a CYP inducer should be a sufficient criterion to consider this molecule as a so-called enzyme inducer.

4.1   Direct receptor-ligand binding assays

Several methods have been used to determine whether the tested molecule is a ligand of a receptor. Basically, a radiolabelled high affinity ligand (used over a large range of concentrations) is incubated with an *in vitro*-translated, partially purified or enriched subcellular fraction of the receptor. The mixture is then fractionated by an appropriate method (centrifugation on sucrose gradient, gel filtration) and the binding to the receptor is monitored by analysis of fraction radioactivity. The presence of increasing concentrations of unlabelled competitive ligand results in a decrease or disappearance of the radioactive peak, thus allowing evaluation of binding parameters. Such experiments have been carried out with AhR and PXR/CAR.[86,178,179] More recently a high throughput *in vitro* screening for PXR activation has been described using a scintillation proximity assay. This assay is based on the use of streptavidin-polyvinyltoluene beads and a biotinylated PXR ligand-binding domain/SRC-1 peptide complex (see Paragraph 4.3.2.), in the presence of tritiated SR12813, a high affinity PXR ligand.[180] Both saturation and competition binding assays can be performed.

4.2 Enhanced green fluorescent protein (EGFP)-based recombinant cell bioassay

This assay has been recently used for the rapid screening of combinatorial chemical libraries for AhR activatory ligands.[37] It consists of a mouse hepatoma cell line (Hepa 1c 1c7) stably transfected with a reporter plasmid harboring the EGFP gene, downstream of a AhR-dependent module consisting of four XREs (see Paragraph 3.1.). Both AhR and ANRT are expressed and functional in this cell line. When cells are treated with dioxin or other AhR agonists, EFGP is induced which is easily detected by monitoring the increased fluorescence in the cells, making the assay suitable for high throughput screening applications. The assay identifies compounds which are able to trigger the transcriptional activity of the AhR but which are not necessarily ligands of the receptor. This bioassay might be adapted to the screening of ligands to human AhR by using human cell lines such as HepG2 or HuH7 for instance, or to other receptors such as PXR or CAR. However, in this last case, cells will have to be transfected with receptor-expression plasmids since neither PXR nor CAR are expressed at a significant level in hepatoma or other cell lines.

4.3 Ligand-induced receptor-coactivator interaction assays

After ligand binding, nuclear receptors undergo a conformational change that results in the recruitment of coactivators such as TRIP2, SRC-1, TIF2, pCIP, and the release of corepressors such as NcoR, and silencing mediator SMRT.[157] Coactivator recruitment is a key step in the assembly of an active transcription complex [181] and can be used as a screening test for inducers. Coactivators utilize α-helical motifs (receptor interaction domains, RIDs) to make direct contact with the conserved ligand-binding domain (LBD) located at the C terminus of the receptor.[182] The LBDs of several nuclear receptors, including PXR,[183] have been crystallized and found to possess a common overall structure, including 12 α-helices (H1 to H12) that enclose the ligand-binding pocket.[184] The ability to discriminate between coactivators and corepressors is determined by the position of H12, which contains the ligand-dependent activation domain (AF2). Several methods have been used to determine whether a tested molecule is able to induce a receptor-coactivator interaction.

4.3.1 Mammalian two hybrid-assay

A mammalian two hybrid-assay has been used to demonstrate functional interaction between the LBD of receptors and a portion of SRC-1 that contains the three canonical receptor interaction domains (RID 1, RID 2, and RID 3). Mammalian cells are co-transfected with three plasmids harboring inserts encoding: (i) a Gal4-SRC-1 hybrid in which the DNA binding domain of Gal4 is fused in frame to SRC-1, (ii) a VP16-LBD receptor hybrid in which the LBD of the nuclear receptor is fused in frame to the transactivating domain of the Herpes Simplex virus protein VP16, and (iii) a reporter construct containing four copies of a Gal4 response element (UAS_Gx4) upstream of the luciferase reporter gene. After transfection, cells are treated with the molecule to be tested. Should the molecule be a ligand, its binding promotes association between the receptor LBD and SRC-1 which eventually leads to the activation of the reporter gene. [185,186]

4.3.2 In vitro protein interaction biochemical assay

The receptor-interacting domain of the coactivator SRC-1 has been mapped to a short motif with the amino acid sequence LXXLL, where L is a leucine and X is any amino acid.[187] Fragments of SRC-1 or short synthetic peptides containing one or more LXXLL motifs bind the LBD of nuclear receptors in a ligand-dependent manner. Hence, the association of LXXLL-containing peptides with a nuclear receptor LBD can be used to obtain a quantitative measure of ligand-receptor binding.[188] This approach has been used in simplified biochemical assays such as *in vitro* protein interaction assays (GST-pull down, or CARLA) or enzyme-linked immunosorbent assay (ELISA). These assays measure the interaction between bacteria-expressed and purified GST-LBD chimeras in which the LBD of the receptor is fused in frame to gluthathione *S* transferase (GST) and a ^{35}S-labelled [6,83,189] or biotin-labelled peptide containing a LXXLL motif of SRC-1.[190]

4.3.3 Fluorescence resonance energy transfer assay

Fluorescence resonance energy transfer (FRET) has been used for the screening of antagonists or agonists of nuclear receptors in high-throughput assays based on the interaction between nuclear receptor LBD and the LXXLL motif of SRC-1.[191,192] In these assays, the nuclear receptor LBD, and the SRC-1 LXXLL motif are labeled with two fluorophores, i.e. europium chelate and allophycocyanin, respectively. The interaction between both peptides, stimulated by the presence of ligand, results in close proximity of the two fluorophores. This eventually leads to energy transfer resulting in fluorescence.[10,193]

4.4   Cell lines for the direct screening of CYP1A inducers

Both the expression and activity of AhR and ARNT are maintained in cell lines such as HepG2 and Caco-2 (as well as its different clones such as TC-7), in contrast to PXR, CAR and GR the expression of which is strongly down-regulated. These cell lines can therefore be used to screen directly for putative CYP1A inducers, by evaluating the expression of endogenous CYP1A1 mRNA either by northern blot, quantitative RT-PCR or even by measuring probe substrate metabolic activity.[33,194] To our knowledge, there is no case in which the behaviour of a CYP1A inducer has been shown to be different in these cell lines compared with its behaviour *in vivo* or in primary cultures of hepatocytes.

4.5   Cell line co-transfections: gene reporter assays

In this type of assay, a cell line (such as HepG2, HuH7 or CV1) is cotransfected with a receptor expression plasmid (for example encoding for PXR) and a reporter plasmid in which the reporter gene (for example luciferase) is placed under the control of a receptor-specific response element (for example CYP3A4-PXRE for PXR, see Paragraph 3.2.2.). Interestingly, the expression of RXR, the partner of this receptor, is maintained in these cell lines. After transfection, cells are treated with the molecule to be tested and reporter gene activity is evaluated.[83,96]

A special comment is necessary for CAR at this point. This receptor exhibits two peculiarities when compared to PXR or AhR. First, CAR exhibits full transcriptional activity in the absence of a ligand, as indicated in Paragraph *3.2.3*. Second, transfection of a CAR-expressing plasmid in cell lines leads to the spontaneous translocation of the receptor into the nucleus, in contrast to the normal situation in which the receptor is sequestered in the cytosol and translocates into the nucleus upon activation by xenobiotics or endogenous activators.[11] Under these conditions, when co-transfection experiments such as those described in the previous paragraph are carried out with CAR (including a CAR-response element such as the PBREM of CYP2B genes for instance) full activity of the reporter gene is measured in the absence of ligand or activator. Therefore this test is inefficient for the screening of ligands or activators. However, this activity is inhibited if cells are treated with inverse agonists such as androstenol or clotrimazole (see Paragraph 3.2.3.). The assay can therefore be used to detect CAR inverse agonists.[86] Interestingly, some CAR activators such as TCPOBOP and estrogens (but not phenobarbital), are able to reverse the inhibition of transcriptional activity promoted by inverse agonists.[86] CAR activators may therefore be screened in the following way. Cells are co-transfected with a CAR expression plasmid and a gene reporter plasmid in which the reporter gene is placed under the control of a CAR-specific response element (for example CYP2B6-PBREM). After transfection, cells are treated with an inverse agonist both in the absence and presence of the molecule to be tested and reporter gene activity is evaluated.

4.6 Primary human hepatocytes for the direct screening of CYP inducers

Ideally, the screening of CYP inducers should be carried out using primary cultures of normal human hepatocytes. Indeed, previous reports from different laboratories have shown that, provided adequate culture conditions are used to allow the expression of the hepatic differentiated phenotype, these cultures constitute the best model to predict with a reasonable confidence whether a compound is an inducer.[195-200] In addition, they enable the direct measurement of enzyme activity considered to be the most relevant parameter to characterize enzyme inducers.[201,202] However, their use is limited for different reasons including: the scarcity of human liver tissue of adequate quality for hepatocyte isolation, difficulties in obtaining robust and reliable primary cultures, and the impossibility of developing high-throughput bioassays. In addition, available data suggest that the normal phenotype of cryopreserved hepatocytes is not fully conserved after thawing and the present status recommendation is to use these cells only for short-term experiments[203] which is not compatible with inducer screening. For the majority of drug companies, these limitations should therefore restrain the use of these cultures to final confirmation of the conclusions drawn from early discovery screen assays described in the previous sections.

When primary human hepatocytes are being used for inducer screening two problems have to be emphasized: (i) the inter-individual (liver donor) variability in the response to inducers; (ii) the *in vivo* significance of the results. A typical example in this respect is the case of eletriptan.[199] This molecule was found to be a CYP3A4 inducer in approximately 60% of hepatocyte cultures tested. The induction was concentration-dependent with significant increases in CYP3A4 protein level and activity at 25 and 50 µM eletriptan, but not detectable for concentrations of 5 µM and lower. Because the maximum blood concentration of eletriptan after a therapeutic dose is 0.5 µM, we predicted that the induction observed *in vitro* in primary hepatocytes should not be clinically significant *in vivo*. This was eventually demonstrated by using the 6β-hydroxycortisol test [204] in patients before and after treatment with this molecule. This case shows that some precautions have to be taken when conducting such studies.

Although a potent inducer (such as rifampicin for example) will easily be recognized, the definition of a "significant clinical response" might require additional information. For instance, the CYP1A-inducing capacity of omeprazole was first reported using human hepatocytes in primary cultures.[32] Based on these *in vitro* data, the CYP1A-induction was further investigated *in vivo* in humans following standard therapeutic doses. Results were controversial. CYP1A-induction was confirmed using the caffeine test following a four week-treatment with omeprazole,[205] but was not observed following standard therapeutic doses.[206] Some additional studies performed in CYP2C19-phenotyped patients (the enzyme responsible for the major part of omeprazole metabolism) further demonstrated that following a 120 mg/day oral administration of omeprazole, *N*-3-demethylation of caffeine was increased by less than 30 % in CYP2C19-extensive metabolizers and by 40 % in CYP2C19-poor metabolizers.[205,207] All these data demonstrate that omeprazole is an inducer of CYP1A *in vivo*, as suggested by *in vitro* studies, only following a high dose administration or in patients deficient in active CYP2C19. On the other hand, the CYP1A induction caused by omeprazole administered at a standard dose of 40 mg/day is unlikely to have clinical consequences.

Rifampicin, a widely used antibiotic for the treatment of tuberculosis, is a potent CYP3A inducer both *in vitro* using primary hepatocytes [196,198,202,208] and *in vivo* as demonstrated by the decreased bioavailability of numerous drugs, all of them being substrates for CYP3A4.[2] The good correlation between both *in vitro* and *in vivo* data with this molecule results from the fact that rifampicin serum concentration after a standard therapeutic dose of 600 mg is in the range of 2-30 μM, that is in a range of concentrations giving a maximum response *in vitro*.[209] These different results emphasize the fact that the induction potential of a drug can be determined using human hepatocytes in primary cultures, but that the pharmacokinetic and therapeutic consequences will be dependent on factors such as therapeutic dose, frequency of administration, plasma and hepatic levels, and genetic status of the patients.

It is generally agreed that induction experiments with primary hepatocytes should be conducted as follows: (i) use prototypical inducers such as omeprazole (20 μM), 3-methylcholanthrene (5 μM), β-naphthoflavone (5 μM) or dioxin (10 nM) for CYP1A-, and rifampicin (10 μM) or phenobarbital (500 μM) for CYP2B6, CYP2C9/8 and CYP3A4-mediated activities, in parallel with the drug to be tested; (ii) use at least a two-log range of concentration of the drug tested, encompassing the actual maximum blood concentration after normal therapeutic dose; (iii) use a 3-5 day treatment period; (iv) use at least three preparations from different liver donors; (v) evaluate enzyme activity directly in the culture dish using CYP-specific substrates.[202]

5   *IN VIVO* EVALUATION OF CYP INDUCTION IN MAN

Induction of CYP isoforms can also be evaluated directly in healthy volunteers or patients, following repeated administration of a putative inducer drug, using different substrate-probes. In such studies, all subjects have to be not taking any medication regularly, including oral contraceptives. Moreover they must abstain from ethanol, caffeine and grapefruit juice for 48 hours before each period of study. In general, CYP1A2 and CYP3A4 enzymes are the most strongly responsive to inducers, although other CYPs notably CYP2C9 are also responsive to the same compounds.[210,211] Therefore, the demonstration that a tested molecule is a CYP1A2 or 3A4 inducer is a sufficient criterion to consider this molecule as a so-called enzyme inducer. However, it must be kept in mind that drug transporters notably MDR1 [98] and MRP2 [100] are likely to be inducible *in vivo* and this may interfere with CYP induction and eventually affect the clinical significance of interactions. In this section, we will focus on *in vivo* methods for evaluating CYP1A2 and 3A4 induction.

5.1 Induction of CYP1A2 isoform

The role of CYP1A2 in the metabolism of drugs is increasingly acknowledged, this isoform being involved in the metabolism of several clinically important drugs including notably psychotropic drugs (clozapine, and olanzapine).[212] CYP1A2 is inducible by smoking, environmental factors and xenobiotics such as omeprazole, carbamazepine and phenobarbital.[213] Hence, all these factors cause variability in psychotropic drug response. The availability of a reliable probe drug providing a measure of CYP1A2 activity is therefore of considerable interest. The high affinity component of phenacetin-*O*-deethylation appears to be entirely attributable to CYP1A2.[214,215] This analgesic drug has been given to healthy volunteers as a probe for CYP1A2 activity. These studies revealed that phenacetin metabolism was increased by cigarette smoking[216] and by consumption of charcoal-broiled meat.[217] Phenacetin, however, is no longer used for CYP1A2 inducer evaluation since it has been withdrawn from the market and cannot be given for ethical reasons.

Among the various metabolic probes which have been used for investigating CYP1A2 activity, caffeine is widely used for numerous reasons such as its ubiquitous consumption, its rapid and complete gastrointestinal absorption, its distribution throughout the total body water, and its low plasma binding. In addition, caffeine has a short half-life, exhibits minimal renal elimination and its biotransformation is virtually confined to the liver, *via* the CYP1A2 isoform. Lelo et al.[218] reported that caffeine *N*-3-demethylation was the most prominent reaction in human subjects and accounted for 83.9 % ± 5.4 % of all its demethylation processes, including *N*-1-demethylation (12.1 % ± 4.1 %) and *N*-7-demethylation (4.1 % ± 1.4 %). Since CYP1A2 is the major enzyme accounting for caffeine metabolism and specially the *N*-3-demethylation (1,3,7-trimethylxanthine) to paraxanthine (1,7-dimethylxanthine),[215] caffeine has become a probe of choice for the quantitative measurement of CYP1A2 *in vivo*[219,220]. Quantification of caffeine partial clearance through production of 1,7-dimethylxanthine, *i.e.* corresponding to the specific *N*-3-demethylation process, offers itself as the primary standard of CYP1A2 activity. This is obtained by determining the area under the curve of caffeine blood levels *versus* time, following a single caffeine oral dose.

A second approach, which is also widely used, is the "caffeine breath test". A precise dose of ^{13}C- or ^{14}C-caffeine (at the 3-methyl group) is administered and the 2-hour cumulative exhalation of labelled carbon dioxide is measured. CYP1A2 activity measured by this test exhibited excellent correlation with the activity evaluated from systemic caffeine clearance. This test revealed high CYP1A2 values in smokers [221] and in subjects exposed to polybrominated biphenyls,[222] as well as after CYP1A2 induction by omeprazole.[207] The major drawback of the test is the requirement for radiolabelled caffeine and the need for specialized equipment for measuring the radiolabelled carbon dioxide in exhaled air.[220]

5.2   Induction of CYP3A4 isoform

Different approaches have been developed to evaluate CYP3A4 induction in man.

The urinary "6β-hydroxycortisol-to-cortisol" ratio has been used as a general indicator of CYP3A induction since cortisol 6β-hydroxylation is catalyzed by CYP3A4.[204] For this purpose, urine is collected over a 0-8 hour period, before and after repeated administration (for 5-7 days) of the drug being tested, and are analyzed for the 6β-hydroxycortisol-to-cortisol ratio. An increase in this ratio after treatment, when compared to the ratio before treatment in the same subject, is indicative of CYP3A4 induction. The advantage of this test is that it does not necessitate the administration of a probe. However, there are two caveats to be borne in mind with this approach. First, besides the liver, kidney appears to be another source of 6β-hydroxycortisol [223] and second, cortisol is a MDR1 substrate.[224] Hence, the 6β-hydroxycortisol-to-cortisol ratio *per se* cannot be used as a "pure" phenotypic marker of liver CYP3A activity.

The urinary "dapsone recovery" ratio has also been used to phenotype patients for CYP3A activity, since dapsone *N*-hydroxylation appears to be mediated by CYP3A4.[225] The dapsone recovery ratio was determined in patients/healthy volunteers, following a single oral dose of 100 mg dapsone. Urine was collected during the 0-8 hour period after drug administration and both dapsone and its *N*-hydroxy metabolites were quantified. The "dapsone hydroxylamine-to-dapsone" ratio was then determined.

The erythromycin breath test (ERBT) is based on the observation that CYP3A-mediated *N*-demethylation of erythromycin leads ultimately to the formation of carbon dioxide.[226] Generally, 2-5 µCi [^{14}C-*N*-methyl]-erythromycin (dissolved in 5 % dextrose solution) were administered by intravenous injection, 24 hours following 5-7 days of repeated oral administration of the drug investigated. Exhaled breath air samples were collected by 60 minute-fractions over 4 hours after the administration of radiolabelled erythromycin. The area under the curve of the [^{14}C]-excretion *versus* time profile, extrapolated to infinity was subsequently estimated by the trapezoidal rule to estimate the total percentage of the administered dose of radioactivity excreted in the breath. Data obtained in untreated healthy volunteers were then statistically analyzed relative to the drug-treated group. Clinical studies were generally performed on 10-12 subjects judged to be healthy on the basis of medical history, physical examination and routine clinical laboratory determinations. Each subject was randomized before the beginning of the study. Although widely used, the ERBT has several practical limitations, including exposure to radioactivity, the intravenous administration of the drug and the need for a period of supervised study. This approach is therefore not suitable for use on subjects in large number. An additional limitation is the requirement for systemic administration, so that intestinal CYP3A activity is not evaluated.

Table I: *In vivo* inducers of CYP isoforms in humans Adapted from reference 236

Inducer drug	Plasma Conc. (µM)	CYP-induced	Drug affected
Omeprazole	1-10	CYP1A1, 1A2	-
Carbamazepine	20 - 40	CYP3A	Praziquantel, Itraconazole, Cyclosporin A
Glucocorticoids (dexamethasone)		CYP3A	Praziquantel, Erythromycin, Cyclosporin A
Phenytoin	> 10	CYP2C CYP3A	Warfarin, Praziquantel, Quinidine, Cyclosporin A
Phenobarbital	40 - 130	CYP2C CYP3A	Warfarin, Cyclosporin A
Rifampin	10	CYP2C	Tolbutamide, Warfarin
Rifampin	10	CYP3A	Triazolam, Oral Contraceptives, Cyclosporin, tamoxifen
Troglitazone	7	CYP3A	Cyclosporin A, Terfenadine

These three tests were evaluated in parallel on healthy subjects. No statistically significant correlation was observed among any of these measurements (dapsone recovery ratio *versus* erythromycin breath test, urinary 6 β-hydroxycortisol/ cortisol ratio *versus* erythromycin breath test, and urinary 6 β-hydroxycortisol/ cortisol ratio *versus* dapsone recovery ratio). This lack of correlation most likely results from the fact that these probes are administered through different routes and are markers of either hepatic or hepatic and extrahepatic metabolism.[227-230] Pharmacokinetic approaches using other CYP3A4 probe-substrates including felodipine, simvastatin,[231] lovastatin[232] and midazolam[233,234] are also being currently used following oral administration to evaluate the CYP3A4 activity *in vivo*. These tests reflect both intestinal and hepatic CYP3A metabolic activity. Terfenadine is no longer used in this respect due to the possibility of severe drug interactions.[235]

Nowadays in many drug companies, the routine analysis to explore the potential of a drug as a putative CYP3A4 inducer is based on the pharmacokinetics of midazolam.[233,234] In this test, healthy volunteers receive first a single intravenous dose of the CYP3A4-probe midazolam and its pharmacokinetic profile is monitored. Next, following a 10-day washout-period, they are treated for 5-10 days (depending on the pharmacokinetic behavior of the drug, especially its elimination half-life) with the putative enzyme inducer and the new pharmacokinetic profile of midazolam is monitored. In the presence of a CYP3A4 inducer, the clearance of intravenous midazolam increases and the area under the curve of midazolam *versus* time decreases. These studies can also be performed according to a double-blind crossover study, using rifampicin as a positive control.

6   CONCLUSION

Recent findings on the function and regulation of xenoreceptors controlling CYP gene induction in response to xenobiotics and drugs have considerably enlarged our understanding of the multiple ways by which the detoxication function may be regulated. New tools and experimental approaches are now available to screen enzyme inducers both *in vitro* and *in vivo*. It appears that many physiopathological factors are able to interfere with the process of enzyme induction through cross-talk, interference, cooperation between these xenoreceptors and other signalling pathways or transcription factors. From a practical point of view this means that conclusions drawn from healthy subjects regarding the induction potential of a drug might be not relevant in patients. Indeed, stimuli such as hypoxia, oxidative stress, cytokines, bile salts accumulation, or glucocorticoids might affect drastically the xenobiotic signalling pathways which control the detoxication function through AhR, PXR and CAR. Conversely, it must be realized that drugs shown to be activator of AhR, PXR and CAR are not only likely to produce drug interactions due to enzyme induction. Indeed, such molecules may also be expected to affect various metabolic pathways through interference between xenoreceptors and the nuclear receptors controlling these pathways. The long story of enzyme induction continues, while that of the toxicity of xenobiotics resulting from interference between signalling pathways is just beginning.

REFERENCES

1. Lin, J. H.; Lu, A. Y. *Clin. Pharmacokinet.* **1998**, *35*, 361-390.
2. McInnes, G. T.; Brodie, M. J. *Drugs* **1988**, *36*, 83-110.
3. Hankinson, O. *Ann. Rev. Pharmacol Toxicol* **1995**, *35*, 307-340.
4. Nebert, D. W.; Jones, J. E. *Int. J. Biochem.* **1989**, *21*, 243-252.
5. Kliewer, S. A.; Moore, J. T.; Wade, L.; Staudinger, J. L.; Watson, M. A.; Jones, S. A.; McKee, D. D.; Oliver, B. B.; Willson, T. M.; Zetterstrom, R. H.; Perlmann, T.; Lehmann, J. M. *Cell* **1998**, *92*, 73-82.
6. Lehmann, J. M.; McKee, D. D.; Watson, M. A.; Willson, T. M.; Moore, J. T.; Kliewer, S. A. *J. Clin. Invest.* **1998**, *102*, 1016-1023.
7. Sueyoshi, T.; Kawamoto, T.; Zelko, I.; Honkakoski, P.; Negishi, M. *J. Biol. Chem.* **1999**, *274*, 6043-6046.
8. Waxman, D. J. *Arch. Biochem. Biophys.* **1999**, *369*, 11-23.
9. Zelko, I.; Negishi, M. *Biochem. Biophys. Res. Commun.* **2000**, *277*, 1-6.
10. Moore, J. T.; Kliewer, S. A. *Toxicology* **2000**, *153*, 1-10.
11. Honkakoski, P.; Zelko, I.; Sueyoshi, T.; Negishi, M. *Mol. Cell Biol.* **1998**, *18*, 5652-5658.
12. Gonzalez, F. J. *Trends Pharmacol. Sci.* **1992**, *13*, 346-352.
13. Shimada, T.; Yamazaki, H.; Mimura, M.; Inui, Y.; Guengerich, F. P. *J. Pharmacol. Exp. Ther.* **1994**, *270*, 414-423.
14. Daniels, N. J.; Dover, J. S.; Schachter, R. K. *Lancet* **1984**, *2*, 639.
15. Campana, C.; Regazzi, M. B.; Buggia, I.; Molinaro, M. *Clin. Pharmacokinet.* **1996**, *30*, 141-179.
16. Modry, D. L.; Stinson, E. B.; Oyer, P. E.; Jamieson, S. W.; Baldwin, J. C.; Shumway, N. E. *Transplantation* **1985**, *39*, 313-314.
17. Slattery, J. T.; Nelson, S. D.; Thummel, K. E. *Clin. Pharmacol. Ther.* **1996**, *60*, 241-246.
18. Shenfield, G. M. *Med. J. Aust.* **1986**, *144*, 205-211.
19. Back, D. J.; Breckenridge, A. M.; Crawford, F.; MacIver, M.; Orme, M. L.; Park, B. K.; Rowe, P. H.; Smith, E. *Eur. J. Clin. Pharmacol.* **1979**, *15*, 193-197.
20. Conney, A. H. *Pharmacol Rev* **1967**, *19*, 317-366.
21. Bonfils, C.; Dalet-Beluche, I.; Maurel, P. *Biochem. Biophys. Res. Commun.* **1982**, *104*, 1011-1017.

22. Elshourbagy, N. A.; Guzelian, P. S. *J. Biol. Chem.* **1980**, *255*, 1279-1285.

23. Lu, A. Y.; Somogyi, A.; West, S.; Kuntzman, R.; Conney, A. H. *Arch. Biochem. Biophys.* **1972**, *152*, 457-462.

24. Gibson, G. G.; Orton, T. C.; Tamburini, P. P. *Biochem. J.* **1982**, *203*, 161-168.

25. Orton, T. C.; Parker, G. L. *Drug Metab. Dispos.* **1982**, *10*, 110-115.

26. Johansson, I.; Ekstrom, G.; Scholte, B.; Puzycki, D.; Jornvall, H.; Ingelman-Sundberg, M. *Biochemistry* **1988**, *27*, 1925-1934.

27. Ryan, D. E.; Ramanathan, L.; Iida, S.; Thomas, P. E.; Haniu, M.; Shively, J. E.; Lieber, C. S.; Levin, W. *J. Biol. Chem.* **1985**, *260*, 6385-6393.

28. Nebert, D. W. *Crit Rev Toxicol* **1989**, *20*, 153-174.

29. Shimada, T.; Inoue, K.; Suzuki, Y.; Kawai, T.; Azuma, E.; Nakajima, T.; Shindo, M.; Kurose, K.; Sugie, A.; Yamagishi, Y.; Fujii-Kuriyama, Y.; Hashimoto, M. *Carcinogenesis* **2002**, *23*, 1199-1207.

30. Whitlock, J. P., Jr. *Annu. Rev. Pharmacol. Toxicol.* **1999**, *39*, 103-125.

31. Sogawa, K.; Fujii-Kuriyama, Y. *J. Biochem. (Tokyo)* **1997**, *122*, 1075-1079.

32. Diaz, D.; Fabre, I.; Daujat, M.; Saint Aubert, B.; Bories, P.; Michel, H.; Maurel, P. *Gastroenterology* **1990**, *99*, 737-747.

33. Daujat, M.; Charrasse, S.; Fabre, I.; Lesca, P.; Jounaidi, Y.; Larroque, C.; Poellinger, L.; Maurel, P. *Eur. J. Biochem.* **1996**, *237*, 642-652.

34. Fontaine, F.; Delescluse, C.; de Sousa, G.; Lesca, P.; Rahmani, R. *Biochem. Pharmacol.* **1999**, *57*, 255-262.

35. Gradelet, S.; Astorg, P.; Pineau, T.; Canivenc, M. C.; Siess, M. H.; Leclerc, J.; Lesca, P. *Biochem. Pharmacol.* **1997**, *54*, 307-315.

36. Ledirac, N.; Delescluse, C.; Lesca, P.; Piechocki, M. P.; Hines, R. N.; de Sousa, G.; Pralavorio, M.; Rahmani, R. *Toxicol. Appl. Pharmacol.* **2000**, *164*, 273-279.

37. Nagy, S. R.; Liu, G.; Lam, K. S.; Denison, M. S. *Biochemistry* **2002**, *41*, 861-868.

38. Dzeletovic, N.; McGuire, J.; Daujat, M.; Tholander, J.; Ema, M.; Fujii-Kuriyama, Y.; Bergman, J.; Maurel, P.; Poellinger, L. *J. Biol. Chem.* **1997**, *272*, 12705-12713.

39. Hahn, M. E. *Comp. Biochem. Physiol. C Pharmacol. Toxicol. Endocrinol.* **1998**, *121*, 23-53.

40. Burbach, K. M.; Poland, A.; Bradfield, C. A. *Proc. Natl. Acad. Sci. U S A* **1992**, *89*, 8185-8189.

41. Kazlauskas, A.; Sundstrom, S.; Poellinger, L.; Pongratz, I. *Mol. Cell Biol.* **2001**, *21*, 2594-2607.

42. Meyer, B. K.; Pray-Grant, M. G.; Vanden Heuvel, J. P.; Perdew, G. H. *Mol. Cell Biol.* **1998**, *18*, 978-988.

43. Ma, Q. *Curr. Drug. Metab.* **2001**, *2*, 149-164.

44. Pollenz, R. S. *Chem. Biol. Interact.* **2002**, *141*, 41-61.

45. Song, Z.; Pollenz, R. S. *Mol. Pharmacol.* **2002**, *62*, 806-816.

46. Beischlag, T. V.; Wang, S.; Rose, D. W.; Torchia, J.; Reisz-Porszasz, S.; Muhammad, K.; Nelson, W. E.; Probst, M. R.; Rosenfeld, M. G.; Hankinson, O. *Mol. Cell Biol.* **2002**, *22*, 4319-4333.

47. Semenza, G. L. *Trends Mol Med* **2001**, *7*, 345-350.

48. Ema, M.; Hirota, K.; Mimura, J.; Abe, H.; Yodoi, J.; Sogawa, K.; Poellinger, L.; Fujii-Kuriyama, Y. *Embo J.* **1999**, *18*, 1905-1914.

49. Gradin, K.; McGuire, J.; Wenger, R. H.; Kvietikova, I.; Whitelaw, M. L.; Toftgard, R.; Tora, L.; Gassmann, M.; Poellinger, L. *Mol. Cell Biol.* **1996**, *16*, 5221-5231.

50. Chan, W. K.; Yao, G.; Gu, Y. Z.; Bradfield, C. A. *J. Biol. Chem.* **1999**, *274*, 12115-12123.

51. Baba, T.; Mimura, J.; Gradin, K.; Kuroiwa, A.; Watanabe, T.; Matsuda, Y.; Inazawa, J.; Sogawa, K.; Fujii-Kuriyama, Y. *J. Biol. Chem.* **2001**, *276*, 33101-33110.

52. Mimura, J.; Ema, M.; Sogawa, K.; Fujii-Kuriyama, Y.; Inazawa, J. *Genes Dev.* **1999**, *13*, 20-25.

53. Israel, D. I.; Estolano, M. G.; Galeazzi, D. R.; Whitlock, J. P., Jr. *J. Biol. Chem.* **1985**, *260*, 5648-5653.

54. Kobayashi, A.; Sogawa, K.; Fujii-Kuriyama, Y. *J. Biol. Chem.* **1996**, *271*, 12310-12316.

55. Scheidereit, C. *Nature* **1998**, *395*, 225-226.

56. Barker, C. W.; Fagan, J. B.; Pasco, D. S. *J. Biol. Chem.* **1992**, *267*, 8050-8055.

57. Muntane-Relat, J.; Ourlin, J. C.; Domergue, J.; Maurel, P. *Hepatology* **1995**, *22*, 1143-1153.

58. Philipsen, S.; Suske, G. *Nucleic Acids Res* **1999**, *27*, 2991-3000.

59. Imataka, H.; Sogawa, K.; Yasumoto, K.; Kikuchi, Y.; Sasano, K.; Kobayashi, A.; Hayami, M.; Fujii-Kuriyama, Y. *Embo J.* **1992**, *11*, 3663-3671.

60. Kaczynski, J. A.; Conley, A. A.; Fernandez Zapico, M.; Delgado, S. M.; Zhang, J. S.; Urrutia, R. *Biochem. J.* **2002**, *366*, 873-882.

61. Jones, K. W.; Whitlock, J. P., Jr.; Chung, I.; Bresnick, E. *Mol. Cell Biol.* **1990**, *10*, 5098-5105.

62. Morel, Y.; Barouki, R. *J. Biol. Chem.* **1998**, *273*, 26969-26976.

63. Morel, Y.; Coumoul, X.; Nalpas, A.; Barouki, R. *Mol. Pharmacol.* **2000**, *58*, 1239-1246.

64. Zhang, J.; Zhang, Q. Y.; Guo, J.; Zhou, Y.; Ding, X. *J. Biol. Chem.* **2000**, *275*, 8895-8902.

65. Wu, X.; Bishopric, N. H.; Discher, D. J.; Murphy, B. J.; Webster, K. A. *Mol Cell Biol* **1996**, *16*, 1035-1046.

66. Corcoran, E. E.; Means, A. R. *J. Biol. Chem.* **2001**, *276*, 2975-2978.

67. Berghard, A.; Gradin, K.; Pongratz, I.; Whitelaw, M.; Poellinger, L. *Mol. Cell Biol.* **1993**, *13*, 677-689.

68. Long, W. P.; Chen, X.; Perdew, G. H. *J. Biol. Chem.* **1999**, *274*, 12391-12400.

69. Le Ferrec, E.; Lagadic-Gossmann, D.; Rauch, C.; Bardiau, C.; Maheo, K.; Massiere, F.; Le Vee, M.; Guillouzo, A.; Morel, F. *J. Biol. Chem.* **2002**, *277*, 24780-24787.

70. Bertilsson, G.; Heidrich, J.; Svensson, K.; Asman, M.; Jendeberg, L.; Sydow-Backman, M.; Ohlsson, R.; Postlind, H.; Blomquist, P.; Berkenstam, A. *Proc Natl Acad Sci U S A* **1998**, *95*, 12208-12213.

71. Blumberg, B.; Sabbagh, W., Jr.; Juguilon, H.; Bolado, J., Jr.; van Meter, C. M.; Ong, E. S.; Evans, R. M. *Genes Dev.* **1998**, *12*, 3195-3205.

72. Park, Y.; Li, H.; Kemper, B. *J. Biol. Chem.* **1996**, *271*, 23725-23728.

73. Trottier, E.; Belzil, A.; Stoltz, C.; Anderson, A. *Gene* **1995**, *158*, 263-268.

74. Handschin, C.; Podvinec, M.; Stockli, J.; Hoffmann, K.; Meyer, U. A. *Mol. Endocrinol.* **2001**, *15*, 1571-1585.

75. Beato, M.; Herrlich, P.; Schutz, G. *Cell* **1995**, *83*, 851-857.

76. Giguere, V. *Endocr. Rev.* **1999**, *20*, 689-725.

77. Mangelsdorf, D. J.; Thummel, C.; Beato, M.; Herrlich, P.; Schutz, G.; Umesono, K.; Blumberg, B.; Kastner, P.; Mark, M.; Chambon, P.; et al. *Cell* **1995**, *83*, 835-839.

78. Lee, J. W.; Lee, Y. C.; Na, S. Y.; Jung, D. J.; Lee, S. K. *Cell Mol Life Sci* **2001**, *58*, 289-297.

79. Tsai, M. J.; O'Malley, B. W. *Annu. Rev. Biochem.* **1994**, *63*, 451-486.

80. Mangelsdorf, D. J.; Evans, R. M. *Cell* **1995**, *83*, 841-850.

81. Aumais, J. P.; Lee, H. S.; DeGannes, C.; Horsford, J.; White, J. H. *J. Biol. Chem.* **1996**, *271*, 12568-12577.

82. Zhou, Z.; Corden, J. L.; Brown, T. R. *J. Biol. Chem.* **1997**, *272*, 8227-8235.

83. Drocourt, L.; Pascussi, J. M.; Assenat, E.; Fabre, J. M.; Maurel, P.; Vilarem, M. J. *Drug Metab. Dispos.* **2001**, *29*, 1325-1331.

84. El-Sankary, W.; Plant, N. J.; Gibson, G. G.; Moore, D. J. *Drug Metab. Dispos.* **2000**, *28*, 493-496.

85. Harvey, J. L.; Paine, A. J.; Maurel, P.; Wright, M. C. *Drug Metab. Dispos.* **2000**, *28*, 96-101.

86. Moore, L. B.; Parks, D. J.; Jones, S. A.; Bledsoe, R. K.; Consler, T. G.; Stimmel, J. B.; Goodwin, B.; Liddle, C.; Blanchard, S. G.; Willson, T. M.; Collins, J. L.; Kliewer, S. A. *J. Biol. Chem.* **2000**, *275*, 15122-15127.

87. Ogg, M. S.; Williams, J. M.; Tarbit, M.; Goldfarb, P. S.; Gray, T. J.; Gibson, G. G. *Xenobiotica* **1999**, *29*, 269-279.

88. Pascussi, J. M.; Drocourt, L.; Gerbal-Chaloin, S.; Fabre, J. M.; Maurel, P.; Vilarem, M. J. *Eur. J. Biochem.* **2001**, *268*, 6346-6358.

89. LeCluyse, E. L. *Chem. Biol. Interact.* **2001**, *134*, 283-289.

90. Watkins, R. E.; Wisely, G. B.; Moore, L. B.; Collins, J. L.; Lambert, M. H.; Williams, S. P.; Willson, T. M.; Kliewer, S. A.; Redinbo, M. R. *Science* **2001**, *292*, 2329-2333.

91. Zhang, J.; Kuehl, P.; Green, E. D.; Touchman, J. W.; Watkins, P. B.; Daly, A.; Hall, S. D.; Maurel, P.; Relling, M.; Brimer, C.; Yasuda, K.; Wrighton, S. A.; Hancock, M.; Kim, R. B.; Strom, S.; Thummel, K.; Russell, C. G.; Hudson, J. R., Jr.; Schuetz, E. G.; Boguski, M. S. *Pharmacogenetics* **2001**, *11*, 555-572.

92. Bertilsson, G.; Berkenstam, A.; Blomquist, P. *Biochem. Biophys. Res. Commun.* **2001**, *280*, 139-144.

93. Pascussi, J. M.; Jounaidi, Y.; Drocourt, L.; Domergue, J.; Balabaud, C.; Maurel, P.; Vilarem, M. J. *Biochem. Biophys. Res. Commun.* **1999**, *260*, 377-381.

94. Goodwin, B.; Moore, L. B.; Stoltz, C. M.; McKee, D. D.; Kliewer, S. A. *Mol. Pharmacol.* **2001**, *60*, 427-431.

95. Gerbal-Chaloin, S.; Pascussi, J. M.; Pichard-Garcia, L.; Daujat, M.; Waechter, F.; Fabre, J. M.; Carrere, N.; Maurel, P. *Drug Metab. Dispos.* **2001**, *29*, 242-251.

96. Gerbal-Chaloin, S.; Daujat, M.; Pascussi, J. M.; Pichard-Garcia, L.; Vilarem, M. J.; Maurel, P. *J. Biol. Chem.* **2002**, *277*, 209-217.

97. Synold, T. W.; Dussault, I.; Forman, B. M. *Nat. Med.* **2001**, *7*, 584-590.

98. Geick, A.; Eichelbaum, M.; Burk, O. *J. Biol. Chem.* **2001**, *276*, 14581-14587.

99. Schuetz, E. G.; Strom, S.; Yasuda, K.; Lecureur, V.; Assem, M.; Brimer, C.; Lamba, J.; Kim, R. B.; Ramachandran, V.; Komoroski, B. J.; Venkataramanan, R.; Cai, H.; Sinal, C. J.; Gonzalez, F. J.; Schuetz, J. D. *J. Biol. Chem.* **2001**, *276*, 39411-39418.

100. Kast, H. R.; Goodwin, B.; Tarr, P. T.; Jones, S. A.; Anisfeld, A. M.; Stoltz, C. M.; Tontonoz, P.; Kliewer, S.; Willson, T. M.; Edwards, P. A. *J. Biol. Chem.* **2002**, *277*, 2908-2915.

101. Goodwin, B.; Hodgson, E.; Liddle, C. *Mol. Pharmacol.* **1999**, *56*, 1329-1339.

102. Staudinger, J.; Liu, Y.; Madan, A.; Habeebu, S.; Klaassen, C. D. *Drug Metab Dispos* **2001**, *29*, 1467-1472.

103. Staudinger, J. L.; Goodwin, B.; Jones, S. A.; Hawkins-Brown, D.; MacKenzie, K. I.; LaTour, A.; Liu, Y.; Klaassen, C. D.; Brown, K. K.; Reinhard, J.; Willson, T. M.; Koller, B. H.; Kliewer, S. A. *Proc. Natl. Acad. Sci. U S A* **2001**, *98*, 3369-3374.

104. Xie, W.; Radominska-Pandya, A.; Shi, Y.; Simon, C. M.; Nelson, M. C.; Ong, E. S.; Waxman, D. J.; Evans, R. M. *Proc. Natl. Acad. Sci. U S A* **2001**, *98*, 3375-3380.

105. Moore, L. B.; Maglich, J. M.; McKee, D. D.; Wisely, B.; Willson, T. M.; Kliewer, S. A.; Lambert, M. H.; Moore, J. T. *Mol. Endocrinol.* **2002**, *16*, 977-986.

106. Baes, M.; Gulick, T.; Choi, H. S.; Martinoli, M. G.; Simha, D.; Moore, D. D. *Mol. Cell Biol.* **1994**, *14*, 1544-1551.

107. Honkakoski, P.; Negishi, M. *Biochem. J.* **1998**, *330*, 889-895.

108. Pascussi, J. M.; Gerbal-Chaloin, S.; Fabre, J. M.; Maurel, P.; Vilarem, M. J. *Mol. Pharmacol.* **2000**, *58*, 1441-1450.

109. Kawamoto, T.; Sueyoshi, T.; Zelko, I.; Moore, R.; Washburn, K.; Negishi, M. *Mol. Cell Biol.* **1999**, *19*, 6318-6322.

110. Sueyoshi, T.; Negishi, M. *Annu. Rev. Pharmacol. Toxicol.* **2001**, *41*, 123-143.

111. Sidhu, J. S.; Omiecinski, C. J. *J. Pharmacol. Exp. Ther.* **1997**, *282*, 1122-1129.

112. Forman, B. M.; Tzameli, I.; Choi, H. S.; Chen, J.; Simha, D.; Seol, W.; Evans, R. M.; Moore, D. D. *Nature* **1998**, *395*, 612-615.

113. Kawamoto, T.; Kakizaki, S.; Yoshinari, K.; Negishi, M. *Mol. Endocrinol.* **2000**, *14*, 1897-1905.

114. Larsen, M. C.; Brake, P. B.; Parmar, D.; Jefcoate, C. R. *Arch. Biochem. Biophys.* **1994**, *315*, 24-34.

115. Yoshinari, K.; Sueyoshi, T.; Moore, R.; Negishi, M. *Mol. Pharmacol.* **2001**, *59*, 278-284.

116. Sugatani, J.; Kojima, H.; Ueda, A.; Kakizaki, S.; Yoshinari, K.; Gong, Q. H.; Owens, I. S.; Negishi, M.; Sueyoshi, T. *Hepatology* **2001**, *33*, 1232-1238.

117. Muangmoonchai, R.; Smirlis, D.; Wong, S. C.; Edwards, M.; Phillips, I. R.; Shephard, E. A. *Biochem. J.* **2001**, *355*, 71-78.

118. Goodwin, B.; Hodgson, E.; D'Costa, D. J.; Robertson, G. R.; Liddle, C. *Mol. Pharmacol* **2002**, *62*, 359-365.

119. Kassam, A.; Winrow, C. J.; Fernandez-Rachubinski, F.; Capone, J. P.; Rachubinski, R. A. *J. Biol. Chem.* **2000**, *275*, 4345-4350.

120. Wei, P.; Zhang, J.; Egan-Hafley, M.; Liang, S.; Moore, D. D. *Nature* **2000**, *407*, 920-923.

121. Wei, P.; Zhang, J.; Dowhan, D. H.; Han, Y.; Moore, D. D. *Pharmacogenomics J* **2002**, *2*, 117-126.

122. Ekins, S.; Mirny, L.; Schuetz, E. G. *Pharm Res* **2002**, *19*, 1788-1800.

123. Smirlis, D.; Muangmoonchai, R.; Edwards, M.; Phillips, I. R.; Shephard, E. A. *J. Biol. Chem.* **2001**, *276*, 12822-12826.

124. Pascussi, J. M.; Busson-Le Coniat, M.; Maurel, P.; Vilarem, M. J. *Mol Endocrinol* **2003**, *17*, 42-55.

125. Pascussi, J. M.; Drocourt, L.; Fabre, J. M.; Maurel, P.; Vilarem, M. J. *Mol. Pharmacol* **2000**, *58*, 361-372.

126. Pascussi, J. M.; Gerbal-Chaloin, S.; Drocourt, L.; Maurel, P.; Vilarem, M. J. *Biochimica Biophysisca Acta* **2003**, *1619*, 243-53.

127. Thummel, K. E.; Brimer, C.; Yasuda, K.; Thottassery, J.; Senn, T.; Lin, Y.; Ishizuka, H.; Kharasch, E.; Schuetz, J.; Schuetz, E. *Mol. Pharmacol.* **2001**, *60*, 1399-1406.

128. Drocourt, L.; Ourlin, J. C.; Pascussi, J. M.; Maurel, P.; Vilarem, M. J. *J. Biol. Chem.* **2002**, *277*, 25125-25132.

129. Kliewer, S. A.; Lehmann, J. M.; Willson, T. M. *Science* **1999**, *284*, 757-760.

130. Kliewer, S. A.; Willson, T. M. *J Lipid Res* **2002**, *43*, 359-364.

131. Schuetz, E. G.; Wrighton, S. A.; Barwick, J. L.; Guzelian, P. S. *J. Biol. Chem.* **1984**, *259*, 1999-2006.

132. Schuetz, E. G.; Guzelian, P. S. *J. Biol. Chem.* **1984**, *259*, 2007-2012.

133. Schuetz, J. D.; Schuetz, E. G.; Thottassery, J. V.; Guzelian, P. S.; Strom, S.; Sun, D. *Mol. Pharmacol.* **1996**, *49*, 63-72.

134. Pereira, T. M.; Carlstedt-Duke, J.; Lechner, M. C.; Gustafsson, J. A. *DNA Cell Biol* **1998**, *17*, 39-49.

135. Wright, M. C.; Wang, X. J.; Pimenta, M.; Ribeiro, V.; Paine, A. J.; Lechner, M. C. *Mol. Pharmacol.* **1996**, *50*, 856-863.

136. Sidhu, J. S.; Omiecinski, C. J. *Pharmacogenetics* **1995**, *5*, 24-36.

137. Waxman, D. J.; Azaroff, L. *Biochem. J.* **1992**, *281*, 577-592.

138. Nemoto, N.; Sakurai, J. *Arch. Biochem. Biophys.* **1995**, *319*, 286-292.

139. Kocarek, T. A.; Schuetz, E. G.; Guzelian, P. S. *Biochem. Pharmacol.***1994**, *48*, 1815-1822.

140. Kremers, P.; Roelandt, L.; Stouvenakers, N.; Goffinet, G.; Thome, J. P. *Cell Biol Toxicol* **1994**, *10*, 117-125.

141. Yamaguchi, S.; Murata, Y.; Nagaya, T.; Hayashi, Y.; Ohmori, S.; Nimura, Y.; Seo, H. *J Mol. Endocrinol.* **1999**, *22*, 81-90.

142. Moochhala, S. M. *Ann Acad Med Singapore* **1991**, *20*, 13-18.

143. Peterson, T. C.; Renton, K. W. *Biochem. Pharmacol.***1986**, *35*, 1491-1497.
144. Morgan, E. T. *Mol. Pharmacol.* **1989**, *36*, 699-707.
145. Abdel-Razzak, Z.; Loyer, P.; Fautrel, A.; Gautier, J. C.; Corcos, L.; Turlin, B.; Beaune, P.; Guillouzo, A. *Mol. Pharmacol.* **1993**, *44*, 707-715.
146. Williams, J. F.; Bement, W. J.; Sinclair, J. F.; Sinclair, P. R. *Biochem Biophys Res Commun* **1991**, *178*, 1049-1055.
147. Pascussi, J. M.; Gerbal-Chaloin, S.; Pichard-Garcia, L.; Daujat, M.; Fabre, J. M.; Maurel, P.; Vilarem, M. J. *Biochem. Biophys. Res. Commun.* **2000**, *274*, 707-713.
148. Fukuda, Y.; Sassa, S. *Biochem. Pharmacol.***1994**, *47*, 1187-1195.
149. Khatsenko, O. G.; Gross, S. S.; Rifkind, A. B.; Vane, J. R. *Proc. Natl. Acad. Sci. U S A* **1993**, *90*, 11147-11151.
150. McKay, L. I.; Cidlowski, J. A. *Mol. Endocrinol.* **1998**, *12*, 45-56.
151. Adcock, I. M.; Caramori, G. *Immunol Cell Biol.* **2001**, *79*, 376-384.
152. Schmiedlin-Ren, P.; Thummel, K. E.; Fisher, J. M.; Paine, M. F.; Lown, K. S.; Watkins, P. B. *Mol. Pharmacol.* **1997**, *51*, 741-754.
153. Kato, S. *J. Biochem. (Tokyo)* **2000**, *127*, 717-722.
154. Haussler, M. R.; Haussler, C. A.; Jurutka, P. W.; Thompson, P. D.; Hsieh, J. C.; Remus, L. S.; Selznick, S. H.; Whitfield, G. K. *J. Endocrinol* **1997**, *154*, S57-73.
155. Toell, A.; Polly, P.; Carlberg, C. *Biochem. J.* **2000**, *352 Pt 2*, 301-309.
156. Freedman, L. P. *J Nutr* **1999**, *129*, 581S-586S.
157. Freedman, L. P. *Cell* **1999**, *97*, 5-8.
158. Brodie, M. J.; Boobis, A. R.; Hillyard, C. J.; Abeyasekera, G.; Stevenson, J. C.; MacIntyre, I.; Park, B. K. *Clin. Pharmacol. Ther.* **1982**, *32*, 525-530.
159. Brodie, M. J.; Hillyard, C. J. *J R Soc Med* **1982**, *75*, 919.
160. Chan, T. Y. *Int J Clin. Pharmacol. Ther.* **1996**, *34*, 533-534.
161. D'Erasmo, E.; Ragno, A.; Raejntroph, N.; Pisani, D. *Recenti Prog Med* **1998**, *89*, 529-533.
162. Zehnder, D.; Hewison, M.; Nemoto, N.; Sakurai, J. *Mol. Cell Endocrinol* **1999**, *151*, 213-220.
163. Bock, H. H.; Lammert, F. *Hepatology* **2002**, *35*, 232-234.
164. Seol, W.; Choi, H. S.; Moore, D. D. *Science* **1996**, *272*, 1336-1339.
165. Seol, W.; Chung, M.; Moore, D. D. *Mol. Cell Biol.* **1997**, *17*, 7126-7131.
166. Goodwin, B.; Jones, S. A.; Price, R. R.; Watson, M. A.; McKee, D. D.; Moore, L. B.; Galardi, C.; Wilson, J. G.; Lewis, M. C.; Roth, M. E.; Maloney, P. R.; Willson, T. M.; Kliewer, S. A. *Mol Cell* **2000**, *6*, 517-526.
167. Gobinet, J.; Auzou, G.; Nicolas, J. C.; Sultan, C.; Jalaguier, S. *Biochemistry* **2001**, *40*, 15369-15377.
168. Johansson, L.; Thomsen, J. S.; Damdimopoulos, A. E.; Spyrou, G.; Gustafsson, J. A.; Treuter, E. *J. Biol. Chem.* **1999**, *274*, 345-353.
169. Klinge, C. M.; Jernigan, S. C.; Risinger, K. E.; Lee, J. E.; Tyulmenkov, V. V.; Falkner, K. C.; Prough, R. A. *Arch. Biochem. Biophys.* **2001**, *390*, 64-70.
170. Lee, Y. K.; Dell, H.; Dowhan, D. H.; Hadzopoulou-Cladaras, M.; Moore, D. D. *Mol. Cell Biol.* **2000**, *20*, 187-195.

171. Lu, T. T.; Makishima, M.; Repa, J. J.; Schoonjans, K.; Kerr, T. A.; Auwerx, J.; Mangelsdorf, D. J. *Mol Cell* **2000**, *6*, 507-515.

172. Seol, W.; Hanstein, B.; Brown, M.; Moore, D. D. *Mol. Endocrinol.* **1998**, *12*, 1551-1557.

173. Brendel, C.; Schoonjans, K.; Botrugno, O. A.; Treuter, E.; Auwerx, J. *Mol. Endocrinol* **2002**, *16*, 2065-2076.

174. Parks, D. J.; Blanchard, S. G.; Bledsoe, R. K.; Chandra, G.; Consler, T. G.; Kliewer, S. A.; Stimmel, J. B.; Willson, T. M.; Zavacki, A. M.; Moore, D. D.; Lehmann, J. M. *Science* **1999**, *284*, 1365-1368.

175. Handschin, C.; Podvinec, M.; Amherd, R.; Looser, R.; Ourlin, J. C.; Meyer, U. A. *J. Biol. Chem.* **2002**, *277*, 29561-29567.

176. Paolini, M.; Pozzetti, L.; Piazza, F.; Guerra, M. C.; Speroni, E.; Cantelli-Forti, G.; Roda, A. *J. Investig. Med.* **2000**, *48*, 49-59.

177. Maurel, P. In *Cytochromes P450 Metabolic and Toxicological Aspects*; Ioannides, C., Ed.; CRC Press: Boca Raton, New York, London, Tokyo, 1996, pp 241-270.

178. Lesca, P.; Peryt, B.; Soues, S.; Maurel, P.; Cravedi, J. P. *Arch. Biochem. Biophys.* **1993**, *303*, 114-124.

179. Wilhelmsson, A.; Cuthill, S.; Denis, M.; Wikstrom, A. C.; Gustafsson, J. A.; Poellinger, L. *Embo J.* **1990**, *9*, 69-76.

180. Jones, S. A.; Moore, L. B.; Wisely, G. B.; Kliewer, S. A. *Methods Enzymol* **2002**, *357*, 161-170.

181. Xu, L.; Glass, C. K.; Rosenfeld, M. G. *Curr. Opin. Genet. Dev.* **1999**, *9*, 140-147.

182. Feng, W.; Ribeiro, R. C.; Wagner, R. L.; Nguyen, H.; Apriletti, J. W.; Fletterick, R. J.; Baxter, J. D.; Kushner, P. J.; West, B. L. *Science* **1998**, *280*, 1747-1749.

183. Ekins, S.; Schuetz, E. *Trends Pharmacol. Sci.* **2002**, *23*, 49-50.

184. Wurtz, J. M.; Bourguet, W.; Renaud, J. P.; Vivat, V.; Chambon, P.; Moras, D.; Gronemeyer, H. *Nat Struct Biol* **1996**, *3*, 87-94.

185. Makishima, M.; Lu, T. T.; Xie, W.; Whitfield, G. K.; Domoto, H.; Evans, R. M.; Haussler, M. R.; Mangelsdorf, D. J. *Science* **2002**, *296*, 1313-1316.

186. Dussault, I.; Lin, M.; Hollister, K.; Fan, M.; Termini, J.; Sherman, M. A.; Forman, B. M. *Mol. Cell Biol.* **2002**, *22*, 5270-5280.

187. Heery, D. M.; Kalkhoven, E.; Hoare, S.; Parker, M. G. *Nature* **1997**, *387*, 733-736.

188. Nishikawa, J.; Saito, K.; Goto, J.; Dakeyama, F.; Matsuo, M.; Nishihara, T. *Toxicol. Appl. Pharmacol.* **1999**, *154*, 76-83.

189. Tzameli, I.; Pissios, P.; Schuetz, E. G.; Moore, D. D. *Mol. Cell Biol.* **2000**, *20*, 2951-2958.

190. Hall, J. M.; McDonnell, D. P.; Korach, K. S. *Mol. Endocrinol.* **2002**, *16*, 469-486.

191. Zhou, G.; Cummings, R.; Li, Y.; Mitra, S.; Wilkinson, H. A.; Elbrecht, A.; Hermes, J. D.; Schaeffer, J. M.; Smith, R. G.; Moller, D. E. *Mol. Endocrinol.* **1998**, *12*, 1594-1604.

192. Llopis, J.; Westin, S.; Ricote, M.; Wang, Z.; Cho, C. Y.; Kurokawa, R.; Mullen, T. M.; Rose, D. W.; Rosenfeld, M. G.; Tsien, R. Y.; Glass, C. K.; Wang, J. *Proc. Natl. Acad. Sci. U S A* **2000**, *97*, 4363-4368.

193. Grant, S. A.; Xu, J.; Bergeron, E. J.; Mroz, J. *Biosens. Bioelectron.* **2001**, *16*, 231-237.

194. Boulenc, X.; Bourrie, M.; Fabre, I.; Roque, C.; Joyeux, H.; Berger, Y.; Fabre, G. *J. Pharmacol. Exp. Ther.* **1992**, *263*, 1471-1478.

195. Li, A. P.; Maurel, P.; Gomez-Lechon, M. J.; Cheng, L. C.; Jurima-Romet, M. *Chem. Biol. Interact.* **1997**, *107*, 5-16.

196. Li, A. P.; Rasmussen, A.; Xu, L.; Kaminski, D. L. *J. Pharmacol. Exp. Ther.* **1995**, *274*, 673-677.

197. Clement, B.; Guguen-Guillouzo, C.; Campion, J. P.; Glaise, D.; Bourel, M.; Guillouzo, A. *Hepatology* **1984**, *4*, 373-380.

198. Pichard, L.; Fabre, I.; Fabre, G.; Domergue, J.; Saint Aubert, B.; Mourad, G.; Maurel, P. *Drug Metab. Dispos.* **1990**, *18*, 595-606.

199. Pichard-Garcia, L.; Hyland, R.; Baulieu, J.; Fabre, J. M.; Milton, A.; Maurel, P. *Drug Metab. Dispos.* **2000**, *28*, 51-57.

200. Meunier, V.; Bourrie, M.; Julian, B.; Marti, E.; Guillou, F.; Berger, Y.; Fabre, G. *Xenobiotica* **2000**, *30*, 589-607.

201. Tucker, G. T.; Houston, J. B.; Huang, S. M. *Eur. J. Pharm. Sci.* **2001**, *13*, 417-428.

202. Pichard-Garcia, L.; Gerbal-Chaloin, S.; Ferrini, J. B.; Fabre, J. M.; Maurel, P. *Methods Enzymol.* **2002**, *357*, 311-321.

203. Li, A. P.; Gorycki, P. D.; Hengstler, J. G.; Kedderis, G. L.; Koebe, H. G.; Rahmani, R.; de Sousas, G.; Silva, J. M.; Skett, P. *Chem. Biol. Interact.* **1999**, *121*, 117-123.

204. Ged, C.; Rouillon, J. M.; Pichard, L.; Combalbert, J.; Bressot, N.; Bories, P.; Michel, H.; Beaune, P.; Maurel, P. *Br. J. Clin. Pharmacol.* **1989**, *28*, 373-387.

205. Rost, K. L.; Brosicke, H.; Heinemeyer, G.; Roots, I. *Hepatology* **1994**, *20*, 1204-1212.

206. Rizzo, N.; Padoin, C.; Palombo, S.; Scherrmann, J. M.; Girre, C. *Eur. J. Clin. Pharmacol.* **1996**, *49*, 491-495.

207. Rost, K. L.; Brosicke, H.; Brockmoller, J.; Scheffler, M.; Helge, H.; Roots, I. *Clin. Pharmacol. Ther.* **1992**, *52*, 170-180.

208. Schuetz, E. G.; Schuetz, J. D.; Strom, S. C.; Thompson, M. T.; Fisher, R. A.; Molowa, D. T.; Li, D.; Guzelian, P. S. *Hepatology* **1993**, *18*, 1254-1262.

209. Acocella, G. *Clin. Pharmacokinet.* **1978**, *3*, 108-127.

210. Niemi, M.; Backman, J. T.; Neuvonen, M.; Neuvonen, P. J.; Kivisto, K. T. *Clin. Pharmacol. Ther.* **2001**, *69*, 400-406.

211. Niemi, M.; Kivisto, K. T.; Backman, J. T.; Neuvonen, P. J. *Br J. Clin. Pharmacol.* **2000**, *50*, 591-595.

212. Carrillo, J. A.; Benitez, J. *Clin. Pharmacokinet.* **2000**, *39*, 127-153.

213. Benitez, J.; Dahl, M. L.; Spina, E.; Carrillo, J. A. In *Interindividual Variability in Human Drug Metabolism*; Pacifici GM, P. O., Ed.; Taylor and Francis: London, New-York, 2001, pp 85-128.

214. Bourrie, M.; Meunier, V.; Berger, Y.; Fabre, G. *J. Pharmacol. Exp. Ther.* **1996**, *277*, 321-332.

215. Butler, M. A.; Iwasaki, M.; Guengerich, F. P.; Kadlubar, F. F. *Proc. Natl. Acad. Sci U S A* **1989**, *86*, 7696-7700.

216. Pantuck, E. J.; Hsiao, K. C.; Maggio, A.; Nakamura, K.; Kuntzman, R.; Conney, A. H. *Clin. Pharmacol. Ther.* **1974**, *15*, 9-17.

217. Conney, A. H.; Pantuck, E. J.; Hsiao, K. C.; Garland, W. A.; Anderson, K. E.; Alvares, A. P.; Kappas, A. *Clin. Pharmacol. Ther.* **1976**, *20*, 633-642.

218. Lelo, A.; Miners, J. O.; Robson, R. A.; Birkett, D. J. *Br. J. Clin. Pharmacol.* **1986**, *22*, 183-186.

219. Carrillo, J. A.; Christensen, M.; Ramos, S. I.; Alm, C.; Dahl, M. L.; Benitez, J.; Bertilsson, L. *Ther. Drug Monit* **2000**, *22*, 409-417.

220. Kalow, W.; Tang, B. K. *Clin. Pharmacol. Ther.* **1993**, *53*, 503-514.

221. Wietholtz H, V. M., Arnaud MJ, Bircher J, Preisig R *Eur. J. Clin. Pharmacol.* **1981**, *21*, 53-59.

222. Lambert, G. H.; Schoeller, D. A.; Humphrey, H. E.; Kotake, A. N.; Lietz, H.; Campbell, M.; Kalow, W.; Spielberg, S. P.; Budd, M. *Environ. Health Perspect.* **1990**, *89*, 175-181.

223. Schuetz, E. G.; Schuetz, J. D.; Grogan, W. M.; Naray-Fejes-Toth, A.; Fejes-Toth, G.; Raucy, J.; Guzelian, P.; Gionela, K.; Watlington, C. O. *Arch. Biochem. Biophys.* **1992**, *294*, 206-214.

224. Ueda, K.; Okamura, N.; Hirai, M.; Tanigawara, Y.; Saeki, T.; Kioka, N.; Komano, T.; Hori, R. *J. Biol. Chem.* **1992**, *267*, 24248-24252.

225. May, D. G.; Porter, J.; Wilkinson, G. R.; Branch, R. A. *Clin. Pharmacol. Ther.* **1994**, *55*, 492-500.

226. Watkins, P. B. *Pharmacogenetics* **1994**, *4*, 171-184.

227. Watkins, P. B.; Turgeon, D. K.; Saenger, P.; Lown, K. S.; Kolars, J. C.; Hamilton, T.; Fishman, K.; Guzelian, P. S.; Voorhees, J. J. *Clin. Pharmacol. Ther.* **1992**, *52*, 265-273.

228. Hunt, C. M.; Watkins, P. B.; Saenger, P.; Stave, G. M.; Barlascini, N.; Watlington, C. O.; Wright, J. T., Jr.; Guzelian, P. S. *Clin. Pharmacol. Ther.* **1992**, *51*, 18-23.

229. Kinirons, M. T.; O'Shea, D.; Downing, T. E.; Fitzwilliam, A. T.; Joellenbeck, L.; Groopman, J. D.; Wilkinson, G. R.; Wood, A. J. *Clin. Pharmacol. Ther.* **1993**, *54*, 621-629.

230. Kinirons, M. T.; O'Shea, D.; Kim, R. B.; Groopman, J. D.; Thummel, K. E.; Wood, A. J.; Wilkinson, G. R. *Clin. Pharmacol. Ther.* **1999**, *66*, 224-231.

231. Prueksaritanont, T.; Gorham, L. M.; Ma, B.; Liu, L.; Yu, X.; Zhao, J. J.; Slaughter, D. E.; Arison, B. H.; Vyas, K. P. *Drug Metab. Dispos.* **1997**, *25*, 1191-1199.

232. Wang, R. W.; Kari, P. H.; Lu, A. Y.; Thomas, P. E.; Guengerich, F. P.; Vyas, K. P. *Arch. Biochem. Biophys.* **1991**, *290*, 355-361.

233. Thummel, K. E.; Shen, D. D.; Podoll, T. D.; Kunze, K. L.; Trager, W. F.; Bacchi, C. E.; Marsh, C. L.; McVicar, J. P.; Barr, D. M.; Perkins, J. D.; et al. *J. Pharmacol. Exp. Ther.* **1994**, *271*, 557-566.

234. Thummel, K. E.; Shen, D. D.; Podoll, T. D.; Kunze, K. L.; Trager, W. F.; Hartwell, P. S.; Raisys, V. A.; Marsh, C. L.; McVicar, J. P.; Barr, D. M.; et al. *J. Pharmacol. Exp.* Ther. **1994**, *271*, 549-556.

235. Honig, P. K.; Wortham, D. C.; Zamani, K.; Conner, D. P.; Mullin, J. C.; Cantilena, L. R. *JAMA* **1993**, *269*, 1513-1518.

236. Masimirembwa, C. M.; Thompson, R.; Andersson, T. B. *Comb. Chem. High Throughput Screen* **2001**, *4*, 245-263.

Chapter 11

Cytochrome P450 Pharmacogenetics

Robert L. Haining, Ph.D.[1] and Aiming Yu, Ph.D.[2]

[1]*West Virginia University, School of Pharmacy,
  Department of Basic Pharmaceutical Sciences, Morgantown, WV*
[2]*National Cancer Institute, Laboratory of Metabolism, Bethesda, MD*

1   INTRODUCTION

1.1   Chapter introduction

Numerous excellent review articles have been published by others in recent months on the subject of P450 pharmacogenetics. It is a field which changes almost daily as new discoveries are made and clinical outcomes improve with our understanding. Thus this chapter was not meant to be all-inclusive nor exhaustive. Rather, I will touch on subjects largely in the manner and proportions to which I have investigated them. My hope is that this work would spark the reader's interest in the subject, and provide direction for further reading rather than serve as a comprehensive manual. Being a protein biochemist by training, my focus is from an enzyme structure/function viewpoint. Much of the experimental detail presented (Section 2.2) was obtained in my laboratory by my co-author, Dr. Aiming Yu, though in writing, I will try to stay in the first person.

1.2   Brief history of pharmacogenetics

Today's rapid advances in our understanding of medicine would not be possible without the contributions and ideas of investigators present and past. Nevertheless, some key ideas defined a new way of thinking. The idea that genetic factors controlling the biochemistry of living cells could contribute to inter-individual variability in the response to toxins and the susceptibility to unwanted drug reactions was suggested by Gorrod and Oxon at the turn of the

20th century.[1] In the 1950s, Arno G. Motulsky expanded on the ideas of Gorrod and Oxon to say that differences in drug efficacy as well as drug toxicity could be traced to individual genetic variation, and only nine years later, Friedrich Vogel coined the term "pharmacogenetics." The full account in these early investigators own words can be found in their later definitive work, *Human Genetics*, published in 1986.[2] The topic quickly gained ground, and in 1977 the first clinically important polymorphism in a cytochrome P450 was described.[3] It was apparently discovered at a lab in London's St. Mary's Hospital, when Bob Smith, the principal investigator, had a hypotensive crisis (he passed out) after a standard dose of debrisoquine. In 1979, Eichelbaum et al. reported that the "poor metaboliser" phenotype (PM) for sparteine metabolism resulted in deficient N-oxidation.[4] Later these findings were both traced to chromosome 22 and deficiencies at the CYP2D6 locus of man.

Since that time, the list of pharmacogenetic polymorphisms has continued to grow. It appears that the rule-of-thumb for nearly every major gene or gene family involved in the adsorption, distribution, metabolism or excretion of a drug, is that there are polymorphisms. Variability is the rule rather than the exception. Phase 2 metabolic enzymes, such as N-acetyl transferases, thiopurine methyltransferases, and glutathione S-transferases, are all subject to such variability among individuals. In current terminology, most investigators use the word pharmacogenetics to mean the study of how a small, defined number of base substitutions in a single gene or related genes affect drug disposition in humans. This is to be distinguished from 'pharmacogenomics,' a term that was once used interchangeably with pharmacogenetics but has since taken on a much more expanded meaning. That is, with the advent of gene micro-array chips, changes in thousands of genetic pathways can be detected simultaneously. Pharmacogenomics then refers to the way that a drug affects major genetic pathways and regulatory mechanisms rather than specific targets.

1.3   Fundamentals of pharmacogenetic mutations

1.3.1 DNA mutations

Most of us were taught the consequences of some major human genetic mutations from our high school biology class days—sickle cell anemia, cystic fibrosis and the predisposition to countless diseases. However, though it is tempting and sometimes even appropriate when speaking of single nucleotide polymorphisms of human drug-metabolic enzymes, one must be cautious with the term 'mutation'. For one thing, not all polymorphisms are deleterious. Drug metabolism mutants are, for the most part, silent. The hallmark of a pure pharmacogenetic trait is that it does not appear until a person is challenged with a chemical drug entity not normally encountered by the body. When a single enzyme is involved in the metabolism of both endogenous chemicals and exogenous xenobiotics, the line between 'inherited genetic trait' and 'mutation' becomes blurred substantially. Hence, 'polymorphism,' 'substitution,' 'variant,' and 'SNP' become overused alternatives. And as the examples included in this Chapter will suggest, some cytochrome P450 isoforms may not merely be the promiscuous oxidants we have been led to believe.

Heritable genetic polymorphisms that alter enzyme activity can fall into several categories. Single nucleotide polymorphisms (SNPs) are the most frequent and perhaps the easiest-to-grasp concept, as they result from the simple exchange of one nucleotide base for another at a particular genetic locus in a subset of the population. In addition to point mutations within an enzyme coding sequence, point mutations in up- or down-stream regulatory elements are also very common causes of variable enzyme expression. If these point mutations are single nucleotide insertions or deletions that result in a frame shift or mRNA splicing defect or some other 'all-or-none' situation, they are more likely to cause observable phenotypic traits. Most point substitutions are silent, after all. However, several instances in which a single nucleotide substitution encodes a highly conserved amino acid and results in drastic changes in enzyme activity are known. The CYP2C9/warfarin phenomenon described in Section 2.1 below is a prime example. Every cell has two copies of each chromosome, and therefore, two copies of each genetic locus. 'Alleles' occur when there are related but distinct heritable genetic sequences within a gene. Homozygotic carriers of a particular allele are those that carry two identical copies of a given gene, while heterozygotic carriers of a particular allele have two distinct versions of the same gene. Most drug metabolic polymorphisms are inherited in an autosomal recessive pattern; they do not occur on the X or Y chromosomes and it often takes more than one 'bad' copy of a gene to observe a given trait. This is an example also of a 'gene dosage' effect, which becomes particularly relevant in those less common cases of gene duplication or amplification (note CYP2D6 and the ultra-rapid or UM metabolizer phenotype).[5]

The frequency of a given genotypic trait being expressed is governed by the Hardy-Weinberg equilibrium, which states that the probability of all possible alleles occurring randomly in a population is equal to one. In symbolic terms, if 'p' represents the frequency of occurrence of the 'good' allele and 'q' represents the frequency of occurrence of the 'bad' allele of a given pharmacogenetic polymorphism, and these are the only two possible alleles, $\mathbf{p + q = 1}$. The frequency of encountering a particular allele combination can then be determined, as $(p + q)^2$ must also be equal to 1, and therefore, $\mathbf{p2 + 2pq + q2 = 1}$. If each allele was encountered at equal frequency (p = q = 0.5), the probability of finding heterozygotic carriers is 2pq or 2(0.5)(0.5) = 0.5. Thus poor metabolizers would be expected 25% of the time in this example. In contrast, compare the situation where the allele frequency of the deleterious allele is only 10 % (q = 0.1). In this case, true poor metabolizers (homozygotic carriers of the 'bad' allele) are only expected with a frequency of $(0.1)^2 = 0.01$, or 1% of the time. Following this up, alleles that occur with a frequency of less than 1% would result in PM status only $(.01)^2 = 0.0001$, or in one-in-10,000 individuals. Genetic traits that occur at allele frequencies below 1% therefore are not strictly considered 'polymorphisms' but are rather termed 'rare genetic traits'.

1.3.2 P450 polymorphism terminology

Experienced P450 investigators and novices alike may sometimes get confused with the genetic terminology and nomenclature used with the P450 superfamily, therefore I have attempted to put together a few simple guidelines. Standardized rules have been adopted for organizing and naming P450 alleles, providing clarity for investigators in the field. The internet web sites maintained by Oscarson and Ingelman-Sundberg, Daly, and Nebert are particularly helpful. With a quick search of the internet (http://www.imm.ki.se/CYPalleles), each allele can be found, along with its nucleotide substitutions, and the combinations of predicted amino acid changes found in full-length transcripts. Briefly, cytochrome P450's are sorted into families based on an amino acid identity of approximately 40% or greater. Sub-families are recognized when they contain over 55% identity and individual genes may have up to 97% identity. Finally, a polymorphism within a gene is defined as such if it occurs at an allele frequency of at least 1%. As one can see, the total number of possible permutations between individuals appears endless.

According to the Human Genome Nomenclature Committee conventions[6] it is recommended that gene and allele symbols be underlined in manuscript form and italicized in print; protein symbols should be represented in standard fonts. Accordingly, the name for the genetic locus of a drug metabolizing enzyme is always italicized, i.e., *CYP2D6*. Allele terminology is organized by the HUGO Mutation Database Initiative (now the Human Genome Variation Society). Alleles should be limited to an optimum of three characters using only capital letters or Arabic numerals and the allele designation should be written on the same line as the gene symbol separated by an asterisk e.g. *CYP2D6*1*, the allele is printed as *1*. This number often approximately corresponds to their order of discovery and often therefore, frequency. In other words, *CYP2D6*1* is the most frequently encountered allele encoding this enzyme, while *CYP2D6*2* comes in a close second in both frequency (and enzyme activity, as it turns out).[7] To distinguish between mRNA, genomic DNA and cDNA the recommended prefix is written in parentheses as in the following examples: (mRNA) *CYP2D6*1*, (gDNA) *CYP2D6*1*, (cDNA) *CYP2D6*1*.

Proteins are referred to with the same gene name as the gene, printed in non-italicized letters, followed by a period/dot and the number of the allele. For example, the enzyme translated from the mRNA that was transcribed from the *CYP2D6*2* allele of the *CYP2D6* gene is referred to as 'CYP2D6.2'. Though individual splice variants in that the same mRNA that may result in two distinct functional proteins have not been demonstrated for a human P450, it is now apparent that many alternately spliced mRNA variants exist.[8] The chances that one or more of these will encode active enzyme may be good given the precedent found in other enzyme systems (tyrosine hydroxylase, for example). The recommended nomenclature in this case would be the gene symbol followed by an underscore and the lower case letter "v" then a consecutive number to denote which variant is which, e.g., TH_v1.

Polymorphism means, literally, many forms. Pharmacogenetic polymorphisms were typically recognized by a bimodal or trimodal population distribution in their pharmacokinetic or pharmacodynamic profile. In other words, when some measure such as drug efficacy, drug toxicity, parent-compound to metabolite ratio, or some other indicator of drug concentration or effect is plotted on a graph versus the number of individuals, two (or more) bell-shaped humps appear. One group may be classified as PMs if they have an unusually high parent/metabolite ratio due to the lack of a particular P450 isoform, while those in the other group are referred to as rapid or 'extensive metabolizers' (EMs). In some cases, notably CYP2D6, there appear phenotypes in addition to these extremes. For example, the *CYP2D6*17* allele encodes for an enzyme with 3 amino acid changes from the 'wild-type' enzyme. The 2D6.17 enzyme has been shown to have reduced but still significant enzymatic activity when compared to 2D6.1, thus constituting a third 'intermediate metabolizer' (IM) phenotype. Further, unusually high levels of enzyme activity may be observed in some individuals, leading to the classification of 'ultra-rapid metabolizers' (UMs). The relatively few pharmacogenetic polymorphisms that have received the bulk of the investigation are those that appear with distinct clarity. However, most phenotypic traits are controlled by more than one genetic locus. This means there are often several different enzymes in the liver that are capable of performing the same reaction. Such traits will rarely fall neatly into discreet metabolizer categories and more likely form a continuum or unimodal distribution curve. So, is the heterozygotic condition, when an individual has one 'good' copy and one 'bad' copy of a gene, equivalent to 'intermediate metabolizer'? The answer is no, it seems. Having one good copy of a gene is generally enough to place an individual within the normal range of 'extensive metabolizers'.

1.4   *In vivo versus in vitro considerations*

Ethical considerations prevent many of the definitive studies one could to do to get answers to P450 pharmacogenetic questions. Therefore, much of the work focuses on making links between genetic polymorphisms and toxic events recorded from accidental exposures or unwanted side effects. P450 research on the other hand has historically been a biochemical venture. One of the goals of pharmacogenetics must be the merger of these disciplines. Now with the advent of advances from the human genome project and the concomitant explosion in DNA technology, the studies of the specific enzymes involved, their amino acid substitutions, gene deletions, or altered transcriptional events that affect the metabolism of drugs have become within reach of the drug metabolism laboratory. There is now a substantial body of literature on how to clone and express P450s. A method is described in Section 2.2.6 that has been very successful in the expression and purification of human CYP2D6 allelic variants. Cell-based methods also offer an attractive alternative to human pharmacogenetic research. For example, a given variant P450 can be over-expressed in a human cell line, which is then challenged with a drug or other xenobiotic to measure the causation or extent of protection from chemical toxicity due to the P450 enzyme activity or lack thereof. It is now well within reach to study the effects of allelic variation of human drug metabolizing enzymes in a non-human whole-animal model, though the extrapolation of these results to humans will undoubtedly remain a daunting task. Nevertheless, the study of human drug metabolizing enzymes in their native environment remains the gold standard. For this reason and despite advances in DNA technology, the study of P450 activities in microsomes will no doubt continue unabated.

1.5   *Key principles of pharmacogenetics*

As reported by Lash et al.,[9] a recent pharmacogenetics symposium/ roundtable session, sponsored by the Division of Toxicology of the American Society for Pharmacology and Experimental Therapeutics, was held at the 2002 Experimental Biology meeting in New Orleans, Louisiana. In their report, the authors summarize some important and emerging principles of pharmacogenetics. First, as mentioned above, polymorphisms that cause pharmacogenetic disorders will be inherited and can occur at high frequency (greater than 1% allele frequency, as defined), but the trait will be expressed phenotypically only at low frequency, i.e., after environmental exposure. Second, not all pharmacogenetic traits are of high toxicological or clinical significance. Factors that come into play include the frequency of exposure, the availability of alternate therapeutic agents, the availability of alternate clearance pathways, and the therapeutic index of the given drug (the difference between therapeutic and toxic concentrations). Third, an individual's ability to metabolize a given drug can change in a temporal fashion. In other words, a gene encoding a given enzyme may be turned on at a specific point in one's life, notably in the transition from neonate to infant. Fourth, although an emerging library of SNPs has and will continue to be invaluable in our search for the causes of human disorders, many genetic traits will depend on numerous genetic loci. The phenomenon of a single nucleotide change that drastically reduces the activity of CYP2C9 and disrupts the clearance of warfarin is not likely to typify most pharmacogenetic SNPs. Instead, most drugs are eliminated by multiple pathways. Herein lies the challenge for pharmacogenomics—to identify the complex haplotypes involved in a given phenotypic disorder. Finally, ethnic variation in pharmacogenetic polymorphisms is frequently observed (references too numerous to mention). Individualized medicine will likely be race-specific before it is truly individual-specific. Many common polymorphisms segregate along racial lines and it is critical to know this when entering into clinical pharmacogenetic studies.

2   P450 PHARMACOGENETICS

2.1   *CYP2C9 and warfarin*

2.1.1 CYP2C9 function and allelic variation

CYP2C9 catalyzes the oxidation of many drugs of note including several non-steroidal anti-inflammatory drugs (NSAIDs), phenytoin, tetrahydrocannabinol, and warfarin. In addition, it catalyzes the oxidation of carcinogens or procarcinogens such as benz[α]anthracene implicated in tobacco toxicity.[10] An identified pharmacophore for 2C9 substrates is that they are weakly acidic,[11] making them very distinct from substrates for 2D6 for example, that frequently are basic. A total of 6 allelic variants of *CYP2C9* have been described at last count (March 2003), including *CYP2C9*1* (wild type), *CYP2C9*2* (R144C), *CYP2C9*3* (I359L), *CYP2C9*4* (I359T), *CYP2C9*5* (D360E), and *CYP2C9*6* (818delA). A recent report from Goldstein lists six more SNPs identified in CYP2C9 that are predicted to result in single amino acid substitutions.[12] Of the resulting enzymes, the only well characterized are CYP2C9.1, CYP2C9.2, and CYP2C9.3. Compared to individuals with the 2C9.1 enzyme, individuals expressing the CYP2C9.2 and CYP2C9.3 variants, found at an allele frequency of 10-15% in the Caucasian population, exhibit reduced metabolism of a number of substrates.[13-16] The CYP2C9.2 variant appears to reduce the V_{max} of substrate turnover but have little effect on K_m.[16] However, in subjects expressing the CYP2C9.3 variant, K_m is increased for all studied substrates whereas V_{max}

is decreased.[15] The *CYP2C9*4* variant was found in a single Japanese epileptic patient who required a lower than normal dose of phenytoin,[17] suggesting that, like the CYP2C9.3 protein which also involves replacement of I359, this enzyme is deficient in substrate turnover. *CYP2C9*5*, which occurs in African-American and African populations with a frequency of 3%, has been associated with reductions in in-vitro intrinsic clearance of a several substrates, including lauric acid, S-warfarin, diclofenac, flurbiprofen, naproxen, and piroxicam,[18, 19] but *in vivo* consequences of this predicted amino acid change are unknown. The *CYP2C9*6* allele was identified in an African-American subject exhibiting phenytoin toxicity[20] and results in a premature stop codon and hence, non-functional protein.

2.1.2 Expression, purification and catalytic activity of CYP2C9 allelic variants

The I359L (*CYP2C9*3*) variant is of particular interest due to the putative location of this amino acid near the active-site as determined by homology modeling with P450$_{(camphor)}$. Updated models based on the rabbit 2C5 crystal structure confirm this prediction (not shown).[21] The intrinsic enzymatic differences between the CYP2C9.1 and CYP2C9.3 isoforms were investigated. An early report using HPLC analysis of CYP2C9-catalyzed warfarin metabolites suggested that the I359L form, when expressed in a yeast microsomal preparation, catalyzed the formation of 4'-OH (S)-warfarin, at a site completely removed from that of 'normal' (7-OH) CYP2C9-catalyzed oxidation (see Figure 1 below). In addition, this report had found that 2C9 also formed significant quantities of (R)-warfarin metabolites. However, subsequent data generated using GC/MS analysis[22] suggested that this was not true.

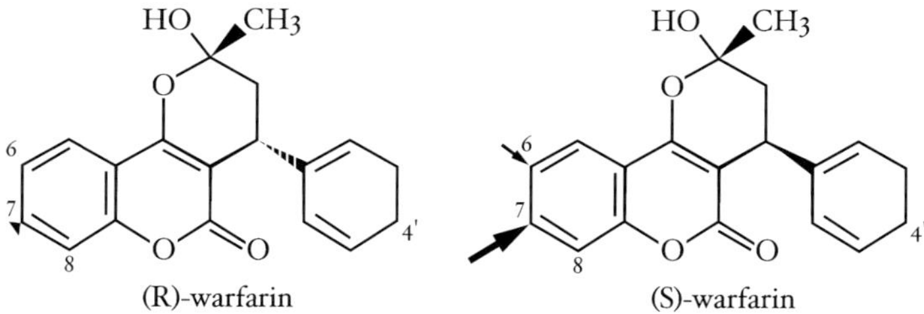

Figure 1. Enantiomers of racemic warfarin used as a probe drug in CYP2C9 allelic variation studies, shown in the ring-closed 11(S) hemi-ketal configuration thought to bind to CYP2C9 (M. He, University of Washington Ph.D. thesis). Arrows indicate major sites of oxidation as determined by a GC/MS assay.

Warfarin is particularly useful as a probe drug for regiospecific and entantiospecific oxidations, since CYP2C9 and CYP2C19 produce distinct 'fingerprint' profiles of metabolites from the (R) and (S) entantiomers. The drug is administered as a racemic mixture, and the fate of the relatively inactive (R)-enantiomer follows a distinctly different path than the active (S)-enantiomer.[14,22-24]

Many expression systems have been used for P450s. The use of *E. coli* for P450 expression is not optimal because N-terminal modifications are generally required in order to keep the expressed protein from aggregating and becoming inactive, presumably because the bacterial membrane is not suitable for insertion of the putative P450 membrane anchor peptides. The baculovirus mediated insect-cell expression offers some advantages in this regard in that, being eukaryotic in nature, the insect cell is able to accept native human protein coding sequences and, in addition, is capable of applying post-translational modifications as necessary.

Starting with a *CYP2C9*3* cDNA provided by Dr. Veronese in Australia, back-mutation to *CYP2C9*1* cDNA was attempted by mutagenesis. However, during preparation of the I359L reverse nucleotide substitution in the *CYP2C9*3* cDNA, inverted repeats and mutational hot spots in the DNA created some complex tertiary structures that interfered with mutagenesis. Sub-cloning a PCR fragment containing the correct sequence into place was more successful than trying to create the nucleotide substitution in the whole cDNA. Once the correct cDNA was prepared, it was inserted into a baculoviral transfer vector for recombination with the viral genome, creating a recombinant baculovirus for amplification and expression. While there are now many simplified commercial expression systems that facilitate baculovirus expression, details of the initial efforts described above have been previously published[13] and a diagram showing an updated procedure later used for CYP2D6 is shown in Figure 6.

Following expression and purification of CYP2C9.1 and CYP2C9.3 (I359L),[13] the 2C9.3 enzyme was found to exhibit a 5-fold increase in K_m (S)-warfarin and a 5-fold decrease in V_{max}, leading to an overall 30-fold decrease in intrinsic clearance (V/K) compared to wild-type 2C9 (CYP2C9.1). Also, purified CYP2C9.3 produced (S)-sulfoxide from the napthyl methyl sulfide in a ratio of 2:1, not markedly different from the ratio obtained from CYP2C9.1. In addition, the phenolic metabolite profile of 2C9.3 towards warfarin was superimposable upon that exhibited by 2D6.1, although the absolute rates of metabolism for the I359L variant were substantially reduced. Therefore no evidence that the orientation of napthyl methyl sulfide or (R)-and (S)- warfarin within the active site of CYP2C9 is altered significantly by mutation of Ile to Leu at position 359. Very similar kinetic changes for other 2C9.3 substrates including tolbutamide[25] and phenytoin[26] have been observed. In other words, all substrates examined had about a five-fold increase in the K_m and a five-fold decrease in V_{max}.

2.1.3 Phenotypic consequences of CYP2C9 genotypic variation.

The kinetic changes induced by the I359L modification are described as *substrate-independent*, caused by some global effect on substrate binding or electron transfer rather than a sterically driven change in active site-orientation. Modeling studies suggest that this residue is near the active-site iron, however, it appears at or below the plane of the heme, making it difficult to envision steric interference. A recent report based on homology modeling of CYP2C9 with the now solved rabbit 2C5 x-ray crystal structure predicts that Leu 362, but not I359, lies in direct contact with the substrates warfarin and diclofenac.[21] In the case of the *CYP2C9*2* variant (R144C), the effects on the inherent catalytic activity of the enzyme toward (S)-warfarin were limited to 60-70% decreases in V_{max} with no change in K_m and hence some 3-fold decrease in V/K. This amino acid appears to be a surface residue, in contact with solvent and/or reductase based on modeling studies, and this observation is backed up by reductase-saturation kinetic studies from the author's laboratory (Haining et al., unpublished results) and others.[27]

Although other drug metabolizing enzymes may also participate in warfarin clearance, notably CYP2C19, the K_m for this reaction as determined in the author's laboratory is significantly greater than physiologically relevant concentrations. From our in-vitro enzyme kinetic data, predictions were that homozygotic carriers of the *CYP2C9*3* allele would be extremely deficient in the metabolism of (S)-warfarin. Heterozygotic *CYP2C9*3* carriers are predicted to be significantly affected, while carriers of the *CYP2C9*2* allele would be only minimally affected. If these predictions were true, warfarin patients who are homozygotic for *CYP2C9*3* should be particularly sensitive to warfarin as reflected in their prothrombin coagulation index ('pro-time'), so a collaborative warfarin clinic was monitored accordingly. L. W., a myocardial infarction patient who was hospitalized for risk of hemorrhage after his pro-time skyrocketed following a standard initial dosing of warfarin, was soon identified. It was subsequently found that L.W. was indeed homozygous for the *CYP2C9*3* allele. In addition, plasma parent drug and urinary metabolite ratios of (R)-and (S)- enantiomers measured were consistent with the hypothesis that this patient should be extremely deficient in (S)-warfarin metabolism, but not (R)-warfarin metabolism. The results of these studies were published recently.[14] In a follow up study conducted by Higashi and others,[28] data from 185 patients from a warfarin clinic were analyzed in relation to CYP2C9 genotype. These workers confirmed an increased risk of adverse bleeding events for carriers of the CYP2C9*3 allele.[28] Other studies find similar results for other CYP2C9 substrates in the clinic, thus constituting a case where an *in vitro* prediction held true *in vivo*.[16]

Substituting an isoleucine for leucine is as probably the most conservative amino acid substitution—zero change in molecular weight or hydrophobicity. It is surprising that such a small change could result in any significant alteration in enzyme activity or substrate preference, such that a thorough characterization to confirm only one amino acid change. Though the vast majority of SNPs in human P450 genes are innocuous, the lesson from the CYP2C9 example is that big changes can accompany small amino acid substitutions.

Interestingly, alternative mechanisms exist for P450s to overcome such deficiencies, and these mechanisms suggest some clinical implications. Recently, a group from Tim Tracy's laboratory reported their findings regarding the CYP2C9 activator, dapsone, in relation to the catalytic activity of allelic variants including CYP2C9.1, 2C9.2, 2C9.3 and 2C9.5. They found that dapsone increased the efficiency (V_{max}/K_m) of flurbiprofen 4β-hydroxylation by CYP2C9.1, CYP2C9.2, CYP2C9.3, and CYP2C9.5 by 690%, 3029%, 4600%, and 2100%, respectively. In similar experiments using the substrate naproxen, dapsone increased the efficiency of naproxen demethylation 591%, 1400%, 1150%, and 2100% in CYP2C9.1, CYP2C9.2, CYP2C9.3, and CYP2C9.5, respectively.[29] By apparently binding to the active site of 2C9, instead of inhibiting the binding of a second substrate, dapsone actually encourages it. Interestingly, the addition of 100 µM dapsone to the 2C9.3 incubation mixture can correct enzyme deficiency to the levels above that seen for 2C9.1 in the absence of dapsone. Although this concentration of dapsone is too high to be clinically feasible, the idea that an small molecule activator could be added to help overcome a deficiency in turnover is intriguing from a drug development perspective.

2.2   The CYP2D6 story

2.2.1 Overview of CYP2D6 and the sparteine/debrisoquine polymorphism:

CYP2D6 was the first generally recognized polymorphic oxidative drug-metabolizing enzyme, associated first more than 20 years ago with the sparteine/debrisoquine polymorphism.[3,30] Two major CYP2D6 phenotypes, poor-metabolizer (PM) and extensive-metabolizer (EM) are generally recognized, with the more recent classification of ultra-rapid metabolizers (UM) and intermediate metabolizers (IM) as a wealth of new information has become available. Normal metabolizers of typical 2D6 substrates have at least one functional copy of *CYP2D6 *1* or *CYP2D6*2*, the most common of the *CYP2D6* alleles and those encoding for the enzymes with the greatest debrisoquine and sparteine hydroxylase activities.[31, 32] Poor metabolizers have deletions or mutations in both alleles that prohibit functional CYP2D6 expression.[33, 34] In the case of intermediate metabolizers, effects may be due to altered orientation within the active site as evidenced by decreased binding affinity, while in others the reductive transfer of electrons from NADPH via cytochrome P450 reductase may be disrupted. Ultra-rapid metabolizers result from a heritable gene amplification, these individuals having multiple copies of functional *CYP2D6*2*.[5] Large interracial variability is observed between the phenotypes,[35] with up to 10% of Caucasians being poor metabolizer[7] while very few PMs are found among Chinese subjects[36-38] and Black Americans.[39]

2.2.2 Genotypic variation of CYP2D6

The CYP2D6 locus has been mapped to chromosome 22q13.1[34] and genotyping studies have proceeded rapidly in recent years. Much is now known about the DNA mutations underlying the clinically observed phenotypes. A standardized nomenclature system has been proposed for the human cytochrome P450 allelic variation and these standards have been adopted here.[40]   Only recently, 17 of the more common CYP2D6 alleles had been identified and named systematically *CYP2D6*1* ('wild-type') through *CYP2D6*17*. However, another report using single strand conformation polymorphism (SSCP) reports another dozen amino acids that are affected by allelic variation in CYP2D6, resulting from a total of some **43** distinct alleles. Much less information regarding allele frequencies and effects of the encoded amino acid substitutions is available for these newly discovered SNPs at this time. Of the first commonly recognized alleles (**1-*17*), **alleles *3, *4, *5, *6, *8, *11, *13, *15, and *16 may be classified as null** due to mutations causing mRNA splicing defects (**4, *11*), premature chain termination including stop codons (**8*), frame shifts (**3, *6, *13, *15, *16*) or gene deletions (**5*) and hence no possible CYP2D6 protein (Daly et al.[41] and references therein). However, the eight other alleles encode full-length, translatable *CYP2D6* mRNA with the following amino acid changes: *CYP2D6*1* (none), **2* (R296C, S486T), **7* (H324P), **9* (K281 deletion), **10* (P34S, S486T), **12* (G42R, R296C, S486T), **14* (P34S, G169R, R296C, S486T) and *CYP2D6*17* (T107I, R296C, S486T). Studies from our laboratory and others indicates that alleles CYP2D6*7, *12 and *14 fold improperly and may never result in protein with heme binding capacity and subsequent CO binding ability (and therefore presumably, O_2 binding and activity). Therefore, five of these allelic proteins show catalytic competence: CYP2D6.1 (wt), 2D6.2 (R296S, S486T), 2D6.9 (ΔK281), 2D6.10 (P34S, S486T), and 2D6.17 (T107I, R296S, S486T).

Figure 2 below shows the results of a western blot analysis of human liver (HL) microsomes from a tissue bank maintained in the University of Washington, Departments of Medicinal Chemistry and Pharmaceutics in Seattle. Liver samples were analyzed for CYP2D6 genotype using a mass-spectral tag procedure described recently.[42] Antibody used to recognize CYP2D6 protein was a polyclonal mixture raised in rabbit against the peptide, and was kindly provided by Steven Leeder. As predicted, individuals homozygotic for the *CYP2D6*4* allele do not show any immuno-recognizable CYP2D6 protein.

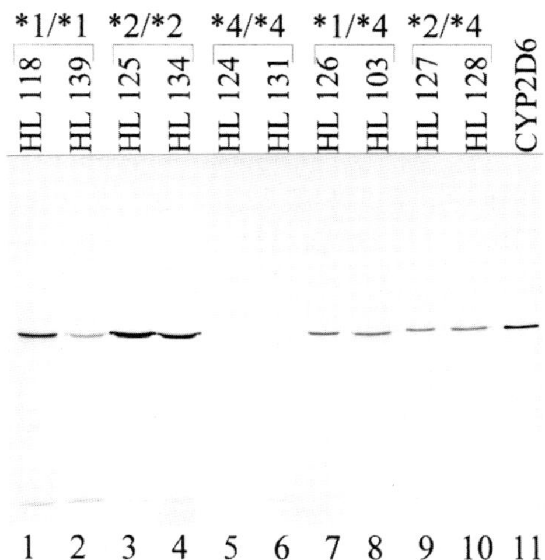

Figure 2. Western blot analysis of CYP2D6 status in human livers genotyped for common allelic variants.

One of the largest correlation studies of CYP2D6 genotypes with 2D6 metabolizer phenotypes included 589 unrelated German subjects.[7] In this population, the frequency of the *CYP2D6*1* allele was 0.364, and that of *CYP2D6*2*, that is associated with slightly reduced activity 0.324. Alleles *9* and *10*, associated with moderately reduced activity and intermediate metabolizer phenotype exhibited allele frequencies of 0.018 and 0.015, respectively. Alleles associated with poor-metabolizer phenotype had frequencies of 0.207 (*4*), 0.020 (*3* and *5*), 0.009 (*6*) and 0.001 (*7*, *15* and *16*). The remaining alleles *8*, *11*, *12*, *13* and *14* were not found in this study.

2.2.3 In vitro studies of 2D6 allelic variant proteins

In addition to the numerous genotyping and *in vivo* phenotyping studies, a number of studies have now been performed specifically to detect the effects of allelic variation on the inherent enzymatic capabilities of CYP2D6 isoforms. *CYP2D6*9* was first noted as a variant with altered electrophoretic mobility in a human liver microsomal preparation with decreased 2D6 marker activities. Subsequent cDNA cloning and genotyping experiments revealed that the gene encoding this protein was lacking an entire codon, that for Lys281. However, when the *CYP2D6*9* protein was produced in HepG2 cells using vaccinia-virus mediated cDNA expression, it displayed K_m values toward bufuralol, debrisoquine and sparteine that were not significantly different from wild type CYP2D6.[43] Despite these observations, this allele has been correlated with poor metabolizer phenotype, and it remains to be determined whether the deletion of this amino acid contributes to other factors such as membrane association or cofactor interactions.

*CYP2D6*7* encodes a full-length protein with the single amino acid substitution H324P. When the corresponding cDNA was expressed in insect cells from recombinant baculovirus, mutant protein amounts comparable to the wild-type enzyme were produced.[44] However, no spectrally detectable P450 was formed, no catalytic activity was detected toward bufuralol and sparteine, and the mutant protein was found almost exclusively in a detergent-insoluble insect cell fraction. These results suggest that the H324P mutation contributes to the *in vivo* poor metabolizer phenotype associated with the *CYP2D6*7* allele by preventing normal protein folding and heme incorporation.

Oscarson et al. recently expressed cDNA corresponding to *CYP2D6*1* and *CYP2D6*17* and also mutants containing all possible combinations of the three amino acids that vary as predicted by these two alleles.[45] Interestingly, *CYP2D6*17* transiently expressed in Cos-1 cells exhibited 20% of the V_{max} compared with *CYP2D6*1* toward bufuralol, while the V_{max} for bufuralol hydroxylase activity in yeast microsomes was unchanged. Both T107I and R296C mutations were required to raise the K_m for this substrate, while the S486T mutation appeared unimportant alone or in combination. However when codeine O-demethylation was examined, the T107I mutation alone was sufficient to increase the K_m toward codeine significantly.

More recently, Zanger et al. investigated the expression of CYP2D6 allelic variants 2, 9 and 10 in an insect cell system.[46] These investigators used crude membrane pellets from *T.ni* expression cultures for their studies and were limited in their ability to repeat kinetic experiments. Nevertheless, they noted decreased expression levels of allelic variants in the insect system and determined the kinetic effects of these variants toward bufuralol.

2.2.4 Hypothesis: Substrate dependent changes in kinetic constants resulting from amino acid substitution

Obviously, a mutation causing in the complete lack of a particular enzyme will result in a blanket effect toward any substrate. Similarly, nucleotide substitutions in up- or downstream DNA regulatory regions may drastically alter the levels of a given enzyme, but would affect all substrates equally. But what about those polymorphisms that instead result in seemingly minor amino acid substitutions? As in the case of P450 CYP2C9 polymorphisms, outcomes are not easily predicted. Despite its putative location in the active site and the drastic effect of the highly conservative substitution (I359L) encoded by *CYP2C9*3*, the kinetic effects of this amino acid substitution were largely substrate-independent. On the other hand, in the case of CYP2D6.2 the amino acid change in question, R144C, does not appear in the active site region, yet does seem to result in a substrate dependent change in kinetics. In this instance, the effect appears to be the result of an allosteric change induced by P450 binding to its redox partner, cytochrome P450 reductase.[27]

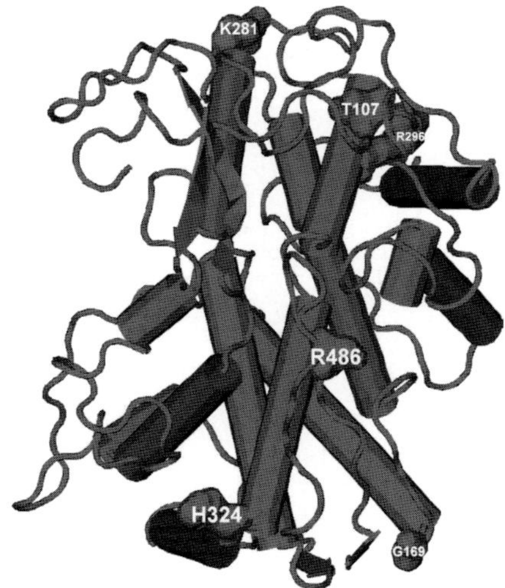

Figure 3. Homology model of CYP2D6 indicating amino acids affected by allelic variation.

Several homology models of CYP2D6 have now been published.[47] Our own appears in Figure 3 above and was based on the rabbit CYP2C5 crystal structure. It was created using the SWISS-MODEL program.[48]

Unlike classical lock and key, single-substrate, single-enzyme systems, P450s have evolved to metabolize broad classes of compounds of often seemingly unrelated structures. Interest in P450 point mutations has greatly increased in recent years following a report[49] that a single amino acid substitution can have profound effects on substrate specificity. It is becoming increasingly clear, as mentioned above, that stereo- and regio-specificity changes in drug metabolic pathways caused by amino acid substitutions of cytochrome P450s can be substrate dependent. Individual substrates may bind in individual orientations and contact different amino acids, so that mutations that abolish the oxidative metabolism of one substrate may leave the activity toward another largely unaffected. For example, the mutation of Ile380 to Phe in rat CYP2D1 differentially decreased activity toward oxidation of bufuralol but not debrisoquine.[50] Likewise, using NMR data Modi et al.[51] have shown that an allosteric effect on the active-site of CYP2D6 mediated by cytochrome P450 reductase was substrate-dependent. Oscarson's report that examined the in-vitro properties of recombinant *CYP2D6*17* (T107I, R296C, S486T) found that the mutation T107I was sufficient to cause a significant increase in the apparent K_m toward codeine, but required an additive effect of the R296C mutation to cause a similar K_m increase toward bufuralol.[45]

Most recently the ability of three separate marker substrates to distinguish 2D6 poor-metabolizer phenotype was examined in three racial populations.[52] This study looked for genotypes *CYP2D6*1, *3, *4, *5, *9, *10,* and **17* correlated with phenotype for sparteine, debrisoquine and dextromethorphan. They found that in Chinese and Caucasian populations, the metabolism of all three drugs correlated positively. However, in Ghanaians, the dextromethorphan metabolic rates did not correlate to either sparteine or debrisoquine. Notably this held true even when only subjects genotyped as **1/*1* were included, suggestive of an inability to distinguish alleles by PCR or perhaps the presence of a novel allele not yet accounted for.

2.2.5 *Expression and characterization of CYP2D6 allelic variants*

The above results are interesting in that they strengthen the hypothesis that changes noted in CYP2D6 enzymes may not affect all substrates equally. During the last 5 years, our laboratory has focused on the CYP2D6 proteins that yield enzymatically competent P450, as based on the standard carbon monoxide binding difference spectra.[53] Our goals have been to understand the inherent capabilities of CYP2D6 variants and the consequences of carrying them. Recombinant cDNA incorporating the allelic changes were successfully created for *CYP2D6*2, *7, *9, *10, *12, *14* and **17* from *CYP2D6*1*. Figure 4 outlines the procedure we used to perform site-directed mutagenesis of cDNA encoding CYP2D6.1 and to create infectious recombinant baculovirus particles for the expression of the resulting 2D6 gene products.

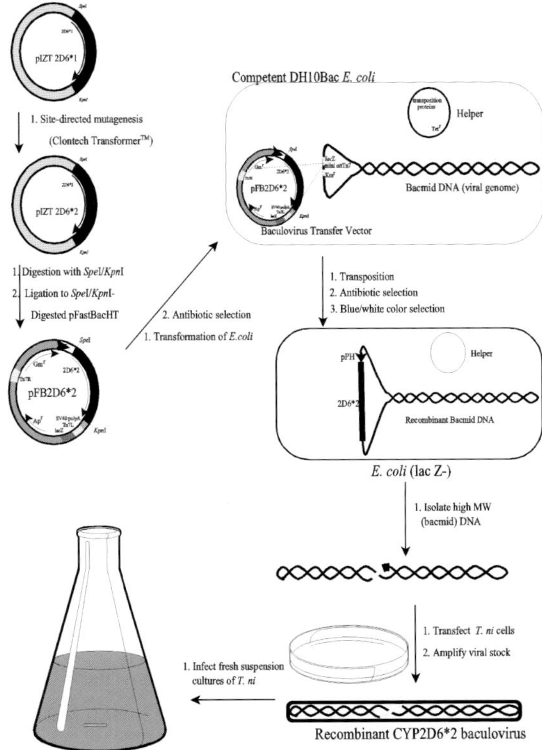

Figure 4. Schematic diagram of recombinant baculovirus creation for the expression of CYP2D6 allelic variants.

Figure 4 shows a schematic diagram of recombinant baculovirus creation for the expression of CYP2D6 allelic variants. Recombinant baculovirus constructs were constructed using the BAC-TO-BAC™ system from the GibcoBRL division of Life Technologies (Gaithersburg, MD).[54] Mutagenesis was carried out in the plasmid pIZT2D6*1 using the Transformer™ kit from BD Biosciences Clontech (Palo Alto, CA). CYP2D6 was excised using *Spe*I and *Kpn*I and ligated into similarly digested pFastBac1. Ligation into pFastBac created the donor plasmid pFB2D6 that places the *CYP2D6* coding sequence behind the viral polyhedrin promoter. Elements of *E. coli* transposon 7 located on the donor plasmid, helper plasmid, and bacmid DNA allowed the site-specific transposition of 2D6 to create a recombinant bacmid. Selection was achieved by simple blue/white screening of bacterial colonies due to the disruption of the lacZα peptide during the recombination process. Bacmid DNA contains the low-copy number mini-F replicon plus a kanamycin resistance marker and thus can propagate in *E. coli*. The bacmid also contains a fully functional baculoviral genome, therefore, it was a simple matter to isolate high molecular weight DNA and transfect insect cells to generate infectious viral particles. Viral stocks were amplified from these transfected cells and used to infect fresh *T. ni* cultures for subsequent expression of 2D6 from the strong polyhedrin promoter.

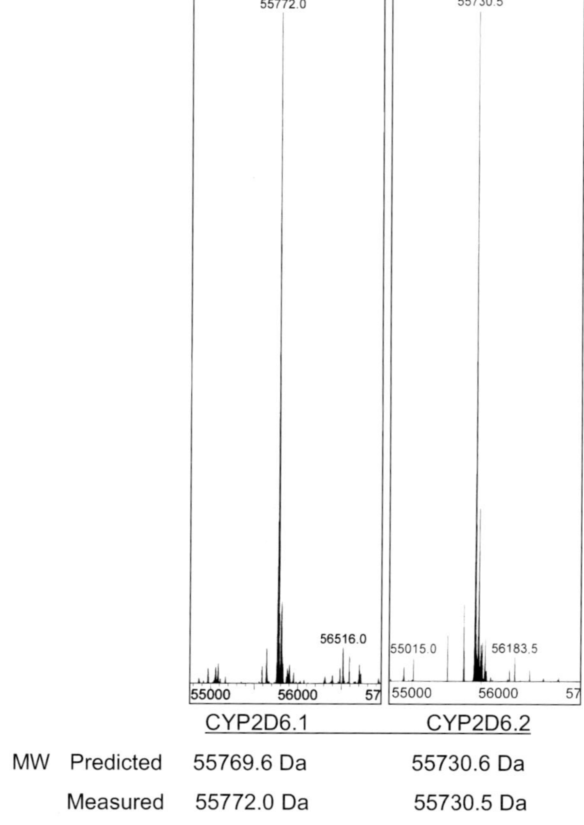

MW	Predicted	55769.6 Da	55730.6 Da
	Measured	55772.0 Da	55730.5 Da

Figure 5. Molecular weight analysis of CYP2D6.1 and CYP2D6.2 by ESI/MS[55].

Of the known active 2D6 variants, recombinant CYP2D6.1, CYP2D6.2, CYP2D6.9, CYP2D6.10, and CYP2D6.17 were over-expressed, and nmole quantities of each were purified to a high degree of homogeneity. ESI/MS analysis of the molecular weight of 2D6.1 and 2D6.2 are shown in Figure 5.

Our results[56] are largely consistent with the observations of Zanger et al. [46] concerning the overall effect of the *2, *9 and *10 alleles on *in vitro* enzyme activity. Other published kinetic analyses indicate that CYP2D6.10[57] and

CYP2D6.17[45] enzymes each have a higher K_m resulting in a lower V_{max}/K_m value than the CYP2D6.1 isoform. Overall, the intrinsic clearance values of DXM O-demethylation as estimated by V_{max}/K_m by CYP2D6.2, 9, 10 and 17 decreased approximately 5-, 50-, 100- and 10-fold compared to CYP2D6.1. Kinetic analysis of these allelic variants toward codeine and fluoxetine showed consistent affinity and catalytic efficiency changes, decreasing in the order CYP2D6.1>CYP2D6.2> CYP2D6.17>CYP2D6.10. In the case of CYP2D6.9, the O-demethylation activity toward DXM places it just in front of 2D6.10 in activity.

2.2.6 *Possible endogenous substrates and a role for CYP2D6 in brain?*

The goal was to investigate the role of CYP2D6 in the causation of prevention of diseased states in humans, specifically in the brain. Its substrate preference for small monoamines is remarkably similar to that exhibited by monoamine oxidase B (MAO-B). Most CYP2D6 substrates contain a basic nitrogen atom located either 5 or 7Å away from the site of oxidation and a planar hydrophobic area nearby.[58, 59] However, several substrates with a distance of approximately 10Å between the basic nitrogen atom and the site of oxidation are also known. More detailed pharmacophore and homology models for CYP2D6 have been developed to predict the interactions between CYP2D6 enzyme and it substrates.[60, 61] Many of the antidepressants, stimulants, antipsychotics—most psychoactive or neuroactive agents, show an interaction with CYP2D6. But the presence of CYP2D6 enzyme in human brain was at one time quite controversial. Its existence has now been convincingly demonstrated using immunoblotting, in situ hybridization, RT-PCR, [62] and turnover of the well-known CYP2D6 probe drug dextromethorphan.[63]

The finding that CYP2D6 mRNA and protein are localized in a cell specific pattern in normal human brain tissue raises the intriguing question of why it is there, given the dogmatic view that most detoxification reactions occur in the liver. Is brain 2D6 an additional site of detoxification for compounds that enter the CNS, or does it rather play a more fundamental role? No endogenous function is yet proven for CYP2D6, as in the case of other P450s such as aromatase that are involved in hormone synthesis pathways. Several investigators have now raised the possibility for 2D6, making the characterization of 2D6 allelic variants a more pressing goal. An observable difference in personalities between extensive metabolizer and poor metabolizer individuals has been reported, suggesting that CYP2D6 may be involved in the regulation of endobiotics critical to CNS function.[64, 65] *In vitro* studies from others indicate that CYP2D6 enzyme catalyzes the biosynthesis of endogenous neurosteroids such as progesterone and monoamine neurotransmitters involved in important signaling pathways.[66-68]

Tyramine is normally produced from tyrosine by aromatic amino acid decarboxylase (AAAD) among other sources (Figure 6). Interestingly, the concentration of tyramine in brain changes directly with dietary protein content. It has been shown that tyramine concentrations rise as much as two to threefold between 0% and 10% dietary protein content. This increase further produces a clear stimulation of the rate of dopamine synthesis, notably in the hypothalamus,[69] where a significant amount of CYP2D6 isoform is observed.[70,71] The difference in the tyramine and dopamine levels regulated by CYP2D6 allelic variants may lead to differences in appetite motivation among CYP2D6 polymorphic individuals. The identification and better understanding of CYP2D6 endogenous substrates in human brain will provide great benefit to CYP2D6 pharmacogenetics, and new insights for the control and therapy of neurological diseases. Therefore, as a logical extension of our work with dextromethorphan, we tested the ability of CYP2D6 allelic variants to metabolize dopamine precursors.

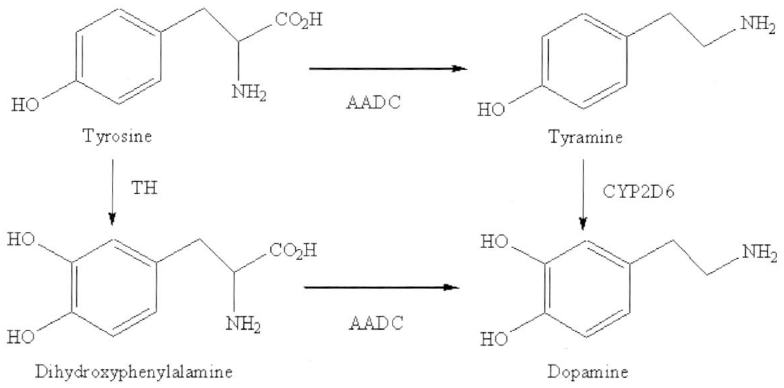

Figure 6. Shown are the biosynthesis pathways of dopamine. TH: tyrosine hydroxylase; AADC: aromatic L-amino acid decarboxylase.

Binding spectra of p-tyramine and m-tyramine with CYP2D6 enzyme: To determine whether tyramine is a good substrate for CYP2D6, we first measured the effect of the substrate on the heme-iron visible absorption spectrum.[53,72] Three classes of binding spectra are usually observed for the interactions between P450 enzymes and chemicals. Compound binding tightly with P450 enzyme forms Type II spectrum, and could be good inhibitor to the P450 isoform. Substrates displace water from the heme in the active site of P450 enzyme leading to the ferric ion from low to high spin transition, thus show Type-I binding spectra. On the contrary, substrates that interact with P450 enzyme resulting to the ferric ion from high to low spin transition indicate reverse Type I binding spectra. Dextromethorphan, a probe drug for CYP2D6,[73] formed a distinct Type I spectrum with CYP2D6 isoform showing a peak at 385nm and a trough at 418nm as expected. However, both *p*-tyramine and *m*-tyramine formed a peculiar Type II binding spectrum with CYP2D6 isoform with the peak around 430nm (Figure 7). These results suggested that *p*-tyramine and *m*-tyramine have similar interactions with CYP2D6 enzyme, but this interaction is different from that found with dextromethorphan.

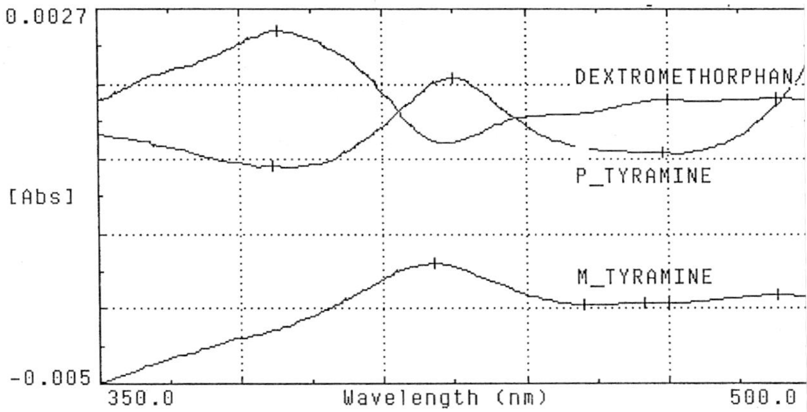

Figure 7. Spectral analysis of dextromethorphan and tyramine binding to CYP2D6

Enzyme reconstitutions and metabolite analysis for tyramine to dopamine conversion: Reconstitution of purified 2D6 enzyme with cytochrome P450 reductase, lipid, NADPH and substrate was performed essentially as described.[56] Without the presence of NADPH, CYP2D6 or P450 reductase, no tyramine was converted to dopamine in the reaction volume. The retention time of dopamine produced from *p*-tyramine and *m*-tyramine was the same, 4.18min, and identical to that of the authentic standard (Figure 8). Dopamine formed in the reaction was further confirmed using ESI/LC-MS analysis. CYP2D6-produced dopamine, absent in the negative control reactions, had the same retention time (3.53min), molecular weight (MH[+], m/z 154), and consistent fragmental pattern (strongest ion m/z 137) as the standard dopamine (spectra not shown).

The CYP2D6.1 and .2 allelic isoforms catalyzed *p*-tyramine and *m*-tyramine hydroxylation to form dopamine with differing enzyme activities as expected. Michaelis-Menten plots of dopamine formation from *p*-tyramine (A), *m*-tyramine (B), and dextrorphan formation from dextromethorphan (C) catalyzed by CYP2D6.1 and 2 allelic isoforms are shown in Figure 8. Kinetic parameters of CYP2D6 allelic isoforms were calculated in order to compare their enzyme efficiency quantitatively for the formation of dopamine from *p*-tyramine and *m*-tyramine. A 5-minute incubation was applied for these reactions since a linear correlation was observed between dopamine formation and the incubation time period ranging from 0 to 10 minutes (data not shown). K_m and V_{max} results are shown in Table 1.

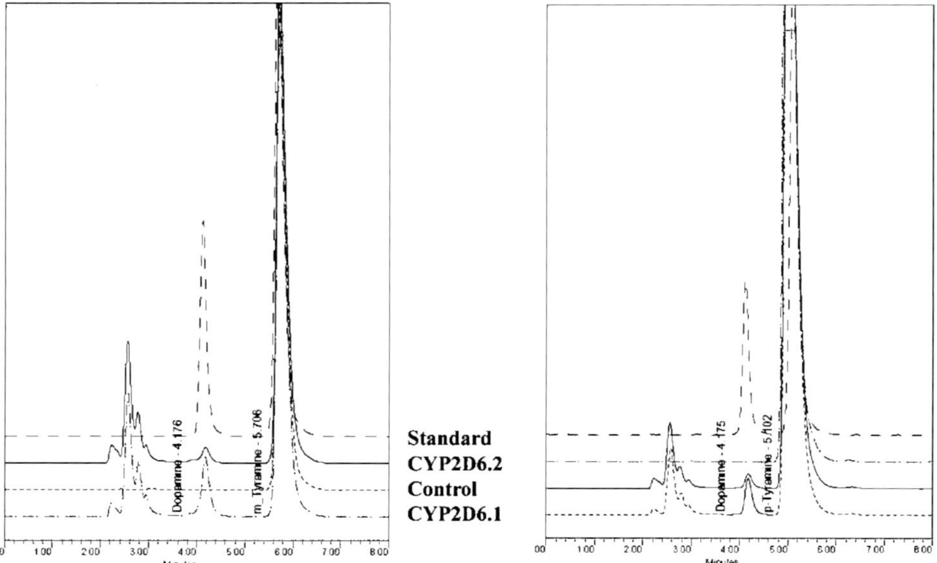

Figure 8. HPLC traces showing dopamine formation from *m*-tyramine (left panel) and *p*-tyramine (right panel) catalyzed by CYP2D6 allelic variants.

The CYP2D6 index reaction, dextromethorphan O-demethylation, was used as a positive control alongside the tyramine incubations. The CYP2D6.1 isozyme has a much higher intrinsic clearance value than the CYP2D6.2 isoform, consistent with our previous results.[56, 74] Dopamine was formed from *p*-tyramine by CYP2D6.1 enzyme with a K_m value of 152 µM, much lower than that (1700 µM) estimated for the CYP2D6.2 isoform, yielding a much higher intrinsic clearance value (49.7 nL/pmol P450/min) for CYP2D6.1 than that for CYP2D6.2 (8.24 nL/pmol P450/min; Table 1). CYP2D6.1 also showed much higher *m*-tyramine hydroxylase activity to produce dopamine than CYP2D6.2 allelic isoform (Figure 10). As shown in Table 2, the K_m values of CYP2D6.1

and CYP2D6.2 allelic variants for dopamine formation from *m*-tyramine were 75.1 µM and 477 µM respectively, and the intrinsic clearance values were 109 nL/pmol P450/min and 17.9 nL/pmol P450/min respectively. These results indicated that CYP2D6.2 allelic isoform had modestly decreased activity compared with the wild type CYP2D6.1 isoform.

Table 1 Estimated kinetic parameters of dopamine formation from *p*- and *m*-tyramine and dextrorphan formation from dextromethorphan as catalyzed by CYP2D6.1 and .2 allelic variants.

Reaction	Allelic Isoform	K_m (µM)	V_{max} (pmol/pmol P450/min)	V_{max}/K_m (nL/pmol P450/min)	V_{max}/K_m Ratio
Dopamine formation from p-Tyramine	CYP2D6.1	152±52.3	7.56±1.23	49.7	1
	CYP2D6.2	1700±341	14.0±1.21	8.24	0.17
Dopamine formation from m-Tyramine	CYP2D6.1	75.1±10.1	8.17±0.40	109	1
	CYP2D6.2	477±174	8.35±0.89	17.5	0.16
Dextrorphan formation from dextromethorphan	CYP2D6.1	2.94±0.361	9.62±0.305	3270	1
	CYP2D6.2	9.27±1.93	5.79±0.273	625	0.19

The CYP2D6 index reaction, dextromethorphan O-demethylation, was used as a positive control alongside the tyramine incubations. The CYP2D6.1 isozyme has a much higher intrinsic clearance value than the CYP2D6.2 isoform, consistent with our previous results.[56, 74] Dopamine was formed from *p*-tyramine by CYP2D6.1 enzyme with a K_m value of 152 µM, much lower than that (1700 µM) estimated for the CYP2D6.2 isoform, yielding a much higher intrinsic clearance value (49.7 nL/pmol P450/min) for CYP2D6.1 than that for CYP2D6.2 (8.24 nL/pmol P450/min; Table 1). CYP2D6.1 also showed much higher *m*-tyramine hydroxylase activity to produce dopamine than CYP2D6.2 allelic isoform (Figure 9). As shown in Table 1, the K_m values of CYP2D6.1 and CYP2D6.2 allelic variants for dopamine formation from *m*-tyramine were 75.1 µM and 477 µM respectively, and the intrinsic clearance values were 109 nL/pmol P450/min and 17.9 nL/pmol P450/min respectively. These results indicated that CYP2D6.2 allelic isoform had modestly decreased activity compared with the wild type CYP2D6.1 isoform.

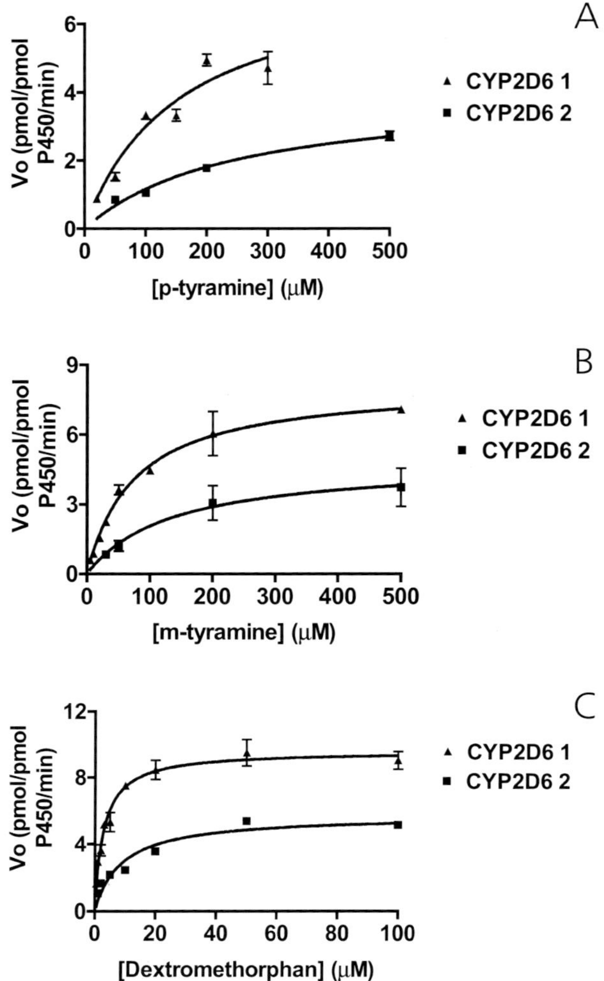

Figure 9.

Functional differences in in vitro modeled CYP2D6 polymorphic subjects: Since the frequency of CYP2D6 heterozygotes in several populations is equal to or even higher than CYP2D6 homozygotes[7, 75-77] we attempted to build an *in vitro* model for the comparison of their drug metabolizing capacity. Recombinant CYP2D6.1, 2, 9, 10, and 17 allelic isoforms were simply mixed in a 1:1 ratio during reconstitution to (crudely) mimic the situation in CYP2D6 heterozygous subjects. At an initial *p*-tyramine concentration of 500 µM, *in vitro* modeled wild type *CYP2D6*1/*1* homozygote showed the highest catalytic heterozygotes (Figure 10 upper panel). For the formation of dopamine from *m*-

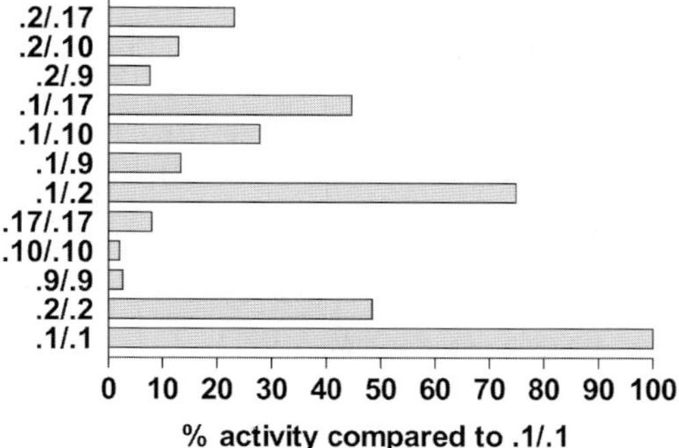

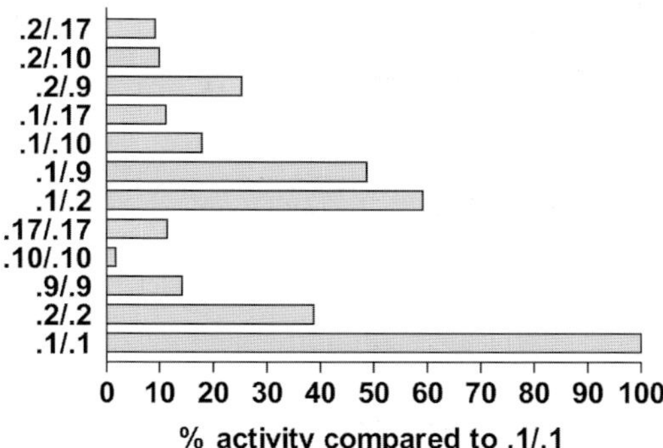

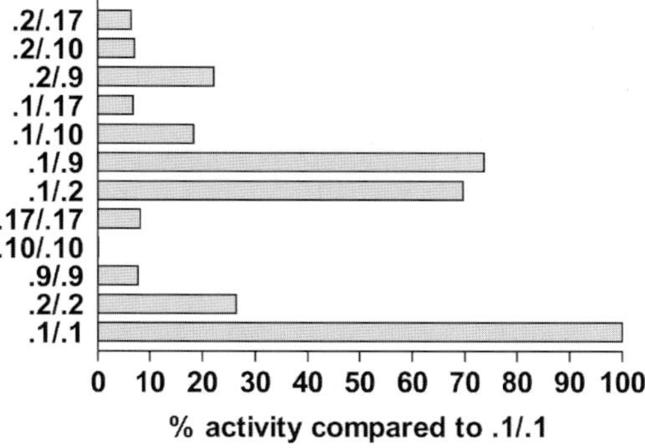

Figure 10. Relative activity of dopamine formation from *p*-tyramine (upper panel), *m*-tyramine (middle panel), and dextrorphan (bottom panel) formation from dextromethorphan catalyzed by *in vitro* modeled CYP2D6 homozygotes and heterozygotes using recombinant CYP2D6 allelic isoforms. Initial substrate concentration was 200 μM for *m*-tyramine, 500 μM for *p*-tyramine, and these reactions were incubated at 37°C for 5 min. Initial dextromethorphan concentration was 10 μM, and incubations were agitated at 37°C for 2 min.

activity toward dopamine formation than any other modeled homozygotes or tyramine at an initial concentration of 200 μM, the same result was obtained, i.e. modeled *CYP2D6*1/*1* homozygous subject possessed the highest activity (Figure 10 middle panel). These results were identical to that obtained from the CYP2D6 index reaction, dextromethorphan O-demethylation (Figure 10 lower panel). Moreover, *p*-tyramine and *m*-tyramine hydroxylase activity for the production of dopamine correlated well with dextromethorphan O-demethylase activity within these modeled CYP2D6 polymorphic subjects (not shown).

2.2.7 The CYP2D6 Story: Discussion and conclusions

Although the initial hypothesis predicted obvious substrate dependent differences in the metabolism of chosen substrates would be observed, the magnitude of the effects could to the opposite conclusion, that the amino acid substitutions in question do not directly affect substrate binding. Only in the case of dextromethorphan N-demethylation reaction, which only appears at high substrate concentrations, does the pattern for allelic variations in activity change. However, in this case, the activity is relatively unaffected and we have made the argument that two substrate molecules occupy the active site simultaneously; therefore, the molecule that binds second and results in the N-demethylated product may not utilize active site binding determinants but rather force its way in while the enzyme is occupied with another molecule.[74] When we directly compared dextromethorphan with the possible endogenous substrates *m*- and *p*-tyramine, again modest decreases are seen in the production of dopamine when 2D6.1 is compared with 2D6.2, similar to findings with other substrates. Therefore, the hypothesis appears to have been refuted, at least for the amino acid changes examined. Thus, for the substrates examined to date, our *in vitro* data predicts that knowledge of the CYP2D6 genotype and the use of a single probe substrate is sufficient to predict metabolizer status of other drugs.

The above study shows that CYP2D6 allelic isoforms catalyze the formation of dopamine from *p*-tyramine and *m*-tyramine with differing efficiencies. Dopamine is a catecholamine neurotransmitter and also a precursor of norepinephrine and epinephrine, and plays a fundamental role in the regulation of behavior and neurodevelopment across all animal species.[78] Dopamine is synthesized from dihydroxyphenylalanine as catalyzed by aromatic L-amino acid decarboxylase (AADC). The latter is derived from L-tyrosine by tyrosine hydroxylase (TH) during the rate-limiting step.[79] Significant differences in the catalytic activity of dopamine formation were observed among CYP2D6 allelic variants, consistent with the results obtained from dextromethorphan O-demethylation, codeine O-demethylation and fluoxetine N-demethylation. Kinetic analysis showed that the wild type CYP2D6.1 isoform had much higher enzyme efficiency than CYP2D6.2 allelic isoform, which contains two amino acid substitutions (R296C and S486T). Using the *in vitro* models, we attempted to compare the functional differences among CYP2D6 polymorphic subjects containing different alleles by making the gross assumption that the production of equal amounts of spectrally active P450 protein arises from each allele in the heterozygote. Therefore, to mimic the heterozygotic condition, proteins were simply mixed in equal (based on CO-difference spectroscopy) parts during the reconstitution step prior to assay. As expected, 'homozygote' samples containing only CYP2D6.1 possessed the highest activity (Figure 10). Samples with 50% of CYP2D6.1 protein and 50% of another active protein had comparatively higher activity than those samples with 100% of variant protein. These results from endogenous compounds and the probe drug dextromethorphan are consistent, and identical to the reported results derived from *in vivo* and *in vitro* studies using other probe drugs.[7,45,57,76,77,80] Based on this observation, the hypothesis is that CYP2D6 polymorphic individuals have different dopamine levels modulated by this pathway, possibly leading to variations in behavior among them as reported.[64, 65] Future studies are needed to further confirm this connection, and to examine the association of CYP2D6 genetic polymorphism with other dopamine regulated behavior.

2.3 *Other human cytochrome P450s of real and/or potential pharmacogenetic interest*

An extremely helpful, interactive and frequently updated guide to human P450 alleles, nomenclature, SNPs, predicted effects of polymorphism, and the references describing their discovery and analysis can be found at http://www.imm.ki.se/CYPalleles/. Due to the rapid accumulation of new human P450-encoding DNA sequences as a result of the human genome sequencing project, a number of alleles have been omitted from the following due to the limited information available. Undoubtedly, there are many cases of clinically relevant P450 polymorphisms yet to be uncovered. On the other hand, one can make the argument that such cases, if common, would already have been recognized. The known and possible endogenous roles for many cytochrome P450s will need to be fully elucidated to fully realize the promise of pharmacogenetics. Therefore, much work remains to be done.

2.3.1 *CYP1A1*

CYP1A1 is found at very low abundance in human livers.[81, 82] For this reason, CYP1A1 is not considered to be a major player in human drug metabolism. In terms of pharmacogenetics and drug metabolism then, it is of little concern despite 10 recognized allelic variants do date.

More importantly, 1A1 is present in human lung tissue and may play a role in the susceptibility to tobacco induced lung cancers. The observation for CYP1A isozymes is that both human members of this family prefer polycyclic aromatic hydrocarbon substrates. *In vitro* enzyme incubations have shown that 1A1 is a potent contributor to carcinogen formation, notably through epoxidation of polycyclic aromatic hydrocarbons such as benzo[α]pyrene. In particular, presence of the 1A1*2 allele, which is rare in Caucasians but common in Asians, has been associated with increased risk for lung cancer. Upon in vitro enzyme expression and investigation of wt and 1A1.2 variant, it was found that this amino acid substitution at position 462 (Ile for Val) probably was not responsible for any link with lung cancer as its catalytic properties toward pro-carcinogens was not significantly different.[83,84] Another variant, in which the amino acid directly next to Val462, Thr461Asp, was found to have no significant allele bias in cancer patients[85] and also not to have a role in lung cancer. On an interesting side note, one of the earliest recognized pharmacogenetic traits was associated with this enzyme. That is, in the early 1970s it was found that approximately 20% of subjects exhibited 'high inducibility' of 1A1-associated phenotype in response to 3-methylcholanthrene. In mitogen-stimulated lymphocytes, a trimodal distribution of CYP1A1 activity was observed.[86]

2.3.2 CYP1A2

Like the related form above, CYP1A2 is inducible by cigarette smoke, polycyclic aromatic hydrocarbons, polychlorinated biphenyls, cooked meat and cruciferous vegetables. Certain allelic variants of 1A2 appear to be more[87] or less[88] inducible in cigarette smokers, again suggesting a link with lung cancer through the aryl hydrocarbon receptor pathway and its resultant up-regulation of CYP1A. However, this isoform, unlike 1A1, is mainly and constitutively expressed in the liver rather than the lung. Notably, both isoforms are located opposite each other while sharing some of the same regulatory elements on chromosome 15.[89] Also unlike 1A1, this isoform has some substrates of clinical and potential pharmacogenetic significance, including theophylline, caffeine, olanzapine, acetaminophen, propranolol and phenacetin. Some 15% of total human liver P450 may be CYP1A2.[90, 91] Despite its abundance in human liver, only one common allelic variant of CYP1A2 that could cause a change in amino acid sequence and thus directly affect enzyme function has been identified. The F21L amino acid change was found in only 1 of 157 Chinese patients, in a heterozygous manner (thus officially making this a rare mutation rather than an allelic variant) but is listed on the allele nomenclature committee table (CYP1A2.2).[92] Chevalier et al. [93] recently reported 5 new allelic variants of CYP1A2 in a French Caucasian population with the following amino acid changes: D348N (CYP1A2.3), I386F (CYP1A2.4), C406Y (CYP1A2.5) and R431W (CYP1A2.6), but the effect of these changes is not known at this time. For conducting in-vitro assays of CYP1A2 function in human liver microsomes, ethoxyresorufin has been used as a specific 1A2 probe[94] and furafylline is a potent and fairly selective mechanism based inactivator.[95] Fluvoxamine has been used as a semi-specific *in vivo* 1A2 inhibitor,[96, 97] but a quick scan of the subject reveals that its use in this regard has been inconsistent and controversial.

2.3.3 CYP1B1

The CYP1B1 isoform has been understudied from a drug metabolism standpoint compared with some of its counterparts due to its relatively low expression in human liver. Thus, it has not historically been considered to be of major concern to drug discovery and pharmacogenetics. However, it plays a role in steroid metabolism, notably in the hydroxylation of 17β-estradiol and is found at fairly high concentrations in steroidogenic tissue.[98] It has also been shown able to activate the procarcinogens benzo[α]pyrene and dimethylbenz[α]anthracene and therefore too, perhaps not surprisingly, has been shown to be inducible through the aryl hydrocarbon receptor pathway.[99, 100]

Shimada and others have conducted an extensive *in vitro* study on the effects of allelic variation on enzyme activity of four active CYP1B1 variants toward 19 different substrates of endogenous or xenobiotic nature. Their results indicate that polymorphisms in the human CYP1B1 gene cause some alterations in catalytic function towards procarcinogens and steroid hormones and thus may make some contribution to the susceptibilities of individuals towards mammary and lung cancers in humans.[101] McLellan et al. [102] also expressed one such variant from the *1B1*2* allele that encodes R48G and A119S and found it had catalytic properties toward 17β-estradiol indistinguishable from CYP1B1.1. Li et al. found that the amino acid changes R48G, A119S (encoded by the *CYP1B1*2* allele) and N453S (encoded by the *CYP1B1*4* allele) had little influence on the V_{max} and the K_m of the CYP1B1 mediated 2- and 4-hydroxylation of estradiol. In contrast, the K_m of these metabolic pathways was increased three-fold by the change L432V (*CYP1B1*3*) or in the L432V and N453S bi-substituted enzyme. However, these alterations had little effect on the kinetic parameters of other CYP1B1 mediated reactions such as the epoxidation of (-)-trans-(7R,8R)-benzo[α]pyrene 7,8-dihydrodiol as determined by (r-7,t-8,t-9,c-10)-benzo[α]pyrene tetraol formation, or such as the O-dealkylation of ethoxyresorufin and the 1'-hydroxylation of bufuralol. Molecular modeling suggests that amino acid residue 432 of CYP1B1 may be involved in the interaction between CYP1B1 and P450 reductase.[99]

Another of the more interesting aspects of this isoform is its reputed role in glaucoma. The CYP1B1 gene has been linked to the GLC3A locus on chromosome 2p21. Stoilov and others[103] found three truncating mutations with a strong correlation in families with glaucoma. They found numerous mutations suggested to affect hinge region or core structure involved in heme binding upon examination with homology modeling.[104] In addition, they used northern blot analysis to show high levels of mRNA expression in regions of the eye involved in regulating intraocular pressure. Aklillu et al studied an Ethiopian population by genomic DNA sequencing to identify a number of distinct 1B1 polymorphisms occurring at various allele frequencies from 0.7-39%. Five different amino acids substitutions appear and occur in varying combinations of : Arg48Gly, Ala119Ser, Leu432Val, Ala443Gly and Asn453Ser. However, only alleles *CYP1B1*6* and **7*, that both result in the R48G, A119S and L432V changes, appeared to be associated with altered kinetics, showing an increase in the K_m and decrease in the V_{max} for in the 4-hydroxylation of 17β-estradiol when expressed in a yeast system. Very little information is available concerning these alleles in other populations. Given the potential for interactions with steroid drugs in humans, much further study needs to elucidate the incidence and possible effect of allelic variation in this gene on pharmacogenetics.[105]

2.3.4 CYP2A6

CYP2A6 is found mainly in human liver where it is principally responsible for the metabolism of nicotine,[106, 107] cotinine,[108] and coumarin.[109] In addition, it has overlapping activities toward substrates such as nitrosamines, disulphuram, halothane and butadiene.[110, 111] A total of 11 distinct alleles and minor variants of 2A6 have been identified to date, of which *2A6*4* is the most common. This deletion mutant occurs at a frequency of some 15-20% in Orientals (21.2% in Japanese)[112] but was rarely found in a sample of Europeans. The *CYP2A6*2* allele, that encodes a protein with a single amino acid change that appears to have no enzyme activity, occurs at a frequency of some 3% in Europeans.[113] Accordingly, studies have shown a link to this allele and susceptibility to nicotine addiction by modification of smoking behavior.[106, 114] Sellers et al. took this a step further and tested a population of smokers using methoxsolen, a potent 2A6 inhibitor.[115] These workers found that those given the drug smoked less tobacco than their control counterparts. Inhibition of CYP2A6 activity apparently leads to increased nicotine concentrations in the blood and less need for smoking to maintain the desired levels. Like 1A1 and 1A2, that are associated with metabolism of the polycyclic aromatic hydrocarbons in tobacco smoke, a link with lung cancer risk and CYP2A6 allelic variation has also been suggested due to its known activation of tobacco smoke carcinogens. One study found that the frequency of homozygotic carriers of the CYP2A6 gene deletion mutation, (*CYP2A6*4*) was lower in lung cancer patients than in healthy control subjects, suggesting that deficient CYP2A6 activity reduces lung cancer risk.[116]

The *CYP2A6*3* allele is interesting from an evolutionary perspective. This gene encodes a protein with 96% predicted amino acid identity with 2A6.1 and appears to have arisen from a gene conversion or splice variant with the nearby CYP2A7 pseudogene. Likewise, the *CYP2A6*12* allele carries an unequal crossover between the CYP2A6 and CYP2A7 genes in intron 2 resulting in a hybrid allele, where the 5' regulatory region and exons 1-2 are of CYP2A7 origin and exons 3-9 are of CYP2A6 origin. It is present at an allele frequency of 2.2% among Spaniards, but is absent in Chinese. In total, 10 amino-acid substitutions are encoded when compared with the *CYP2A6*1* allele. *In vivo* phenotyping with coumarin as well as *in vitro* expression in COS-1 cells confirms that this enzyme retains only 60% of the activity of the wild-type enzyme.[117] Xu et al. examined the effects of amino acid substitutions encoded by alleles *CYP2A6*7* (I471T), **8* (R485L) and **10* (I471T, R485L) on enzyme activity. These alleles occur at higher frequencies among Oriental groups but are less frequent among Caucasians. They found that the CYP2A6.8 enzyme activity was relatively unaffected, the activity of CYP2A6.7 was moderately affected, but the combined substitution caused a drastic decrease in enzyme activity toward nicotine and coumarin.[118] In summary, there is significant potential for a role for this enzyme in the pharmacogenetic drug discovery process.

2.3.5 CYP2B6

Another comparatively underappreciated P450 isoform, CYP2B6, is coming into its own as a player in the pharmacogenetic arena. It catalyzes the biotransformation of buproprion and cyclophosphamide as well as xenobiotic nitrosamines found in cigarette smoke. In addition, Hidestrand et al. have suggested that CYP2B6, not CYP2D6 as others have suggested, is a major enzyme in the metabolism of selegiline.[119] It has been thought that levels of this isoform constitute less than one percent of total human liver P450. However, this expression has been recently reported to be highly variable and there may be much more present in liver than has been previously appreciated.[120] Much progress has been made on the identification of selective substrates such as 7-ethoxy-4-triflouromethylcoumarin (7-EFC; O-deethylation),[121] S-mephenytoin (N-demethylation), buproprion (hydroxylation),[122] and S-mephobarbitol,[123] However, no suitable selective inhibitors have emerged to help clarify the significance of this isoform in human liver.

Lang et al. identified a total of nine point mutations in the human CYP2B6 gene, five of which result in amino the acid substitutions R22C, Q172H, S259R, K262R and R487C while the others are silent for a total of 6 alleles. In human liver samples, significantly reduced CYP2B6 protein expression and S-mephenytoin N-demethylase activity were found in carriers of alleles *5 and *7 (both of which encode for the R487C amino acid substitution).[124] These alleles occur at a combined frequency of some 14% in Europeans and are thus the two most likely to be of major significance from a drug development perspective. Ariyoshi and coworkers found a single nucleotide polymorphism (SNP) in the CYP2B6 gene resulting in the substitution Q172H (*CYP2B6*9*). The frequency of the variant allele was found to be 19.9% in a Japanese population. The mutant- and the wild-type enzymes were expressed in *E. coli*, and the effects of the single amino acid substitution on the catalytic activity were examined by investigating the kinetic profiles of 7-ethoxycoumarin O-deethylase activity. Interestingly, the wild-type enzyme showed typical Michaelis-Menten kinetics, while the mutant-type enzyme exhibited sigmoidal kinetics, with a higher V_{max} value compared to that of the wild-type enzyme. These workers suggest this SNP enhances the catalytic activity of CYP2B6 through homotropic cooperativity.[125]

2.3.6 CYP2C8

CYP2C8 is a relative newcomer on the pharmacogenetic list, mainly due to confusion over its exact identity. Some 20-30% of total human liver P450 is from the CYP2C family, however, the similarity of 2C8 with 2C18, 2C9 and 2C19 historically made it difficult to distinguish between isoforms. In addition, no good specific substrates or inhibitors have arisen to help distinguish between 2C isoforms in human liver microsomes. However, several drugs of clinical significance appear to be largely metabolized by 2C8, including paclitaxel (Taxol™),[126] all trans retinoic acid,[127] amodiaquine,[128] and rosiglitazone.[129] CYP2C8 is also the predominant P450 responsible for the metabolism of arachidonic acid to biologically active epoxyeicosatrienoic acids (EETs) in human liver and kidney.[130, 131] Three allelic variants of note that encode for amino acid changes, all of which are associated with decreased enzyme activity, have been described as follows: *CYP2C8*2* (I269F) with an allele frequency of 0.18 in African Americans and rare in Europeans; *CYP2C8*3* (R139K, K399R), with a frequency of 0.02 in African Americans and 0.13 in Europeans; and *CYP2C8*4* (I264M) found rarely in African Americans and at a frequency of 0.075 in Europeans.[132] In addition to their genotyping study, these investigators also expressed CYP2C8.1, CYP2C8.2 and CYP2C8.3 in *E. coli* for comparative kinetic analysis. They found that the turnover number of CYP2C8.3 for paclitaxel was 15% of that exhibited by CYP2C8.1, while CYP2C8.2 had a two-fold higher K_m and two-fold lower intrinsic clearance for paclitaxel than CYP2C8.1. CYP2C8.3 was also markedly defective in the conversion of arachidonic acid to 11,12- and 14,15-EET, with turnover numbers ranging from 35-40% of that observed with CYP2C8.1.[132] This polymorphism therefore appears to have important clinical and physiological implications in individuals homozygous for this allele. The effect of these variants on the clinical usage of substrates affected by this polymorphism is yet to be determined. Notably, Bahadur et al. found that human liver microsome samples heterozygous for *CYP2C8*3* showed significantly lower paclitaxel 6α-hydroxylase activity when compared with wild-type samples.[133]

2.3.7 CYP2C18

The study of CYP2C18 has also lagged behind other human P450s, such that no drug or physiological substrates of significance have been identified to date and it is not listed on the human P450 allele nomenclature site. As is seemingly universally true among P450s, this gene is also subject to polymorphism, however, and cannot be discounted for drug discovery purposes. At least one allelic variant cDNA cloned resulting in a predicted Thr/Met substitution at position 385 has been identified.[134] However, no data on the enzymatic activity of this substitution is available. Another predicted null allele has also been described,[135] as have some upstream regulatory region polymorphisms that co-segregate with the *2C19*2* polymorphism.[136]

2.3.8 CYP2C19 and the mephenytoin story

Following the awareness of the debrisoquine polymorphism in the late 1970s (covered in Section 2.2), many studies identified additional examples of P450-related pharmacogenetic phenomena. In 1984, investigators at Vanderbilt University were testing the pharmacokinetics of racemic mephenytoin (MP), an anticonvulsant medication that acts as a sedative at high concentrations. One individual in the study population apparently complained of excessive sedation upon ingestion of a normal dose. A follow-up of this person's immediate family revealed a similar phenotypic trait in other members, thus establishing a genetic basis for the anomaly, not dietary or other unknown influences.[137] From here, the search for the defective gene and its gene product (or lack thereof) began. It was found that liver samples from poor mephenytoin metabolizers lacked (S)-mephenytoin 4-hydroxylase activity, while extensive metabolizers did not. Therefore, an attempt was made to purify the mephenytoin-4-hydroxylase enzyme from human EM liver samples by Meyer et al. as well as Shimada and Guengerich.[137, 138] These investigators managed to purify an enzyme with weak

MP 4-hydroxylase activity and proceeded to develop a polyclonal antibody against it in rabbit, in the hopes of using this antibody to probe for the presence of MP-4-OH enzyme in human liver samples. However, they instead found that it identified a strong protein both in mephenytoin PMs and EMs liver samples—no apparent correlation was found between their enzyme and the enzyme activity in question. Perplexed, they also probed a human cDNA expression library (in which human gene products are expressed in a bacterial host and can be recognized by their respective antibodies) and found several cDNAs that resulted in immunoreactive protein products. Eventually the entire CYP2C family was elucidated in this way.[134] As it turns out, Dr. Shimada probably mistakenly purified CYP2C9, now known to constitute a high percentage of the total hepatic P450, that has only weak MP 4-hydroxylase activity. Being related to CYP2C19, the actual enzyme responsible for this elimination pathway, the polyclonal antibody his lab made recognized enough epitopes in common so as to be indiscriminate between the two.

The end result of the above story was that it is polymorphism of the CYP2C19 gene that was found responsible for poor mephenytoin 4-hydroxylation. A total of 7 variants, two common and 5 rarer ones, have been described that result in deficient enzyme activity and together account for the PM phenotype.[139-143] Alleles *2C19*9-*15* are rather predicted to result in two amino acid changes each, except 2C19.12 that contains an extra 26 amino acids in addition to the two substitutions. These are the most recently described and appear to be specific to persons of black African descent.[144] In addition to mephenytoin, other drug substrates that are affected by this polymorphism include diazepam,[145] and proguanil, an antimalarial medication that is activated partly by CYP2C19 but not solely.[146] Omeprazole, the popular proton pump inhibitor, is the most widely used selective inhibitor, though it has been shown to have some cross-inhibition with other P450 isoforms.[147] Most interestingly, CYP2C19 PMs were shown to have a better response to peptic ulcers treated with Omeprazole,[148] opening up a whole new arena in co-medication therapy. Recently, (+)-*N*-3-benzyl nirvanol and (-)-*N*-3-benzyl-phenobarbitol have been described as new potent and selective *in vitro* inhibitors of CYP2C19.[149]

2.3.9 CYP2E1

CYP2E1 has received considerable attention over the years despite the relatively few clinically important drug substrates it metabolizes. Though few in number, they include such common drugs as acetaminophen,[150] caffeine,[151] and ethanol, as well as chlorzoxazone,[152] enflurane,[153] halothane,[154, 155] benzene, acetone,[156] and some nitrosamines and azocarcinogens found in cigarette smoke.[157] CYP2E1 also affects numerous endogenous reactions including lipid peroxidation,[158] and metabolism of fatty acids such as arachidonic and linoleic,[159] and as such its relative conservation among P450s has been postulated to be the result of such selective pressure.[160] One of the most important and potentially toxic reactions catalyzed by this enzyme is that observed upon overdoses of acetaminophen. In these cases, hepatic glutathione pools, that normally detoxify the small amount of the NAPQI quinone formed from acetaminophen by 2E1, may become depleted. Thus overdoses result in the covalent binding of NAPQI to cellular constituents and resulting damage.[161] CYP2E1 is widely known as the 'ethanol-inducible' form, as alcohol and fasting have been shown to up-regulate its expression,[162] further aggravating the potential for acetaminophen overdoses particularly in cases of concomitant opiate addiction (consider Tylenol™ with codeine for example). Notably, up to 20-fold differences in inter-individual enzyme expression in human liver have suggested, though phenotyping with the muscle relaxant chlorzoxazone in Caucasians suggests this variation is only 2-3 fold.[160] Pyridine appears to be a fairly specific *in vitro* inhibitor of 2E1 at low concentrations.[163]

In terms of pharmacogenetics, most polymorphisms to date appear to be in intronic or upstream sequences and are not significant. *CYP2E1*2* encodes for enzyme with a single deleterious amino acid change, Arg76 to His,[160] and is found rarely in Chinese. Two other amino acid changes found in 2E1 allelic variants appear to result in fully functional enzyme activity.[160] *CYP2E1*5* encodes for an allele with one nucleotide substitution upstream in an HNF-1 regulatory-element site. *In vitro* studies suggest that this substitution could result in up to 10-fold variation in expression levels, though the significance of this *in vivo* has yet to be determined. This allele occurs at a frequency of 0.24 in Japanese and 0.02 in Caucasians.[164] Importantly, a study of Taiwanese patients showed a strong association of this allele with nasopharyngeal cancer[165] and there is some evidence for an increased frequency of this allele among alcoholic liver-disease patients.[166]

2.3.10 CYP2J2

CYP2J2 is a recently described human cytochrome P450 that is abundant in cardiovascular tissue and is active in the metabolism of arachidonic acid to eicosanoids that possess potent anti- inflammatory, vasodilatory, and fibrinolytic properties.[167] King et al. found a variety of polymorphisms in this gene, five of which resulted in predicted amino acid coding changes (R158C, I192N, D342N, N404Y, and T143A). Each of these substitutions was created *in vitro* and expressed using baculovirus mediated insect cell expression. The R158C, N404Y, and T143A variants showed significantly reduced metabolism of both arachidonic acid and linoleic acid, while the I192N variant showed significantly reduced activity only toward arachidonic acid and enzyme activity of the D342N variant was relatively unaffected.[167] The possibility that this polymorphism could affect cardiovascular therapeutics will undoubtedly be examined further.

2.3.11 CYP3A4

Members of the CYP3A family are some of the most commonly involved in drug-drug interactions due to their broad, overlapping substrate specificity. The ability to metabolize large substrates is their observed pharmacophore and they may possess the ability to accommodate more than one substrate per active-site simultaneously.[168, 169] As such, they can exhibit non-Michaelis-Menten type kinetics, often showing a 'low K_m' component as well as a 'high K_m' component to the metabolism of substrates, or more commonly a 'hooked' Eadie-Hofstee plot, even in the purified state. This fact undoubtedly led to much consternation over the years as investigators attempted to elucidate the various P450 isozymes involved in a given elimination pathway. Ketoconozole is a relatively selective inhibitor of CYP3A at 1 μM[170] and Gestodene (a progesterone analog) has long been known as a 3A4 suicide inactivator.[171] But only with the advent of recombinant DNA technologies are some of the kinetic puzzles being to be unraveled. It has been suggested that exclusion of water may be a key component of the enzymatic mechanism of CYP3A4.[172] There are at least three members of the 3A family in humans, most notably 3A4, 3A5 and 3A7. CYP3A4 typically constitutes some 30% of total human liver cytochrome P450 in adults.[173] However, Guengerich et al. have reported up to 40-fold inter-individual variability in hepatic 3A4 content, suggestive of a pharmacogenetic link.[174]

Indeed, many SNPs resulting in amino acid substitutions are reported for 3A4, some of which result in altered catalytic activities. Sata and others identified a 5'-non-coding substitution of unknown effect (*CYP3A4*1B*) that occurred in 4.2% of Caucasians and 66.7% of blacks but was absent in Chinese subjects. *CYP3A4*2*, encoding the amino acid change S222P, was found in 2.7% of whites but was absent in blacks and Chinese subjects. Using an *in vitro* expression system, these workers report that the enzyme activity of the CYP3A4.2 variant was decreased toward nifedipine but unchanged toward testosterone 6β-hydroxylation. *CYP3A4*3* (M445T) was found in a single Chinese subject. No information concerning this isozyme's activity is available, but homology modeling places this residue in the highly conserved P450 heme pocket making this protein probably inactive.[175] Hsieh et al. then reported 3 other alleles, *CYP3A4*4* (I118V) that occurred in 3% of Chinese subjects, *CYP3A4*5* (P218R; 2%) and *CYP3A4*6* (1%), that causes a frame shift in exon nine (thus

CYP3A4.6 could be minimally truncated and enzymatically active). Following the 6β–hydroxylation of cortisol as a urinary marker, these investigators found phenotypic evidence for a change in activity in subjects carrying these alleles.[176] No *in vitro* enzyme activity data is yet available for these variants however.

Eiselt and coworkers identified 7 allelic variants of CYP3A4 in a population of 213 Europeans, each of which is predicted to result in a single amino acid change. At least one of these alleles occurred in 7.5% of the population examined. Expressed *in vitro*, the CYP3A4.8 (R130Q) and CYP3A4.13 (P416L) appeared to be unstable, while CYP3A4.11 (T363M) expressed at reduced levels. CYP3A4.7 (G56D), CYP3A4.9 (V170I), CYP3A4.10 (D174H) and the previously identified CYP3A4.3 (M445T) were indistinguishable from wild-type CYP3A4 in expression or steroid hydroxylase activity. However, the CYP3A4.12 (L373F) was significantly altered in terms of metabolite profile toward testosterone and kinetics of midazolam hydroxylation.[177] Dai et al. subsequently identified four new allelic variants predicted to result in amino acid substitutions: *CYP3A4*15* (R162Q), *CYP3A4*17* (F189S), *CYP3A4*18* (L293P), and *CYP3A4*19* (P467S). The *CYP3A4*15* allele was found only in black populations with a frequency of 4%, while *CYP3A4*17* and *CYP3A4*3* (M445T; previously identified) were found in Caucasians with allelic frequencies 2% and 4%, respectively. *CYP3A4*18* and *CYP3A4*19* appeared be Asian specific, found at an allele frequency of 2% in this population. These workers then expressed cDNAs corresponding to *3A4*3, *17, *18* and *19 in vitro* and used testosterone and chlorpyrifos (an insecticide) as substrates to measure enzymatic activity of the resulting proteins. They found that CYP3A4.17 was impaired in the turnover of both substrates, the activities of 3A4.3 and 3A4.19 were unchanged, but interestingly, the activity of the 3A4.18 isoform appeared to be increased.[178] Most recently, Lamba et al. analyzed DNA from 82 individuals who had been phenotyped for CYP3A4 activity and attempted to correlate reduced activity of CYP3A4 substrates or lower hepatic expression with the 28 SNPs they examined. However, none of them appeared to correlate well.[179]

Confounding the correlation measurements mentioned above is the fact that many classic inducers of CYP3A activity are known that affect the high variable expression, including rifampicin, dexamethasone, carbamezapine, and phenobarbitol.[123] In addition, polymorphisms in the PXR nuclear transcriptional regulator locus explain much of the observed CYP3A variability.[180] Recently, it was announced (Oct. 28, 2002) that the x-ray crystal structure of CYP3A4 has been solved (by Astex Technology, whose prelude to this announcement was the x-ray crystal structure of CYP2C9). This will surely spur many more analyses leading to a better understanding of the pharmacogenetic significance of 3A4 polymorphisms.

2.4.12 CYP3A5, CYP3A7 and CYP3A43

CYP3A5 is closely related to CYP3A4 and the two have historically been confused. 3A5 is widely found in human intestine,[181] lung,[182] and fetal liver,[183] but is only found in a percentage of adult human livers (60% of African Americans and 33% of Caucasians.[183,184] Interestingly, CYP3A5 mRNA also has been identified by reverse transcriptase-polymerase chain reaction in human pituitary gland and shown by immuno-histochemistry to be localized to growth hormone containing cells of the anterior pituitary gland.[185] Several alleles have been identified that account for the lack of hepatic 3A5. Of these, *CYP3A5*3* is more common, resulting in a cryptic splice site and premature chain termination.[186-188] The frequencies of the variant that allows for normal splicing and thus expression of CYP3A5 transcripts are were found to be 5% in Caucasians, 29% in Japanese, 27% in Chinese, 30% in Koreans and 73% in African-Americans,[187] indicating wide inter-racial variability in the expression of this isoform. CYP3A5 may be the most important genetic contributor to interindividual and interracial differences in CYP3A-dependent drug clearance and in responses to many medicines.[188] CYP3A7 is mainly restricted to fetal liver, though it has been reported in a minority of adult liver sample.[183] Finally, the CYP3A43 gene has recently been reported but nothing yet is known about the catalytic properties of its protein product.[189]

3   PHARMACOGENETICS: PROMISE VERSUS PRACTICE

The dawn of the genetic era has been heralded as a new age in medicine. Our increasing understanding of the underlying causes of some adverse drug reactions and drug-drug interactions may lead to avoiding them entirely. Individualized medicine based on pharmacogenetic information may become common prescribing practice. Many hold a vision that patients may be genotyped for drug metabolizing enzymes and drug targets in the physician's office, and that patients' genotypes become part of their medical record to aid physicians in prescribing medications (in both the selection of which medication to prescribe and what dosing regimen should be applied). While such a scenario is futuristic, those involved in the discovery and development of new medications and those in the medical profession must pay attention to these developments. A recent article raises some very important considerations, many of which are summarized in the following paragraphs.[190]

To this point, the few known and well-characterized clinically important pharmacogenetic polymorphisms were initially discovered because of their occurrence at a high frequency in the population and because the expression of these polymorphisms resulted in clinically relevant effects. It was only after the clinical observations were made that the polymorphisms were traced back to their genetic roots. However, the technologies that exist today have the capabilities to seek out the genetic polymorphisms before the ramifications of these polymorphisms, if any, are identified or understood. Oligonucleotide microarrays, or "gene chips" are being miniaturized, and it will soon be possible to screen for millions of SNPs simultaneously. As an alternative approach, robotic high-throughput DNA sequencing technologies also exist, thanks to the human genome sequencing project. Thus, the process is now reversed: genetic polymorphisms can be found before possessing an understanding of the meaning of the polymorphisms to human disease and clinical practice. So with the technology in hand, the question arises as to which SNPs or other DNA biomarkers should be sought after as most SNPs are silent. Many attractive, yet incorrect hypotheses have been proven wrong when a specific trait has been attributed to a polymorphism. Also, part of the reason that the polymorphisms described in this chapter were discovered is because they resulted in discreet bimodal or multimodal population groups. On the other hand, multigenic traits, those that arise due to changes in multiple genetic loci, are not as distinct, particularly if they affect two or more pathways, i.e., metabolism via P450 polymorphism and distribution via transporter polymorphism. Ultimately, medicines may be group-specific, as identified by allele frequencies and polymorphisms, but not, by and large, individual-specific.

Another promise of pharmacogenetics is the reduced clinical trial costs that might be achieved by testing only genetic sub-populations expected to respond to the test drug. However, it is not clear whether such streamlined trials will satisfy FDA requirements. Additionally, recruitment for such studies will be difficult if the genetic trait is rare. For example, if a clinical trial were to require subjects homozygous for an allele with a frequency of 10%, then just 1% of the population will satisfy this requirement and a very large number of prospective study subjects will need to be screened in order to obtain a large enough number of subjects for the study. This benefit would need to be counterbalanced with the notion that a smaller number of subjects may be adequate for the study in order to demonstrate the effect with statistical significance, since the subject population will have been specifically enriched with those anticipated to show a positive response. However, this changes the paradigm of drug discovery from blockbusters that are able to be prescribed to virtually anyone to agents that can only be prescribed to specific subpopulations, and many companies express concerns of market fragmentation in this paradigm. Newer drugs such as Herceptin, an anti-breast-tumor antibody and Glivec, an anti-tumor specific antibody that recognizes tumors containing the Philadelphia chromosomal aberration, have shown that it is possible to market a drug to such a subpopulation. Concerns over market fragmentation may be alleviated by the example of Glivec, which originally fragmented the leukemia drug market, but was then found to be useful in other disorders containing the same chromosomal abnormality.

The use of genetic information in prescription practice could reduce the frequencies of adverse drug reactions, however there are many other potential factors that can contribute to adverse drug reactions. Environmental factors, such as diet, alcohol, tobacco use, concomitant medications, as well as other non-genetic factors such as age, sex, body weight, disease, etc. can also affect response to drugs. Some of these non-genetic factors can play a more substantial role in adverse drug reactions.   Thus, genetic information alone will not resolve the challenge of attaining complete efficacy and complete lack of toxicity with pharmacotherapy. Additionally, a large obstacle to the realization of the use of pharmacogenetics in drug therapy is the negative public perception of genotyping. Bioethics and informed consent are major obstacles. "SNP prints" have been argued to be a way to distinguish between genotyping for disease susceptibility and genotyping for polymorphisms in genes involved in drug metabolism and drug disposition.[190] Herceptest™, which identifies candidates for Herceptin™ treatment, showed that this approach was feasible. Education appears to be the answer in getting the public to accept and embrace these medicines that are utilized with genetic information. A good review of the ethical, legal and social issues at stake has been presented recently by Renegar.[191]

The Pharmacogenetics Working Group (www.pharmgkb.org) was spearheaded by GlaxoSmithKline and has 12 other members in the pharmaceutical industry. Their goal is to standardize the terminology, protocols and consent forms in order that such genetic testing can be done smoothly and uniformly.[192] Support may also be found from the Pharmacogenetics Research Network, sponsored by the National Institute of General Medical Sciences. The SNP consortium is yet another pharmacogenetic working group. This is a non-profit organization created by the Wellcome Trust and includes at least 11 other members. It is dedicated to making information on SNP biomarkers publicly available and can be found on the internet at http://snp.cshl.org/.

4　　CONCLUSIONS

The discovery of polymorphism in the human cytochrome P450 genes presents a Pandora's box of both promise and difficulties. Undiscovered polymorphisms are almost certain to arise, given the sheer number of human P450s and their SNPs. It appears to be likely that SNP printing will be used to choose safer medications in the future. Our knowledge of pharmacogenetics will also lead us to a better understanding of many disease processes, since many polymorphisms involving drug substrates overlap with endogenous metabolic pathways. However, many hurdles have yet to be overcome, largely in the arena of public opinion, before these goals can become a reality.

ACKNOWLEDGEMENT

The experimental work presented in Section 2.2 was kindly funded by NIH grant ES09894.

REFERENCES

1. Gorrod, A.E. and M.D. Oxon, *Lancet*, **1902**, 1616-1620.
2. Vogel, F. and A.G. Motulsky, *Human Genetics: Problems and Approaches*. 2nd ed. 1986, Berlin; New York: Springer-Verlag. 807.
3. Mahgoub, A., *et al.*, *Lancet*, **1977,** *2,* 584-586.
4. Eichelbaum, M., *et al.*, *Eur. J. Clin. Pharmacol.*, **1979,** *16,* 183-7.
5. Johansson, I., *et al.*, *Proceedings of the National Academy of Sciences of the United States of America*, **1993,** *90,* 11825-9.
6. Wain, H.M., *et al.*, *Genomics*, **2002,** *79,* 464-70.
7. Sachse, C., *et al.*, *American Journal of Human Genetics*, **1997,** *60,* 284-95.
8. Huang, Z., *et al.*, *Cancer Res*, **1997,** *57,* 2589-92.
9. Lash, L.H., *et al.*, *J. Pharmacol. Exp. Ther.*, **2003,** *24,* 24.
10. London, S.J., *et al.*, *Pharmacogenetics*, **1996,** *6,* 527-33.
11. Leemann, T., C. Transon, and P. Dayer, *Life Sci*, **1993,** *52,* 29-34.
12. Goldstein, J.A., *Br. J. Clin. Pharmacol.*, **2001,** *52,* 349-55.
13. Haining, R.L., *et al.*, *Archives of Biochemistry and Biophysics*, **1996,** *333,* 447-58.
14. Steward, D.J., *et al.*, *Pharmacogenetics*, **1997,** *7,* 361-7.
15. Takanashi, K., *et al.*, *Pharmacogenetics*, **2000,** *10,* 95-104.
16. Yamazaki, H., *et al.*, *Biochem. Pharmacol.*, **1998,** *56,* 243-51.
17. Ieiri, I., *et al.*, *Ther Drug Monit*, **2000,** *22,* 237-44.
18. Dickmann, L.J., *et al.*, *Mol. Pharmacol.*, **2001,** *60,* 382-7.
19. Tracy, T.S., *et al.*, *Drug Metab. Dispos.*, **2002,** *30,* 385-90.
20. Kidd, R.S., *et al.*, *Pharmacogenetics*, **2001,** *11,* 803-8.
21. Afzelius, L., *et al.*, *Mol. Pharmacol.*, **2001,** *59,* 909-19.
22. Rettie, A.E., *et al.*, *Chemical Research in Toxicology*, **1992,** *5,* 54-9.
23. Rettie, A.E., *et al.*, *Pharmacogenetics*, **1994,** *4,* 39-42.
24. Wienkers, L.C., *et al.*, *Drug Metabolism and Disposition*, **1996,** *24,* 610-4.
25. Sullivan, K.-T.H., *et al.*, *Pharmacogenetics*, **1996,** *6,* 341-9.
26. Rettie, A.E., *et al.*, *Epilepsy Res.*, **1999,** *35,* 253-5.
27. Crespi, C.L. and V.P. Miller, *Pharmacogenetics*, **1997,** *7,* 203-10.
28. Higashi, M.K., *et al.*, *JAMA*, **2002,** *287,* 1690-8.
29. Hutzler, J.M., *et al.*, *Drug Metab. Dispos.*, **2002,** *30,* 1194-200.
30. Idle and Smith, *Drug Metab. Rev.*, **1979,** *9,* 301-317.
31. Mura, C., *et al.*, *Human Genetics*, **1993,** *92,* 367-72.

32. Panserat, S., *et al.*, *Human Genetics*, **1994**, *94,* 401-6.
33. Gonzalez, F.J., *et al.*, *Nature*, **1988**, *331,* 442-6.
34. Gough, A.C., *et al.*, *Nature*, **1990**, *347,* 773-6.
35. Bertilsson, L., *Clinical Pharmacokinetics*, **1995**, *29,* 192-209.
36. Johansson, I., *et al.*, *European Journal of Clinical Pharmacology*, **1991**, *40,* 553-6.
37. Lee, E.J. and K. Jeyaseelan, *British Journal of Clinical Pharmacology*, **1994**, *37,* 605-7.
38. Lane, H.Y., *et al.*, *Clinical Pharmacology and Therapeutics*, **1996**, *60,* 696-8.
39. Evans, W.E., *et al.*, *Journal of Clinical Investigation*, **1993**, *91,* 2150-4.
40. Daly, A.K., *et al.*, *Pharmacogenetics*, **1996**, *6,* 193-201.
41. Daly, A.K., K.S. Fairbrother, and J. Smart, *Toxicol Lett*, **1998**, *102-103,* 143-7.
42. Kokoris, M., *et al.*, *Mol Diagn*, **2000**, *5,* 329-40.
43. Tyndale, R., *et al.*, *Pharmacogenetics*, **1991**, *1,* 26-32.
44. Evert, B., E.U. Griese, and M. Eichelbaum, *Naunyn Schmiedebergs Archives of Pharmacology*, **1994**, *350,* 434-9.
45. Oscarson, M., *et al.*, *Molecular Pharmacology*, **1997**, *52,* 1034-40.
46. Zanger, U.M., *et al.*, *Pharmacogenetics*, **2001**, *11,* 573-85.
47. de, G.-M.J., *et al.*, *Chemical Research in Toxicology*, **1996**, *9,* 1079-91.
48. Guex, N. and M.C. Peitsch, *Electrophoresis*, **1997**, *18,* 2714-23.
49. Lindberg, R.L. and M. Negishi, *Nature*, **1989**, *339,* 632-4.
50. Matsunaga, E., *et al.*, *Journal of Biological Chemistry*, **1990**, *265,* 17197-201.
51. Modi, S., *et al.*, *Biochemistry*, **1996**, *35,* 4540-50.
52. Droll, K., *et al.*, *Pharmacogenetics*, **1998**, *8,* 325-33.
53. Omura, T., *et al.*, *Fed. Proc.*, **1965**, *24,* 1181-9.
54. Luckow, V.A., *et al.*, *Journal of Virology*, **1993**, *67,* 4566-79.
55. Koenigs, L.L., *et al.*, *Biochemistry*, **1999**, *38,* 2312-9.
56. Yu, A., *et al.*, *J. Pharmacol. Exp. Ther.*, **2002**, *303,* 1291-300.
57. Fukuda, T., *et al.*, *Arch. Biochem. Biophys.*, **2000**, *380,* 303-8.
58. Strobl, G.R., *et al.*, *Journal of Medicinal Chemistry*, **1993**, *36,* 1136-45.
59. de, G.-M.J., *et al.*, *Chemical Research in Toxicology*, **1997**, *10,* 41-8.
60. de Groot, M.J., *et al.*, *J. Med. Chem.*, **1999**, *42,* 4062-70.
61. Ekins, S., M.J. de Groot, and J.P. Jones, *Drug Metab. Dispos.*, **2001**, *29,* 936-44.
62. McFayden, M.C., W.T. Melvin, and G.I. Murray, *Biochem. Pharmacol.*, **1998**, *55,* 825-30.
63. Voirol, P., *et al.*, *Brain Res.*, **2000**, *855,* 235-43..
64. Bertilsson, L., *et al.*, *Lancet*, **1989**, *1,* 555.
65. Llerena, A., *et al.*, *Acta Psychiatrica Scandinavica*, **1993**, *87,* 23-8.
66. Mart'inez, C., *et al.*, *Pharmacogenetics*, **1997**, *7,* 85-93.
67. Hiroi, T., *et al.*, *Biochemical Pharmacology*, **1997**, *53,* 1937-9.
68. Hiroi, T., *et al.*, *Endocrinology*, **2001**, *142,* 3901-8.
69. Fernstrom, J.D. and M.H. Fernstrom, *Nestle Nutr. Workshop Ser. Clin. Perform. Programme*, **2001**, *5,* 117-31.
70. Rosel, P., *et al.*, *J. Neural. Transm.*, **1997**, *104,* 89-96.
71. Siegle, I., *et al.*, *Pharmacogenetics*, **2001**, *11,* 237-45.
72. Jefcoate, C.R., *J. Biol. Chem.*, **1975**, *250,* 4663-70.
73. Yu, A. and R.L. Haining, *Drug Metab. Dispos.*, **2001**, *29,* 1514-20.

74. Yu, A., *et al.*, *Drug Metab. Dispos.*, **2001**, *29*, 1362-5.

75. Marez, D., *et al.*, *Pharmacogenetics*, **1997**, *7*, 193-202.

76. Panserat, S., *et al.*, *Br. J. Clin. Pharmacol.*, **1999**, *47*, 121-4.

77. Shimada, T., *et al.*, *Pharmacogenetics*, **2001**, *11*, 143-56.

78. Seeman, P. and H.H. Van Tol, *Curr. Opin. Neurol. Neurosurg.*, **1993**, *6*, 602-8.

79. Udenfriend, S., *Pharmacol. Rev.*, **1966**, *18*, 43-51.

80. Johansson, I., *et al.*, *Molecular Pharmacology*, **1994**, *46*, 452-9.

81. Shimada, T. and F.P. Guengerich, *Cancer Res.*, **1991**, *51*, 5284-91.

82. McKinnon, R.A., *et al.*, *Hepatology*, **1991**, *14*, 848-56.

83. Zhang, Z.Y., *et al.*, *Cancer Res.*, **1996**, *56*, 3926-33.

84. Persson, I., I. Johansson, and M. Ingelman-Sundberg, *Biochem. Biophys. Res. Commun.*, **1997**, *231*, 227-30.

85. Cascorbi, I., J. Brockmoller, and I. Roots, *Cancer Res.*, **1996**, *56*, 4965-9.

86. Kellermann, G., M. Luyten-Kellermann, and C.R. Shaw, *Am. J. Hum. Genet.*, **1973**, *25*, 327-31.

87. Sachse, C., *et al.*, *Br. J. Clin. Pharmacol.*, **1999**, *47*, 445-9.

88. Nakajima, M., *et al.*, *J. Biochem. (Tokyo)*, **1999**, *125*, 803-8.

89. Corchero, J., *et al.*, *Pharmacogenetics*, **2001**, *11*, 1-6.

90. Yamazaki, H., *et al.*, *Mol. Pharmacol.*, **1994**, *46*, 568-77.

91. Hakkola, J., *et al.*, *Biochemical Pharmacology*, **1994**, *48*, 59-64.

92. Huang, J.D., *et al.*, *Drug Metab. Dispos.*, **1999**, *27*, 98-101.

93. Chevalier, D., *et al.*, *Hum. Mutat.*, **2001**, *17*, 355-6.

94. Burke, M.D., *et al.*, *Biochem. Pharmacol.*, **1985**, *34*, 3337-45.

95. Sesardic, D., *et al.*, *Br. J. Clin. Pharmacol.*, **1990**, *29*, 651-63.

96. Brosen, K., *et al.*, *Biochem. Pharmacol.*, **1993**, *45*, 1211-4.

97. Rasmussen, B.B., *et al.*, *Br. J. Clin. Pharmacol.*, **1995**, *39*, 151-9.

98. Bailey, L.R., *et al.*, *Cancer Res.*, **1998**, *58*, 5038-41.

99. Li, D.N., *et al.*, *Pharmacogenetics*, **2000**, *10*, 343-53.

100. Sutter, T.R., *et al.*, *J. Biol. Chem.*, **1994**, *269*, 13092-9.

101. Shimada, T., *et al.*, *Carcinogenesis*, **1999**, *20*, 1607-13.

102. McLellan, R.A., *et al.*, *Arch. Biochem. Biophys.*, **2000**, *378*, 175-81.

103. Stoilov, I., A.N. Akarsu, and M. Sarfarazi, *Hum Mol Genet*, **1997**, *6*, 641-7.

104. Stoilov, I., *et al.*, *Am. J. Hum. Genet.*, **1998**, *62*, 573-84.

105. Aklillu, E., *et al.*, *Mol. Pharmacol.*, **2002**, *61*, 586-94.

106. Rao, Y., *et al.*, *Mol. Pharmacol.*, **2000**, *58*, 747-55.

107. Cashman, J.R., *et al.*, *Chem. Res. Toxicol.*, **1992**, *5*, 639-46.

108. Nakajima, M., *et al.*, *J. Pharmacol. Exp. Ther.*, **1996**, *277*, 1010-5.

109. Yamano, S., J. Tatsuno, and F.J. Gonzalez, *Biochemistry*, **1990**, *29*, 1322-9.

110. Nesnow, S., *et al.*, *Mutat. Res.*, **1994**, *324*, 93-102.

111. Duescher, R.J. and A.A. Elfarra, *Arch. Biochem. Biophys.*, **1994**, *311*, 342-9.

112. Yokoi, T., *et al.*, *Pharmacogenetics*, **1996**, *6*, 395-401.

113. Oscarson, M., *et al.*, *FEBS Lett.*, **1998**, *438*, 201-5.

114. Pianezza, M.L., E.M. Sellers, and R.F. Tyndale, *Nature*, **1998**, *393*, 750.

115. Sellers, E.M., H.L. Kaplan, and R.F. Tyndale, *Clin. Pharmacol. Ther.*, **2000**, *68*, 35-43.

116. Miyamoto, M., *et al.*, *Biochem. Biophys. Res. Commun.*, **1999**, *261*, 658-60.
117. Oscarson, M., *et al.*, *Hum. Mutat.*, **2002**, *20*, 275-83.
118. Xu, C., *et al.*, *Biochem. Biophys. Res. Commun.*, **2002**, *290*, 318-24.
119. Hidestrand, M., *et al.*, *Drug Metab. Dispos.*, **2001**, *29*, 1480-4.
120. Code, E.L., *et al.*, *Drug Metab. Dispos.*, **1997**, *25*, 985-93.
121. Ekins, S. and S.A. Wrighton, *Drug Metab Rev*, **1999**, *31*, 719-54.
122. Faucette, S.R., *et al.*, *Drug Metab. Dispos.*, **2000**, *28*, 1222-30.
123. Pelkonen, O., *et al.*, *Xenobiotica*, **1998**, *28*, 1203-53.
124. Lang, T., *et al.*, *Pharmacogenetics*, **2001**, *11*, 399-415.
125. Ariyoshi, N., *et al.*, *Biochem. Biophys. Res. Commun.*, **2001**, *281*, 1256-60.
126. Sonnichsen, D.S., *et al.*, *J. Pharmacol. Exp. Ther.*, **1995**, *275*, 566-75.
127. Nadin, L. and M. Murray, *Biochem. Pharmacol.*, **1999**, *58*, 1201-8.
128. Li, X.Q., *et al.*, *J. Pharmacol. Exp. Ther.*, **2002**, *300*, 399-407.
129. Baldwin, S.J., S.E. Clarke, and R.J. Chenery, *Br. J. Clin. Pharmacol.*, **1999**, *48*, 424-32.
130. Rifkind, A.B., *et al.*, *Arch. Biochem. Biophys.*, **1995**, *320*, 380-9.
131. Zeldin, D.C., *et al.*, *Arch. Biochem. Biophys.*, **1996**, *330*, 87-96.
132. Dai, D., *et al.*, *Pharmacogenetics*, **2001**, *11*, 597-607.
133. Bahadur, N., *et al.*, *Biochem. Pharmacol.*, **2002**, *64*, 1579-89.
134. Romkes, M., *et al.*, *Biochemistry*, **1991**, *30*, 3247-55.
135. Komai, K., *et al.*, *Pharmacogenetics*, **1996**, *6*, 117-9.
136. Inoue, K., H. Yamazaki, and T. Shimada, *Xenobiotica*, **1998**, *28*, 403-11.
137. Meyer, U.A., *et al.*, *Xenobiotica*, **1986**, *16*, 449-64.
138. Shimada, T., K.S. Misono, and F.P. Guengerich, *J. Biol. Chem.*, **1986**, *261*, 909-21.
139. De Morais, S.M., *et al.*, *Mol. Pharmacol.*, **1994**, *46*, 594-8.
140. Ibeanu, G.C., *et al.*, *Pharmacogenetics*, **1998**, *8*, 129-35.
141. Ibeanu, G.C., *et al.*, *J. Pharmacol. Exp. Ther.*, **1998**, *286*, 1490-5.
142. Ferguson, R.J., *et al.*, *J. Pharmacol. Exp. Ther.*, **1998**, *284*, 356-61.
143. Ibeanu, G.C., *et al.*, *J. Pharmacol. Exp. Ther.*, **1999**, *290*, 635-40.
144. Blaisdell, J., *et al.*, *Pharmacogenetics*, **2002**, *12*, 703-11.
145. Andersson, T., *et al.*, *Br. J. Clin. Pharmacol.*, **1994**, *38*, 131-7.
146. Wright, J.D., N.A. Helsby, and S.A. Ward, *Br. J. Clin. Pharmacol.*, **1995**, *39*, 441-444.
147. Funck-Brentano, C., *et al.*, *J. Pharmacol. Exp. Ther.*, **1997**, *280*, 730-8.
148. Furuta, T., *et al.*, *Ann Intern Med*, **1998**, *129*, 1027-30.
149. Suzuki, H., *et al.*, *Drug Metab. Dispos.*, **2002**, *30*, 235-9.
150. Patten, C.J., *et al.*, *Chem. Res. Toxicol.*, **1993**, *6*, 511-8.
151. Gu, L., *et al.*, *Pharmacogenetics*, **1992**, *2*, 73-7.
152. Peter, R., *et al.*, *Chem. Res. Toxicol.*, **1990**, *3*, 566-73.
153. Kharasch, E.D., *et al.*, *Clin. Pharmacol. Ther.*, **1994**, *55*, 434-40.
154. Spracklin, D.K., *et al.*, *J. Pharmacol. Exp. Ther.*, **1997**, *281*, 400-11.
155. Kharasch, E.D., *et al.*, *Lancet*, **1996**, *347*, 1367-71.
156. Tierney, D.J., A.L. Haas, and D.R. Koop, *Arch. Biochem. Biophys.*, **1992**, *293*, 9-16.

157. Guengerich, F.P., D.H. Kim, and M. Iwasaki, *Chem. Res. Toxicol.*, **1991**, *4*, 168-79.
158. French, S.W., *et al.*, *Exp. Mol. Pathol.*, **1993**, *58*, 61-75.
159. Laethem, R.M., *et al.*, *J. Biol. Chem.*, **1993**, *268*, 12912-8.
160. Hu, Y., *et al.*, *Mol. Pharmacol.*, **1997**, *51*, 370-6.
161. Manyike, P.T., *et al.*, *Clin. Pharmacol. Ther.*, **2000**, *67*, 275-82.
162. Johansson, I., *et al.*, *Biochem. Biophys. Res. Commun.*, **1990**, *173*, 331-8.
163. Hargreaves, M.B., *et al.*, *Drug Metab. Dispos.*, **1994**, *22*, 806-10.
164. Kato, S., *et al.*, *Cancer Res.*, **1992**, *52*, 6712-5.
165. Hildesheim, A., *et al.*, *J. Natl. Cancer Inst.*, **1997**, *89*, 1207-12.
166. Ingelman, S.-M., *et al.*, *Exs.*, **1994**, *71*, 197-207.
167. King, L.M., *et al.*, *Mol. Pharmacol.*, **2002**, *61*, 840-52.
168. Shou, M., *et al.*, *Biochemistry*, **1994**, *33*, 6450-5.
169. Korzekwa, K.R., *et al.*, *Biochemistry*, **1998**, *37*, 4137-47.
170. Baldwin, S.J., *et al.*, *Xenobiotica*, **1995**, *25*, 261-70.
171. Back, D.J., *et al.*, *J. Steroid Biochem. Mol. Biol.*, **1991**, *38*, 219-25.
172. Shou, M., *et al.*, *Curr. Drug Metab.*, **2001**, *2*, 17-36.
173. Thummel, K.E., *et al.*, *J. Pharmacol. Exp. Ther.*, **1994**, *271*, 557-66.
174. Guengerich, F.P., *Am. J. Clin. Nutr.*, **1995**, *61*, 651S-658S.
175. Sata, F., *et al.*, *Clin. Pharmacol. Ther.*, **2000**, *67*, 48-56.
176. Hsieh, K.P., *et al.*, *Drug Metab. Dispos.*, **2001**, *29*, 268-73.
177. Eiselt, R., *et al.*, *Pharmacogenetics*, **2001**, *11*, 447-58.
178. Dai, D., *et al.*, *J. Pharmacol. Exp. Ther.*, **2001**, *299*, 825-31.
179. Lamba, J.K., *et al.*, *Pharmacogenetics*, **2002**, *12*, 121-32.
180. Zhang, J., *et al.*, *Pharmacogenetics*, **2001**, *11*, 555-72.
181. Gibbs, M.A., *et al.*, *Drug Metab. Dispos.*, **1999**, *27*, 180-7.
182. Kivisto, K.T., *et al.*, *Pharmacogenetics*, **1997**, *7*, 295-302.
183. Schuetz, J.D., D.L. Beach, and P.S. Guzelian, *Pharmacogenetics*, **1994**, *4*, 11-20.
184. Aoyama, T., *et al.*, *J. Biol. Chem.*, **1989**, *264*, 10388-95.
185. Murray, G.I., *et al.*, *FEBS Lett*, **1995**, *364*, 79-82.
186. Chou, F.C., S.J. Tzeng, and J.D. Huang, *Drug Metab. Dispos.*, **2001**, *29*, 1205-9.
187. Hustert, E., *et al.*, *Pharmacogenetics*, **2001**, *11*, 773-9.
188. Kuehl, P., *et al.*, *Nat Genet*, **2001**, *27*, 383-91.
189. Domanski, T.L., *et al.*, *Mol. Pharmacol.*, **2001**, *59*, 386-92.
190. Bromley, N., *Pharmafocus*, **2002,**
191. Renegar, G., P. Rieser, and P. Manasco, *Expert Rev. Mol. Diagn.*, **2001**, *1*, 255-63.
192. Hewett, M., *et al.*, *Nucleic Acids Res.*, **2002**, *30*, 163-5.

Chapter 12

Role of Intestinal Cytochromes P450 in Drug Disposition

Mary F. Paine[1] and Kenneth E. Thummel[2]

[1]*Division of Pharmacotherapy, School of Pharmacy, University of North Carolina, Durham, NC*
[2]*Department of Pharmaceutics, University of Washington, Seattle, WA*

ABSTRACT

Drugs are prime examples of ingested xenobiotics. Following their disintegration and dissolution in the gastrointestinal (GI) lumen, ensuing drug molecules must make an obligatory passage through a series of organs before reaching the systemic circulation. The first in the series, the GI tract, contains significant levels of numerous drug biotransformation enzymes, including the cytochromes P450 (CYPs). Although it has been known for well over a decade that CYP3A (CYP3A4 + CYP3A5) is the predominant CYP subfamily expressed in the adult small intestinal mucosa, the role of intestinal CYP3A in human drug metabolism and its potential impact on the oral bioavailability of a drug, and hence therapeutic outcome, has only recently become appreciated. Indeed, through the use of direct (duodenal instillation into anhepatic transplant patients) and indirect (intravenous vs. oral administration) methods, it appears that the intestinal contribution to the overall first-pass extraction of the CYP3A substrates midazolam, nifedipine, and verapamil rivals the hepatic contribution. Although other CYP isoforms have been identified and characterized in the human intestine (CYP2D6, CYP2C9, CYP2C19, CYP1A1, CYP2J2), their importance in drug disposition remains to be determined. In vitro studies, however, indicate that, as with CYP3A, all exhibit large interindividual variation in specific protein content and catalytic activity. Also like CYP3A, the expression and activity of

CYP2D6, CYP2C9, and CYP2C19 appear to vary along the length of the small intestine; that is, protein and activity tend to increase slightly from duodenum to mid-jejunum then decline to distal ileum, suggesting that the site of absorption could contribute to the large interindividual variation often observed in the extent of the systemic exposure of orally administered drugs. Although the liver and intestine express several of the same CYPs, exhibiting similar kinetic properties, some appear not to be constitutively expressed in the intestine (CYP1A2, CYP2A6, CYP2B6), while only one appears not to be constitutively expressed in the liver (CYP1A1). Similarly, at least for CYP3A4, while several modulators (rifampin, St. John's wort, ketoconazole, clarithromycin) can alter both intestinal and hepatic enzyme function, others appear to modulate only the hepatic form (phenobarbital, dexamethasone, clotrimazole) or only the intestinal form (grapefruit juice, Seville orange juice). Moreover, although the pregnane X receptor may be important in the regulation of constitutive hepatic CYP3A4, other factors (e.g., the vitamin D receptor) are likely involved in the regulation of constitutive intestinal CYP3A4. Finally, it has become apparent that for some drugs, P-glycoprotein-mediated efflux at the lumenal membrane of the intestinal mucosa may modulate the efficiency of intestinal CYP3A-dependent first-pass metabolism, greatly complicating in vitro to in vivo predictions of drug disposition and drug-drug interactions. Collectively, these intestine-liver 'disconnects' support the contention that intestinal and hepatic CYPs are under different regulatory controls and intimate that hepatic models of CYP-mediated metabolism will not necessarily predict the behavior of a drug in the intestine. Further refinement of human intestinal cell culture models and/or the identification of an appropriate animal model should improve our understanding of the unique nature of intestinal CYPs and allow us to not only better predict the impact of the intestine on the overall first-pass extraction of existing drugs, but also to improve the drug selection process at the preclinical level.

1   INTRODUCTION

1.1   *Absorption and the First-Pass Effect*

The oral route continues to be the most popular, convenient, and generally safest means for the administration of systemically acting drugs. It is not the most efficient, however, due to the several anatomic and physiologic obstacles drugs encounter during *absorption* – that is, from the time of ingestion until the time of appearance in the systemic circulation. For some drugs, these obstacles can be so great as to obviate their use as oral agents. Isoproterenol, dihydroergotamine, lidocaine, nitroglycerin, fentanyl, and naloxone are such examples.[1] All of these agents exhibit a high *first-pass effect*, which is defined as the loss of drug as the dose passes, for the first time, through sites of elimination during absorption.[1] Therefore, before orally administered drugs reach the systemic circulation and exert their effects in the target tissues, loss can occur in any one or a combination of the following obligatory series of organs: the gastrointestinal (GI) tract, the liver, the right heart, the lungs, and the left heart. Processes than can lead to significant loss of active drug during first-pass include incomplete release from the dosage form, degradation in the GI lumen, poor permeation across the GI epithelial barrier, active efflux into the GI lumen, excretion into the bile, and metabolism. Of these, only metabolism can occur in all of the aforementioned first-pass eliminating organs.

Metabolism is the major means by which the body eliminates drugs.[1] Enzymatic alteration of the drug usually results in inactive metabolites and reduced pharmacological activity. Therefore, extensive first-pass metabolism can lead to systemic concentrations too low to achieve the desired effect. Of the first-pass eliminating organs, the liver and GI tract (specifically, the small intestine) are the most commonly implicated, as these organs contain by far the highest levels of drug biotransformation enzymes. While the role of such enzymes in the liver has been well established for some time, relatively less is known about the role of the complement of enzymes in the small intestine. Nevertheless, much progress has been made in the last decade regarding the identification of different subfamilies and individual isoforms. As a consequence, the potential impact of intestinal enzymes on drug disposition has become increasingly recognized. Before discussing these enzymes and their clinical relevance, however, an understanding of the absorption process is warranted.

1.2   Drug Movement Through the Gastrointestinal Barrier

The GI tract represents the first barrier to the complete absorption of orally administered drugs. As the majority are given as solid dosage forms, disintegration and then dissolution of the dosage form must occur in the lumen before drug molecules can be absorbed into the systemic circulation. Several physicochemical factors related to both the drug and the GI environment govern the rate of dissolution. Drug factors include solubility, particle size, salt form, complexation, and crystal form.[2] Environmental factors include pH and lumenal stirring. Following dissolution of the dosage form, ensuing drug molecules must then traverse the contents of the lumen (e.g., water, food, bacteria) and a mucous layer before they can be absorbed across the GI wall into the lamina propria, which contain the capillaries that eventually lead to the portal vein. Most drugs are weak acids or weak bases. According to the pH partition hypothesis, which assumes that only unionized drug can traverse biological membranes, weak acids are predicted to be more rapidly absorbed from the stomach (pH 1 to 3) than from the small intestine (pH 6 to 8) or large intestine (pH 6 to 7),[1,3] and the converse is predicted for weak bases. However, the vast majority of drugs, whether a weak acid or weak base, and whether predominantly in the unionized or ionized form, are largely absorbed in the small intestine. These apparent anomalies can be explained by the small intestine being favored over the stomach and large intestine in terms of surface area and permeability, two key anatomic/physiologic factors that govern the rate and extent of drug absorption.

1.3   Anatomic and Physiologic Considerations

The adult small intestine, with an anatomical length (i.e., length at autopsy or after surgical removal) that measures approximately 500 cm,[4] is divided into three segments: the duodenum, jejunum, and ileum. The duodenum, 25 to 30 cm in length, begins just distal to the pyloric sphincter of the stomach and ends at the ligament of Treitz. While the ligament of Treitz distinguishes the duodenum from the jejunum, no such anatomical landmark distinguishes the jejunum from the ileum; however, it is generally assumed that the jejunum represents the proximal two-fifths, and the ileum represents the distal three-fifths of the remainder of the small intestine (approximately 190 and 280 cm, respectively). In comparison, the length of the entire large intestine is approximately 160 cm.[4]

The primary function of the small intestine is to absorb nutrients. The unique morphology of the mucosal surface facilitates this task. Illustrated in Figure 1 is a cross-section of the small intestine showing extensive folding (folds of Kerckering). The folds of Kerckering, or plicae circulares, are circular folds created by mucosal/submucosal invaginations into the lumen. Predominant in the duodenum and jejunum and essentially absent by middle ileum, these circular folds are lined with finger-like projections (villi). At the base of the villi are the crypts of Lieberkuehn, which consist of undifferentiated (crypt) cells. Over the course of two to three days, crypt cells migrate to the villous tips while maturing into various cell types, including mucus-secreting goblet cells, enteroendocrine cells, and absorptive cells (enterocytes); after three more days, the mature cells of the villous tips are shed into the lumen.[5] Enterocytes, the predominant cell types that compose the single layer of absorptive columnar epithelial cells lining the villi, are further lined with microvilli. This combination of circular folds, villi, and microvilli thus creates a tremendous absorptive surface area. In adults, this area has been estimated to be 200 m^2, which is roughly the size of a tennis court. In comparison, the surface area of the stomach is approximately 1 m^2,[1] and that of the large intestine is approximately 0.3 m^2.[4]

In addition to having a much greater surface area than the stomach and large intestine, the small intestine is more permeable, due in part to having both a relatively thin epithelial layer and a lower electrical resistance.[1] Therefore, as with nutrients, the small intestine is the prime site for the absorption of drugs. Moreover, because surface area and permeability decline from proximal to distal regions, it can be argued that, at least for immediate-release formulations, drug absorption occurs primarily in the duodenum and jejunum.

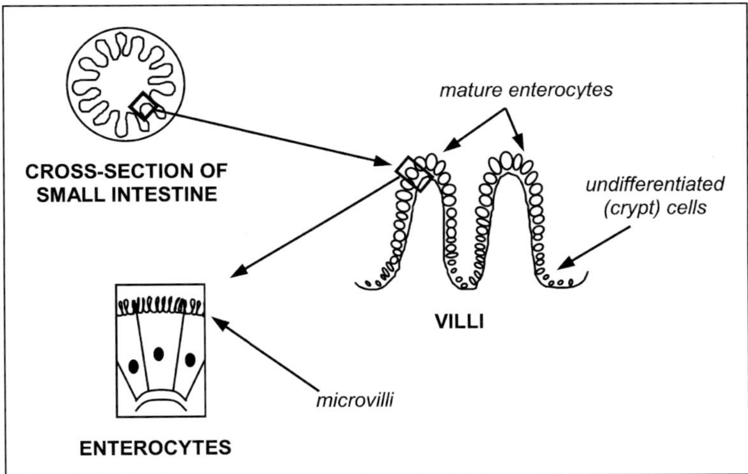

Figure 1. Morphology of the small intestine. Invaginations in the lumen are lined with villi, which are lined with a single layer of absorptive columnar epithelial cells (enterocytes), which are further lined with microvilli (adapted from reference 5).

Absorption of drug molecules across the small intestinal epithelial layer occurs via paracellular or transcellular mechanisms. That is, the molecule can either passively diffuse between enterocytes (paracellular), or it can passively diffuse or be actively transported across the apical (lumenal) and basolateral membranes of enterocytes (transcellular). While most hydrophilic, polar drugs are absorbed by the paracellular route (exceptions are polar drugs that are actively transported across the epithelial cell), most lipophilic, nonpolar drugs are absorbed by the transcellular route.[6] For the latter, the opportunity exists for first-pass metabolism, as the relevant enzymes are located intracellularly, predominantly in the endoplasmic reticulum or cytosol of enterocytes. Indeed, evidence continues to accumulate that the small intestine can represent a major contributor to the overall first-pass metabolism of drugs, the extent of which can have clinical ramifications.

1.4   Clinical Relevance of Intestinal First-Pass Metabolism

Many commonly prescribed drugs undergo extensive first-pass metabolism following oral ingestion. Such examples are listed in Table 1. For all of these, at least 45% of the original dose is lost, on average, by the time it reaches the systemic circulation. That is, all exhibit a low average *oral bioavailability* (F_{oral}), which is the fraction of the oral dose that reaches the systemic circulation in the unchanged form. Because metabolism is frequently the major source of first-pass loss, F_{oral} is often used as a measure of the extent of first-pass metabolism. F_{oral} is typically calculated from the ratio of the area under the blood or plasma concentration-time curve following oral administration (AUC_{oral}) to that following intravenous administration (AUC_{iv}) and correcting for dose:

$$F_{oral} = \frac{AUC_{oral}}{AUC_{iv}} \times \frac{Dose_{iv}}{Dose_{oral}}$$

Alternatively, for most drugs, F_{oral} may be viewed as the product of the fractions of the dose that escape first-pass elimination in the intestine and liver:

$$F_{oral} = F_{abs} \times F_I \times F_L$$

where F_{abs} is the fraction of an oral dose absorbed intact across the apical membrane of the epithelial layer, and F_I and F_L are the fractions that escape elimination by the intestine and liver, respectively. In terms of extraction ratios (the fractions that do not escape first-pass metabolism), the above equation can be rewritten as follows:

$$F_{oral} = F_{abs} \times (1 - E_I) \times (1 - E_L)$$

where E_I and E_L are the extraction ratios across the intestine and liver, respectively. These simple equations clearly demonstrate the impact of a second presystemic site of metabolism on F_{oral}. For example, if a drug is completely absorbed from the intestinal lumen ($F_{abs} = 1$), has a F_I of 0.6, and has a F_L of 0.5, an F_{oral} of 30% is predicted. If only hepatic first-pass metabolism is considered, an F_{oral} of 50% is predicted. Omission of the intestinal component would therefore lead to an overestimation of F_{oral}, which could potentially lead to suboptimal dosing and ineffective concentrations at the site(s) of action. For those drugs that have a wide therapeutic window, this can simply be remedied by increasing the dose. However, for those that have a narrow therapeutic window and/or exhibit a large interindividual variation in F_{oral}, optimization of oral drug therapy becomes more challenging. Knowledge of the degree and variation in the expression of the

Table 1. Selected drugs with low and variable oral bioavailability believed to be due in part to intestinal first-pass metabolism.

Drug	Intestinal Enzyme(s)	Oral Bioavailability (%) (Average ± SD)
Amiodarone	CYP3A	46 ± 22
Cyclosporine	CYP3A	28 ± 18
Diclofenac	CYP2C9	54 ± 2
Dihydroergotamine	CYP3A[8]	0.5 ± 0.1[9]
Diltiazem	CYP3A, esterases	38 ± 11
Erythromycin	CYP3A	35 ± 25
Ethinyl estradiol	CYP3A, SULT1E1, UGT1A1[10-12] 2A1, UGT1A1[10-12]	51 ± 9[10]
Felodipine	CYP3A	15 ± 8
Irinotecan	CYP3A, esterases[13,14]	8[14]
Isoproterenol	SULT1A3[10,11]	28[10]
Lidocaine	CYP3A	35 ± 11
Losartan	CYP2C9, CYP3A	36 ± 16
Lovastatin	CYP3A	≤5
Midazolam	CYP3A	44 ± 17
Nicardipine	CYP3A	18 ± 11
Nifedipine	CYP3A	50 ± 13
Omeprazole	CYP2C19, CYP3A	53 ± 29
Raloxifene	UGT1A1, UGT1A8, UGT1A10[15]	2
Saquinavir	CYP3A	4-13
Sirolimus	CYP3A	15
Tacrolimus	CYP3A	25 ± 10
Terbutaline	SULT1A3[11]	14 ± 2
Triazolam	CYP3A	44
Verapamil	CYP3A, CYP2C9	22 ± 8

Enzyme(s) and bioavailability values are from reference 16 unless indicated otherwise.

major drug metabolizing enzymes of the human intestine is therefore essential, as not only can such enzymes represent a key determinant of the extent of first-pass elimination, but also the interindividual variation in F_{oral} and the probability and magnitude of drug-drug (or food-drug) interactions.[7] Recognition of these potential consequences of intestinal metabolism should further aid in the prediction of oral drug disposition during both the clinical and preclinical stages of development.

1.5   *Overview of Intestinal Drug Metabolizing Enzymes*

Compared to drug metabolizing enzymes in the human liver, research on the expression and catalytic properties of corresponding enzymes in the intestine has been lagging, largely due to a limited supply of high quality tissue. Over the past few years, however, several types of human intestinal tissue preparations (e.g., microsomes, cytosol, tissue slices, modified cell lines) (reviewed by Thummel and Wilkinson[17]) have become commercially available. Therefore, through the application of the same molecular biological techniques as for liver enzymes, along with the ongoing identification of selective catalytic probe substrates and inhibitors, a thorough characterization of the various intestinal enzymes is becoming more feasible.

Several of the drug metabolizing enzymes expressed in hepatocytes are also expressed in the enterocytes lining the small intestinal mucosa, including both phase I (oxidative and reductive) and phase II (conjugative) enzymes. Of the phase I enzymes, the cytochromes P450 (CYPs) are the most prominent and are discussed in great detail in the remaining sections of this chapter. Other phase I enzymes include esterases (e.g., carboxylesterase, amidase) and epoxide hydrolases. Intestinal carboxylesterase, for example, may contribute to the first-pass conversion of the chemotherapeutic agent and prodrug irinotecan to the active metabolite, SN-38.[13,14]

The UDP-glucuronosyl transferases (UGTs) constitute a major family of phase II enzymes that are ubiquitous in a number of extrahepatic tissues.[18,19] Gastrointestinal UGTs (e.g., UGT1A1, UGT1A7-10) may contribute to the low oral bioavailability of ethinyl estradiol[12] and raloxifene[15,20,21] (Table 1) and to the first-pass glucuronidation of SN-38[14,22-24] and troglitazone.[25] The extensive first-pass loss of ethinyl estradiol is also attributed to sulfotransferases (SULTs) in the intestine (SULT1E1, SULT2A1)[11,26,27] (Table 1). Likewise, intestinal SULT1A3 has been implicated in the low oral bioavailability of the β-adrenergic agents isoproterenol and terbutaline[11,28,29] (Table 1). Other phase II enzymes that have been identified in the intestinal mucosa include those belonging to the *N*-acetyltransferase (NAT) and glutathione *S*-transferase (GST) families.[30-33] 5-Aminosalicylic acid, used to treat inflammatory bowel disease, exhibits a low F_{oral},[34] and intestinal NAT (most likely NAT1) is believed to play a significant role in the first-pass metabolism of this agent.[35] The GSTs are commonly implicated in the detoxification or bioactivation of environmental toxins and carcinogens and some chemotherapeutic agents. Intestinal GSTα has been shown to conjugate busulfan with an intrinsic clearance (V_{max}/K_m, µL/min/mg cytosolic protein) comparable to that for the liver.[32]

As with enzymes in the hepatocyte, enzymes in the enterocyte generally reside either in the microsomal (CYPs, UGTs, esterases, epoxide hydrolases) or cytosolic (SULTs, NATs, GSTs) fraction. On a per mg microsomal or cytosolic protein basis, catalytic activities and/or levels of expression are generally highest in the duodenum/proximal jejunum then progressively decline toward distal ileum. This pattern of expression indicates that the extent of intestinal first-pass metabolism may depend on the site of absorption. In contrast, enzymes that appear to be more uniformly distributed along the small intestinal tract include NAT1,[30] GSTα,[32,33] and GSTπ.[31]

As the small intestine represents the first in the series of eliminating organs, the numerous drug metabolizing enzymes expressed in the mucosa are well suited for the task of first-pass xenobiotic metabolism. Indeed, one CYP subfamily, namely CYP3A, has been shown conclusively to be a significant determinant of the magnitude and interindividual variation in the F_{oral} of some drugs in humans. Strong evidence also exists for intestinal metabolism as a major determinant of the low F_{oral} of ethinyl estradiol.[26] Although sulfotransferases were implicated, other intestinal enzymes (e.g., UGTs, CYP3A) cannot be excluded. For all of these studies, the contribution of the intestine to first-pass metabolic extraction was investigated under rare conditions (i.e., during the anhepatic phase of a liver transplant operation or during a situation in which the hepatic portal vein was cannulated). Through the identification of specific probe substrates and appropriate in vitro and in vivo models of human intestinal metabolism, the importance of other intestinal enzymes to drug disposition should become more easily determined.

2 INTESTINAL CYTOCHROMES P450

The existence of CYP protein and associated monooxygenase activity (7-ethoxycoumarin *O*-deethylation) in the human small intestine was first reported by Hoensch and coworkers in 1979.[36] Total CYP content, as measured by carbon monoxide difference spectra in a small number of surgical samples, declined from proximal to distal regions. Average (± SD) total CYP content ranged from 93 ± 19 (proximal) to 35 ± 4 (distal) pmol/mg microsomal protein; 7-ethoxycoumarin *O*-deethylase activity paralleled this pattern. Similar values were later reported by others.[37,38] Total CYP content in the average small intestine therefore ranges from approximately 10 to 30% of that in the average human liver, suggesting a role for CYP isoform(s) in the first-pass metabolism of xenobiotics, including drugs.

2.1   Individual Subfamilies/Isoforms

2.1.1 CYP3A

Eight years following the report by Hoensch et al., Watkins and coworkers identified and characterized CYP3A as the major CYP subfamily expressed in the human small intestine. Ensuing in vitro and in vivo studies later established intestinal CYP3A as a major determinant of the extent of first-pass extraction, and hence F_{oral}, of several CYP3A substrates.

2.1.1.1   In Vitro Studies

In mucosae isolated from jejunal sections from four surgical patients, Watkins et al. found a CYP isoform and associated mRNA to be selectively recognized by an anti-CYP3A1 monoclonal antibody (which detected all human CYP3A forms) and HLp (CYP3A4) cDNA, respectively.[39] Using purified HLp as the reference standard, average (± SD) CYP3A protein content in microsomes prepared from the jejunal specimens was comparable to that for liver microsomes prepared from four separate organ donors/surgical patients (70 ± 20 vs. 65 ± 20 pmol/mg). Moreover, the average CYP3A-catalyzed rate of erythromycin N-demethylation in jejunal microsomes was comparable to that for liver microsomes and was inhibited by anti-CYP3A1. Three years later, de Waziers and coworkers[31] quantified, by Western blot analysis, the levels of various CYP isoforms/subfamilies in microsomes prepared from the following human extrahepatic tissues: esophagus, stomach, duodenum, jejunum, ileum, colon, and kidney. These investigators found that, next to the liver, the duodenum contained the highest amounts of CYP3A protein, followed by the jejunum, followed by the ileum. Average duodenal, jejunal, and ileal CYP3A content represented approximately 50%, 30%, and 10% of average hepatic CYP3A content, respectively. Corresponding values for the remaining extrahepatic organs were less than 5%. Consistent with the report by Watkins et al., CYP3A was the dominant CYP expressed in all three regions of the small intestine. Collectively, these in vitro results supported the growing belief that the small intestine may contribute to the first-pass metabolism of some CYP3A substrates. The in vivo significance of intestinal CYP3A to first-pass metabolism of a drug substrate was subsequently demonstrated.

2.1.1.2 In Vivo Studies

The widely used immunosuppressant cyclosporine is notorious for exhibiting large interindividual variation in its oral bioavailability, which has been reported to range from 5 to 89% for the conventional formulation (Sandimunne)[40] and from 21 to 73% for the microemulsion formulation (Neoral).[41-44] This property, coupled with a narrow therapeutic window, can lead to an under- or overdosing of the patient, which in turn can lead to graft rejection or toxicity. It was initially believed that the low and unpredictable oral bioavailability of cyclosporine was due to erratic absorption across the intestinal lumen and variable hepatic first-pass extraction (i.e., a low and variable F_{abs} and F_L). Kolars et al.,[45] however, after instilling cyclosporine into the duodenum of two patients during the anhepatic phase of their liver transplant operations, measured appreciable concentrations of two CYP3A-mediated primary metabolites in hepatic portal and systemic blood. The investigators concluded that the extrahepatic site of metabolism was the gut as (1) tissues other than the gut (e.g., the kidney and lung) express low levels of CYP3A and (2) portal metabolite concentrations exceeded systemic concentrations at the end of the anhepatic phase. These findings provided the first direct evidence that the small intestine can play a significant role in the first-pass metabolism of a CYP3A substrate. A subsequent pharmacokinetic analysis of cyclosporine AUCs following oral and intravenous administration indicated that the intestine, rather than the liver, was largely responsible for the first-pass extraction of cyclosporine.[46] However, cyclosporine is now known to be a substrate for the apically-directed efflux pump P-glycoprotein (P-gp). Thus this indirect approach would not distinguish between intestinal CYP3A-mediated metabolism and P-gp-mediated efflux.

The sedative-hypnotic and CYP3A substrate midazolam has been shown not to be a substrate for P-gp[47] and therefore might serve as a 'cleaner' in vivo CYP3A probe. Using a study design similar to that for the cyclosporine study, the disposition of midazolam and its primary metabolite, 1'-hydroxymidazolam, were examined in a larger group of anhepatic transplant patients following either intravenous (n = 5) or intraduodenal (n = 5) administration.[48] Multiple blood samples were collected simultaneously from the hepatic portal vein and a peripheral artery during the approximately one-hour anhepatic phase. From the difference between the arterial and hepatic portal venous midazolam AUCs (intravenous) or between the hepatic portal venous and arterial midazolam and 1'-hydroxymidazolam AUCs (intraduodenal), an average (± SD) extraction fraction of 8 ± 12% and 43 ± 18% was calculated for subjects who received midazolam by the intravenous and intraduodenal route, respectively. The low and variable extraction fraction following intravenous administration indicated that the intestine contributed some to the systemic extraction of midazolam. The fivefold greater value following intraduodenal administration, however, was identical to the average intestinal extraction ratio estimated in healthy volunteers (43 ± 24%).[49] Moreover, these values were essentially identical to the average hepatic extraction ratio estimated in the healthy volunteer study (44 ± 14%).[49] These findings strongly indicated that the small intestine is a major determinant of the overall extent of the first-pass extraction of midazolam, and it can even rival the liver.

By the indirect approach, intestinal CYP3A has also been shown to contribute significantly to the first-pass extraction of the calcium channel blockers nifedipine[50] and verapamil.[51] The comparable contributions of the intestine and liver (mean extraction ratio of 49% and 48%, respectively) to the overall first-pass elimination of verapamil was later confirmed using a newly developed method involving a multilumen intestinal perfusion technique and stable isotope-labeled drug.[52] In addition, active intestinal secretion was shown to be as important as biliary excretion for the elimination of verapamil metabolites.

Interestingly, for all of the aforementioned drugs, significant intestinal first-pass extraction occurred despite that total CYP3A content of the entire gut mucosa has been estimated to be much less than total hepatic CYP3A (70 nmol vs. 5490 nmol).[37] Of apparently more importance than total enzyme mass is the comparable intracellular enzyme concentration (enterocyte vs. hepatocyte) and the obligatory nature of drug passage through the enterocyte (if transcellular absorption is operative). Thus, a more appropriate comparison might be microsomal intrinsic activities. Indeed, mean CYP3A-mediated rates of erythromycin *N*-demethylation,[39] tacrolimus *O*-demethylation,[53] midazolam 1'-hydroxylation,[37] and testosterone 6β-hydroxylation[54] in intestinal (duodenal or jejunal) microsomes were 45-120% of corresponding metabolic rates in hepatic microsomes. Based on these reports, mean intestinal mucosal intrinsic clearances may be within two- to threefold of corresponding hepatic intrinsic clearances. Whether there will be a similarity in vivo for a given drug is more difficult to predict, as total oral dose, enzyme saturability, and absorption rate become relevant. Should the dose be large enough, and the affinity of the drug for the enzyme active site be high enough (or, the K_m be low enough), it is possible that the majority of absorbed drug could escape intestinal first-pass metabolism. Such may be the case with some of the HIV protease inhibitors (e.g., indinavir, saquinavir, and ritonavir).

2.1.1.3 CYP3A4 vs. CYP3A5

As in the liver, CYP3A4 protein appears to be constitutively expressed in the small intestine of all individuals, while CYP3A5 protein is polymorphically expressed. Immunoreactive CYP3A5 protein was readily detected at a frequency of 20-30% in intestinal tissue obtained from adult Caucasians.[37,55] Moreover, if detected, it was expressed along the length of the small intestinal tract.[37] As has been shown for the liver,[56] the frequency of CYP3A5 expression in the small intestine likely varies among different ethnic groups.

With the advent of specific, commercially available antibodies and suitable reference standards for immunoblot analysis, CYP3A5 was shown to constitute from 30 to 80% of total intestinal CYP3A (CYP3A4+CYP3A5) protein content (range, 5.4-34.1 pmol/mg jejunal homogenate protein) in individuals that carried at least one reference (wild-type) allele of the *CYP3A5* gene (*CYP3A5*1*).[57] Therefore, as has been reported for hepatic CYP3A5,[56] intestinal CYP3A5 may play a larger role in the first-pass metabolism of drugs than previously thought. Identification of a selective in vivo CYP3A5 probe substrate is needed to test this hypothesis.

Interestingly, significant expression of CYP3A4 in the gastrointestinal tract appears to be restricted to the small intestine. In mucosa of the stomach and colon, CYP3A5 mRNA and protein were more prominent than corresponding CYP3A4 measures.[58,59] Consistent with these findings, in two full-length human donor small intestines that were CYP3A5-positive, the ratio of CYP3A5 to CYP3A4 immunoreactive protein decreased from duodenum to jejunum, then increased in ileum to values comparable to or greater than those observed for the duodenum.[37] Finally, Gervot et al.[60] detected CYP3A5 protein, but not CYP3A4 protein, in colonic mucosa from 40 different uninduced tissue donors. These authors suggested that any CYP3A4 in colonic tissue is likely to be a consequence of prior treatment of the donor with an enzyme inducer.

2.1.1.4 Localization

CYP3A expression along the small intestine is not uniform. It appears to be greatest within proximal regions (duodenum and jejunum) and lowest in the distal region (ileum)[31,37,38] (Figure 2). In microsomes prepared from mucosal scrapings obtained from 20 donor small intestines, median CYP3A content declined from 31 to 23 to 17 pmol/mg protein in duodenum, jejunum, and ileum, respectively.[37] CYP3A-catalyzed midazolam 1'-hydroxlation activity displayed a similar pattern (Figure 2). Median microsomal intrinsic clearance (V_{max}/K_m) declined from 157 to 85 to 14 mL/min/mg.[37] Likewise, erythromycin *N*-demethylase activity declined from proximal to distal regions.[38] These findings suggest that the site of absorption can represent another source in the large interindividual variation in the oral bioavailabilies of CYP3A substrates.

CYP3A expression from the crypt to the tip of the small intestinal villus is also not uniform. By immunohistochemical analysis, CYP3A protein was not detected in the crypt cells or goblet cells but was readily detected in enterocytes, with the most intense staining evident in the mature enterocytes lining the villous tips.[58,61] By in situ hybridization, McKinnon et al.[59] reported a similar pattern for CYP3A mRNA. Within the enterocyte, CYP3A protein was largely located at the apex of the cell, adjacent to the microvillous border.[58] The strategic location of CYP3A in the mature enterocytes further highlights the small intestine as uniquely suited for the task of first-pass xenobiotic metabolism.

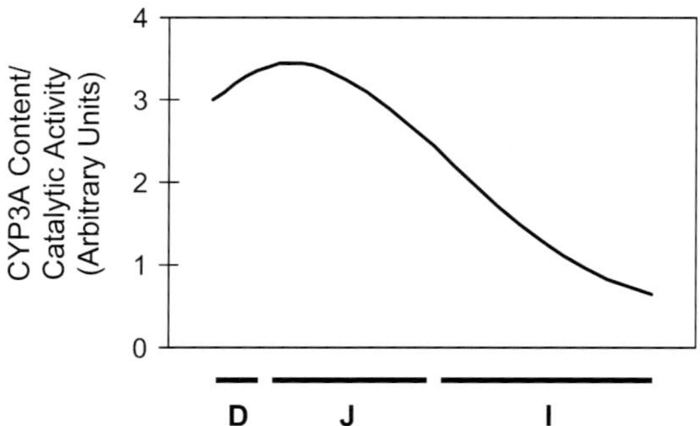

Figure 2. General pattern of CYP3A expression/catalytic activity along the human small intestine. D, duodenum; J, jejunum; I, ileum.

2.1.1.5 Gene Regulation

Localization of CYP3A within only the mature enterocytes of the small intestinal mucosa is consistent with a wider pattern of differentiation of cell function as cells formed within the crypts migrate towards the villus tip and are eventually sloughed off. Total CYP3A content even within a defined region of small bowel varies considerably. CYP3A mRNA and protein content were reported to vary 8- and 11-fold, respectively, among duodenal pinch biopsies obtained from 20 healthy volunteers.[55] Even greater variability (>30-fold) was reported for CYP3A protein content and catalytic activity in duodenal, jejunal, and ileal mucosal scrapings obtained from 20 organ donors.[37] Although some of the extreme variability in the latter study could have been due to events preceding organ procurement (e.g., reduced nutritional intake, antibiotic administration, brain death, and ischemia), it does suggest a remarkably dynamic system of enzyme expression that responds to a variety of dietary, therapeutic, pathophysiologic, and endogenous signaling factors. Several studies in humans support this contention.

2.1.1.5.1 Dietary Factors

The most widely studied dietary component in terms of CYP3A-mediated drug metabolism is grapefruit juice. It has been shown to markedly elevate blood levels of a variety of drugs by inhibiting intestinal, but not hepatic,

CYP3A during first passage of the drug from the intestinal lumen to the systemic circulation.[62] Juice prepared from a related citrus fruit, the Seville ('sour') orange, exhibited a similar effect on the CYP3A probe substrate felodipine.[63] The lack of effect of either juice on hepatic CYP3A may simply be due to dilution of the active ingredients in portal blood to concentrations below their effective K_i's, or, as was recently proposed, to avid binding of the active ingredients to plasma and/or cellular proteins in portal blood.[64] The active ingredients appear to be furocoumarins, several of which are potent inhibitors of CYP3A catalytic activity.[65-67] In addition, the pioneering study by Lown et al.[68] showed that grapefruit juice significantly reduced average intestinal CYP3A4 immunoreactive protein (measured in duodenal pinch biopsies) by 60% in 10 healthy men; the lack of a decrease in corresponding mRNA suggested a post-transcriptional mechanism. In vitro studies later confirmed that two such furocoumarins (bergamottin and 6',7'-dihydroxybergamottin) reduced CYP3A4 protein by accelerating its degradation.[69,70] Other citrus fruits that contain appreciable concentrations of furocoumarins include the pomelo and sweetie.[67]

2.1.1.5.2 Therapeutic agents

Therapeutic agents that have been shown to inhibit intestinal CYP3A in vivo include the azole antifungals ketoconazole[71-73] and fluconazole,[74] the macrolide antibiotics erythromycin[75] and clarithromycin,[76] and the HIV protease inhibitor saquinavir,[77] as evidenced by a greater decrease in the oral clearance (or E_I) compared to the decrease in systemic clearance (or E_L) or increase in elimination half-life of the CYP3A substrate (cyclosporine, tirilazad, midazolam). Caution is warranted in this interpretation for cyclosporine, however, as ketoconazole has also been shown to be an inhibitor of P-gp.[78]

Exposure of human subjects to the known hepatic enzyme inducer rifampin (7-10 days) and to the popular herbal medicine St. John's wort (14 days) increased average duodenal CYP3A protein content by ≥ 4- and 1.5-fold, respectively, relative to baseline.[61,79,80] Moreover, a comparison of the effect of rifampin on the systemic and oral clearance of midazolam,[81,82] verapamil,[51] nifedipine,[50] and triazolam[83] suggested that the inducer increased intestinal enzyme levels to a greater extent than hepatic levels. Interestingly, while short-term treatment (two days) with rifampin significantly increased average enterocyte CYP3A4 content, other known hepatic CYP3A4 inducers (phenobarbital, dexamethasone, clotrimazole) given for the same period of time had no effect.[84] The greater effect of enzyme inhibitors and inducers on intestinal CYP3A4 activity compared to hepatic CYP3A4 activity may simply be the consequence of higher intracellular concentrations and greater receptor occupancy in the enterocyte that occurs during absorption of the modifying agent. Unlike CYP3A4, intestinal CYP3A5 expression is not induced by rifampin.[60,61] This is consistent with the apparent lack of inducibility of hepatic CYP3A5.[85,86]

2.1.1.5.3 Pathophysiologic Conditions

Compared to hepatic CYP3A, less is known about the effect of disease on intestinal CYP3A. However, Lang et al.[87] reported that patients with Celiac disease had reduced levels of jejunal mucosal CYP3A protein as a consequence of widespread epithelial cell destruction. Treatment with a gluten-free diet reversed this aberration. In addition, Chalasani et al.[88] recently compared the disposition of midazolam among cirrhotic patients, cirrhotic patients with transjugular intrahepatic portosystemic shunts (TIPS), and healthy volunteers. The significantly higher mean F_{oral} in the cirrhotic patients with TIPS compared with the cirrhotic controls and healthy controls (0.76 vs. 0.27 and 0.30) was largely due to the significantly higher F_I in the TIPS patients compared with the cirrhotic and healthy subjects (0.83 vs. 0.32 and 0.42). The markedly lower extent of first-pass metabolism of midazolam in the TIPS patients was concluded to be a result of diminished intestinal CYP3A activity. In support of this interpretation, duodenal CYP3A4 expression was found to be markedly more variable in patients with compensated cirrhosis than in healthy volunteers, with a significant fraction of the cirrhotic patients exhibiting a specific content that was 50-75% of the lowest expression level in the control subjects (K.E. Thummel, unpublished observation).

2.1.1.5.4 Endogenous Signaling Factors

Despite years of effort, there is no definitive understanding of the regulation of constitutive CYP3A4 expression in humans. Studies in rats[89,90] indicated repression of CYP3A2 by a male pulsatile pattern of growth hormone secretion from the pituitary. Experiments with primary human hepatocytes also demonstrated transcriptional activation and increased CYP3A4 expression following continuous exposure to physiologically relevant concentrations of growth hormone.[91] More recently, replacement of growth hormone in deficient patients led to alterations in the erythromycin breath test, consistent with changes in hepatic CYP3A activity.[92] A female pattern of continuous growth hormone replacement resulted in increased activity, whereas a pulsatile male pattern of replacement led to decreased activity. These effects are in keeping with the apparent higher level of hepatic CYP3A activity in women compared to men.[76]

The mechanism for the stimulatory effects of growth hormone on hepatic CYP (2C and 3A) expression have been elucidated[93] and involves binding of growth hormone to hepatocyte surface receptors and activation of the JAK/ STAT5 signaling pathway. Interestingly, the liver is more enriched in growth hormone receptors than other tissues,[94] which is consistent with the possibility of preferential effects on hepatic rather than intestinal CYP3A4 expression. Moreover, recent studies suggested a regulatory role of intestinal CYP3A4 by a different hormone, vitamin D.

Treatment of the human intestinal cell lines Caco-2 and LS-180 with the fully hydroxylated and active form of vitamin D, 1α,25-dihydroxy vitamin D_3 (1α,25-$(OH)_2$-D_3), stimulated CYP3A4 expression and associated midazolam 1'-hydroxylation activity from an essentially deficient baseline.[95,96] The mechanism of enhanced enzyme expression involves binding of 1α,25-$(OH)_2$-D_3 to the vitamin D receptor (VDR), formation of a heterodimer with the retinoic acid receptor (RXRα), and binding of the complex to ER-6 and DR-3 enhancer motifs in the proximal and distal 5'-flanking regions of the *CYP3A4* gene,[96] the same regions that mediate CYP3A4 induction by ligands of the pregnane X receptor (PXR).[97] Although experiments showed that transcriptional activation of CYP3A4 by 1α,25-$(OH)_2$-D_3 does not involve the PXR, work from Mangelsdorf and colleagues[98] demonstrated that other endogenous steroids (e.g., 3-keto lithocholic acid) can bind to the VDR and activate CYP3A transcription in mice both in vitro and in vivo, albeit at ligand concentrations that were several-fold higher than those associated with 1α,25-$(OH)_2$-D_3 effects.

1α,25-$(OH)_2$-D_3 exerts important physiological roles in calcium homeostasis, including effects on enterocytes of the small intestine.[99] Binding of the hormone to a specific intracellular receptor (VDR) triggers a cascade of events that promote calcium absorption. VDR expression in the gut is greater in the proximal small intestine compared to ileum and colon.[100] Thus, delivery of the fully active hormone, which is produced in the kidney (1α-hydroxylation of hepatically generated 25-hydroxyvitamin D_3), and interaction with vitamin D_3 receptors within the small intestine may explain the preferential localization of CYP3A4 in the proximal small intestine.[37,58] Alternatively, the secretion of keto lithocholic acid into bile and delivery to the duodenal lumen may contribute to preferential CYP3A4 expression in the proximal small intestine.[98]

2.1.1.5.5 Intestinal vs. Hepatic CYP3A

In light of the many differing responses between intestinal and hepatic CYP3A to the various aforementioned exogenous and endogenous regulatory factors, it is not surprising that intestinal and hepatic CYP3A appear to be independently regulated. This was first demonstrated by Lown and coworkers,[55] who found that neither duodenal CYP3A protein content or catalytic activity correlated with hepatic CYP3A activity (as measured by the erythromycin breath test) in 20 healthy subjects. Likewise, other investigators found no rank order correlation between intestinal and hepatic CYP3A protein content or midazolam 1'-hydroxylation activity in microsomes prepared from eight matched intestine-liver donor pairs.[37] Finally, Thummel et al.[49] and Gorski et al.[76] found no correlation between the F_I and F_L of midazolam in 20 and 16 healthy subjects, respectively. This noncoordinate regulation between intestinal and hepatic CYP3A therefore indicates that a measure of one cannot be used to predict the other. However, it does not exclude the possibility of overlapping mechanisms of constitutive and inducible CYP3A4 expression. Indeed, Maurel and colleagues[101] recently reported that treatment of primary human hepatocytes and HepG2 cells with $1\alpha,25\text{-}(OH)_2\text{-}D_3$ activated CYP3A4 transcription. Thus, constitutive expression of CYP3A4 in the liver may involve multiple hormone (i.e., growth hormone and vitamin D) signaling pathways, whereas expression in the intestine may involve a single cell receptor (VDR) but multiple ligands ($1\alpha,25\text{-}(OH)_2\text{-}D_3$ and keto bile acids).

Interestingly, in a recent study involving 31 jejuna and 60 livers from Caucasian donors, for those livers with the *CYP3A5*1/*3* genotype (n = 13), CYP3A4 protein content strongly correlated with CYP3A5 protein content (r = 0.93), suggesting that hepatic *CYP3A4* and *CYP3A5* genes share a common regulatory pathway for constitutive expression.[57] Due to the limited number of donor intestines with the *CYP3A5*1/CYP3A5*3* genotype, the relationship between CYP3A4 and CYP3A5 expression for this tissue could not be assessed. Moreover, until a selective probe substrate for CYP3A5 is identified, the significance of these relationships in vivo remains unknown.

2.1.2 CYP2D6

Extrahepatic CYP2D6 expression is not uncommon. The enzyme has been identified in mucosal epithelium of the bladder,[102] lymphocytes,[103] breast tissue,[104,105] certain cells of the brain,[106-108] and the small intestine.[31,109,110] Intestinal CYP2D6 protein was first identified by de Waziers et al.[31] Like CYP3A4, it was found to be localized within mucosal enterocytes and most concentrated within the proximal small intestine. The mean specific protein content of CYP2D6 in duodenal and jejunal microsomes were both approximately 20% of hepatic CYP2D6 microsomal content. The enzyme was not detected in ileum or colon.

Other investigators later confirmed the expression of CYP2D6 in the human small intestine. In separate studies, Prueksaritanont and coworkers detected CYP2D6 protein in mucosal microsomes from two[109] and five[110] human donors. Likewise, CYP2D6-catalyzed (+)-bufuralol 1'-hydroxylation activity was measurable in all preparations. Moreover, catalytic activity was largely inhibited by the CYP2D6 inhibitors quinidine and ajmaline, as well as anti-CYP2D6 IgG.[109] Zhang et al.[38] subsequently demonstrated the presence of CYP2D6 mRNA in human enterocytes. In a more comprehensive study, Madani et al.[111] quantified CYP2D6 protein in microsomes prepared from 19 human jejuna and 31 human livers. As expected, the enzyme was not detected in a small percentage of these predominantly Caucasian donors (one jejunum and two livers). Median jejunal content was less than 8% of the hepatic content (0.9 *vs.* 12.8 pmol/mg); this lower percentage compared to that reported by de Waziers et al.[31] was attributed to the smaller number of tissues examined (eight intestines and five livers) and the pooling of duodena and jejuna in the earlier study. Even excluding the CYP2D6-negative organs, Madani et al.[111] found extensive interindividual variation in protein content for both tissues (100- and 13-fold for jejunum and liver, respectively). These investigators also characterized the catalytic activity of the same jejunal microsomes toward the CYP2D6 substrate metoprolol and found that the α-hydroxylation rate correlated significantly with CYP2D6 protein content (r = 0.75). Finally, similar to CYP3A, both CYP2D6 protein content and metoprolol α-hydroxylase activity measured along the length of four donor small intestines tended to peak in the proximal jejunal sections and decline toward distal ileum. Intraintestinal variation (< twofold), however, was much less than that for CYP3A protein content and catalytic activity.

Although several CYP2D6 substrates undergo extensive first-pass metabolism, it is unlikely that the gut wall contributes significantly to the elimination process. For example, the jejunal microsomal intrinsic clearance for metoprolol oxidation was reported to be only a fraction of the hepatic intrinsic clearance (0.7 and 19.7 μL/min/mg, respectively).[111] Likewise, the predicted average E_I for metoprolol was a minute fraction of the predicted average E_L (0.01 and 0.48, respectively). Based on these findings, the authors concluded that, unless a CYP2D6 substrate has a long residence time in the intestinal mucosa or undergoes futile cycling via an efflux transporter, intestinal metabolism is expected to contribute minimally to its first-pass removal. However, the presence of CYP2D6 in the intestine may become clinically relevant if it mediates the formation of a cytotoxic metabolite that could lead to mucosal damage.[111]

2.1.3 CYP2C

Although CYP2C mRNAs have been detected in a number of human extrahepatic tissues (e.g., kidney, testes, adrenal gland, prostate, brain, duodenum), significant protein expression appears to be limited to the small intestinal tract.[31,112]   de Waziers et al.[31] first detected what was described as 'CYP2C8-10' in mucosal microsomes, which, like CYP3A and CYP2D6, was found predominantly in the proximal region. Other investigators later confirmed the descending pattern of expression of a CYP2C isoform along the length of the small intestine.[38] However, in both studies, it was not clear which isoform was detected (CYP2C8, CYP2C9, or CYP2C19), as the CYP2Cs can be easily separated by 1-dimensional gel electrophoresis.[112] Based on the relative amount of each CYP2C isoform in human liver, the intestinal form identified was likely to have been CYP2C9. Indeed, in our analysis of 12 different proximal (duodenal or jejunal) microsomal preparations, two proteins were detected that reacted with a CYP2C-selective anti-CYP2C19 antibody and that comigrated with authentic cDNA-expressed CYP2C9 and CYP2C19 protein standards (Figure 3). CYP2C9 protein content was estimated to vary approximately 5-fold among the different preparations, with a median specific content (7.6 pmol/mg) that was approximately 20% of the median specific content determined for 29 hepatic microsomal preparations (38 pmol/mg protein). Intestinal CYP2C19 content was equally variable, with a median specific content (2.7 pmol/mg) near that measured in five liver microsomal preparations (4.1 pmol/mg).

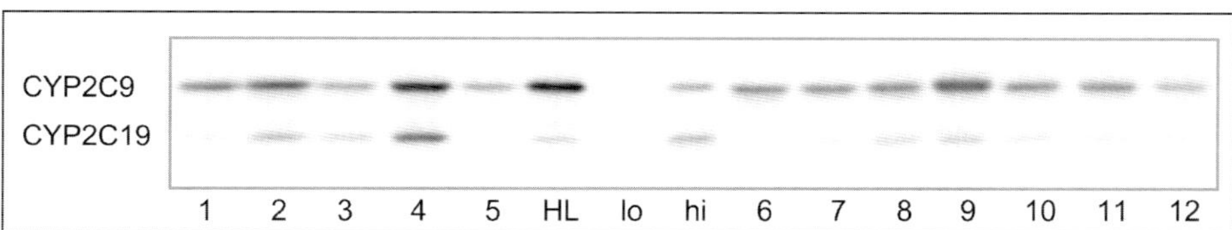

Figure 3. Western blot of 12 human jejunal microsomal preparations (30 μg protein/lane) showing the presence of CYP2C9 and CYP2C19. HL, human liver microsomes (5 μg); lo, low standard (100 fmol cDNA-expressed CYP2C9 and CYP2C19); hi, high standard (300 fmol cDNA-expressed CYP2C9 and CYP2C19).

With respect to intestinal CYP2C catalytic activity, Prueksaritanont et al.[110] reported a >20-fold variation in tolbutamide methylhydroxylase activity (<0.5-9.8 pmol/min/mg) for five duodenal/jejunal microsomal preparations that was ascribed to 'CYP2C8-10'; average (± SD) activity (5.1 ± 3.8 pmol/min/mg) was at least an order of magnitude lower than the hepatic counterpart. Other investigators later reported a similar variation in CYP2C9-catalyzed diclofenac 4'-hydroxylase activity (7.3-129 pmol/min/mg) for 10 human jejunal microsomal preparations; median activity was 55 pmol/min/mg.[54] Likewise, we have found a 22-fold variation in CYP2C9-catalyzed *S*-warfarin 7-hydroxylase activity for 11 proximal intestinal preparations (Table 2). Similar to results reported by Prueksaritanont et al.,[110] median activity was an order of magnitude lower than that measured in liver microsomes (n = 27). As for intestinal CYP2C19 catalytic activity, we (Table 2) and others[54] have measured similar variations in *S*-mephenytoin 4'-hydroxylase activity (0.50-19.6 and 0.8-13.1 pmol/min/mg, respectively) in the corresponding preparations analyzed for CYP2C9 activity. Consistent with our findings for CYP2C19 protein content, median intestinal CYP2C19 activity was nearly identical to that measured in 27 liver microsomal preparations (Table 2). Collectively, based on the appreciable median intestinal CYP2C9 and CYP2C19 protein contents and catalytic activities relative to hepatic measures (10-20% and >60%, respectively), the incomplete oral bioavailabilities of some CYP2C substrates may in part be attributed to intestinal first-pass metabolism. Such drugs include the CYP2C9 substrates verapamil, losartan, and diclofenac, and the CYP2C19 substrate omeprazole (Table 1).

Table 2. CYP2C9-catalyzed *S*-warfarin 7-hydroxylation and CYP2C19-catalyzed *S*-mephenytoin 4'-hydroxylation activity in human proximal small intestinal and hepatic microsomes.

	Intestine	*Liver*
S-warfarin 7-hydroxylation rate (pmol/min/mg):		
Median	0.7	6.6
Range	0.14 – 3.1	1.9 – 15.7
n	11	27
S-mephenytoin 4'-hydroxylation rate (pmol/min/mg):		
Median	5.5	5.7
Range	0.5 – 19.6	0 - 58
n	12	27

Interestingly, the pattern of CYP2C9 and CYP2C19 protein expression we measured along the length of a human donor small intestine was similar to that reported for CYP3A, with high specific contents in the duodenum and jejunum compared to ileum (Figure 4, upper panels). This corresponded to heterogeneous patterns of CYP2C9-catalyzed *S*-warfarin 7-hydroxylase and CYP2C19-catalyzed *S*-mephenytoin 4'-hydroxylation activity, respectively (Figure 4, lower panels). The concordance between patterns of intestinal CYP3A and CYP2C expression suggest a common constitutive regulatory pathway. Indeed, Maurel and colleagues[101] showed that the *CYP2C9* gene, which contains a DR-4 response element, can be transcriptionally activated by the VDR. Thus, one can speculate that the VDR mediates the expression of both CYP3A4 and CYP2C9/19 in the small intestine.

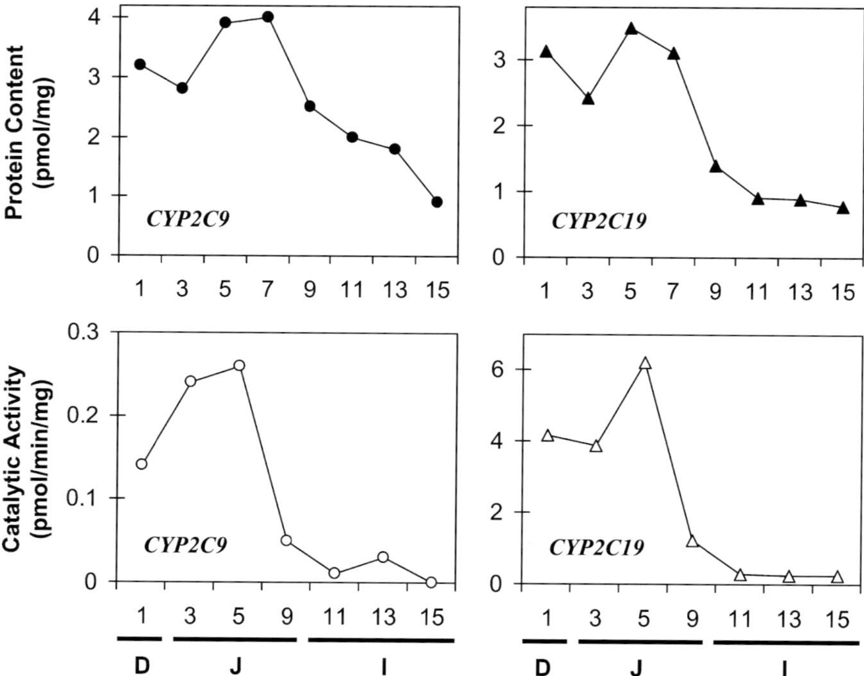

Figure 4. CYP2C immunoreactive protein content (upper panels) and catalytic activity (*S*-warfarin 7-hydroxylation and *S*-mephenytoin 4'-hydroxylation for CYP2C9 and CYP2C19, respectively) (lower panels) along the length of a human donor small intestine. D, duodenum; J, jejunum; I, ileum.

2.1.4 CYP1A1

CYP1A1 is expressed predominantly in extrahepatic tissues, including the lungs,[113-115] placenta,[116,117] stomach, and small intestine.[110,118-121] In two independent investigations in which duodenal biopsies were obtained from healthy volunteers, CYP1A1 mRNA was constitutively expressed in all specimens; as with other CYP isoforms, there was large interindividual variation among the specimens, at least 6-fold.[119,122] CYP1A1 protein and/or catalytic (ethoxyresorufin *O*-deethylase, or EROD) activity were undetectable or low. Following treatment with the CYP1A inducers omeprazole[119] or the polycyclic aromatic hydrocarbons and heterocyclic amines found in chargrilled meat,[122] CYP1A1 protein and catalytic activity became readily detectable. Similarly, median duodenal EROD activity was significantly higher in smokers and omeprazole-treated patients compared to non-smoking control subjects (2.1 vs. 1.1 vs. 0.5 pmol/min/mg homogenate protein).[120]

Characterization of a bank of 18 human small intestinal microsomal preparations revealed measurable rates of ethoxyresorufin *O*-deethylation in one-third of the preparations, with a median and range (23.7 and 1.4-124 pmol/min/mg, respectively)[121] comparable to those reported for CYP1A2-catalyzed EROD activity in human liver microsomes (39.4 and 10.1-224 pmol/min/mg, respectively).[123] Using cDNA-expressed human CYP1A1 as the reference standard, median CYP1A1 immunoreactive protein content for three CYP1A1-expressing human intestinal microsomal preparations (3.7 pmol/mg) (MF Paine, unpublished observations) was 9% of average CYP1A2 protein content reported for human liver microsomes (41 pmol/mg).[124] The differing CYP1A protein contents despite comparable EROD activities between intestine and liver can be attributed to CYP1A1 being more efficient than CYP1A2 in catalyzing the *O*-deethylation of ethoxyresorufin, as evidenced by cDNA-expressed CYP1A1 having both a lower K_m and a higher turnover number compared to cDNA-expressed CYP1A2 (87 nM and 7.6 min^{-1} vs. 240 nM and 1.9 min^{-1}).[125] A greater metabolic activity for CYP1A1 compared to CYP1A2 has also been demonstrated for ethoxycoumarin *O*-deethylation and benzo(a)pyrene hydroxylation.[126] The catalytic activity of CYP1A1 towards the CYP1A drug substrates caffeine, theophylline, phenacetin, and warfarin, however, is lower than that compared to CYP1A2.[126]

Interestingly, intestinal, but not hepatic, EROD activity was potently inhibited by ketoconazole, with a mean K_i (40 nM) near the low end of the range reported for inhibition of CYP3A4 (15-100 nM).[121] Therefore, ketoconazole should not be considered a selective CYP3A4 inhibitor for in vitro screening studies involving intestinal tissue, unless the individual is known not to express CYP1A1. In addition, these observations inferred that in vivo drug-drug interactions involving ketoconazole could result from CYP1A1 inhibition in the intestine in some individuals.

2.1.5 CYP2J2

While screening a human kidney cDNA library in an attempt to identify the CYP isoform(s) involved in the metabolism of arachidonic acid, Zeldin and coworkers[127] identified a novel cDNA that encoded a 57-kDa polypeptide that was 80% identical to rabbit CYP2J1 and was 40-43% identical to other members of the CYP2 family. This same cDNA was also found in screens of human heart, liver, and intestinal cDNA libraries. Based on its deduced amino acid sequence, this new human CYP was designated CYP2J2. Although most abundant in the heart, CYP2J2 is also expressed at significant levels (both mRNA and immunoreactive protein) in a number of other tissues, including the liver, kidney, stomach, and small intestine.[127] Due to its abundance in the heart, it has been postulated that CYP2J2 plays a role in the cardioprotective effects of epoxyeicosatrienoic acids (EETs), the major CYP2J2-mediated metabolites of arachidonic acid.[127] The biological role of CYP2J2 in other tissues is largely unknown. Likewise, the role of this enzyme in xenobiotic metabolism is largely unknown. Recent in vitro studies, however, have suggested that the intestinal form contributes to the first-pass metabolism of the non-sedating antihistamines astemizole and ebastine.

In both human intestinal and liver microsomes, Matsumoto et al.[128] found *O*-demethylation to be the primary metabolic pathway for astemizole, with the average ($\pm$ SD) rate in intestinal microsomes being approximately one-third that in liver microsomes (171 ± 57 vs. 478 ± 88 pmol/min/mg). Using recombinant CYP2J2 as the reference standard, immunoreactive CYP2J2 protein in microsomes prepared from five human small intestines averaged 2.1 ($\pm$ 0.6) pmol/mg. A protein band electrophoretically similar to that observed in intestinal microsomes and recombinant enzyme was also observed in liver microsomes. However, because this band was not completely separated from a band with lower mobility, CYP2J2 protein content in human liver could not be ascertained. That CYP2J2 was the major astemizole *O*-demethylase in human intestine was supported by (1) the excellent correlation between CYP2J2 protein content and *O*-demethylastemizole formation rate in intestinal microsomes ($r = 0.90$, $p < 0.05$) and (2) the CYP2J2 substrates ebastine and arachidonic acid strongly inhibited *O*-demethylastemizole formation in both intestinal microsomes and recombinant CYP2J2. Using similar strategies, along with an inhibitory anti-CYP2J2 antibody, Hashizume et al.[129] reported CYP2J2 to be the major ebastine hydroxylase in human intestinal microsomes. It will be interesting to learn whether these intriguing findings translate to the *in vivo* situation.

2.1.6 Other CYP Isoforms

CYP4F12 is another newly identified CYP isoform that was shown to be expressed, by RT-PCR, in human liver and small intestine.[130] This enzyme, when expressed in *Saccaromyces cerevisiae*, was also capable of catalyzing the hydroxylation of ebastine, suggesting that CYP4F12 in the small intestine (and liver) may play a role in the first-pass metabolism of this drug. However, as later reported by this same group of investigators, although CYP4F12 contributed some to the hydroxylation of ebastine in human intestinal microsomes, CYP2J2 was the predominate enzyme involved in this pathway.[129] In addition to CYP4F12, other CYP isoforms shown to be present in human small intestine at the mRNA level include CYP1A2, but only after treatment with omeprazole,[119] CYP1B1,[38] and CYP2C8 and CYP2C18.[112] The significance of these forms in vivo remains to be determined. By Western blot analysis, and with prolonged exposure, CYP isoforms shown to be expressed in only trace amounts include CYP2E1,[31] CYP2B6,[131] and CYP2A6 (M.F. Paine, unpublished observations). The role of these enzymes in intestinal first-pass metabolism is likely to be negligible.

3   SUMMARY AND PERSPECTIVE

Of the extrahepatic organs in humans, the small intestine contains the highest levels of total CYP protein. CYP3A (CYP3A4 + CYP3A5), the most widely studied, represents at least 50% of spectrally determined total CYP content in the small intestine. Its catalytic activity and immunoreactive protein content in the most proximal region (duodenum to mid-jejunum) are within the range reported for the hepatic counterpart. These in vitro findings are consistent with in vivo studies showing that the intestinal contribution to the low and variable F_{oral} of some CYP3A substrates can rival the hepatic contribution. However, because intestinal and hepatic CYP3A appear to be under different regulatory controls, and thus do not correlate, CYP3A activity in one organ will not necessarily predict CYP3A activity in the other. Thus the challenge remains to develop an in vivo method capable of delineating intestinal from hepatic first-pass metabolism, and which would also delineate CYP3A-mediated metabolism from P-gp-mediated efflux. This is currently an active area of research, as the successful prediction of intestinal first-pass metabolism could aid in the management of drugs with a low and variable F_{oral}, particularly those with a narrow therapeutic window, e.g., some immunosuppressive and chemotherapeutic agents.

With the increasing availability of quality human intestinal tissue and molecular biological techniques for the isolation and identification of xenobiotic metabolizing enzymes, along with the ongoing identification of selective probe substrates and inhibitors, characterization of other CYP isoforms has become more feasible. Based on median (or average) immunoreactive protein contents and/or catalytic activities relative to the hepatic counterparts, CYP2C9, CYP2C19, and CYP2J2 may play a role in the intestinal first-pass metabolism, and hence low F_{oral}, of some drugs. Although CYP2D6 and CYP1A1 are present and catalytically active, their low levels relative to hepatic CYP2D6 and CYP1A2 render them unlikely to contribute significantly to the first-pass metabolism of drugs. However, a role in the bioactivation or detoxification of other xenobiotics ingested in trace amounts, e.g., environmental contaminants, cannot be excluded. Clearly, further research is necessary to clarify the putative roles of these non-CYP3A isoforms in vivo. Meanwhile, knowledge of the differences between the intestinal and hepatic forms should be taken into account during the drug selection process.

REFERENCES

1. Rowland, M.; Tozer, T.N. Clinical Pharmacokinetics. Concepts and Applications. Philadelphia: Lippincott Williams & Wilkins, **1995**, 119.
2. Aungst, B.; Shen, D.D. Gastrointestinal absorption of toxic agents. In: Rozman, K.; Hanninen, O., eds. Gastrointestinal Toxicology. Amsterdam: Elsevier Science Publishers B.V., **1986**, 29.
3. Nugent, S.G.; Kumar, D.; Rampton, D.S.; Evans, D.F. *Gut* **2001**, *48*, 571.
4. Snyder, W.S.; Cook, M.J.; Karhausen, L.R.; Nasset, E.S.; Howells, G.P.; Tipton, I.H. Report of the Task Group on Reference Man. Oxford: Pergamon Press, **1975**, 134.
5. Rubin, D.C. Small intestine: anatomy and structural anomalies. In: Yamada, T., ed. Textbook of Gastroenterology. Vol. 2. Philadelphia: Lippincott Williams & Wilkins, **1999**, 1561.
6. Meunier, V.; Bourrie, M.; Berger, Y.; Fabre, G. *Cell. Biol. Toxicol.* **1995**, *11*, 187.
7. George, C.F. *Clin. Pharmacokinet.* **1981**, *6*, 259.
8. Delaforge, M.; Riviere, R.; Sartori, E.; Doignon, J.L.; Grognet, J.M. *Xenobiotica* **1989**, *19*, 1285.
9. Little, P.J.; Jennings, G.L.; Skews, H.; Bobik, A. *Br. J. Clin. Pharmacol.* **1982**, *13*, 785.
10. Thummel, K.E.; Kunze, K.L.; Shen, D.D. *Adv. Drug Deliv. Rev.* **1997**, *27*, 99.
11. Glatt, H.; Boeing, H.; Engelke, C.E.; Ma, L.; Kuhlow, A.; Pabel, U.; Pomplun, D.; Teubner, W.; Meinl, W. *Mutat. Res.* **2001**, *482*, 27.
12. Ebner, T.; Remmel, R.P.; Burchell, B. *Mol. Pharmacol.* **1993**, *43*, 649.
13. Khanna, R.; Morton, C.L.; Danks, M.K.; Potter, P.M. *Cancer Res.* **2000**, *60*, 4725.
14. Garcia-Carbonero, R.; Supko, J.G. *Clin. Cancer Res.* **2002**, *8*, 641.
15. Kemp, D.C.; Fan, P.W.; Stevens, J.C. *Drug Metab. Dispos.* **2002**, *30*, 694.
16. Thummel, K.E.; Shen, D.D. Design and optimization of dosage regimens; pharmacokinetic data. In: Hardman, J.G.; Limbird, L.E.; Goodman Gilman, A.,

eds. Goodman & Gilman's The Pharmacological Basis of Therapeutics. New York: McGraw-Hill, **2001**, 1917.

17. Thummel, K.E.; Wilkinson, G.R. *Annu. Rev. Pharmacol. Toxicol.* **1998**, *38*, 389.

18. Tukey, R.H.; Strassburg, C.P. *Mol. Pharmacol.* **2001**, *59*, 405.

19. Fisher, M.B.; Paine, M.F.; Strelevitz, T.J.; Wrighton, S.A. *Drug Metab. Rev.* **2001**, *33*, 273.

20. Hochner-Celnikier, D. *Eur. J. Obstet. Gynecol. Reprod. Biol.* **1999**, *85*, 23.

21. Snyder, K.R.; Sparano, N.; Malinowski, J.M. *Am. J. Health Syst. Pharm.* **2000**, *57*, 1669.

22. Iyer, L.; King, C.D.; Whitington, P.F.; Green, M.D.; Roy, S.K.; Tephly, T.R.; Coffman, B.L.; Ratain, M.J. *J. Clin. Invest.* **1998**, *101*, 847.

23. Ciotti, M.; Basu, N.; Brangi, M.; Owens, I.S. *Biochem. Biophys. Res. Commun.* **1999**, *260*, 199.

24. Gagne, J.F.; Montminy, V.; Belanger, P.; Journault, K.; Gaucher, G.; Guillemette, C. *Mol. Pharmacol.* **2002**, *62*, 608.

25. Watanabe, Y.; Nakajima, M.; Yokoi, T. *Drug Metab. Dispos.* **2002**, *30*, 1462.

26. Back, D.J.; Breckenridge, A.M.; MacIver, M.; Orme, M.; Purba, H.S.; Rowe, P.H.; Taylor, I. *Br. J. Clin. Pharmacol.* **1982**, *13*, 325.

27. Her, C.; Szumlanski, C.; Aksoy, I.A.; Weinshilboum, R.M. *Drug Metab. Dispos.* **1996**, *24*, 1328.

28. Hochhaus, G.; Mollmann, H. *Int. J. Clin. Pharmacol. Ther. Toxicol.* **1992**, *30*, 342.

29. Hartman, A.P.; Wilson, A.A.; Wilson, H.M.; Aberg, G.; Falany, C.N.; Walle, T. *Chirality* **1998**, *10*, 800.

30. Hickman, D.; Pope, J.; Patil, S.D.; Fakis, G.; Smelt, V.; Stanley, L.A.; Payton, M.; Unadkat, J.D.; Sim, E. *Gut* **1998**, *42*, 402.

31. de Waziers, I.; Cugnenc, P.H.; Yang, C.S.; Leroux, J.P.; Beaune, P.H. *J. Pharmacol. Exper. Ther.* **1990**, *253*, 387.

32. Gibbs, J.P.; Yang, J.S.; Slattery, J.T. *Drug Metab. Dispos.* **1998**, *26*, 52.

33. Coles, B.F.; Chen, G.; Kadlubar, F.F.; Radominska-Pandya, A. *Arch. Biochem. Biophys.* **2002**, *403*, 270.

34. Klotz, U. *Clin. Pharmacokinet.* **1985**, *10*, 285.

35. Vree, T.B.; Dammers, E.; Exler, P.S.; Sorgel, F.; Bondesen, S.; Maes, R.A. *Int. J. Clin. Pharmacol. Ther.* **2000**, *38*, 514.

36. Hoensch, H.P.; Hutt, R.; Hartmann, F. *Environ. Health Perspect.* **1979**, *33*, 71.

37. Paine, M.F.; Khalighi, M.; Fisher, J.M.; Shen, D.D.; Kunze, K.L.; Marsh, C.L.; Perkins, J.D.; Thummel, K.E. *J. Pharmacol. Exper. Ther.* **1997**, *283*, 1552.

38. Zhang, Q.Y.; Dunbar, D.; Ostrowska, A.; Zeisloft, S.; Yang, J.; Kaminsky, L.S. *Drug Metab. Dispos.* **1999**, *27*, 804.

39. Watkins, P.B.; Wrighton, S.A.; Schuetz, E.G.; Molowa, D.T.; Guzelian, P.S. *J. Clin. Invest.* **1987**, *80*, 1029.

40. Ptachcinski, R.J.; Venkataramanan, R.; Burckart, G.J. *Clin. Pharmacokinet.* **1986**, *11*, 107.

41. Wallemacq, P.E.; Reding, R.; Sokal, E.M.; de Ville de Goyet, J.; Clement de Clety, S.; Van Leeuw, V.; De Backer, M.; Otte, J.B. *Transpl. Int.* **1997**, *10*, 466.

42. Chueh, S.C.; Kahan, B.D. *J. Am. Soc. Nephrol.* **1998**, *9*, 297.

43. Ku, Y.M.; Min, D.I.; Flanigan, M. *J. Clin. Pharmacol.* **1998**, *38*, 959.

44. Lee, M.; Min, D.I.; Ku, Y.M.; Flanigan, M. *J. Clin. Pharmacol.* **2001**, *41*, 317.

45. Kolars, J.C.; Awni, W.M.; Merion, R.M.; Watkins, P.B. *Lancet* **1991**, *338*, 1488.

46. Hebert, M.F.; Roberts, J.P.; Prueksaritanont, T.; Benet, L.Z. *Clin. Pharmacol. Ther.* **1992**, *52*, 453.

47. Kim, R.B.; Wandel, C.; Leake, B.; Cvetkovic, M.; Fromm, M.F.; Dempsey, P.J.; Roden, M.M.; Belas, F.; Chaudhary, A.K.; Roden, D.M.; Wood, A.J.; Wilkinson, G.R. *Pharm. Res.* **1999**, *16*, 408.

48. Paine, M.F.; Shen, D.D.; Kunze, K.L.; Perkins, J.D.; Marsh, C.L.; McVicar, J.P.; Barr, D.M.; Gillies, B.S.; Thummel, K.E. *Clin. Pharmacol. Ther.* **1996**, *60*, 14.

49. Thummel, K.E.; O'Shea, D.; Paine, M.F.; Shen, D.D.; Kunze, K.L.; Perkins, J.D.; Wilkinson, G.R. *Clin. Pharmacol. Ther.* **1996**, *59*, 491.

50. Holtbecker, N.; Fromm, M.F.; Kroemer, H.K.; Ohnhaus, E.E.; Heidemann, H. *Drug Metab. Dispos.* **1996**, *24*, 1121.

51. Fromm, M.F.; Busse, D.; Kroemer, H.K.; Eichelbaum, M. *Hepatology* **1996**, *24*, 796.

52. von Richter, O.; Greiner, B.; Fromm, M.F.; Fraser, R.; Omari, T.; Barclay, M.L.; Dent, J.; Somogyi, A.A.; Eichelbaum, M. *Clin. Pharmacol. Ther.* **2001**, *70*, 217.

53. Lampen, A.; Christians, U.; Guengerich, F.P.; Watkins, P.B.; Kolars, J.C.; Bader, A.; Gonschior, A.K.; Dralle, H.; Hackbarth, I.; Sewing, K.F. *Drug Metab. Dispos.* **1995**, *23*, 1315.

54. Obach, R.S.; Zhang, Q.Y.; Dunbar, D.; Kaminsky, L.S. *Drug Metab. Dispos.* **2001**, *29*, 347.

55. Lown, K.S.; Kolars, J.C.; Thummel, K.E.; Barnett, J.L.; Kunze, K.L.; Wrighton, S.A.; Watkins, P.B. *Drug Metab. Dispos.* **1994**, *22*, 947.

56. Kuehl, P.; Zhang, J.; Lin, Y.; Lamba, J.; Assem, M.; Schuetz, J.; Watkins, P.B.; Daly, A.; Wrighton, S.A.; Hall, S.D.; Maurel, P.; Relling, M.; Brimer, C.; Yasuda, K.; Venkataramanan, R.; Strom, S.; Thummel, K.; Boguski, M.S.; Schuetz, E. *Nat. Genet.* **2001**, *27*, 383.

57. Lin, Y.S.; Dowling, A.L.; Quigley, S.D.; Farin, F.M.; Zhang, J.; Lamba, J.; Schuetz, E.G.; Thummel, K.E. *Mol. Pharmacol.* **2002**, *62*, 162.

58. Kolars, J.C.; Lown, K.S.; Schmiedlin-Ren, P.; Ghosh, M.; Fang, C.; Wrighton, S.A.; Merion, R.M.; Watkins, P.B. *Pharmacogenetics* **1994**, *4*, 247.

59. McKinnon, R.A.; Burgess, W.M.; Hall, P.M.; Roberts-Thomson, S.J.; Gonzalez, F.J.; McManus, M.E. *Gut* **1995**, *36*, 259.

60. Gervot, L.; Carriere, V.; Costet, P.; Cugnenc, P.H.; Berger, A.; Beaune, P.H.; de Waziers, I. *Environ. Toxicol. Pharmacol.* **1996**, *2*, 381.

61. Kolars, J.C.; Schmiedlin-Ren, P.; Schuetz, J.D.; Fang, C.; Watkins, P.B. *J. Clin. Invest.* **1992**, *90*, 1871.

62. Bailey, D.G.; Arnold, J.M.O.; Spence, J.D. *Br. J. Clin. Pharmacol.* **1998**, *46*, 101.

63. Malhotra, S.; Bailey, D.G.; Paine, M.F.; Watkins, P.B. *Clin. Pharmacol. Ther.* **2001**, *69*, 14.

64. Criss, A.B.; Paine, M.F.; Watkins, P.B. *Drug Metab. Rev.* **2002**, *34*, 169.

65. Schmiedlin-Ren, P.; Edwards, D.J.; Fitzsimmons, M.E.; He, K.; Lown, K.S.; Woster, P.M.; Rahman, A.; Thummel, K.E.; Fisher, J.M.; Hollenberg, P.F.; Watkins, P.B. *Drug Metab. Dispos.* **1997**, *25*, 1228.

66. Fukuda, K.; Ohta, T.; Oshima, Y.; Ohashi, N.; Yoshikawa, M.; Yamazoe, Y. *Pharmacogenetics* **1997**, *7*, 391.

67. Guo, L.Q.; Fukuda, K.; Ohta, T.; Yamazoe, Y. *Drug Metab. Dispos.* **2000**, *28*, 766.

68. Lown, K.S.; Bailey, D.G.; Fontana, R.J.; Janardan, S.K.; Adair, C.H.; Fortlage, L.A.; Brown, M.B.; Guo, W.; Watkins, P.B. *J. Clin. Invest.* **1997**, *99*, 2545.

69. Malhotra, S.; Schmiedlin-Ren, P.; Paine, M.F.; Criss, A.B.; Watkins, P.B. *Drug Metab. Rev.* **2001**, *2001*, 97.

70. Paine, M.F.; Criss, A.B.; Malhotra, S.; Watkins, P.B. *Drug Metab. Rev.* **2002**, *34*, 176.

71. Gomez, D.Y.; Wacher, V.J.; Tomlanovich, S.J.; Hebert, M.F.; Benet, L.Z. *Clin. Pharmacol. Ther.* **1995**, *58*, 15.

72. Wienkers, L.C.; Steenwyk, R.C.; Sanders, P.E.; Pearson, P.G. *J. Pharmacol. Exper. Ther.* **1996**, *277*, 982.

73. Tsunoda, S.M.; Velez, R.L.; von Moltke, L.L.; Greenblatt, D.J. *Clin. Pharmacol. Ther.* **1999**, *66*, 461.

74. Ahonen, J.; Olkkola, K.T.; Neuvonen, P.J. *Eur. J. Clin. Pharmacol.* **1997**, *51*, 415.

75. Olkkola, K.T.; Aranko, K.; Luurila, H.; Hiller, A.; Saarnivaara, L.; Himberg, J.J.; Neuvonen, P.J. *Clin. Pharmacol. Ther.* **1993**, *53*, 298.

76. Gorski, J.C.; Jones, D.R.; Haehner-Daniels, B.D.; Hamman, M.A.; O'Mara, E.M., Jr.; Hall, S.D. *Clin. Pharmacol. Ther.* **1998**, *64*, 133.

77. Palkama, V.J.; Ahonen, J.; Neuvonen, P.J.; Olkkola, K.T. *Clin. Pharmacol. Ther.* **1999**, *66*, 33.

78. Siegsmund, M.J.; Cardarelli, C.; Aksentijevich, I.; Sugimoto, Y.; Pastan, I.; Gottesman, M.M. *J. Urol.* **1994**, *151*, 485.

79. Greiner, B.; Eichelbaum, M.; Fritz, P.; Kreichgauer, H.P.; von Richter, O.; Zundler, J.; Kroemer, H.K. *J. Clin. Invest.* **1999**, *104*, 147.

80. Dürr, D.; Stieger, B.; Kullak-Ublick, G.A.; Rentsch, K.M.; Steinert, H.C.; Meier, P.J.; Fattinger, K. *Clin. Pharmacol. Ther.* **2000**, *68*, 598.

81. Backman, J.T.; Olkkola, K.T.; Neuvonen, P.J. *Clin. Pharmacol. Ther.* **1996**, *59*, 7.

82. Kharasch, E.D.; Russell, M.; Mautz, D.; Thummel, K.E.; Kunze, K.L.; Bowdle, A.; Cox, K. *Anesthesiology* **1997**, *87*, 36.

83. Villikka, K.; Kivisto, K.T.; Backman, J.T.; Olkkola, K.T.; Neuvonen, P.J. *Clin. Pharmacol. Ther.* **1997**, *61*, 8.

84. Wille, R.T.; Lown, K.S.; Huszczo, U.R.; Schmiedlin-Ren, P.; Watkins, P.B. *Gastroenterology* **1997**, *112*, A419.

85. Wrighton, S.A.; Ring, B.J.; Watkins, P.B.; VandenBranden, M. *Mol. Pharmacol.* **1989**, *36*, 97.

86. Schuetz, E.G.; Schuetz, J.D.; Strom, S.C.; Thompson, M.T.; Fisher, R.A.; Molowa, D.T.; Li, D.; Guzelian, P.S. *Hepatology* **1993**, *18*, 1254.

87. Lang, C.C.; Brown, R.M.; Kinirons, M.T.; Deathridge, M.A.; Guengerich, F.P.; Kelleher, D.; O'Briain, D.S.; Ghishan, F.K.; Wood, A.J. *Clin. Pharmacol. Ther.* **1996**, *59*, 41.

88. Chalasani, N.; Gorski, J.C.; Patel, N.H.; Hall, S.D.; Galinsky, R.E. *Hepatology* **2001**, *34*, 1103.

89. Agrawal, A.K.; Shapiro, B.H. *J. Pharmacol. Exper. Ther.* **2000**, *292*, 228.

90. Kawai, M.; Bandiera, S.M.; Chang, T.K.; Bellward, G.D. *Biochem. Pharmacol.* **2000**, *59*, 1277.

91. Liddle, C.; Goodwin, B.J.; George, J.; Tapner, M.; Farrell, G.C. *J. Clin. Endocrinol. Metab.* **1998**, *83*, 2411.

92. Jaffe, C.A.; Turgeon, D.K.; Lown, K.; Demott-Friberg, R.; Watkins, P.B. *Am. J. Physiol. Endocrinol. Metab.* **2002**, *283*, E1008.

93. Davey, H.W.; Wilkins, R.J.; Waxman, D.J. *Am. J. Hum. Genet.* **1999**, *65*, 959.

94. Bauman, G. Growth hormone and its disorders. In: Becker, K.L., ed. Principles and Practice of Endocrinology and Metabolism. Philadelphia: Lippincott Williams & Wilkins, **2001**, 129.

95. Schmiedlin-Ren, P.; Thummel, K.E.; Fisher, J.M.; Paine, M.F.; Lown, K.S.; Watkins, P.B. *Mol. Pharmacol.* **1997**, *51*, 741.

96. Thummel, K.E.; Brimer, C.; Yasuda, K.; Thottassery, J.; Senn, T.; Lin, Y.; Ishizuka, H.; Kharasch, E.; Schuetz, J.; Schuetz, E. *Mol. Pharmacol.* **2001**, *60*, 1399.

97. Goodwin, B.; Redinbo, M.R.; Kliewer, S.A. *Annu. Rev. Pharmacol. Toxicol.* **2002**, *42*, 1.

98. Makishima, M.; Lu, T.T.; Xie, W.; Whitfield, G.K.; Domoto, H.; Evans, R.M.; Haussler, M.R.; Mangelsdorf, D.J. *Science* **2002**, *296*, 1313.

99. Kumar, R. *J. Am. Soc. Nephrol.* **1990**, *1*, 30.

100. Feldman, D.; McCain, T.A.; Hirst, M.A.; Chen, T.L.; Colston, K.W. *J. Biol. Chem.* **1979**, *254*, 10378.

101. Drocourt, L.; Ourlin, J.C.; Pascussi, J.M.; Maurel, P.; Vilarem, M.J. *J. Biol. Chem.* **2002**, *277*, 25125.

102. Romkes-Sparks, M.; Mnuskin, A.; Chern, H.D.; Persad, R.; Fleming, C.; Sibley, G.N.; Smith, P.; Wilkinson, G.R.; Branch, R.A. *Carcinogenesis* **1994**, *15*, 1955.

103. Carcillo, J.A.; Parise, R.A.; Adedoyin, A.; Frye, R.; Branch, R.A.; Romkes, M. *Res. Commun. Mol. Pathol. Pharmacol.* **1996**, *91*, 149.

104. Huang, Z.; Fasco, M.J.; Figge, H.L.; Keyomarsi, K.; Kaminsky, L.S. *Drug Metab. Dispos.* **1996**, *24*, 899.

105. Huang, Z.; Fasco, M.J.; Spivack, S.; Kaminsky, L.S. *Cancer Res.* **1997**, *57*, 2589.

106. Gilham, D.E.; Cairns, W.; Paine, M.J.; Modi, S.; Poulsom, R.; Roberts, G.C.; Wolf, C.R. *Xenobiotica* **1997**, *27*, 111.

107. Voirol, P.; Jonzier-Perey, M.; Porchet, F.; Reymond, M.J.; Janzer, R.C.; Bouras, C.; Strobel, H.W.; Kosel, M.; Eap, C.B.; Baumann, P. *Brain Res.* **2000**, *855*, 235.

108. Siegle, I.; Fritz, P.; Eckhardt, K.; Zanger, U.M.; Eichelbaum, M. *Pharmacogenetics* **2001**, *11*, 237.

109. Prueksaritanont, T.; Dwyer, L.M.; Cribb, A.E. *Biochem. Pharmacol.* **1995**, *50*, 1521.

110. Prueksaritanont, T.; Gorham, L.M.; Hochman, J.H.; Tran, L.O.; Vyas, K.P. *Drug Metab. Dispos.* **1996**, *24*, 634.

111. Madani, S.; Paine, M.F.; Lewis, L.; Thummel, K.E.; Shen, D.D. *Pharm. Res.* **1999**, *16*, 1199.

112. Klose, T.S.; Blaisdell, J.A.; Goldstein, J.A. *J. Biochem. Mol. Toxicol.* **1999**, *13*, 289.

113. Wheeler, C.W.; Park, S.S.; Guenthner, T.M. *Mol. Pharmacol.* **1990**, *38*, 634.

114. Shimada, T.; Yun, C.H.; Yamazaki, H.; Gautier, J.C.; Beaune, P.H.; Guengerich, F.P. *Mol. Pharmacol.* **1992**, *41*, 856.

115. Shimada, T.; Yamazaki, H.; Mimura, M.; Wakamiya, N.; Ueng, Y.F.; Guengerich, F.P.; Inui, Y. *Drug Metab. Dispos.* **1996**, *24*, 515.

116. Sesardic, D.; Pasanen, M.; Pelkonen, O.; Boobis, A.R. *Carcinogenesis* **1990**, *11*, 1183.

117. Hakkola, J.; Pasanen, M.; Hukkanen, J.; Pelkonen, O.; Maenpaa, J.; Edwards, R.J.; Boobis, A.R.; Raunio, H. *Biochem. Pharmacol.* **1996**, *51*, 403.

118. Peters, W.H.; Kremers, P.G. *Biochem. Pharmacol.* **1989**, *38*, 1535.

119. McDonnell, W.M.; Scheiman, J.M.; Traber, P.G. *Gastroenterology* **1992**, *103*, 1509.

120. Buchthal, J.; Grund, K.E.; Buchmann, A.; Schrenk, D.; Beaune, P.; Bock, K.W. *Eur. J. Clin. Pharmacol.* **1995**, *47*, 431.

121. Paine, M.F.; Schmiedlin-Ren, P.; Watkins, P.B. *Drug Metab. Dispos.* **1999**, *27*, 360.

122. Fontana, R.J.; Lown, K.S.; Paine, M.F.; Fortlage, L.; Santella, R.M.; Felton, J.S.; Knize, M.G.; Greenberg, A.; Watkins, P.B. *Gastroenterology* **1999**, *117*, 89.

123. Pelkonen, O.; Pasanen, M.; Kuha, H.; Gachalyi, B.; Kairaluoma, M.; Sotaniemi, E.A.; Park, S.S.; Friedman, F.K.; Gelboin, H.V. *Br. J. Clin. Pharmacol.* **1986**, *22*, 125.

124. Shimada, T.; Yamazaki, H.; Mimura, M.; Inui, Y.; Guengerich, F.P. *J. Pharmacol. Exper. Ther.* **1994**, *270*, 414.

125. Penman, B.W.; Chen, L.; Gelboin, H.V.; Gonzalez, F.J.; Crespi, C.L. *Carcinogenesis* **1994**, *15*, 1931.

126. Miners, J.O.; McKinnon, R.A. CYP1A. In: Levy, R.H.; Thummel, K.E.; Trager, W.F.; Hansten, P.D.; Eichelbaum, M., eds. Metabolic Drug Interactions. Philadelphia: Lippincott Williams & Wilkins, **2000**, 61.

127. Scarborough, P.E.; Ma, J.; Qu, W.; Zeldin, D.C. *Drug Metab. Rev.* **1999**, *31*, 205.

128. Matsumoto, S.; Hirama, T.; Matsubara, T.; Nagata, K.; Yamazoe, Y. *Drug Metab. Dispos.* **2002**, *30*, 1240.

129. Hashizume, T.; Imaoka, S.; Mise, M.; Terauchi, Y.; Fujii, T.; Miyazaki, H.; Kamataki, T.; Funae, Y. *J. Pharmacol. Exper. Ther.* **2002**, *300*, 298.

130. Hashizume, T.; Imaoka, S.; Hiroi, T.; Terauchi, Y.; Fujii, T.; Miyazaki, H.; Kamataki, T.; Funae, Y. *Biochem. Biophys. Res. Commun.* **2001**, *280*, 1135.

131. Mouly, S.; Lown, K.S.; Kornhauser, D.; Joseph, J.L.; Fiske, W.D.; Benedek, I.H.; Watkins, P.B. *Clin. Pharmacol. Ther.* **2002**, *72*, 1.

Chapter 13

Prediction of Hepatic Clearance in Humans from Experimental Animals and *In Vitro* Data

Masato Chiba[1], Yoshihiro Shibata[1], Hiroyuki Takahashi[1], Yasuyuki Ishii[1] and Yuichi Sugiyama[2]

[1]*Drug Metabolism, Tsukuba Research Institute, Banyu Pharmaceutical Co. Ltd., Tsukuba, Japan*

[2]*Department of Pharmaceutics, Faculty of Pharmaceutical Sciences, University of Tokyo, Tokyo, Japan*

ABSTRACT

It has become critically important to discover more bioavailable drug candidates in many therapeutic targets where the drug would be preferably given per oral to the patients. At the early discovery stage in the development of new drug, *in vitro* metabolic stability is more routinely examined than pharmacokinetics in experimental animals. The rationale of this strategy is that *in vitro* metabolic stability reasonably well predicts *in vivo* clearance. The first section summarizes recent progress in the prediction strategy for *in vivo* clearance in humans. The predictions were based on *in vivo* pharmacokinetics data in experimental animals and/or *in vitro* metabolic data obtained from both pre-clinical animals and humans. Prediction methods were first classified into two categories: (1) allometric scaling methods and (2) *in vitro-in vivo* scaling methods. Allometric scaling methods involve several approaches with different correction factors such as brain weight, *MLP* (maximum life-span potential) and *in vitro* metabolism data in experimental animals and humans. *In vitro- in vivo* scaling methods can be further classified into the following three different approaches, depending on the scaling factor used for the extrapolation of *in vitro* data to the *in vivo* value: (1) Biologically based scaling factors such as mg-microsomal protein/kg body weight and cells/kg body weight have been used in successful predictions in both humans and animals. (2) The scaling factor was also empirically obtained from *in vitro-in vivo* relationship of metabolic clearance for the target drug candidate in multiple pre-clinical animals. (3) Another successful prediction was achieved by using *in vitro* data from cryopreserved preparations of human hepatocytes with an aid of the scaling factor empirically obtained from the *in vitro-in vivo* calibration with the reference drugs in humans. It was found to be important to correct/compensate the potential inter-individual variation and loss of metabolic activity when cryopreserved human hepatocytes were used for the prediction of clearance in humans. The second section of this chapter describes the structure-activity relationship (SAR) between metabolic stability, CYP3A4 inhibitory potency and type of substrate-induced P450 binding spectra in human liver microsomes. The SAR successfully guided structure modifications to yield metabolically stable candidates for further development.

1   INTRODUCTION

Total body clearance (*CL*) is one of the most important pharmacokinetic parameters because it directly relates to the drug elimination and bioavailability, which determine the AUC after drugs are orally given to human subjects. The half-life at the terminal log-linear phase ($t_{1/2}\beta$ when drugs exhibit 2-compartment model) in the drug concentration-time profile after the drug is intravenously given to the subjects, which relates to the dosing regimen, is a function of volume of distribution (*Vdβ*) and total body clearance (*CL*) as follows:

$$t_{1/2}\beta = 0.693 \times \frac{Vd\beta}{CL} \tag{1}$$

The value of *CL* is the sum of all clearances in the organs participating in the elimination of drugs. Generally, the liver and kidney are the two major organs for elimination: the liver has a high expression level of drug metabolizing enzymes and its location is anatomically important for the elimination between the site of absorption (intestine) and systemic circulation; and the kidney has a large physical capacity for the elimination of hydrophilic xenobiotics. If both organs are involved in elimination, the total body clearance is described by the following equation using clearances in the liver [hepatic clearance (*CL$_H$*)] and kidney [renal clearance (*CL$_R$*)]:

$$CL = CL_H + CL_R \tag{2}$$

For the drug undergoing exclusive elimination by the liver, the total body clearance becomes equal to the hepatic clearance:

$$CL = CL_H \tag{3}$$

In this case, the hepatic clearance directly determines the hepatic availability (F_H), the ratio of drug concentration leaving the liver ($C_{out,liver}$) to that entering the same organ ($C_{in,liver}$) as follows:[1]

$$F_H = \frac{C_{out,liver}}{C_{in,liver}} = 1 - \frac{CL_H}{Q_H} \tag{4}$$

where Q_H is the hepatic blood flow rate. The oral bioavailability (F) is further affected by the fraction of absorption (f_{abs}) and the intestinal availability (F_G) in addition to the F_H after the drug orally dosed to subjects as follows:[2]

$$F = f_{abs} \times F_G \times F_H = f_{abs} \times F_G \times \left(1 - \frac{CL_H}{Q_H}\right) \tag{5}$$

Therefore, at the discovery stage, much attention is generally focused to identify the drug candidate with a lower value of CL and/or CL_H, which more likely leads to the pharmacokinetics having a longer half-life ($t_{1/2}\beta$) (Equation 1) and higher bioavailability (F) (Equation 5), though the other parameters such as absorption/intestinal availability ($f_{abs} \times F_G$) and volume of distribution ($Vd\beta$) are also involved in the overall pharmacokinetics. The example of empirical prediction of absorption/intesinal availability ($f_{abs} \times F_G$) in humans based on the data in experimental animals is described in the later Section 3.3. Recently, an *in vitro* approach to the predictions for tissue-to-plasma partition coefficients ($P_{t:p}$) which directly relate to the volume of distribution has been proposed.[3-6] The values of $P_{t:p}$ were calculated only based on the *in vitro* physicochemical values such as partition coefficients between octanol:water (and olive oil:water) and unbound fraction in plasma. The concentrations (in tissues as well as plasma)-time profiles of drugs in rats were subsequently predicted using PBPK (physiologically-based pharmacokinetic) model incorporating both parameters of $P_{t:p}$ and metabolic (intrinsic) clearance obtained from isolated rat hepatocytes which had been scaled-up to the *in vivo* value as described in Section of 3.2. The predictions were very successful, suggesting that the volume of distribution (or the partition of drugs to each major distribution organ) is reasonably dependent on the physicochemical property of the drug of interest, and this approach can be a potential tool to identify more promising drug candidate before an *in vivo* pharmacokinetic study is initiated.

Hepatic clearance (CL_H) is pharmacokinetically described by different equations, depending on the mathematical model for the disposition of compounds in the liver (for review article, reference 7). Under the assumption that the drug (or chemical entity) is mixed infinitely well inside the liver, the "well-stirred model" is applicable to the hepatic clearance:[8]

$$CL_H = \frac{Q_H \times CL_{int}}{Q_H + CL_{int}} \tag{6}$$

where Q_H and CL_{int} represent the hepatic blood flow and total intrinsic clearance in the liver ($CL_{int} = CL'_{int} \times f_u$; see below). In the other extreme case, the drug is mixed only in the infinitely small section along the flow from input to output of the liver, "parallel-tube model" is applicable:[8]

$$CL_H = Q_H \times (1 - e^{\frac{-CL_{int}}{Q_H}}) \tag{7}$$

In addition to the aforementioned two extreme cases, based on the analysis on the dilution curve of non-extractable tracer after the bolus injection into the liver, the "dispersion model" was introduced as an alternative mathematical model:[9]

$$CL_H = Q_H \times \left[1 - \left(\frac{4a}{(1+a)^2 \exp[(a-1)/2D_N] - (1-a)^2 \exp[-(a+1)/2D_N]} \right) \right] \tag{8}$$

where $a = (1 + 4R_N D_N)^{1/2}$ and R_N is CL_{int}/Q_H. D_N is the dispersion number which determines the extent of dispersion of drug in the liver: the "dispersion model" becomes both "well-stirred model" (Equation 6) and "parallel-tube model" (Equation 7) when the D_N approaches infinite (infinitely mixed condition) and zero (no mixing condition), respectively. Figure 1 shows the relation between CL_H and CL_{int} by each mathematical model. Simulation results clearly indicate that with the increase of CL_{int}, CL_H increases and approaches a maximum value equal to hepatic blood flow rate (Q_H) regardless of the model used. Therefore, the drug design efforts would reasonably be directed to the decrease of CL_{int} in order to improve the pharmacokinetics by lowering hepatic clearance.

The intrinsic clearance (CL_{int}) is further defined using unbound fractions in blood (f_B) and intrinsic clearance by metabolism (if metabolism is the exclusive process for the drug elimination) for unbound drugs ($CL_{u,int}$) as follows:

$$CL_{int} = f_B \times CL_{u,int} \qquad (9)$$

where the $CL_{u,int}$ relates to the enzymatic parameters (V_{max} and K_m) under the condition that the drug concentration is much smaller than K_m value:

$$CL_{u,int} = \frac{V_{max}}{K_m} \qquad (10)$$

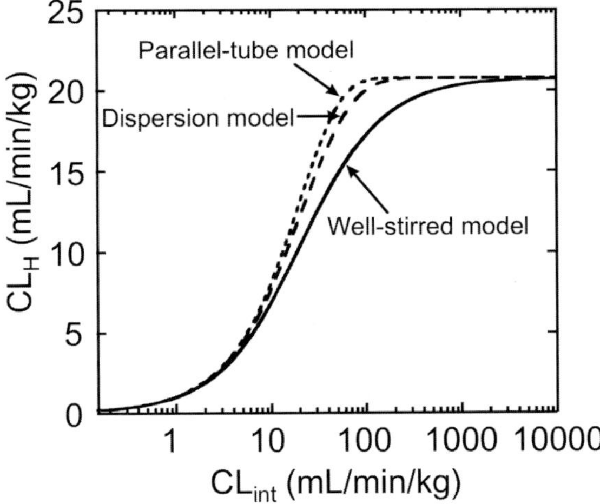

Figure 1. Relation between hepatic clearance (CL_H) and intrinsic clearance (CL_{int}) by various mathematical models. Simulations were carried out by "Well-stirred model" (Equ.6), "Parallel-tube model" (Equ. 7) and "Dispersion model" (Equ. 8). Q_H, f_B and D_N are 21 (mL/min/kg), 1.0 and 0.17, respectively.

Therefore, in order to reduce hepatic clearance, pharmaceutical scientists pay attention to the structural modification/optimization of the compound which leads to the decrease of metabolic rate (or increase of metabolic stability) and/or to the increase of plasma protein binding (which results decrease of f_B). However, the attempt to increase plasma protein binding is less straightforward than metabolic stability from the point of view of the structure-activity relationship, and the consequence would be negative for the development of drugs: it likely reduces the pharmacological potential of the drug candidate as unbound drug concentrations generally relate to the *in vivo* pharmacological efficacy.

Metabolic stability is often evaluated in a higher throughput screening (HTS) manner using subcellular fractions in the liver (for review article, reference 10). The last section presents successful examples for optimizing chemical structures in the light of metabolic stability and/or inhibitory potency, based on the information on the structure-activity relationship between metabolic stability, inhibitory potency and type of substrate-induced P450 (CYP3A4) binding spectra.

2   STRATEGY FOR THE PREDICTION OF HEPATIC CLEARANCE IN HUMANS

New drug candidates can be eliminated from the body by numerous processes including excretion of unchanged compound by renal and/or biliary routes as well as hepatic (and extra-hepatic) metabolism. There are clear advantages in allometric scaling methods, in that they only require the *in vivo* pharmacokinetic data in experimental animals and knowledge of mechanism(s) underlying the elimination route(s) of the particular compound of interest is not necessary for the prediction of clearance in humans. In contrast, prediction methods using *in-vitro* data obtained from hepatocytes and/or liver subcellular fractions are limited only to the NCEs predominantly undergoing hepatic metabolism in humans. Therefore, at the discovery and early development stages, compounds are often evaluated through an iterative validation process to identify their major route(s) of elimination in experimental animals. These data assure that any *in vitro* data using hepatic derived systems would be meaningful for quantitative prediction in humans. *In vitro* methods can account for inter-species differences in metabolism and clearance mechanisms, and may therefore be considered more superior to allometric scaling methods, which are empirical in nature. However, *in vitro* systems do not possess all determinants of in vivo clearance, and since specific activities may be lost or reduced in the preparation of these systems, some clearance mechanisms may be underpredicted or overlooked altogether. Therefore, pharmaceutical scientists need to employ multiple strategies available for the prediction of clearance in humans.

2.1   *Allometric Scaling Methods to predict Clearance in Humans*

Many different physical and/or physiological parameters are known to be a function of body weight. Evidence has been accumulated to encourage scientists to apply the pharmacokinetic parameters obtained in the experimental animals to other mammalian species including humans.[11] The usefulness of allometric scaling has been demonstrated especially for the drugs which mainly undergo physical elimination processes such as biliary and renal excretions if the plasma protein binding was appropriately corrected in different species.[12-17] In contrast, the prediction of clearance by allometric scaling methods for the drugs mainly undergoing enzymatic degradations such as P450-catalyzed metabolism appeared to be less straightforward, considering the fact that the species difference in the metabolic activities depends on the species-dependent expression of different isoform(s) of P450 and differences between individual specific activities of animal vs. human P450 enzymes for any given compound. Any *a priori* knowledge of species differences in the P450 isoform(s) involved and in *in vitro* metabolic rates between humans and experimental animals may or may not be helpful to improve predictability of the allometric scaling methods.[18-20] This section describes the allometric scaling methods for predicting clearance in humans with and without the aid of *in vitro* metabolism data.

The allometric approach is based on the power function, as the body weights from several animal species are plotted against the pharmacokinetic parameter of interest. The power function is written as follows:

$$Y = a \times W^b \tag{11}$$

where Y and W represent the pharmacokinetic parameter of interest (for example, clearance) and body weight, respectively. The coefficient and exponent of the allometric equation are described by a and b, respectively. The log transformation of Equation 11 is represented as follows:

$$\log Y = \log a + b \times \log W \tag{12}$$

where $\log a$ is the y-intercept, and b is the slope. For simple allometric scaling, Y (in Equation 11) is then replaced by the clearance (CL) as follows:

$$CL = a \times W^b \tag{13}$$

Maximum life-span potential (MLP),[21] the maximum documented longevity for a species, has been used as a correction factor in the allometric scaling for predicting clearance in humans.[11,21-24] MLP (in years) was calculated from both brain weight (BW) and body weight (W), in kilogram, as follows:[11]

$$MLP = 185.4 \times (BW)^{0.636} \times (W)^{-0.225} \tag{14}$$

Table 1 summarizes the MLP values of different species. The ratio of MLP values in humans to rats is 20, while the species difference between both species in brain weight is more than 800. The rank order of species difference in parameters is brain weight > body weight $\gg$ *MLP*. The clearance value obtained in the experimental animal is first multiplied by the *MLP* of the corresponding animal (Equation 15) and plotted as a function of body weight on a log-log scale. By using the *a* and *b* in the following equation, the *CL•MLP* value for humans is extrapolated from the body weight of humans. The value was then divided by the *MLP* for humans (8.18 x 10^5 hr) to calculate the predicted clearance in humans:

$$CL \times MLP = a \times W^b \tag{15}$$

Table 1. Body weight, brain weight and maximum life-span potential (MLP) in experimental animals and humans.

	Body Weight in kg		Brain Weight in kg		Body Weight %	MLP in years	
Mouse	0.023	(0.1)	0.000334	(0.18)	1.45	2.7	(0.57)
Rat	0.25	(1.0)	0.00188	(1.0)	0.75	4.7	(1.0)
Dog	14.2	(57)	0.0754	(40)	0.53	19.7	(4.2)
Monkey	4.7	(19)	0.0620	(33)	1.32	22.3	(4.8)
Human	70	(280)	1.53	(814)	2.19	93.4	(20)

Number in parenthesis represents the value normalized by that in rats.

Similarly, the product of brain weight (*BW*) and clearance is also used where the following equation (Equation 16) predicted the clearance in humans at the body weight of humans:

$$CL \times BW = a \times W^b \tag{16}$$

Mahmood *et al.* extensively examined the predictability of allometric scaling methods and found that the "rule of exponents" improved the prediction by helping to choose a more appropriate correction factor for the particular compound of interest.[22,25] Total body clearance of more than 40 drugs in at least three animal species were first fitted by simple allometric scaling method (Equation 13) to evaluate the exponent value (*b*). Drugs were then classified into four categories based on the *b* value (Figure 2): (1) 0.55 < *b* < 0.70, the clearance was well predicted by the simple allometric scaling method (Equation 13), as shown by closed circles on panel A, (2) 0.71 < *b* < 1.0, the *CL x MLP* (Equation

15) reasonably well predicted clearance values in humans compared to other allometric scaling methods, as shown by open circles on panel B, (3) $b > 1.0$, the product of *BW* and *CL* (Equation 16) predicted clearance in humans better than simple and *CL x MLP* scaling methods, as shown by open squares on panel C, and (4) $b < 0.55$ or $b > 1.3$, the clearance in humans was not predicable by any allometric scaling methods. The exponents of allometry have no physiological meaning and the normalization by *MLP* or brain weight is only a mathematical manipulation which may not be associated with any physiological relevance.[25] The value of the exponent in the simple allometric scaling method appeared to be more affected by the observed clearance in the animals with larger body weight. Therefore, the clearance of drug with large value of *b* is more reasonably corrected by a larger (correction) factor (*BW* >> *MLP*, Table 2) especially in larger animal species (humans). The corrections may compensate more than disproportionally lower metabolic clearance in humans than the value corrected only by body weight.

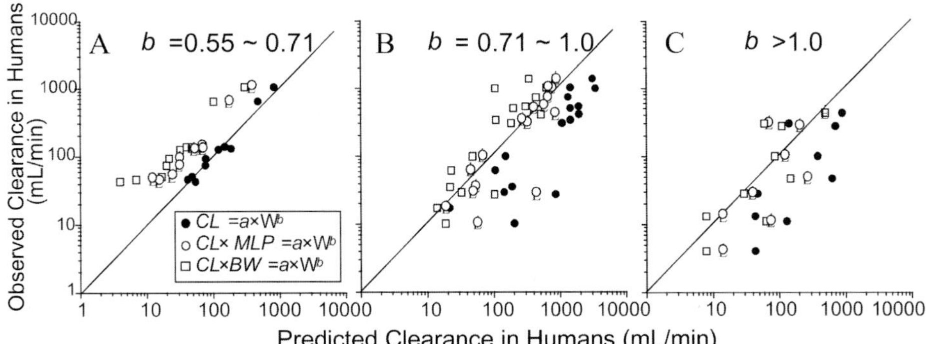

Figure 2. Prediction of clearance in humans using different allometric models. Data were taken from ref. 22.

The successful corrections by these correction factors (*BW* and *MLP*) to improve the predictability of clearance in humans clearly implied that the clearance and metabolic activity in humans, both normalized by the body weight, generally were lower than the experimental animals. The extent of species difference in metabolic activities (experimental animals > humans) likely depend on the drug of interest or the metabolizing enzyme(s) involved. Instead of using the allometrically (or empirically) selected correction factor based on the "rule of exponents", the clearance can be also corrected by the ratio of *in vitro* (metabolic or intrinsic) clearance in both experimental animals and humans. The corrected clearance is then applied to the simple allometric equation as follows:[19]

$$CL_{animal} \times \frac{CL_{human(hepatocytes)}}{CL_{animal(hepatocytes)}} = a \times W^b \tag{17}$$

where $CL_{human(hepatocytes)}$ and $CL_{animal(hepatocytes)}$ represent the *in vitro* metabolic clearance obtained in the preparations of hepatocytes from humans and experimental animals, respectively. Based on the Equation 17, the predictions were carried out for 10 extensively metabolized drugs with at least three animal species.[19,26] The corrections by *in vitro* metabolic data improved the predictions of human clearance compared to the simple allometric approach based only on the body weight. However, the relative predictability between allometric scaling approaches, one with the empirically selected correction factor (*BW* or *MLP*) and the other with an experimentally evaluated correction factor (from hepatocytes studies), was reported comparable[20] although it was emphasized that the integration of *in vitro* data would provide a more rational basis for predicting metabolic clearance in humans.[26]

It has been well recognized that the plasma protein binding of drugs is also subject to species-differences. The effect of incorporating plasma protein binding (as an unbound fraction in the blood) of the target drug into the allometric scaling approaches on the predictability of *in vivo* data in experimental animals was examined for the prediction of clearance in humans.[19,27] Both reports consistently concluded that the replacement of total clearance (*CL*) in the simple allometric calculation (Equation 13) by unbound clearance (*CL/f_B* where f_B is the unbound fraction in blood) resulted in better predictions than either simple allometric scaling excluding plasma protein binding or allometric scaling method corrected by *MLP* with taking plasma protein binding into account: the successful rate for the prediction (predicted value/observed value < 2-fold error of observed value) was 69% (nine of 13 predictions were successful);[27] with the same criteria, the rate was 60% in another set of 15 compounds.[19] Inclusion of plasma protein binding in allometric scaling, which was corrected by *MLP* or *BW*, resulted in poor predictions.[19,27] In addition, the combination of plasma protein binding and allometric scaling with the correction factor of *in vitro* metabolism in animals and humans (Equation 17) decreased the predictability of this allometric scaling method by a successful rate of 82% (without f_B) to 70% (with f_B).[19] The usefulness of plasma protein binding data was demonstrated only in the simple allometric scaling method and appeared to depend on the target compound.[28]

2.2 Direct Prediction of Hepatic Clearance in Humans and Experimental Animals from In-vitro Liver Microsomal Data

Quatitative predictions of hepatic clearance in humans based on *in-vitro* liver microsomal data have been comprehensively reviewed.[7,29-31] Combination of Equations 9 and 10 resulted the following equation:

$$CL_{int} = f_B \times CL_{u,int} = f_B \times \frac{V_{max}}{K_m} \tag{18}$$

where $CL_{u,int}$ represents the *in vitro* metabolic clearance, which can be directly estimated from *in vitro* kinetic data (V_{max} and K_m). In the direct prediction method for the hepatic clearance discussed in this section, the metabolic clearance ($CL_{u,int}$) thus obtained from *in vitro* kinetic experiments is directly scaled up to an *in vivo* value ($CL_{u,int,in\ vivo}$) by using a scaling factor (*SF*) as follows:

$$CL_{u,\ int,\ in\ vivo} = CL_{u,\ int} \times SF \tag{19}$$

where *SF* represents the scaling factor for the *in vitro-to-in vivo* scaling up of the *in vitro* V_{max} value (usually expressed in nmol/min/mg microsomal protein) in humans. The *SF* value (1050, mg microsomal protein/kg body weight) for humans is calculated from (52.5 mg microsomal protein/g liver[29]) x (20 g liver/ kg body weight[32]). The calculated *in vivo* intrinsic clearance ($CL_{u,int,in\ vivo}$) then can replace CL_{int} in the appropriate mathematical model (X-6, X-7 or X-8) in order to predict hepatic clearance in humans.

Successful examples of the *in vitro-in vivo* correlations in the hepatic clearance in humans as well as experimental animals are shown in Figure 3. Both *Indinavir* (HIV protease inhibitor) and *Montelukast* (Leukotriene D$_4$ receptor antagonist) demonstrated that the *in vitro* metabolic clearance obtained from microsomal kinetic data (as V_{max}/K_m) was well consistent with *in vivo* metabolic clearance calculated from AUC values after intravenous administrations to experimental animals and humans (in-house data). Even when nonlinear metabolism is involved in the *in vivo* pharmacokinetics, *in vitro* microsomal data quantitatively well predicted nonlinear hepatic clearance and first-pass metabolism in rats, dogs and humans for *YM796*.[33,34]

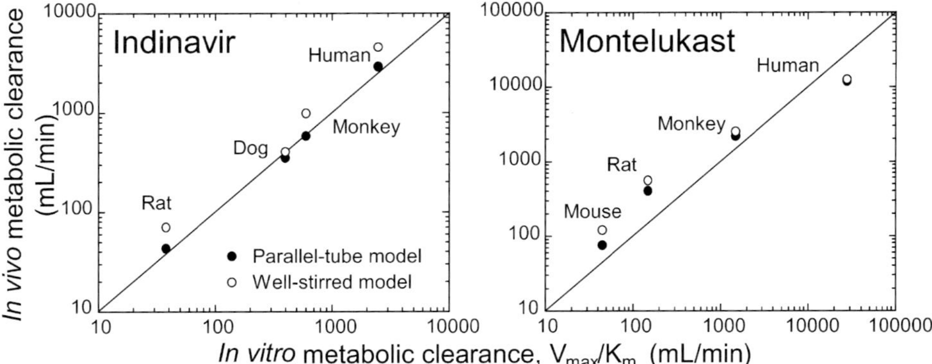

Figure 3. *In vitro- in vivo* correlation of metabolic (intrinsic) clearance in experimental animals and humans for *indinavir* and *montelukast*. The metabolic intrinsic clearance was calculated from *in vivo* clearance based on either "Well-stirred model" (Equ. 6) or "Parallel-tube model"(Equ. 7). The values were compared with *in vitro* metabolic clearance (Equ. 10) obtained from *in vitro* kinetics studies using liver microsomes.

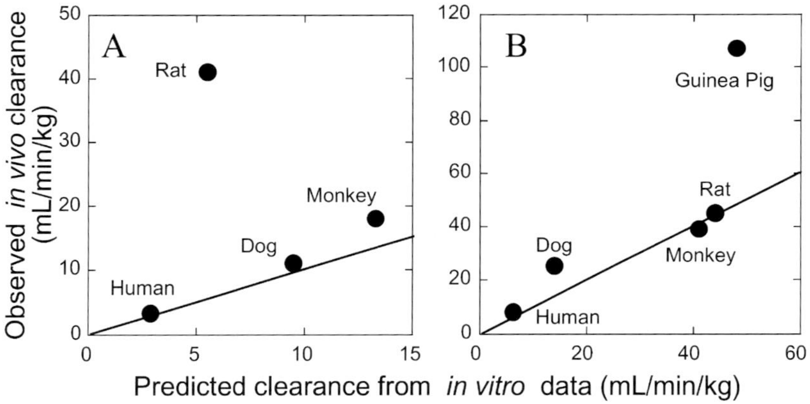

Figure 4. *In vitro- in vivo* correlation of hepatic clearance in experimental animals and humans for *ras FPTase inhibitor I* (A) and *Elzopitant* (B). Data on panels A and B were taken from refs. 45 and 46, respectively. The predictions were carried out by "Well-stirred model" (see Equ. 6) with the *in vitro* intrinsic clearance (Equations 9 and 10). Nonspecific binding to microsomal fraction was corrected for the prediction of *Elzopitant*.

However, the limitations have been also evident for the direct prediction using *in vitro* metabolism data obtained from liver microsomes. In case of *ras FPTase (farnesyl-protein transferase) inhibitor I* (Figure 4A),[35] the *in vivo* pharmacokinetics study in rats showed more than 10-fold higher clearance than that predicted from *in vitro* microsomal data, presumably due to the extensive extra-hepatic metabolism in this species, while *in vitro-in vivo* correlations were relatively good in dogs, monkeys and humans. For the other example, the observed values of clearance in rats, dogs, monkeys and humans were well predicted from *in vitro* microsomal kinetic data for *Ezlopitant* (substance P receptor antagonist) except with that in Guinea Pigs.

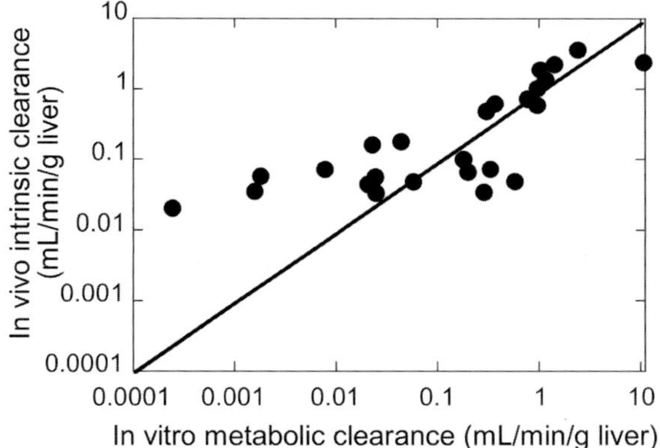

Figure 5. Correlation between *in vivo* intrinsic clearance and *in vitro* metabolic clearance for 25 drugs in humans. Data were taken from ref. 29.

Comparison of *in vitro* and *in vivo* metabolic clearance in humans for 25 drugs also clearly demonstrated the limited predictability of *in vitro* metabolism data (Figure 5).[7,29,30] For the 25 drugs examined, greater than 5-fold differences between predicted and observed metabolic clearance were observed for six. For most outliers, *in vivo* intrinsic clearance was 7~80-fold larger than that predicted from *in vitro* data. Several possible reasons for these discrepancies included:[7] metabolism in the extra-hepatic tissues[37]; incorrect assumption of rapid equilibrium of drugs between blood and hepatocytes[38]; presence of active transporter(s) through the sinusoidal membrane[39-41]; and inter-individual variation.[11,42,43] Furthermore, the misunderstanding of the rate-determination step in the hepatic elimination of drugs can be another potential reason for this discrepancy. The intrinsic clearance for overall elimination ($CL_{int,all}$) of drugs in the liver consists of three distinct processes (Equation 20): [30]

$$CL_{int,all} = \frac{PS_{inf} \times CL_{int}}{PS_{eff} + CL_{int}} \tag{20}$$

where, in addition to the intrinsic clearance (CL_{int}) by the metabolic degradation if the metabolism exclusively eliminates the drug as described by Equation 9, the clearance for the diffusion (including transporter-mediated transport (if any) through the sinusoidal membrane) in the directions of sinusoid→hepatocytes (influx, PS_{inf}) and hepatocytes→sinusoid (efflux, PS_{eff}) are taken into account. In the case that the membrane permeability (or clearance for the diffusion through membrane in the direction of hepatocytes→sinusoid) of drugs is much less than the metabolic intrinsic clearance ($PS_{eff} \ll CL_{int}$), $CL_{int,all}$ becomes equal to PS_{inf} (Equation 21):

$$CL_{int,all} = PS_{inf} \tag{21}$$

This indicates that the intrinsic clearance calculated from *in vivo* data may not necessarily represent the metabolic intrinsic clearance. The *in vivo* intrinsic clearance can be much larger than the predicted intrinsic clearance from *in vitro* metabolism studies if the PS_{inf} is larger than CL_{int}. This is more likely affecting the *in vivo* intrinsic clearance when anactive transporter(s) is (are) involved in the influx process especially for less lipophilic drugs which showed larger discrepancies between *in vivo* (overall) intrinsic clearance and the *in vitro* intrinsic clearance predicted from metabolic data in Figure 5.

In addition to the aforementioned factors underlying the *in vitro*-and-*in vivo* discrepancy, the nonspecific binding of drugs to the subcellular fractions such as microsomal lipid and protein has been reported to hamper the predictability of *in vitro* data.[44-47] Kinetically, it is an important assumption that the only unbound drug can undergo the free exchange and reach rapid equilibrium between outside and inside cells, which can be subject to the metabolism. Therefore, both protein binding in blood and nonspecific binding to the *in vitro* incubation matrix (microsomes etc.) were often examined and have been used to correct enzyme kinetic value for the unbound drug for the purpose of quantitative predictions of pharmacokinetics in rats.[44-47] Contrary to this pharmacokinetic premise, it was recently reported that the hepatic clearance in humans was more accurately predicted by the model disregarding plasma protein binding (as f_B) than that incorporating the same parameter.[48] This was more evident for lipophilic drugs which more extensively bound to the plasma protein. Twenty-nine drugs with wide ranges of structure and physicochemical property were examined for the predictability of *in vitro* metabolic data obtained from human liver microsomes.[27,49,50] Comprehensive analysis highlighted the potential pitfall in *in vitro* experimental systems. This included (1) inaccuracy to measure the unbound fraction in plasma especially for the tightly bound drugs, and/or (2) significant nonspecific bindings of drugs to the *in vitro* reaction matrix, leading to the underestimation of intrinsic clearance for the unbound drugs. Predictions were carried out with "well-stirred model" either disregarding both unbound fractions in the blood (f_B) and in the microsomal reaction mixture (f_m) (Equation 22), incorporating only f_B (Equation 23), or incorporating both f_B and f_m (Equation 24).

$$CL = \frac{Q_H \times CL_{u,int}}{Q_H + CL_{u,int}} \tag{22}$$

$$CL = \frac{Q_H \times f_B \times CL_{u,int}}{Q_H + f_B \times CL_{u,int}} \tag{23}$$

$$CL = \frac{Q_H \times f_B \times \dfrac{CL_{u,int}}{f_m}}{Q_H + f_B \times \dfrac{CL_{u,int}}{f_m}} \tag{24}$$

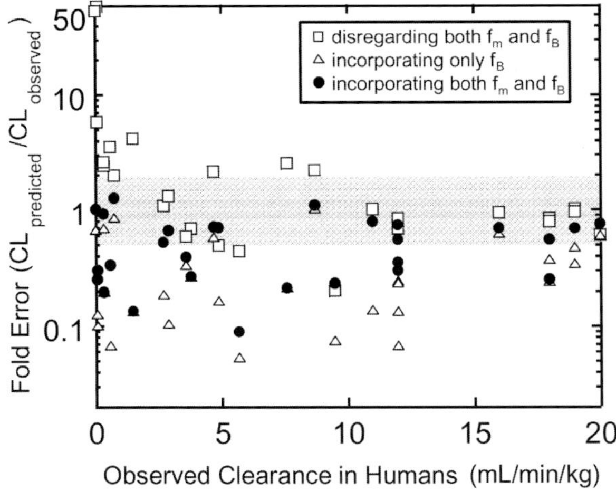

Figure 6. Accuracy of human clearance values predicted from *in vitro* intrinsic clearance data versus observed human clearance. Data were taken from ref. 50. Data in the shaded area are considered to be successful (0.5<fold error <2).

As shown in Figure 6, predictions based on the *in vitro* values reasonably well agreed with human clearance when both bindings were completely disregarded (Equation 22) or both were taken into considerations (Equation 24). For these drugs, the poorest predictions were obtained if only unbound fraction in the blood was included in the equation (Equation 23). Results indicated that the incorporation of both bindings (f_B and f_U) would provide better predictions of human clearance by using *in vitro* microsomal data regardless of structure and physicochemical property of drugs.

2.3 Empirical Prediction of Hepatic Clearance in Humans by In vitro-In vivo Relationship in Experimental Animals

At the pre-development stage of new drug candidates before safety and clinical studies, both *in vitro* metabolism and *in vivo* pharmacokinetic studies in two or three different experimental animals would have been already conducted. The pre-clinical information on the absorption and/or systemic exposure not only supports safety studies, but also helps go/no-go decisions for continued development of the candidate compound by providing preliminary predictions of oral bioavailability and half-life for clinical studies. This section review two examples of the empirical use of *in vivo* pharmacokinetics data in pre-clinical animals for the prediction of human clearance and oral bioavailability. Both predictions were successful with an aid of *in vitro* metabolic data obtained from experimental animals and humans.

One of the practical (empirical) approaches to overcome the potential discrepancy between *in vitro* and *in vivo* intrinsic clearance, which is often observed when the biologically-based scaling factor is used for scaling-up of *in vitro* data to *in vivo* (see section 2.2), has been recently proposed by Naritomi *et al.*[51] In this method, the *in vitro-to-in vivo* scaling factor is calculated as follows:

$$\text{Scaling factor (animal)} = \frac{CL_{int, in\ vivo, animal}}{CL_{int, in\ vitro, animal}} \tag{25}$$

where $CL_{int, in\ vivo,\ animal}$ and $CL_{int, in\ vitro, animal}$ represent the intrinsic clearance obtained from *in vivo* pharmacokinetics and *in vitro* experiments using the same experimental animal, respectively. Then, the *in vivo* intrinsic clearance for humans ($CL_{int,\ in\ vivo,\ humans}$) is calculated by:

$$CL_{int, in\ vivo, humans} = CL_{int, in\ vitro, humans} \times \text{Scaling factor(animal)} \tag{26}$$

Comparisons of $CL_{int, in\ vivo}$ and $CL_{int, in\ vitro}$ in rats, dogs and humans for eight model compounds indicated that (1) values of scaling factor were similar among animals for each compound; (2) values of scaling factor differed between compounds; and (3) the *in vivo* intrinsic clearance for humans calculated from

Equation 26 accurately predicted the observed clearance in humans. This empirical prediction method is based on the assumption that any discrepancy between *in vitro* and *in vivo* in experimental animals is applicable to the humans. Though the mechanism of discrepancy would not likely be fully understood by the time the new drug candidate enters the pre-development stage, the method would be helpful for the preliminary prediction of clearance in humans.

The oral bioavailability is defined as the fraction of oral dose which reaches the systemic circulation after the passages through the gut wall and liver where the drug may undergo the first-pass metabolism. The oral bioavailability (F) can be described as shown before (Equation 5):

$$F = f_{abs} \times F_G \times F_H = f_{abs} \times F_G \times \left(1 - \frac{CL_H}{Q_H}\right) \tag{5}$$

where f_{abs} is the fraction of oral dose absorbed from the gastro-intestinal lumen. F_G and F_H represent the fractions of drug reaching the outflow of the intestine and liver as an intact drug, respectively.[2] A marked species difference was observed in the pharmacokinetics of indinavir between rats, dogs and monkeys (Table 2). Inter-species difference (and the rank order) in the hepatic clearance observed *in vivo* (CL_H) was quantitatively consistent with that ($CL_{H, in vitro}$) predicted from V_{max} and K_m values obtained in the liver microsomes. Species difference appeared to be further amplified by the variations of absorption/intestinal availability ($f_{abs} \times F_G$). Extrapolation of the predicted hepatic clearance from *in vitro* values was validated by good *in vitro-in vivo* correlations among experimental animals (Figure 3 and Table 2). The predicted oral bioavailability in humans [26~75% $= 0.75 \times (0.35~1.0) \times 100$] was reasonably consistent with the clinical data (60-65%).[53] Pharmacokinetic modeling for the intestinal extraction of drugs appeared to be much complicated than that for the liver due to the anatomical complexity of the intestine including localization of metabolic site and gradient of drug in the lumen.[54] However, Thummel *et al.* proposed the quantitative approach to the prediction of intestinal first-pass extraction/metabolism based on the modified well-stirred model.[54] This model can be a potential tool for the prediction of the average extraction and maximum first-pass availability of drugs in the human intestine. Nonetheless, the empirical approach adopted for the prediction of oral bioavailability of indinavir in humans may provide the preliminary information at the stage of pre-development enough to make a go/no-go decision for further development.

In summary, the empirical methods described herein would be useful approaches to obtain pre-clinical information needed for predictions of hepatic clearance and oral bioavailalibility in humans before the new drug candidate enters clinical studies. However, it should be noted that these methods are applicable to prediction only when the scaling factor for the *in vitro-to-in vivo* correlation is relatively constant among multiple pre-clinical animals or the *in vivo* hepatic clearance is reasonably well predicted from *in vitro* data in multiple experimental animals. Any outlier may discount the value of application unless the outlying animal(s) is (are) known to be pharmacokinetically less representative of humans than others. These premises may limit their versatility for the accurate predictions in humans.

Table 2. Species difference in pharmacokinetics and *in vitro-in vivo* correlation of *indinavir*.

	Observed *In Vivo* Pharmacokinetics					Predicted from In Vitro
	$CL_H{}^a$ (mL/min/kg)	$Q_H{}^b$	$Vdss^a$ (L/kg)	F^a (%)	$f_{abs} \times F_G{}^c$ (%)	$CL_{H,in\ vitro}{}^d$ (mL/min/kg)
Rat	47	65	2.2	24	87	33
Dog	15	45	1.4	72	100	8
Monkey	27	45	2.3	14	35	29
Human		20				5

[a] In-house data; [b] ref. 42; [c] from Equ. 5; [d] Hepatic clearance predicted from Equations 6, 9 and 10 using *in vitro* V_{max} and K_m values (ref. 52), unbound fractions in blood (ref. 52) and Q_H.

2.4 *Empirical Prediction of Hepatic Clearance in Humans by In vitro-In vivo Calibration of Human Hepatocytes Data*

It has been long recognized that the isolated hepatocytes generally provide better *in vitro* information for quantitative/qualitative predictions of *in vivo* metabolism/degradation than liver slices and subcellular fractions.[55-61] The superior predictions yielded by isolated rat hepatocytes vs liver microsomes was more evident for highly cleared drugs.[57] Isolated rat hepatocytes accurately predicted *in vivo* metabolic activities in the presence or absence of P450 inhibitors and inducers (P450 modifiers). Successful predictions included: *in vivo* metabolism of ethoxycoumarin in control and inducer-pretreated rats[55]; product inhibition observed in the diazepam metabolism in rats[62]; the pharmacokinetics of caffeine,[63] diazepam,[59] ethoxycoumarin,[56] ondansetron,[64] tolbutamide and phenytoin[60] in control rats; and the inhibitory effect of omeprazole on diazepam disposition in rats.[58]

Based on the evidence of good predictability of isolated rat hepatocytes, freshly isolated human hepatocytes appeared to be one of the best preparations for the prediction of *in vivo* metabolism in humans. Indeed, isolated human hepatocytes reasonably well predicted hepatic clearance (or hepatic extraction ratio) for the drugs with wide range of metabolic clearance undergoing human UGT-mediated phase II metabolism[65] as well as P450-catalyzed Phase I metabolism.[66]

As mentioned in the previous Section 3.2, intrinsic (metabolic) clearance (CL_{int}) in the equation of hepatic clearance (CL, Equation 24) consists of three independent parameters (Equation 27): $CL_{u,int}$, intrinsic (metabolic) clearance for the unbound (free) drugs; f_B, unbound fraction in blood (or unbound fraction in plasma multiplied by the blood-to-plasma partition ratio); and f_m or f_U (unbound fraction in *in vitro* incubation matrix such as subcellular fraction or hepatocytes).

$$CL_{int} = f_B \times \frac{CL_{u,int}}{f_u} \tag{27}$$

The importance of incorporating unbound fractions in blood and/or in incubation mixture has been extensively discussed in the previous section and reports mentioned therein. Although the unbound fractions in plasma are routinely measured by ultra-filtration or equilibrium dialysis techniques, many compounds adsorb onto the membrane and/or apparatus, which would hamper the accuracy of estimated values especially for the drugs tightly bound to proteins.[67,68] At early discovery or screening stages, it is impractical to measure both unbound fractions in plasma and *in vitro* incubation matrix as well as *in vitro* intrinsic clearance for unbound drugs for every drug candidate. In order to avoid complexity and improve the accuracy for predicting *in vivo* metabolic clearance, a novel and convenient *in vitro* method for predicting *in vivo* metabolic clearance has been established by using freshly isolated rat hepatocytes suspended in 100% rat serum.[69] Metabolic (intrinsic) clearance for sixteen widely used compounds, directly evaluated from the incubation with freshly isolated rat hepatocytes suspended in 100% rat serum, was extrapolated to the *in vivo* value using biologically based scaling factor [6×10^9 cells/kg = 45 (g/kg of body weight) $\times$ 1.35×10^8 (cells/g of liver)[61]]. *In vitro* metabolic clearance obtained from a single incubation accurately predicted both hepatic clearance and oral bioavailability without separate evaluation of unbound fractions in the plasma and/or incubation medium (Figure 7).

Data showing that freshly isolated rat hepatocytes provided accurate predictions of hepatic clearance in rats was encouraging, and the application of this system was extended to the prediction of clearance in humans. Freshly isolated human hepatocytes appeared to be a superior *in vitro* tool for the prediction of clearance in humans, however the availability of such material is inconsistent, precluding

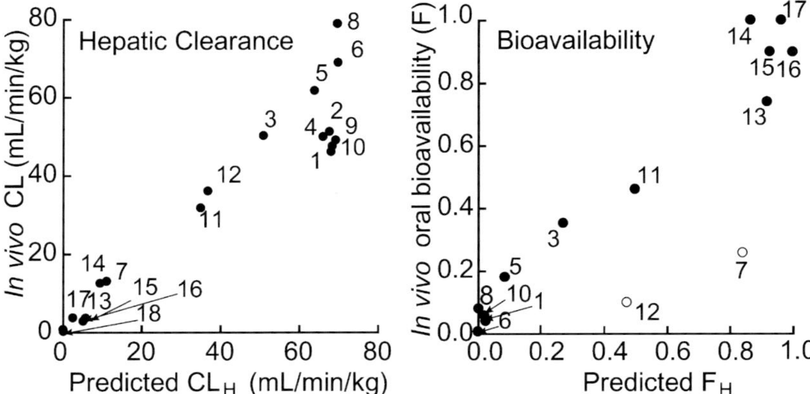

Figure 7. Observed versus predicted hepatic clearance (left) and oral bioavailability (right) from *in vitro* data in isolated rat hepatocytes suspended in 100% serum. Data were taken from ref. 69. 1,alprenolol; 2,chlorpromazine; 3,diazepam; 4,7-ethoxycoumarin; 5,hexobarbital; 6,lidocaine; 7,phenytoin; 8,propranolol; 9,quinidine; 10,verapamil; 11,compound X; 12,compound Y; 13, antipyrine; 14, caffeine; 15, ibuprofen; 16,tolbutamide; 17, theophylline; 18, (S)-warfarin. Note: Poor intestinal availability was known for compounds 7 and 12.

routine application of this *in vitro* system. Cryopreserved human hepatocytes have instead became more prevalent and widely used for routine application because they have been reported to qualitatively retain most of the metabolic activities observed in hepatocytes from fresh liver.[70-72] *In vitro* metabolic clearance for seven clinically used drugs was measured in the cryopreserved preparations of human hepatocytes suspended in 100% human serum and scaled up to the *in vivo* clearance values by using either biologically based or empirical scaling factor (Figure 8).[73] Values were generally underpredicted when the biologically based scaling factor [3.1×10^9 cells/kg $= 26$ (g/kg of body weight) $\times 1.2 \times 10^8$ (cells/g of liver)[29,74]] was used. In contrast, the empirical scaling factor (calculated from the best-fitting between observed and predicted intrinsic clearance values in Figure 8, 8.5×10^9 cells/kg body weight, note: the number depends on the preparation used), which was approximately three-fold larger than the biologically based scaling factor, improved predictability of cryopreserved preparation of human hepatocytes. The larger scaling factor, empirically thus obtained, may compensate the potential loss of metabolic capacity during cryopreservation or the fact that the hepatocytes are from only a few donors which may not represent a good average. Both hepatic clearance and oral bioavailability were reasonably well predicted from the *in vitro* metabolic clearance with the empirical scaling factor for 14 widely used drugs (Figure 9). The method will be helpful at the early discovery stage to identify more promising candidates which have lower hepatic clearance and higher oral bioavailability in humans.

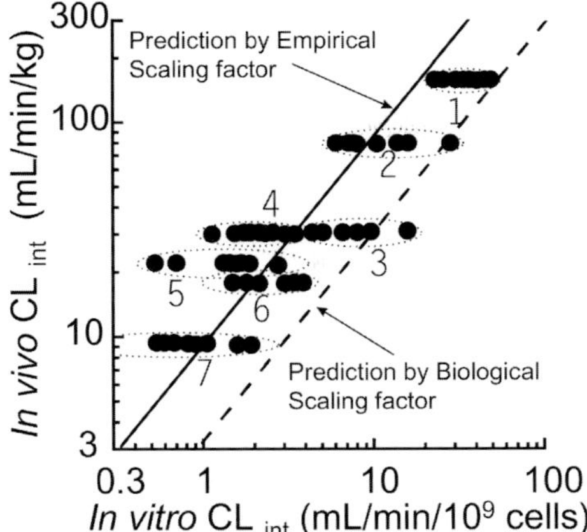

Figure 8. Prediction of hepatic intrinsic clearance from *in vitro* data in ten cryopreserved preparations of human hepatocytes suspended in 100% serum. Data were taken from ref. 67. Predictability was compared with different scaling factors: empirically calculated scaling factor (8.5×10^9 cells/kg body weight) versus biologically-based (reported) scaling factor (3.1×10^9 cells/kg). Each symbol represents the mean of three determinations obtained in each preparation (total=10): 1.naloxone; 2, buspirone; 3.verapamil; 4.lidocaine; 5.imipramine; 6.metoprolol; 7.timolol.

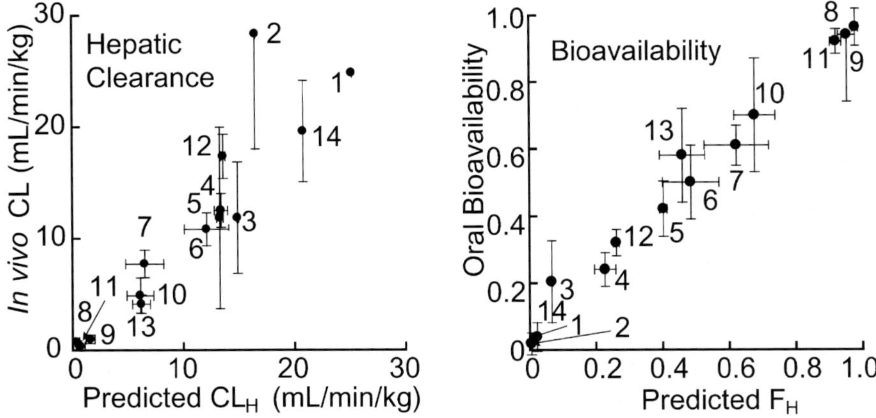

Figure 9. Prediction of hepatic clearance (left) and bioavailability (right) for 14 standard drugs by pooled preparation of human hepatocytes suspended in 100% serum. Data were taken from ref. 73. Prediction was carried out using empirially obtained scaling for the pooled preparation used (10.8×10^9 cells/kg). 1.naloxone; 2, buspirone; 3.verapamil; 4.lidocaine; 5.imipramine; 6.metoprolol; 7.timolol; 8.antipyrine; 9.diazepam; 10.quinidine; 11.caffeine; 12.propranolol; 13.diclofenac; 14. phenacetin.

3   OPTIMIZATION OF METABOLIC CLEARANCE BY STRUCTURE MODIFICATION

Relation between Metabolic Stability, Inhibitory Potency and P450-Substrate Binding Spectra

Despite CYP3A4 being one of the most prevalent isoforms in the human liver[75] and its importance in the metabolism of pharmaceuticals[76] the understanding of active site structure is rather limited.[77,78] In a series of HIV (human immunodeficiency virus) protease inhibitor candidates sharing a similar template structure, there was a positive relationship between metabolic stability (as intrinsic clearance, $CL_{u,int}$) and inhibitory potency (as IC_{50}) for CYP3A4-mediated marker metabolism (Figure 10);[79] namely, the more metabolically stable HIV protease inhibitor had a greater potency for inhibition of CYP3A4. In addition, The drug candidates formed two clusters defined by the distinct type of P450 binding spectra: the compounds with type II binding spectra generally were more stable and potent for CYP3A4 inhibition than those with type I binding spectra. The comparison between compounds having the same structural template clearly demonstrated that minor structural modifications dramatically changed the CYP3A4 inhibitory potency as well as the *in vitro* metabolic stability. Examples are shown in Figure 11. The *gem*-dimethyl modification to block *N*-dealkylation (BM→compound A) did not change the type of P450 binding spectra (both type II) and had little effect on IC_{50} and metabolic stability, while one-methyl substitutions on the pyridine ring (compound A→B or C) switched the binding spectra from type II to I. This substitution decreased both CYP3A4 inhibitory potency and metabolic stability by a factor of >10 compared with BM. Also, the position of the nitrogen atom on the pyridine ring appears to be critical: compound D (nitrogen *ortho* to the linkage) did not produce type II binding spectrum and the stability/inhibitory potency was much lower than compound A (nitrogen *meta* to a linkage). These results confirmed that the nitrogen atom on the pyridine moiety can coordinate with a ferric form of P450 heme iron. Substitutions on the nitrogen-containing heterocycles and the change of position of nitrogen clearly affect the manner/magnitude of coordination between the ligand and P450 heme, presumably due to the change in the steric environment for the interaction between substrate and P450 active site. Replacements on the benzene ring (compounds E~G) had little effect on either the type of spectra, inhibitory potency or metabolic stability. This suggested that the functional group at the position of benzene may not be involved in the stabilization of the nitrogen (on the pyridine)-P450 heme coordination.

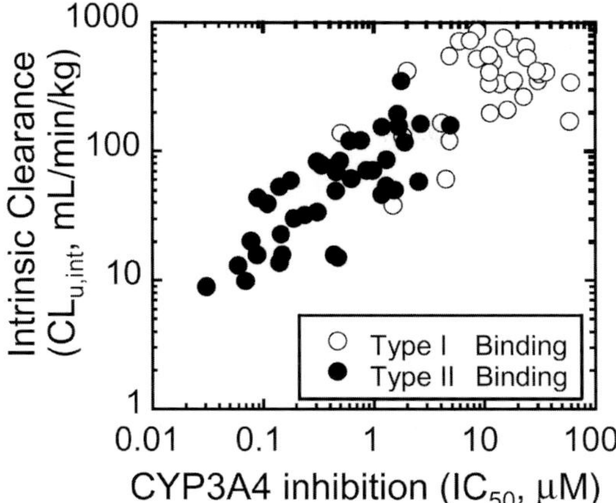

Figure 10. Correlation between metabolic stability (intrinsic clearance) and CYP3A4 inhibitory potency (IC$_{50}$) in HIV protease inhibitor candidates in pooled human liver microsomes. Open and closed circles represent the candidates showing type I and type II binding spectra induced by substrates with P450(CYP3A4), respectively.

	X	Y	P450 SP	CL$_{u,int}$	IC$_{50}$
BM			I I	75	0.45
A			I I	71	0.87
B			I	763	15
C			I	854	8.7
D			I	636	19
E		H	I I	54	1.3
F			I I	71	0.99
G			I I	61	0.63

Figure 11. Effect of structure modifications on the type of P450 binding spectra (SP), intrinsic clearance (CL$_{u,int}$, mL/min/kg), and CYP3A4 inhibitory potency (IC$_{50}$, µM). BM represents "Bench Mark".

A similar relationship between metabolic stability, inhibitory potency and P450 (CYP3A4) binding spectra was also obtained in the series of ras FTI (farnesyl-protein transferase inhibitor) candidates.[78] The methyl-substitution on the imidazole ring dramatically reduced the inhibitory potency of all non-substituted FTI candidates for the marker metabolism catalyzed by CYP3A4 and, to a lesser extent, by CYP2C9 (Table 3). In contrast, this substitution had little effect on the inhibitory potency (IC_{50} ~ 20 μM) for the CYP2D6-catalyzed metabolism. Non-substituted FTI candidates (A1, B1, and C1) all demonstrated concentration-dependent type II binding spectra (max, ~425 nm; min, ~390 nm) with P450, presumably CYP3A4. One methyl-substitution on the imidazole ring at the 2 position dramatically increased V_{max} as well as K_m values. This appears to support the hypothesis that the steric factors play an important role in the interaction between the substrate and active oxygen at the CYP3A4 heme and affect the reactivity (turnover) in CYP3A4-catalyzed metabolism. The substitution on the imidazole ring successfully improved the drug metabolism profiles of FTI candidates by reducing the potential to inhibit CYP3A4-mediated metabolism without affecting in vitro metabolic clearance or intrinsic clearance (V_{max}/K_m).

Compounds that can bind simultaneously to both the lipophilic region of the P450 protein and to the prosthetic heme iron are known to be inherently more potent inhibitors than those depending on only one of these binding interactions.[77,78,81,82] The inhibitory potency of such compounds is governed not only by their hydrophobic character but also by the strength of the bond between their heteroatomic lone pair electrons and the prosthetic heme iron. The binding of inhibitors that are strong iron ligands gives rise to a type II difference spectrum. Strong ligands such as the heterocyclic nitrogen displace weak ligands (water) from the hexacoordinated heme of P450. This leads to the subsequent coordination to the pentacoordinated heme, resulting in the P450 shift from its high-spin to low-spin dominant form. This spin state change is accompanied by an increase in the redox potential of the P450, which makes P450 reduction (by NADPH P450 reductase) more difficult.[83,84] Therefore, both the change in redox potential and the physical occupation of the sixth coordination site for oxygen by the strong iron ligand of drug candidates may be responsible for more potent CYP3A4 inhibition (IC_{50} and/or K_m) and lower turnover rates (V_{max}) observed in the compounds with type II binding spectra than those with type I. These information on the structure-activity relationship help medicinal chemists to design metabolic stability and/or CYP3A4 inhibitory potency of drug candidates for further development.

Table 3. P450 interaction and metabolic kinetics of imidazole- and 2-methylimidazole-containing FTIs in human liver microsomes. Reproduced from ref. 80. K_s, dissociation constant, μM; K_m, μM; V_{max}, nmol/min/mg protein; V_{max}/K_m, mL/min/mg protein.

		P450 Binding	P450 Inhibition (IC_{50}, μM)			Metabolic Kinetics		
	type	Ks	3A4	2C9	2D6	K_m	V_{max}	V_{max}/K_m
A1	II	0.0845	0.10	2.38	25.2	0.256	0.0916	0.358
A2	I	4.03	8.09	13.2	16.0	3.22	0.943	0.293
(ratio)		(48)	(81)	(5.6)	(0.64)	(13)	(10)	(0.82)
B1	II	0.0451	0.25	3.89	15.3	0.144	0.0757	0.526
B2	I	3.79	9.64	18.2	20.2	2.87	1.20	0.418
(ratio)		(84)	(39)	(4.7)	(1.3)	(20)	(16)	(0.80)
C1	II	0.0675	0.24	5.85	14.9	0.169	0.0430	0.254
C2	I	10.5	41.8	163	18.8	7.76	1.46	0.188
(ratio)		(156)	(174)	(28)	(1.3)	(46)	(34)	(0.74)

A1(R=H);**A2**(R=CH$_3$)   **B1**(R=H);**B2**(R=CH$_3$) 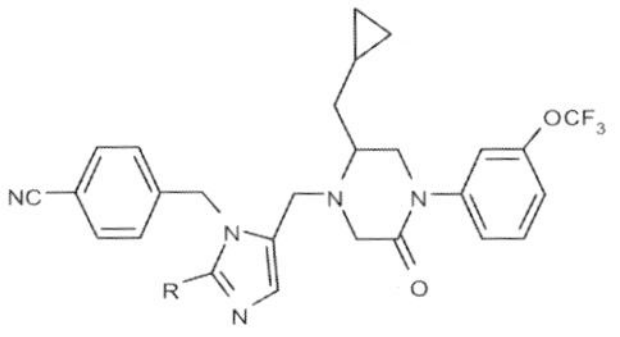   **C1**(R=H);**C2**(R=CH3)

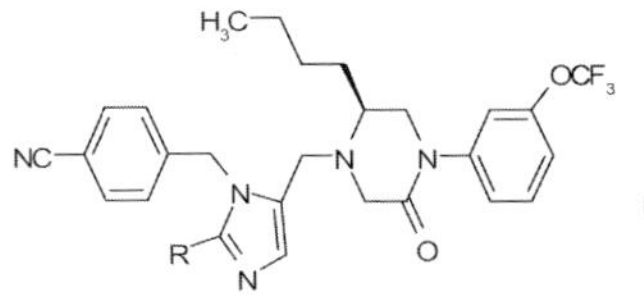

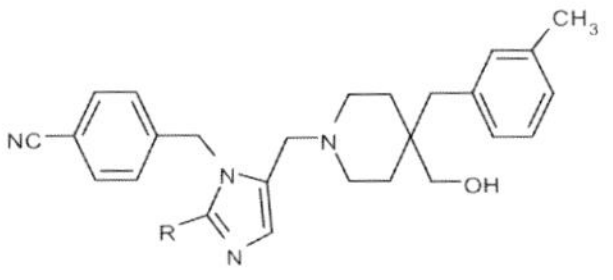

4   CONCLUSION

At discovery and pre-development stages, the time and resource that drug metabolism scientists can spend on each drug candidate are generally very limited. In order to facilitate the drug design process in such constrained environment, many pharmaceutical companies have already set up the HTS-based assay systems for metabolic liability. Though our knowledge and understanding of three-dimensional structure active site modeling and pharmacophores for metabolism by P450 isoforms have increased enormously, the prospective application for the quantitative prediction of P450-substrate binding (and metabolism) remains to be further refined as a practical tool in drug discovery. The examples of SAR between metabolic stability (by CYP3A4),

CYP3A4 inhibitory potency and substrate-induced P450 (CYP3A4) binding spectra might be a useful probe for the effect of structure modification on the manner of substrate-P450 interaction and provide a practical guide for the direction of structure optimization.

Quantitative predictions of hepatic clearance in humans have become more important at earlier stages of discovery and development in order to speed up the structure optimization process which would lead to a better pharmacokinetic profile in humans. Once a chemical entity is identified as a promising new drug candidate, *in vivo* pharmacokinetic studies are usually initiated using multiple pre-clinical animal species. Pharmacokinetic data in animals then are useful for the extrapolation to humans if the allometric scaling method is applied with or without correction factors such as brain weight and *MLP* (maximum life-span potential). The data of metabolic clearance obtained from *in vitro* experiments using subcellular fractions (and hepatocytes) in both experimental animals and humans further expand the range of applicable methodology for predictions: (1) allometric scaling method corrected by the ratio of *in vitro* metabolic clearance in humans to that in animals; (2) direct extrapolation of *in vitro* metabolic clearance in humans to the *in vivo* value using biologically based scaling factors; and (3) empirical extrapolation of *in vitro* metabolic clearance in humans to the *in vivo* value using scaling factor calculated from the discrepancy between *in vivo* and *in vitro* metabolic clearance in experimental animals. The *in vitro* method using cryopreserved preparation of human hepatocytes suspended in 100% serum with an aid of the scaling factor calibrated by standard compounds would be an alternative to the existing methods. Although many approaches are available for the prediction of clearance in humans, pharmaceutical scientists need to judge which approach(es) can be most relevant and accurate with the limited data at hand at the stages of discovery and pre-development of new drugs.

REFERENCES

1. Wilkinson, G.R. *Pharmacol. Rev.* **1987**, *39*, 1.
2. Lin, J.H. and Lu, A.Y.H. *Pharmacol. Rev.* **1997**, *49*, 403.
3. Paulin, P. and Theil, F.P. *J.Pharm. Sci.* **2000**, *89*,16.
4. Paulin, P., Schoenlein, K. and Theil, F.P. *J.Pharm.Sci.* **2001**, *90*, 436.
5. Paulin, P. and Theil, F.P. *J.Pharm.Sci.* **2002**, *91*, 129.
6. Paulin, P. and Theil, F.P. *J.Pharm.Sci.* **2002**, *91*, 1358.
7. Iwatsubo, T., Hirota, N., Ooie, T., Suzuki, H., Shimada, N., Chiba, K., Ishizaki, T., Green, C.E., and Tyson, C.A. *Pharmacol. Ther.* **1997**, *73*, 147.
8. Pang, K.S. and Rowland, M. *J. Pharmacokinet. Biopharm.* **1977**, *5*, 625.
9. Roberts, M.S. and Rowland, M. *J. Pharmacokinet. Biopharm.* **1986**, *14*, 227.
10. Masimirembwa, C.M., Thompson, R., and Andersson, T.B. *Comb.Chem High Throughput.Screen.* **2001**, *4*, 245.

11. Boxenbaum, H. *J. Pharmacokinet. Biopharm.* **1982**, *10*, 201.

12. Mahmood, I. *Life Sci.* **1998**, *63*, 2365.

13. Lin, J.H. *Drug Metab. Dispos.* **1995**, *23*, 1008.

14. Dedrick, R.L., Forrester, D.D., Cannon, J.N., el Dareer, S.M., and Mellett, L.B. *Biochem. Pharmacol.* **1973**, *22*, 2405.

15. Mordenti, J. *J. Pharm. Sci.* **1986**, *75*, 1028.

16. Dedrick, R.L. *J. Pharmacokinet. Biopharm.* **1973**, *1*, 435.

17. Sawada, Y., Hanano, M., Sugiyama, Y., and Iga, T. *J. Pharmacokinet. Biopharm.* **1984**, *12*, 241.

18. Mahmood, I. *Drug Metabolism and Drug Interaction* **2001**, *18*, 135.

19. Lave, T., Schmitt, C., Chou, R.C., Jaeck, D., and Coassolo, P. *J. Pharm. Sci.* **2002**, *86*, 584.

20. Mahmood, I. *J.Pharm.Sci.* **1998**, *87*, 527.

21. McNamara PJ, Interspecies scaling in pharmacokinetics. In: *Pharmaceutical Bioequivalence*, (Eds. Welling PG, Tse FLS, and Dighe SV). Macel Dekker Inc., New York, **1991**.

22. Mahmood, I. and Balian, J.D. *Xenobiotica* **1996**, *26*, 887.

23. Boxenbaum, H. and Fertig, J.B. *Eur. J. Drug Metab. Pharmacokinet.* **1984**, *9*, 177.

24. Mahmood, I. and Balian, J.D. *J. Pharm. Sci.* **1996**, *85*, 411.

25. Mahmood, I. *J.Pharm.Sci.* **1999**, *88*, 1101.

26. Lave, T., Coassolo, P., and Reigner, B. *Clin.Pharmacokin.* **1999**, *36*, 211.

27. Obach, R.S., Baxter, J.G., Liston, T.E., Silber, B.M., Jones, B.C., MacIntyre, F., Rance, D.J., and Wastall, P. *J. Pharmacol. Exp. Ther.* **1997**, *283*, 46.

28. Mahmood, I. *J.Clin.Pharmacol.* **2000**, *40*, 1439.

29. Iwatsubo, T., Hirota, N., Ooie, T., Suzuki, H., Shimada, N., Chiba, K., Ishizaki, T., Green, C.E., and Tyson, C.A. *Pharmacol. Ther.* **1997**, *73*, 147.

30. Ito, K., Iwatsubo, T., Kanamitsu, S., Nakajima, Y., and Sugiyama, Y. *Annu. Rev. Pharmacol. Toxicol.* **1998**, *38*, 461.

31. Obach, R.S. *Curr. Opin. Drug Discov. Devel.* **2001**, *4*, 36.

32. Boxenbaum, H. *J. Pharmcokinet. Biopharm.* **1980**, *8*, 165.

33. Iwatsubo, T., Suzuki, H., Shimada, N., Chiba, K., Ishizaki, T., Green, E., Tyson, C.A., Yokoi, T., and Kamataki, T.S.Y. *J. Pharmacol. Exp. Ther.* **1997**, *282*, 909.

34. Iwatsubo, T., Hisaka, A., Suzuki, H., and Sugiyama, Y. *J. Pharmacol. Exp. Ther.* **1998**, *286*, 122.

35. Singh, R., Chen, I., Jin, L., Silva, M.V., Arison, B.H., Lin, J.H., and Wong, B.K. *Drug Metab. Dispos.* **2001**, *29*, 1578.

36. Obach, R.S. *Drug Metab. Dispos.* **2000**, *28*, 1069.

37. De Waziers, I., Cugnenc, P.H., Yang, C.S., Leroux, J.-P., and Beaune, P.H. *J. Pharmacol. Exp. Ther.* **2002**, *253*, 387.

38. Miyauchi, S., Sawada, Y., Iga, T., Hanano, M., and Sugiyama, Y. *Pharm. Res.* **1993**, *10*, 434.

39. Yamazaki, M., Suzuki, H., and Sugiyama, Y. *Pharm. Res.* **1996**, *13*, 497.

40. Kusuhara, H., Suzuki, H. and Sugiyama, Y. *J.Pharm.Sci.* **1998**, *87*, 1025.

41. Kusuhara, H. and Sugiyama, Y. *J. Control. Rel. (Review)* **2002**, *78*, 43.

42. Thummel, K.E., Shen, D.D., Podoll, T., Kunze, K.L., Trager, W.F., Hartwell, P.,

Raisys, V.A., Marsh, C.L., McVicar, J.P., Barr, D.M., Perkins, J.D., and Carithers, R.L.Jr. *J. Pharmacol. Exp. Ther.* **1994**, *271*, 549.

43. Imaoka, S., Enomoto, K., Oda, Y., Asada, A., Fujimori, M., Shimada, T., Fujita, S., Guengerich F.P., and Funae, Y. *J. Pharmacol. Exp. Ther.* **2002**, *255*, 1385.

44. Lin, J.H., Sugiyama, Y., Awazu, S., and Hanano, M. *J. Pharmacokinet. Biopharm.* **1982**, 10, *649*.

45. Chiba, M., Fujita, S., and Suzuki, T. *J. Pharm. Sci.* **1990**, *79*, 281.

46. Lin, J.H., Sugiyama, Y., Awazu, S., and Hanano, M. *Biochem. Pharmacol.* **1980**, *29*, 2825.

47. Lin, J.H., Hayashi, M., Awazu, S., and Hanano, M. *J. Pharmacokinet. Biopharm.* **1978**, *6*, 327.

48. Obach, R.S. *Drug Metab. Dispos.* **1996**, *24*, 1047.

49. Obach, R.S. *Drug Metab. Dispos.* **1997**, *25*, 1359.

50. Obach, R.S. *Drug Metab. Dispos.* **1999**, *27*, 1350.

51. Naritomi, Y., Terashita, S., Kimura, S., Suzuki, A., Kagayama, A., and Sugiyama, Y. *Drug Metab. Dispos.* **2001**, *29*, 1316.

52. Lin, J.H., Chiba, M., Balani, S.K., Chen, I.W., Kwei, G.Y., Vastag, K.J., and Nishime, J.A. *Drug Metab. Dispos.* **1996**, *24*, 1111.

53. Yeh, K.C., Stone, J.A., Carides, A.D., Rolan, P., Woolf, E., and Ju, W.D. *J. Pharm. Sci.* **1999**, *88*, 568.

54. Lin, J.H., Chiba, M., and Baillie, T.A. *Pharmacol. Rev.* **1999**, *51*, 135. 54-1. Thummel, K.E., Kunze, K.L., and Shen, D.D. *Adv. Drug Del. Rev.* **1997**, 27, 99.

55. Carlile, D.J., Hakooz, N., and Houston, J.B. *Drug Metab Dispos.* **1999**, *27*, 526.

56. Carlile, D.J., Stevens, A.J., Ashforth, E.I., Waghela, D., and Houston, J.B. *Drug Metab. Dispos.* **1998**, *26*, 216.

57. Houston, J.B. and Carlile, D.J. *Drug Metab. Rev.* **1997**, *29*, 891.

58. Zomorodi, K., Carlile, D.J., and Houston, J.B. *Xenobiotica* **1995**, *25*, 907.

59. Zomorodi, K. and Houston, J.B. *Pharm. Res.* **1995**, *12*, 1642.

60. Ashforth, E.I., Carlile, D.J., Chenery, R., and Houston, J.B. *J. Pharmacol. Exp. Ther.* **1995**, *274*, 761.

61. Houston, J.B. *Biochem. Pharmacol.* **1994**, *47*, 1469.

62. Carlile, D.J., Zomorodi, K., and Houston, J.B. *Drug Metab. Dispos.* **1997**, *25*, 903.

63. Hayes, K.A., Brennan, B., Chenery, R., and Houston, J.B. *Drug Metab. Dispos.* **1995**, *23*, 349.

64. Worboys, P.D., Brennan, B., Bradbury, A., and Houston, J.B. *Xenobiotica* **1996**, *26*, 897.

65. Soars, M.G., Burchell, B., and Riley, R.J. *J. Pharmacol. Exp. Ther.* **2002**, *301*, 382.

66. Lave, T., Dupin, S., Schmitt, C., Valles, B., Ubeaud, G., Chou, R.C., Jaeck, D., and Coassolo, P. *Pharm. Res.* **1997**, *14*, 152.

67. Bertilsson, L., Braithwaite, R., Tybring, G., Garle, M., and Borga, O. *Clin. Pharmacol. Ther.* **1979**, *26*, 265.

68. Desoye, G. *J. Biochem. Biophys. Methods* **1988**, *17*, 3.

69. Shibata, Y., Takahashi, H., and Ishii, Y. *Drug Metab. Dispos.* **2000**, *28*, 1518.

70. Li, A.P., Gorycki, P.D., Hengstler, G.L., Kedderis, G.L., Koebe, H.G., Rahmani, R., de Sousas, G., Silva, J.M., and Skett, P. *Chem.-Biol.Inter.* **1999**, *121*, 117.

71. Li, A.P., Lu, C., Brent, J.A., Pham, C., Fackett, A., Ruegg, C.E., and Silber, P.M. *Chem.-Biol.Inter.* **1999**, *121*, 17.

72. Li, A.P. *Chem.-Biol.Inter.* **1999**, *121*, 1.

73. Shibata, Y., Takahashi, H., Chiba, M., and Ishii, Y. *Drug Metab. Dispos.* **2002**, *30*, 892.

74. Davis, B. and Morris, T. *Pharm. Res.* **1993**, *10*, 1093.

75. Shimada, T., Yamazaki, H., Mimura, M., Inui, Y., and Guengerich, F.P. *J. Pharmacol. Exp. Ther.* **1994**, *270*, 414.

76. Li, A.P., Kaminski, D.L., and Rasmussen, A. *Toxicology* **2002**, *15*, 1.

77. Smith, D.A., Ackland, M.J., and Jones, B.C. *Drug Discovery Today* **1997**, *2*, 406.

78. Smith, D.A., Ackland, M.J., and Jones, B.C. *Drug Discovery Today* **1997**, *2*, 479.

79. Chiba, M., Jin, L., Neway, W., Vacca, J.P., Tata, J.R., Chapman, K., and Lin, J.H. *Drug Metab. Dispos.* **2001**, *29*, 1.

80. Chiba, M., Tang, C., Neway, W.E., Williams, T.M., Desolms, S.J., Dinsmore, C.J., Wai, J.S., and Lin, J.H. *Biochem. Pharmacol.* **2001**, *62*, 773.

81. Schenkman, J.B. *Pharmacol. Ther.* **81**, *12*, 43.

82. Smith, D.A. and Jones, B.C. *Biochem. Pharmacol.* **92**, *44*, 2089.

83. Blanck, J., Rein, H., Sommer, M., Ristau, O., Smettan, G., and Ruckpaul, K. *Biochem. Pharmacol.* 1983, *32*, 1683.

84. Sligar, S.G., Cinti, D.L., Gibson, G.G., and Schenkman, J.B. *Biochem. Biophys. Res. Comm.* **1979**, *90*, 925.

Chapter 14

Non-P450 Mediated Oxidative Metabolism of Xenobiotics

Dieter Lang[1] and Amit S. Kalgutkar[2]

[1]*Drug Metabolism and Isotope Chemistry, Bayer AG, 42096 Wuppertal, Germany*

[2]*Pharmacokinetics, Dynamics, and Metabolism Department, Pfizer Global Research and Development, Groton, Connecticut 06340*

ABSTRACT

In addition to cytochrome P450, oxidative metabolism of drugs and several other xenobiotics can also be mediated by non-P450 enzymes, the most significant of which are the flavoproteins monoamine oxidases (MAOs) and flavin-containing monooxygenases (FMOs), the molybdenum-containing oxidoreductases aldehyde oxidase and xanthine oxidase and the NAD^+-dependent alcohol and aldehyde dehydrogenases. The review article highlights the importance of these non-P450 enzymes in drug metabolism. A brief introduction to each of the non-P450 oxidizing enzymes (including structure-function relationships, tissue distribution, catalysis, species differences, and structure-activity relationship analysis with substrates and inhibitors) is provided and the oxidative pathways for drug elimination and bioactivation have been illustrated with clinically relevant examples. Oxidative metabolism of xenobiotics by FMOs or MAOs often times generates the same metabolites as those generated by P450. As a consequence, elucidation of the enzyme(s) responsible for the biotransformation of small molecule xenobiotics (P450s versus FMOs and MAOs) requires a clear knowledge of the underlying enzymology and mechanisms. In contrast, oxidations mediated by aldehyde oxidase and xanthine oxidase afford different metabolites than those resulting from P450 hydroxylation. Based on literature evidence on the dramatic species differences, e.g., in substrate/inhibitor specificities or tissue distribution demonstrated by these non-P450 enzymes, it is concluded that there is still much to learn about factors affecting the metabolism of drugs by these enzymes in a clinically relevant situation.

1   FLAVIN-CONTAINING MONOOXYGENASES (FMOs)

Flavin-containing monooxygenases (EC 1.14.13.8) is a family of FAD-containing enzymes located in the endoplasmatic reticulum. At least five gene products are known to exist in mammals. The catalytic function of these microsomal enzymes involves the oxygenation of soft nucleophilic heteroatoms, including nitrogen, sulfur, phosphorous, and selenium (Equation 1). The catalysis is NADPH- and oxygen-dependent. FMO was discovered by Ziegler (Ziegler's enzyme) in 1964[1] and purified for the first time from hog liver in 1972.[2] For many years FMO was thought to exist as a single enzyme and all of the fundamental studies on the catalytic mechanism and substrate specificity were conducted with purified hog liver FMO1. The extensive use of molecular biology techniques over the last 15 years has resulted in the identification of a family of structurally related isoforms.

$$X + O_2 + NADPH/H^+ \rightarrow$$
$$XO + H_2O + NADP^+ \ (X = N, S, P, Se \ in \ xenobiotic) \tag{1}$$

1.1   *Nomenclature and Molecular Characteristics of FMOs*

Enough evidence now exists which suggests that the mammalian FMO gene family is comprised of a minimum of five distinct forms, and appears to be expressed in all mammalian species yet examined. Recently, a sixth isoform (FMO6) was identified on human chromosome 1 (ENTREZ accession AL 021026). However, detailed investigations suggests that at the moment human FMO6 should be considered a pseudogene.[3] The isoforms are distinguished by Arabic numerals and simply termed: FMO1, FMO2, FMO3, FMO4 and FMO5. A systematic nomenclature[4] has replaced trivial names used previously, which are listed below.

FMO1: human FMO1, pig liver FMO, rabbit FMO 1A1, rabbit FMO form 1,
    and rat RFMO1
FMO2: rabbit FMO 1B1, rabbit lung FMO, and guinea pig 1B1
FMO3: human FMO form II, rabbit FMO form 2 and rabbit 1D1
FMO4: human FMO2, rabbit FMO 1E1
FMO5: rabbit 1C1 and FMO form 3

The primary amino acid sequence is about 50 – 58 % identical across the five isoforms, and more than 80 % identical to the corresponding mammalian orthologues.[5] The molecular masses are about 60 kDa with the exception of FMO4, which consists of additional 25 amino acids and has a molecular mass of ~ 63 kDa. The molecular properties of the human FMO isoforms[5] are summarized in Table 1.

Table 1. Properties of Human FMOs

FMO	Amino acids	Molecular mass [kDa]	pI
FMO1	532	60.3	6.9
FMO2	535	60.9	8.9
FMO3	532	60.0	8.3
FMO4	558	63.3	9.1
FMO5	533	60.2	8.6

Two conserved glycine rich motifs (GxGxxG), essential for FAD and NADPH binding are present at identical positions in all human FMO isoforms. The putative FAD- and NADPH-binding sites are located between residues 9 – 14 and 191 – 196, respectively.[6,7] Several other motifs including a FATGY motif at residues 327 - 331 are common in all isoforms, but their function is yet unknown.

1.2   *Tissue Distribution and Species Differences of FMOs*

FMOs are expressed in almost all tissues examined and several studies (mRNA-, protein analysis and metabolic reactions with selective substrates) indicate tissue and/or species specific expression in addition to developmental controlled expression.

1.2.1 *Humans*

In humans, FMO1 mRNA is present in fetal liver, kidney and in some brain tissues examined, as well as in adult kidney, but not in adult liver.[8] Functional protein has only been detected in fetal liver, adult kidney and intestine by polyacrylamide gel electrophoresis using isoform specific antibodies and by specific catalytic activities.[9,10,11,12] The expression of functional full length FMO2 protein has only been documented in some African-Americans, but not in Caucasians and Koreans due to a premature stop codon in the DNA sequence of the latter two ethnic groups.[13, 14] FMO3 mRNA has been detected in low abundance in fetal liver and lung and in adult lung and kidney, but in much greater abundance in adult liver.[8] Functional FMO3 protein, however, has only been detected in large quantities, (80 ± 40 pmol/mg[15] or 12.5 - 117 pmol/ml microsomal protein[16]) in adult liver.[15,9,12] Although FMO4 mRNA is present in low abundance in several tissues and seems to be expressed constitutively,[8] functional protein has not been detected thus far. FMO5 is expressed in adult liver (3.5 - 34 pmol/mg microsomal protein),[17] albeit at a lower extent than FMO3. In addition, mRNA of some FMOs has also been detected in the skin of some individuals,[18] and the expression of an unidentified but catalytically active FMO isoform(s) has been observed in human brain.[19]

1.2.2 Other Mammalian Species

The most detailed analysis of the tissue distribution of FMOs in animal species has been conducted in rabbits.[20,21,22] FMO1 mRNA expression is highest in liver and intestine, FMO2 mRNA highest in lung, and variable levels of mRNA encoding FMO3, FMO4, and FMO5 are present in liver and kidney. Similar to rabbit, FMO1 seems to be the most abundant form in pig[23,24] and male rat liver.[25, 26] Hepatic FMO3 expression levels in adult rats and rabbits are unaffected by gender[27, 26] but there is an age-dependent change in FMO3 expression in male rats[28] (see Section 1.3). In dog, FMO1 and FMO3 are expressed in liver and lung,[29, 27] with small amounts of FMO3 expression in kidney. A significant gender difference in hepatic expression of FMO isoforms exists in mice.[30, 26] In females, FMO1 and FMO3 are expressed in almost equal amounts. Compared to females, FMO3 is not expressed in males and the expression of FMO1 is ~ a third to a half of that expressed in females. FMO5 expression is low but equivalent in both sexes. Enzymatic FMO activity was found in brain microsomes of several species.[31, 32, 33] In general, besides FMO3[26, 27] and FMO5,[26] FMO1 is the predominant form in the liver of many experimental animal species, but is apparently absent in adult human liver, where FMO3 is the major isoform. Due to variations in substrate specificities of the various FMO isoforms (especially FMO1 and FMO3), extrapolation of liver microsomal data for FMO-catalyzed reactions from preclinical species to humans must be executed with caution.

1.3 Regulation of FMO Expression

FMOs are not induced by prototypical cytochrome P450 inducers like barbiturates or polycyclic hydrocarbons. Less is known about the inducibility by other xenobiotics, except for some evidence suggesting that dietary constituents can influence levels of FMO expression. However, there is plenty of evidence suggesting that FMOs are subject to hormonal and developmental regulation. An obvious example of developmental control includes the selective expression of FMO1 but not FMO3 in human fetal liver, whereas in adult human liver the reverse is true. These observations have been substantiated with a variety of biochemical approaches including mRNA analysis,[8, 5] protein analysis with isoform specific antibodies,[10, 34] and specific FMO catalyzed reactions.[10, 15, 9, 34] In a recent study, the developmental expression pattern of human hepatic FMO1 and FMO3 was investigated by Western blot analysis of microsomal protein from a bank of 240 human liver samples representing ages from 8 weeks gestation to 18 years.[12] Highest level of FMO1 expression was observed at the earliest gestational ages less than 15 weeks and declined in the period up to 40 weeks. After birth FMO1 suppression occurs in the first 3 days of life. In contrast, no expression of FMO3 was observed between 15 and 40 weeks gestation. After birth the onset of FMO3 expression was generally observed during the first year of life followed by moderate expression up to 11 years of age. A second phase of FMO3 expression occurs from puberty reaching the maximum level of expression in adulthood. These data indicate that suppression of FMO1 is tightly linked to the birth process and that birth is necessary but not sufficient for the onset of FMO3 expression. Similar developmental isoform-specific FMO expression has been observed in male and female mice.[26] Hepatic expression of FMO3 is not detectable in any fetal samples, but appears 2 weeks after birth equally in males and females. After puberty (6 weeks), the gender difference described in Section 1.2.2 becomes apparent with undetectable FMO3 levels in male mice. FMO1 expression occurs during the late gestation period and similar to FMO3, the gender difference begins to manifest itself after 4 weeks. In rats, a gender-dependent change in hepatic and renal expression of FMO1 versus FMO3 was observed in an age period from 3 to 11 weeks, which corresponds to the acquisition of sexual maturity.[28] Expression of FMO1 remains unchanged in male rat liver and kidney and female kidney, but a decrease in expression in female rat liver has been observed. In contrast expression of FMO3 remains unchanged in female rat liver and male kidney and significantly increases in both male rat liver and female rat kidney. The total FMO activity decreases in female rats and increases in male rats.[28, 35] FMO1 and FMO3 expression in mouse liver[36] and FMO expression in rat liver[37, 38] appear to be regulated by sex steroids, including testosterone or estradiol, which may be linked to developmental expression.[26, 28]

FMO2 mRNA and protein expression varies in the rabbit lung during pregnancy. Higher levels in late gestation appear to correlate with elevated progesterone concentrations in the plasma.[39, 40] In contrast, FMO2 expression in the rabbit kidney seems to be closely linked with glucocorticoid levels.[41] Dietary factors have been shown to exert pronounced effects on liver FMO activity.[42, 43] Overall, experimental data suggests that depending on the diet administered, certain dietary constituents induce or inhibit FMO expression. Indole-3-carbinol, a plant alkaloid found in cruciferous vegetables was found to dramatically down-regulate FMO1 expression in rat liver and intestine,[44, 45] but not in guinea pig or mouse liver[46] Indole-3-carbinol mediated induction of P450s (mainly CYP1A1) was also observed in all species investigated,[45, 44, 46] resulting in a dramatic change in the FMO/P450 ratio, especially in rats.

1.4 Catalytic Mechanism

NADPH and molecular oxygen are required for the catalytic function of FMOs. In contrast to P450 reactions, NADPH directly binds to a glycine-rich region near the N-terminus of the enzyme, and therefore no coenzymes are necessary. To elucidate the catalytic mechanism, extensive spectroscopic measurements, steady state and stopped flow kinetic studies have been performed with purified hog liver FMO1[47] and the mechanistic interpretations are generally considered to reflect the function of all other isoforms. The widely accepted catalytic cycle consistent with 4 spectroscopic distinguishable forms of the enzyme,[47] involves the sequential binding of NADPH and oxygen. The catalytic cycle is depicted in Figure 1.

NADPH binds to the enzyme and reduces FAD_{OX}. Oxygen binds in the next step to this reduced form of the enzyme to generate an FAD C-4α-hydroperoxy derivative considered to function as the reactive intermediate in the overall pathway. The soft nucleophilic heteroatom (N or S) of the substrate attacks the distal oxygen of this hydroperoxide resulting in oxygen transfer and formation of the corresponding N-oxides, sulfoxides, and a hydroxyflavin species of the enzyme. Subsequent loss of water from the hydroxyflavin intermediate regenerates FAD_{OX}. The decomposition of the hydroxyflavin is generally considered as the rate-limiting step in FMO catalysis. As a consequence, the observation that this rate-limiting step occurs after the irreversible oxygenation of the substrate and in contrast to P450, no substrate binding is required to trigger the formation of the reactive hydroperoxide intermediate, suggests, that the hydroperoxide intermediate can be considered as resting state of the enzyme, and therefore all FMO catalyzed reactions should exhibit the same V_{max}. This proposal is also based on several model pig liver FMO1 mediated reactions. In addition the K_m, usually a parameter of substrate affinity for the enzyme, seems to be more a parameter that characterizes the accessibility of the substrate to the active site of the enzyme, which could in some cases reflect the rate-limiting step.

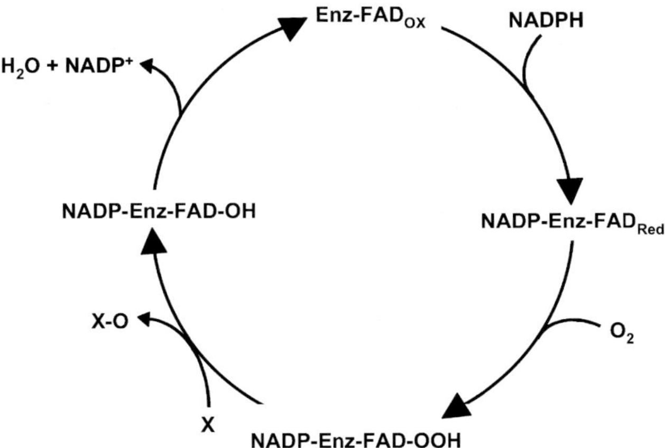

Figure 1. Catalytic Cycle of FMOs.

1.5   Characteristics of FMO Catalyzed Reactions

FMOs catalyze the biotransformation of a wide variety of xenobiotics including several drugs, pesticides, plant alkaloids, dietary constituents and endogenous compounds. These pathways have been the subject of several past reviews.[48, 49, 50, 51] Most FMO substrates are lipophilic sulfur- or nitrogen-containing compounds which are oxygenated to the corresponding hydrophilic sulfoxides or N-oxides. The number of transformations described for organophosphorus[52] or Se-containing[53, 54, 55] compounds are relatively limited.

On the basis of structure-activity relationship (SAR) studies with purified hog liver FMO1, it appears that any xenobiotic containing a soft nucleophilic heteroatom and which gains access to the FMO active site can undergo oxygenation. The structural diversity of pig liver FMO1 substrates cannot be extrapolated to all FMO isoforms, since FMO1 has apparently the broadest substrate specificity.[56] In addition, most of the SAR analysis is the result of indirect measurement such as NADPH- or O_2-consumption which may not always lead to metabolite formation. In general, the major determinants of FMO substrates are electronic and steric factors associated with the heteroatom within the xenobiotic. Compared with the oxidizing equivalents generated during P450 catalysis, the active 4α-hydroperoxyflavin oxidant in FMO is a milder oxidizing agent and thus only soft nucleophiles are oxygenated. FMOs can also be considered as an example of "enzymatic hydrogen peroxide". In addition, steric constraints are also considered to be important for FMO substrate specificity since many FMO catalyzed reactions are stereoselective.

In general, FMO catalyzed reactions are also potential metabolic pathways catalyzed by cytochrome P450s, and available methodologies that distinguish between P450 and FMO contribution in metabolism are discussed in Section 1.6. The following discussion will focus only on N- and S-oxygenation with a selection of substrates relevant to drug metabolism.

1.5.1 Nitrogen-oxygenation

Nitrogen-containing functionalities are common in many xenobiotics. Considering the weak oxidation potential of the 4α-hydroperoxyflavin, only sp^3-hybridized electron-rich nitrogen atoms serve as good candidates for oxygenation by FMOs (Figure 2). FMOs are not involved in the oxidation of heteroaromatics or amides. Although it is reported, that FMOs can catalyze the oxidation of primary and secondary amines[57] the majority of good FMO substrates are tertiary acyclic and cyclic amines such as trimethylamine,[9] dimethylaniline, imipramine,[11, 37, 58] nicotine,[59] clozapine,[60] itopride,[61] benzydamine,[62, 63] xanomeline,[64] moclobemide,[65, 66] and tamoxifen[67] (Figure 2). The oxidation of these tertiary amines results in the formation of stable N-oxide metabolites, many of which have been detected upon analysis of plasma or urine samples after *in vivo* administration. A few FMO catalyzed stereoselective N-oxidations are described for prochiral tertiary amines such as (*S*)-Nicotine, ABT-418,[68] and N-Ethyl-N-methylaniline.[69, 70]

1.5.2 Sulfur-oxygenation

S-Oxygenation reactions are also limited to electron rich soft nucleophilic centers for instance thioethers (Figure 3). Sulfoxide formation is easily accomplished by either FMO or P450. FMOs are rather slow or incapable of catalyzing the further oxidation of the sulfoxide to the sulfone metabolite, an outcome easily catalyzed by P450 enzymes. As with nitrogen-containing compounds, sulfur-containing compounds in which the sulfur has poor electron density associated with it (adjacent electron-withdrawing groups or heteroaromatic functionalities) are generally poor FMO substrates or virtually devoid of FMO substrate properties. For instance, due to the strong electron-withdrawing effect of the s-triazine ring, sulfoxide formation in the two methylthiotriazines ametryne and terbutryne is exclusively catalyzed by P450- and not by FMO.[71] Prototypical substrates include methimazole, thiobenzamide, or methyl-p-tolyl sulfide (Figure 3). A wide variety of drugs including sulindac sulfide,[72] albendazole,[73] cimetidine[59] (Figure 3), or thioether containing pesticides also undergo FMO catalyzed S-oxygenation. Oxidation of prochiral thioethers generates two possible sulfoxide enantiomers, and stereoselectivity in FMO catalyzed sulfoxidation have been extensively studied.[74, 75, 15, 76] Some examples of stereoselectivity are described in more detail in the subsequent chapter on substrate specificities.

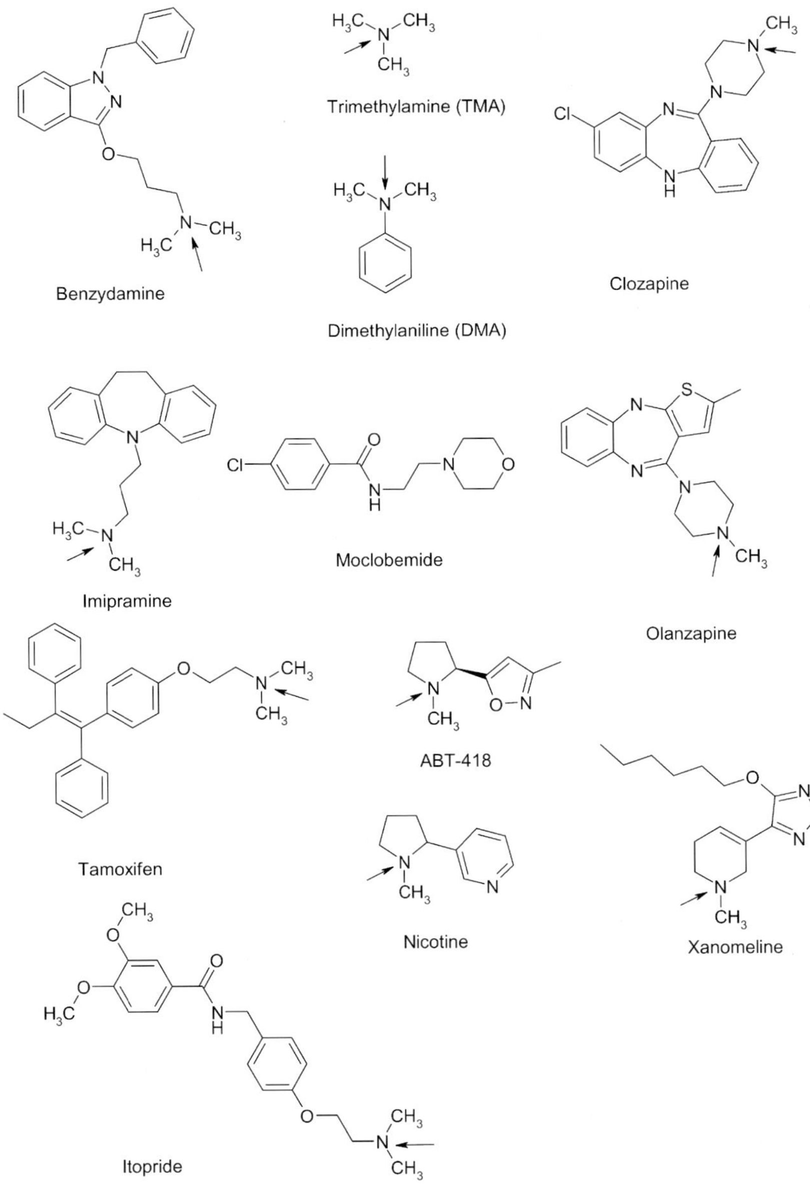

Figure 2. Examples of FMO Catalyzed N-Oxygenation Reactions. Arrows indicate the site of N-oxygenation.

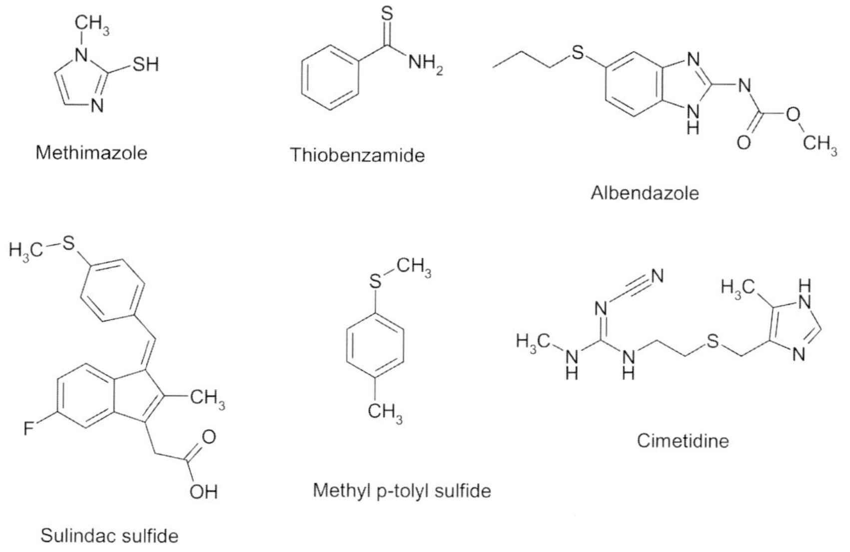

Methimazole

Thiobenzamide

Albendazole

Sulindac sulfide

Methyl p-tolyl sulfide

Cimetidine

Figure 3. FMO Catalyzed S-Oxygenation Reactions.

1.5.3 Substrate Specificities of Individual FMO Isoforms

In recent years, the availability of individual FMO isoforms from several species has led to detailed investigations designed to unearth FMO-specific substrates and recently some examples have begun to emerge. Although several studies[75, 77] have indicated that steric constraints or molecular shape/size are important for the specificity towards a certain FMO isoform as well as for the observed stereoselectivity in sulfoxidations, additional factors which determine substrate selectivity must also exist. For instance, trimethylamine (TMA), the smallest possible tertiary amine which, theoretically, should be a nonspecific FMO substrate, is selectively oxygenated by human FMO3.[9] Overall substrate specificities based on some SAR analysis on the individual FMO isozymes can be summarized as follows:

FMO1

Generally, FMO1 seems to possess the broadest substrate acceptance when considering steric constraints. The human isoform is more restricted compared to the pig and guinea pig orthologues.[56] Bulky non polar lipophilic compounds are generally good FMO1 substrates. For example, imipramine or orphenadrine are selectively N-oxygenated at pH 7.4 by recombinant human FMO1 but not

FMO3, FMO4, and FMO5.[11] Thus, imipramine and orphenadrine may be utilized as marker substrates of human FMO1 reactions in the kidney, since this enzyme is not expressed in the adult human liver.[11, 12] In many instances, FMO1 also demonstrates highly stereoselective sulfoxidation of prochiral thioethers.[74, 78, 10]

FMO2

SAR studies on pulmonary rabbit FMO2 revealed that from a series of 10-(N,N-dimethylalkyl)-2-(trifluoromethyl)phenothiazines with different methylene spacers (n = 2 – 10) between the phenothiazine moiety and the potential site of oxygenation (dimethylamino group), only derivatives with n > 5 underwent N-oxygenation suggesting that the substrate channel leading to the hydroperoxyflavin is about 6 - 8Å across in its longest axis.[77] In contrast to FMO2, porcine FMO1 catalyzed the N-oxidation of all derivatives. In another study using a series of n-alkyl p-tolylsulfides with n-alkyl ranging from methyl to n-heptyl, stereoselectivity of the sulfoxidation continuously changed from 99 % R (methyl) to 96 % S for the n-hexyl derivative.[75]

FMO3

The active site of human FMO3 appears to be smaller and more restricted compared to human FMO1. Human FMO3 has a unique property in selectively metabolizing small amines including TMA, the smallest possible tertiary amine, which is a highly selective FMO3 probe substrate.[9] This phenomenon remains poorly understood and there is still a lot to learn about substrate specificities of FMO isoforms. FMO3 mediated sulfoxidation of prochiral thioethers is often times not stereoselective. For instance, sulfoxidation of methyl-p-tolyl sulfide results in an *R/S* ratio of about 55/45.[15] In contrast, S-oxygenation of sulindac sulfide to the (*R*)-sulfoxide is stereoselectively catalyzed by recombinant human FMO3.[72]

FMO4

Little is known about the substrate specificity for FMO4. In addition to prototypical FMO substrates, thioethers containing a carboxylic acid seem to be good FMO4 substrates.[11]

FMO5

FMO5 exhibits a very poor catalytic activity towards conventional FMO substrates. However, that the enzyme is indeed active was evident from the highly stereoselective sulfoxidation of short-chain arylalkyl sulfides. Interestingly, the *S*-enantiomer predominates in the FMO5 mediated reactions as opposed to the *R*-sulfoxide metabolite usually observed with all other FMO isoforms.[76]

1.6 Diagnostic Tools to Distinguish P450- and FMO-Catalyzed Reactions

For in vitro-in vivo predictions or for a better understanding of potential drug-drug interactions, the elucidation of the drug metabolizing enzymes involved in the metabolism of drugs is very important in drug discovery or development. Since both FMOs and P450s can catalyze N- or S-oxygenations leading to identical metabolites, biochemical tools that allow differentiation between FMO or P450 involvement in drug metabolism are very useful. In this context, N- or S-oxygenation of soft nucleophilic N- and S-centers are potentially catalyzed by FMOs and P450s, whereas N- or S-dealkylated metabolites generally result from P450 catalyzed reactions. Procedures described below are not proven to work quantitatively similar with all FMO isoforms in all species. These are more general ways, based on principal differences of enzyme characteristics or behaviour.

1.6.1 Thermolability, pH-Profile, and Use of Detergent

Compared with P450s, FMOs are relatively more thermolabile in the absence of NADPH. Upon heating at 45 – 50 °C for 1 – 3 min in the absences of NADPH, most FMO isoforms, except FMO2,[79] demonstrate a substantial irreversible loss of FMO activity.[80, 81, 79] Under these experimental conditions P450 activities are only slightly reduced.[81] Monitoring the effects using diagnostic P450 and FMO substrates as positive controls is always recommended. Due to this thermal lability of FMOs, their activity in liver microsomes can vary, dependent on the procurement and initial storage of the tissues and the procedure employed for the preparation of microsomes. Soon after the death of an animal, NADPH-levels drop dramatically, which also may influence FMO stability. Thus, especially commercially available human liver microsomes often exhibit quite low FMO3 activity, and as a result FMO3 contribution to the metabolism of a xenobiotic could be underestimated.

The maximal velocity of an FMO catalyzed reaction is for most isoforms around or above pH 9.0.[9, 80] In contrast, cytochrome P450 activity is maximal around pH 7.4. Therefore, enhanced activity at elevated pH is a hint for a potential FMO involvement in metabolism. In many laboratories buffers between pH 8.5 – 9.0 are routinely utilized for *in vitro* FMO related studies. However, for assessing potential FMO involvement in metabolism, the activity at physiologically relevant pH is more decisive rather than a forced activity at elevated pH ranges.

The addition of detergent (0.1 - 0.2 % Lubrol) to liver microsomal incubations at elevated pH (pH 8.5) has been shown to abrogate all P450 activity.[34] An explanation for this phenomenon may be the lack of P450-oxidoreductase interaction, which is disturbed by solubilization and may attenuate the flow of electrons from the reductase to P450. In contrast, FMO catalysis does not require coenzymes because NADPH directly binds to FMO, and consequently alteration of FMO activity via addition of detergent is insignificant.

1.6.2 Inhibitors

Currently, no specific reversible, and unfortunately, no irreversible chemical inhibitor for FMOs are available. Several stilbene derivatives bearing N,N-dimethylamino moieties were described as FMO inhibitors, but the observed inhibition appeared to be caused by an influence on NADPH-binding rather than inhibition of enzyme activity.[82] Consequently, typical FMO substrates are often used as competitive inhibitors of alternative substrate. Methimazole, a prototypical FMO substrate, is also the most widely used FMO inhibitor, but it also inhibits CYP2B6, CYP2C9, and CYP3A4[83] at concentrations below 100 µM. An alternative approach is the abrogation of total P450 activity by using the general P450 suicide inactivator, aminobenzotriazole,[34] and subsequently monitoring N- or S-oxygenation for the compound of interest. As a more selective approach, FMO isoform specific antibodies may be used for inhibition of FMO activity. Alternately, indirect inactivation of total P450 activity can also be achieved through inhibition of the P450 oxidoreductase using selective antibodies against this protein. FMO activity should remain unaffected, since NADPH directly binds to FMOs.

1.6.3 Heterologously cDNA-Expressed FMO Isoforms

A valuable approach to distinguish between FMO- versus P450-catalyzed reactions is the subsequent incubation of the xenobiotic with cDNA-expressed individual FMO isoforms, some of which are commercially available. In recent years, many FMOs from different species have been cloned and expressed for instance in yeast, E. coli, and baculovirus expression systems. A major drawback, however, is that a quantitative assessment of enzymes involved is rather difficult with recombinant proteins, although the capability of an FMO isozyme in catalyzing N- or S-oxygenation preferentially under physiological conditions can easily be answered.

1.7 Diagnostic FMO Substrates *In Vitro* and *In Vivo*

1.7.1 *In Vitro*

Structurally diverse compounds are used *in vitro* as FMO substrates. The FMO catalyzed conversion of the tertiary amine N,N-dimethylaniline (DMA) to its N-oxide was historically one of the most widely used reactions for FMO activity. Monitoring FMO activity in microsomal preparations or purified FMOs using DMA is often conducted indirectly by NADPH or oxygen consumption, since direct DMA-N-oxide analysis is difficult. Methimazole, thiobenzamide, and methyl-p-tolyl sulfide are commonly used as sulfur-containing FMO substrates, and, as with N-containing compounds, NADPH or oxygen consumption is often monitored. Analysis of methimazole disappearance[84] can be used as marker of FMO activity. A common disadvantage of most FMO-substrates discussed above is the lack of FMO selectivity (versus P450) under more physiological conditions (pH 7.4); a situation which is further complicated by difficulties in the analytical methodologies for the respective metabolites. However, of interest are the observations, that the anti-inflammatory drug benzydamine is a good substrate for several human FMO isoforms.[63] In microsomal incubations the major N-oxide metabolite is easily and selectively formed by FMOs at pH 7.4. Furthermore, since benzydamine and its N-oxide are fluorescent, a simple and highly sensitive fluorometric assay for general FMO activity is available.[62]

FMO specific and isoform selective substrates are rare. FMO1 seems to selectively catalyze imipramine or orphenadrine N-oxygenation.[11] As mentioned earlier, human FMO3 selectively catalyzes the N-oxidation of TMA. In contrast, N-oxygenation by other FMOs or P450s is minimal at best.[9] Although TMA is highly selective as an FMO3 substrate, the lack of a facile analytical method for the assessment of TMA-N-oxide diminishes its utility as a selective FMO3 probe substrate.[9, 85, 86] The nicotine N-oxide pathway seems to be another FMO specific reaction, but the selectivity as an FMO3 specific substrate has not yet been conclusively determined.

1.7.2 *In Vivo*

The identification and validation of drugs as *in vivo* probes of drug metabolizing enzymes is important for conducting studies on the pharmacogenetic mechanisms underlying population variability in drug response and the evaluation of their clinical consequences. Drugs with the greatest utility as *in vivo* probes should be easily administered to human populations, the metabolic pathway of interest should be isoform-specific and dominate the *in vivo* clearance of the probe, and finally a simple analysis would be preferred.

A general problem in the quantification of metabolites from FMO-mediated reactions, namely N-oxides and sulfoxides, is their ability to undergo redox cycling *in vivo*. For instance, the pharmacological activity of the nonsteroidal anti-inflammatory drug sulindac sulfide relies on the metabolic reduction of the corresponding sulindac sulfoxide prodrug. Following its reduction, sulindac sulfide is capable of undergoing reoxidation to the sulfoxide in a reaction mediated by FMO. Furthermore, sulfoxides are also prone to further "irreversible" oxidation to the corresponding sulfones in the presence of P450 enzymes. Therefore, an ideal *in vivo* FMO probe substrate should be one that upon FMO mediated oxygenation results in the formation of a stable metabolite without further oxidation. Consequently, the formation of an N-oxide metabolite as stable endpoint is a more suitable metabolic pathway for a marker substrate of FMO activity *in vivo*, compared to the formation of a sulfoxide metabolite.

Considering the current knowledge of FMOs in humans, the FMO3 isoform is of greatest interest, since it is expressed in large quantities in adult liver as a highly active protein. In contrast, FMO5, which is also expressed in liver in considerably amounts, is almost inactive against prototypical FMO substrates. FMO1, also highly active, is only expressed in small quantities in extrahepatic tissues. Thus, the overall contribution of FMO1 to total FMO activity *in vivo* may be quite low. For instance, imipramine N-oxide which is selectively formed by recombinant human FMO1 *in vitro*, is virtually undetectable in plasma[87] and extremely low levels are observed in the urine (0.6 – 3 % of the dose).

Several amine-containing drugs, e.g. benzydamine,[63] clozapine,[88] ranitidine,[89] nicotine,[90] caffeine,[88] or trimethylamine have been postulated or used as *in vivo* probes for FMO3. Both nicotine[17] and ranitidine are low affinity FMO substrates. More than 40 % of the dose of ranitidine is excreted unchanged in the urine and only ~ 6 % as ranitidine N-oxide.[91] In the case of nicotine, ~ 4 % of the dose is excreted as nicotine N-oxide in the urine.[90] Although, clozapine and caffeine[88] have been used in the analysis of FMO3 variations in the human population, the usefulness of caffeine as an FMO probe is more than questionable,[92, 93] and the affinity of the contribution of FMO3 towards clozapine metabolism appears to be too low to be clinically relevant.[88] Benzydamine, a good *in vitro* FMO substrate, might be a reasonable *in vivo* probe, because benzydamine N-oxide is a major metabolite in humans with 12 - 17 % of the dose excreted in urine,[94] and provides a simple and highly sensitive analysis. Ratios of benzydamine to its N-oxide metabolite in human plasma could be a suitable parameter for assessing FMO3 activity in the human liver.[92]

Even more promising in this aspect is itopride, a gastroprokinetic agent which undergoes extensive hepatic metabolism in humans to the corresponding N-oxide (75.4 % of the administered dose in the urine).[95] Since it has been shown that the N-oxygenation pathway is FMO-mediated *in vitro,*[61] itopride-N-oxygenation may serve as an *in vivo* marker of FMO activity. The probably best *in vivo* probe for the assessment of human FMO3 activity is TMA N-oxidation. In humans, TMA is exclusively metabolized (>95 %) to its stable N-oxide[96] in a reaction selectively mediated by human FMO3. As a dietary constituent with an uptake of up to 50 mg/day, or derived via the decomposition of other dietary constituents, ratios of TMA/TMA-N-oxide can be easily analyzed in the urine of individuals without additional administration of TMA. However, sensitive bioanalytical methodology remains a most critical issue.

1.8 Polymorphism of Human FMOs and Relevance of FMOs to Drug Metabolism

Trimethylaminuria (fish-odour syndrome) is a genetic disorder, that results from an inability to form odourless TMA N-oxide from malodourous TMA, which is then excreted unchanged. A first direct causative relationship between a defective FMO3 gene (P153L mutation) and trimethylaminuria has been unequivocally demonstrated.[97] Several other reports have followed, describing mutations in the FMO3 gene which result in mild to severe trimethylaminuria.[98, 99, 100, 85, 101] A TMA/TMA-N-oxide ratio of >9/1 is considered as normal.[100, 98] Although rare, severe cases of trimethylaminuria related to mutations in the FMO3 gene with ratios of up to 92 % unchanged TMA excreted in the urine have been described.[101] Mild trimethylaminuria has not clearly been shown to be associated with defects in the FMO3 gene. For a better evaluation, a larger set of data is necessary considering all other physiological parameters which may cause mild trimethylaminuria such as hepatic dysfunction, age-dependent low hepatic expression of FMO3 (see Section 1.3). Therefore, based on the current knowledge, it is not very likely that FMO3 polymorphism causes significant problems in patient populations. To date, FMO polymorphisms, other than for human FMO3, have not been reported. That human FMOs are able to metabolize a variety of CNS-active tertiary amines, including nicotine, olanzapine, clozapine, xanomeline, or tricyclic antidepressants, has resulted in a considerable interest in the indentification of the nature of human brain FMO isoforms which may modulate the pharmacological activity of these tertiary amines at their sites of action within the CNS.

In conclusion, based on the current state-of-the-art, it does not appear that human FMOs are involved in the primary metabolism of many drugs *in vivo.* As a result, severe drug-drug interactions caused by inhibition of FMO mediated metabolic pathways do not appear to be a major concern, at the present time, but the possibility cannot be excluded.

2   MONOAMINE OXIDASES

The monoamine oxidases (MAO) A and B (EC 1.4.3.4; amine-oxygen oxido-reductases) are flavoenzymes that catalyze the oxidation of structurally diverse amines including endogenous amines such as dopamine, norephinephrine, serotonin (5-HT), tyramine, 2-phenylethylamine (PEA) and exogenous amines including the triptan class of anti-migraine drugs and the parkinsonian neuro-toxin 1-methyl-4-phenyl-1,2,3,6-tetrahydropyridine (MPTP) (Figure 4).[102, 103]

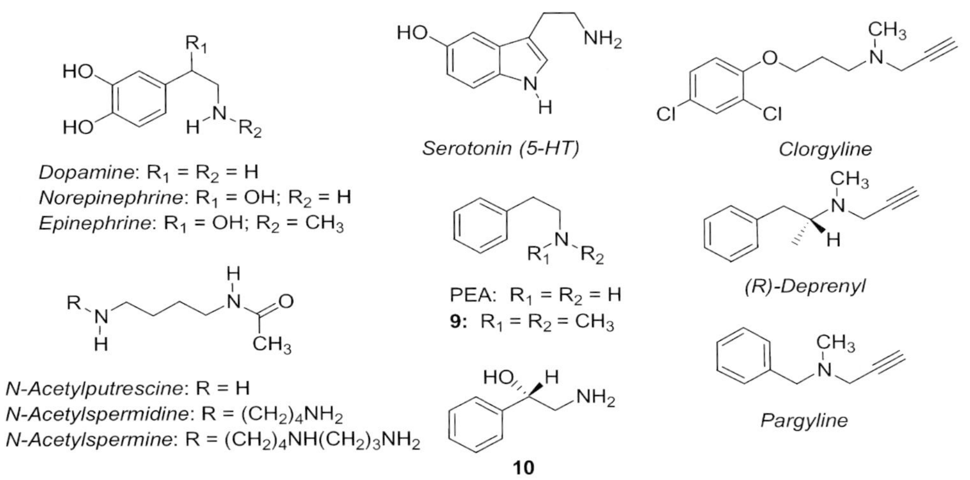

Figure 4. Structures of MAO Substrates and Irreversible Inactivators.

2.1   *Multiplicity, Tissue Distribution, and Species Differences for the Mammalian Isozymes*

MAO is an integral protein of the mitochondrial outer membrane. Two isozymes, termed MAO-A and MAO-B, are known. These isozymes are con-veniently distinguished in mitochondrial preparations by their differences in substrate and mechanism-based inactivator specificities.[104, 105, 106] MAO-A is inhibited by low concentrations of clorgyline (see Figure 4) and preferentially catalyzes the oxidation of 5-HT and norephinephrine; MAO-B selectively catalyzes the oxidation of PEA and benzyl amine and is inhibited by nanomolar concentrations of (*R*)-deprenyl. Tyramine, dopamine and tryptamine appear to be substrates for both subtypes. Primary structures deduced from cDNA clones indicate that the two isozymes consists of 526 (MAO-A) and 520 (MAO-B) amino acid residues with molecular weights of 59,700 and 58,800, respectively, and have 70 % sequence identity.[107, 108, 109, 110] Rat and bovine MAO-B share ~88 % and 86 % sequence identity with human MAO-B,[111] whereas bovine MAO-B displays ~ 81 % identity to rat MAO-B.[112]

Although MAO is expressed in most mammalian tissues, species dependent differences in MAO activity have been observed by studying the inactivation properties of clorgyline, (*R*)-deprenyl, and the mixed inhibitor pargyline (see Figure 4) on the oxidation of MAO substrates[113] and by sensitivity of interspecies preparations towards proteolytic cleavage.[114] For instance, MAO-A preparations in brain and peripheral tissues of human, dog, rat, and rabbit are inactivated by trypsin to an extent that is species dependent, but independent of tissue source. In contrast, MAO-B activity in human, dog, and rat tissues is unaffected by trypsin, but rabbit MAO-B is. Furthermore, Denny *et al.* have reported that a monoclonal antibody to human platelet MAO-B does not interact with mouse liver MAO-B, although it does recognize human liver MAO-B.[115] Comparison of the MAO-B substrate and inhibitor properties of the anticonvulsant milacemide in bovine and rat liver mitochondria reveals greater affinity towards bovine MAO-B.[116] Likewise, the inhibitory potency of several oxadiazolones and oxadiazolethiones (selective, slow, tight-binding MAO-B inhibitors) varies by 3- to 4-orders of magnitude between rat and bovine liver MAO-B, whereas inhibitory potency against rat and human liver enzymes is relatively similar.[117] More recently, Inoue and co-workers have characterized species and organ differences in MAO activity by studying the rates of α-carbon oxidation of MPTP and analogs in the brain and liver mitochondria of rabbits, dogs, monkeys and baboons.[118] The most significant finding is the greater similarity in MAO activity profiles in humans versus rodents compared to humans versus subhuman primates.

In humans, most tissues, except the blood platelets (MAO-B), express both isozymes.[119] The highest MAO levels are present in the liver and the placenta and the lowest in spleen. The preponderance of MAO-A in human placenta, lung and the small intestine and MAO-B in the myocardium has been established.[120, 121, 122] Rodriguez and co-workers have studied the localization of MAO-A and -B in human heart, liver, duodenum, blood vessels, kidney, lung, spleen, pancreas, thyroid and the adrenal glands using monoclonal anti-human MAO-A (6G11/E1) and anti-human MAO-B (3F12/G10/2E3) antibodies.[123] The results of these studies indicate that both enzymes have a widespread distribution in the human body with a matching pattern in many, but not all, tissues and with strong differences from the pattern of distribution in rodents. MAO isozymes are also present in most areas of the human brain. MAO-B appears to be predominantly localized in serotonergic neurons whereas dopaminergic neurons contain MAO-A.[124, 125]

2.2   Crystal Structure

The crystal structure of native and inhibitor-bound human MAO-B has recently been determined at 3-Å resolution.[126] Analysis of the crystal structure indicates that the enzyme is dimeric and binds to the membrane through a C-terminal transmembrane helix. The flavinadenine dinucleotide (FAD) moiety is covalently linked via its 8α-methyl group to Cys397, and electron density maps reveal that the mechanism-based inactivator pargyline binds covalently to the flavin N5 atom. The substrate binding site is formed by a flat cavity with a volume of 420 $Å^3$ and is lined with a number of hydrophobic amino acid residues, as predicted by structure-activity relationship (SAR) and quantitative structure-activity relationship studies on numerous MAO-B substrates and inhibitors.[127, 128] Subsequent docking studies with benzyl amine suggest that the substrate carbon undergoing flavin-dependent oxidation binds in a highly conserved position in front of the flavin N5-C4a locus. On this basis, the benzyl amine methylene carbon is positioned 3.6 Å from the flavin N5 and the nitrogen atom binds to the phenolic side chain of Tyr398 and Tyr435 (7 Å away).

Adjacent to the substrate cavity is a smaller hydrophobic cavity (situated between the active site and the protein surface and termed as the entrance cavity). Tyr326, Ile199, Leu171, and Phe168 are residues that separate the active site from the entrance cavity. Minor structural changes in these amino acid residues can result in the fusion of the two cavities to form a single layer that could accommodate larger substrates and inhibitors as observed with MAO-A. These hypotheses are also consistent with recent site directed mutagenesis data that indicates that smaller Tyr326Ile MAO-B mutant exhibits substrate/inhibitor specificity similar to MAO-A.[129] Overall, the crystal structure suggests that variations in substrate/inhibitor specificities between the two isozymes mainly result from the fine-tuning of the size and shape of the hydrophobic substrate cavity and its steric relation with the entrance cavity.

2.3   Catalysis

Although both isozymes differ in their substrate and inhibitor selectivities, they both are able to catalyze the oxidation of 1°, 2° and 3° amines. The FAD moiety is functional in the oxidative deamination of amine substrates and is involved in the transfer of electrons from the amine nitrogen to molecular oxygen (Figure 5). In the process, the amine **1** is converted to a protonated iminium species **2** that is released from the active site and hydrolyzed to the corresponding aldehyde **3** and ammonium ion. The most important structural feature for MAO substrate properties is the presence of the two hydrogen atoms on the α-carbon in amines. Furthermore, Yu and co-workers have demonstrated that both MAO isozymes exhibit the same stereospecificity by abstracting the pro-*R*-hydrogen exclusively from the prochiral methylene group in amine substrates.[130]

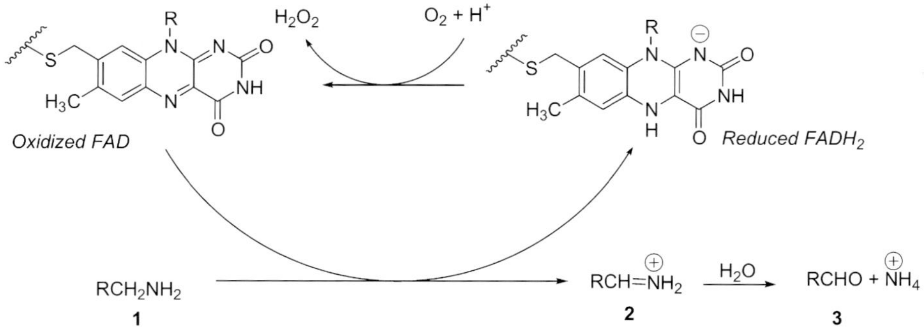

Figure 5. Oxidation of amines by MAO-bound FAD.

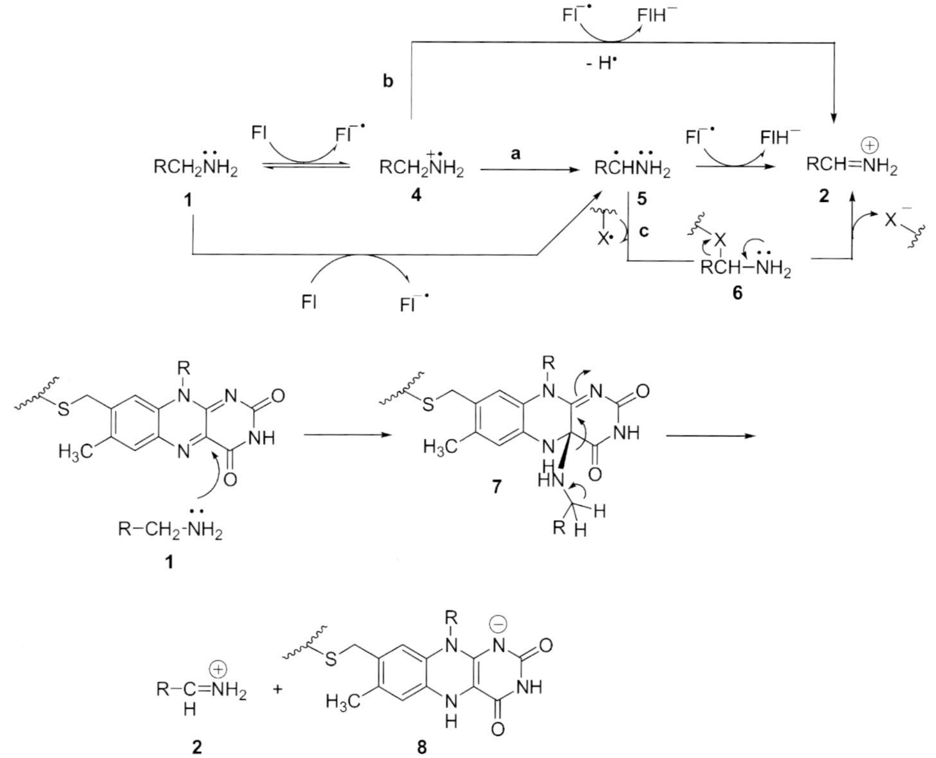

Figure 6. Proposed Mechanistic Pathways for the MAO-B Catalyzed Oxidation of Amines.

Three pathways for the MAO catalyzed oxidation of amines have been described (Figure 6). These are: (1) the single electron transfer (SET) pathway,[131, 132] (2) a hydrogen atom transfer (HAT) pathway,[133] and (3) a nucleophilic or polar mechanism.[134] The SET pathway proceeds by an initial electron transfer step from the nitrogen lone pair of the amine substrate **1** resulting in the aminyl radical cation (**4**) and flavin semiquinone (Fl$^{\bullet-}$). Proton loss from **4** results in the carbon-centered radical **5** which can transfer the second electron to the flavin semiquinone (pathway a) to yield the iminium species (**2**) and reduced flavin (FADH$_2$). Direct α-hydrogen atom loss from the radical cation **4** leading to **2** also has been considered.[135] Alternatively, the carbon-centered radical **5** could undergo radical combination with an active site radical (pathway c) to give a covalent adduct **6** which could yield the iminium species **2** by β-elimination. According to the polar mechanism, the amine substrate forms an adduct (**7**) with FAD. Subsequent cleavage of **7** yields the iminium metabolite **2** and reduced FADH$_2$ (**8**). Experimental evidence for[136, 137] and against[138, 139, 140] each of these mechanistic pathways has been presented.

2.4   SAR Analysis of MAO Substrates

2.4.1 Acyclic Amines

Besides neurotransmitters, several acetylated endogenous polyamines including N-acetyl-putrescine, -spermidine and -spermine are MAO substrates (see Figure 4).[141] Although many amines are metabolized by a specific MAO isozyme, none is a "true isozyme-specific substrate". For instance, at 10 μM, PEA is preferentially metabolized by MAO-B, whereas at 1000 μM, the amine serves as a common substrate for both MAO isozymes.[142] Optimal substrate and inhibitor properties of amines are discernible with compounds that bear an aryl group 1 – 3 carbons from the amine nitrogen. Although 1° and 2° amines are deaminated indiscriminately by both isozymes, 3° amine substrates are relatively MAO-B selective.[143] For instance, the MAO-B-selectivity for PEA is enhanced further by conversion to the corresponding N,N-dimethyl derivative **9** (see Figure 4).[144] Besides arylalkylamines, MAO also catalyzes the oxidation of simple aliphatic alkyl amines.[145] Thus, amines tethered to medium length alkyl chains (C5-C10) are excellent MAO substrates, whereas those tethered to short chain alkyl groups (C1-C4), or alkyl groups with chain lengths longer than C10, are poor MAO substrates.

Analogs where the hydrogen atom on the α-carbon is substituted with a methyl group results in compounds that are devoid of MAO substrate properties. The resulting compounds, however, are reversible, competitive inhibitors.[130] Modification of β-position also influences substrate recognition. The *R*-enantiomer **10** of 2-phenylethanolamine is a selective MAO-A substrate whereas both enantiomers of 2-phenylethanolamine are substrates for MAO-B.[146] Overall, these observations suggest that the MAO-A active site may present steric constraints at the position corresponding to the β-carbon in arylalkylamines that do not exist for MAO-B. In contrast, MAO-B displays low affinity towards α-substituted amines as judged from the poor MAO-B substrate properties of α-methyl monoamines.

Noteworthy examples of MAO's involvement in drug metabolism include the recent reports of MAO catalyzed oxidation of the selective serotonin reuptake inhibitor citalopram and the triptan class of anti-migraine drugs. Testa *et al.* noted that citalopram (**11**) and its N-dealkylated metabolites **12** and **13** undergo MAO catalyzed oxidation in human hepatic tissue and whole blood to yield the corresponding aldehyde **14** that upon further aldehyde oxidase mediated oxidation affords the carboxylic acid **15** (Figure 7).[147, 148] That MAO and not P450 catalyzed these oxidative deaminations was evident from the observations that the formation of **15** occurred in the absence of NADPH and was prevented following co-incubation with MAO-A and -B inhibitors. Although the extent to which MAO participates in the clearance of citalopram *in vivo* is not clear, several cases of the 'serotonin syndrome' have been reported following overdoses of citalopram when given with the selective MAO-A inhibitor moclobemide. It is tempting to speculate that MAO inhibitors may enhance the citalopram-induced side effects by inhibiting its metabolism.

MAO plays a dominant role in the clearance of the triptan class of anti-migraine drugs (Figure 8). Members of this group of drugs include acyclic 3° amines such as sumatriptan, zolmitriptan, rizatriptan, and almotriptan and cyclic 3° amines such as eletriptan and naratriptan. MAO-A is the principal enzyme responsible for the clearance of sumatriptan and almotriptan leading to the formation of the corresponding carboxylic acid analogs **16** and **17**, respectively, as the exclusive metabolites of these two agents in humans (Figure 8).[149, 150] The principal involvement of MAO-A in sumatriptan metabolism in rat, rabbit, dog, and humans and the similarities in protein binding across these species has also resulted in successful interspecies allometric scaling of pharmacokinetic parameters.[151] Rizatriptan and its N-desmethyl metabolite also undergo MAO-A catalyzed oxidation to the indole acetic acid metabolite **18** in humans (35 % to 57 % of the dose).[152] The structurally related zolmitriptan is cleared in humans primarily by metabolism (N-demethylation and N-oxidation) mediated by P450 1A2.[153] The N-demethylated metabolite of zolmitriptan undergoes selective MAO-A catalyzed oxidation leading to the indole acetic acid metabolite **19**. Since the *in vivo* clearance of the triptans is primarily dependent on MAO, interactions with drugs that induce or inhibit MAO-A are possible. For instance, moclobemide, a selective MAO-A inhibitor, produces a 3-fold increase in the C_{max} and the AUC of N-desmethylzolmitriptan when co-administered with zolmitriptan.[154] Despite their structural similarity to sumatriptan, zolmitriptan, and rizatriptan, the cyclic 3^0 amines, eletriptan and naratriptan, are not MAO substrates.

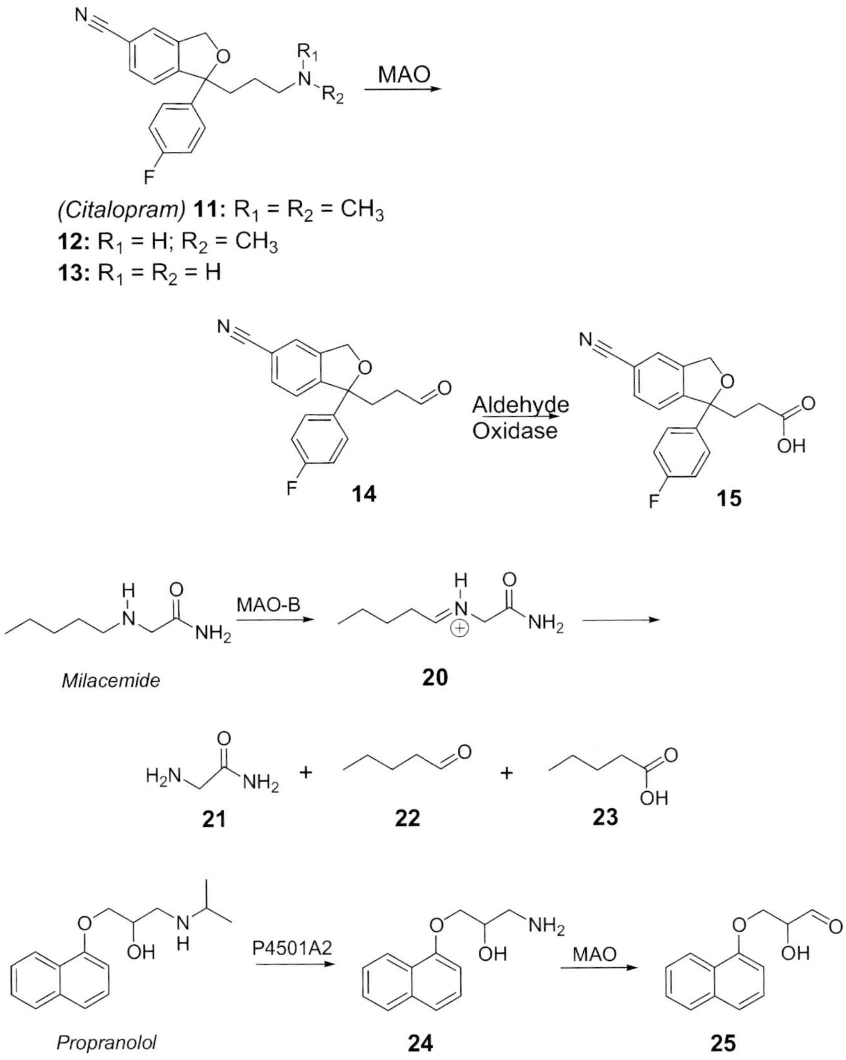

Figure 7. Examples of Drugs Metabolized by MAO Isozymes.

Besides citalopram and the triptans, the involvement of MAO in the metabolism of the anticonvulsant agent milacemide and the β-blocker propranolol has also been documented. That MAO-B is responsible for the metabolism of milacemide to glycine in rats was confirmed when the formation of glycine from milacemide was prevented by pretreatment with the MAO-B inhibitor (*R*)-deprenyl.[155] Biochemical studies revealed that the formation of glycinamide (**21**) from milacemide is mediated through the selective MAO-B catalyzed oxidation at the α-methylene carbon of the pentyl group via the intermediate iminium species **20**. Other oxidative products include pentanal (**22**) and pentanoic acid (**23**) (Figure 7).[156] Evidence for MAO's involvement in the metabolism of propranolol has been obtained in rat hepatocytes where oxidative deamination of the 1° amine metabolite **24** of propranolol results in the formation of the corresponding stable aldehyde **25**.[157]

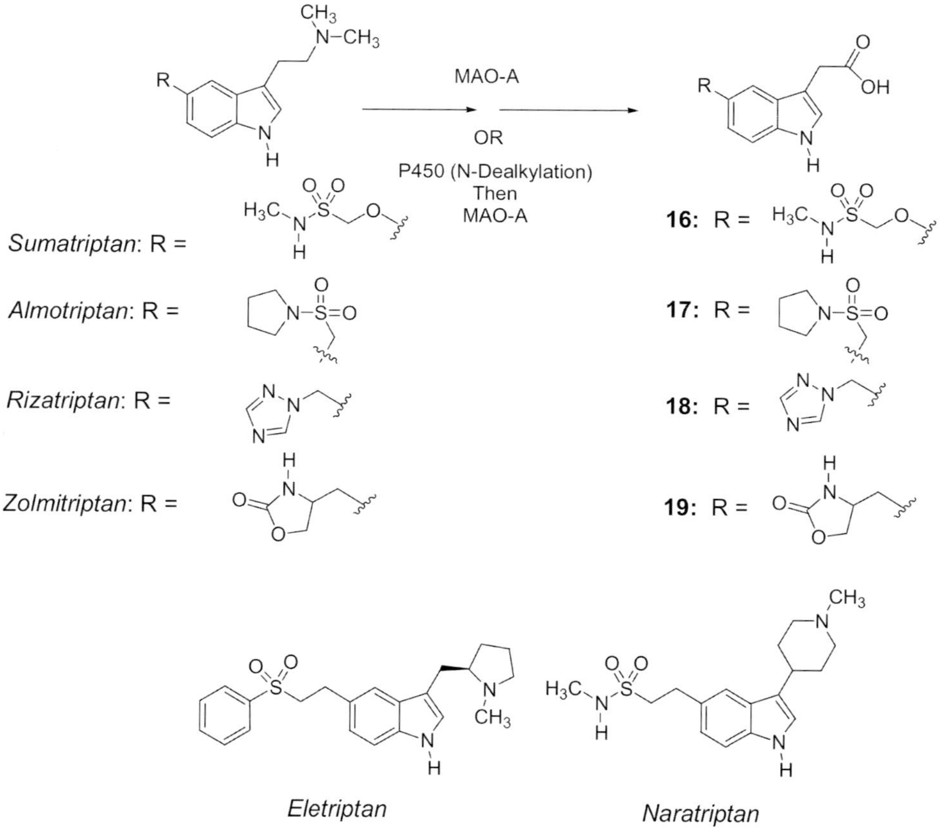

Figure 8. MAO-A Catalyzed Oxidative Deamination of the Triptan Class of Anti-migraine Drugs.

2.4.2 Cyclic Tertiary Amines

Extensive metabolic, biochemical and toxicological studies have established that the neurodegenerative properties of the parkinsonian inducing neurotoxin MPTP (Figure 9) are mediated by MAO-B.[158] The cascade of events leading to MPTP's selective destruction of nigrostriatal dopaminergic neurons can be summarized as follows: (1) Selective MAO-B catalyzed extraneuronal oxidation of MPTP generates the pyridinium species MPP$^+$ via the intermediate dihydropyridinium MPDP$^+$ (see Figure 9).[159] (2) Active transport of MPP$^+$ into dopaminergic nerve terminals by the dopamine uptake system follows.[160] (3) MPP$^+$ localizes in the mitochondrial matrix where it inhibits complex I of the mitochondrial electron transport chain leading to cessation of oxidative phosphorylation and ATP depletion.[161] The neurotoxicological importance of this oxidation was demonstrated in experiments in which irreversible, selective MAO-B inhibitors protected monkeys and mice against MPTP's neurotoxicity.[162, 163] MPDP$^+$ is chemically unstable under physiological conditions. Amongst other oxidative pathways, it undergoes a bimolecular redox reaction with its conjugate base **26** to form stoichiometric amounts of MPTP and MPP$^+$ (see Figure 9).[164] The principal fate of MPDP$^+$ in brain homogenates is its further two-electron oxidation to MPP$^+$, an event that is also thought to be MAO-B catalyzed. Extensive SAR studies on MPTP analogs have been undertaken in an attempt to understand the unique structural features of these cyclic tertiary allylamines that lead to their unexpected MAO substrate properties and neurotoxicity and these have been recently reviewed.[165]

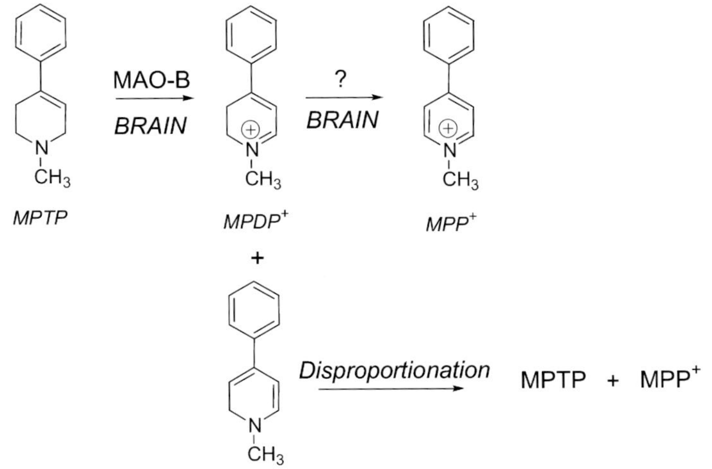

Figure 9. Metabolic Fate of the Nigrostriatal Neurotoxin, 1-Methyl-4-phenyl-1,2,3,6-tetrahydropyridine (MPTP).

2.5 MAO Inhibition

The past 30 years has witnessed the discovery of numerous competitive and slow, tight-binding monoamine oxidase inhibitors (MAOIs) that preferentially inhibit one isozyme over the other. MAOIs typically are classified into reversible (competitive or slow, tight-binding) or irreversible (affinity labeling agents or mechanism-based inactivators). Compounds belonging to the first generation of MAOIs were mechanism-based inactivators that acted via formation of reactive electrophilic intermediates that covalently modified the protein. Most of these compounds also inactivated the P450s, a consequence that led to hepatotoxic side effects.[166] A second, more serious side effect with irreversible MAOIs is the occurrence of the so-called cheese effect, a phenomenon that led to severe hypertensive crises induced by elevated dietary tyramine levels resulting from MAO inhibition.[167] Such hepatotoxic and hypertensive liabilities led to a sharp decline in the popularity of MAOIs as anti-depressants and to efforts directed towards the identification of reversible, isozyme-selective MAOIs.

2.5.1 Reversible MAOIs

We have recently reviewed the structural diversity of reversible (competitive and slow, tight-binding) isozyme-selective MAOIs[165] including α-methyl monoamines, aryloxazolidinones, hydroquinones, 5H-indeno[1,2-c]pyridazines,

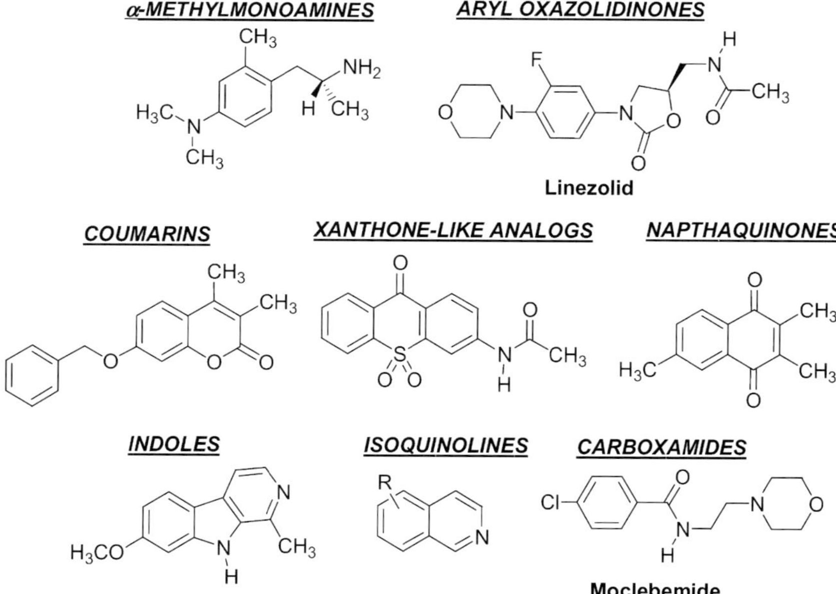

Figure 10. Structural Classes of Reversible MAO Inhibitors.

coumarins, xanthones, menadione analogs, tetrahydropyridines, indoles, isoquinolines, and carboxamides such as moclobemide. Representative examples are highlighted in Figure 10.

2.5.2 Irreversible MAOIs

A general strategy that allows for the design of selective, irreversible MAOIs involves the incorporation of substituents on the nitrogen atom of isozyme selective substrates which undergo MAO catalyzed conversion to electrophilic intermediates that alkylate the active site of the enzyme. N-substituents that are capable of imparting irreversible inhibition properties to MAO substrates include amino (hydrazines), allyl, propargyl, cyclopropyl, cyclobutyl, trialkylsilanyl, oxazolidinonyl, and furanoyl groups (Figure 11). For example, the MAO-B-selective substrate benzyl amine is converted to the MAO-B-selective and irreversible inhibitor pargyline by incorporating a propargyl group on the nitrogen atom that is susceptible to enzyme mediated oxidation. Furthermore, extending the distance between this electron rich center and the aromatic ring, or incorporating a bulky group on the aromatic ring of irreversible MAO-B selective inhibitors, affords irreversible MAO-A selective inhibitors. This is apparent from structural differences between (*R*)-deprenyl (MAO-B selective) and clorgyline (MAO-A selective).

The mechanism of MAO inactivation by hydrazine derivatives such as phenyl hydrazine (**27**) (Figure 11) is thought to involve an initial MAO catalyzed dehydrogenation to a corresponding phenyldiazine **28** intermediate which in turn loses a H atom and N_2 to afford the phenyl radical **29**. Based on results obtained from model studies, covalent attachment of **29** has been proposed to occur at the 4a- on the flavin group in MAO-B.[168] Allyl- and propargylamines undergo MAO catalyzed oxidation to the corresponding dieniminium and eyniminium species, **30** and **31**, respectively. These highly electrophilic Michael acceptors then inactivate the enzyme by bonding covalently to an active site residue or the flavin group as demonstrated in the pargyline: MAO-B cocrystal structure.[169,170] The pathway(s) responsible for the mechanism-based inhibition by cyclopropylamines remains the subject of intense debate.[131, 136] One of the most widely accepted proposals involves the initial single electron transfer from cyclopropyl amines such as *N*-cyclopropyl-α-methyl benzyl amine (**32**) to the flavin co-factor to generate the aminium radical cation **33**. Spontaneous ring opening of the cyclopropyl group in **33** generates the primary carbon-centered **34** that alkylates an active site residue leading to covalent inactivation. Peptide-mapping studies followed by mass spectral analysis revealed that **32** labels Cys365 in bovine liver MAO-B, which corresponds to Cys374 and Cys365 in human MAO-A and MAO-B, respectively.[171] Not all cyclopropylamines, however, are mechanism-based inactivators of MAO. Certain MPTP analogs bearing N-cyclopropyl substituents have been reported to be excellent MAO-B substrates that do not inactivate MAO.[138, 139] These compounds undergo α-carbon oxidation to the corresponding dihydropyridinium species, the same pathway observed with MPTP. As noted earlier, irreversible inhibition by aryl oxazolidinones such as MD 780236 (**35**) is dependent on the presence of the amino methyl group, since replacement of nitrogen with other heteroatoms alters the mode of inhibition from irreversible to reversible. The proposed mechanism of inhibition involves initial α-carbon oxidation of the amino methyl group to give the iminium species **36** that covalently modifies MAO by reacting with an active site residue to generate adduct **37**.[172]

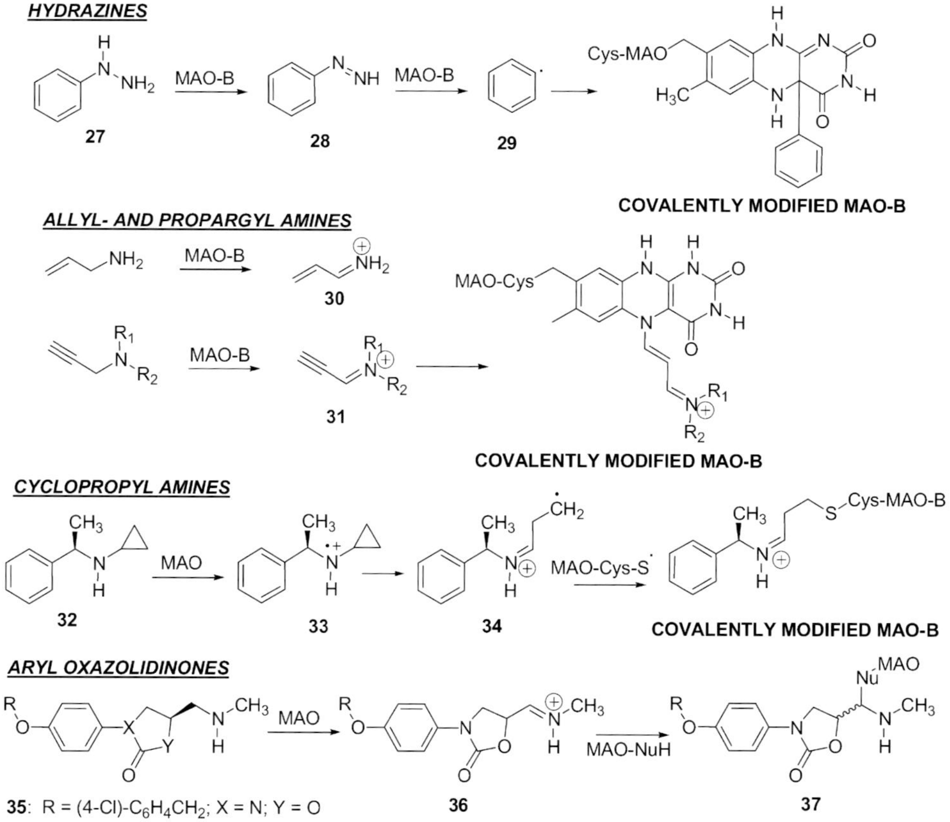

Figure 11. Proposed Mechanism of Covalent Inactivation of MAO by Irreversible Inhibitor Classes.

2.6   MAO and Drug Metabolism

The last decade has witnessed a resurgence of interest in MAO, particularly due to its involvement drug metabolism. A dramatic illustration is the role of MAO-A as the principal phase I enzyme involved in the clearance of the several anti-migraine drugs from the triptan series in humans. The MAO-A substrate properties of sumatriptan and the related drugs, zolmitriptan and rizatriptan, are not surprising, given their structural similarity to 5-HT, the preferred MAO-A substrate. Overall, a better understanding of the tissue distribution of the MAOs may help to address questions of the possible roles of MAO in first pass metabolism and the exposure to amine xenobiotics, especially those that are inhaled or administered orally. Furthermore, the finding that several amine-containing drugs including the triptans, citalopram, primaquine, milacemide, and diltiazem are MAO substrates, raises the possibility of potentially important drug-drug interactions that may lead to side-effects in patients on MAOIs as has been demonstrated with citalopram. Given the structural diversity of MAO substrates, any small molecule drug containing a $RCH_2NR_1R_2$ functionality may be metabolized by MAOs. Likewise, given the structural diversity of MAOIs, any small molecule xenobiotic bearing resemblance to known or hitherto unknown structural class may function as a MAOI, as observed with the anti-bacterial agent linezolid. Thus, inhibition and modulation of MAOs can result in undesirable elevations in plasma concentrations of MAO substrates and in some instances cytochrome P450 substrates.

3   MOLYBDENUM HYDROXYLASES

Aldehyde oxidase (EC 1.2.3.1; aldehyde-oxygen oxidoreductase, AO), xanthine oxidase (EC 1.2.3.2; xanthine-oxygen oxidoreductase, XO) and xanthine dehydrogenase (EC 1.1.1.204; xanthine: NAD^+ oxidoreductase, XOD) are members of the molybdopterin family of proteins collectively referred to as molybdenum hydroxylases. XO/XOD are two forms of the same enzyme and most mammalian forms exist as the dehydrogenase form *in vivo*.[173] Despite the parallel in XO/XOD nomenclature, AO is not an alternative form of aldehdye dehydrogenase. AO exists solely in its oxidase form such that electron transfer from the reducing substrate is utilized by the flavin to reduce molecular oxygen instead of NAD^+ (see Section 3.3). Attempts to convert AO back to a dehydrogenase form, such that it will reduce NAD^+ instead of molecular oxygen as is observed with XO have been unsuccessful. Overall, these enzymes are responsible for the α- or γ-carbon oxidation of imines ($R_1R_2C=NR_3$) in nitrogen-containing aromatic heterocyclic rings (azaarenes) resulting in the formation of the corresponding lactam (α-carbon oxidation) or aminoenone

(γ-carbon oxidation) derivatives, respectively. For instance, the conversion of 1-methylxanthine, a secondary metabolite of caffeine, to 1-methyluric acid is catalyzed by XO (Figure 12), and the urinary ratio of these two metabolites represents an *in vivo* index of XO activity.[174, 175] AO is also responsible for the oxidation of structurally diverse aldehydes to the corresponding carboxylic acids. The reaction catalyzed by molybdenum hydroxylases is unusual in comparison to other oxidases in that water rather than molecular oxygen is the ultimate source of the oxygen atom incorporated in the product and the reaction generates rather than consumes reducing equivalents. It is noteworthy to emphasize on the "crucial" requirement of the nitrogen atom in molybdenum hydroxylase substrates. While naphthalene is a specific P450 substrate, 1,3,5,8-tetraazanaphthalene (pteridine) is a selective substrate of molybdenum hydroxylases, and 1,3-diazanaphthalene (quinazoline) demonstrates similar affinity towards P450 as well as molybdenum hydroxylases as a substrate (see Figure 12).[176, 177, 178] Of some interest is

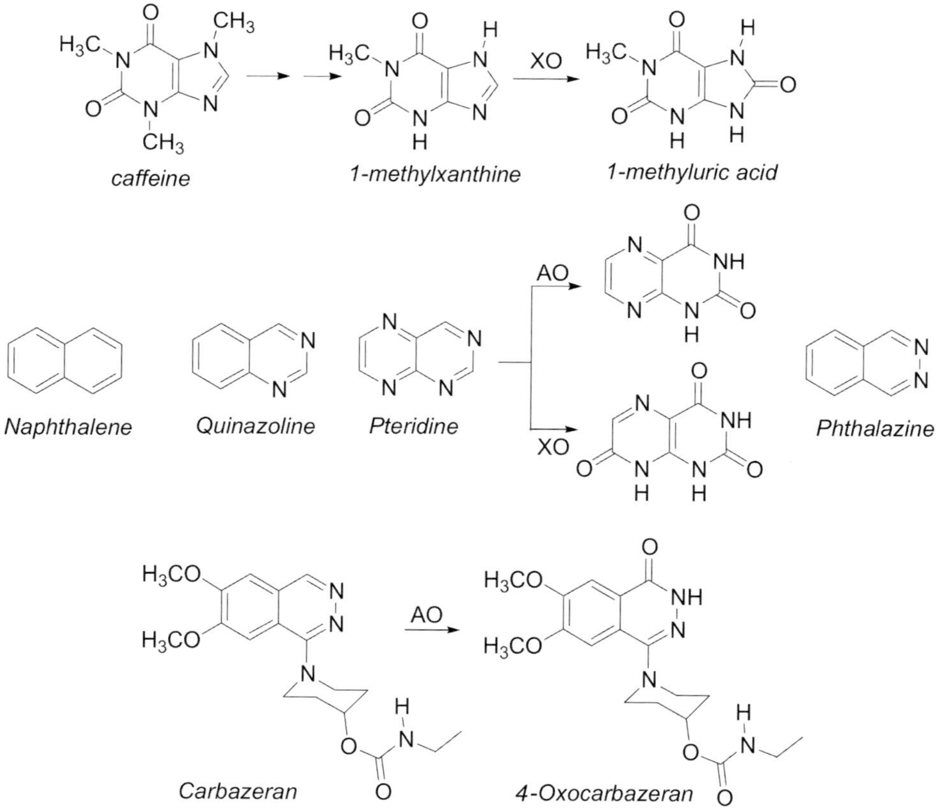

Figure 12. Crucial Requirement of the Nitrogen Atom in Azaarene Substrates of Molybdenum Hydroxylases.

the regiospecificity of pteridine hydroxylation in that XO converts it sequentially to pteridin-2,4,7-trione, whereas AO oxidizes it to pteridin-2,4-dione (see Figure 12).[178] Since molybdenum hydroxylase mediated oxidations involves the nucleophilic addition of oxygen to the most electrophilic site in the azaarene skeleton, it is not surprising that pteridine (which has 4 electron-deficient carbon atoms as a result of 4 imine bonds) functions as a selective substrate for molybdenum hydroxylases. In contrast to the mechanism of oxidation by molybdenum hydroxylases, P450 mediated oxidations involves the formation of an electrophilic oxo species (derived from molecular oxygen) which then generally reacts with the most nucleophilic site within the xenobiotic; a feature that does not exist in the strongly electron-deficient pteridine ring system.

3.1   Tissue Distribution

Both AO and XO are homodimers of 300 kDa and each 150-kDa subunit contains two iron-sulfur (Fe_2S_2) centers, an FAD, and the molybdenum-pterin cofactor (MoCo). The FeS, FAD, and MoCo cofactors comprise an internal electron transfer chain in which electrons are passed from the active site molybdenum center to the FeS centers and finally to FAD. A major portion of the molybdenum hydroxylase activity is found in the cytosol. The highest concentration of XO in mammals is found in the milk and lactating mammary gland.[179] However, there seems to be no correlation between the amount of XO present in milk and liver of the same species. For instance, felines which exhibit the highest levels of hepatic XO activity compared to other preclinical species and humans, possess very low milk XO activity.[179, 180] Although there are no reports of any milk AO activity, liver contains the highest levels in all mammalian tissues studied, with other organs such as the lung, kidney, and small intestine containing less than 50 % of the hepatic AO activity.[181, 182, 183] In contrast, whole brain homogenates from preclinical species have been found to contain very low levels or to be devoid of either molybdenum hydroxylases.[184] Studies on the cellular localization of molybdenum hydroxylases reveals that XO is localized exclusively in sinusoid and capillary endothelial cells of hepatic, cardiac, pulmonary and adipose tissue as well as in the lactating epithelial cells in the mammary gland.[185, 186] Likewise, AO is widespread in respiratory, digestive, urogenital, and endocrine tissues.[187] One of the problems associated in considering tissue distribution of AO activity is the plethora of substrates utilized to measure enzyme activity, many of which are known to possess diverse tissue or species specificity. In contrast, studies on the tissue and species distribution of XO are easier since only one substrate xanthine is generally used to measure enzyme activity.

3.2 Species Distribution

XOs from different tissues and species demonstrate similar substrate specificities. For instance, K_m values for endogenous substrates (e.g. xanthine) or xenobiotics (e.g. theophylline) do not vary markedly between human liver and bovine milk XO.[188] Despite the established role of XO in endogenous purine metabolism, lower concentrations of XO activity have been detected in human liver than in guinea pig, rat, or rabbit[189] and *in vitro* and *in vivo* studies have revealed a wide variation in human XO activity.[190] A sex difference in XO activity also exists in rats, with mature females possessing half the XO activity of males.[191] In contrast, studies in humans performed either with liver biopsy samples or using caffeine metabolite ratios to assess XO levels did not reveal a substantial sex difference in human XO activity.[192] AO activity across species reveals marked variations and as a consequence, extrapolation of enzyme activity in preclinical species to humans is difficult. For instance, dog liver apparently contains very little AO[193] and dramatic intra-species variation in AO activity has been noted in certain in-bred rat and mouse strains, particularly Sprague-Dawley rats.[194] An example of the marked species difference in AO activity is demonstrated in the metabolism of the inotropic agent, carbazeran. Although carbazeran demonstrates excellent pharmacological activity in the dog, its pharmacological effect in humans is negligible due to its extensive AO mediated metabolism. While carbazeran is stable in dog liver cytosol, it undergoes rapid metabolism to 4-oxocarbazeran (see Figure 12) upon incubation with human liver cytosol or partially purified AO. These *in vitro* species differences in carbazeran metabolism were consistent with those *in vivo*; these showed that following an oral dose to man and baboon, carbazeran was almost completely cleared via pre-systemic 4-hydroxylation, whereas in the dog, this metabolic route appeared unimportant.[195] Rabbit liver contains high levels of AO and lower XO activity than rat liver.[194] Consequently, rabbit liver preparations are often times used as model for human AO, even though the rabbit enzyme has a distinct substrate specificity from human liver AO.[196] The binding site of rabbit AO appears to be more restricted than human or guinea pig liver AO in that bulkier phthalazine derivatives such as carbazeran are not substrates, whereas similar kinetics of oxidation are observed with unsubstituted phthalazine itself (Figure 12).[196] In contrast, guinea pig AO activity correlates well with human liver enzyme in terms of substrate and inhibitor specificity and seems to be a better predictor of human AO substrate specificity.[197] As with XO, both mice and rats express kinetically distinct forms of AO, which are thought to arise via differences in redox states of the enzyme and not as a result of expression from two independent genes.[198, 199] Additional details on the species differences in AO activity are discussed in Section 3.4 on the structural diversity of AO/XO substrates.

3.3   Catalysis

Although extensive mechanistic studies have not been conducted in AO, the catalytic mechanism described for XO is most likely operative for AO as well, since spectroscopic analysis (UV-VIS, EPR) has revealed a remarkable similarity and identity in the cofactor and apoprotein environment of the two enzymes.[173] Substrate hydroxylation takes place at the molybdenum center that undergoes reduction from Mo^{VI} to Mo^{IV}.[200] The reducing equivalents introduced at the molybdenum center are subsequently passed via intramolecular electron transfer to FAD (AO and XO) resulting in the formation of $FADH_2$. $FADH_2$ returns to its resting state (FAD) via the reduction of either NAD^+ (XO/XDH) or molecular oxygen (AO), which results in the formation of peroxide or superoxide, depending on the level of enzyme reduction.[201] The observation that XO/XDH or AO can utilize an oxidized substrate instead of NAD^+ or molecular O_2, and act essentially as an oxidoreductase, has considerably clarified the substrate specificities attributed to these enzymes. Compounds that interact with AO and XO/XDH can be categorized as either reducing (electron donor at the Mo center) or oxidizing (electron acceptor from $FADH_2$). The general reaction catalyzed by molybdenum hydroxylases is depicted in Equation 2 (RH is the reducing substrate and ROH is the hydroxylated metabolite).

$$RH \quad + \quad H_2O \quad \longrightarrow \quad ROH \quad + \quad 2e^- \quad + \quad 2H^+ \tag{2}$$

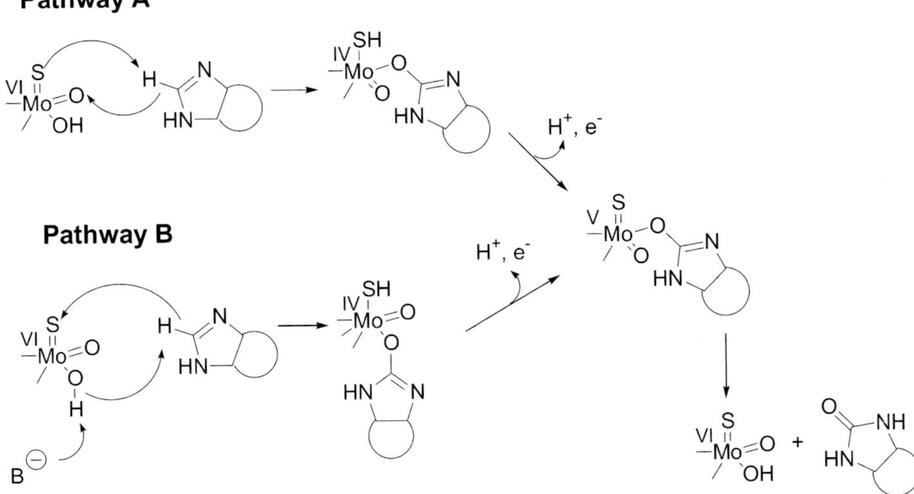

Figure 13. Potential Reaction Mechanisms for Molybdenum Hydroxylases.

The pH dependence of molybdenum hydroxylase mediated metabolism indicates that hydroxylation of substrates occurs via a base-catalyzed mechanism, and that the nitrogen atom in the substrate must be protonated for the hydroxylation to occur.[202] Although the oxygen atom incorporated into the product is ultimately derived from water,[203] it has been known for some time that the proximal oxygen atom donor is a catalytically labile site on the enzymes, which, in the course of a single turnover, transfers its oxygen to substrate, to be regenerated subsequently by oxygen derived from solvent prior to a second turnover.[204] Speculations on the catalytically labile site on the enzymes include the Mo=O group or a metal-coordinated hydroxide.[205, 206] The latter proposal is consistent with (1) crystallographic data on the structure of AO from *D. gigas,*[207] (2) single turnover experiments involving incorporation of ^{17}O into the molybdenum center,[208] and (3) with EPR studies on model inorganic molybdenum compounds.[209] The two potential mechanisms of catalysis involving either the Mo=O (pathway A) or Mo-OH (pathway B) as the catalytically labile site are depicted in Figure 13. The initial $Mo^{IV}O(SH)(OR)$ intermediate of the catalytic sequence [210] may be arrived at either by proton abstraction from the carbon atom α to the nitrogen in the substrate followed by nucleophilic attack on the electron-deficient Mo^{VI}=O unit, (Figure 13A) or by base-assisted nucleophilic attack on the carbon atom α to the nitrogen in the substrate by the metal-bound hydroxide followed by hydride transfer to the Mo=S group (Figure 13B).

3.4   Substrate Specificity of Molybdenum hydroxylases

3.4.1 Oxidation of Aldehydes to Carboxylic Acids

Although aliphatic and aromatic aldehydes are subject to AO and XO mediated oxidation, the K_m values are much lower with AO than XO.[211, 212] However, these K_m values are some 100-fold higher than the corresponding values with aldehyde dehydrogenase.[213] Optimal activity towards AO occurs with a 3- or 4-carbon unbranched chain.[214] Compared with aliphatic aldehydes, aromatic aldehydes are better AO substrates. Rabbit liver retinal oxidase, which is the enzyme responsible for the biosynthesis of retinoic acid, has been found to be identical to rabbit liver AO.[215] Thus, tissue AO levels can play a major role in the regulation of cellular growth, differentiation, and morphogenesis through the modulation of retinoic acid levels. Some notable examples of AO involvement in drug metabolism include the oxidation of the aldehyde metabolites of citalopram (see Section 2.4.1) and the nonsteroidal antiestrogen tamoxifen to the corresponding carboxylic acid derivatives (Figure 14).[216]

3.4.2 Oxidation of Nitrogen-Containing Aromatic Heterocycles (Azaarenes)

AO and XO catalyze the α-carbon oxidation of structurally diverse azaarenes and although the enzymes demonstrate similar structure-function relationships, their substrate specificities are quite different, with regard to both rates and the regiochemistry of oxidation (Figure 14). XO generally has a narrower substrate specificity than AO and is mainly involved in the oxidation of purine derivatives. The conversion of hypoxanthine to uric acid via xanthine, in the final step in purine catabolism, is mediated by XO. Although azaarenes besides purines are also subject to XO mediated oxidation, the overall rates of oxidation are much lower than those observed with AO.

3.4.2.1 Effect of Physiochemical Parameters

The electron density at the carbon undergoing oxidation does appear to have a greater influence on the regiochemistry of AO oxidations than on that of XO. Thus, cinnoline is oxidized at the β-carbon (the most electron-deficient carbon) by AO but is refractory to oxidation by XO (see Figure 14).[217] In the case of XO, the relative position of the electronegative nitrogen atoms, perhaps to bind to a complimentary amino acid at the XO active site, seems to be more significant. For instance, the C8 position in purine is the most electron-deficient and is also the position of AO attack, whereas the compound is oxidized at the C6 position by XO.[218] Incorporation of additional nitrogen atoms in azarene derivatives further increases selectivity for molybdenum hydroxylases as was indicated earlier with pteridine. Likewise, charged nitrogen compounds are better AO substrates as compared with the corresponding 3° nitrogen-containing analogs. Thus quaternization of a 3° ring nitrogen activates the α- as well as the γ-carbon atoms in that ring towards nucleophilic attack, and AO can catalyze simultaneous oxidations at either of these positions. The product ratio (α-hydroxylation versus γ-hydroxylation) is dependent on electronic as well as steric factors. Thus, pyridinium salts such as **38** undergo exclusive α-carbon oxidation to the corresponding α-pyridone derivatives (compd. **39**).[219] In contrast, quinolinium salts (compd. **40**) are more susceptible to steric effects, presumably because of their larger size, and in most cases the 4-quinolone (compd. **41**) derivative constitutes the major product of oxidation (see Figure 14).[220]

A further complication in considering AO substrate specificity is that the product ratio also varies with species. Thus, upon incubation with guinea pig AO, N-phenylquinolinium perchlorate (**42**) undergoes regioselective oxidation to the corresponding 1-phenyl-4-quinolone derivative **43**, whereas 1-phenyl-2-quinolone (**44**) is the major oxidation product with rabbit liver AO (see Figure 14).[221]

Although quaternary azaarene salts are excellent AO substrates, they are refractory to oxidation by XO under physiological conditions; however some oxidation is discernible at higher pH values (pH ~ 9.6).[222] Besides the activating effect of the positively charged azaarenes on the AO substrate specificity, lipophilicity also plays an important role. Thus, N-methylpyridinium is not an AO substrate, but the more lipophilic long chain N-alkylpyridinium derivatives are excellent substrates.[223] The influence of lipophilicity and the number of carboxylic acid groups on the rate of hydroxylation of the classical folate antagonist methotrexate (MTX) (see Figure 14) by AO has also been studied.[224] 7-HydroxyMTX is the major metabolite of MTX in humans, the formation of which is thought to be AO catalyzed.[225] Thus, decreasing the number of carboxylic acid groups in MTX from two to one or increasing the lipophilicity of MTX via incorporation of long chain alkyl groups led to enhanced catalytic efficiency, involving both a decrease in K_m and an increase in V_{max}.

3.4.2.2 Influence of Steric Effects

Studies with quinoline analogs containing an additional benzene ring fused to the quinoline nucleus have provided some insight on the steric constraints for AO substrates. Thus, phenanthridine is one of the most efficient AO substrates (K_m < 1 μM), whereas 5,6-benzoquinoline and 7,8-benzoquinoline are stable towards AO-mediated hydroxylation (see Figure 14).[226, 227] SAR studies on the interaction of substituted quinazolines and phthalazine derivatives with partially purified human, rabbit, guinea pig, and baboon liver AO have also been conducted in attempts to gain insight into the species differences in AO substrate specificity and the regiospecificity of oxidation.[197] The results of this analysis indicates that unsubstituted quinazolines were the most optimal AO substrates in the series giving rise to 2- and 4-oxoquinazolines (see Figure 14).

Marked differences in substrate specificity were also discernible between human liver AO and AO from preclinical species with substrate size being the differentiating factor. The molecular size of the substrates estimated using calculated molar refractivities, ranked the size of the binding site of AO in the order rabbit < guinea pig < baboon < human. Thus, the phthalazine derivative, carbazeran, which undergoes extensive AO mediated first pass metabolism in humans was not a rabbit liver AO substrate. Likewise, marked species difference in AO mediated metabolism has also been observed with the antivasospasm agent, **45** following intravenous administration of the radiolabeled drug to rats and rabbits.[228] In rabbits, the major urinary metabolites of **45** were 6- or 6'-monopyridone (24% of the dose) whereas analysis of rat urine revealed that only 5 % of the radioactivity was due to metabolites, mainly the N-oxide derivatives of **45** (Figure 15). Consistent with the *in vivo* findings, formation of the monopyridone metabolites by rabbit liver cytosol (menadione sensitive) was much higher than in rats. An additional example of species differences in AO mediated metabolism has also been documented with the α-adrenergic agent, brimonidone.[229] Human, rabbit, rat, and monkey liver fractions catalyzed the *in vitro* metabolism of brimonidone to the corresponding 2- and 3-oxo and the 2,3-dioxobrimonidone metabolites in a reaction that was menadione sensitive implicating AO involvement (Figure 15). In contrast, dog liver fractions converted brimonidone to the corresponding 4',5'-dehydrobrimonidine and guanidine metabolites in a NADPH dependent fashion implicating the involvement of P450 (see Figure 15). The species difference in the hepatic metabolism of brimonidone were attributed to the low AO activity of dog liver as opposed to poor brimonidone substrate properties for dog AO and overall, these observations are consistent with the species differences also discernible in the metabolism of carbazeran.[195]

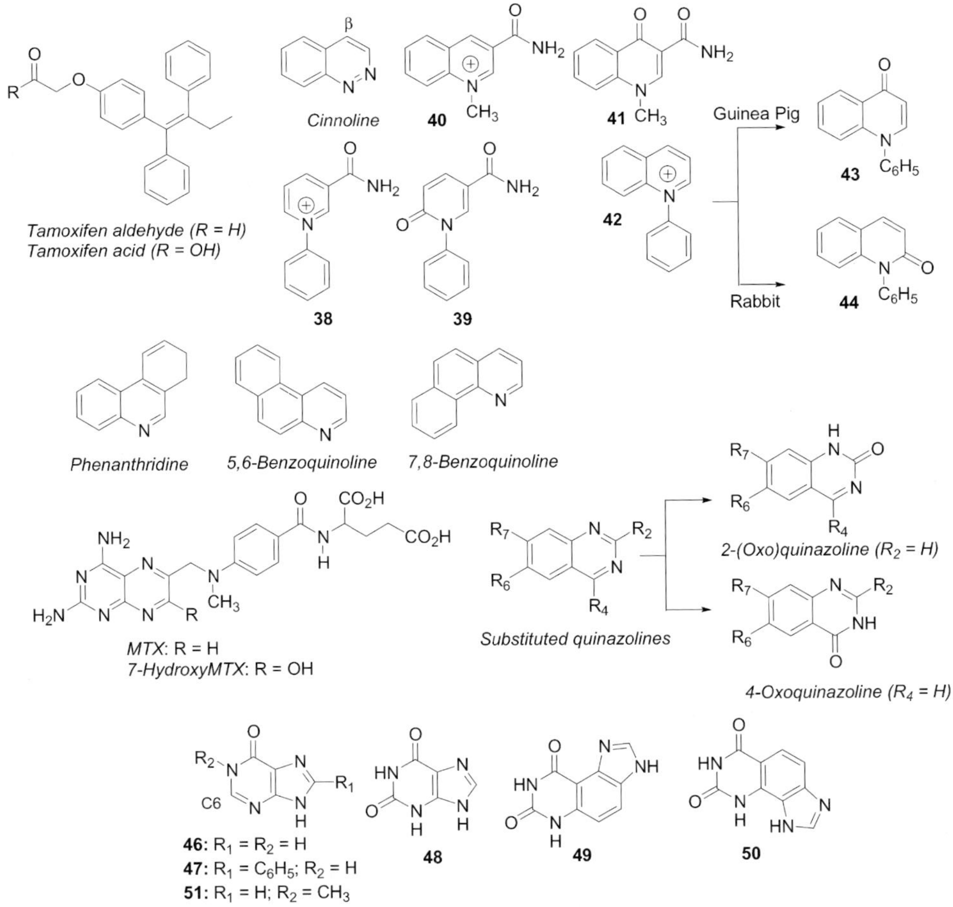

Figure 14. Probing the Effect of Electronics and Sterics on Molybdenum Hydroxylase Mediated Oxidations.

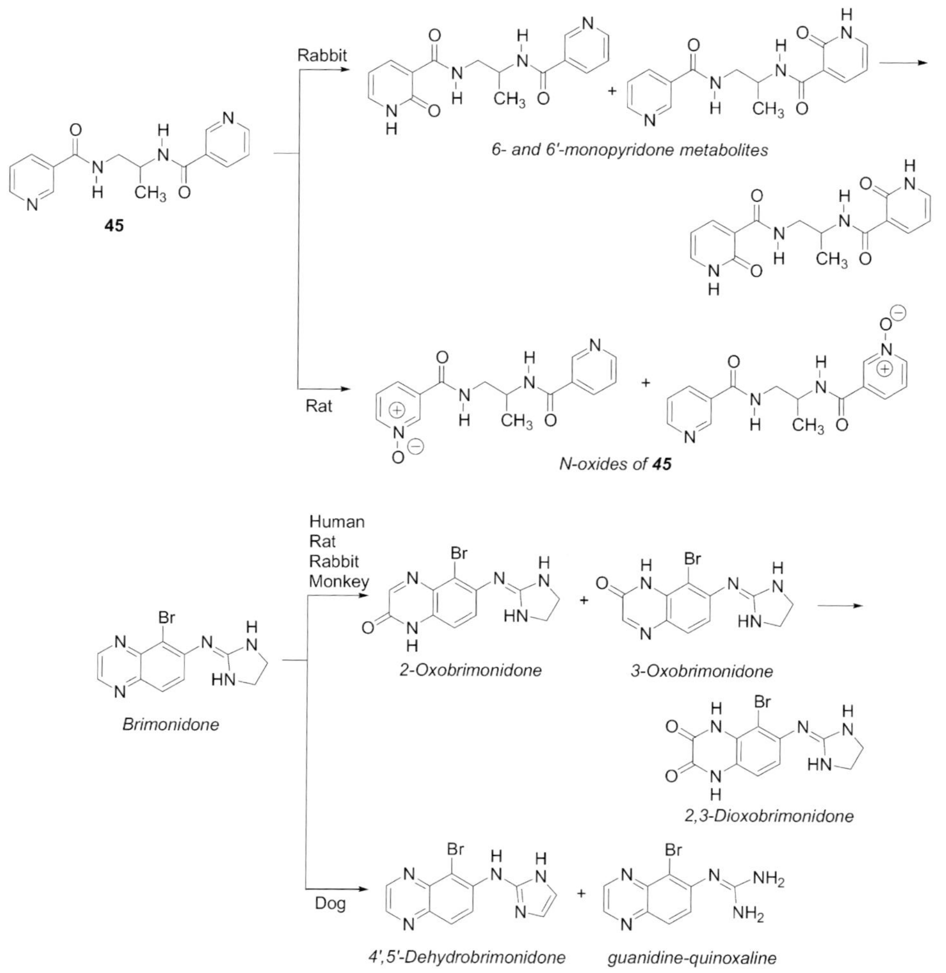

Figure 15. Species Differences in the AO Mediated Metabolism of **45** and Brimonidone.

Novel substrates based on the purine nucleus have also been utilized to probe SAR requirements for XO substrates (Figure 14). Conventional XO substrates based on the purine skeleton are known to undergo XO mediated oxidation in both heterocyclic rings at the C-2, C-6, and/or C-8 positions. Introduction of an aryl group at the C-8 position (analog **47**) in hypoxanthine (**46**) hinders C-6 oxidation.[230] Furthermore, lipophilic analogs of hypoxanthine or xanthine (**48**) such as benzohypoxanthine (**49**) are readily oxidized by XO, but the K_m value is 5 – 10-fold higher than that observed for hypoxanthine.[231] In contrast, binding of angular benzopurines at the active site of XO appears to be sterically hindered to a greater extent; thus the proximal angular benzopurine analog **49** is oxidized by XO at much slower rates than xanthine and the distal angular benzopurine analog **50** is not a XO substrate.[231] Furthermore, when tautomerism in the pyrimidine ring is blocked, as in 1-methylhypoxanthine (**51**), the compound is not a XO substrate suggesting that the N-methyl group sterically hinders C2 oxidation.[232]

3.5   Antiviral Prodrug Activation by AO and XO

6-Deoxyaciclovir was synthesized as an XO-activated prodrug of aciclovir for the treatment of chronic type B hepatitis (Figure 16). Aciclovir was the first orally active antiviral drug to show selectivity against herpes viruses. Aciclovir suffers from slow, variable, and incomplete absorption upon oral administration in humans; thus high doses, up to five times daily, have to be given particularly against less sensitive viruses. *In vitro* studies with human liver cytosol indicate that oxidation of deoxyaciclovir to aciclovir is indeed catalyzed by XO (C6 oxidation) whereas the prodrug is inactivated to 8-oxo-6-deoxyaciclovir by rabbit liver AO.[233] Consistent with the *in vitro* results, urinary excretion of aciclovir in human volunteers accounted for 65 – 70 % of the dose following oral or intravenous administration which indicates that 6-deoxyaciclovir is readily absorbed and extensively bioactivated (rather than deactivated) via oxidation.[234] Toxicological consequences in preclinical species, however, led to its discontinuation from further development.

Like aciclovir, penciclovir is also active *in vitro* against the herpes virus but suffers from poor oral absorption in humans. However, diacetyl-6-deoxypenciclovir (famciclovir) is rapidly absorbed and metabolized to penciclovir via the 6-deoxypenciclovir intermediate (see Figure 16). Rapid first pass metabolism of famciclovir occurs in the intestinal wall (ester bond cleavage) and liver (oxidation of 6-deoxyfamciclovir to penciclovir).[235, 236] Although initial studies hinted at the involvement of XO in the oxidation of 6-dexoyfamciclovir to penciclovir, coadministration of famciclovir and the XO inhibitor allopurinol to human volunteers did not result in significant pharmacokinetic interactions with famciclovir.[237] Recent *in vitro* studies have established that it is AO and not XO that catalyzes the conversion of 6-deoxypenciclovir to penciclovir in human liver cytosol.[238] In fact, famciclovir is not a XO substrate and 6-deoxypenciclovir is oxidized very slowly.[239] Pharmacokinetic studies indicate that the intrinsic capacity of human liver AO towards famciclovir is very high.[240] Peak plasma levels of penciclovir are discernible in 1 hour following famciclovir administration to humans.[241] This has important implications since *in vitro* 6-deoxypenciclovir is a modest AO substrate with a relatively low V_{max} and high K_m compared with other N-heterocyclic or aldehyde substrates. Penciclovir and 6-deoxypenciclovir account for ~ 60 % and ~ 5 % of the administered dose, respectively, in urine and feces.[240] The wide substrate specificity of AO suggests that its utility as a prodrug activator is not restricted to oxidations on the purine ring. For instance, 5-iodo-2-pyrimidinone-2'-deoxyribose, a pyrimidine nucleoside is converted to a clinical radiosensitizer, 5-iodo-deoxyuridine by AO (Figure 16).[242]

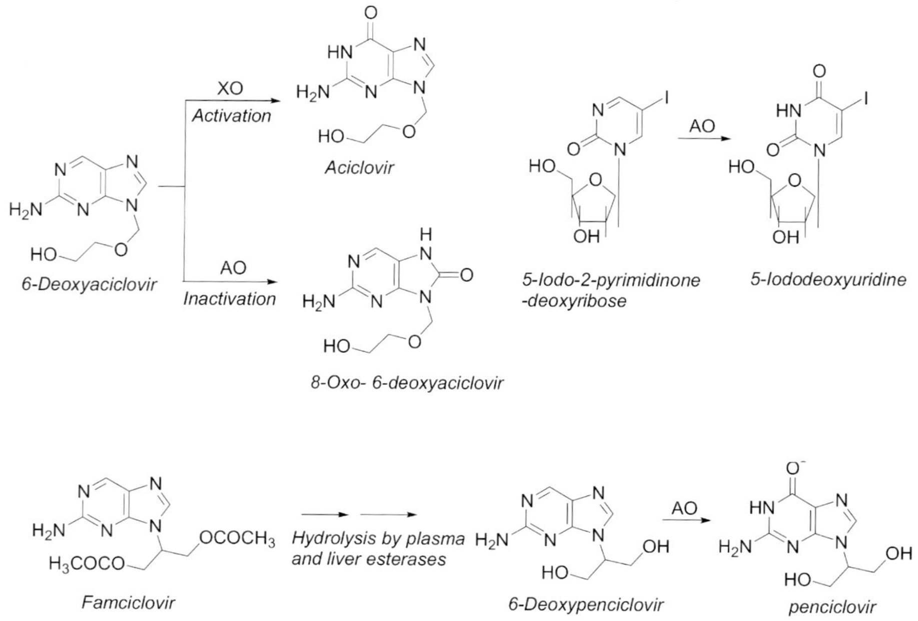

Figure 16. Antiviral Prodrug Activation by AO and XO.

3.6 Molybdenum Hydroxylase Mediated Reductions

Besides oxidative transformations, molybdenum hydroxylases are also responsible for reductive metabolism reactions involving electron transfer from $FADH_2$ to the oxidized xenobiotic rather than to molecular oxygen or NAD^+. *In vitro*, these reductions are facilitated under anaerobic conditions and in the presence of an electron donor. Common electron donors include azaarenes such as 2-hydroxypyrimidine and N-methylnicotinamide and aldehydes such as benz-aldehyde. NADPH or NADH are inefficient as electron donors in the reductive metabolism by molybdenum hydroxylases. A variety of substrates (Figure 17) are subject to molybdenum hydroxylase reduction including N-oxides (brucine-N-oxide → brucine),[243] sulfoxides (sulindac sulfoxide → sulindac sulfide),[244, 245] hydroxamic acids (Nicotinohydroxamic acid → nicotinic acid),[246, 247] and epoxides (benzo[a]pyrene-4,5-oxide → benzo[a]pyrene).[248] As observed in the oxidation reactions, substrate affinity towards AO or XO mediated reductions is also species dependent. An example of AO mediated reductive bioactivation has also been documented with the antitrypanosomiasis drug benznidazole.[249] Thus, under anaerobic conditions and in the presence of appropriate electron donors, both rat liver microsomes and cytosol are capable of reducing the nitro group in benznidazole. Inhibition of this pathway in liver microsomes and cytosol by carbon monoxide, and by menadione and/or allopurinol, respectively, suggests the involvement of P450 as well as AO/XO in the reduction. Subsequent studies with [^{14}C]-benznidazole indicate that under the experimental conditions leading to enzymatic reduction of the nitro group, reactive metabolites are also produced as judged by the incorporation of radioactivity in the microsomal and cytosolic protein. Whether benznidazole functions as a suicide inactivator of P450 and/or AO/XO remains unclear.

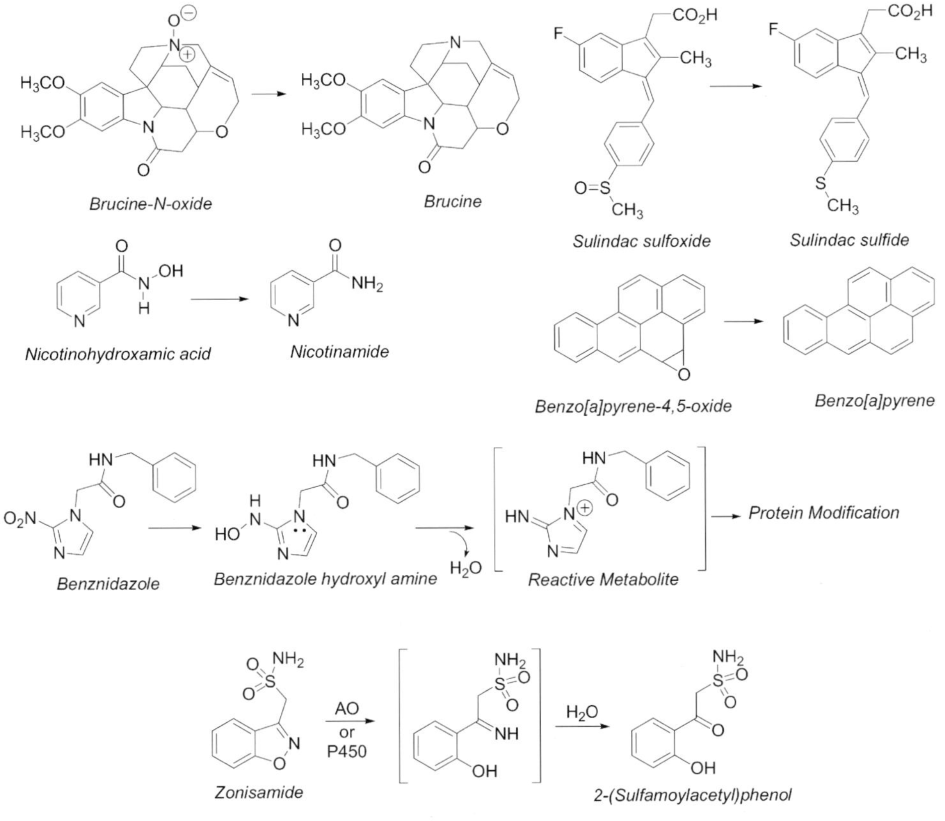

Figure 17. Molybdenum Hydroxylase Mediated Reduction Reactions.

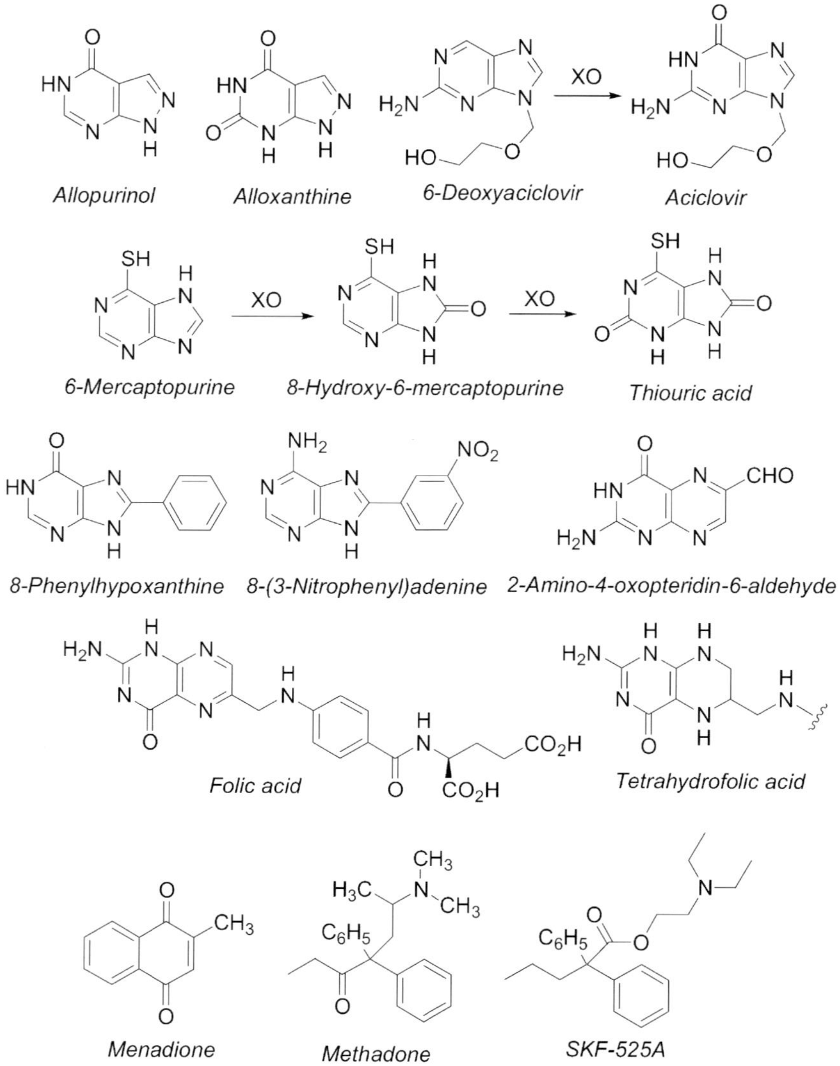

Figure 18. Molybdenum Hydroxylase Inhibitors.

An example of AO mediated reductive heterocyclic ring scission has been demonstrated in the metabolism of the anticonvulsant agent zonisamide (see Figure 17).[250] Under anaerobic conditions, rabbit or rat liver cytosolic AO in the presence of 2-hydroxypyrimidine or N-methylnicotinamide as electron donors catalyzes the two-electron reduction of zonisamide to 2-sulfamoylacetylphenol and ammonia in a 1:1 stoichiometry (presumably via the intermediate imine derivative). In the absence of an electron donor however, no reduction is discernible in rat liver cytosol, which initially led Nakasa *et al.* to propose the exclusive involvement of CYP 3A in the reductive metabolism of zonisamide in rat[251] and human[252] liver microsomes. Comparative studies in rabbit, hamster, mouse, and guinea pig liver cytosol and microsomes indicate that cytosolic reductase activity in the presence of 2-hydroxypyrimidine is significantly higher than microsomal reductase activity in the presence of NADPH.

3.7   Inhibitors of Molybdenum Hydroxylases

Both XO and AO metabolize allopurinol (Figure 18), a drug administered to patients suffering from gout, to alloxanthine (oxipurinol), which is a potent, and selective, tight-binding inhibitor of XO.[253, 254] Allopurinol- or alloxanthine-dependent XO inhibition can be effectively used to evaluate XO-dependent metabolism *in vitro* as well as *in vivo*. For instance, allopurinol attenuates the XO mediated oxidative activation of the antiviral prodrug 6-deoxyaciclovir to acyclovir (see Figure 18) in rat liver cytosol as well as in studies involving intraperitoneal co-administration of allopurinol and desciclovir to rats. Furthermore, the observation that allopurinol decreases the ratio of aciclovir to desciclovir 10- to 15-fold in the rat implies that XO is the major enzyme involved in the oxidative activation of the prodrug to aciclovir.[255] Likewise, a 400 – 500 % increase in peak plasma concentrations and AUC of the orally administered anticancer agent 6-mercaptopurine in primates and humans was discernible following pre-administration with allopurinol.[256] The observation that allopurinol pretreatment did not affect intravenous 6-mercaptopurine pharmacokinetics suggested that increases in 6-mercaptopurine levels were due to inhibition of first-pass metabolism of oral 6-mercaptopurine mediated by liver or intestinal XO. These results also confirm the involvement of XO as the principal enzyme in humans responsible for the sequential metabolism of 6-mercaptopurine to 8-hydroxy-6-mercaptopurine and thiouric acid (see Figure 18).[257] In general, the potency of XO inhibitors such as allopurinol can be dramatically improved by increasing the lipophilicity. Thus, 8-phenylhypoxanthine and 8-(3-nitrophenyl)adenine (see Figure 18) are superior to allopurinol as XO inhibitors with IC_{50} values 10 – 50-fold lower than allopurinol.[258]

Besides purine-like structures, Lewis *et al.* have also shown that folic acid and its coenzyme derivative tetrahydrofolic acid are several fold more potent than allopurinol as XO inhibitors (see Figure 18). In fact, a comparison of the inhibition constants (K_i) for folic acid ($K_i = 0.5$ μM), tetrahydrofolic acid ($K_i = 1$ μM), and allopurinol ($K_i = 4$ μM) suggests that the folate compounds are the most potent inhibitors of mammalian XO described to date.[259] There are a few reports of aldehydes acting solely as molybdenum hydroxylase inhibitors. For instance, 2-amino-4-oxopteridin-6-aldehyde is a potent XO inhibitor that progressively inactivates the enzyme and shows a high affinity for it.[260] Menadione and hydralazine are potent inhibitors of AO mediated oxidations (see Figure 18). Hydralazine has been used *in vivo* to implicate AO involvement in the human first-pass metabolism of carbazeran.[261] Menadione is perhaps the most widely used inhibitor of AO activity *in vitro*, and together with allopurinol, is conveniently utilized to distinguish between AO- and XO-catalyzed reactions.[262] The latter study also demonstrated that the rat is not a good model for human AO mediated reactions, since the C6 oxidation of antiviral deoxyguanine prodrugs was catalyzed exclusively by XO in rat liver, but by AO in human liver. Methadone has also been reported to be an extremely potent inhibitor of rat liver AO with an IC_{50} value of 0.3 μM.[263] Interestingly, the methadone analog proadifen (SKF-525A), generally considered to be a broad spectrum P450 inhibitor, also inhibits AO activity (IC_{50} = 2.6 μM).[264, 265] Kinetic analysis indicated the mode of inhibition to be mixed non-competitive type for the two compounds.

3.8   *Molybdenum Hydroxylases and Drug Metabolism*

Despite indications to the contrary, AO and XO/XDH play a major role in the metabolism of azaarenes and aromatic aldehydes that can be derived as intermediates in amine metabolism. AO substrate specificity varies markedly between species whereas that of XO appears to be more consistent. Due to the species differences in AO mediated metabolism of xenobiotics, extrapolation of metabolism data from preclinical species to humans is difficult. Based on SAR studies, guinea pig liver AO has similar kinetic properties to the human liver ortholog for structurally diverse AO substrates and appears to be a reasonable model, both *in vitro* and *in vivo*, for human AO. Furthermore, *in vitro* measurements indicate that human liver AO and XO are labile but *in vivo* enzyme activity is high, as demonstrated with carbazeran and this phenomenon has been utilized quite successfully in the activation of antiviral prodrugs. Given the medicinal chemist's passion for incorporating nitrogen-containing heterocyclic rings in drug candidates and wide substrate specificity of AO, any small molecule N-heterocycle can serve as a molybdenum hydroxylase substrate in addition to its P450 substrate properties. Although the relative *in vivo* contribution of human P450 isozymes in drug metabolism can easily be assessed in liver microsomes, that by AO often times is difficult due to the instability of the enzyme in surgically excised and post-mortem human hepatic tissue.

4   CARBONYL REDUCTASES AND ALCOHOL- AND ALDEHYDE DEHYDROGENASES

P450 mediated hydroxylation of aliphatic 1° or 2° carbon atoms to the corresponding 1° or 2° alcohols followed by sequential oxidation of the 1° alcohols via intermediate aldehydes to carboxylic acids constitutes a common metabolic fate of many drugs. Likewise, many amines undergo P450 or MAO catalyzed oxidations to the corresponding iminium species, spontaneous hydrolysis of which leads to the intermediate aldehyde followed by oxidation to the corresponding carboxylic acid derivatives. Alcohols, aldehydes, and ketones are oxidized or reduced by a number of enzymes including alcohol dehydrogenase (ADH; E.C.1.1.1.1), aldehyde dehydrogenase (ADLH; EC1.2.1.3), and carbonyl reductases. NAD^+ is the preferred cofactor for both ADH and ADLH.

4.1   Alcohol Dehydrogenases

ADH is a zinc-containing cytosolic enzyme present in several mammalian tissues including the liver, which has the highest levels, the kidney, the lung, and the gastric mucosa.[266] Human ADH is a dimeric protein consisting of two 40-kDa subunits. The subunits (α, β, γ, π, and χ) are encoded by five distinct gene loci. However, because there are three allelic variants of the β subunit (β_1, β_2, and β_3) and two allelic variants of the gamma subunit (γ_1 and γ_2), the human ADH enzymes are comprised of eight subunits, which can combine as homo- or heterodimers.[267] These different molecular forms of ADH are divided into three major classes: Class I contains ADH_1, ADH_2, and ADH_3. The class I enzymes are responsible for the oxidation of ethanol and other small, aliphatic alcohols, and they are strongly inhibited by zinc chelators such as pyrazole.[268] In contrast, class II (ADH_4) and class III (ADH_5) enzymes preferentially oxidize aromatic or medium to long alkyl chain aliphatic alcohols and are resistant to pyrazole mediated inhibition. Although many 1° alcohols undergo facile oxidation via the cytosolic ADHs *in vitro*, it is generally difficult to establish the identity of the particular alcohol-oxidizing enzyme in a clinical situation as several other enzymes including P450 and steroid dehydrogenases are also efficient catalysts of the same reaction.[269]

4.2   Aldehyde Dehydrogenases

Oxidation of aldehydes is considered to be a general detoxification process due to the electrophilic nature of aldehyde metabolites, and polymorphisms of several human ALDHs are associated with altered drug metabolism and disease phenotypes or pathophysiological consequences.[270] As with alcohol oxidation, aldehyde oxidation is usually attributed to cytosolic or mitochondrial ALDHs, although P450s and molybdenum hydroxylases are also capable of catalyzing aldehyde dehydrogenations. Sixteen ALDH genes and three pseudogenes have been identified so far in the human genome with distinct chromosomal locations.[271] An example of ALDH contribution in drug metabolism is highlighted in the detoxification of the cytosolic drug, cyclophosphamide, and other related oxazaphosphorines by cytosolic ALDH.[272] The prodrug, cyclophosphamide, undergoes initial P450 mediated oxidation to 4-hydroxycyclophosphamide which exists in equilibrium as the corresponding ring opened aminoaldehyde (Figure 19).[273] In tumorigenic cells, under relatively anaerobic conditions, the aminoaldehyde decomposes to acrolein and phosphoramide mustard, the ultimate cytotoxic metabolite. Alternately, the aminoaldehyde can be deactivated via ALDH ($ALDH_1$ or $ALDH_3$) mediated oxidation to carboxyphosphamide (see Figure 19). As a consequence, elevated ALDH levels in tumor cells can lead to acquired resistance to cyclophosphamide.[274]

Alcohol and aldehyde dehydrogenases both show marked genetic variations.[275] Genetic deficiencies in individual isozymes are found with varying frequencies in different ethnic groups who may metabolize alcohols and aldehydes at widely different rates. For instance, polymorphism in $ALDH_2$ is characterized by decreased acetaldehyde metabolism, low risk of alcoholism and increased risk of ethanol-induced cancers.[276] On the basis of ethanol metabolism, it has been suggested that genetic polymorphism in alcohol or aldehyde dehydrogenase is the underlying factor in racial or ethnic variations in ethanol metabolism.[277, 278] Although the polymorphism of alcohol and aldehyde dehydrogenase clearly alters ethanol and acetaldehyde metabolism it is not clear whether metabolism of other substrates is also affected. In this context, it is interesting to note that the conversion of 5-HT to 5-hydroxyindoleacetic acid, which involves both alcohol and aldehyde dehydrogenase, is not significantly different between subjects with varying alcohol and aldehyde genotypes.[279]

4.3 Carbonyl Reductases

The reduction of certain aldehydes to 1° alcohols and of ketones to 2° alcohols is catalyzed by alcohol dehydrogenase and by a family of carbonyl reductases (CR; EC 1.1.1.184). CRs belong to a class of oxidoreductase proteins that are part of the family of short chain dehydrogenases/reductases.[280] Generally, CRs are low molecular weight, monomeric, cytosolic, NADPH-dependent enzymes that reduce aldehyde and keto groups of prostaglandins, steroids, pterins, biogenic amines, and quinones derived from polycyclic aromatic hydrocarbons.[281] CR's are ubiquitously expressed in most mammalian tissues including blood and the cytosolic fraction of the liver, kidney, brain, and other tissues.[282] The major circulating metabolite of the antipsychotic agent haloperidol, is reduced-haloperidol formed by carbonyl reductases in the blood and liver cytosol (Figure 19).[283] Other carbonyl substrates of noteworthy mention include daunorubicin, warfarin, and menadione.[284] The activity of low- and high-affinity carbonyl reductases in human liver cytosol varies ~ 10-fold among individuals.[285]

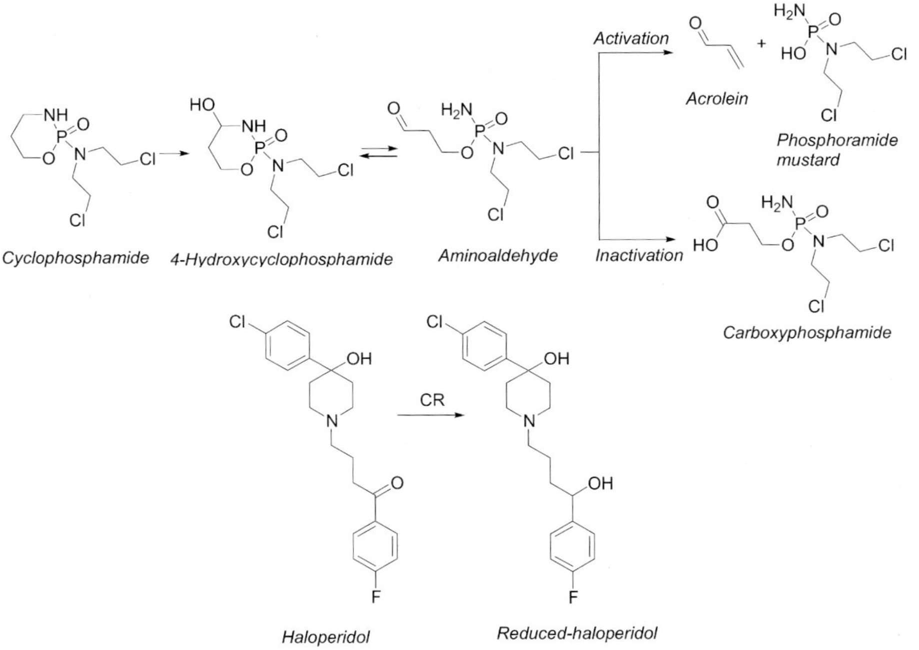

Figure 19. Reduction of Carbonyl Compounds.

REFERENCES

1. Pettit, F. H.; Orme-Johnson, W.; Ziegler, D. M. *Biochem. Biophys. Res. Commun.* **1964**, *16*, 444.
2. Ziegler, D. M; Mitchell, C. H. *Arch. Biochem. Biophys.* **1972**, *150*, 116.
3. Hines, R. N.; Hopp, K. A.; Franco, J.; Saeian, K.; Begun, F. P. *Mol. Pharmacol.* **2002**, *62*, 320.
4. Lawton, M. P.; Cashman, J. R.; Cresteil, T.; Dolphin, C.; Elfarra, A.; Hines, R. N.; Hodgson, E.; Kimuar, T.; Ozols, J.; Phillips, I.; Philpot, R. M.; Poulsen, L. L.; Rettie, A. E.; Williams, D. E.; Ziegler, D. M. *Arch. Biochem. Biophys.* **1994**, *308*, 254.
5. Phillips, I. R.; Dolphin, C. T.; Clair, P.; Hadley, M. R.; Hut, A. J.; McCombie, R. R.; Smith, R. L.; Shephard, E. A. *Chemico-Biological Interactions* **1995**, *96*, 17.

6. Kubo, A.; Itoh, S.; Itoh, K.; Kamataki, T. *Arch. Biochem. Biophys.* **1997**, *345*, 271.

7. Lawton, M. P.; Philpot, R. M. *J. Biol. Chem.* **1993**, *268*, 5728.

8. Dolphin, C. T.; Cullingford, T. E.; Shephard, E. A.; Smith, R. L.; Phillips, I. R. *Eur. J. Biochem.* **1996**, *235*, 683.

9. Lang, D. H.; Yeung, C. K.; Peter, R. M.; Ibarra, C.; Gasser, R.; Itagaki, K.; Philpot, R. M.; Rettie, A. E. *Biochem. Pharmacol.* **1998**, *56*, 1005.

10. Yeung, C. K.; Lang, D. H.; Thummel, K. E.; Rettie, A. E. *Drug Metab. Dispos.* **2000**, *28*, 1107.

11. Lang, D. H.; Rettie, A. E. *5th Int. ISSX Meeting*, **1998**, *13*, 185.

12. Koukouritaki, S. B.; Simpson, P.; Yeung, C. K.; Rettie, A. E.; Hines R. N. *Pediatric Research* **2002**, *51*, 236.

13. Whetstine, J. R.; Yueh, M. F.; McCarver, D. G.; Williams, D. E.; Park, C.S.; Kang, J. H.; Cha, Y. N.; Dolphin, C. T.; Shephard, E. A.; Phillips, I. R.; Hines, R. N. *Toxicol. Appl. Pharmacol.* **2000**, *168*, 216.

14. Krueger, S. K.; Martin, S. R.; Yueh, M. F.; Pereira, C. B.; Williams, D. E. *Drug Metab. Dispos.* **2002**, *30*, 34.

15. Haining, R. L.; Hunter, A. P.; Sadeque, A. J.; Philpot, R. M.; Rettie, A. E. *Drug Metab. Dispos.* **1997**, *25*, 790.

16. Cashman, J. R.; Zhang, J.; Leushner, J.; Braun, A. *Drug Metab. Dispos.* **2001**, *29*, 1629.

17. Overby, L. H.; Carver, G. C.; Philpot, R. M. *Chem. Biol. Interact.* **1997**, *106*, 29.

18. Janmohamed, A.; Dolphin, C. T.; Phillips, I.R.; Shephard, E. A. *Biochem. Pharmacol.* **2001**, *62*, 777.

19. Bhagwat, S. V.; Bhamre, S.; Boyd, M. R.; Ravindranath, V. *Neuropsycho-pharmacology.* **1996**, *15*, 133.

20. Atta-Asafo-Adjei, E.; Lawton, M. P.; Philpot, R. M. *J. Biol. Chem.* **1993**, *268*, 9681.

21. Shehin-Johnson, S. E.; Williams, D. E.; Larsen-Su, S.; Stresser, D. M.; Hines, R. N. *J. Pharmacol. Exp. Ther.* **1995**, *272*, 1293.

22. Burnett, V. L.; Lawton, M. P.; Philpot, R. M. *J. Biol. Chem.* **1994**, *269*, 14314.

23. Gasser, R.; Tynes, R. E.; Lawton, M. P.; Korsmeyer, K. K.; Ziegler, D. M.; Philpot, R. M. *Biochemistry* **1990**, *29*, 119.

24. Sabourin, P. J.; Smyser, B. P.; Hodgson, E. *Int. J. Biochem.* **1984**, *16*, 713.

25. Itoh, K.; Kimura, T.; Yokoi, T.; Itoh, S.; Kamataki, T. *Biochim. Biophys. Acta* **1993**, *1173*, 165.

26. Cherrington, N. Y.; Cao, Y.; Cherrington, J. W.; Rose, R. L.; Hodgson, E. *Xenobiotica* **1998**, *28*, 673.

27. Ripp, S. L.; Itagaki, K.; Philpot, R. M.; Elfarra, A. A. *Drug Metab. Dispos.* **1999**, *27*, 46.

28. Lattard, V.; Lachuer, J.; Buronfosse, T.; Garnier, F.; Benoit, E. *Biochem. Pharmacol.* **2002**, *63*, 1453.

29. Lattard, V.; Longin-Sauvageon, C.; Lachuer, J.; Delatour, R.; Benoit, E. *Drug Metab. Dispos.* **2002**, *30*, 119.

30. Falls, J. G.; Blake, B. L.; Cao, Y.; Levi, P. E.; Hodgson, E. *J. Biochem. Toxicology.* **1995**, *10*, 171.

31. Kawaji, A.; Isobe, M.; Takabatake, E. *Biol. Pharm. Bull.* **1997**, *20*, 917.

32. Fang, J. *Eur J Drug Metab Pharmacokinet* **2000**, *25*, 109.

33. Kawaji, A.; Ohara, K.; Takabatake, E. *Biol. Pharm. Bull.* **1994**, *17*, 603.

34. Rettie, A. E.; Meier, G. P.; Sadeque, A. J. M. *Chemico-Biological Interactions* **1995**, *96*, 3.

35. Das, M.L.; Ziegler, D.M. *Arch. Biochem. Biophys.* **1970**, *140*, 300.

36. Falls, J. G.; Ryu, D. Y.; Cao, Y.; Levi, P. E.; Hodgson, E. *Arch. Biochem. Biophys.* **1997**, *342*, 212.

37. Lemoine, A.; Williams, D. E.; Cresteil, T.; Leroux, J. P. *Mol. Pharmacol.* **1991**, *40*, 211.

38. Dannan, G.A.; Guengerich, F.P.; Waxman, D.J. *J. Biol. Chem.* **1986**, *261*, 10728.

39. Williams, D. E.; Hale, S. E.; Muerhoff, A. S.; Masters, B. S. S. *Mol. Pharmacol.* **1985**, *28*, 381.

40. Lee, M. Y.; Clark, J. E.; Williams, D. E. *Arch. Biochem. Biophys.* **1993**, *302*, 332.

41. Lee, M. Y.; Smiley, S.; Kadkhodyan, S.; Hines, R. N.; Williams, D. E. *Chem. Biol. Interact.* **1995**, *96*, 75.

42. Kaderlik, R. K.; Weser, E.; Ziegler, D. M. *Prog. Pharmacol. Clin. Pharmacol.* **1991**, *3*, 95.

43. Nnane, L. P.; Damani, L. A.; Hutt, A. J. *Eur. J. Drug Metab. Pharmacokinet.* **2001**, *26*, 17.

44. Katchamart, S.; Stresser, D. M.; Dehal, S. S.; Kupfer, D.; Williams, D. E. *Drug Metab. Dispos.* **2000**, *28*, 930.

45. Larsen-Su, S.; Williams, D. E. *Drug Metab. Dispos.* **1996**, *24*, 927.

46. Katchamart, S.; Williams, D. E. *Comp Biochem. Physiol. C Toxicol. Pharmacol.* **2001**, *129*, 377.

47. Poulsen, L. L.; Ziegler, D. M. *J. Biol. Chem.* **1979**, *254*, 6449.

48. Ziegler, D. M. *Drug Metab. Rev.* **1988**, *19*, 1.

49. Ziegler, D. M. *Annu. Rev. Pharmacol. Toxicol.* **1993**, *33*, 179.

50. Cashman, J. R. *Chem. Res. Toxicol.* **1995**, *8*, 165.

51. Hodgson, E.; Levi, P. E. *Xenobiotica* **1992**, *22*, 1175.

52. Jokanovic, M. *Toxicology* **2001**, *166*, 139 Review.

53. Rooseboom, M.; Commandeur, J. N.; Floor, G. C.; Rettie, A. E.; Vermeulen, N. P. *Chem. Res. Toxicol.* **2001**, *14*, 127.

54. Goeger, D. E.; Ganther, H. E. *Arch. Biochem. Biophys.* **1994**, *310*, 448.

55. Chen, G. P.; Ziegler, D. M. *Arch. Biochem. Biophys.* **1994**, *212*, 566.

56. Kim, Y. M.; Ziegler, D. M. *Drug Metab. Dispos.* **2000**, *28*, 1003.

57. Lin, J.; Berkman, C. E.; Cashman, J. *Chem. Res. Toxicol.* **1996**, *9*, 1183.

58. Narimatsu, S.; Yamamoto, S.; Kato, R.; Masubuchi, Y.; Horie, T. *Biol. Pharm. Bull.* **1999**, *22*, 567.

59. Cashman, J. R.; Park, S. B.; Berkman, C. E.; Cashman, L. E. *Chemico-Biological Interactions* **1995**, *96*, 33.

60. Tugnait, M.; Hawes, E. M.; McKay, G.; Rettie, A. E.; Haining, R. L.; Midha, K. K. *Drug Metab. Dispos.* **1997**, *25*, 524.

61. Mushiroda, T.; Douya, R.; Takahara, E.; Nagata, O. *Drug Metab. Dispos.* **2000**, *28*, 1231.

62. Kawaji, A.; Ohara, K.; Takabatake, E. *Anal. Biochem.* **1993**, *214*, 409.
63. Lang, D. H.; Rettie, A. E. *Br. J. Clin. Pharmacol.* **2000**, *50*, 311.
64. Ring, B. J.; Wrighton, S. A.; Aldridge, S. L.; Hansen, K.; Haehner, B.; Shipley, L. A. *Drug Metab. Dispos.* **1999**, *27*, 1099.
65. Lang, D. H.; Rettie, A. E. *5th Int. ISSX Meeting*, **1998**, *13*, 186.
66. Hoskins, J.; Shenfield, G.; Murray, M.; Gross, A. *Xenobiotica* **2001**, *31*, 387.
67. Hodgson, E.; Rose, R. L.; Cao, Y.; Dehal, S. S.; Kupfer, D. *J. Biochem. Molecular Toxicology* **2000**, *14*, 118.
68. Rodrigues, A. D.; Kukulka, M. J.; Ferrero, J. L.; Cashman, J. R. *Drug Metab. Dispos.* **1995**, *23*, 1143.
69. Hadley, M. R.; Oldham, H. G.; Camilleri, P.; Damani, L. A.; Hutt, A. J. *Chirality* **1996**, *8*, 430.
70. Hadley, M. R.; Oldham, H. G.; Damani, L. A.; Hutt, A. J. *Chirality* **1994**, *68*, 98.
71. Lang, D. H.; Rettie, A. E.; Böcker, R. H. *Chem. Res. Toxicol.* **1997**, *10*, 1037.
72. Hamman, M. A.; Haehner-Daniels, D. B.; Wrighton, S. A.; Rettie, A. E.; Hall, S. D. *Biochem. Pharmacol.* **2000**, *60*, 7.
73. Rawden, H. C.; Kokwaro, G. O.; Ward, S. A.; Edwards, G. *Br. J. Clin. Pharmacol.* **2000**, *49*, 313.
74. Rettie, A. E.; Lawton, M. P.; Sadeque, A. J.; Meier, G. P. *Arch. Biochem. Biophys.* **1994**, *311*, 369.
75. Fisher, M. B.; Rettie, A. E. *Tetrahedron Asymmetry* **1997**, *8*, 613.
76. Fisher, M. B.; Lawton, M. P.; Atta-Asafo-Adjei, E.; Philpot, R. M.; Rettie, A. E. *Drug Metab. Dispos.* **1995**, *23*, 1431.
77. Nagata, T.; Williams, D. E.; Ziegler, D. M. *Chem. Res. Toxicol.* **1990**, *3*, 372.
78. Light, D. R.; Waxman, D. J.; Walsh, C. *Biochemistry*, **1982**, *231*, 2490.
79. Lawton, M. P.; Kronbach, T.; Johnson, E. F.; Philpot, R. M. *Mol. Pharmacol.* **1991**, *660*, 692.
80. Itagaki, K.; Carver, G. T.; Philpot, R. M. *J. Biol. Chem.* **1996**, *271*, 20102.
81. Grothusen, A.; Hardt, J.; Brautigam, L.; Lang, D.; Bocker, R. *Arch. Toxicol.* **1996**, *71*, 64.
82. Clement, B.; Weide, M.; Ziegler, D. M. *Chem. Res. Toxicol.* **1996**, *9*, 599.
83. Guo, Z.; Raeissi, S.; White, R. B.; Stevens, J. C. *Drug Metab. Dispos.* **1997**, *25*, 390.
84. Grothusen, A.; Hardt, J.; Brautigam, L.; Lang, D.; Bocker, R. *Arch. Toxicol.* **1996**, *71*, 64.
85. Lambert, D. M.; Mamer, O. A.; Akerman, B. R.; Choiniere, L.; Gaudet, D.; Hamet, P.; Treacy, E. P. *Mo.l Genet. Metab.* **2001**, *73*, 224.
86. Mayatepek, E.; Kohlmüller, D. *Acta Paediatr.* **1998**, *87*, 1205.
87. Koyama, E.; Kikuchi, Y., Echizen, H.; Chiba, K.; Ishizaki, T. *Therapeutic Drug Monitoring.* **1993**, *15*, 224.
88. Sachse, C.; Ruschen, S.; Dettling, M.; Schley, J.; Bauer, S.; Müller-Oerlinghausen, B.; Roots, I.; Brockmöller, J. *Clin. Pharmacol. Ther.* **1999**, *66*, 431.
89. Kang, J. H.; Chung, W. G.; Lee, K. H.; Park, C. S.; Kang, J. S.; Shin, I. C.; Roh, H. K.; Dong, M. S., Baek, H. M.; Cha, Y. N. *Pharmacogenetics* **2000**, *10*, 67.
90. Berkman, C. E.; Park, S. B.; Wrighton, S. A.; Cashman, J. R. *Biochem. Pharmacol.* **1995**, *50*, 565.

91. Prueksaritanout, T.; Sittichai, N. *J. Chromatogr.* **1989**, *490*, 175.

92. Lang, D. H.; Rettie, A. E. *Br. J. Clin. Pharmacol.* **2000**, *50*, 311.

93. Rettie, A. E.; Lang, D. H. *Pharmacogenetics* **2000**, *10*, 1.

94. Baldock, G. A.; Brodie, R. R.; Chasseaud, L. F.; Taylor, T. *J. Chromatogr.* **1990**, *529*, 113.

95. Nakashima, M.; Uematsu, T.; Nakajima, S.; Nagata, O.; Yamaguti, T. *Jpn. Pharmacol. Ther.* **1993**, *21*, 4157.

96. Al-Waiz, M.; Mitchell, S. C.; Idle, J. R.; Smith, R. L. *Xenobiotica.* **1987**, *17*, 551.

97. Dolphin, C. T.; Janmohamed, A.; Smith, R. L.; Shephard, E. A.; Phillips, I. R. *Nature Genet.* **1997**, *17*, 491.

98. Zschocke, J.; Kohlmueller, D.; Quak, E.; Meissner, T.; Hoffmann, G. F.; Mayatepek, E. *Lancet* **1999**, *354*, 834.

99. Zschocke, J.; Mayatepek, E. *J. Inherit. Metab. Dis.* **2000**, *23*, 378.

100. Lee, C.W.; Tomlinson, B.; Yeung, J. H.; Lin, G.; Damani, L. A. *Pharmacogenetics* **2000**, *10*, 829.

101. Akerman, B. R.; Lemass, H.; Chow, L. M.; Lambert, D. M.; Greenberg, C.; Bibeau, C.; Mamer, O. A.; Treacy, E. P. Mol. Genet. Metab. **1999**, *68*, 24.

102. Singer, T. P. Chem. Biochem. Flavoenzymes **1991**, *2*, 437.

103. Dostert, P.; Strolin Benedetti, M.; Tipton, K. F. *Med. Res. Rev.* **1989**, *9*, 45.

104. Johnston, J. P. (1968) *Biochem. Pharmacol.* **1968**, *17*, 1285.

105. Knoll, J.; Magyar, K. *Adv. Biochem.* **1972**, *5*, 393.

106. Tipton, K. F. *Biochem. Soc. Trans.* **1994**, *22*, 764.

107. Bach, A. W. J.; Lan, N.C.; Johnson, D. L.; Abell, C. W.; Benmbenek, M. E.; Kwan, S-W.; Seeburg, P. H.; Shih J. C. *Proc. Natl. Acad. Sci. USA* **1988**, *85*, 4934.

108. Hsu, Y-P. P; Weyler, W.; Chen, S.; Sims, K. B.; Rinehart, W. B.; Utterback, M.; Powell, J. F.; Breakefield, X. O. *J. Neurochem.* **1988**, *51*, 1321.

109. Powell, J. F.; Hsu, Y. -P.P.; Weyler, W.; Chen, S.; Salach, J.; Andrilopoulos, K.; Mallet, J.; Breakefield, X. O. *Biochem. J.* **1989**, *259*, 407.

110. Weyler, W.; Hsu, Y. P.; Breakefield, X. O. Biochemistry and genetics of monoamine oxidase. In *Pharmacogenetics of drug metabolism.* (Kalow, W., Ed.), **1992**, pp 333.

111. Ito, A.; Kuwahara, T.; Inadome, S.; Sagara, Y. *Biochem. Biophys. Res. Commun.* **1988**, *157*, 970.

112. Hsu, Y.-P.; Powell, J. F.; Sims, K. B.; Breakefield, X. O. *J. Neurochem.* **1989**, *53*, 12.

113. Squires, R. F. *Adv. Biochem. Pharmacol.* **1972**, *5*, 355.

114. White, H. L.; Stine, D. K. *Life Sci.* **1984**, *35*, 827.

115. Denny, R. M.; Patel, N. T.; Fritz, R. R.; Abell, C. W. *Mol. Pharmacol.* **1982**, *22*, 500.

116. O'Brien, E.; Dostert, P.; Tipton, K. F. *Biochem. Pharmacol.* **1995**, *50*, 317.

117. Krueger, M. G.; Mazouz, F.; Ramsay, R. R.; Milcent, R.; Singer, T. P. *Biochem. Biophys. Res. Commun.* **1995**, *206*, 556.

118. Innoue, H.; Castagnoli, K.; Cornelis, V.D.S.; Mabic, S.; Igarashi, K.; Castagnoli, N.; Jr. *J. Pharmacol. Exp. Ther.* **1999**, *291*, 856.

119. O'Carroll, A. M.; Anderson, M. C.; Tobbia, I.; Phillips, J. P.; Tipton, K. F. *Biochem. Pharmacol.* **1989**, *38*, 901.

120. Saura, J.; Nadal, E.; Van den Berg, B.; Vila, M.; Bombi, J. A.; and Mahy, N. *Life Sci.* **1996**, *59*, 1341.

121. Shih, J. C.; Grimsby, J.; Chen, K. *J. Neural Transm. (Suppl.)* **1990**, *32*, 41.

122. Grimsby, J.; Lan, N. C.; Neve, R.; Chen, K.; Shih, J C. *J. Neurochem.* **1990**, *55*, 1166.

123. Rodriguez, M. Z.; Saura, J.; Billett, E. E.; Finch, C. C.; Mahy, N. *Cell. Tissue Res.* **2001**, *304*, 215.

124. Thorpe, L. W.; Westlund, K. N.; Kochersperger, L. M.; Abell, C. W.; Denney, R. M. *J. Histochem. Cytochem.* **1987**, *35*, 23.

125. Westlund, K. N. *Neurol. Dis. Ther.* **1994**, *21*, 1.

126. Binda, C.; Newton-Vinson, P.; Hubalek, F.; Edmondson, D. E.; Mattevi, A. *Nature Struct. Biol.* **2002**, *9*, 22.

127. Mabic, S.; Castagnoli, Jr.; N. *J. Med. Chem.* **1996**, *39*, 3694.

128. Miller, J. R.; Edmondson, D. E. (1999) *Biochemistry* **1999**, *38*, 13670.

129. Geha, R. M.; Rebrin, I.; Chen, K.; Shih, J. C. *J. Biol. Chem.* **2001**, *276*, 9877.

130. Yu, P. H.; Davis, A. B. *Int. J. Biochem.* **1988**, *20*, 1197.

131. Silverman, R. B. *Acc. Chem. Res.* **1995**, *28*, 335.

132. Silverman, R. B.; Hoffman, S. J.; Catus, W. B. *J. Am. Chem. Soc.* **1980**, *102*, 7126.

133. Walker, M. C.; Edmondson D. E. *Biochemistry* **1994**, *33*, 7088.

134. Kim, J.–M.; Hoegy, S. E.; Mariano, P. S. *J. Am. Chem. Soc.* **1995**, *117*, 100.

135. Silverman, R. B.; Zelechonok, Y. *J. Org. Chem.* **1992**, *57*, 6379.

136. Silverman, R. B. *Biochem. Soc. Trans.* **1991**, *19*, 201.

137. Miller, R. J.; Edmondson, D. E.; Grissom, C. *J. Am. Chem. Soc.* **1995**, *117*, 7830.

138. Kuttab, S.; Kalgutkar, A.; Castagnoli, N.; Jr. *Chem. Res. Toxicol* . **1994**, *7*, 740.

139. Anderson, A. H.; Kuttab, S.; Castagnoli, N.; Jr. *Biochemistry* **1996**, *35*, 3335.

140. Edmondson, D. E. *Xenobiotica* **1995**, *25*, 735.

141. Youdim, M. B. H.; Harshak, N.; Yoshioka, M.; Araki, H.; Mukai, Y.; Gotto, G. *Biochem. Soc. Trans.* **1991**, *19*, 224.

142. Suzuki, O.; Katsumata, Y.; Oya, M. *J. Neurochem.* **1981**, *36*, 1298.

143. Blaschko, H. *J. Pharm. Pharmacol.* **1989**, *41*, 664.

144. Barwell, C. J.; Basma, A. N.; Canham, C. A.; Williams, C. *Pharmacol. Res. Commun.* **1989**, *20*, 101.

145. Yu, P. H. *J. Pharm. Pharmacol.* **1989**, *41*, 205.

146. Williams, C. H. *Biochem. Soc. Trans* **1977**, *5*, 1770.

147. Rochat, B.; Kosel, M.; Boss, G.; Testa, B.; Gillet, M.; Baumann, P. *Biochem. Pharmacol.* **1998**, *56*, 15.

148. Kosel, M.; Amey, M.; Aubert, A.-C.; Baumann, P. *Eur. Neuropsychopharmacol.* **2001**, *11* , 75.

149. Dixon, C. M.; Park, G. R.; Tarbit, M. H. *Biochem. Pharmacol.* **1994**, *47*, 1253.

150. Fleishaker, J. C.; Ryan, K. K.; Jansat, J. M.; Carel, B. J.; Bell, D. J.; Burke, M. T.; Azie, N. E. *Br. J. Clin. Pharmacol.* **2001**, *51*, 437.

151. Cosson, V. F.; Fuseau, E.; Efthymiopoulos, C.; Bye, A. *J. Pharmacokinet. Biopharmaceutics* **1997**, *25*, 149.

152. Wild, M. J.; McKillop, D.; Butters, C. J. *Xenobiotica* **1999,** *29*, 847.

153. Rolan, P. *Cephalalgia* **1997**, *17(suppl. 18)*, 21.

154. Vyas, K. P.; Halpin, R. A.; Geer, L. A.; Ellis, J. A.; Liu, L.; Cheng, H.; Chavez-Eng, C.; Matuszewski, B. K.; Varga, S. L.; Guiblin, A. R.; Rogers, J. D. *Drug Metab. Dispos.* **2000**, *28*, 89.

155. Christophe, J.; Kutzner, R.; Hguyen-Bui, N. D.; Damien, C.; Chatelain, P.; Gillet, L. *Life Sci.* **1983**, *33*, 533.

156. Silverman, R. B.; Nishimura, K., Lu, X. *J. Am. Chem. Soc.* **1993**, *115*, 4949.

157. Wu, X.; Noda, A.; Imamura, Y. *Comp. Biochem. Physiol. Part (C)* **2001**, *129*, 361.

158. Trevor, A. J.; Castagnoli, Jr. N.; Singer, T. P. *Toxicology* **1988**, *49*, 513.

159. Chiba, K.; Trevor, A.; Castagnoli, Jr., N. *Biochem. Biophys. Res. Commun.* **1984**, *120*, 574.

160. Dungigan, C. D.; Shamoo, A. E. *Neuroscience* **1996**, *75*, 37.

161. Nicklas, W. J.; Vyas, I.; Heikkila, R. E. *Life Sci.* **1985**, *36*, 2503.

162. Youngster, S. K.; Sonsalla, P. K.; Sieber, B.-A.; Heikkila, R. E. *J. Pharmacol. Exp. Ther.* **1989**, *249*, 820.

163. Langston, J. W.; Irwin, I.; Langston, E. B.; Forno, L. S. *Science* **1984**, *225*, 1480-1482.

164. Peterson, L. A.; Caldera, P.; Trevor, A.; Chiba, A.; Castagnoli, Jr.; N. *J. Med. Chem.* **1985**, *28*, 1432.

165. Kalgutkar, A. S.; Dalvie, D. K.; Castagnoli, N. Jr.; Taylor, T. J. *Chem. Res. Toxicol.* **2001**, *14*, 1139.

166. De Master E. G.; Sumner, H. W.; Kaplan, E.; Shirota, F. N.; Nagasawa, H. T. *Toxicol. Appl. Pharmacol.* **1982**, *65*, 390.

167. Davies, B.; Bannister, R.; Sever, P. *Lancet* **1978**, *1*, 172.

168. Patek, D.; Hellerman, L. *J. Biol. Chem.* **1974**, *249*, 2373.

169. Silverman, R. B.; Hiebert, C. K.; Vazquez, M. L. *J. Biol. Chem.* **1985**, *260*, 14648.

170. Maycock, A. L.; Abeles, R. H.; Salach, J. I.; Singer, T. P. *Biochemistry* **1976**, *15*, 114.

171. Zhong, B.; Silverman, R. B. *J. Am. Chem. Soc.* **1997**, *119*, 6690.

172. Gates, K. S.; Silverman, R. B. *J. Am. Chem. Soc.* **1990**, *112*, 9364.

173. Turner, N. A.; Doyle, W. A.; Ventom, A. M.; Bray, R. C. *Eur. J. Biochem.* **1995**, *232*, 646.

174. Miners, J. O.; Birkett, D. J. *Gen. Pharmacol.* **1996**, *27*, 245.

175. Grant, D. M.; Tang, B. K.; Campbell, M. E.; Kalow, W. *Br. J. Clin. Pharmacol.* **1986**, *21*, 454.

176. Beedham, C. *Drug Metab. Rev.* **1985**, *16*, 119.

177. Stubley, C.; Stell, J. G. P.; Mathieson, D. W. M. *Xenobiotica* **1978**, *9*, 474.

178. Hodnett, C. M.; McCormack, J. S.; Sabean, J. A. *J. Pharm. Sci.* **1976**, *68*, 1150.

179. Krenitsky, T. A.; Tuttle, J. V.; Cattau, E. L.; Wang, P. *Comp. Biochem. Physiol.* **1974**, *49B*, 687.

180. Beedham, C. Molybdenum hydroxylases: biological distribution and substrate-inhibitor specificity. In *Progress in Medicinal Chemistry* (Ellis, G. P.; West, G. B.; Ed.), **1987**, Vol. 24, pp 85.

181. Chen, M.-L.; Chiou, W. L. *Drug Metab. Dispos.* **1982**, *10*, 706.
182. Holmes, R. S. *Biochem. Physiol.* **1978**, *61B*, 339.
183. Schoutsen, B.; de Tombe, P.; Harmsen, E.; Keijzer, E.; de Jong, J. W. *Adv. Exp. Med. Biol.* **1984**, *165*, 497.
184. Westerfield, W. W.; Richter, D. A. *Proc. Soc. Exp. Biol. Med.* **1949**, *71*, 181.
185. Bruder, G.; Heid, H.; Jarasch, E.-D.; Keenan, T. W.; Mather, I. H. *Biochim. Biophys. Acta* **1982**, *701*, 357.
186. Pritsos, C. A. *Chemico-Biol. Interact.* **2000**, *129*, 195.
187. Moriwaki, Y.; Yamamoto, T.; Takahashi, S.; Tsutsumi, Z.; Hada, T. *Histol. Hisotopathol.* **2001**, *16*, 745.
188. Krenitsky, T. A.; Spector, T.; Hall, W. W. *Arch. Biochem. Biophys.* **1986**, *247*, 108.
189. Krenitsky, T. A. *Biochem. Pharmacol.* **1978**, *27*, 2763.
190. Guerciolini, R.; Szumlanski, C.; Weinshilboum, R. M. *Clin. Pharmacol Ther.* **1991**, *50*, 663-672.
191. Levinson, D. J. and Chalker, D. *Arthritis Rheum.* **1980**, *23*, 77.
192. Relling, M. V.; Lin, J. S.; Ayers, G. D.; Evans, W. E. *Clin. Pharmacol. Ther.* **1992**, *52*, 643.
193. Rodrigues, A. D. *Biochem. Pharmacol.* **1994**, *48*, 197.
194. Beedham, C.; Bruce, S. E.; Critchley, D. J.; al-Tayib, Y.; Rance, D. J. *Eur. J. Drug Metab. Dispos.* **1987**, *12*, 307.
195. Kaye, B.; Rance, D. J.; Waring, L. *Xenobiotica* **1985**, *15*, 237.
196. Beedham, C.; Bruce, S. E.; Critchley, D. J.; Rance, D. J. *Biochem. Pharmacol.* **1990**, *39*, 1213.
197. Beedham, C.; Critchley, D. J.; Rance, D. J. *Arch. Biochem. Biophys.* **1995**, *319*, 481.
198. Wright, R. M.; Clayton, D. A.; Riley, M. G.; McManaman, J. L.; Repine, J. E. *J. Biol. Chem.* **1999**, *274*, 3878.
199. Yoshihara, S.; Tatsumi, K. *Arch. Biochem. Biophys.* **1997**, *338*, 29.
200. Bray, R. C.; Palmer, G.; Beinert, H. *J. Biol. Chem.* **1964**, *239*, 2667.
201. Porras, A. G.; Olson, J. S.; Palmer, G. *J. Biol. Chem.* **1981**, *256*, 9096.
202. Morpeth, F. F. *Biochim. Biophys. Acta* **1983**, *744*, 328.
203. Hille, R. *Chem. Rev.* **1996**, *96*, 2757.
204. Hille, R.; Sprecher, H. *J. Biol. Chem.* **1987**, *262*, 10914.
205. Howes, B. D.; Bray, R. C.; Richards, R. L.; Turner, N. A.; Bennett, B.; Lowe, D. J. *Biochemistry* **1996**, *35*, 1432.
206. Huber, R.; Hof, P.; Duarte, R. O.; Moura, J. J. G.; Moura, I.; Liu, M.-Y.; LeGall, J.; Hille, R.; Archer, M.; Romabo, M. J. *Proc. Natl. Acad. Sci. U.S.A.* **1996**, *93*, 8846.
207. Romabo, M. J.; Archer, M.; Moura, I.; Moura, J. J. G.; LeGall, J.; Engh, R.; Schneider, M.; Hof, P.; Huber, R. *Science* **1995**, *270*, 1170.
208. Xia, M.; Dempski, R.; Hille, R. *J. Biol. Chem.* **1999**, *274*, 3323.
209. Greenwood, R. J.; Wilson, G. L.; Pilbrow, J. R.; Wedd, A. G. *J. Am. Chem. Soc.* **1993**, *115*, 5385.
210. McWhirter, R. B.; Hille, R. *J. Biol. Chem.* **1991**, *266*, 23724.

211. Booth, V. H. *Biochem. J.* **1938**, *32*, 503.

212. Pelsey, G.; Klibanov, A. M. *Biochim. Biophys. Acta* **1983**, *742*, 352.

213. Feldman, R. J.; Weiner, H. *J. Biol. Chem.* **1972**, *247*, 260.

214. Palmer, G. *Biochim. Biophys. Acta* **1962**, *56*, 444.

215. Tomita, S.; Tsujita, M.; Ichikawa, Y. *FEBS Lett.* **1993**, *336*, 272.

216. Ruenitz, P. C.; Bai, X. *Drug Metab. Dispos.* **1995** *23*, 993.

217. Stubley, C.; Stell, J. G. P.; Mathieson, D. W. *Xenobiotica* **1979**, *9*, 475.

218. Nieman, Z. *Isr. J. Chem.* **1972**, *10*, 819.

219. Angelino, S. A. G. F.; Buurman, D. J.; van der Plas, H. C.; Muller, F. *Recl. Trav. Chim. Pays-Bas.* **1982**, *98*, 342.

220. Angelino, S. A. G. F.; van Valkengoed, B. H.; Buurman, D. J.; van der Plas, H. C.; Muller, F. *J. Heterocycl. Chem.* **1984**, *21*, 1057.

221. Taylor, S. M.; Stubley-Beedham, C.; Stell, J. G. P. *Biochem. J.* **1984**, *220*, 67.

222. Bunting, J. W.; Laderoute, K. R.; Norris, D. J. *Can. J. biochem.* **1980**, *58*, 49.

223. Rajgopalan, K. V.; Handler, P. *J. Biol. Chem.* **1964**, *239*, 2027.

224. Rosowsky, A.; Wright, J. E.; Holden, S. A.; Waxman, D. J. *Biochem. Pharmacol.* **1990**, *40*, 851.

225. Jordan, C. G. M.; Rashidi, M. R.; Laljee, H.; Clarke, S. E.; Brown, J. E.; Beedham, C. *J. Pharm. Pharmacol.* **1999**, *51*, 411.

226. Stubley, C.; Stell, J. G. P. *J. Pharm. Pharmacol.* **1980**, *32*, 51P.

227. Moder, K. P.; Leonard, N. J. *J. Am. Chem. Soc.* **1982**, *104*, 2613.

228. Ishigai, M.; Ishitani, Y.; Orikasa, Y.; Kamiyama, H.; Kumaki, K. *Arzneim.-Forsch./ Drug Res.* **1998**, *48(I)*, 429.

229. Acheampong, A. A.; Chien, D.-S.; Lam, S.; Vekich, S.; Breau, A.; Usansky, J.; Harcourt, D.; Munk, S. A.; Nguyen, H.; Garst, M.; Tang-Liu, D. *Xenobiotica* **1996**, *26*, 1035.

230. Bergmann, F.; Levene, L.; Govrin, H. *Biochim. Biophys. Acta* **1977**, *484*, 275.

231. Leonard, N. J.; Sprecker, M. A.; Morrice, A. G. *J. Am. Chem. Soc.* **1976**, *98*, 3987.

232. Bergmann, F.; Kwietny, H. W.; Levin, G.; Brown, D. J. *J. Am. Chem. Soc.* **1960**, *82*, 598.

233. Krenitsky, T. A.; Hall, M.; De Miranda, P.; Beauchamp, L. M.; Schaeffer, H. J.; Whiteman, P. D. *Proc. Natl. Acad. Sci. U. S. A.* **1984**, *81*, 3209.

234. Whiteman, P. D.; Bye, A.; Fowle, A. S. E.; Jeal, S.; Land, G.; Posner, J. *Eur. J. Clin. Pharmacol.* **1984**, *27*, 471.

235. Vere Hodge, R. A.; Sutton, D.; Boyd, M. R.; Harnden, M. R.; Jarvest, R. L. *Antimicrob. Agents Chemother.* **1989**, *33*, 1765.

236. Filer, C. W.; Allen, G. D.; Brown, T. A.; Fowles, S. E.; Hollis, F. J.; Mort, E. E.; Prince, W. T.; Ramji, J. V. *Xenobiotica* **1994**, *24*, 357.

237. Daniels, S.; Schentag, J. *J. Antiviral Chem. Chemother.* **1993**, *4*, 57.

238. Clarke, S. E.; Harrell, A. W.; Chenery, R. J. *Drug Metab. Dispos.* **1995**, *23*, 251.

239. Rashidi, M. R.; Smith, J. A.; Clarke, S. E.; Beedham, C. *Drug Metab. Dispos.* **1997**, *25*, 805.

240. Filer, C. W.; Allen, G. D.; Brown, T. A.; Fowles, S. E.; Hollis, F. J.; Mort, E. E.; Prince, W. T.; Ramji, J. V. *Xenobiotica* **1994**, *24*, 357.

241. Pue, M. A.; Pratt, S. K.; Fairless, A. J.; Fowles, S.; Laroche, J.; Giorgiou, P.; Prince, W. *J. Antimicrob. Chemother.* **1994**, *33*, 119.

242. Guo, X.; Lerner-Tung, M.; Chen, H.-X.; Chang, C.-N.; Zhu, J.-I.; Chang C.-P. *Biochem. Pharmacol.* **1995**, *46*, 1111.

243. Takekawa, K.; Sugihara, K.; Kitamura, S.; Ohta, S. *Xenobiotica* **2001**, *31*, 769.

244. Lee, S. C.; Renwick, A. G. *Biochem. Pharmacol.* **1995**, *49*, 1557.

245. Ratnayake, J. H.; Hanna, P. E.; Anders, M. W.; Duggan, D. E. *Drug Metab. Dispos.* **1981**, *9*, 85.

246. Sugihara, K.; Tatsumi, K. *Arch. Biochem. Biophys.* **1986**, *247*, 289.

247. Tatsumi, K.; Ishigai, M. *Arch. Biochem. Biophys.* **1987**, *253*, 413.

248. Hirao, Y.; Kitamura, S.; Tatsumi, K. *Carcinogenesis* **1994**, *15*, 739.

249. Masana, M.; de Toranzo, E. G. D.; Castro, J. A. *Biochem. Pharmacol.* **1984**, *33*, 1041

250. Sugihara, K.; Kitamura, S.; Tatsumi, K. *Drug Metab. Dispos.* **1996**, *24*, 199.

251. Nakasa, H.; Komiya, M.; Ohmori, S.; Kitada, M.; Rikihisa, T.; Kanakubo, Y. *Drug Metab. Dispos.* **1993**, *21*, 777.

252. Nakasa, H.; Koniya, M.; Ohmori, S.; Rikihisa, T.; Kiuchi, M.; Kitada, M. *Mol. Pharmacol.* **1993**, *44*, 216.

253. Massey, V.; Komai, H.; Palmer, G.; Elion, G. B. *J. Biol. Chem.* **1970**, *245*, 2837.

254. Spector, T.; Hall, W. W.; Krenitsky, T. A. *Biochem. Pharmacol.* **1986**, *35*, 3109.

255. Krasny, H. C.; Krenitsky, T. A. *Biochem. Pharmacol.* **1986**, *35*, 4339.

256. Zimm, S.; Collins, J. M.; O'Neill, D.; Chabner, B. A.; Poplack, D. G. *Clin. Pharmacol. Ther.* **1983**, *34*, 811.

257. Elion, G. B. *Fed. Proc.* **1967**, *26*, 898.

258. Baker, B. R.; Wood, W. F.; Kozma, J. A. *J. Med. Chem.* **1968**, *11*, 661.

259. Lewis, A. S.; Murphy, L.; McCalla, C.; Fleary, M.; Purcell, S. *J. Biol. Chem.* **1984**, *259*, 12.

260. Spector, T.; Ferone, R. *J. Biol. Chem.* **1984**, *259*, 10784.

261. Critchley, D. J. P.; Rance, D. J.; Beedham, C. *Xenobiotica* **1994**, *24*, 37.

262. Harrell, A. W.; Wheeler, S. M.; East, P.; Clarke, S. E.; Chenery, R. J. *Drug Metab. Dispos.* **1994**, *22*, 189.

263. Robertson, I. G. C.; Gamange, R. S. K. A. *Biochem. Pharmacol.* **1994**, *47*, 584.

264. Robertson, I. G. C.; Bland, T. J. *Biochem. Pharmacol.* **1993**, *45*, 2159.

265. Robertson, I. G. C.; Gamange, R. S. K. A. *Biochem. Pharmacol.* **1994**, *47*, 584.

266. Yin, S. J.; *Alcohol Alcohol Suppl.* **1994**, *2*, 113.

267. Jornvall, H.; Hoog, J. O.; Persson, B.; Pares, X. *Pharmacology* **2000**, *61*, 184.

268. Brent, J. *Drugs* **2001**, *61*, 979.

269. Cholerton, S.; Daly, A. K.; Idle, J. R. *Trends Pharmacol. Sci.* **1992**, *13*, 434.

270. Vasiliou, V.; Pappa, A. *Pharmacology* **2000**, *61*, 192.

271. Vasiliou, V.; Pappa, A.; Petersen, D. R. *Chemico.-Biol. Interact.* **2000**, *129*, 1.

272. Lindahl, R. *Crit. Rev. Biochem. Mol. Biol.* **1992**, *27*, 283.

273. Chang, T. K. H.; Weber, G. F.; Crespi, C. L.; Waxman, D. J. *Cancer Res.* **1993**, *45*, 1685.

274. Rekha, G. K.; Sreerama, L.; Sladel, N. E. *Biochem. Pharmacol.* **1994**, *48*, 1943.

275. Agarwal, D. P. *Pathol. Biol.* **2001**, *49*, 703.

276. Yamauchi, M.; Maezawa, Y.; Mizauhara, Y.; Ohata, M.; Hirakawa, J.; Nakajima, H.; Toda, G. *Hepatology* **1995**, *22*, 1136.

277. Agarwal, D. P.; Goedde, H. W. *Pharmogen* **1992**, *2*, 48.

278. Thomasson, H. R.; Crabb, D. W.; Edenberg, H. J.; Li, T.-K. *Behav. Gen.* **1993**, *23*, 131.

279. Helander, A.; Walzer, C.; Beck, O.; Balant, L.; Borg, S.; von Wartburg, J.-P. *Life Sci.* **1994**, *55*, 359.

280. Forrest, G. L.; Gonzalez, B. *Chemico.-Biol. Interact.* **2000**, *129*, 21.

281. Ahmed, N. K.; Felsted, R. L.; Bachur, N. R. *Xenobiotica* **1981**, *11*, 131.

282. Wirth, H.; Wermuth, B. *J. Histochem. Cytochem.* **1992**, *40*, 1857.

283. Kudo, S.; Ishizaki, T. *Clin. Pharmacokinet.* **1999**, *37*, 435.

284. Bachur, N. R. *Science* **1976**, *193*, 595.

285. Wong, J. M. Y.; Kalow, W.; Kadar, D. et al. *Pharmacogenetics* **1993**, *3*, 110.

Chapter 15

The Role of Sulfotransferases (SULTs) and UDP-Glucuronosyltransferases (UGTs) in Human Drug Clearance and Bioactivation

Michael W.H. Coughtrie[1] and Michael B. Fisher[2]

[1]*Department of Molecular & Cellular Pathology, University of Dundee, Ninewells Hospital & Medical School, Dundee DD1 9SY, Scotland, UK*

[2]*Discovery Pharmacokinetics, Dynamics, and Metabolism, Pfizer Inc., Eastern Point Rd, Groton, CT 06340*

GENERAL INTRODUCTION

The bulk of this book has focused on the biochemistry, mechanism, and tools pertaining to the cytochrome P450 superfamily of enzymes. This is appropriate, since this enzyme system is responsible for the metabolism of approximately 80% of pharmaceutical products. However, other enzymatic processes significantly affect the ADME of marketed drugs. Two of these, the sulfotransferases (SULT) and UDP-glucuronosyltransferases (UGT), will be discussed in this chapter. While parallels will be made between these enzymes and the CYP superfamily, it is valuable to keep in mind that significant differences exist. Important similarities and dissimilarities will be emphasized.

INTRODUCTION TO THE SULFOTRANSFERASE ENZYME FAMILY

Sulfation occurs widely in nature and this chapter concerns those sulfation reactions carried out by members of the cytosolic sulfotransferase (SULT) enzyme family. Baumann first discovered xenobiotic sulfation in the 1870s, by isolating phenyl sulfate from someone treated with carbolic acid. PAPS (3'-phosphoadenosine 5'-phosphosulfate) is the sulfuryl donor for all these cytosolic sulfation reactions (Figure 1).

The cytosolic SULTs are derived from a large gene superfamily. Almost 60 mammalian and avian cytosolic SULTs cDNAs have now been cloned and sequenced, and based on amino acid sequence analysis can be subdivided into 6 families.[1] A new nomenclature system for the cytosolic SULTs has recently been devised[1], and this system is used here. Application of heterologous expression technologies, particularly in bacteria such as *E. coli* and *S. typhimurium*, has facilitated the detailed examination of the properties of a number of SULT proteins and has provided the raw materials for antibody production and characterization.[2,3] These studies confirm that the majority of the SULT enzymes sulfate xenobiotic small molecules and/or important endogenous chemicals including many hormones and neurotransmitters.[4] The SULT1 and, to a lesser extent, SULT2 families are the most important for xenobiotic metabolism. In humans the SULT enzyme family comprises 11 isoforms of the SULT1, SULT2 and SULT4 families that are coded for by 10 distinct genes. Some properties of the various human SULTs are summarized in Table 1. Expression of SULTs in humans is regulated with respect to tissue type, development and hormonal influences, supporting proposed functional roles for some of the enzymes. In humans SULTs 1C2 and 1C4 are almost exclusively expressed in fetal tissues, although the function of these enzymes remains unclear. SULT1A3 (known as the catecholamine sulfotransferase) is absent from the adult human liver, but the major site of expression is the gastrointestinal tract.[5] This has implications for the metabolism of a number of drugs, for example salbutamol is almost exclusively sulfated in humans. Another important enzyme is estrogen sulfotransferase (SULT1E1), which has a very high affinity (low nM) for its natural substrate 17β-estradiol and for a major drug substrate 17α-ethinylestradiol. SULT1E1 is important in modulating estrogen action. For instance the enzyme is expressed in the endometrium and is very tightly regulated during the menstrual cycle, where is believed to moderate estrogenic stimulation of the endometrium at the time of implantation.[6] Recently a potentially important toxicological mechanism involving estrogen sulfation has been identified. A number of hydroxylated polychlorinated biphenyls are extremely potent inhibitors (K_i in the pM range) of SULT1E1, pointing to a mechanism through which these chemicals exert their well-known endocrine disrupting effects[7]. These data also raise the possibility that other drugs that inhibit this enzyme may manifest endocrine-disrupting side effects.

A problem currently taxing investigators in the field is to uncover the function of the most recently discovered SULT family, SULT4A. SULT4A1 cDNAs have been isolated so far from rat, mouse and human and the predicted proteins share a remarkable degree of similarity. The mouse and the rat proteins have identical amino acid sequences, and the human has only 6 amino acid differences out of

284, whilst sharing only around 30 % amino acid sequence identity with any other member of the family. The SULT4A1 proteins seem to be expressed only in the brain,[8] and despite significant effort no natural or xenobiotic substrate (or ligand) has yet been identified. These observations strongly implicate a specific and important function for this protein.

1   ASSAY OF SULFOTRANSFERASE ACTIVITY

Various methods and approaches may be applied to assess SULT enzyme activity. However it is the availability of a rapid, simple and substrate-independent assay that has greatly facilitated the study of these enzymes.[9] This assay is based on the fact that all cytosolic SULTs utilize PAPS as co-substrate and sulfuryl donor. PAPS is available commercially in the [^{35}S]-labeled form, and following incubation with substrate, enzyme source and unlabeled PAPS in a suitable buffer the unreacted PAPS can be precipitated using a combination of barium and zinc salts, leaving the labeled sulfate conjugate which can be quantified by liquid scintillation spectrometry.

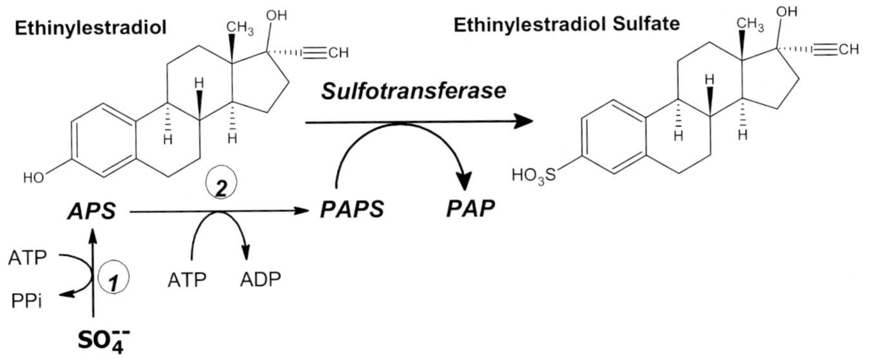

Figure 1. The Sulfotransferase Reaction

All sulfotransferase reactions utilise the co-substrate PAPS as sulfuryl donor. PAPS synthesis is carried out by a single bifunctional enzyme called PAPS synthetase (PAPSS) that contains both the ATP sulfurylase (1) and APS kinase (2) catalytic activities. In humans and mice, 2 isoforms of PAPSS exist that are the products of the *PAPSS1* and *PAPSS2* genes. It is believed that interindividual variation in the ability to synthesise PAPS, arising from genetic and/or environmental effects on PAPSS, may influence sulfation capacity.

Table 1. Some Properties of the Human Sulfotransferase Enzyme Family

SULT Isoform (Old Name)	Probe Substrates	Known Drug Substrates	Chromosomal Location	Major Sites of Expression
SULT1A1 (P-PST)	4-Nitrophenol	Acetaminophen, Troglitazone, Minoxidil, 4-OH Tamoxifen, Apomorphine	16p12.1-11.2	Adult Liver, Adult GI Tract, Adult Platelets, Placenta
SULT1A2	No Selective Substrate Known	-	16p12.1-11.2	?
SULT1A3 (M-PST)	Dopamine	Salbutamol, Dobutamine	16p11.2	Adult GI Tract, Adult Platelets, Adult Brain, Placenta, Fetal Liver
SULT1B1	No Selective Substrate Known	-	4q11-13	Adult Liver, Adult GI Tract, Fetal GI Tract
SULT1C2 (ST1C1)	No Selective Substrate Known	-	2q11.2	Fetal Kidney, Fetal Lung, Fetal GI Tract
SULT1C4 (ST1C2)	No Selective Substrate Known	-	2q11.2	Fetal Kidney, Fetal Lung
SULT1E1 (EST)	17β-Estradiol	17α-Ethinylestradiol	4q13	Fetal Liver, Fetal Lung, Fetal Kidney, Adult Liver, Endometrium
SULT2A1 (HST, DHEA-ST)	Dehydroepiandrosterone	Budenoside, Dehydroepiandrosterone, Pregnenolone	19q13.3	Fetal Adrenal, Fetal Liver, Adult Liver, Adult Adrenal
SULT2B1	Cholesterol (2B1b), Pregnenolone (2B1b)	-	19q13.3	Adult Skin, Prostate, Placenta
SULT4A1	No Substrate Known	-	22q13.1-13.2	Brain

Although this PAP[35]S assay is extremely useful and broadly applicable, it does suffer a number of drawbacks. First, it is not particularly sensitive - in our hands the limit of detection is approximately 10 pmol reaction product. Second, PAP[35]S is expensive, compounded by the fact that ^{35}S is a relatively short-lived ($t_{1/2}$ = 87.4 days) isotope. Third, it has been reported that the assay is not suitable for acids,[10] although in our hands this did not appear to be a problem.[11]

It is also possible to measure SULT activity with radioactive acceptor substrates. This is most often used for steroid substrates such as dehydroepiandrosterone, 17β-estradiol and progesterone, although certain xenobiotics such as paracetamol, 4-nitrophenol and 1-naphthol have also been used.[12,13] In these cases, extraction of the unreacted (more lipophilic) acceptor substrate into a suitable organic solvent provides a simple means of separating the radioactive sulfate conjugate.

As working with radioactive material becomes more difficult, there is increasing pressure to develop non-radioactive, higher-throughput assays. A recently published method utilizes microtitre plates to assay SULT1A activity towards 2-naphthol.[14] The assay detects the synthesis of 2-naphthylsulfate from 2-naphthol and PAPS, and the addition of 4-nitrophenylsulfate to the assay provides a PAPS-regenerating system. Compared with other SULT assay methods, the second reaction product (PAP, a SULT inhibitor) does not accumulate, although the method is rather limited in sensitivity. The relative affordability of mass spectrometers as detection systems means that sulfation assays directly detecting individual sulfate conjugates can be readily devised, and quantitation is possible where suitable standards are available. There are many fluorescent compounds that are substrates for SULTs, and it is possible to devise disappearance assays based on these compounds. However, for accurate and reliable quantification of activity in tissue samples it is of course necessary to use substrates that are isoform selective. The (difficult) challenge will therefore be to identify or design compounds that are (a) highly fluorescent, (b) that are rapidly sulfated and (c) that are highly selective for individual SULT isoforms.

2   MOLECULAR TOOLS AVAILABLE TO STUDY SULFOTRANSFERASES

Understanding of the role of SULTs in drug metabolism *in vitro* has been advanced through the application of many experimental approaches. The closest systems to the *in vivo* situation are isolated perfused organs,[15-17] tissue slices[18-20] and freshly isolated cells (e.g. hepatocytes),[21-25] and all of these have been applied in studying the sulfation of drugs and other compounds. However, the most significant advances have been made over the last decade since the availability of SULT cDNA clones became widespread. cDNAs representing more than 50 SULTs from avian and mammalian species have been cloned, sequenced and in many cases characterized. The SULT1 and SULT2 groupings make up the major families, and are the most important for xenobiotic metabolism. The human SULT enzyme family is the best characterised to date, and comprises 11 different isoforms from families 1, 2 and 4, coded for by 10 distinct genes (SULTs 2B1a and 2B1b are generated by alternate splicing of the first exon of SULT2B1) that share many common structural features (see ref. 1,26). The properties of the various human SULTs are summarised in Table 1. In addition to facilitating the classification of the various SULTs, the cDNA clones have been vital for the detailed characterization of these enzymes. Heterologous expression systems, essentially devoid of endogenous SULT activity, have been developed for all the human SULTs. Early work using transient transfection of cDNAs in e.g. COS cells (e.g., ref. 27), or mammalian cell lines (e.g. V79) expressing SULTs[28,29] allowed confirmation of substrate specificities of the recombinant proteins predicted from studies with tissue cytosols or with purified enzymes. As soluble cytosolic enzymes, the SULTs are readily expressed at high levels in prokaryotic systems such as *E. coli* (e.g., ref. 30) and *S. typhimurium* (e.g., ref. 31), and the recombinant proteins appear to faithfully represent the enzymes that can be studied in tissue samples and in eukaryotic expression systems.[3,32] These expression systems have provided a wealth of information on the substrate specificities of the various SULTs, and it is now readily possible to take a compound that is known to be sulfated and screen it against a bank of recombinant SULTs in order to determine the isoform(s) most likely responsible for its sulfation. Such studies are facilitated by the commercial availability of recombinant human SULTs.

In addition to facilitating enzymatic studies, the expression of SULTs in *E. coli* has provided the raw material for the production and characterization of antibodies against various SULT isoforms. The enzymes may be readily purified in large quantities from *E. coli* using e.g. a combination of ion exchange chromatography and affinity chromatography on ADP-agarose[3] and can be used as immunogens for antibody production (e.g., refs. 33,34). However, the high degree of amino acid sequence identity between various SULTs presents a significant problem for the production of truly isoform-specific antibodies, and most antibody preparations produced from immunization with the whole purified recombinant proteins cross-react (to varying degrees) with other SULTs. For example, the 3 SULT1A enzymes in humans, SULTs 1A1, 1A2 and 1A3, share at least 93% amino acid sequence identity making it extremely difficult to produce molecular probes of any sort (especially antibodies) to distinguish between them. Various approaches have been taken to try and circumvent this problem, including the production of anti-peptide antibodies[5,35] or adsorption of antibodies with cross-reacting enzymes,[36] but the lack of availability of high quality, high-specificity antibodies is often a limiting factor in studying SULT expression. Nevertheless, such reagents have provided researchers in the field with the opportunity to study many facets of sulfotransferase biology. For example, expression patterns of SULT isoforms in different human tissues involved in drug metabolism have been (qualitatively) assessed. From these studies it is clear that in the liver, the major SULT enzyme is likely to be SULT1A1. From a functional perspective, this makes sense: SULT1A1 is a very broad specificity enzyme, with a low K_m towards a range of phenolic substrates.[37] It is therefore an important part of the detoxification process, in particular "mopping up" many Phase 1 metabolites as well as providing front-line chemical defence against drugs (e.g. acetaminophen (paracetamol), troglitazone) as well as dietary and environmental chemicals including phenols and hydroxylated aromatic amines. The other major member of the phenol sulfotransferase (SULT1A) family, SULT1A3, is not expressed at any significant level in the adult human liver. However, this enzyme is one of the major SULTs expressed in the upper gastrointestinal tract and is therefore important for the first pass metabolism of drugs administered orally. SULT1A3 displays considerable specificity towards catecholamines and related chemicals,[37-39] and is involved in the metabolism of a number of clinically important drugs in humans such as salbutamol and dobutamine. The enzyme SULT1B1 is also expressed at high levels in the gastrointestinal tract and this is probably its major site of expression. No specific substrate is known for this enzyme. The liver expresses the steroid metabolising SULTs 1E1 and 2A1, which are also expressed in the gut. Intestinal first pass sulfation (probably by SULT1E1) is believed to be the major factor in the low bioavailability of 17α-ethinylestradiol, the estrogenic component of the combined oral contraceptive pill.[40]

Studies on the expression of various SULT enzymes during human development have shown that some of the enzymes are expressed quite differently early in life compared with the adult, and this may have implications for drug metabolism in infants and children as well as during pregnancy. The most significant difference identified so far is that the SULT1A3 enzyme is expressed at high levels in the human fetal liver, and its expression appears to be switched off around the time of birth.[33] This is in sharp contrast to the adult liver, where this enzyme is not expressed to any significant extent. The estrogen sulfotransferase, SULT1E1, is also expressed preferentially in human fetal liver, at about 4 times the level of the adult[41].

The rapidly developing techniques of structural biology and computational chemistry have recently been applied to the sulfotransferase enzymes. These studies have dramatically increased our appreciation of the mechanism of the sulfation reaction, and have provided important clues to the substrate specificities of various SULT enzymes. The first X-ray crystal structure to be solved was of the monomeric mouse estrogen sulfotransferase, SULT1E1.[42] Since then, structures of 3 other (dimeric) cytosolic SULTs (human isoforms 1A3,[37,43] 2A1[44] and 1E1[45]) have been solved, along with the related insect enzyme retinol dehydratase[46] and the sulfotransferase domain of the Golgi enzyme human heparan sulfate *N*-deacetylase/*N*-sulfotransferase 1.[47] Together, these studies clearly demonstrate the highly conserved nature of the sulfotransferase structure, which is characterized by a 5-stranded beta-sheet forming the core of the PAPS binding domain and active site. The active site is designed to bind the common sulfuryl donor PAPS and acceptor substrate in appropriate proximity to facilitate the SN_2–type reaction mechanism.[48,49] The regions of the SULT molecules involved in the binding of PAPS had been largely demonstrated before the availability of the X-ray crystal structures, particularly by the elegant studies from Strott and colleagues (e.g. refs. 50-52) and by mutational analysis that confirmed the role of the absolutely conserved catalytic centre of the enzyme (a histidine residue at position 108 (in the human SULT1A enzymes) that acts as the catalytic base).[48]

This newly-available structural information has greatly advanced our understanding of the molecular basis of SULT enzyme specificity. For example, mutagenesis experiments had revealed that a variable region within the SULT amino acid sequence (amino acids 143-151 in the SULT1A sequences) strongly influenced the specificity of the SULT1A enzymes,[39,53,54] with Glu_{146} playing a key role in the high degree of selectivity for catecholamines demonstrated by SULT1A3. It seems that the influence of Glu_{146} is more wide-reaching however, as it appears to influence the orientation of many compounds in the SULT1A3 active site, thereby providing an explanation for the selectivity of SULT1A3 for catechols over phenols.[37] The only other major report of the molecular basis of SULT specificity indicates that specificity of mouse SULT1E1 for 17β–estradiol (in preference to DHEA) arises from a "gating" phenomenon. Petrotchenko *et al.*[55] showed that Tyr_{81} and Phe_{142} together form a narrow gate permitting entry of 17β-estradiol to the active site whilst preventing access by DHEA (with its 19-methyl group). Phe_{142} is absolutely conserved amongst SULTs, and Tyr_{81} alone was found to regulate the gating phenomenon.

3   ROLE OF SULFATION IN BIOACTIVATION OF DRUGS AND OTHER XENOBIOTICS

Although sulfation, along with other Phase 2 metabolism, is generally perceived as a predominantly detoxification pathway, there are a number of important instances where sulfate conjugates are more biologically active than the parent molecules.

For some drugs, sulfation is an essential step in their pharmacology, and for many environmental chemicals sulfation catalyses the terminal step in their activation to mutagens and/or carcinogens. Minoxidil (2,4-diamino-6-piperidinylpyrimidine 3-oxide) was originally developed as an antihypertensive drug, but was soon found to have hair growth-promoting properties.[56] Both activities require metabolic activation by sulfation to reveal the pharmacological effects, which include smooth muscle relaxation.[57] Minoxidil is metabolised by a number of human SULTs, including 1A1[58], 1A2[59] and 1A3[60] isoforms, and sulfation takes place in the scalp.[61,62] It is noteworthy that some individuals refractory to the hypertrichotic effects of topically-applied minoxidil have low levels of minoxidil sulfotransferase activity in the scalp. Responders have significantly higher levels of activity than non-responders.[63] Other examples of drugs bioactivated by sulfation are the diuretic/anti-hypertensive CRE 1094 (2-(*p*-fluorophenoxy), 1-(*o*-hydroxyphenyl)-ethane)[64] and the natriuretic cicletanine.[65]

The pioneering work of the Millers (reviewed in ref. 66) laid the foundations for our understanding of the role of sulfation in chemical carcinogenesis. The requirement for sulfation in the mutagenicity and/or carcinogenicity of dietary and environmental chemicals has been demonstrated both *in vivo* and *in vitro*. For example, safrole, and related compounds such as estragole, are known rodent hepatocarcinogens. With these and similar chemicals, activation occurs firstly via metabolism at the 1' allylic-benzylic position of the allyl group, forming 1'-hydroxy metabolites, which are more potent carcinogens than the parent molecule and which are metabolised by sulfation.[67] Treatment of mice with pentachlorophenol (a potent inhibitor of phenol SULT activity) prior to administration of 1'-hydroxysafrole results in a substantial reduction in DNA adduct formation and subsequent tumorigenesis.[68] Similarly the brachymorphic mouse, which is genetically defective in PAPS synthesis,[69] is resistant to tumour formation when fed compounds for which sulfuric acid ester metabolites are the ultimate carcinogens.[68] Other evidence includes the observation that male rats are far more susceptible to hepatocarcinogenesis following exposure to aromatic amines than female rats, because male rat livers express much higher levels of expression of the phenol SULTs which sulfate hydroxyarylamines and arylhydroxamic acids (SULT1A1, SULT1C1).[70,71]

In addition to these *in vivo* studies, there is a host of evidence from experiments *in vitro* to implicate sulfation and particular SULT isoforms in the metabolism and bioactivation of chemical procarcinogens and promutagens. Liver cytosol and purified human liver phenol sulfotransferases can metabolise a number of *N*-hydroxyarylamines, *N*-hydroxy-heterocyclic amines and arylhydroxamic acids,[72] and the use of cDNA-expressed human SULTs has demonstrated the metabolism and bioactivation of dietary mutagens such as *N*-OH-PhIP (preferentially by SULT1A1 and SULT1A2)[73] and benzylic alcohols of polycyclic aromatic hydrocarbons (preferentially by SULT2A1).[74] These and other studies have been extensively reviewed recently.[2,75,76] A very sensitive and specific *in vitro* model system for assessing SULT-dependent activation has been develop by Glatt's group, where the cloning of human and rat SULT cDNAs directly into *Salmonella typhimurium* allows the (highly sensitive) detection of the genotoxicity of unstable sulfuric acid esters generated directly in the indicator cell, by a modified Ames test.[77] The use of these and other *in vitro* systems, such as the expression of SULTs in V79 cells[29,77,78] (also a standard cell line for genotoxicity testing), has been of considerable value in furthering our appreciation of the importance of individual SULTs in bioactivating dietary and environmental chemicals.

4   SULFOTRANSFERASE PHARMACOGENETICS

The activity of SULT enzymes expressed in human tissues and sulfation of xenobiotics *in vivo* varies widely in the human population. The pioneering early work mainly concerned members of the phenol SULT enzyme subfamily, and was conducted mostly on the platelet enzymes, and has benefited from considerable attention elsewhere (e.g. refs. 1,79-81).

SULT1A1 (or "thermostable" phenol SULT) is probably most significant and widely studied polymorphic SULT, since it is involved in the metabolism and detoxification of many drugs and other xenobiotics as well being responsible for the bioactivation of a host of dietary and environmental promutagens/procarcinogens.[2] Variation in this enzyme activity was shown to be regulated by inheritance,[82] and was associated with sulfation of drugs such as paracetamol.[83,84] It was also discovered that the thermal stability of this enzyme activity was under the control of a genetic polymorphism.[85] The availability of antibodies recognising specific human SULTs allowed demonstration that variability in SULT1A1 activity strongly correlates with the level of enzyme protein.[34,86] Cloning of human SULT1A1 cDNAs provided the first evidence that variant alleles (in the coding region at least) of the *SULT1A1* gene existed in the population. The GenBank/EMBL sequence databases contain at least 8 SULT1A1 cDNA sequences, mainly representative of the 2 most common *SULT1A1* alleles, *SULT1A1*1* (allele frequency in Caucasians approximately 0.66) and *SULT1A1*2* (allele frequency in Caucasians approximately 0.32).[87,88]

Important studies from Weinshilboum's laboratory clearly demonstrated that platelet phenol sulfotransferase enzyme activity and thermal stability of the protein were closely related to *SULT1A1* genotype. These data clearly demonstrated that individuals homozygous for *SULT1A1*2* possess about 15% of the platelet phenol sulfotransferase activity of *1A1*1/1A1*2* heterozygotes or of individuals homozygous for *1A1*1*, and that SULT1A1 thermal stability was associated with *SULT1A1* genotype.[87,88] Another study confirmed the reduced platelet SULT activity observed in *1A1*2* homozygotes,[89] although these authors showed only a 50% reduction in activity. It is still not entirely clear how the single amino acid change ($Arg_{213}His$) responsible for SULT1A1*2 causes a reduction in protein expression, enzyme activity and thermal stability, although the thermal stability of SULT1A1*2 is less than that of the wild-type SULT1A1*1 enzyme,[88,90,91] and the 1A1*2 protein appears to have a shorter biological half-life than SULT1A1*1.[92]

A number of recent studies have examined the molecular epidemiology of the *SULT1A1* polymorphism. Frame *et al.*[93] showed that low levels of platelet phenol SULT(1A1) activity was more prevalent in patients with colorectal cancer, and another study indicated that *SULT1A1*1* homozygosity is associated with reduced risk of colorectal cancer.[94] In contrast, however, a large study in breast cancer showed no overall effect of *SULT1A1* genotype on disease risk.[95] Another breast cancer patient cohort showed increased risk associated with *SULT1A1*1* genotype when combined with consumption of well-cooked meat, a major source of heterocyclic amines,[96] many of which require bioactivation by the action of SULTs including SULT1A1.[2,97] This same study,[96] however, found that overall risk of breast cancer was associated with the *SULT1A1*2* allele.

Sequence variants of a number of other human SULTs have been identified, but little or no functional or epidemiological information on these exists.

5   INTRODUCTION TO THE UDP-GLUCURONOSYLTRANSFERASE ENZYME FAMILY

Improvements in molecular techniques and the relative increase in importance of phase 2 metabolic processes have led to a renaissance in our understanding of UGT biochemistry and enzymology. The goal of this chapter is to incorporate as much information as possible that has been learned about the UGT superfamily of enzymes to facilitate its application to drug discovery and development. This information has been the result of several decades of intense research, and several excellent review articles have summarized this information in a few dozen succinct pages. Some recent outstanding reviews include Radominska,[100] King,[101] Tukey,[102] Burchell,[103] Fisher,[104] and should be consulted for more in depth information about topics in this chapter.

The UGTs catalyze the addition of a glucuronic acid moiety to a nucleophilic substrate using UDP-glucuronic acid as cofactor.[189] As with the CYPs, the

UGT superfamily possesses a large multiplicity of forms, each with a broad and overlapping substrate selectivity, and wide tissue distribution.[104] This multiplicity, while probably resulting from evolutionary pressure to metabolize the array of xenobiotics that an organism may be exposed to, also resulted in a very confusing nomenclature. Fortunately, a rational nomenclature system has since been established.[105] Three subfamilies exist to date: UGT1A, 2B, and 2A, with the latter localized in olfactory tissue and probably not involved in drug metabolism to a substantial degree.

Interesting differences between subfamilies exist in terms of genetic multiplicity. In the UGT2B subfamily, a separate gene appears to code for each human enzyme (UGT2B4, 7, 10, 11, 15, and 17), similarly to the CYPs. However, multiplicity in the UGT1A subfamily (UGT1A1, 3, 4, 5, 6, 7, 8, 9, and 10) appears to occur through a concept called exon sharing. Each isoform in this subfamily share the same last 4 exons (exon 2-5), and therefore are identical in their C-terminal 245 amino acids. Only unique first exons differentiate them, meaning that diversity in amino acid sequence exists only in the N-terminal half of the protein.[102] Each first exon contains a unique 5' regulatory region, and unique mRNA expression occurs via splicing of the 3' end of the appropriate exon 1 to the 5' end of shared exon 2. The identical C-terminal half of the UGT1As shares high homology with the C-terminal half of all other UGTs, suggesting that this region is responsible for binding of UDP-glucuronic acid while the N-terminus determines nucleophilic substrate selectivity. The implications of this will be discussed.

6   REACTION MECHANISM/STRUCTURE

The UGTs are reported to be distantly related to vertebrate and invertebrate proteins that catalyze the addition of glycosyl groups from UDP-sugars to small molecules.[102] The transfer of glucuronic acid from UDPGA to the nucleophilic substrate occurs with inversion of stereochemistry to yield a glucuronide in the β configuration (Figure 2). Indeed, this carbon atom that has undergone inversion contains the anomeric proton, which has been used diagnostically for structural confirmation by NMR.[106] Evidence to date examining a series of phenols of varying nucleophilicity supports an SN2 mechanism of addition.[100] Specific amino acid modifiers and mutagenesis have implicated histidine, arginine, cysteine, and either a glutamate or aspartate in the catalytic mechanism.[107,190,191] Further work is necessary to conclusively identify the catalytic amino acids.

Any atom with sufficient nucleophilicity is capable of being glucuronidated. For example, heteroatoms such as oxygen in alcohols, phenols, and acids; nitrogen in primary, secondary, and tertiary amines (the latter results in a quaternary ammonium glucuronide); and nucleophilic carbon atoms, such as in phenylbutazone.[104] A wide range in substrate size has been observed, from ethanol[108] to tacrolimus and cyclosporin,[109] a molecular weight range of nearly 2 orders of magnitude. Contrary to most CYP-generated metabolites, glucuronide formation is almost always reversible. The enzyme β-glucuronidase catalyzes the reverse reaction.[110]

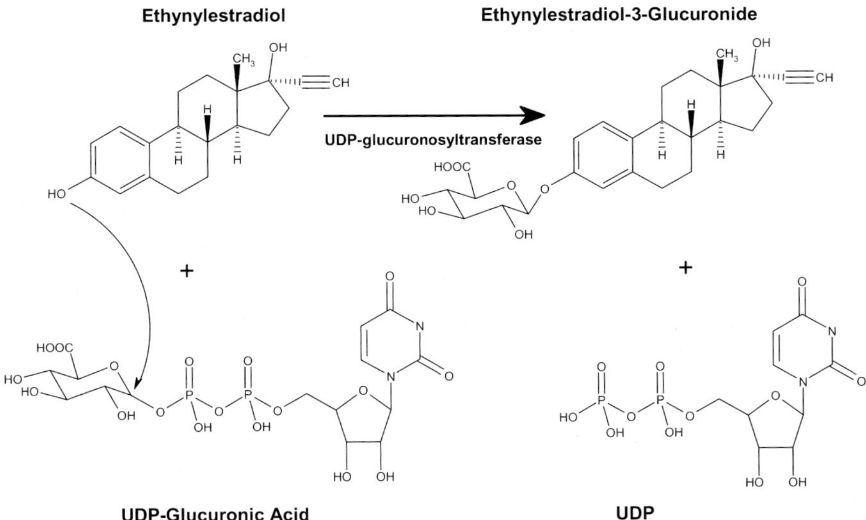

Figure 2. The reaction mechanism of the UDP-glucuronosyltransferases.

One important difference between UGTs and CYPs is their subcellular localization. While both are endoplasmic reticulum membrane associated proteins, the CYP active site is exposed on the cytosolic side of the ER while the UGTs are lumenal. This has important implications for assay methodologies, as discussed below.[111] Sequence features that impart topological characteristics to the protein are discussed by Tukey and Strassburg, including an N-terminal signal sequence and a C-terminal membrane spanning region.[102]

A second distinction of the UGTs is recent evidence that these enzymes may exist as dimers, both hetero- and homodimers, as evidenced by chemical cross-linking experiments, immunoprecipitation, and other observations.[112,113,114] Mutagenesis experiments have shown that the N-terminus plays an important role in dimerization, and that these dimers are catalytically active. In some cases, the catalytic activity of the dimers is different than that observed in either monomer alone. For example, when guinea pig UGT2B21 and 22 were coexpressed, significant stimulation of morphine-6-glucuronide was observed relative to each form alone. Also, coexpression of wild type and inactive mutated UGT1A1 resulted in significantly lower than expected enzyme activity, indicating that heterodimerization in this instance can result in a dominant negative effect of some mutant forms, such as what would be observed in a heterozygotic individual. An important implication of this phenomenon pertains to the use of recombinant UGT enzymes. While a systematic study has not yet been performed, the above data suggests that a single expressed UGT may not reflect the rate and/or metabolite profile that is observed for that enzyme in a mixture of different UGTs. Therefore, care must be taken in extrapolating the data from expressed enzymes.

7   ENZYMOLOGY AND METHODOLOGIES

As stated above, the UGTs possess a lumenal localization. Since the active site is in the ER lumen, the ER membrane imparts a diffusional barrier to substrate and cofactor access and metabolite and by-product removal. In vivo, specific transporters probably mediate the trafficking of these components. However, in a microsomal incubation, this diffusional barrier must be overcome in order to more closely mimic the in vivo situation. This is important, since reaction byproducts, such as UDP,[115,189] have been shown to inhibit glucuronidation. Classical methods of overcoming this diffusional barrier include the addition of detergents, albumin, or lipids.[111,192] The increase in activity observed with these modifiers has been described as unmasking a "latency" in activity. The disadvantage of detergents and lipid treatment is that CYP activity is usually inhibited, either because the additive is a substrate for CYP or that P450 oxidoreductase: CYP interactions are disrupted. Additionally, evidence suggests that detergent effects are a balance between activity stimulation and inhibition, probably due to membrane and protein perturbations. Finally, different isoforms, and even different substrates of an isoform, often require an optimization of detergent type and amount.[193]

Recently, my lab and others have begun using the pore-forming peptide alamethicin instead of detergents to overcome this diffusional barrier.[111] Alamethicin is a commercially available, 20-amino acid peptide that is known to form well-defined pores in membrane bilayers. This method offers distinct advantages over detergents in that conditions appear to be isoform independent and analytically show less interference than detergent; and since alamethicin has been shown to not inhibit CYP activity, it can be used to examine coupled CYP:

UGT metabolism.[121] The characteristics of this methodology have recently been reviewed.[104]

Similarly to the CYPs, the UGTs maintain a wide tissue distribution, with liver possessing the highest levels of activity, in general.[102] Kidney and small intestine may also contain high levels of certain isoforms.[116] Forms exist that possess only extrahepatic localization, such as UGT1A8, 1A7, 1A10.[117] Table 2 shows a qualitative summary of the present understanding of tissue distribution of UGT isoforms. An important feature of the extrahepatic expression of UGTs is that the intestinal expression is extremely polymorphic, with some individuals lacking a certain form entirely.[117] This genetic variability will be discussed later.

Table 2. Some Properties of the Human UDP-glucuronosyltransferase Enzyme Family

UGT Isoform	Probe substrates	Major sites of expression
UGT1A1	Bilirubin β-estradiol (3-gluc) 17α-ethynylestradiol (3-gluc)	Liver, intestine
UGT1A3	Hyodeoxycholic acid (COOH-gluc)?	Liver
UGT1A4	Imipramine, trifluoperazine	Liver
UGT1A5	No substrates known	Unknown
UGT1A6	Serotonin	Liver, brain
UGT1A7	Benzo[a]pyrene metabolites	Stomach
UGT1A8	No selective substrate known	Intestine
UGT1A9	propofol	Liver, kidney
UGT1A10	No selective substrate known	Stomach, intestine
UGT2A1	No selective substrate known	Olfactory epithelium
UGT2B4	Hyodeoxycholic acid (6-gluc)	Liver
UGT2B7	AZT, morphine	Liver, kidney, intestine
UGT2B10	No selective substrate known	Liver, prostate, mammary
UGT2B11	No selective substrate known	Liver, kidney, prostate, adrenal
UGT2B15	(S)-oxazepam	Liver, prostate
UGT2B17	Dihydrotestosterone?	Prostate

In addition to the ER lumenal localization, evidence exists that nuclear membranes contain significant UGT activity. Immunohistochemical and electron microscopy analysis showed the presence of enzyme in the ER and nuclear envelopes.[118] Recently, Radominska-Pandya and coworkers have performed a further characterization of the subcellular localization of UGT1A6 and 2B7. This data convincingly shows that both human liver and UGT expressed in HK293 cells possess localization in the nuclear membrane, and that, although less than ER, the nuclear membrane and whole nuclei contain significant activity toward fatty acid derivatives. The possible significance of this localization, such as for cellular defense or to regulate levels of nuclear receptor ligands, is discussed.[119]

A list of commonly used substrates for the various isoforms is shown in Table 2. Many endogenous and xenobiotic compounds are able to be glucuronidated by the UGTs. While many of these are substrates for more than one UGT isoform, a partial list of selective substrates has begun to evolve. For example, bilirubin and estradiol-3-glucuronidation are selective for UGT1A1,[101,111] propofol glucuronidation appears to be catalyzed by UGT1A9,[120] and AZT and morphine glucuronidation is selective for UGT2B7.[119] The glucuronidation of endogenous substrates has been shown to be physiologically significant for compounds such as bilirubin, estradiol and other steroids, and some fatty acids.[119]

A typical reaction "recipe" includes a substrate (dissolved in organic solvent and added at $\geq 1:100$ dilution), enzyme source (tissue microsomes or recombinant enzyme), a latency-removing agent (although this is sometimes omitted), buffer (pH 7-7.5), and cofactor (UDPGA 2.5-5 mM).[104] Alamethicin is usually added at a ratio of 50 µg/mg microsomal protein, although lower amounts have been used. Detergent type and amount should be empirically determined. Physiologically relevant buffer pH for the ER lumen is 7.1, but classical pH of 7.4 appears to yield qualitatively similar data.[111] A pH "optimum" may exist outside this range, but relevance to in vivo conditions should outweigh the use of non-physiological conditions to generate increased reaction rates. Our recent work has shown that, at least for morphine glucuronidation in human liver microsomes, UGTs are somewhat insensitive to organic concentration up to 2%. Sometimes, a β-glucuronidase inhibitor, saccharolactone (saccharic acid- 1,4-lactone, 2-5 mM), is added in order to measure the true "intrinsic" activity of the UGT without the confounding factor of hydrolysis.[104,111] For example, if one was trying to generate V_{max} values for an enzymatic process, it may be desirable to inhibit glucuronide hydrolysis. However, if one is attempting an *in vitro-in vivo* correlation, it is intuitively most appropriate to allow formation and hydrolysis to occur at their physiologically relevant rates and measure a net formation.

One source of enzyme that can be used "as is" for determining UGT activity is human hepatocytes. Hepatocytes are a whole cell system, and therefore the requisite transport systems and cofactor catabolic processes are presumably active. The availability of cryopreserved human hepatocytes has made hepatic metabolite generation very convenient, albeit expensive. More detailed kinetic and enzymological studies still require microsomal preparations.

In some cases, synthetic standards for the glucuronide metabolites are commercially available, such as for metabolites of morphine, codeine, estradiol, and testosterone, among others. However, often this is not the case, and other detection methods must be utilized to either qualitatively or quantitatively analyze UGT activities. Quantitation can be performed using either scintillation counting or HPLC linked to a radiomatic detector, if radiolabeled UDPGA or substrate is used. HPLC-MS is a convenient method for identifying glucuronides. An increase of 176 amu in M+H is indicative of a glucuronide, as is the presence of aglycone due to mass spectral source-induced loss of the conjugate. Constant neutral loss of m/z 176 is also a generic, useful means for detecting glucuronides. Identification of the site of attachment by LCMS is more difficult, since the most labile fragmentation is usually also loss of the conjugate to give aglycone. NMR is frequently utilized, since the anomeric proton is diagnostic of a glucuronide and the protons from the site of attachment will be either lost and/or shifted.[121]

Immunological techniques have been extremely valuable in the CYP and SULT fields as described elsewhere in this chapter and book, both in terms of immunoinhibition as a mechanism for selective inhibition of a CYP isoform and quantitation by SDS-PAGE and immunoblotting with selective antibodies. These techniques are, by comparison, less straightforward for UGTs. The lumenal localization of the UGTs may limit access of inhibitory antibodies to the enzyme active site. A few reports exist of immunoinhibition of UGTs,[122,194] although it is unclear how universal this method will be. Generating selective antibodies for a UGT isoform has been complicated by the high degree of homology between the forms, especially the UGT1A subfamily, which shares the same 245 C-terminal amino acid sequence. This feature has been utilized to generate an antibody against the entire subfamily. More recently, selective antibodies against UGT1A1, 1A6, and 2B7 have become commercially available, and used for isoform identification in human tissues.[123]

8   REGULATION

Information concerning the regulation of specific UGT isoforms has lagged behind that of the CYPs. However, clinically significant drug-drug interactions (see below) have supported the notion that UGTs are indeed inducible by other xenobiotics. More recently, in vitro evidence has been gathered to begin to identify inducers for specific forms and the mechanisms of regulation. For example, inducers such as chrysin, t-butylhydroquinone, and tetrachlorodibenzodioxin have been shown to induce UGT1A6, 1A9, and 2B7 in Caco-2 cells.[124] Flavonoids, such as chrysin, also induce UGT1A1, and a distinct structure-activity relationship for this effect has been identified.[125] UGT1A1 also appears to be inducible in human hepatic systems by phenobarbital, oltipraz, 3-methylcholanthrene, and phenytoin.[126,127,195] In fact, the phenobarbital response element was identified in human UGT1A1, and was shown to be activated by the human constitutive active receptor (CAR).[128] This induction of UGT1A1 by phenobarbital explains the efficacy of this barbiturate in the treatment of hyperbilirubinemia resulting from Gilbert's syndrome, since bilirubin is cleared by UGT1A1-mediated metabolism.

The ontogenic regulation of UGTs has recently been reviewed, with general trends suggesting postnatal and developmental increases.[129] The low levels of UGT1A1 immediately after birth explain neonatal jaundice, since bilirubin clearance is catalyzed by this isoform. The endobiotic metabolizing UGT2B15 and 2B17 are differentially regulated by growth factors and cytokines.[130] Similarly, hepatic inflammation, but not fibrosis, has been shown to correlate with a decrease in mRNA levels for UGT1A4, 2B4, and 2B7.[131]

The regulation of UGT2B7 has been one of the most studied mechanistically. Recent analysis of the 5' flanking region has identified several putative transcription factor binding sites that may explain the tissue specificity of expression, including HNF-1, CDX-2, AP-1, and CREBP.[119] Indeed, variations in levels of the transcription factor HNF-1 have been identified as a factor modulating the significant interindividual differences in UGT2B7 expression,[132] and Oct-1 appears to be a transcription enhancer.[133]

The posttranscriptional regulation of UGTs has not been as well studied. One regulatory mechanism identified in rat is acylation. Acyl-CoA was shown to inhibit UGT activity in a manner consistent with acylation.[134] This occurred with a range of acyl-CoA analogs and inhibited a range of UGT activities. Interestingly, the inhibition was accompanied by a conformational change in the protein structure, and was slowly reversible. It remains to be seen if this mechanism is operative for human UGTs.

9   PHARMACOGENETICS

The effect of pharmacogenetic variability in drug metabolism on drug efficacy and toxicity has been known for decades. The significant genetic variability in CYP was first appreciated for debrisoquine hydroxylase, and in fact CYP2D6 poor metabolizers remain the most noteworthy example of pharmacogenetic variability in this gene family. Subsequently, important genetic variability in the population has also been identified in CYP2C19, 2A6, 2C9, and 3A5.[135]

The UGT isoform with the largest human genetic variability is clearly UGT1A1. More than 50 genetic variants of this form have been identified.[136] The consequence of variability in this isoform is genetic unconjugated hyperbilirubinemia, namely Crigler-Najjar (rare) and Gilbert's (more common) syndromes, resulting from deficient or reduced capacity to glucuronidate bilirubin. These conditions, affecting at least 10% of the population and possessing interethnic differences, can have considerable effects on drug metabolism. This has led to this isoform being called "the CYP2D6 of the UGTs".[104]

Two types of Crigler-Najjar syndrome exist, type 1 and 2. Type 1 results in complete lack of the ability to glucuronidate bilirubin, and is caused by coding region mutations in exons 1-5 that result in inactive enzyme. While type 1 mutations in exon 1 only affect the production of UGT1A1, type 1 deficiencies located in the shared exons 2-5 will affect the entire UGT1A subfamily. Type 1 is typically fatal. Crigler-Najjar type 2 results in decreased bilirubin conjugation, and appears to be caused by mutations in exon 1 that lead to impaired enzyme activity.[137] Clearly, individuals with Crigler-Najjar will possess a significant impaired ability to glucuronidate drugs that are substrates either for UGT1A1 or the entire UGT1A subfamily.

The more common syndrome, Gilbert's, is characterized by moderately reduced bilirubin conjugation and is generally caused by promoter region mutations 5' to exon 1. Wild type individuals contain a TATA promoter element with six TA repeats (TA_6). Individuals with Gilbert's syndrome often possess a TA insertion mutation, which results in a less active promoter (TA_7), and subsequently decreased production of active enzyme.[137-139,196,197] Individuals with Gilbert's syndrome will have a moderately impaired ability to metabolize drugs that are substrates for UGT1A1 that could be significant for low therapeutic index (TI) drugs, although the inducibility of this isoform may partially mask the contribution of genetics to the observed population variability in this isoform. A more important outcome for Gilbert's individuals is their increased susceptibility for hyperbilirubinemia from inhibition of UGT1A1 by xenobiotics (see below).

One well-characterized example of the influence of UGT1A1 activity on drug toxicity is the low TI drug irinotecan. This antitumor agent is actually a prodrug for the active topoisomerase I inhibitor SN-38, which is a substrate for UGT1A1. Patients with Gilbert's syndrome have been reported to exhibit toxicity related to accumulation of SN-38. A significant correlation was also observed between the glucuronidation of bilirubin and SN-38 in liver microsomal samples prepared from wild type and Gilbert's individuals.[140] A pharmacogenetic analysis of patients taking irinotecan identified two common Gilbert's mutations as major risk factors for severe toxicity.[141] Clearly, for low TI drugs that are UGT1A1 substrates, some diagnostic test for impaired capacity for glucuronidation, such as serum bilirubin or 1A1 genotyping, is warranted.

The pharmacogenetics of other UGT isoforms have not been as well studied, although examples exists of polymorphisms having been identified and evaluated. For example, Strassburg, Tukey, and coworkers identified a tissue-specific polymorphism in the expression of UGT1A forms.[117] Specifically, while a common battery of hepatic expression was identified, complete absence of at least one form was observed in some individuals. The polymorphism for one isoform was independent of that for another. The genetic lesions leading to this polymorphism is unclear. Apparently, variability in intestinal glucuronidation may be considerable as a result of extreme tissue-specific regulation.

Additionally, allelic variants have been identified in other UGT isoforms, and a few of these have been phenotypically characterized. For example, several promoter region and coding region single nucleotide polymorphisms (SNPs) within the UGT2B7 gene were identified recently in a large Norwegian population. However, none of the haplotypes resulting from these SNPs caused a significant alteration in morphine metabolite serum ratios.[142] One of these SNPs, a common coding region mutation possessing either a tyrosine or histidine at position 268, had been observed previously. Although up to 50% of the population may display differences at this position, no significant catalytic differences have been identified to date resulting from this change. Coding region mutations in UGT1A6 (T181A, R184S, or Y486D), which are found in the general population in both unique and shared exons and display ethnic variability in allele frequency, were found to result in impaired glucuronidation capacity.[143,144] Analysis of the UGT1A8 exon 1 identified a mutation (C277Y) that, although present at a low allele frequency, resulted in a dramatic decrease in catalytic activity.[145] Since this isoform is found mostly in gastric tissues, this UGT1A8 polymorphism may affect the first pass metabolism of UGT1A8 substrates. Recently, the UGT1A7*3 allele, defined by a combination of mutations W208R, N129K, and R131K, was associated with a significant risk factor for hepatocellular carcinoma.[146] Interestingly, this mutated gene codes for a protein with low detoxification potential, since this enzyme normally metabolizes carcinogens to inactive metabolites, and that >15% of the population appears to be homozygous for UGT1A7*3. Clearly, pharmacogenetic variability is fairly common among UGT genes, but our understanding of the clinical outcomes of this variability must await a more thorough analysis of all human UGT genes.

10   ROLE IN DRUG AND XENOBIOTIC METABOLISM

A cursory examination of the literature combined with experience in metabolite structure elucidation of NCEs in pharmaceutical industry reveals that, although CYPs are quantitatively the most important enzyme system in drug metabolism, an extensive number of pharmaceutical agents exists that are at least partially metabolized via glucuronidation in humans. Clarke and Burchell have published a partial list of drugs possessing glucuronidation as a significant portion of their clearance.[147] This list is by no means complete, but instead serves to demonstrate that glucuronidation occurs for a substantial number of drugs. Some very recent and interesting examples of xenobiotic glucuronidation include the estrogen receptor modulating drug raloxifene,[148] the anticonvulsant GI265080,[149] the anticancer agent flavopiridol,[150] the cholesterol-lowering agent simvastatin,[151] the lung carcinogen NNAL (4-(methylnitrosamino)-1-pyridyl)-1-butanol),[152] and the antipsychotic olanzapine.[153] While some of these biotransformations are fairly common, such as phenol glucuronidation, these also include formation of quaternary ammonium glucuronides, acyl glucuronides, and O-glucuronidation on an N-oxide.

A rare biotransformation involving amines deserves some mention due to its uniqueness and its analytical difficulty. Carbamoyl glucuronides were first reported in 1980 for tocainide.[154] Apparently, the addition of CO_2 to an available amine can be subsequently conjugated by UGTs. Since tocainide, other compounds have been implicated in this unique biotransformation, including sertraline, rimantadine, carvedilol, mofegiline, and the desmethyl metabolites of several other drugs. A couple of aspects of these metabolites can be problematic. Firstly, carbamoyl glucuronides are extremely labile, and can undergo in-source breakdown in the mass spectrometer to yield just parent, and thus can artificially enhance the quantitated levels of parent unless chromatographic separation is achieved. While this is an issue with all glucuronides, a comparison of different glucuronides showed that carbamoyl glucuronides are more labile than others, such as acyl glucuronides. Secondly, a search for metabolites that are increased by 176 amu compared to parent may overlook these carbamoyl glucuronides, as they are parent + CO_2 + glucuronic acid (parent + 220).

10.1 Bioactivation.

Drug biotransformation is generally considered to be an inactivation or elimination pathway. Some examples exist of glucuronide conjugation resulting in a net bioactivation event, such as increased biological potency or toxicity.[155] The glucuronidation of morphine is the most notorious example of this phenomenon. In humans, morphine is cleared by metabolism to two regioisomeric glucuronides: morphine-3- and 6-glucuronide (M3G and M6G). The major metabolite, M3G, is inactive and in fact antagonizes morphine analgesia. M6G, the minor metabolite, is more potent than morphine at inducing analgesia. An analogous situation occurs with codeine and codeine-6-glucuronide.[156] Other lesser known examples of UGT-mediated bioactivation also exist. For example, steroid D-ring glucuronides of testosterone, dihydrotestosterone, estradiol, 17α-ethynylestradiol, and estriol appear to be cholestatic. Interestingly, the corresponding A-ring conjugates of estrogens can have the opposite effect and increase bile flow. Glucuronides of retinoic acid and its analogs have been reported to have effects on cell growth and differentiation. It has become apparent that the process of glucuronidation can be more complex than that of simple bioinactivation.[155]

Glucuronidation has also been shown to bioactivate some functional groups into a potentially more toxic species via the rearrangement to reactive, electrophilic intermediates. Specifically, acyl glucuronides, including zomepirac and diclofenac conjugates, can either covalently bind directly via nucleophilic displacement of the glucuronic acid moiety, rearrange into non-hydrolyzable conjugates or into electrophilic species that can covalently bind to protein lysine side chains. Drugs such as zomepirac, tolmetin, and diflunisal have been shown to form acyl glucuronides that result in covalent binding and subsequent toxicity.[157] The acyl glucuronide of the immunosuppressant mycophenolic acid has been linked with gastrointestinal side effects.[158] Additionally, N-O-glucuronides of hydroxamic acids such as N-hydroxy-2-acetylaminofluorine (N-hydroxy-2AAF) can form an arylnitrenium ion and react with cellular nucleophiles.[155]

The issue of acyl glucuronides deserves further discussion. Many acidic drugs are metabolized to acyl glucuronides. The rearrangement of these metabolites, and the toxicological implications, has been extensively studied. However, while several examples exist of toxicity purportedly due to this bioactivation pathway, other examples of benign outcomes also exist.[157,159] For example, ibuprofen undergoes significant acyl glucuronidation, but is very safe. Valproic acid, although hepatotoxic in rare instances, is reportedly bioactivated by other mechanisms and detoxified by glucuronidation. Also, in some instances, the targets of acyl glucuronide covalent binding are other hepatic and serum proteins, while in other cases the covalent binding is to the UGT that generated the metabolite and results in enzyme inactivation (ketoprofen).[160]

Recent reports have begun to elucidate an SAR for acyl glucuronide rearrangement, covalent binding, and toxicity. Acyl glucuronides unsubstituted at the α position appear to be the most reactive and give the highest covalent binding, presumably due to the lack of steric hindrance for rearrangement and nucleophilic attack.[161]   The reactivity appears to decrease with increasing substitution, although exceptions to this relationship, such as valproic acid, exist. Bolze et al.[157] have developed a model for predicting the reactivity toward human serum albumin as a model protein target. Their model showed an excellent correlation between the maximal amounts of bound drug, expressed as percentage of acyl glucuronide present in the incubation medium, and the aglycone appearance rate constant weighted by the percentage of rearrangement. Although preliminary, this approach incorporates the features of carboxylic acid bioactivation that are expected to be important determinants of amount of covalent binding, namely amount of glucuronide formed and rearrangement rate.

10.2 Structure-biotransformation relationships.

With the availability of recombinant UGTs, the relationship between the structure or physico-chemical properties of compounds and their ability to be metabolized by certain isoforms has begun to evolve. For example, UGT1A1, which metabolizes bilirubin, also is active in the 3-glucuronidation (A ring) of some estrogens (β-estradiol and 17α-ethynylestradiol).[101,116] UGT1A3 and 1A4 are apparently the most active in the N-glucuronidation of many amine substrates, including ketotifen and amitriptyline, sometimes to quaternary N-glucuronides.[162] UGT1A6 is active in the glucuronidation of simple planar phenols.[163] UGT1A9, interestingly, has a wider SAR for phenols, appearing to prefer sterically hindered substrates such as propofol (2,6-di-*t*-butylphenol).[164] UGT2B7 appears to be a very important and broad selectivity enzyme in xenobiotic metabolism, accepting steroids, opioids, nucleoside analogs, benzodiazepines, fatty acids, and others.[119] Endogenous and xenobiotic steroids are low K_m substrates for UGT2B15 and 2B17.[165]

Although a large number of substrate enzyme kinetics have not been determined with single isoforms, some QSAR modeling has been attempted for UGTs with encouraging results. Radominska-Pandya et al. have recently reviewed this,[100] and data from several sources has shown a trend toward increased rates and catalytic efficiency with higher lipophilicity. Recently, enzyme kinetic parameters using expressed UGT1A6 and 1A9 has been determined with a series of simple phenols, and the data used to generate MSWHIM (molecular surface-weighted holistic invariant molecular) descriptors.[163] This data showed a clear trend between the size of the *p*-substituent and lower V_{max} for UGT1A6, and showed differences in substrate acceptance between these 2 isoforms that have considerable overlap in substrate specificity. Although this work is limited, the results encourage more QSAR generation over a more diverse range of substrate characteristics and for more isoforms.

10.3 *Kinetics and scaling.*

One of the utilities of in vitro systems to study drug metabolism is the ability to kinetically characterize specific biotransformations. This has been especially true for the CYPs.[166] Enzyme kinetic parameters, usually K_m and V_{max}, are typically generated in liver microsomes and have been applied to the problems of identifying conditions where substrate metabolism is isoform selective, selecting isoform-selective inhibitor concentrations, investigating the enzyme multiplicity of a biotransformation, and predicting hepatic clearance.

Recently, examples have accumulated of non-Michaelis-Menten behavior for some CYP biotransformations.[167] The chapter from Houston and coworkers in this book should be consulted for more detail. While the exact biochemical cause for atypical kinetics is not known, the most common explanations are the presence of multiple binding sites or conformations for a CYP isoform. This manifests itself as a sigmoidal v vs. [S] curve, or a hooked Eadie-Hofstee plot. The latter is a more sensitive indicator of atypical kinetics. The kinetic model used to characterize this behavior is most commonly the Hill equation. While the *n* value generated from the Hill equation indicates the degree of receptor occupancy for allosteric oligomeric receptors, it is unknown what the meaning of *n* is for biotransformation enzymes besides a degree of deviation from classical Michaelis-Menten behavior.

Since the UGT field has lagged behind that of the P450s, so has the application of *in vitro* enzyme kinetics. This has at least partially been a result of the difficulty in working with UGTs in vitro (see above). Over the last several years, the kinetic characterization of UGT activities has become more common. Glucuronidation is considered a low affinity high capacity process, although K_m values cover a range from sub µM to >mM. Atypical kinetics have been observed for some glucuronidations, including heterotropic activation, but the mechanistic meaning remains indistinguishable.[104,168]

Predicting clearance due to hepatic metabolism has become fairly routine due to the pioneering work of Sugiyama,[169] Houston,[170] Obach,[171] and others.[172] The chapter from Dr. Sugiyama discusses this in more detail. Briefly, the intrinsic clearance activity (CL_{int}) of the liver preparation is determined, and this CL_{int} value is scaled-up, or extrapolated, to determine the CL_{int} of the whole liver, or CL'_{int}. The in vivo blood clearance is predicted through the application of a flow model, usually the well-stirred equation. Two main experimental methods exist to predict whole organ CL_{int}: V/K and in vitro half-life. In the former, the kinetic parameters V_{max} and K_m are determined, and their ratio is the microsomal CL_{int}, which is multiplied by known scaling factors to calculate a whole organ CL'_{int}. In the latter, the disappearance half-life of substrate is determined under conditions where it is assumed that $[S] \ll K_m$, and whole organ CL'_{int} is calculated from this value. For drugs known to be cleared largely by hepatic CYP metabolism, predicted blood clearance correlate fairly well with observed in vivo clearances.

Mistry and Houston first tried to scale *in vitro* glucuronidation over a decade ago.[173] For 3 drugs, rat *in vitro* clearance underpredicted the *in vivo* clearance by more than 10-fold. Recent reports have begun to evaluate the *in vitro-in vivo* scaling of human glucuronidation reactions. Soars et al.[174] reported an underprediction of clearance using human liver microsomes for several drugs, while human hepatocytes were much more accurate. Interestingly, the underprediction in human liver microsomes was similar to that reported for rat, and occurred for all but the extremely low clearance substrates (naproxen and valproate). Work in one author's lab (MBF) have extended this observation for AZT, mycophenolate, and diclofenac. Additionally, the underprediction is probably even worse, since many drugs that are primarily hepatically glucuronidated are also recirculated, suggesting that hepatic metabolism should overpredict the overall clearance. It was found that consideration of protein binding in plasma and in microsomes was important for many substrates. A very recent publication[175] found that assay conditions could have a significant effect on the relative success of clearance predictions from microsomal glucuronidation, with the degree of underprediction of AZT clearance minimized by reducing the concentration of phosphate buffer in the incubation. The reasons for this underprediction are presently unknown, but nonetheless care must be taken in selecting the proper *in vitro* system for quantitative predictions of glucuronidation.

10.4 First-pass metabolism.

As stated above, the UGTs possess a significant extrahepatic presence, with some forms existing in only extrahepatic organs. One organ that has been identified as having a significant effect on the pharmacokinetics of pharmaceuticals is the small intestine. The UGT enzyme activities in intestine could play a significant role in the bioavailability of orally administered drugs.[176] Examination of the rates of estrogen conjugation in human intestine and liver microsomes demonstrates that intestinal glucuronidation can be more important than liver on a per milligram of protein basis.[116,148,177] One can also envision a scenario where a drug can be metabolically stable in liver microsomes, but still be significantly glucuronidated in intestine if the drug is a substrate for UGT1A8, 1A10, or some other intestinal-specific isoform.

DDI. While drug-drug interactions are a difficult and recurring problem with the CYPs (see Chapter 9), there have been relatively few drug-drug interactions reported with UGTs. A series of studies have examined drug interactions *in vivo* and *in vitro* with AZT.[178] Four drugs often used in combination with AZT- atovaquone, fluconazole, methadone, and valproic acid- were found to increase the AUC for AZT in patients, often accompanied by an indication of decreased glucuronidation, such as a decrease in the AUC for, or decreased urinary excretion of, the glucuronide metabolite. *In vitro* data in human liver microsomes supported these observations, although inhibition should have been more significant with fluconazole and valproic acid. A potent interaction was predicted from *in vitro* data between diclofenac and codeine glucuronidation, and this interaction may explain the increase in efficacy of opioids with concomitant NSAID therapy.[179] The immunosuppressant tacrolimus has been shown to enhance the bioavailability of mycophenolate mofetil with concurrent reduction of mycophenolate glucuronidation, and this interaction was confirmed in liver microsomes at clinically relevant drug concentrations.[180] Other known drug interactions involving decreased glucuronide formation include gemfibrozil-statins,[181] acetaminophen and lorazepam-probenecid,[182] CPT-11-FK506,[183] and others.[184] Also, although it's an older reference, Miners and Mackenzie generated a useful list of known pharmacokinetic drug interactions involving glucuronidated drugs.[185]

A drug interaction phenomenon that has been more commonly observed is a drug-endobiotic interaction. As stated above, UGT1A1 is responsible for the clearance of bilirubin. Drugs have now been identified that interfere with this process, leading to hyperbilirubinemia in susceptible patients. For example, lorazepam, buprenorphine, and ethynylestradiol have been suggested to be prone to interactions with UGT1A1. *In vitro* evidence supports this with in vitro K_i values being similar to steady state plasma concentrations.[186] Recently, hyperbilirubinemia was observed with indinavir therapy or indinavir/ritonavir regimen for HIV protease inhibition, and this increase in serum bilirubin, while not due to hepatic damage, was found to be significantly greater in those patients that were heterozygous or homozygous for the Gilbert's TATA box mutation.[187] Interestingly, saquinavir, which also has been shown to competitively inhibit UGT1A1 activity, has not been associated with hyperbilirubinemia, possibly due to its higher K_i for the enzyme compared to indinavir (360 vs 183 μM). Recently, a UGT1A1 inhibition fluorescent screen was described, which should facilitate detection of drugs that may be prone to causing this inhibition.[188]

11 CONCLUSIONS

Many advances in our understanding of the enzymology of the sulfotransferases and UDP-glucuronosyltransferases have been made in recent years, and this trend will continue. It is likely to be driven, at least in part, by the needs of the Pharmaceutical Industry. The fact that many adverse drug reactions are directly related to the major cytochromes P450 (particularly CYP2D6 and the CYP3A family)[98] is leading the industry, as well as the regulatory authorities, to re-evaluate drug discovery and development priorities away from compounds for which oxidative metabolism predominates.[99] The consequence of this is that many drugs coming to market in the future will be metabolised primarily by other pathways, in particular the conjugating enzymes such as UDP-glucuronosyltransferases and SULTs. It will therefore be crucial to understand the biology and genetics of these enzymes if we are to be able to predict adverse reactions involving compounds metabolised by conjugation. Studies of *in vivo* sulfation and glucuronidation in humans will be of key importance in assessing the functional consequences of interindividual variation in SULTs and UGTs. Hopefully this will be an area where significant advances can be made in the future.

ACKNOWLEDGEMENTS

M.W.H.C.: I am grateful to all the members of my laboratory (past and present), and to the numerous valued collaborators, who have contributed to the work described here. Current research in the laboratory is supported by the Medical Research Council, the Commission of the European Communities (QLG3-CT-2000-00930), Tenovus Scotland, the Scottish Office Chief Scientist Office and the Human Drug Conjugation Consortium.

M.B.F.: I would like to acknowledge past and present collaborators and current members of my laboratory, including, but not exclusively, Tim Strelevitz, Kwansik Yoon, Rob Foti, Barbara Ring, Kristina Campanale, Brad Ackermann, Mark VandenBranden, Steve Wrighton, Ken Thummel, Mary Paine, and Brian Burchell. I would also like to thank Drs. Jae Lee, Mike Schrag, Jennifer Liras, and Cliff Fisher for many fruitful discussions.

REFERENCES

1. Coughtrie, M. W. H. *Pharmacogenomics J.* **2002**, *2*, 297.
2. Glatt, H.; Engelke, C. E. H.; Pabel, U.; Teubner, W.; Jones, A. L.; Coughtrie, M. W. H.; Andrae, U.; Falany, C. N.; Meinl, W. *Toxicol. Lett.* **2000**, *112*, 341.
3. Dajani, R.; Sharp, S.; Graham, S.; Bethell, S. S.; Cooke, R. M.; Jamieson, D. J.; Coughtrie, M. W. H. *Prot. Expression Purif.* **1999**, *16*, 11.
4. Falany, C. N. *FASEB J.* **1997**, *11*, 206.
5. Rubin, G. L.; Sharp, S.; Jones, A. L.; Glatt, H.; Mills, J. A.; Coughtrie, M. W. H. *Xenobiotica* **1996**, *26*, 1113.

6. Rubin, G. L.; Harrold, A. J.; Mills, J. A.; Falany, C. N.; Coughtrie, M. W. H. *Mol. Human Reprod.* **1999**, *5*, 995.

7. Kester, M. H.; Bulduk, S.; van, T.; Tibboel, D.; Meinl, W.; Glatt, H.; Falany, C. N.; Coughtrie, M. W.; Schuur, A. G.; Brouwer, A.; Visser, T. J. *J. Clin. Endocrinol. Metab.* **2002**, *87*, 1142.

8. Falany, C. N.; Xie, X. W.; Wang, J.; Ferrer, J.; Falany, J. L. *Biochem. J.* **2000**, *346*, 857.

9. Foldes, A.; Meek, J. L. *Biochim. Biophys. Acta* **1973**, *327*, 365.

10. Carter, S. B.; Glover, V.; Sandler, M. *Biochem. Pharmacol.* **1986**, *35*, 1205.

11. Jones, A. L., PhD Thesis. University of Dundee **1995**.

12. Borthwick, E. B.; Burchell, A.; Coughtrie, M. W. H. *Biochem. J.* **1993**, *289*, 719.

13. Coughtrie, M. W. H.; Sharp, S. *Biochem. Pharmacol.* **1990**, *40*, 2305.

14. Frame, L. T.; Ozawa, S.; Nowell, S. A.; Chou, H. C.; Delongchamp, R. R.; Doerge, D. R.; Lang, N. P.; Kadlubar, F. F. *Drug Metab. Dispos.* **2000**, *28*, 1063.

15. Pang, K. S.; Koster, H.; Halsema, I. C. M.; Scholtens, E.; Mulder, G. J.; Stillwell, R. N. *J. Pharmacol. Exp. Ther.* **1981**, *224*, 647.

16. Yang, C. M.; Carlson, G. P. *Pharmacology* **1991**, *42*, 28.

17. Ring, J. A.; Ghabrial, H.; Ching, M. S.; Shulkes, A.; Smallwood, R. A.; Morgan, D. J. *Drug Metab. Dispos.* **1996**, *24*, 1378.

18. Oddy, E. A.; Manchee, G. R.; Coughtrie, M. W. H. *Xenobiotica* **1997**, *27*, 369.

19. Evdokimova, E.; Taper, H.; Calderon, P. B. *Toxicology In Vitro* **2001**, *15*, 683.

20. Kuhn, U. D.; Rost, M.; Muller, D. *Exp. Toxicol. Pathol.* **2001**, *53*, 81.

21. Tosh, D.; Borthwick, E. B.; Sharp, S.; Burchell, A.; Burchell, B.; Coughtrie, M. W. H. *Biochem. Pharmacol.* **1996**, *51*, 369.

22. Li, A. P.; Lu, C.; Brent, J. A.; Pham, C.; Fackett, A.; Ruegg, C. E.; Silber, P. M. *Chem. -Biol. Interact.* **1999**, *121*, 17.

23. Tan, E.; Pang, K. S. *Drug Metab. Dispos.* **2001**, *29*, 335.

24. Slaus, K.; Coughtrie, M. W.; Sharp, S.; Vanhaecke, T.; Vercruysse, A.; Rogiers, V. *Biochem. Pharmacol.* **2001**, *61*, 1107.

25. Duanmu, Z. B.; Locke, D.; Smigelski, J.; Wu, W.; Dahn, M. S.; Falany, C. N.; Kocarek, T. A.; Runge-Morris, M. *Drug Metab. Dispos.* **2002**, *30* , 997.

26. Weinshilboum, R. M.; Otterness, D. M.; Aksoy, I. A.; Wood, T. C.; Her, C.; Raftogianis, R. B. *FASEB J.* **1997**, *11*, 3.

27. Otterness, D. M.; Wieben, E. D.; Wood, T. C.; Watson, R. W. G.; Madden, B. J.; McCormick, D. J.; Weinshilboum, R. M. *Mol. Pharmacol.* **1992**, *41*, 865.

28. Glatt, H.; Pauly, K.; Pieestaffa, A.; Seidel, A.; Hornhardt, S.; Czich, A. *Toxicol. Lett.* **1994**, *72*, 13.

29. Jones, A. L.; Hagen, M.; Coughtrie, M. W. H.; Roberts, R. C.; Glatt, H. *Biochem. Biophys. Res. Commun.* **1995**, *208*, 855.

30. Falany, C. N.; Wheeler, J.; Oh, T. S.; Falany, J. L. *J. Steroid Biochem. Mol. Biol.* **1994**, *48*, 369.

31. Glatt, H. R.; Christoph, S.; Czich, A.; Pauly, K.; Schwierzok, A.; Seidel, A.; Coughtrie, M. W. H.; Doehmer, J.; Falany, C. N.; Phillips, D. H.; Yamazoe, Y.; Bartsch, I. In *Control Mechanisms of Carcinogenesis*; Oesch, F.; Hengstler, J. G. Eds.; Tieme: Meissen, Germany, **1996**; pp. 98.

32. Ganguly, T. C.; Krasnykh, V.; Falany, C. N. *Drug Metab. Dispos.* **1995**, *23*, 945.

33. Richard, K.; Hume, R.; Kaptein, E.; Stanley, E. L.; Visser, T. J.; Coughtrie, M. W. H. *J. Clin. Endocrinol. Metab.* **2001**, *86*, 2734.

34. Stanley, E. L.; Hume, R.; Visser, T. J.; Coughtrie, M. W. H. *J. Clin. Endocrinol. Metab.* **1-12-2001**, *86*, 5944.

35. Sharp, S.; Coughtrie, M. W. H.; Forbes, K. J.; Hume, R. *J. Pharmacol. Toxicol. Methods* **1995**, *34*, 89.

36. Wang, J.; Falany, J. L.; Falany, C. N. *Mol. Pharmacol.* **1998**, *53*, 274.

37. Dajani, R.; Cleasby, A.; Neu, M.; Wonacott, A. J.; Jhoti, H.; Hood, A. M.; Modi, S.; Hersey, A.; Taskinen, J.; Cooke, R. M.; Manchee, G. R.; Coughtrie, M. W. H. *J. Biol. Chem.* **1999**, *274*, 37862.

38. Reiter, C.; Mwaluko, G.; Dunnette, J.; Van Loon, J.; Weinshilboum, R. *Naunyn-Schmiedeberg Arch. Pharmacol.* **1983**, *324*, 140.

39. Dajani, R.; Hood, A. M.; Coughtrie, M. W. H. *Mol. Pharmacol.* **1998**, *54*, 942.

40. Back, D. J.; Breckenridge, A. M.; MacIver, M.; Orme, M.; Purba, H. S.; Rowe, P. H.; Taylor, I. *Br. J. Clin. Pharmacol.* **1982**, *13*, 325.

41. Stanley, E. L.; Coughtrie, M. W. H.; Hume, R. *The Pharmacologist* **2002**, *44 (Suppl1)*, A177.

42. Kakuta, Y.; Pedersen, L. G.; Carter, C. W.; Negishi, M.; Pedersen, L. C. *Nature Struct. Biol.* **1997**, *4*, 904.

43. Bidwell, L. M.; McManus, M. E.; Gaedigk, A.; Kakuta, Y.; Negishi, M.; Pedersen, L.; Martin, J. L. *J. Mol. Biol.* **1999**, *293*, 521.

44. Pedersen, L. C.; Petrotchenko, E. V.; Negishi, M. *FEBS Lett.* **2000**, *475*, 61.

45. Pedersen, L. C.; Petrotchenko, E.; Shevtsov, S.; Negishi, M. *J. Biol. Chem.* **10-5-2002**, *277*, 17928.

46. Pakhomova, S.; Kobayashi, M.; Buck, J.; Newcomer, M. E. *Nat. Struct. Biol.* **2001**, *8*, 447.

47. Kakuta, Y.; Sueyoshi, T.; Negishi, M.; Pedersen, L. C. *J. Biol. Chem.* **16-4-1999**, *274*, 10673.

48. Kakuta, Y.; Petrotchenko, E. V.; Pedersen, L. C.; Negishi, M. *J. Biol. Chem.* **1998**, *273*, 27325.

49. Negishi, M.; Pedersen, L. G.; Petrotchenko, E.; Shevtsov, S.; Gorokhov, A.; Kakuta, Y.; Pedersen, L. C. *Arch. Biochem. Biophys.* **2001**, *390*, 149.

50. Komatsu, K.; Driscoll, W. J.; Koh, Y. C.; Strott, C. A. *Biochem. Biophys. Res. Commun.* **1994**, *204*, 1178.

51. Driscoll, W. J.; Komatsu, K.; Strott, C. A. *Proc. Natl. Acad. Sci. USA* **1995**, *92*, 12328.

52. Chiba, H.; Komatsu, K.; Lee, Y. C.; Tomizuka, T.; Strott, C. A. *Proc. Natl. Acad. Sci. USA* **1995**, *92*, 8176.

53. Sakakibara, Y.; Takami, Y.; Nakayama, T.; Suiko, M.; Liu, M.-C. *J. Biol. Chem.* **1998**, *273*, 6242.

54. Brix, L. A.; Duggleby, R. G.; Gaedigk, A.; McManus, M. E. *Biochem. J.* **1999**, *337*, 337.

55. Petrotchenko, E. V.; Doerflein, M. E.; Kakuta, Y.; Pedersen, L. C.; Negishi, M. *J. Biol. Chem.* **1999**, *274*, 30019.

56. Campese, V. M. *Drugs* **1981**, *22*, 257.

57. Meisheri, K. D.; Johnson, G. A.; Puddington, L. *Biochem. Pharmacol.* **1993**, *45*, 271.

58. Falany, C. N.; Kerl, E. A. *Biochem. Pharmacol.* **1990**, *40*, 1027.

59. Ozawa, S.; Nagata, K.; Shimada, M.; Ueda, M.; Tsuzuki, T.; Yamazoe, Y.; Kato, R. *Pharmacogenetics* **1995**, *5*, S135-S140.

60. Kudlacek, P. E.; Clemens, D. L.; Anderson, R. J. *Biochem. Biophys. Res. Commun.* **1995**, *210*, 363.

61. Dooley, T. P.; Walker, C. J.; Hirshey, S. J.; Falany, C. N.; Diani, A. R. *J. Invest. Dermatol.* **1991**, *96*, 65.

62. Baker, C. A.; Uno, H.; Johnson, G. A. *Skin Pharmacology* **1994**, *7*, 335.

63. Buhl, A. E.; Baker, C. A.; Dietz, A. J.; Murray, F. T.; Johnson, G. A. *J. Invest. Dermatol.* **1994**, *102*, 534.

64. Garay, R. P.; Labaune, J.-P.; Mesangeau, D.; Nazaret, C.; Imbert, T.; Moinet, G. *J. Pharmacol. Exp. Ther.* **1990**, *255*, 415.

65. Garay, R. P.; Rosati, C.; Fanous, K.; Allard, M.; Morin, E.; Lamiable, D.; Vistelle, R. *Eur. J. Pharmacol.* **1995**, *274*, 175.

66. Miller, J. A. *Chem. -Biol. Interact.* **1994**, *92*, 329.

67. Wakazono, H.; Gardner, I.; Eliasson, E.; Coughtrie, M. W. H.; Kenna, J. G.; Caldwell, J. *Chem. Res. Toxicol.* **1998**, *11*, 863.

68. Boberg, E. W.; Miller, E. C.; Miller, J. A.; Poland, A.; Liem, A. *Cancer Res.* **1983**, *43*, 5163.

69. Sugahara, K.; Schwartz, N. B. *Proc. Natl. Acad. Sci. USA* **1979**, *76*, 6615.

70. Meerman, J. H. N.; Ringer, D. P.; Coughtrie, M. W. H.; Bamforth, K. J.; Gilissen, R. A. H. J. *Chem. -Biol. Interact.* **1994**, *92*, 321.

71. Nagata, K.; Ozawa, S.; Miyata, M.; Shimada, M.; Gong, D.-W.; Yamazoe, Y.; Kato, R. *J. Biol. Chem.* **1993**, *268*, 24720.

72. Gilissen, R. A. H. J.; Bamforth, K. J.; Stavenuiter, J. F. C.; Coughtrie, M. W. H.; Meerman, J. H. N. *Carcinogenesis* **1994**, *15*, 39.

73. Ozawa, S.; Chou, H. C.; Kadlubar, F. F.; Yamazoe, Y.; Kato, R. *Jpn. J. Cancer Res.* **1994**, *85*, 1220.

74. Glatt, H.; Pauly, K.; Czich, A.; Falany, J. L.; Falany, C. N. *Eur. J. Pharmacol.* **1995**, *293*, 173.

75. Glatt, H. R. *Biochem. Soc. Trans.* **2000**, *28*, 1.

76. Glatt, H. *Chem. -Biol. Interact.* **2000**, *129*, 141.

77. Glatt, H.; Bartsch, I.; Czich, A.; Seidel, A.; Falany, C. N. *Toxicol. Lett.* **1995**, *82-3*, 829.

78. Forbes, K. J.; Hagen, M.; Glatt, H.; Hume, R.; Coughtrie, M. W. H. *Mol. Cell. Endocrinol.* **1995**, *112*, 53.

79. Weinshilboum, R. *Cell. Mol. Neurobiol.* **1988**, *8*, 27.

80. Weinshilboum, R. *Pharmac. Ther.* **1990**, *45*, 93.

81. Weinshilboum, R.; Aksoy, I. *Chem. -Biol. Interact.* **1994**, *92*, 233.

82. Price, R. A.; Spielman, R. S.; Lucena, A. L.; van Loon, J. A.; Baidak, B. L.; Weinshilboum, R. M. *Genetics* **1989**, *122*, 905.

83. Reiter, C.; Weinshilboum, R. *Clin. Pharmacol. Ther.* **1982**, *32*, 612.

84. Bonham-Carter, S. M.; Rein, G.; Glover, V.; Sandler, M.; Caldwell, J. *Br. J. Clin. Pharmacol.* **1983**, *15*, 323.

85. van Loon, J. A.; Weinshilboum, R. M. *Biochem. Genet.* **1984**, *22*, 997.

86. Jones, A. L.; Roberts, R. C.; Coughtrie, M. W. H. *Biochem. J.* **1993**, *296*, 287.

87. Raftogianis, R. B.; Wood, T. C.; Otterness, D. M.; van Loon, J. A.; Weinshilboum, R. M. *Biochem. Biophys. Res. Commun.* **1997**, *239*, 298.

88. Raftogianis, R. B.; Wood, T. C.; Weinshilboum, R. M. *Biochem. Pharmacol.* **1999**, *58*, 605.

89. Nowell, S.; Ambrosone, C. B.; Ozawa, S.; MacLeod, S. L.; Mrackova, G.; Williams, S.; Plaxco, J.; Kadlubar, F. F.; Lang, N. P. *Pharmacogenetics* **2000**, *10*, 789.

90. Ozawa, S.; Shimizu, M.; Katoh, T.; Miyajima, A.; Ohno, Y.; Matsumoto, Y.; Fukuoka, M.; Tang, Y.-M.; Lang, N. P.; Kadlubar, F. F. *J. Biochem.* **1999**, *126*, 271.

91. Li, X.; Clemens, D. L.; Cole, J. R.; Anderson, R. J. *J. Endocrinol.* **2001**, *171*, 525.

92. Raftogianis, R. B. *Drug Metab. Rev.* **2001**, *33 (Suppl 1)*, 10.

93. Frame, L. T.; Gatlin, T. L.; Kadlubar, F. F.; Lang, N. P. *Environ. Toxicol. Pharmacol.* **1997**, *4*, 277.

94. Bamber, D. E.; Fryer, A. A.; Strange, R. C.; Elder, J. B.; Deakin, M.; Rajagopal, R.; Fawole, A.; Gilissen, R. A. H. J.; Campbell, F. C.; Coughtrie, M. W. H. *Pharmacogenetics* **2001**, *11*, 679.

95. Seth, P.; Lunetta, K. L.; Bell, D. W.; Gray, H.; Nasser, S. M.; Rhei, E.; Kaelin, C. M.; Iglehart, D. J.; Marks, J. R.; Garber, J. E.; Haber, D. A.; Polyak, K. *Cancer Res.* **2000**, *60*, 6859.

96. Zheng, W.; Xie, D. W.; Cerhan, J. R.; Sellers, T. A.; Wen, W. Q.; Folsom, A. R. *Cancer Epidem. Biomar.* **2001**, *10*, 89.

97. Chou, H. C.; Lang, N. P.; Kadlubar, F. F. *Carcinogenesis* **1995**, *16*, 413.

98. Ingelman-Sundberg, M. *Drug Metab. Dispos.* **2001**, *29*, 570.

99. Hodgson, J. *Nat. Biotechnol.* **2001**, *19*, 722.

100. Radominska-Pandya, A.; Czernik, P.J.; Little, J.M.; Battaglia, E.; Mackenzie, P.I. *Drug Metab. Rev.* **1999**, *31*, 817.

101. King, C.D.; Rios, G.R.; Green, M.D.; Tephly, T.R. *Curr. Drug Metab.* **2000**, *1*, 143.

102. Tukey, R.H.; Strassburg, C.P. *Annu. Rev. Pharmacol. Toxicol.* **2000**, *40*, 581.

103. Burchell, B. In *Handbook of Drug Metabolism*; Woolf, T.F. Ed.; Marcel Dekker: New York, **1999**; pp. 153.

104. Fisher, M.B.; Paine, M.F.; Strelevitz, T.J.; Wrighton, S.A. *Drug Metab. Rev.* **2001**, *33*, 273.

105. Mackenzie, P.I.; Owens, I.S.; Burchell, B.; Bock, K.W. *Pharmacogenetics* **1997**, *7*, 255.

106. Bradow, G.; Kan, L.S.; Fenselau, C. *Chem. Res. Toxicol.* **1989**, *5*, 316.

107. Ouzzine, M.; Antonio, L.; Burchell, B.; Netter, P.; Fournel-Gigleux, S.; Magdalou, J. *Mol. Pharmacol.* **2000**, *58*, 1609.

108. Stephanson, N.; Dahl, H.; Helander, A.; Beck, O. *Ther. Drug Monitor.* **2002**, *24*, 646.

109. Strassburg, C.P.; Barut, A.; Obermayer-Straub, P.; Li, Q.; Nguyen, N.; Tukey, R.H.; Manns, M.P. *J. Hepatol.* **2001**, *34*, 865.

110. Sperker, B.; Backman, J.T.; Kroemer, H.K. *Clin. Pharmacokinet.* **1997**, *33*, 18.

111. Fisher, M.B.; Campanale, K.; Ackermann, B.L.; VandenBranden, M.; Wrighton, S.A. *Drug Metab. Dispos.* **2000**, *28*, 560.

112. Meech, R.; Mackenzie, P.I. *J. Biol. Chem.* **1997**, *272*, 26913.

113. Ishii, Y.; Miyoshi, A.; Watanabe, R.; Tsuruda, K.; Tsuda, M.; Yamaguchi-Nagamatsu, Y.; Yoshisue, K.; Tanaka, M.; Maji, D.; Ohgiya, S.; Oguri, K. *Mol. Pharmacol.* **2001**, *60*, 1040.

114. Ghosh, S.S.; Sappal, B.S.; Kalpana, G.V.; Lee, S.W.; Chowdhury, J.R.; Chowdhury, N.R. *J. Biol. Chem.* **2001**, *45*, 42108.

115. Yokota, H.; Ando, F.; Iwano, H.; Yuasa, A. *Life Sci.* **1998**, *63*, 1693.

116. Fisher, M.B.; VandenBranden, M.; Findlay, K.; Burchell, B.; Thummel, K.E.; Hall, S.D.; Wrighton, S.A. *Pharmacogenetics* **2000**, *10*, 727.

117. Tukey, R.H.; Strassburg, C.P. *Mol. Pharmacol.* **2001**, *59*, 405.

118. Radominska-Pandya, A.; Pokrovskaya, I.D.; Little, J.M.; Jude, A.R.; Kurten, R.C.; Czernik, P.J. *Arch. Biochem. Biophys.* **2002**, *399*, 37.

119. Radominska-Pandya, A.; Little, J.M.; Czernik, P.J. *Curr. Drug Metab.* **2001**, *2*, 283.

120. Court, M.H.; Duan, S.X.; Von Moltke, L.L.; Greenblatt, D.J.; Patten, C.J.; Miners, J.O.; Mackenzie, P.I. *J. Pharmacol. Exp. Ther.* **2001**, *299*, 998.

121. Fisher, M.B.; Jackson, D.; Kaerner, A.; Wrighton, S.A.; Borel, A.G. *Drug Metab. Dispos.* **2002**, *30*, 270.

122. Nowell, S.A.; Massengill, J.S.; Williams, S.; Radominska-Pandya, A.; Tephly, T.R.; Cheng, Z.; Strassburg, C.P.; Tukey, R.H.; MacLeod, S.L.; Lang, N.P.; Kadlubar, F.F. *Carcinogenesis* **1999**, *20*, 1107.

123. Paine, M.F.; Fisher, M.B. *Biochem. Biophys. Res. Commun.* **1998**, *1394*, 199.

124. Munzel, P.A.; Schmohl, S.; Heel, H.; Kalberer, K.; Bock-Henning, B.S.; Bock, K.W. *Drug Metab. Dispos.* **1999**, *27*, 569.

125. Walle, U.K.; Walle, T. *Drug Metab. Dispos.* **2002**, *30*, 564.

126. Ritter, J.K.; Kessler, F.K.; Thompson, M.T.; Grove, A.D.; Auyeung, D.J.; Fisher, R.A. *Hepatology* **1999**, *30*, 476.

127. Brierley, C.H.; Senafi, S.B.; Clarke, D.; Hsu, M.-H.; Johnson, E.F.; Burchell, B. *Adv. Enzyme Regul.* **1996**, *36*, 85.

128. Sugatani, J.; Kojima, H.; Ueda, A.; Kakizaki, S.; Yoshinari, K.; Gong, Q.H.; Owens, I.S.; Negishi, M.; Sueyoshi, T. *Hepatology* **2001**, *33*, 1232.

129. McCarver, D.G.; Hines, R.N. *J. Pharmacol. Exp. Ther.* **2002**, *300*, 361.

130. Levesque E.; Beaulieu M.; Guillemette C.; Hum D.W.; Belanger A. *Endocrinology* **1998**, *139*, 2375.

131. Congiu, M.; Mashford, M.L.; Slavin, J.L.; Desmond, P.V. *Drug Metab. Dispos.* **2002**, *30*, 129.

132. Toide, K.; Takahashi, Y.; Yamazaki, H.; Terauchi, Y.; Fujii, T.; Parkinson, A.; Kamataki, T. *Drug Metab. Dispos.* **2002**, *30*, 613.

133. Ishii Y.; Hansen A.J.; Mackenzie P.I. *Mol. Pharmacol.* **2000**, *57*, 940.

134. Yamashita, A.; Nagatsuka, T.; Watanabe, M.; Kondo, H.; Sugira, T.; Waku, K. *Biochem. Pharmacol.* **1997**, *53*, 561.

135. Daly, A.K. *Fund. Clin. Pharmacol.* **2003**, *17*, 27.

136. Tukey, R.H.; Strassburg, C.P.; Mackenzie, P.I. *Mol. Pharmacol.* **2002**, *62*, 446.

137. Kadakol, A.; Ghosh, S.S.; Sappal, B.S.; Sharma, G.; Chowdhury, J.R.; Chowdhury, N.R. *Human Mutation* **2000**, *16*, 297.

138. Mackenzie, P.I.; Miners, J.O.; McKinnon, R.A. *Clin. Chem. Lab. Med.* **2000**, *38*, 889.

139. Burchell, B.; Coughtrie, M.W.H. *Environ. Health Perspect.* **1997**, *105*(Suppl. 4), 739.

140. Iyer, L.; Hall, D.; Das, D.; Mortell, M.A.; Ramirez, J.; Kim, S.; Di Rienzo, A.; Ratain, M.J. *Clin. Pharmacol. Ther.* **1999**, *65*, 576.

141. Ando, Y.; Saka, H.; Ando, M.; Sawa, T.; Muro, K.; Ueoka, H.; Yokoyama, A.; Saitoh, S.; Shimokata, K.; Hasegawa, Y. *Cancer Res.* **2000**, *60*, 6921.

142. Holthe, M.; Rakvåg, T.N.; Klepstad, P.; Idle, J.R.; Kaasa, S.; Krokan, H.E.; Skorpen, F. *The Pharmacogenomics Journal* **2003**, *3*, 17.

143. Lampe J.W.; Bigler J.; Horner N.K.; Potter J.D. *Pharmacogenetics* **1999**, *9*, 341.

144. Ito, M.; Yamamoto, K.; Maruo, Y.; Sato, H.; Fujiyama, Y.; Bamba, T. *Eur. J. Clin. Pharmacol.* **2002**, *58*, 11.

145. Huang, Y.-H.; Galijatovic, A.; Nguyen, N.; Geske, D.; Beaton, D.; Green, J.; Green, M.; Peters, W.H.; Tukeu, R.H. *Pharmacogenetics* **2002**, *12*, 287.

146. Vogel, A.; Kneip, S.; Barut, A.; Ehmer, U.; Tukey, R.H.; Manns, M.P.; Strassburg, C.P. *Gastroenterology* **2001**, *121*, 1136.

147. Clarke D.J.; Burchell, B. *Handbook of Experimental Pharmacology* **1994**, *112*, 3

148. Kemp, D.C.; Fan, P.W.; Stevens, J.C. *Drug Metab. Dispos.* **2002**, *30*, 694.

149. Ismail, I.M.; Dear, G.J.; Roberts, A.D.; Plumb, R.S.; Ayrtont, J.; Sweatman, B.C.; Bowers G.D. *Xenobiotica* **2002**, *32*; 29

150. Innocenti, F.; Stadler, W.M.; Iyer, L.; Ramirez, J.; Vokes, E.E.; Ratain, M.J. *Clin. Cancer Res.* **2000**, *6*, 3400

151. Prueksaritanont, T.; Subramanian, R.; Fang, X.; Ma, Bennett; Qiu, Y.; Lin, J.H.; Pearson, P.G.; Baillie, T.A. *Drug Metab. Dispos.* **2002**, *30*, 505.

152. Ren, Q.; Murphy, S.E.; Zheng, Z.; Lazarus, P. *Drug Metab. Dispos.* **2000**, *28*, 1352.

153. Kassahun, K.; Mattiuz, E.; Franklin, R.; Gillespie, T. *Drug Metab. Dispos.* **1998**, *26*, 848.

154. Liu, D.Q.; Pereira, T. *Rapid Commun. Mass Spectrom.* **2002**, *16*, 142.

155. Ritter, J.K. *Chem. Biol. Interact.* **2000**, *129*, 171.

156. Vree, T.B.; van Dongen, R.T.M.; Koopman-Kimenai, P.M. *IJCP* **2000**, *54*, 395.

157. Bolze, S.; Bromet, N.; Gay-Feutry, C.; Massiere, F.; Boulieu, R.; Hulot, T. *Drug Metab. Dispos.* **2002**, *30*, 404.

158. Weiland, E.; Shipkova, M.; Schellhaas, U.; Schutz, E.; Niedmann, P.D.; Armstrong, V.W.; Oellerich, M. *Clin. Biochem.* **2000**, *33*, 107.

159. Bailey, M.J.; Dickinson, R.G. *Chem. Res. Toxicol.* **1996**, *9*, 659.

160. Terrier, N.; Benoit, E.; Senay, C.; Lapicque, F.; Radominska-Pandya, A.; Magdalou, J.; Fournel-Gigleux, S. *Mol. Pharmacol.* **1999**, *56*, 226.

161. Benet, L.Z.; Spahn-Langguth, H.; Iwakawa, S.; Volland, C.; Mizuma, T.; Mayer, S.; Mutschler, E.; Lin, E.T. *Life Sci.* **1993**, *53*, 141.

162. Breyer-Pfaff, U.; Mey, U.; Green, M.D.; Tephly, T.R. *Drug Metab. Dispos.* **2000**, *28*, 869.

163. Ethell, B.T.; Ekins, S.; Wang, J.; Burchell, B. *Drug Metab. Dispos.* **2002**, *30*, 734.

164. Shimizu, M.; Matsumoto, Y.; Tatsuno, M.; Fukuoka, M. *Biol. Pharm. Bull.* **2003**, *26*, 216.

165. Turgeon, D.; Carrier, J.S.; Levesque, E.; Hum, D.W.; Belanger, A. *Endocrinology* **2001**, *142*, 778.

166. Ekins, S.; Ring, B.J.; Grace, J.; McRobie-Belle, D.J.; Wrighton, S.A. *J. Pharmacol. Toxicol. Methods* **2000**, *44*, 313.

167. Ekins, S.; Ring, B.J.; Binkley, S.N.; Hall, S.D.; Wrighton, S.A. *Int. J. Clin. Pharmacol. Ther.* **1998**, *36*, 642.

168. Williams, J.A.; Ring, B.J.; Cantrell, V.E.; Campanale, K.; Jones, D.R.; Hall, S.D.; Wrighton, S.A. *Drug Metab. Dispos.* **2002**, *30*, 1266.

169. Iwatsubo, T.; Hirota, N.; Ooie, T.; Suzuki, H.; Shimada, N.; Chiba, K.; Ishizaki, T.; Green, C.E.; Tyson, C.A.; Sugiyama, Y. *Pharmacol. Ther.* **1997**, *73*, 147.

170. Houston, J.B. *Biochem. Pharmacol.* **1994**, *47*, 1469.

171. Obach, R.S. *Curr. Opin. Drug Disc. Dev.* **2001**, *4*, 36-44.

172. Pang, K.S.; Rowland, M. *J. Pharmacokinet. Biopharm.* **1977**, *5*, 625.

173. Mistry, M.; Houston, J.B. *Drug Metab. Dispos.* **1987**, *15*, 710.

174. Soars, M.G.; Burchell, B.; Riley, R.J. *J. Pharm. Exp. Ther.* **2002**, *301*, 382.

175. Boase, S.; Miners, J.O. *Br. J. Clin. Pharmacol.* **2002**, *54*, 493.

176. Hall S.D.; Thummel, K.E.; Watkins, P.B.; Lown, K.S.; Benet, L.Z.; Paine, M.F.; Mayo, R.R.; Turgeon, D.K.; Bailey, D.G.; Fontana, R.J.; Wrighton, S.A. *Drug Metab. Dispos.* **1999**, *27*, 161.

177. Czernik, P.J.; Little, J.M.; Barone, G.W.; Raufman, J.P.; Radominska-Pandya, A. *Drug Metab. Dispos.* **2000**, *28*, 1210.

178. Trapnell, C.B.; Klecker, R.W.; Jamis-Dow, C.; Collins, J.M. *Antimicrob. Agents Chemother.* **1998**, *42*, 1592.

179. Ammon, S.; von Richter, O.; Hofman, U.; Thon, K.-P.; Eichelbaum, M.; Mikus, G. *Drug Metab. Dispos.* **2000**, *28*, 1149.

180. Zucker, K.; Tsaroucha, A.; Olson, L.; Esquenazi, V.; Tzakis, A.; Miller, J. *Ther. Drug Monit.* **1999**, *21*, 35.

181. Prueksaritanont, T.; Zhao, J.J.; Ma, B.; Roadcap, B.A.; Tang, C.; Qiu, Y.; Liu, L.; Lin, J.H.; Pearson, P.G.; Baillie, T.A. *J. Pharmacol. Exp. Ther.* **2002**, *301*, 1042.

182. von Moltke, L.L.; Manis, M.; Harmatz, J.S.; Poorman, R.; Greenblatt, D.J. *Biopharm. Drug Dispos.* **1993**, *14*, 119.

183. Gornet, J.M.; Lokiec, F.; Duclos-Vallee, J.-C.; Azoulay, D.; Goldwasser, F. *Anticancer Res.* **2001**, *21*, 4203.

184. Liston, H.L.; Markowitz, J.S.; DeVane, C.L. *J. Clin. Psychopharmacol.* **2001**, *21*, 500.

185. Miners, J.O.; Mackenzie, P.I. *Pharmacol. Ther.* **1991**, *51*, 347.

186. Burchell, B.; Soars, M.; Monaghan, G.; Cassidy, A.; Smith, D.; Ethell, B. *Toxicol. Lett.* **2000**, *112-113*, 333.

187. Zucker, S.D.; Qin, X.; Rouster, S.D.; Yu, F.; Green, R.M.; Keshavan, P.; Feinberg, J.; Sherman, K.E. *Proc. Natl. Acad. Sci. USA* **2001**, *98*, 12671.

188. Broudy, M.I.; Crespi, C.L.; Patten, C.J. *AAPSPharmSci* **2001**, *3*, abstract.

189. Dutton, G.J. In *Glucuronic acid- free and combined*; Dutton, G.J., Ed.; Academic Press: NY, **1966**; pp. 349.

190. Ouzzine, M.; Gulberti, S.; Levoin, N.; Netter, P.; Magdalou, J.; Fournel-Gigleux, S. *J. Biol. Chem.* **2002**, 277, 25439..

191. Senay, C.; Jedlitschky, G.; Terrier, N.; Burchell, B.; Magdalou, J.; Fournel-Gigleux, S. *Biochim. Biophys. Acta* **2002**, 1597, 90.

192. Dutton, G.J.; Burchell, B. *Prog. Drug Metab.* **1977**, 2, 1.

193. Dutton, G.J. In *Glucuronidation of drugs and other compounds*; CRC Press: Boca Raton, FL., **1980**; pp. 268.

194. Coughtrie, M.W.; Burchell, B.; Sheperd, I.M.; Bend, J.R. *Mol. Pharmacol.* **1987**, 31, 585.

195. Sutherland, L.; Ebner, T.; Burchell, B. *Biochem. Pharmacol.* **1993**, 45, 295.

196. Bosma, P.J.; Chowdhury, J.R.; Bakker, C.; Gantla, S.; de Boer, A.; Oostra, B.A.; Lindhout, D.; Tytgat, G.N.; Jansen, P.L.; Oude Elferink, R.P., et al. *N. Engl. J. Med.* **1995**, 333, 1171.

197. Monaghan, G.; Ryan, M.; Seddon, R.; Hume, R.; Burchell, B. *Lancet* **1996**, 347, 578.

Index

acebutolol, 101
acetaminophen, 35, 54-58, 60-61, 88, 110, 148, 150, 198, 258-259, 338, 402, 408-409, 547, 566
acetylation, 94, 97, 105
acetylene, 64
aciclovir, 521, 525
acrolein, 96, 119, 123, 137, 528
acyl glucuronides, 94, 107-108, 561-562
ADH, 527
adverse drug reactions, 87-88, 104, 107, 139, 311, 412, 414, 567
agranulocytosis, 92-93, 97-100, 102, 105-107, 109, 112-113, 115, 117, 120, 126, 128, 133
AIDS, 92, 143
alclofenac, 122-123
aldehyde, 11, 13, 16, 68, 119, 136, 170-173, 339, 483, 501, 504, 506, 511, 516, 521, 527-528
aldehyde dehydrogenases, 483, 527-528
alkyl halides, 96
alleles, 259, 331, 377-378, 386-388, 390, 392, 401, 404-406, 408, 410-412, 550
allometric scaling, 453, 458-462, 478, 504
alprazolam, 175, 249
amadori rearrangement, 108-109
ames test, 550
amiloride, 100
aminoglutethimide, 100, 175
aminopyrine, 106-107, 133
amodiaquine, 112, 259, 406

analytical techniques, 35-36, 39, 62, 71, 83
anastrozole, 175
angiotensin, 126, 173
antibodies, 89-90, 98, 100, 112, 124, 127, 157, 159, 162, 261, 281, 294, 297-299, 408, 433, 485, 487, 495, 500, 546-547, 550, 557
antibody inhibition, 261, 295-296, 298-299
anticancer, 124, 130, 169-170, 188, 314, 525, 561, 574
antidepressants, 158-159, 314, 393, 498, 508
antiestrogen, 95, 115, 516
antipsychotics, 393
arene oxides, 122, 124, 133
aromatase, 11, 13, 16, 100, 175, 393
aromatic amines, 97, 100-101, 135, 321, 547, 549
aromatic nitro drugs, 101
artemisinin, 170
aryl amine, 97-101, 110, 121
aspartic acid, 158
autoactivation, 211, 220, 228-230, 232, 234, 241, 249
azaarenes, 511, 517-518, 522, 526

benoxaprofen, 110
benzodiazepines, 167, 563
benzozapepinone, 172
benzydamine, 490, 496-497
benzydamine, 490, 496-497
benzylamine, 71, 75, 77-78, 80

betaxolol, 168
bile, 36-39, 44, 48, 51-52, 55, 57-58,
    60-62, 65, 68-69, 71, 73-78, 80, 82,
    119, 136, 200, 338, 347, 350,
    352-354, 364, 423, 437-438, 562
bioactivation, 77-78, 80, 87-90, 93-94,
    98, 104, 107, 112-121, 123, 125,
    128-131, 134-135, 137-138, 154,
    189, 197-198, 429, 446, 483, 523,
    541, 548-551, 561-563
bioavailability, 168, 260, 315, 328, 330,
    332, 360, 421, 426-428, 430-431,
    454-455, 468-469, 471-473, 547,
    565-566
biological matrices, 36, 39, 41, 51
biotransformation, 34-36, 179-180, 185,
    199, 201, 255-256, 258, 260, 273,
    277, 285, 295, 298-299, 317-320,
    322, 361, 405, 421, 423, 483, 489,
    561, 564
body weight, 182-184, 414, 453,
    458-463, 471-473
brain weight, 183-184, 453, 459-461,
    478

caffeine, 198, 258, 327, 359-362, 402,
    408, 443, 470, 472-473, 497, 512,
    514
cancer, 30, 100, 115-116, 124, 139, 141,
    175, 196, 203-204, 207-209, 302,
    306, 375, 402, 404-405, 415, 417,
    419, 446, 450, 538, 551, 570-571,
    573
captopril, 126
CAR, 193-195, 199-200, 337-338,
    344-358, 365, 558
carbamazepine, 113, 123, 250, 353, 361
carbocations, 10, 104
carbonyl reductases, 527-529
carcinogenesis, 88, 198, 203, 207, 307,
    366, 417, 450-451, 538, 549, 568,
    570-572
catalysis, 10, 29, 167, 214, 483-484,
    488-489, 495, 501, 515-516

celecoxib, 99
cell stress, 97, 132, 138
chemical carcinogenesis, 549
chemical inhibition, 280, 285, 289,
    292-295
chloramphenicol, 102, 288
chlorpropamide, 169
cimetidine, 285, 288, 325
clearance, 34, 163, 166, 168-171, 175,
    179, 181-185, 191, 195, 198,
    211-212, 215-216, 220, 222-223,
    229, 233, 250, 255-257, 259-260,
    263, 266, 284, 300, 315, 319-320,
    322, 325, 327-328, 332, 362, 364,
    380-384, 393, 396-397, 407, 412,
    429, 433, 435, 439, 453-476, 478,
    496, 504, 511, 541, 558, 561,
    564-566
clozapine, 106-107, 117, 133, 361, 490,
    497-498
coa, 107-110, 120
coa esters, 107-109
colon, 346, 430, 433, 437-438
competitive inhibition, 162, 223, 240,
    242-243, 247, 286, 289, 319, 329
correlation analysis, 255, 260, 274-275,
    277, 279, 281, 283, 294, 297-299
covalent binding, 91-93, 97-99, 104,
    106, 108-109, 113, 115, 124, 126,
    132, 137-138, 147-150, 152-153,
    409, 562-563
COX 2 inhibitors, 102
Crigler-Najjar syndrome, 559
crystal structure, 5-6, 14, 282, 346, 382,
    384, 389, 411, 501, 548
cubane, 10
cumene, 13
cyclooctene, 22
cyclosporine, 165, 248, 328, 430-431,
    435
CYP, 1-7, 10-13, 15-18, 20, 22, 24-25,
    27, 117, 165, 173, 175, 187, 198,
    211-212, 215-216, 218, 227, 229,
    233, 251, 256, 258, 262-264, 266,
    274, 280, 284, 294, 312-317, 319-

327, 329, 331, 337-339, 349-351,
354-355, 359-361, 363-364, 421,
429-430, 436, 443-446, 525, 541,
553-554, 557-558, 564-565
CYP induction, 262, 317, 338, 354,
360-361
CYP1A, 185, 192-193, 198, 330, 339,
342, 355, 357-358, 360, 402, 443,
451
CYP1A1, 185, 192, 198-199, 339-340,
342-344, 358, 401-402, 421-422,
442-443, 446, 488
CYP1A2, 90, 159, 175, 185-186,
198-199, 258, 272, 280-281, 283,
289, 291-292, 313, 315, 327-329,
338-339, 361-362, 402-403, 422,
443, 445-446
CYP1A2 deficient mouse, 198
CYP2A, 186-187
CYP2B, 187-188, 193-194, 199-200,
339, 344, 347-350, 352, 358
CYP2C, 188-190, 258, 283, 338-339,
349-350, 406, 408, 439-442
CYP2C9, 90, 159, 164, 169, 182, 239,
250, 257-259, 263, 282-283, 289,
295-297, 313, 316, 328, 331, 346,
348-350, 352-354, 360-361, 377,
381-385, 389, 408, 411, 421-422,
440-442, 446, 476, 495
CYP2D, 189, 259
CYP2D6, 157-159, 162, 166-168, 176,
181, 184, 189, 201, 249, 256,
258-259, 262, 267, 279-282,
288-289, 292, 313-314, 327-328,
331-332, 376-380, 383, 385-401,
405, 421-422, 438-440, 446, 558-
559, 567
CYP2E, 189, 198
CYP2E1, 17, 20, 22-23, 90, 160,
189-190, 198, 258, 289, 294, 327,
330, 338-339, 408-409, 445
CYP2J2, 409, 421, 444-446
CYP3A, 164, 182, 188, 190-191,
194-196, 199-201, 214, 249-250,
259, 283, 293, 314-315, 325, 328-

330, 337-339, 344, 346-350, 353-
355, 360, 362-364, 410-411, 421,
429-442, 445, 567
CYP3A4, 16, 90, 159-167, 173-176,
184, 190-191, 195, 211-212,
214-215, 218, 227, 229-230,
232-234, 236-239, 246, 248-250,
258-260, 263, 265-266, 274, 277,
279, 282-283, 288-289, 292-297,
299, 313, 315-316, 325, 327-329,
346-349, 352-354, 359-362, 364,
410-412, 421-422, 430, 432-433,
435-438, 442-443, 445, 454, 458,
474-478
CYP4A, 191-192, 196-197, 199, 339
cytochrome, 1-2, 29, 93, 145, 148, 151,
155-158, 162, 170-171, 173-175,
179, 182-187, 191-192, 197-198,
200-204, 214, 223, 249, 255-258,
282-285, 290, 293, 296, 299, 308,
315, 331, 334, 375-378, 386,
389-390, 395, 401, 409-410, 414,
483, 486, 490, 494, 511, 541
cytochrome b5, 214, 282-285, 293
cytochrome P450, 1-2, 29, 93, 145,
155-158, 170-171, 173-175, 179,
182-187, 191-192, 197-198, 200-204,
223, 255-258, 296, 299, 308, 315,
331, 334, 375-378, 386, 389-390,
395, 409-410, 414, 483, 486, 494,
511, 541
cytosol, 340, 345, 347, 354, 358, 425,
428, 513-514, 519, 521, 523, 525,
529, 549
cytotoxicity, 106, 111, 138

danger hypothesis, 91-92
dantrolene, 102
dapsone, 97, 99, 258, 325, 327, 362,
364, 385
death, 88, 198, 434, 494
debrisoquine, 156-157, 189, 201, 376,
385-386, 388, 390, 407, 558
detergent, 494-495, 554, 556

development, 33-35, 81-82, 87, 113,
    117, 136, 138-139, 147-151, 153,
    155, 162-163, 173, 175-176, 181,
    187, 198-199, 202, 215, 223, 254,
    257, 281, 285, 298-300, 312, 314,
    317, 319-322, 331-332, 337, 385,
    406, 412, 428, 453-454, 457-458,
    468-469, 476, 478, 483, 494, 521,
    542, 547, 551, 567
diacetylhydrazine, 105
dibromoethane, 94
diclofenac, 108-109, 114-115, 182, 237,
    249-250, 283, 298, 382, 384,
    440-441, 473, 562, 565-566
diet, 414, 436, 488
dietary factors, 325, 434, 488
dihydralazine, 90, 105, 134
diltiazem, 159, 172, 260, 287, 293-294,
    322, 511
dimers, 329, 553-554
dimethylaniline, 490
dioxygen, 1, 4, 18, 24
dipyrone, 106
disease model, 181
dismutase, 24-25
dispersion model, 456-457
DMA, 496
dog, 153, 184-185, 187-191, 344, 486,
    500, 504, 514, 519
dosage, 377, 423, 446
drug discovery, 33, 35, 82, 147, 150,
    155, 163, 176, 181, 259, 281, 285,
    299, 403, 405, 407, 413, 477, 481,
    494, 551, 567
drug interaction, 162, 248, 254, 257,
    312, 315, 321, 323, 338-339, 479,
    566
duodenum, 422, 424-425, 429-431,
    433-434, 439, 441-442, 445, 500

EadieHofstee plots, 222, 227, 229, 244,
    273-274
electrophiles, 94-96, 108, 110, 118, 129,
    132, 137, 147-149
encainide, 157

endoplasmic reticulum, 2, 98, 425, 553
enterocyte, 429, 432-433, 435
enterocytes, 424-425, 428, 433-434,
    437-439
enterohepatic cycling, 94
enzyme, 2-4, 7, 9, 14, 17, 19-20, 23, 34,
    52, 72, 90, 92-93, 95, 126, 135, 137,
    148-150, 155, 158-160, 162-163, 165,
    167, 173, 175, 179-180, 184-185,
    188-191, 197-198, 201, 211-224,
    227-229, 231-232, 234-235, 240-241,
    245-247, 252, 255-266, 268, 270-275,
    277, 279, 281-289, 292-295, 297-299,
    301, 305, 313, 315-322, 325, 327,
    331-332, 334, 338-339, 347, 353-355,
    359-361, 364-365, 375, 377-385, 388-
    389, 392-397, 400-411, 422, 427,
    432-435, 437-439, 444-445, 461, 466,
    483-484, 488, 493-495, 501, 504,
    509, 511, 513-516, 525-527, 541-544,
    546-548, 550-557, 559-566, 572
enzyme kinetics, 188, 211, 215-216,
    220, 252, 260, 263-264, 266, 268,
    270-271, 273, 285, 293, 305, 317,
    334, 563-564
epithelial, 315, 423-426, 436, 513
epoxides, 96, 121-122, 125, 523
erythromycin, 98, 159-162, 237, 248,
    287, 314-315, 322, 325, 329, 362-
    364, 430, 432-433, 436, 438
estradiol, 159, 194, 278, 348, 403-404,
    428-429, 487, 542, 545, 548,
    556-557, 562-563
estrogens, 11, 115-116, 131, 175, 194,
    338, 345-346, 348, 358, 562-563
ethinylestradiol, 165, 542, 547
ethylmorphine, 156, 159
expressed P450, 258, 263, 282-284, 288
expression, 160, 176, 186, 190, 192,
    194-201, 211, 213, 215, 218, 249,
    255-256, 258-259, 263, 274, 281-285,
    293, 298, 300, 327-328, 341-344,
    349-355, 357-359, 377, 380, 382-383,
    386, 388, 390-392, 402-406, 408-412,
    421, 426, 428-429, 432-434, 436-442,

454, 459, 485-488, 495, 498, 514,
542, 546-547, 549-552, 555, 558,
560, 567
extraction, 37, 39, 50, 73, 421-422, 426,
429-432, 469, 471, 545
extrapolation, 119, 179, 181, 185, 212,
222, 249, 322, 380, 453, 469, 478,
486, 514, 526

fadrozole, 175
famciclovir, 521
FDA Guidance for Industry, 254, 323
felbamate, 91, 94, 118-119, 136-137, 250
felodipine, 175, 237, 242-244, 247-248,
364, 435
Flavin-containing monooxygenase, 93,
128
flavoprotein, 2-3
fluperlapine, 117
FMO, 484-498
fraction of absorption, 455
free radical, 28, 96, 105, 107, 132-135
furans, 125, 135
furosemide, 125

gene knockout, 197
genetic polymorphism, 94, 137, 157,
192, 401, 528, 550
genotype, 105, 124, 201, 331, 384, 387,
400, 438, 550-551
genotyping, 262, 386, 388, 406, 414,
560
gestodene, 162, 288, 290, 292, 410
Gilbert's syndrome, 558-559
glaucoma, 404
glitazones, 130
glucuronidase, 69, 553, 556
glucuronides, 67-69, 94, 107-108, 256,
557, 561-562
glutathione, 34, 40, 55, 64-65, 67, 82,
92, 94-96, 98, 106, 110-111, 116, 119,
122, 124, 129-130, 133, 135-137,
148, 152, 339, 343, 376, 409, 428
glutathione conjugates, 64, 110, 136,
152

glycine, 58, 73, 77, 485, 506
GST, 357, 428-429

halflife, 34, 103, 117, 168-171, 173,
181, 198, 250, 315, 329, 342, 361,
364, 435, 454-455, 468, 551, 564
halothane, 89-90, 96-97, 102, 404, 408
hamster, 186-187, 192, 525
hapten, 89, 92-93, 105, 132
Hardy-Weinberg equilibrium, 377
hepatic availability, 455
hepatic clearance, 453-458, 463-464,
466, 468-473, 478, 564
hepatic necrosis, 148, 150
hepatitis, 90, 94, 111, 521
hepatocytes, 33-34, 52, 136-137, 149-
150, 181, 193, 195, 211, 213, 249,
316, 327, 347, 350-352, 354, 358-
360, 428, 436, 438, 454-455, 458,
462, 466, 470-473, 478, 506, 545,
556-557, 565
heteroactivation, 211, 227, 235-236,
238-240, 249-250
heterozygote, 401
heterozygotes, 398-399, 551
high risk, 316, 321
histidine, 3, 5, 137, 548, 552, 560
homozygote, 398, 401
human, 1, 47, 52, 81, 137, 148, 153,
155-160, 162-167, 170, 175, 177,
179-182, 184-190, 192, 194-197,
200-202, 212-216, 218, 227, 232,
234, 249, 256, 258-263, 266, 268-
270, 272-274, 277-279, 281-290,
292-300, 315, 318, 322, 328-331,
333, 338, 343, 346-348, 350-352,
355-356, 359-361, 373, 376, 378-
381, 383, 385-388, 393-394, 401-
410, 412-416, 421-422, 427-430,
433-445, 454, 459, 462, 466-475,
477-478, 484-487, 492-494, 496-498,
500-501, 504, 510, 514, 518-519,
521, 526-529, 541-542, 544, 546-
552, 555-560, 563, 565-568, 573
humanized CYP2D6 mouse, 201

humanized mouse, 200-201
hydrazines, 104, 134-135, 509
hydroxylamine, 24, 73, 75, 77, 97-102,
    104, 121
hypersensitivity reactions, 89, 98, 149
hypertension, 1, 124, 177

idiosyncratic, 87-106, 108-113, 119,
    121-124, 127, 132-139, 148
idiosyncratic drug reactions, 87, 89-94,
    97, 99, 101-103, 108-109, 112-113,
    121, 123-124, 132, 135-139
idiosyncratic reaction, 88, 91, 106, 108,
    119, 132
ileum, 422, 424, 429-430, 433-434, 437-
    439, 441-442
imidazoles, 174
imidoquinone, 88, 110, 113
immunoquantitation, 157, 284, 294-295,
    298-299
in vitro, 33-34, 36, 51-52, 77, 81-82, 91,
    98, 106, 123, 136-138, 149-150, 152,
    162, 170, 174-176, 179-181, 184,
    187, 201, 211-216, 218, 222-225,
    227, 229-230, 233, 239, 246, 248-
    251, 254-256, 258, 260, 263-264,
    266, 275, 284-285, 287, 294, 298,
    312, 316-317, 319-323, 327, 330,
    332-333, 338, 348, 354-355, 357,
    359-360, 364, 380, 385, 388, 392-
    393, 398-403, 405, 408-411, 421-
    422, 429-430, 435, 437, 443-445,
    453-455, 458-459, 461-474, 476,
    478, 494, 496-498, 514, 519, 521-
    522, 525-527, 545, 549-550, 557,
    564-566, 568
in vivo, 36, 52, 75, 77, 80, 98, 106, 108,
    136-138, 149-150, 152-153, 171,
    174-175, 180-181, 184, 187, 197-
    198, 211-212, 216, 218, 222-223,
    240, 248-250, 258, 260, 263, 266,
    284-285, 287, 294, 312, 316-317,
    319-325, 330-333, 338, 347-348,
    354-355, 358-361, 363-364, 380,
    382, 385, 388, 401, 403, 405, 409,

422, 429-433, 435, 437-438, 443-
    446, 453, 455, 457-458, 462-466,
    468-472, 478, 490, 496-498, 504,
    511-512, 514, 518, 525-526, 545,
    549-550, 554, 556, 564-567
inhibitor, 51-52, 54, 62, 100, 113,
    122-123, 126-127, 150, 161, 164,
    170-171, 175, 211, 215, 223-225,
    235, 240-242, 244-248, 258, 263,
    271-272, 280-282, 285-288, 292,
    294, 297, 316, 318-322, 324-325,
    327-329, 342, 344, 347, 350, 394,
    403-404, 408-410, 435, 443, 463-
    464, 474-476, 484, 495, 500-501,
    503-506, 509-510, 514, 521, 525-
    526, 545, 549, 556, 559, 564
interspecies scaling, 182, 184, 479
intestinal availability, 455, 469, 472
intestine, 109, 259, 346-347, 412, 421-
    434, 437-446, 454, 469, 485-486,
    488, 500, 513, 555, 565
intrinsic clearance, 182-184, 211, 216,
    220, 222-223, 263, 266, 382-383,
    393, 396-397, 407, 429, 433, 439,
    455-457, 462-468, 471-476, 564
iodosobenzene, 13
iproniazid, 134
ipso adduct, 111
irbesartan, 172-173, 288
Irinotecan, 428, 559
ironsulfur protein, 2-3
ischemia, 434
isocyanate, 78, 117, 129-130
isoniazid, 105, 189, 331
itopride., 490, 497

jejunum, 424-425, 429-430, 433-434,
    439, 441-442

ketoconazole, 161-162, 174-175, 286,
    289-290, 292, 314, 325, 422, 435, 443
kidney, 52, 192, 196-197, 199, 325, 346,
    362, 406, 430-431, 437, 439, 444,
    454, 485-488, 493, 500, 513, 527,
    529, 555

kinetics, 188, 211-213, 215-218, 220-221, 227-232, 234-235, 243-244, 247-249, 252, 260, 263-264, 266, 268, 270-271, 273, 285, 293, 305, 312, 317, 323, 329-330, 332, 334, 389, 404, 406, 410-411, 464, 477, 514, 563-564

K_m, 158, 182, 188-190, 211, 213, 215-216, 218-220, 223-225, 229, 232, 234, 238, 240-241, 250, 263-264, 266, 268, 270, 282-284, 317, 319-320, 381, 383-385, 388, 390, 393, 396-397, 403-404, 407, 410, 429, 432-433, 443, 457, 463, 469-470, 476-477, 488, 514, 516, 518, 520-521, 547, 563-565

laboratory animals, 201
lamotrigine, 100
latency, 554
leflunomide, 121
letrozole, 175
leucine, 357, 385
lidocaine, 159, 422, 472-473
lifespan potential, 183, 453, 459-460, 478
Lineweaverburk Plots, 268, 270
liver, 22, 33-34, 52, 77, 80, 90, 92-94, 97, 100, 102-103, 106, 109-110, 113, 115, 117-121, 123, 125, 127-128, 130, 133-134, 137, 147-150, 152-153, 156-159, 161-162, 167, 180, 182, 185, 187-190, 192-193, 196-199, 214, 216, 218, 221-223, 230, 232, 257-264, 266, 268, 270, 272-274, 277-279, 281-284, 286-290, 292-299, 316, 320, 322, 338, 343, 346-348, 351-352, 354, 359-362, 379, 387-388, 393, 402-410, 412, 422-423, 426, 428-432, 437-441, 443-445, 454-456, 458, 463-466, 469-472, 475, 477, 484-489, 493-495, 497, 500, 509, 513-514, 516-519, 521, 523-527, 529, 542, 547, 549, 555-556, 559, 564-566

liver damage, 90, 148, 338
liver microsomes, 34, 80, 150, 153, 157, 159, 167, 188-190, 214, 216, 218, 223, 230, 232, 257, 260-262, 264, 270, 272-274, 278, 281-284, 286-290, 292-296, 298, 322, 354, 403, 406, 430, 440-441, 443-444, 454, 464, 466, 469-470, 475, 477, 494, 523, 525-526, 556, 564-566
liver toxicity, 90, 92, 97, 100, 102-103, 106, 109-110, 117-121, 125, 128, 130, 134, 198
losartan, 172-173, 441
lovastatin, 175, 191, 323
low risk, 316, 321, 528
lupus, 90, 93-94, 98-99, 101, 104-105, 108

mammals, 1-2, 181, 185, 339, 344, 484, 513
MAO, 499-501, 503-511, 527
marker activity, 189-190, 281
mechanismbased inhibition, 131, 322, 329, 509
medium risk, 316, 321
mephenytoin, 188, 258-259, 274, 286, 289, 316, 327, 407-408
metabolic activation, 88-89, 93, 124, 126, 128, 137-138, 147-153, 160, 258, 321, 549
metabolites, 22, 33-48, 50-52, 54-55, 57-58, 62, 68-69, 71, 73, 77-78, 80-83, 87-97, 99-102, 104-105, 109-113, 115-117, 121, 123-124, 128, 130-139, 147-148, 150-151, 153, 161-162, 169-170, 179, 181, 187, 198, 212, 214, 218, 250, 255-256, 314, 327, 338, 346, 362, 382-383, 423, 431-432, 444, 483-484, 490, 494, 496-497, 504, 512, 516, 518-519, 523, 527, 547, 549, 553, 557, 560-562
methemoglobinemia, 35, 97, 99, 130
methimazole, 93, 128, 490, 495-496
3methylcholanthrene, 155, 192, 360, 402, 558

methylenedioxyphenyl, 131
metoclopramide, 99
metronidazole, 103-104
mianserin, 120
michael acceptors, 95, 118, 125, 135, 509
microsomes, 31, 34, 52, 80, 98-99, 117, 123, 137, 149-150, 153, 156-157, 159, 167, 180, 188-190, 211, 213-214, 216, 218, 223, 230, 232, 257, 260-262, 264, 269-270, 272-274, 278, 281-284, 286-290, 292-296, 298, 322, 327, 354, 380, 387-388, 403, 406, 428, 430, 432-433, 438-441, 443-445, 454, 464, 466, 469-470, 475, 477, 486, 494, 523, 525-526, 556, 564-566
midazolam, 167, 175, 191, 236-237, 242-244, 246-249, 260, 266, 277-278, 287-288, 325, 327-328, 364, 411, 421, 431-433, 435-438
minocycline, 117-118
minoxidil, 35, 549
mitochondria, 2, 500
modeling, 15, 285, 288, 382, 384, 403-404, 410, 469, 477, 563
molecular weights, 39, 41, 44, 55, 73, 259, 499
molybdenum hydroxylase, 512-513, 516, 519, 522-526
monkey, 153, 180, 182, 184-191, 299, 344, 519
monoamine oxidase, 134, 393, 508, 533
monoclonal antibody, 159, 296, 298, 430, 500
monooxygenase, 93, 128, 354, 429
moricizine, 164
mouse, 128, 164, 184-201, 343, 348, 355-356, 487-488, 500, 514, 525, 542, 548-549
mucosa, 421-422, 428-429, 432-434, 439, 527
multiple binding, 211, 229, 240, 564
mutation, 14-15, 18-20, 282, 301, 307, 343, 376-378, 383, 388-390, 402, 405, 498, 559-560, 566, 573

myeloperoxidase, 90, 93, 113
myocarditis, 106

NADPH, 2, 93, 149, 216-217, 230, 287-288, 292-294, 322, 339, 386, 395, 476, 484-485, 488, 494-496, 504, 519, 522, 525
NAT, 30-31, 206, 369, 372, 419, 428, 448, 569, 571
NCE, 33-35, 256, 281, 284, 294-295, 298, 317-322
neurotoxicity, 507
neurotransmitters, 393, 503, 542
neutrophils, 93, 97, 99, 105-107, 112, 115, 120, 126, 128, 136-137
nicotine, 404-405, 490, 496-498
nifedipine, 103, 158-159, 162, 175, 234, 237-238, 242-245, 248, 260, 288-289, 346, 410, 421, 435
NIH shift, 23, 122
nimesulide, 102
nitro, 94, 101-104, 132, 135, 523
nitrofurantoin, 104
nitrogen, 1-2, 38, 50, 68, 73, 95-96, 100, 104-105, 110, 115, 117, 121, 124, 134, 136-137, 158-161, 167, 171-172, 174-175, 262, 286, 393, 474, 476, 484, 490, 501, 503, 509-510, 512, 516-517, 552
nitroimidazoles, 103
4nitrophenol, 545
nomifensine, 100
Non-P450 enzymes, 483-484
norcarane, 10
N-oxide, 490, 496-498, 518, 561
N-oxygenation, 491, 493, 496, 498
NSAID, 109-110, 115, 566
nucleophile, 24, 34, 82, 95-96, 149

olanzapine, 106-107, 361, 402, 498, 561
olefins, 13, 16, 19, 23, 25-27
omeprazole, 127, 283, 327, 355, 359-362, 408, 441, 445, 470
oral bioavailability, 330, 421, 426-428, 430-431, 455, 468-469, 471-472

oxidation, 1, 3-4, 6-8, 11, 19-20, 22,
  24-26, 63, 77, 90, 93-94, 97, 101,
  104-107, 110, 112-113, 117-122, 125,
  128-130, 133-137, 148,
  151-152, 156-162, 167-168, 170,
  255, 257, 259, 261, 273, 285, 287,
  289, 312, 314, 329, 343, 353,
  381-382, 390, 393, 439, 490, 497,
  499-504, 506-507, 509-514, 516-518,
  520-521, 523, 526-528

P450, 1-2, 6, 29, 63, 67, 77, 90, 92-94,
  96, 105, 109, 122, 128, 131,
  133-137, 145, 148, 155-164, 166-
  168, 170-171, 173-176, 179-180,
  182-187, 189, 191-192, 197-198,
  200-204, 214, 223, 230, 255-263,
  266, 268, 270-278, 280-289, 291-
  299, 308, 311-315, 331, 334, 337,
  339, 349, 372, 375-383, 385-386,
  388-390, 394-397, 401-403, 405-410,
  413-414, 421, 428-429, 454,
  458-459, 470, 474-478, 483-484,
  486, 488-490, 494-497, 504,
  511-513, 519, 523, 526-528, 541,
  554, 567
P450 binding spectra, 454, 474-475
P450 reductase, 214, 230, 284, 296,
  386, 389-390, 395, 403, 476
pancytopenia, 105
PAPS, 541, 543, 545, 548-549
PAPS, 541, 543, 545, 548-549
paralleltube model, 456-457, 464
paroxetine, 132
penciclovir, 521
penicillamine, 126-127
penicillin, 89, 124
peroxisome, 165, 191, 196
peroxy shunt, 7-8
p-glycoprotein, 164, 200, 325, 328, 330,
  431
pharmacogenetics, 140, 203, 206, 208,
  251, 256, 299-300, 302-303,
  307-310, 331, 335-336, 368, 370,
  374, 375-376, 380-381, 394, 401,

403-404, 409, 412-419, 448-450,
  532-533, 539, 550, 558, 560, 570-573
phenelzine, 134
phenobarbital, 123, 167, 183, 193, 199-
  200, 346-350, 352-354, 358, 360-
  361, 422, 435, 558
phenobarbitone, 155, 339
phenotype, 157, 201, 259, 325-327, 332,
  359, 362, 376-377, 379, 387-388,
  390, 402, 408
phenoxybenzamine, 124
phenytoin, 113, 123, 134, 164, 183, 259,
  315-316, 381-382, 472, 558
pinacidil, 16
piperazine, 121
placenta, 442, 500
plasma, 50-51, 81, 136, 149, 152-153,
  162, 165, 170, 194, 215, 250,
  315-316, 318, 320-321, 325,
  327-328, 360-361, 384, 426, 435,
  455, 457-458, 462, 466, 471, 487,
  490, 497, 511, 521, 525, 565-566
polyamines, 503
polycyclic hydrocarbons, 486
polymorphic, 88, 97, 123, 256, 259,
  262, 293, 331, 385, 394, 398,
  400-401, 550, 555
polymorphism, 94, 137, 157, 192, 201,
  376-379, 385-386, 401, 406-408,
  410, 413-414, 498, 528, 550-551, 560
polypharmacy, 332
practolol, 101
pregnancy, 487, 547
probe, 9, 24, 39, 47-48, 50, 158, 163,
  201, 271, 325-327, 331, 358,
  361-362, 382-383, 393, 395, 400-
  401, 408, 428-429, 431, 433, 435,
  438, 446, 478, 493, 496-498, 520
procainamide, 91, 97, 99-100
progesterone, 12, 16, 187-188, 190, 194,
  229, 237, 249, 265, 338, 345-346,
  348-349, 393, 410, 487, 545
propranolol, 157, 280-281, 402, 472-
  473, 506
propylthiouracil, 128

prostaglandin, 93, 108-109, 134
protein adducts, 148, 150
protein binding, 183, 215, 263, 268, 283,
    320, 457, 459, 462, 466, 504, 565
protein interaction, 158, 357
pteridine, 512-513, 517
puberty, 487
pyrimethamine, 100

quasiirreversible inhibition, 321
quinone, 96, 100, 110-112, 115-118,
    130-131, 409
quinone methides, 110, 116
quinones, 110-111, 115, 118, 132-133,
    135, 528

rabbit, 13, 22, 184-192, 195, 282, 344,
    382, 384, 387, 389, 408, 444, 484,
    486-488, 493, 500, 504, 514, 516-
    519, 521, 524-525
radical clock, 11, 23, 27
raloxifene, 561
rat, 44, 48, 51-52, 55, 57-58, 60-63, 65,
    71, 77, 81, 137, 153, 157-159, 180,
    182, 184-197, 230, 249, 257, 288,
    298-299, 346, 349-350, 355, 390,
    455, 470-472, 484, 486-488, 500,
    504, 506, 514, 518-519, 523-526,
    542, 549-550, 558, 565
reaction phenotyping, 185, 255-257,
    260-262, 266, 270, 281, 283-285,
    289, 294-295, 297-299
reactive metabolites, 34, 44, 52, 78, 82,
    87-97, 101-102, 104-105, 109-113,
    116, 124, 128, 130, 132-139, 147-
    148, 338, 523
reduction, 3, 50, 94, 101-104, 107, 111,
    168, 175, 213, 217, 248, 266, 283,
    338, 354, 476, 497, 515, 522-524,
    528-529, 549, 551, 566
regulation, 1, 179, 192-199, 201-202,
    277, 301, 322, 339, 342-343, 346,
    354, 364, 393, 400, 422, 434, 436,

438, 486-487, 516, 557-558, 560
relative activity factor, 284
reversible inhibition, 223, 226, 265
rifampicin, 164, 195, 200-201, 338-339,
    346, 349-350, 352-355, 359-360,
    364, 411
ring strain, 96, 121, 124
rodents, 34, 119, 155-156, 179, 181,
    186, 188-189, 191, 194, 197, 344,
    347, 350, 500
rosiglitazone, 130-131, 165, 406
rule of exponents, 184-185, 460-461

saccharolactone, 556
safrole, 132, 549
serotonin, 499, 504
simvastatin, 175, 237, 323, 329, 364,
    561
small intestine, 259, 346, 422-425,
    429-434, 437-438, 440-442, 444-446,
    500, 513, 555, 565
smoking, 201, 361, 404
SNP, 377, 406, 414
S-oxygenation, 490, 492-495
sparteine, 157, 376, 385-386, 388, 390
species differences, 34, 51, 82, 160,
    180-182, 185-186, 192, 194-197,
    200-202, 250, 459, 483-485, 499,
    514, 518-520, 526
SR12813, 164, 355
stable isotopes, 33
steric effects, 160, 517-518
stimulants, 393
structure elucidation, 46, 561
substrate, 3-4, 6-10, 14-15, 18-20, 22,
    25, 41-45, 55, 77, 93, 97, 102, 156,
    158-160, 173, 185, 187-189, 191-192,
    197-198, 211-214, 216-221, 223-225,
    227-229, 231-236, 238, 240-242,
    244-248, 250, 257, 259-260, 262-264,
    266-268, 270, 275-277, 280-288, 294,
    299, 316-320, 323, 325, 330, 349,
    358, 362, 381-382, 384-385, 388-390,

393-395, 399-400, 410, 430-431, 433, 435, 438-439, 441, 474, 476, 484, 486, 488-493, 495-497, 499-501, 503-504, 507, 509, 511-519, 521-523, 526, 542-543, 545-548, 551-554, 556-557, 559, 563-565
substrate selectivity, 156, 492, 552
succinimide, 134
sulfamethoxazole, 98, 116
sulfasalazine, 101
sulfation, 36, 52, 94, 541-543, 545-550, 567
sulfinamide, 97
sulfisoxazole, 98
sulfonamides, 92, 97-99
sulfotransferase, 541-544, 547-551
sulfoxide, 383, 490, 493, 497, 522
sulindac sulfide, 490, 493, 497, 522
sulofenur, 130
SULT1A1, 547, 549-551
superoxide, 4, 24-25, 93, 515

tacrine, 117-118, 258
tacrolimus, 175, 432, 553, 566
tamoxifen, 95, 115, 159, 516
terbinafine, 119, 135-136
terminology, 294, 376, 378, 414
tetracyclines, 117
theophylline, 159, 258, 327, 402, 443, 472, 514
thioacetamide, 127
thiobenzamide, 490, 496
thiono sulfur compounds, 127
thiophenes, 125, 135
thiourea, 127, 129
thr252, 5-6, 14-15
thrombocytopenia, 92, 100, 126
ticlopidine, 125-126
ticrynafen, 125-126
tienilic acid, 92, 125
TMA, 492-493, 496, 498
tolbutamide, 129, 168-169, 279, 289-290, 440, 470, 472
toxicity, 35, 51-52, 54, 83, 88-93, 96-97, 100, 102-106, 109-111, 115, 117-121,

123, 125-128, 130-131, 133-134, 138, 148, 153, 179, 181, 192, 195, 197-200, 218, 285, 338, 365, 376, 379-381, 414, 431, 558-559, 561-562
toxicology, 140, 142, 153-154, 179-181, 185, 197, 282, 285, 301-304, 306, 308-310, 365, 380, 415-416, 446, 481, 530-532, 535, 568
triamterene, 100
triazolam, 175, 249-250
trimethoprim, 98, 101, 116
trimethylamine, 490, 492, 497
trimethylaminuria, 498
troglitazone, 117, 130-131, 165-166, 195, 428, 547
tuberculosis, 105, 336, 360

UDPglucuronic acid, 551-552
UGT1A, 552, 557, 559-560
ultra-rapid metabolizers, 379, 385-386
unusual metabolites, 54-55, 69, 78, 80
urine, 36-39, 47, 50-52, 55, 62, 71, 77, 80-82, 156, 362, 490, 497-498, 518, 522

valproic acid, 108, 120, 562-563, 566
variant, 288, 377, 379-384, 388, 401-403, 405-407, 410, 412, 550
vesnarinone, 113-114, 136-137
victim drug, 325
vincristine, 175
V_{max}, 182, 188-190, 211, 213, 215-216, 219-220, 223-224, 229, 232, 234, 240-243, 246, 263, 266, 268, 272, 277, 284, 317, 381, 383-385, 388, 393, 396, 403-404, 406, 429, 433, 457, 463, 469-470, 476-477, 488, 518, 521, 556, 563-564
volume of distribution, 454-455
vorozole, 175
warfarin, 237, 249, 257, 259, 274, 276, 282, 286, 377, 381-384, 443, 472, 529
wellstirred model, 182, 456-457, 464,

467, 469
western blot, 295, 387, 430, 440, 445,
    487

xanthine, 94, 198, 483, 511, 513-514,
    517, 520-521
xanthine oxidase, 94, 483, 511
xenobiotics, 1, 15, 33-34, 83, 94, 147,
    156, 192, 195-199, 257, 285, 337,
    349, 351, 353-355, 358, 361, 364-
365, 377, 421, 430, 446, 454, 483,
    486, 489-490, 511, 514, 526, 545,
    548, 550, 552, 557, 559
xray crystal structure, 384, 411, 548

zomepirac, 110, 562